THE
HERMETIC AND ALCHEMICAL WRITINGS

OF

AUREOLUS PHILIPPUS THEOPHRASTUS BOMBAST,

OF HOHENHEIM, CALLED

PARACELSUS THE GREAT.

NOW FOR THE FIRST TIME FAITHFULLY TRANSLATED INTO ENGLISH.

EDITED WITH A BIOGRAPHICAL PREFACE, ELUCIDATORY NOTES, A COPIOUS HERMETIC
VOCABULARY, AND INDEX,

By ARTHUR EDWARD WAITE.

IN TWO VOLUMES.

VOLS. I & II.

HERMETIC CHEMISTRY.

HERMETIC MEDICINE & HERMETIC PHILOSOPHY.

THE
HERMETIC AND ALCHEMICAL
WRITINGS

OF

AUREOLUS PHILIPPUS THEOPHRASTUS BOMBAST,

OF HOHENHEIM, CALLED

PARACELSUS THE GREAT.

NOW FOR THE FIRST TIME FAITHFULLY TRANSLATED INTO ENGLISH.

EDITED WITH A BIOGRAPHICAL PREFACE, ELUCIDATORY NOTES, A COPIOUS HERMETIC
VOCABULARY, AND INDEX,

By ARTHUR EDWARD WAITE.

IN TWO VOLUMES.

VOL. I.

HERMETIC CHEMISTRY.

London:
JAMES ELLIOTT AND CO.,
TEMPLE CHAMBERS, FALCON COURT, FLEET STREET, E.C.
1894.

Copyright © 2015 Merchant Books

ISBN 978-1-60386-696-5

DIGITALIZED BY
WATCHMAKER PUBLISHING
ALL RIGHTS RESERVED

TABLE OF CONTENTS

VOLUME I

                              PAGE

PREFACE TO THE ENGLISH TRANSLATION   xvii

PART I

HERMETIC CHEMISTRY

THE CŒLUM PHILOSOPHORUM, or Book of Vexations, concerning the Science and Nature of Alchemy, and what opinion should be formed thereof. Regulated by the Seven Rules or Fundamental Canons according to the Seven commonly known Metals; and containing a Preface, with certain Treatises and Appendices   3

> The Preface of Theophrastus Paracelsus to all Alchemists and readers of this book. The First Canon: concerning the nature and properties of Mercury. The Second Canon: concerning the nature and properties of Jupiter. The Third Canon: concerning Mars and his properties. The Fourth Canon: concerning Venus and her properties. The Fifth Canon: concerning the nature and properties of Saturn. The Sixth Canon: concerning Luna and the properties thereof. The Seventh Canon: concerning the nature of Sol and its properties. Certain treatises and appendices arising out of the Seven Canons. God and Nature do nothing in vain. Note on Mercurius Vivus. What is to be thought concerning the congelation of Mercury. Concerning the Recipes of Alchemy. How to conjure the crystal so that all things may be seen in it. Concerning the heat of Mercury. What materials and instruments are required in Alchemy. The method of seeking Minerals. What Alchemy is.

THE BOOK CONCERNING THE TINCTURE OF THE PHILOSOPHERS, written against those Sophists born since the Deluge, in the life of our Lord Jesus Christ, the Son of God   19

> The Preface. Chapter I.: concerning the Arcanum and Quintessence. Chapter II.: concerning the definition of the Subject and Matter of the Tincture of the Philosophers. Chapter III.: concerning the Process of the Ancients for the Tincture of the Philosophers, and a more compendious method by Paracelsus. Chapter IV.: concerning the Process for the Tincture of the Philosophers, as it is shortened by Paracelsus. Chapter V.: concerning the conclusion of the Process of the Ancients, made by Paracelsus. Chapter VI.: concerning the Transmutation of Metals by the Perfection of Medicine. Chapter VII.: concerning the Renovation of Men.

THE GRADATIONS OF METALS   31

> Preface. The First Gradation. The Second Gradation. The Third Gradation. The Fourth Gradation. The Fifth Gradation. The Sixth Gradation. The Seventh Gradation. The Eighth Gradation. The Ninth Gradation. The Tenth Gradation. The Eleventh Gradation. The Twelfth Gradation. The Thirteenth Gradation. The Fourteenth Gradation.

	PAGE
THE TREASURE OF TREASURES FOR ALCHEMISTS	36

Concerning the Sulphur of Cinnabar. Concerning the Red Lion. Concerning the Green Lion.

CONCERNING THE TRANSMUTATIONS OF METALS AND OF CEMENTS — 41

Concerning the First or Royal Cement. Concerning the Second Cement. Concerning the Third Cement. The Fourth Cement. The Fifth Cement. The Sixth Cement. Conclusion.

THE AURORA OF THE PHILOSOPHERS, by Theophrastus Paracelsus, which he otherwise calls his Monarchia — 48

Chapter I. : concerning the Origin of the Philosophers' Stone. Chapter II. : wherein is declared that the Greeks drew a large part of their learning from the Egyptians; and how it came from them to us. Chapter III. : what was taught in the Schools of the Egyptians. Chapter IV. : what Magi the Chaldæans, Persians, and Egyptians were. Chapter V. : concerning the chief and supreme Essence of Things. Chapter VI. : concerning the chief errors as to its discovery and knowledge. Chapter VII. : concerning the errors of those who seek the Stone in Vegetables. Chapter VIII. : concerning those who have sought the Stone in Animals. Chapter IX. : concerning those who have sought the Stone in Minerals. Chapter X. : concerning those who have sought the Stone, and also Particulars, in Minerals. Chapter XI. : concerning the true and perfect special Arcanum of Arsenic for the White Tincture. Chapter XII. : General Instruction concerning the Arcanum of Vitriol and the Red Tincture to be extracted from it. Chapter XIII. : Special Instructions concerning the process of Vitriol for the Red Tincture. Chapter XIV. : concerning the Secrets and Arcana of Antimony, for the Red Tincture, with a view to Transmutation. Chapter XV. : concerning the Projection to be made by the Mystery and Arcanum of Antimony. Chapter XVI. : concerning the Universal Matter of the Philosophers' Stone. Chapter XVII. : concerning the Preparation of the Matter for the Philosophic Stone. Chapter XVIII. : concerning Instruments and the Philosophic Vessel. Chapter XIX. : concerning the Secret Fire of the Philosophers. Chapter XX. : concerning the Ferment of the Philosophers, and the Weight.

CONCERNING THE SPIRITS OF THE PLANETS — 72

Prologue. The First Treatise.—Chapter I. : concerning Simple Fire. Chapter II. : concerning the multiplicity of Fire from whence spring the varieties of Metals. Chapter III. : concerning the Spirit or Tincture of Sol. Chapter IV. : concerning the Spirit or Tincture of Luna. Chapter V. : concerning the Spirit of Venus. Chapter VI. : concerning the Spirit of Mars. Chapter VII. : concerning the Spirit of Jupiter. Chapter VIII. : concerning the Spirit of Saturn. Chapter IX. : concerning the gross Spirit of Mercury. The Second Treatise.—Concerning the Philosophers' Mercury, and the Medium of Tinctures. Chapter I. : from what Tinctures and Leavens are made. Chapter II. : concerning the Conjunction of the Man with the Woman. Chapter III. : concerning the Form of the Glass Instruments. Chapter IV. : concerning the Properties of Fire. Chapter V. : concerning the Signs which appear in the Union of Conjunction. Chapter VI. : concerning the Knowledge of the Perfect Tincture. Chapter VII. : concerning the Augmentation or the Multiplying of Tinctures. The Third Treatise.—Chapter I. : concerning the Building of the Furnace with the Fire. Chapter II. : concerning the Conjunction of the Man with the Woman. Chapter III. : concerning the Copulation of the Man with the Woman. Chapter IV. : concerning the Philosophic Coition of the Husband with his Wife. Chapter V. : concerning the Black Colour. Chapter VI. : concerning the Bud appearing in the Glass. Chapter VII. : concerning the Red Colour. Chapter VIII. : concerning Increase and Multiplication. Conclusion.

THE ECONOMY OF MINERALS, elsewhere called the Genealogy of Minerals — 89

Preface to the Reader. Chapter I. : concerning the Generations of Minerals. Chapter II. : concerning the Ultimate and Primal Matter of Minerals. Chapter III. : concerning the Field, the Roots, and the Trees of Minerals. Chapter IV. : concerning the Fruits and the Harvest of Minerals. Chapter V. : concerning the Death of the Elements, especially of Water. Chapter VI. : concerning the Death of the Tree of Minerals. Chapter VII. : concerning the variation of the Primal Matter of Minerals, in proportion to the different Species and Individuals thereof: also concerning the various colours, etc. Chapter VIII. : concerning the Natural Dispenser of Minerals, and his Ministers. Chapter IX. : concerning

	PAGE

the Virtues and Properties of Salts in Alchemy and in Medicine. Chapter X.: concerning Mumia. Chapter XI.: concerning Dry Salt. Chapter XII.: concerning Salt Nitre. Chapter XIII.: concerning the Ill Effects of Nutrimental Salt. Chapter XIV.: concerning Vitriol. Chapter XV.: concerning the Species of Vitriol and the Tests of it. Chapter XVI.: concerning the Virtues of Vitriol, crude or calcined, in Medicine. Chapter XVII.: concerning the Threefold Sulphur of Minerals. Chapter XVIII.: concerning Arsenic used for Alchemy. Chapter XIX.: concerning Quicksilver. Chapter XX.: concerning Cachimiæ and Imperfect Bodies. Conclusion.

THE COMPOSITION OF METALS — 114

CONCERNING THE NATURE OF THINGS.

    BOOK THE FIRST: concerning the Generation of Natural Things — 120

    BOOK THE SECOND: concerning the Growth of Natural Things — 128

    BOOK THE THIRD: concerning the Preservation of Natural Things — 130

    BOOK THE FOURTH: concerning the Life of Natural Things — 135

    BOOK THE FIFTH: concerning the Death of Natural Things — 138

    BOOK THE SIXTH: concerning the Resuscitation of Natural Things — 146

    BOOK THE SEVENTH: concerning the Transmutation of Natural Objects — 151

    BOOK THE EIGHTH: concerning the Separation of Natural Things — 160

        Concerning the Separation of Metals from their Minerals. Concerning the Separation of Minerals. Concerning the Separation of Vegetables. Concerning the Separation of Animals.

    BOOK THE NINTH: concerning the Signature of Natural Things — 171

        Concerning Monstrous Signs in Men. Concerning the Astral Signs in the Physiognomy of Man. Concerning the Astral Signs of Chiromancy. Concerning Mineral Signs. Concerning certain particular Signs of Natural and Supernatural Things

THE PARACELSIC METHOD OF EXTRACTING MERCURY FROM ALL THE METALS — 195

THE SULPHUR OF THE METALS — 197

THE CROCUS OF THE METALS, OR THE TINCTURE — 199

THE PHILOSOPHY OF THEOPHRASTUS CONCERNING THE GENERATIONS OF THE ELEMENTS.

    BOOK THE FIRST: concerning the Element of Air — 201

    BOOK THE SECOND: concerning the Element of Fire — 210

        Treatise the First: concerning the Separation of Air and Fire. Treatise the Second: concerning the Sun, Light, Darkness, and Night. Concerning Winds. Concerning the Temperate Stars. Concerning Nebulæ. Concerning Metals, Minerals, and Stones from the Upper Regions. Concerning Metals. Concerning Stones from Above. Concerning Crystals and Beryls.

    BOOK THE THIRD: concerning the Element of Earth — 226

    BOOK THE FOURTH: concerning the Element of Water, with its Fruits — 231

APPENDICES

APPENDIX I.: a Book about Minerals — 237

    Concerning Silver. Concerning Jove. Concerning Saturn. Concerning Iron and Steel. Concerning Venus. Note.—Of Mixed Metals. Concerning Spurious Metals. Concerning Zinc. Concerning Cobalt. Concerning Granates. Note concerning Gems. Concerning Quicksilver. Note concerning Cachimiæ, that is, the Three Imperfect Bodies. General Recapitulation concerning Generation. Of the Generation of Marcasites. Autograph Schedule by Paracelsus.

APPENDIX II.: concerning Salt and Substances comprehended under Salt — 257

    Correction and Addition on the Subject of a second time correcting and reducing Dry Salt.

APPENDIX III.: concerning Sulphur — 265

    Concerning the Kinds of Sulphur. Concerning Embryonated Sulphur. Concerning Mineral Sulphur. Concerning Metallic Sulphur, that is, Sulphurs prepared from the entire Metals. Concerning the Alchemical Virtues of Sulphur, and first concerning Embryonated Sulphur. Concerning Mineral Sulphur. Concerning the Use of Sulphur of the Metals in Alchemy.

APPENDIX IV.: the Mercuries of the Metals — 278

    A Little Book concerning the Mercuries of the Metals, by the Great Theophrastus Paracelsus, most excellent Philosopher and Doctor of both Faculties. Mercury of the Sun. Mercury of the Moon. Mercury out of Venus. Mercury out of Mars. Mercury of Jupiter. Mercury of Saturn.

APPENDIX V.: De Transmutationibus Metallorum — 283

    Concerning the Visible and Local Instruments: and first of all concerning the Spagyric Uterus. The Phœnix of the Philosophers. A Very Brief Process for attaining the Stone.

APPENDIX VI.: the Vatican Manuscript of Paracelsus. A Short Catechism of Alchemy — 288

APPENDIX VII.: the Manual of Paracelsus — 306

    The Work on Mercury for Luna and Sol. The Work of Sulphur. The Fixation of Spirits. A Cement of Part with Part. The Solution of Gold by Marcasites. A Great Secret. Method of calcining Mercury. Digestion of the Moon. For the White and Red. For Multiplication. Red Oil which fixes Luna and Sol. The Gradation of Luna. The Oil of the Philosophers. Corporal Mercury. Mercury of all the Metals. The Foundation of the Philosophers. Mercury of Saturn. Fixed Augment. Mercury of Jupiter. Mercury of the Moon. To convert Metals into Mercury. Augment in Luna. Mercury of Sol or Luna. Oil of Arcanum. Water of Mercury. Elixir at the White. Concerning Luna and Venus. Notable Elixir. Rubification. Sal Ammoniac. Sal Borax. Cinabrium. Factitious Corals. Pearls from Chalk. Ruby. Aqua Ardens. Calcination of Sol and Luna. Sublimation and Fixation of Sulphur. Oil of Vitriol. Sal Borax of the Philosophers. Fixation of Arsenic. Coagulation of Mercury. Glorious Oil of Sol. Lac Virginis. The Process of Sulphur. Operation for Sol. To make Precious Stones. Water which makes Luna into Sol. Fixation of Sulphur. How every Stone can be transmuted into a clear one. The Adepts' Fire. Sol produced with Pars cum Parte. Concerning Cements. Method of making Luna. Water of Luna. True Albatio. Rubification of Mercury. Oil of Mercury and the Sun. Quintessence of Luna, etc. Fixation of Orpiment. Spirits of Water. Augmentation for Sol. Fixed Luna. Secret Philosophical Water. The Hermetic Bird. Attinkar of Venus. Cement Regal. Philosophic Water.

PREFACE TO THE ENGLISH TRANSLATION

THERE are many respects in which Paracelsus at the present day seems to be little more than a name. Even among professed mystics the knowledge concerning him, very meagre and very indefinite, is knowledge that has been obtained at second hand, in most cases from Eliphas Levi, who in his *Dogme et Rituel de la Haute Magie*, and again in his *Histoire de la Magie*, has delivered an intuitive judgment upon the German "Monarch of Arcana," expressed epigramatically, after the best manner of a Frenchman.* But, whencesoever derived, the knowledge is thin and phantasmal. Paracelsus is indeed cited as an authority in occult science, as a great alchemist, a great magician, a great doctor; he is somehow supposed to be standing evidence of the "wisdom of a spoliated past," and to offer a peculiar instance of malignity on the part of the enemies of Hermetic philosophy, because such persons have presumed to pronounce him an impostor. Thus there is a very strong opinion concerning him, which occultists and mystics of all schools have derived from a species of mystical tradition, and this represents one side of modern thought concerning him. It is not altogether a satisfactory side, because it is not obtained at first hand. In this respect, however, it may compare, without suffering by comparison, with the alternative opinion which

---

* The cure of Paracelsus were miraculous, and he deserved that there should be added to his name of Philippus Theophrastus Bombast that of Aureolus Paracelsus, with the addition of the epithet of divine.—*Dogme de la Haute Magie*, c. 1. Paracelsus, that reformer in magic, who has surpassed all other initiates by his unassisted practical success.—*Ibid.*, c. 5. Paracelsus, the most sublime of the Christian magi.—*Ibid.*, c. 16. Paracelsus was a man of inspiration and of miracles, but he exhausted his life with his devouring activity, or, rather, he rapidly outwore and destroyed its vestment: for men like Paracelsus can both use and abuse fearlessly; they well know that they can no more die than grow old here below.—*Rituel de la Haute Magie*, c. 2. Paracelsus was naturally aggressive and combative; his familiar, he said, was concealed in the pommel of his great sword, which he never put aside. His life was incessant warfare; he travelled, he disputed, he wrote, he taught. He was more attentive to physical results than to moral conquests; so he was the first of practical magicians and the last of wise adepts. His philosophy was wholly sagacity, and he himself called it *Philosophia Sagax*. He has divined more than anyone without ever completely understanding anything. There is nothing to equal his intuitions unless it be the temerity of his commentaries. He was a man of bold experiences; he was drunk of his opinions and his talk; he even got drunk otherwise, if we are to believe his biographers. The writings which he has left behind him are precious for science, but they must be read with caution; he may be called the divine Paracelsus, if the term be understood in the sense of a diviner; he is an oracle, but not invariably a true master. He is great as a physician above all, for he had discovered the Universal Medicine; yet he could not preserve his own life, and he died while still young, worn out by his toil and excesses, leaving a name of fantastic and doubtful glory, based on discoveries by which his contemporaries did not profit. He died without having uttered his last message, and he is one of those mysterious personages of whom one may affirm, as of Enoch and S. John: he is not dead, and he will revisit the earth before the last day.—*Histoire de la Magie*, Liv. V., c. 5. His success was prodigious, and never has any physician approached Paracelsus in the multitude of his marvellous cures.—*Dogme de la Haute Magie*, c. 16.

obtains among non-mystics, namely, that Paracelsus was a great charlatan, though at the same time it is true that he was a great physician, at least for the period in which he lived. This judgment as little, perhaps less than the other, is derived from any solid knowledge concerning the man or his writings.* At the same time it is noticeable that even hearsay condemnations admit that Paracelsus performed notable cures. How it comes about that the application of what would be termed a distracted theory both in medicine and physics enabled its inventor to astound his age by what seemed miracles of the healing art would be a crux for such criticism if the criticism knew anything about it. It is not a crux for the mystics, because by these it would be replied that Paracelsus was a veritable adept, that his Hermetic teachings require to be interpreted, and that the key to their meaning would lay open for those who possess it an abundant treasure of sapience to which the literal significance is only a *bizarre* veil. Between these views it is unnecessary to make choice here. It is proposed to enable those who are interested in either to judge this matter for themselves by placing completely before them, for the first time, and in an English dress, the Hermetic writings of Paracelsus. It is proposed, also, by way of a brief introduction, to notify a few facts in connection with the life of the author, which may be useful at the beginning of an inquiry.

There are, however, many debateable points in connection with the life of Paracelsus to which a reference in this place scarcely requires to be made. What proportion of his long designation belonged to him by birth or baptism, to what countries he actually extended his travels during incessant wanderings which terminated only with his life, under what circumstances he died and what was the precise manner of his death, all these are points about which there is considerable uncertainty, and they are at this day not likely to be settled. Theophrastus and Bombast seem to have been assumed names, to one of which an unfortunate, and in some respects an undeserved, significance has been since attracted. The surname of Paracelsus was conferred by his father in alchemy, and it signified that he was greater than Celsus, the physician of ancient fame. To the style of Hohenheim it is believed that he had only a doubtful right. His alternative designation of Eremite suggests the monastic state, but the reference is simply to his birthplace, Maria Einsiedeln, or Notre Dame des Eremites, a short distance from Zurich. He appears to have been christened Philippus Aureolus, and in his writings he indifferently

---

* M. Louis Figuier, the French scientist, who otherwise might perhaps be regarded as exhibiting more than Gallic accuracy, may be cited in this connection. Referring to the fact that Paracelsus has laid some stress upon an opinion not uncommon among alchemists, namely, that astrology and magic are collaterally a help to the seeker after the Great Work, he goes on to affirm that the writings of Paracelsus are filled with foolish invocations to the invisible world, while, as a fact, there is not a single treatise comprised in the great Geneva folio, nor is there any other extant work attributed to Paracelsus, and known to the present editor, which contains any invocations at all. M. Louis Figuier subsequently states, apparently on the sole authority of his intuition as a Frenchman and a man of parts, that the fuliginous Swiss physician enjoys only a contested authority among alchemists, which is only partially true; and adds that he was a theoretical writer who did not apply himself manually to the accomplishment of the *Magnum Opus*, which, so far as it is possible to judge, is not true at all.

describes himself as a Helvetian and a German. He was born in the year 1493, following the tradition which is most generally accepted, but other dates have been indicated, the earliest being 1490. According to one account he was descended from the ancient and honourable family of Bombast, which had abode during many generations at the castle of Hohenheim, near Stuttgart, Würtemberg, but this is most probably romancing. His father was a physician of repute, who is said to have been in possession of a large collection of curious books, and has also been described as a grand master of the Teutonic order, but the precise meaning attaching to this high-sounding dignity is uncertain and the authority is suspicious. His mother is variously identified as the matron of a hospital and "superintendent of the university of Einsiedeln." Paracelsus is reputed to have been their only child, born one year after marriage, but it has also been hinted that his parents were not married, and that the great master of Hermetic medicine was a natural son. He is said also to have been emasculated in his childhood, but there is reason to suppose that this was merely a hypothetical explanation to account for his beardless and somewhat feminine appearance, and for that hatred of women which seems to have been one of his social characteristics, and can be traced indirectly, but with sufficient distinctness, in his writings.* About 1502 the family removed to Carinthia, and there the father continued to practise medicine till his death in 1534. From him Paracelsus is supposed to have received the first rudiments of education, and he entered the university of Basle at the age of sixteen, when he betook himself to the study of alchemy, surgery, and medicine. To the first of these sciences he had previously had some introduction through the works of Isaac the Hollander, which are said to have inflamed him with the ambition of curing diseases by medicine superior to the *materia* at that time in use. It was from the same source that he derived the cardinal principle which is enunciated everywhere in his writings, namely, that salt, sulphur, and mercury are the three elementary constituents of all substances. This doctrine, however, by no means originated with the first alchemist of Holland, and Isaac himself was a follower of Geber, Morien, and Arnold.† The actual initiation of Paracelsus into the mysteries of alchemy is, however, supposed to have been attained under the

---

* So free was Paracelsus of all amourous weaknesses, that he made even his sex seem doubtful.—*Dogme de la Haute Magie*, c. 11.

† Contemporary with Basilius Valentinus were Isaac the Hollander and his son, who are supposed to have worked with success. They were the first alchemists of Holland, and their operations were highly esteemed by Paracelsus, Boyle, and Kunckel. In practical chemistry they followed the traditions of Geber, and their alchemical experiments are the most plain and explicit in the whole range of Hermetic literature. They worked principally in metals, describing minutely the particulars of every process. Their lives are almost unknown. . . . They are placed in the fifteenth century by conjecture, from the fact that they do not cite any philosophers subsequent to that period. They speak of Geber, Dastin, Morien, and Arnold, but not of more modern authorities, while, on the other hand, their references to aquafortis and aqua regia, which were discovered in the fourteenth century, prevent us from assigning their labours to an anterior epoch. The two Isaacs were particularly skilful in the manufacture of enamels and artificial gem-stones. They taught that the Grand Magisterium could convert a million times its own weight into gold, and declared that any person taking weekly a small portion of the philosophical stone will be ever preserved in perfect health, and his life will be prolonged to the very last hour which God has assigned to him.—*Lives of Alchemystical Philosophers*.

tuition of the Abbot Trithemius,* who is regarded as an adept of a high order, and had been previously the instructor of the more celebrated, though less illustrious, Henry Cornelius Agrippa.† From this mysterious ecclesiastic, who at the present day, in so far as he is remembered at all, is best known by his treatises on cryptic writing, he is supposed to have acquired "the Kabbalah of the spiritual, astral, and material worlds." About 1516 he is still found at Basle pursuing his researches in mineralogy, medicine, surgery, and chemistry, under the guidance of Sigismund Fugger, a wealthy physician of that city. Subsequently, having got into some trouble with the authorities, he fled, and commenced his nomadic life, which an apparently inaccurate tradition represents to have begun at the age of twenty years. Though his father was still alive he appears to have been without any means of subsistence, and supported himself, like many distressed students of that period, by psalm-singing, astrological predictions, chiromantic soothsaying, and, it is even said, by necromantic practices. He wandered through Germany, Hungary, Italy, France, the Netherlands, Denmark, Sweden, and Russia. In the last mentioned country, if it be true that he ever reached it, he is reported to have been made prisoner by the Tartars, to have been brought before "the Great Cham," to have become a favourite at the court of that potentate, and to have accompanied his son on an embassy from China to Constantinople. In spite of the tuition of Trithemius, he had apparently something to learn, and that was nothing less than "the supreme secret of alchemistry," the "universal dissolvent," the Azoth, alcahest, or sophic fire. This was imparted to him by a generous Arabian, about whom no other particulars are forthcoming. It is easy to see that the greater part of this nomadic legend is purely fabulous, and so also, in all probability, is his subsequent journey to India and Egypt. It is not an unusual device to account for obscure periods in the lives of Hermetic philosophers by extensive eastern travellings. However this may be, Paracelsus ultimately returned to Europe, and passed along the Danube into Italy, where he appears as an army surgeon, and where also his wonderful cures began. He is said to have re-entered Germany in 1526, at the age of thirty-two, and if this be accepted the date 1516, when he is supposed to have been at work with Sigismund Fugger, will be found approximately correct. It is to the period immediately succeeding his return that most of his biographers assign his induction into a professorship of physics, medicine, and surgery, at the university he entered

---

* Trithemius was a monk of the Benedictine order, who began life as a mendicant child setting forth in search of knowledge. He was received into a convent at Trèves, where he made astounding progress in his studies. Having exhausted the possibilities of his teachers, he betook himself to Louvain, thence to Heidelberg, and subsequently to Mayence. He became familiar with oriental languages, pagan and Christian philosophy, astronomy, and alchemy. He was a theologian, a poet, an astronomer, and a necromancer. He took monastic vows in 1482, and in the year following he became the abbot of a convent at Spanheim, which he transformed speedily into a sanctuary of art and the sciences. He subsequently was made superior of an abbey at Wurzbourg, and there it would appear he remained till his death in the year 1516. His works are chiefly historical, but, as above indicated, there are some upon secret writing which are exceedingly curious, and one, *Chronologia Mystica*, is of a magical character.

† Agrippa, who seems to have divided with Paracelsus the reputation of the Trismegistus of his time, was born in 1486 and died in 1535.

as a youth. This was a position of some importance, and it was offered him at the instance of Erasmus and Ecolampidus. "There, in his lectures, he professed internal medicine, denounced the antiquated systems of Galen and other authorities, and began his instruction by burning the works of these masters in a brass pan with sulphur and nitre. He created innumerable enemies by his arrogance and his innovations, but the value of his mineral medicines was proved by the cures which he performed.* These cures only increased the hatred of his persecutors, and Paracelsus, with characteristic defiance, invited the faculty to a lecture, in which he promised to teach the greatest secret of medicine. He began by uncovering a dish which contained excrement. The doctors, indignant at the insult, departed precipitately, Paracelsus shouting after them: 'If you will not hear the mysteries of putrefactive fermentation, you are unworthy of the name of physicians.'" It will be easily understood that the Hermetic doctor did not long retain his professorship at Basle. He came into conflict with the municipal authorities, and a second time he was forced to flee the place. He betook himself once more to a wandering mode of life. In 1528 he proceeded to Colmar; in 1530 he is found at Nuremburg, in embroilment, as usual, with the medical faculty, by whom he was denounced as an impostor, but the tables were turned on his opponents after his successful treatment of several aggravated cases of elephantiasis. For the ten years succeeding this date there are no certain records of his movements; he commonly lodged at inns and other public places, still performing cures which were astonishing for the period, and, according to the accusations of his enemies, also drinking to excess.† The testimony of Oporinus on this point is very clear, though it has been indignantly repudiated by some of his later defenders. In 1541 Paracelsus was invited by Archbishop Ernst to settle at Salzburg, and there, according to one account, he died on September 24 of the same year, but the manner of his death, like that of his birth, has been the subject of contradictory recitals.‡ By an alternative statement it occurred on a bench at the kitchen fire in a Strasburg hostelry. One writer supposes the event to have been accelerated by a scuffle with assassins in the pay of the orthodox medical faculty.

There can be no doubt that Paracelsus obtained a wide, though not altogether a happy, reputation during the brief period of his turbulent life, and there is also no doubt that this was immeasurably increased after death.

---

* Paracelsus, who was the first who made known zinc, has obtained an immense and deserved reputation by introducing into medicine the use of chemical compounds furnished by metals. To the old therapeutics of the Galenists, abounding in complicated and often inoperative preparations, he substituted the simple medicaments furnished by chemistry, and was the first to open the audacious path to the application of this science to human physiology and pathology.—Louis Figuier, *L'Alchimie et les Alchimistes, troisième édition*, pp. 99, 100.

† Marvellous Paracelsus, always drunk and always lucid, like the heroes of Rabelais.—*Dogme de la Haute Magie*, Introduction.

‡ He proceeded to Maehren, Kaernthen, Krain, and Hungary, and finally landed in Salzburg, to which place he was invited by the Prince Palatine, Duke Ernst of Bavaria, who was a great lover of the secret arts. In that place Paracelsus obtained at last the fruits of his long labours and of a widespread fame. But he was not destined to enjoy a long time the rest he so richly deserved. . . . He died, after a short sickness (at the age of forty-eight years), in a small room of the White Horse Inn, near the quay, and his body was buried in the graveyard of St. Sebastian.—Hartmann's *Paracelsus*.

It is in no sense inexact to affirm that he founded a new school both in medicine and in alchemy. The commentaries on his medical system became a literature which, in extent, at least, is formidable; out of the mystic physics of his alchemical teachings the Rosicrucian doctrines developed in the first part of the following century. The works of Benedictus Figulus are evidence that he was idolized by his disciples. He was termed the noble and beloved monarch, the German Hermes, the Philosopher Trismegistus, our dear preceptor and King of Arts, Theophrastus of blessed memory and immortal fame. The collection of his genuine writings was made with devout care, and as a consequence of his celebrity many fictitious treatises were in due course ascribed to him.* Students attracted by his doctrines travelled far in search of like-minded persons to compare observations thereon, and to sift the mystery of his instruction. In the course of these inquiries it seems to have become evident, from the experience of his followers, that his prescriptions in many cases were not to be literally understood, even when they were apparently the ordinary formulæ and concerned with the known *materia* of medicine. It will scarcely be necessary to add that in things alchemical the letter of his teachings was found still more in need of interpretation. The very curious influence exercised by Paracelsus for something like two hundred years over certain sections of restless experiment and speculation is still unwritten, and it would be interesting to trace here, were it possible within the limits of a preface. A task so ambitious is, however, outside those limits, and will perhaps be more wisely surrendered to other hands, for it is, in the main, part of the history of medicine, and demands an expert in the medical literature and medical knowledge of the past. The translations which follow are concerned only with the Hermetic writings of Paracelsus, to the exclusion of many formidable treatises on surgical science, and on the causes and cure of disease. They comprise what Paracelsus would himself have comprised in a collection of his alchemical writings, and this in itself is much more than is ordinarily understood to be within the significance of the term. With Paracelsus the province of alchemy was not limited to the transmutation of metals. It was, broadly speaking, the development of hidden possibilities or virtues in any substance, whether by God, or man, or

---

* More especially, dear friends, have we to complain of the devilish cunning way in which the works of Theophrastus have hitherto been suppressed, only a few of which (and those to be reckoned the very worst) having appeared in print. For although they have been collected together from all countries in which Theophrastus has lived and travelled—the books he has written in Astronomy, Philosophy, Chemistry, Cabala, and Theology, numbering some thousand volumes—yet the same has only been done from avarice to get riches. For, having been trafficked in and sold for great sums, they have become scattered among the courts of princes and nobles, while Christendom at large, for whose use and benefit Theophrastus wrote, has no part in them. Particularly his theological works (because they annihilate the godless, and do not suit children of this world—belly-servers, deceived by the devil) have hitherto been totally suppressed. . . . But, at the Last Day, I, together with all true Sons of the Doctrine, shall demand an account of them for having . . . shut Truth away in boxes, walls, and vaults, and behind locks and bolts. Now, these precious and revered writings were ordered by God in our latter times, through Theophrastus, for the use and weal of the whole of Christendom. As regards our dear, highly favoured Monarch and Preceptor, . . . we, for our part, will not suppress his Life, his well-merited praise, . . . given him by God, the Angels, and the whole Firmament, but will heartily defend his honour and teaching to the very end of our life.—Benedictus Figulus, *A Golden and Blessed Casket of Nature's Marvels.*

Nature. Thus it included the philosophy of creation, and dealt with the first matter as developed into the universe by Divine Power. It included also the natural evolution which takes place round us, whether in the formation of metals within the earth, or the formation of animals in the matrix. Finally, it included the development by man's skill and art of whatsoever was capable of improvement in the products of Nature. Thus the Hermetic and Alchemical writings of Paracelsus have a wider scope than might at first be inferred from the title. The purpose of this translation is altogether of an unpretentious kind. It aims at providing, and that for the first time, a complete and faithful text of all that Paracelsus is known or supposed to have written on the subject of alchemy and Hermetic medicine. It does not attempt to distinguish between the works which have been falsely attributed to him; concerning this question there are no satisfactory canons of criticism, for those which have been indicated by the old author of the *Onomastikon* are of an arbitrary and unpractical kind. A careful reader will probably regard with suspicion the "Aurora of the Philosophers," at least in its present state, and he will possibly reject altogether the treatise "Concerning the Spirits of the Planets." There is nothing else in these volumes, except the uncertain "Manual," which from internal evidence is unlikely to have been the work of Paracelsus, and it is unnecessary to enter into the question which has been raised by some of his biographers as to his employment of scribes under him, who reduced his dictations to writing and have possibly maltreated their master. The text which has been adopted for translation is the Geneva folio, in four volumes, 1658, in Latin. The *bizarre* mixture of Latin and old German in which Paracelsus originally wrote presents many difficulties with which it is unnecessary to grapple, as the Latin collected edition appears to represent in a very satisfactory manner both the letter and the spirit of the originals.

It seems also desirable to state that a comparison of the medical and chemical knowledge possessed by Paracelsus with the chemistry and medicine of to-day is outside the purpose of this edition, because it is outside possibility within the limits of two volumes. There is no doubt that it would be an interesting as well as an important task to establish the exact position of Paracelsus, not only as regards modern knowledge, but as regards the science which preceded him, and it is hoped that such a work will be ultimately performed. Should this translation be regarded as final by students, or at least as a satisfactory foundation for a full and complete comprehension of the sage of Hohenheim, and should the encouragement which is indispensable to an undertaking so long and costly be adequately given, it is proposed, after a reasonable interval, that these two volumes of uncriticised text should be followed by one other which will contain all the materials requisite for understanding Paracelsus, and will further trace, methodically and historically, the development of alchemical symbolism, with the growth of chemical knowledge from the Byzantine period to the dawn of the Reformation. It is anticipated that this inquiry will fix for the first time the true objects of Hermetic physics, and the place which must be assigned to Paracelsus in connection therewith. The less ambitious but indispensable preliminary of this inquiry has been alone attempted here, and the simple provision of a text, as intelligible as the circumstances will allow, has been truly no light undertaking, nor should it be regarded as the exclusive work of one hand. The editor has accomplished his task with the collaboration of other specialists, and is responsible only for certain portions of the actual translation, and for its general revision and collation. The work, as it stands, consists of (*a*) the large body of literature, entire and unabridged, attributed to Paracelsus, and treating directly of alchemy, and the transcendental doctrines and physics of the *Magnum Opus* ; (*b*) The whole Paracelsian literature of the Great Elixir and the Universal Medicine ; (*c*) So much of the Hermetic philosophy and cosmogony of Paracelsus as has been judged necessary to illustrate his alchemical teachings ; (*d*) One important treatise illustrating the application by Paracelsus of metallic and mineral substances to the treatment of diseases; (*e*) An exhaustive collection of alchemical references scattered through the chirurgical works of Paracelsus. Thus, the present edition is practically inclusive of everything except the exoteric medicine of Paracelsus, which, it is thought, is of inferior importance to the modern student.

PART I.

HERMETIC CHEMISTRY.

THE CŒLUM PHILOSOPHORUM,

OR BOOK OF VEXATIONS;

By PHILIPPUS THEOPHRASTUS PARACELSUS

The Science and Nature of Alchemy, and what Opinion should be formed thereof

Regulated by the Seven Rules or Fundamental Canons according to the seven commonly known Metals; and containing a Preface with certain Treatises and Appendices

THE PREFACE

OF THEOPHRASTUS PARACELSUS TO ALL ALCHEMISTS AND READERS OF THIS BOOK

YOU who are skilled in Alchemy, and as many others as promise yourselves great riches or chiefly desire to make gold and silver, which Alchemy in different ways promises and teaches; equally, too, you who willingly undergo toil and vexations, and wish not to be freed from them, until you have attained your rewards, and the fulfilment of the promises made to you; experience teaches this every day, that out of thousands of you not even one accomplishes his desire. Is this a failure of Nature or of Art? I say, no; but it is rather the fault of fate, or of the unskilfulness of the operator.

Since, therefore, the characters of the signs, of the stars and planets of heaven, together with the other names, inverted words, receipts, materials, and instruments are thoroughly well known to such as are acquainted with this art, it would be altogether superfluous to recur to these same subjects in the present book, although the use of such signs, names, and characters at the proper time is by no means without advantage.

But herein will be noticed another way of treating Alchemy different from the previous method, and deduced by Seven Canons from the sevenfold series of the metals. This, indeed, will not give scope for a pompous parade of words, but, nevertheless, in the consideration of those Canons everything which should be separated from Alchemy will be treated at sufficient length, and, moreover, many secrets of other things are herein contained. Hence, too, result certain marvellous speculations and new operations which frequently differ from the writings and opinions of ancient operators and natural philosophers, but have been discovered and confirmed by full proof and experimentation.

Moreover, in this Art nothing is more true than this, though it be little known and gains small confidence. All the fault and cause of difficulty in Alchemy, whereby very many persons are reduced to poverty, and others labour in vain, is wholly and solely lack of skill in the operator, and the defect or excess of materials, whether in quantity or quality, whence it ensues that, in the course of operation, things are wasted or reduced to nothing. If the true process shall have been found, the substance itself while transmuting approaches daily more and more towards perfection. The straight road is easy, but it is found by very few.

Sometimes it may happen that a speculative artist may, by his own eccentricity, think out for himself some new method in Alchemy, be the consequence anything or nothing. He need do nought in order to reduce something into nothing, and again bring back something out of nothing. Yet this proverb of the incredulous is not wholly false. Destruction perfects that which is good; for the good cannot appear on account of that which conceals it. The good is least good whilst it is thus concealed. The concealment must be removed that so the good may be able freely to appear in its own brightness. For example, the mountain, the sand, the earth, or the stone in which a metal has grown is such a concealment. Each one of the visible metals is a concealment of the other six metals.

By the element of fire all that is imperfect is destroyed and taken away, as, for instance, the five metals, Mercury, Jupiter, Mars, Venus, and Saturn.* On the other hand, the perfect metals, Sol and Luna, are not consumed in that same fire. They remain in the fire: and at the same time, out of the other imperfect ones which are destroyed, they assume their own body and become visible to the eyes. How, and by what method, this comes about can be gathered from the Seven Canons. Hence it may be learnt what are the nature and property of each metal, what it effects with the other metals, and what are its powers in commixture with them.

But this should be noted in the very first place: that these Seven Canons cannot be perfectly understood by every cursory reader at a first glance or a single reading. An inferior intelligence does not easily perceive occult and abstruse subjects. Each one of these Canons demands no slight discussion. Many persons, puffed up with pride, fancy they can easily comprehend all which this book comprises. Thus they set down its contents as useless and futile, thinking they have something far better of their own, and that therefore they can afford to despise what is here contained.

---

* The three prime substances are proved only by fire, which manifests them pure, naked, clean, and simple. In the absence of all ordeal by fire, there is no proving of a substance possible. For fire tests everything, and when the impure matter is separated the three pure substances are displayed.—*De Origine Morborum ex Tribus Primis Substantiis-Paramirum*, Lib. I., c. 1. Fire separates that which is constant or fixed from that which is fugitive or volatile.—*De Morbis Metallicis*, Lib. II., Tract I. Fire is the father or active principle of separation.—"Third Fragment on Tartar," from the *Fragmenta Medica*.

THE CŒLUM PHILOSOPHORUM

PART I
THE SEVEN CANONS OF THE METALS

THE FIRST CANON
Concerning the Nature and Properties of Mercury *

ALL things are concealed in all. One of them all is the concealer of the rest—their corporeal vessel, external, visible, and movable. All liquefactions are manifested in that vessel. For the vessel is a living and corporeal spirit, and so all coagulations or congelations enclosed in it, when prevented from flowing and surrounded, are not therewith content. No name can be found for this liquefaction, by which it may be designated; still less can it be found for its origin. And since no heat is so strong as to be equalised therewith, it should be compared to the fire of Gehenna. A liquefaction of this kind has no sort of connection with others made by the heat of natural fire, or congelated or coagulated by natural cold. These congelations, through their weakness, are unable to obtain in Mercury, and therefore, on that account, he altogether contemns them. Hence one may gather that elementary powers, in their process of destruction, can add nothing to, nor take away anything from, celestial powers (which are called Quintessence or its elements), nor have they any capacity for operating. Celestial and infernal powers do not obey the four elements, whether they be dry, moist, hot, or cold. No one of them has the faculty of acting against a Quintessence; but each one contains within itself its own powers and means of action.†

---

* By the mediation of Vulcan, or fire, any metal can be generated from Mercury. At the same time, Mercury is imperfect as a metal; it is semi-generated and wanting in coagulation, which is the end of all metals. Up to the half-way point of their generation all metals are Mercury. Gold, for example, is Mercury; but it loses the Mercurial nature by coagulation, and although the properties of Mercury are present in it, they are dead, for their vitality is destroyed by coagulation.—*De Morbis Metallicis*, Lib. III., Tract II., c. 2. The essences and arcanas which are latent in all the six metals are to be found in the substance of Mercury.—*Ibid.*, c. 3. There are two genera of Mercury, the fixed Mercury of earth and another kind which descends from the daily constellation.—*Ibid.*, Lib. I., Tract II., c. 4. As there is a red and white Sulphur of Marcasites, a yellow, red, and black Sulphur of Talc, a purple and black Sulphur of the Cachimiæ, a Sulphur of Cinnabar, and, in like manner, of marble, amethyst, etc., so is there a special Mercury of Copper, Plumbago, Zinc, Arsenic, etc.—*Ibid.* Mercury is not Quicksilver, for Mercury is dead, while Quicksilver is living.—*De Hydropisi.*

† Nothing of true value is located in the body of a substance, but in the virtue. And this is the principle of the Quintessence, which reduces, say, 20lbs. into a single ounce, and that ounce far exceeds the entire 20lbs. in potency. Hence the less there is of body, the more in proportion is the virtue.—*De Origine Morborum Invisibilium*, Lib. IV.

THE SECOND CANON

Concerning the Nature and Properties of Jupiter

In that which is manifest (that is to say, the body of Jupiter) the other six corporeal metals are spiritually concealed, but one more deeply and more tenaciously than another. Jupiter has nothing of a Quintessence in his composition, but is of the nature of the four elementaries. On this account his liquefaction is brought about by the application of a moderate fire, and, in like manner, he is coagulated by moderate cold. He has affinity with the liquefactions of all the other metals. For the more like he is to some other nature, the more easily he is united thereto by conjunction. For the operation of those nearly allied is easier and more natural than of those which are remote. The remote body does not press upon the other. At the same time, it is not feared, though it may be very powerful. Hence it happens that men do not aspire to the superior orders of creation, because they are far distant from them, and do not see their glory. In like manner, they do not much fear those of an inferior order, because they are remote, and none of the living knows their condition or has experienced the misery of their punishment. For this cause an infernal spirit is accounted as nothing. For more remote objects are on that account held more cheaply and occupy a lower place, since according to the propriety of its position each object turns out better, or is transmuted. This can be proved by various examples.

The more remote, therefore, Jupiter is found to be from Mars and Venus, and the nearer Sol and Luna, the more "goldness" or "silveriness," if I may so say, it contains in its body, and the greater, stronger, more visible, more tangible, more amiable, more acceptable, more distinguished, and more true it is found than in some remote body. Again, the more remote a thing is, of the less account is it esteemed in all the respects aforesaid, since what is present is always preferred before what is absent. In proportion as the nearer is clear the more remote is occult. This, therefore, is a point which you, as an Alchemist, must seriouly debate with yourself, how you can relegate Jupiter to a remote and abstruse place, which Sol and Luna occupy, and how, in turn, you can summon Sol and Luna from remote positions to a near place, where Jupiter is corporeally posited; so that, in the same way, Sol and Luna also may really be present there corporeally before your eyes. For the transmutation of metals from imperfection to perfection there are several practical receipts. Mix the one with the other. Then again separate the one pure from the other. This is nothing else but the process of permutation, set in order by perfect alchemical labour. Note that Jupiter has much gold and not a little silver. Let Saturn and Luna be imposed on him, and of the rest Luna will be augmented.*

---

* Tin, or Jupiter, is pure Mercury coagulated with a small quantity of Salt, but combined with a larger proportion of white Sulphur. It derives its colours, white, yellow, or red, from its Mercury. Its sublimation is also by Mercury, and its resolution by Salt, and it is sublimed and resolved by these.—*De Elemento Aquæ*, Tract III., c. 6.

THE THIRD CANON
Concerning Mars and His Properties

The six occult metals have expelled the seventh from them, and have made it corporeal, leaving it little efficacy, and imposing on it great hardness and weight. This being the case, they have shaken off all their own strength of coagulation and hardness, which they manifest in this other body. On the contrary, they have retained in themselves their colour and liquefaction, together with their nobility. It is very difficult and laborious for a prince or a king to be produced out of an unfit and common man. But Mars acquires dominion with strong and pugnacious hand, and seizes on the position of king. He should, however, be on his guard against snares, that he be not led captive suddenly and unexpectedly. It must also be considered by what method Mars may be able to take the place of king, and Sol and Luna, with Saturn, hold the place of Mars.*

THE FOURTH CANON
Concerning Venus and Its Properties

The other six metals have rendered Venus an extrinsical body by means of all their colour and method of liquefaction. It may be necessary, in order to understand this, that we should show, by some examples, how a manifest thing may be rendered occult, and an occult thing rendered materially manifest by means of fire. Whatever is combustible can be naturally transmuted by fire from one form into another, namely, into lime, soot, ashes, glass, colours, stones, and earth. This last can again be reduced to many new metallic bodies. If a metal, too, be burnt, or rendered fragile by old rust, it can again acquire malleability by applications of fire.†

THE FIFTH CANON
Concerning the Nature and Properties of Saturn

Of his own nature Saturn speaks thus: The other six have cast me out as their examiner. They have thrust me forth from them and from a spiritual place. They have also added a corruptible body as a place of abode, so that I may be what they neither are nor desire to become. My six brothers are spiritual, and thence it ensues that so often as I am put in the fire they penetrate my body and, together with me, perish in the fire, Sol and Luna

---

* In the generation of Iron there is a larger proportion of Salt and Mercury, while the red Sulphur from which copper proceeds is present in a smaller quantity. It contains also a cuprine salt, but not in equal proportion with Mercury. Its constituents are its own body, which preponderates; then comes Salt, afterwards Mercury, and, lastly, Sulphur. When there is more Salt than the composition of Sulphur requires, the metal can in no wise be made, for it depends upon an equal weight of each. For fluxibility proceeds from Mercury and coagulation from Salt. Accordingly, if there be too much Salt it becomes too hard.—*De Elemento Aquæ*, Lib. IV., Tract III., c. 4.

† Venus is the first metal generated by the Archeus of Nature from the three prime principles after the marcasites and cachimiæ have been separated from these. It is formed of the gross redness which is purged off from the primal Sulphur, of the light red expelled in like manner from the Mercury, and of the deep yellow separated in the purification of the prime Salt by this same Archeus.—*Ibid.*, c. 3.

excepted. These are purified and ennobled in my water. My spirit is a water softening the rigid and congelated bodies of my brothers. Yet my body is inclined to the earth. Whatever is received into me becomes conformed thereto, and by means of us is converted into one body. It would be of little use to the world if it should learn, or at least believe, what lies hid in me, and what I am able to effect. It would be more profitable it should ascertain what I am able to do with myself. Deserting all the methods of the Alchemists, it would then use only that which is in me and can be done by me. The stone of cold is in me. This is a water by means of which I make the spirits of the six metals congeal into the essence of the seventh, and this is to promote Sol with Luna.*

Two kinds of Antimony are found : one the common black, by which Sol is purified when liquefied therein. This has the closest affinity with Saturn. The other kind is the white, which is also called Magnesia and Bismuth. It has great affinity with Jupiter, and when mixed with the other Antimony it augments Luna.

THE SIXTH CANON
Concerning Luna and the Properties thereof

The endeavour to make Saturn or Mars out of Luna involves no lighter or easier work than to make Luna, with great gain, out of Mercury, Jupiter, Mars, Venus, or Saturn. It is not useful to transmute what is perfect into what is imperfect, but the latter into the former. Nevertheless, it is well to know what is the material of Luna, or whence it proceeds. Whoever is not able to consider or find this out will neither be able to make Luna. It will be asked, What is Luna? It is among the seven metals which are spiritually concealed, itself the seventh, external, corporeal, and material. For this seventh always contains the six metals spiritually hidden in itself. And the six spiritual metals do not exist without one external and material metal. So also no corporeal metal can have place or essence without those six spiritual ones. The seven corporeal metals mix easily by means of liquefaction, but this mixture is not useful for making Sol or Luna. For in that mixture each metal remains in its own nature, or fixed in the fire, or flies from it. For example, mix, in any way you can, Mercury, Jupiter, Saturn, Mars, Venus, Sol, and Luna. It will not thence result that Sol and Luna will so change the other five that, by the agency of Sol and Luna, these will become Sol and Luna. For though all be liquefied into a single mass, nevertheless each remains in its nature whatever it is. This is the judgment which must be passed on corporeal mixture. But concerning spiritual mixture and

---

* Lead is the blackness of the three first principles, which, however, is by no means a superfluity, but a peculiar metallic nature in them existing. For all metals are latent in Mercury, and they are all only Mercury. The same is to be concluded concerning Salt and Sulphur. Thus, as copper is the abundant redness of the three principles, so Lead is their blackness; but, at the same time, there are four colours concealed therein – the blackness, purged off from the three principles; redness, which contains a precipitate out of Mercury; whiteness, from the calcination of Mercury; and a certain yellowness, derived from Mercury. Thus the grossness and the colours are alike due to Mercury, and Lead is, in fact, a black Mercury. – *Ibid.*, c. 5.

communion of the metals, it should be known that no separation or mortification is spiritual, because such spirits can never exist without bodies. Though the body should be taken away from them and mortified a hundred times in one hour, nevertheless, they would always acquire another much more noble than the former. And this is the transposition of the metals from one death to another, that is to say, from a lesser degree into one greater and higher, namely, into Luna; and from a better into the best and most most perfect, that is, into Sol, the brilliant and altogether royal metal. It is most true, then, as frequently said above, that the six metals always generate a seventh, or produce it from themselves clear in its *esse*.

A question may arise: If it be true that Luna and every metal derives its origin and is generated from the other six, what is then its property and its nature? To this we reply: From Saturn, Mercury, Jupiter, Mars, Venus, and Sol, nothing and no other metal than Luna could be made. The cause is that each metal has two good virtues of the other six, of which altogether there are twelve. These are the spirit of Luna, which thus in a few words may be made known. Luna is composed of the six spiritual metals and their virtues, whereof each possesses two. Altogether, therefore, twelve are thus posited in one corporeal metal, which are compared to the seven planets and the twelve celestial signs. Luna has from the planet Mercury, and from Aquarius and Pisces, its liquidity and bright white colour. ☿, ♒, and ♓. So Luna has from Jupiter, with ♐ (Sagittarius) and Taurus, its white colour and its great firmness in fire. ♃, ♐, ♉. Luna has from Mars, with Cancer and Aries, its hardness and its clear sound. ♂, ♋, and ♈. Luna has from Venus, with Gemini and Libra, its measure of coagulation and its malleability. ♀, ♊, and Libra. From Saturn, with Virgo and ♏ (Scorpio), its homogeneous body, with gravity. ♄, ♍, and ♏. From Sol, with Leo and Virgo, its spotless purity and great constancy against the power of fire. ☉, ♌, and ♍. Such is the knowledge of the natural exaltation and of the course of the spirit and body of Luna, with its composite nature and wisdom briefly summarised.

Furthermore, it should be pointed out what kind of a body such metallic spirits acquire in their primitive generation by means of celestial influx. For the metal-digger, when he has crushed the stone, contemptible as it is in appearance, liquefies it, corrupts it, and altogether mortifies it with fire. Then this metallic spirit, in such a process of mortification, receives a better and more noble body, not friable but malleable. Then comes the Alchemist, who again corrupts, mortifies, and artificially prepares such a metallic body. Thus once more that spirit of the metal assumes a more noble and more perfect body, putting itself forward clearly into the light, except it be Sol or Luna. Then at last the metallic spirit and body are perfectly united, are safe from the corruption of elementary fire, and also incorruptible.*

---

* When the three prime principles have been purged of their superfluities, and from the said superfluities the imperfect metals have been generated, there remains nothing gross or crude, either in colour or substance, but only a very subtle nature of a white and purple hue. This is the most pure quality of Mercury, Salt, and Sulphur, most clear

THE SEVENTH CANON

Concerning the Nature of Sol and its Properties

The seventh after the six spiritual metals is corporeally Sol, which in itself is nothing but pure fire. What in outward appearance is more beautiful, more brilliant, more clear and perceptible, a heavier, colder, or more homogeneous body to see? And it is easy to perceive the cause of this, namely, that it contains in itself the congelations of the other six metals, out of which it is made externally into one most compact body. Its liquefaction proceeds from elementary fire, or is caused by the liquations of Mercury, with Pisces and Aquarius, concealed spiritually within it. The most manifest proof of this is that Mercury is easily mingled corporeally with the Sun as in an embrace. But for Sol, when the heat is withdrawn and the cold supervenes after liquefaction, to coagulate and to become hard and solid, there is need of the other five metals, whose nature it embraces in itself—Jupiter, Saturn, Mars, Venus, Luna. In these five metals the cold abodes with their regimens are especially found. Hence it happens that Sol can with difficulty be liquefied without the heat of fire, on account of the cold whereof mention has been made. For Mercury cannot assist with his natural heat or liquefaction, or defend himself against the cold of the five metals, because the heat of Mercury is not sufficient to retain Sol in a state of liquefaction. Wherefore Sol has to obey the five metals rather than Mercury alone. Mercury itself has no office of itself save always to flow. Hence it happens that in coagulations of the other metals it can effect nothing, since its nature is not to make anything hard or solid, but liquid. To render fluid is the nature of heat and life, but cold has the nature of hardness, consolidation, and immobility, which is compared to death. For example, the six cold metals, Jupiter, Venus, Saturn, Mars, Venus, Luna, if they are to be liquefied must be brought to that condition by the heat of fire. Snow or ice, which are cold, will not produce this effect, but rather will harden. As soon as ever the metal liquefied by fire is removed therefrom, the cold, seizing upon it, renders it hard, congelated, and immovable of itself. But in order that Mercury may remain fluid and alive continually, say, I pray you, whether this will be affected with heat or cold? Whoever answers that this is brought about by a cold and damp nature, and that it has its life from cold—the promulgator of this opinion, having no knowledge of Nature, is led away by the vulgar. For the vulgar man judges only falsely, and always holds firmly on to his error. So then let him who loves truth withdraw therefrom. Mercury, in fact, lives not at all from cold but from a warm and fiery nature.

---

and excellent in form, substance, essence, and colour. These two essences, namely, the white and the purple, are separated by the Archeus, and out of the first, fixed and coagulated, is formed silver, while from the purple there is generated gold, which is the most noble Sulphur, Salt, and Mercury, separated from all other colours, and consisting of purple alone. Its clayey or yellow appearance is accounted for by the subtlety and clearness of the metal, because all the dull colours are removed. In Silver the most prevalent colours are green and blue, which are respectively derived from the Mercury and the Salt, the Sulphur contributing nothing in the matter of colouring, On the other hand, in gold the purple colour is derived from Salt, the pellucid redness from Sulphur, and the yellow from Mercury.
—*Ibid.*, c. 8.

Whatever lives is fire, because heat is life, but cold the occasion of death. The fire of Sol is of itself pure, not indeed alive, but hard, and so far shews the colour of sulphur in that yellow and red are mixed therein in due proportion. The five cold metals are Jupiter, Mars, Saturn, Venus, and Luna, which assign to Sol their virtues; according to cold, the body itself; according to fire, colour; according to dryness, solidity; according to humidity, weight; and out of brightness, sound. But that gold is not burned in the element of terrestrial fire, nor is even corrupted, is effected by the firmness of Sol. For one fire cannot burn another, or even consume it; but rather if fire be added to fire it is increased, and becomes more powerful in its operations. The celestial fire which flows to us on the earth from the Sun is not such a fire as there is in heaven, neither is it like that which exists upon the earth, but that celestial fire with us is cold and congealed, and it is the body of the Sun. Wherefore the Sun can in no way be overcome by our fire. This only happens, that it is liquefied, like snow or ice, by that same celestial Sun. Fire, therefore, has not the power of burning fire, because the Sun is fire, which, dissolved in heaven, is coagulated with us.

| Gold is in its Essence three-fold | 1 Celestial
2 Elementary
3 Metallic | and is | Dissolved
Fluid
Corporeal. |

THE END OF THE SEVEN CANONS

THE CŒLUM PHILOSOPHORUM

PART II

CERTAIN TREATISES AND APPENDICES ARISING OUT OF THE SEVEN CANONS

God and Nature do Nothing in Vain

THE eternal position of all things, independent of time, without beginning or end, operates everywhere. It works essentially where otherwise there is no hope. It accomplishes that which is deemed impossible. What appears beyond belief or hope emerges into truth after a wonderful fashion.

Note on Mercurius Vivus

Whatever tinges with a white colour has the nature of life, and the properties and power of light, which causally produces life. Whatever, on the other hand, tinges with blackness, or produces black, has a nature in common with death, the properties of darkness, and forces productive of death. The earth with its frigidity is a coagulation and fixation of this kind of hardness. For the house is always dead; but he who inhabits the house lives. If you can discover the force of this illustration you have conquered.

Tested liquefactive powder.
Burn fat verbena.*

Recipe.—Salt nitre, four ounces; a moiety of sulphur; tartar, one ounce. Mix and liquefy.

What is to be thought concerning the Congelation of Mercury.

To mortify or congeal Mercury, and afterwards seek to turn it into Luna, and to sublimate it with great labour, is labour in vain, since it involves a dissipation of Sol and Luna existing therein. There is another method, far different and much more concise, whereby, with little waste of Mercury and less expenditure of toil, it is transmuted into Luna without congelation. Any one can at pleasure learn this Art in Alchemy, since it is so simple and easy; and by it, in a short time, he could make any quantity of silver and

---

* Verbenas adole pingues, et mascula tura.—Virg., Ecl. viii, 65.

gold. It is tedious to read long descriptions, and everybody wishes to be advised in straightforward words. Do this, then; proceed as follows, and you will have Sol and Luna, by help whereof you will turn out a very rich man. Wait awhile, I beg, while this process is described to you in few words, and keep these words well digested, so that out of Saturn, Mercury, and Jupiter you may make Sol and Luna. There is not, nor ever will be, any art so easy to find out and practise, and so effective in itself. The method of making Sol and Luna by Alchemy is so prompt that there is no more need of books, or of elaborate instruction, than there would be if one wished to write about last year's snow.

Concerning the Receipts of Alchemy

What, then, shall we say about the receipts of Alchemy, and about the diversity of its vessels and instruments? These are furnaces, glasses, jars, waters, oils, limes, sulphurs, salts, saltpetres, alums, vitriols, chrysocollæ, copper-greens, atraments, auri-pigments, fel vitri, ceruse, red earth, thucia, wax, lutum sapientiæ, pounded glass, verdigris, soot, testæ ovorum, crocus of Mars, soap, crystal, chalk, arsenic, antimony, minium, elixir, lazurium, gold-leaf, salt-nitre, sal ammoniac, calamine stone, magnesia, bolus armenus, and many other things. Moreover, concerning preparations, putrefactions, digestions, probations, solutions, cementings, filtrations, reverberations, calcinations, graduations, rectifications, amalgamations, purgations, etc., with these alchemical books are crammed. Then, again, concerning herbs, roots, seeds, woods, stones, animals, worms, bone dust, snail shells, other shells, and pitch. These and the like, whereof there are some very far-fetched in Alchemy, are mere incumbrances of work; since even if Sol and Luna could be made by them they rather hinder and delay than further one's purpose. But it is not from these—to say the truth—that the Art of making Sol and Luna is to be learnt. So, then, all these things should be passed by, because they have no effect with the five metals, so far as Sol and Luna are concerned. Someone may ask, What, then, is this short and easy way, which involves no difficulty, and yet whereby Sol and Luna can be made? Our answer is, this has been fully and openly explained in the Seven Canons. It would be lost labour should one seek further to instruct one who does not understand these. It would be impossible to convince such a person that these matters could be so easily understood, but in an occult rather than in an open sense.

The Art is this: After you have made heaven, or the sphere of Saturn, with its life to run over the earth, place on it all the planets, or such, one or more, as you wish, so that the portion of Luna may be the smallest. Let all run, until heaven, or Saturn, has entirely disappeared. Then all those planets will remain dead with their old corruptible bodies, having meanwhile obtained another new, perfect, and incorruptible body.

That body is the spirit of heaven. From it these planets again receive a body and life, and live as before. Take this body from the life and the earth,

Keep it. It is Sol and Luna. Here you have the Art altogether, clear and entire. If you do not yet understand it, or are not practised therein, it is well. It is better that it should be kept concealed, and not made public.

How to Conjure the Crystal so that all Things may be Seen in it

To conjure is nothing else than to observe anything rightly, to know and to understand what it is. The crystal is a figure of the air. Whatever appears in the air, movable or immovable, the same appears also in the speculum or crystal as a wave. For the air, the water, and the crystal, so far as vision is concerned, are one, like a mirror in which an inverted copy of an object is seen.

Concerning the Heat of Mercury

Those who think that Mercury is of a moist and cold nature are plainly in error, because it is by its nature in the highest degree warm and moist, which is the cause of its being in a constant state of fluidity. If it were of a moist and cold nature it would have the appearance of frozen water, and be always hard and solid, so that it would be necessary to liquefy it by the heat of fire, as in the case of the other metals. But it does not require this, since it has liquidity and flux from its own heat naturally inborn in it, which keeps it in a state of perpetual fluidity and renders it "quick," so that it can neither die, nor be coagulated, nor congealed. And this is well worth noticing, that the spirits of the seven metals, or as many of them as have been commingled, as soon as they come into the fire, contend with one another, especially Mercury, so that each may put forth its powers and virtues in the endeavour to get the mastery in the way of liquefying and transmuting. One seizes on the virtue, life, and form of another, and assigns some other nature and form to this one. So then the spirits or vapours of the metals are stirred up by the heat to operate mutually one upon the other, and transmute from one virtue to another, until perfection and purity are attained.

But what must be done besides to Mercury in order that its moisture and heat may be taken away, and in their place such an extreme cold introduced as to congeal, consolidate, and altogether mortify the Mercury? Do what follows in the sentence subjoined: Take pure Mercury closely shut up in a silver pixis. Fill a jar with fragments of lead, in the midst of which place the pixis. Let it melt for twenty-four hours, that is, for a natural day. This takes away from Mercury his occult heat, adds an external heat, and contributes the internal coldness of Saturn and Luna (which are both planets of a cold nature), whence and whereby the Mercury is compelled to congeal, consolidate, and harden.

Note also that the coldness (which Mercury needs in its consolidation and mortification) is not perceptible by the external sense, as the cold of snow or of ice is, but rather, externally, there is a certain amount of apparent heat. Just in the same way is it with the heat of Mercury, which is the cause of its fluidity. It is not an external heat, perceptible in the same way as one of our qualities. Nay, externally a sort of coldness is perceptible. Whence the Sophists (a race which has more talk than true wisdom) falsely assert that Mercury is cold and of a moist nature, so that they go on and advise us to congeal it by means of heat ; whereas heat only renders it more fluid, as they daily find out to their own loss rather than gain.

True Alchemy which alone, by its unique Art, teaches how to fabricate Sol and Luna from the five imperfect metals, allows no other receipt than this, which well and truly says : Only from metals, in metals, by metals, and with metals, are perfect metals made, for in some things is Luna and in other metals is Sol.

What Materials and Instruments are required in Alchemy

There is need of nothing else but a foundry, bellows, tongs, hammers, cauldrons, jars, and cupels made from beechen ashes. Afterwards, lay on Saturn, Jupiter, Mars, Sol, Venus, Mercury, and Luna. Let them operate finally up to Saturn.

The Method of Seeking Minerals

The hope of finding these in the earth and in stones is most uncertain, and the labour very great. However, since this is the first mode of getting them, it is in no way to be despised, but greatly commended. Such a desire or appetite ought no more to be done away with than the lawful inclination of young people, and those in the prime of life, to matrimony. As the bees long for roses and other flowers for the purpose of making honey and wax, so, too, men—apart from avarice or their own aggrandisement—should seek to extract metal from the earth. He who does not seek it is not likely to find it. God dowers men not only with gold or silver, but also with poverty, squalor, and misery. He has given to some a singular knowledge of metals and minerals, whereby they have obtained an easier and shorter method of fabricating gold and silver, without digging and smelting them, than they were commonly accustomed to, by extracting them from their primitive bodies. And this is the case not only with subterranean substances, but by certain arts and knowledge they have extracted them from the five metals generally (that is to say, from metals excocted from minerals which are imperfect and called metals), viz., from Mercury, Jupiter, Saturn, Mars and Venus, from all of which, and from each of them separately, Sol and Luna can be made, but from one more easily than from another. Note, that Sol and Luna can be made easily from Mercury, Saturn, and Jupiter, but from Mars and Venus with difficulty. It is possible to make them, however, but with the addition of Sol and Luna. Out of Magnesium and Saturn comes Luna, and out of Jupiter and Cinnabar pure Sol takes its rise. The skilful artist, however (how well I remember !), will be able by diligent consideration to prepare metals so that, led by a true method of reasoning, he can promote the perfection of metallic transformation more than do the courses of the twelve signs and the seven planets. In such matters it is quite superfluous to watch these courses, as also their aspects, good or bad days or hours, the prosperous or unlucky condition of this or that planet, for these matters can do no good, and much less can they do harm in the art of natural Alchemy. If otherwise, and you have a feasible process, operate when you please. If, however, there be anything wanting in you or your mode of working, or your understanding, the planets and the stars of heaven will fail you in your work.

If metals remain buried long enough in the earth, not only are they consumed by rust, but by long continuance they are even transmuted into natural stones, and there are a great many of these; but this is known to few. For there is found in the earth old stone money of the heathens, printed with their different figures. These coins were originally metallic, but through the transmutation brought about by Nature, they were turned into stone.

What Alchemy is

Alchemy is nothing else but the set purpose, intention, and subtle endeavour to transmute the kinds of the metals from one to another.* According to this, each person, by his own mental grasp, can choose out for himself a better way and Art, and therein find truth, for the man who follows a thing up more intently does find the truth. It is highly necessary to have a correct estimation of stars and of stones, because the star is the informing spirit of all stones. For the Sol and Luna of all the celestial stars are nothing but one stone in itself; and the terrestrial stone has come forth from the celestial stone; through the same fire, coals, ashes, the same expulsions and repurgations as that celestial stone, it has been separated and brought, clear and pure in its brightness. The whole ball of the earth is only something thrown off, concrete, mixed, corrupted, ground, and again coagulated, and gradually liquefied into one mass, into a stony work, which has its seat and its rest in the midst of the firmamental sphere.

Further it is to be remarked that those precious stones which shall forthwith be set down have the nearest place to the heavenly or sidereal ones in point of perfection, purity, beauty, brightness, virtue, power of withstanding fire, and incorruptibility, and they have been fixed with other stones in the earth.†

They have, therefore, the greatest affinity with heavenly stones and with the stars, because their natures are derived from these. They are found by

---

* Alchemy is, so to speak, a kind of lower heaven, by which the sun is separated from the moon, day from night, medicine from poison, what is useful from what is refuse.—*De Colica.* Therefore learn Alchemy, which is otherwise called Spagyria. This teaches you to discern between the true and the false. Such a Light of Nature is it that it is a mode of proof in all things, and walks in light. From this light of Nature we ought to know and speak, not from mere phantasy, whence nothing is begotten save the four humours and their compounds, augmentation, stagnation, and decrease, with other trifles of this kind. These proceed, not from the clear intellect, that full treasure-house of a good man, but rather are based on a fictitious and insecure foundation.—*Paramirum*, Lib. I., c. 3.

† When the occult dispenser of Nature in the prime principles, that is to say, the potency called Ares, has produced the gross and rough genera of stones, and no further grossness remains, a diaphanous and subtle substance remains, out of which the Archeus of Nature generates the precious stones or gems.—*De Elemento Aquæ*, Lib. IV., Tract IV., c. 10.

men in a rude environment, and the common herd (whose property it is to take false views of things) believe that they were produced in the same place where they are found, and that they were afterwards polished, carried around, and sold, and accounted to be great riches, on account of their colours, beauty, and other virtues. A brief description of them follows :—

The Emerald. This is a green transparent stone. It does good to the eyes and the memory. It defends chastity; and if this be violated by him who carries it, the stone itself does not remain perfect.*

The Adamant. A black crystal called Adamant or else Evax, on account of the joy which it is effectual in impressing on those who carry it. It is of an obscure and transparent blackness, the colour of iron. It is the hardest of all; but is dissolved in the blood of a goat. Its size at the largest does not exceed that of a hazel nut.†

The Magnet Is an iron stone, and so attracts iron to itself.‡

The Pearl. The Pearl is not a stone, because it is produced in sea shells. It is of a white colour. Seeing that it grows in animated beings, in men or in fishes, it is not properly of a stony nature, but properly a depraved (otherwise a transmuted) nature supervening upon a perfect work.§

The Jacinth Is a yellow, transparent stone. There is a flower of the same name which, according to the fable of the poets, is said to have been a man.‖

The Sapphire Is a stone of a celestial colour and a heavenly nature.¶

The Ruby Shines with an intensely red nature.**

The Carbuncle. A solar stone, shining by its own nature like the sun.††

The Coral Is a white or red stone, not transparent. It grows in the sea, out of the nature of the water and the air, into the form of wood or a shrub; it hardens in the air, and is not capable of being destroyed in fire.‡‡

---

* The body of the Emerald is derived from a kind of petrine Mercury. It receives from the same its colour, coagulated with spirit of Salt.—*Ibid.*, c. 12.

† The most concentrated hardness of all stones combines for the generation of the adamant. The white adamant has its body from Mercury, and its coagulation from the spirit of Salt.—*Ibid.*, c. 14.

‡ Fortified by experience, which is the mistress of all things, and by mature theory, based on experience, I affirm that the Magnet is a stone which not only undeniably attracts steel and iron, but has also the same power over the matter of all diseases in the whole body of man.—*De Corallis.* See *Herbarius Theophrasti.*

§ The Pearl is a seed of moisture. It generates milk abundantly in women if they are deficient therein.—*De Aridura.*

‖ The Jacinth, or Hyacinth, is a gem of the same genus as the Carbuncle, but is inferior thereto in its nature.—*De Elemento Aquæ,* Lib. IV., Tract IV., c. 11.

¶ In the matter of body and colour the Sapphire is generated from Mercury (the prime principle). It is formed over white Sulphur and white Salt from a pallid petrine Mercury. Hence white Sapphires frequently occur because a white Mercury concurs in the formation. In like manner a lute-coloured Mercury sometimes produces a clay-like hue.—*Ibid.*, c. 15.

** The Ruby and similar gems possessing a ruddy hue are generated from the red of Sulphur, and their body is of petrine Mercury. For Mercury is the body of every precious stone.—*Ibid.*, c. 13.

†† The Carbuncle is formed of the most transparent matter which is conserved in the three principles. Mercury is the body and Sulphur the colouring thereof, with a modicum of the spirit of Salt, on account of the coagulation. All light abounds therein, because Sulphur contains in itself a clear quality of light, as the art of its transmutation demonstrates.—*Ibid.*, c. 11.

‡‡ There are two species of red Corals—one a dull red, which varies between sub-purple and semi-black; the other a resplendent and brilliant red. As the colours differ, so also do the virtues. There is also a whitish species which is almost destitute of efficacy. In a word, as the Coral diminishes in redness, so it weakens in its qualities.—*Herbarius Theophrasti; De Corallis.*

The Chalcedony Is a stone made up of different colours, occupying a middle place between obscurity and transparency, mixed also with cloudiness, and liver coloured. It is the lowest of all the precious stones.*

The Topaz Is a stone shining by night. It is found among rocks.†

The Amethyst Is a stone of a purple and blood colour.‡

The Chrysoprasus Is a stone which appears like fire by night, and like gold by day.

The Crystal Is a white stone, transparent, and very like ice. It is sublimated, extracted, and produced from other stones.§

As a pledge and firm foundation of this matter, note the following conclusion. If anyone intelligently and reasonably takes care to exercise himself in learning about the metals, what they are, and whence they are produced : he may know that our metals are nothing else than the best part and the spirit of common stones, that is, pitch, grease, fat, oil, and stone. But this is least pure, uncontaminated, and perfect, so long as it remains hidden or mixed with the stones. It should therefore be sought and found in the stones, be recognised in them, and extracted from them, that is, forcibly drawn out and liquefied. For then it is no longer a stone, but an elaborate and perfect metal, comparable to the stars of heaven, which are themselves, as it were, stones separated from those of earth.

Whoever, therefore, studies minerals and metals must be furnished with such reason and intelligence that he shall not regard only those common and known metals which are found in the depth of the mountains alone. For there is often found at the very surface of the earth such a metal as is not met with at all, or not equally good, in the depths. And so every stone which comes to our view, be it great or small, flint or simple rock, should be carefully investigated and weighed with a true balance, according to its nature and properties. Very often a common stone, thrown away and despised, is worth more than a cow. Regard must not always be had to the place of digging from which this stone came forth; for here the influence of the sky prevails. Everywhere there is presented to us earth, or dust, or sand, which often contain much gold or silver, and this you will mark.

HERE ENDS THE CŒLUM PHILOSOPHORUM

---

* The gem Chalcedony is extracted from Salt.—*Chirurgia Magna; De Tumoribus, etc., Morbi Gallici*, Lib. III., c. 6.

† The Topaz is an extract from the minera of Mars, and is a transplanted Iron.—*Ibid.*

‡ The Amethyst is an extract of Salt, while Marble and Chalcedony are extracted from the same principle through the Amethyst.—*Ibid.*

§ The origin of Crystals is to be referred to water. They contain within them a spirit of coagulation whereby they are coagulated, as water by the freezing and glacial stars.—*Lib. Meteorum*, c. 7.

THE BOOK CONCERNING THE TINCTURE OF THE PHILOSOPHERS

WRITTEN AGAINST THOSE SOPHISTS BORN SINCE THE DELUGE, IN THE AGE OF OUR LORD JESUS CHRIST, THE SON OF GOD;

BY PH. THEOPHRASTUS BOMBAST, OF HOHENHEIM,

Philosopher of the Monarchia, Prince of Spagyrists, Chief Astronomer, Surpassing Physician, and Trismegistus of Mechanical Arcana

---

PREFACE

SINCE you, O Sophist, everywhere abuse me with such fatuous and mendacious words, on the ground that being sprung from rude Helvetia I can understand and know nothing: and also because being a duly qualified physician I still wander from one district to another; therefore I have proposed by means of this treatise to disclose to the ignorant and inexperienced: what good arts existed in the first age; what my art avails against you and yours against me; what should be thought of each, and how my posterity in this age of grace will imitate me. Look at Hermes, Archelaus, and others in the first age: see what Spagyrists and what Philosophers then existed. By this they testify that their enemies, who are your patrons, O Sophist, at the present time are but mere empty forms and idols. Although this would not be attested by those who are falsely considered your authentic fathers and saints, yet the ancient Emerald Table shews more art and experience in Philosophy, Alchemy, Magic, and the like, than could ever be taught by you and your crowd of followers. If you do not yet understand, from the aforesaid facts, what and how great treasures these are, tell me why no prince or king was ever able to subdue the Egyptians. Then tell me why the Emperor Diocletian ordered all the Spagyric books to be burnt (so far as he could lay his hands upon them). Unless the contents of those books had been known, they would have been obliged to bear still his intolerable yoke,—a yoke, O Sophist, which shall one day be put upon the neck of yourself and your colleagues.

From the middle of this age the Monarchy of all the Arts has been at length derived and conferred on me, Theophrastus Paracelsus, Prince of Philosophy and of Medicine. For this purpose I have been chosen by God to extinguish and blot out all the phantasies of elaborate and false works, of delusive and presumptuous words, be they the words of Aristotle, Galen, Avicenna, Mesva, or the dogmas of any among their followers. My theory, proceeding as it does from the light of Nature, can never, through its consistency, pass away or be changed: but in the fifty-eighth year after its millennium and a half it will then begin to flourish. The practice at the same time following upon the theory will be proved by wonderful and incredible signs, so as to be open to mechanics and common people, and they will thoroughly understand how firm and immovable is that Paracelsic Art against the triflings of the Sophists: though meanwhile that sophistical science has to have its ineptitude propped up and fortified by papal and imperial privileges. In that I am esteemed by you a mendicant and vagabond sophist, the Danube and the Rhine will answer that accusation, though I hold my tongue. Those calumnies of yours falsely devised against me have often displeased many courts and princes, many imperial cities, the knightly order, and the nobility. I have a treasure hidden in a certain city called Weinden, belonging to Forum Julii, at an inn,—a treasure which neither you, Leo of Rome, nor you, Charles the German, could purchase with all your substance. Although the signed star has been applied to the arcanum of your names, it is known to none but the sons of the divine Spagyric Art. So then, you wormy and lousy Sophist, since you deem the monarch of arcana a mere ignorant, fatuous, and prodigal quack, now, in this mid age, I determine in my present treatise to disclose the honourable course of procedure in these matters, the virtues and preparation of the celebrated Tincture of the Philosophers for the use and honour of all who love the truth, and in order that all who despise the true arts may be reduced to poverty. By this arcanum the last age shall be illuminated clearly and compensated for all its losses by the gift of grace and the reward of the spirit of truth, so that since the beginning of the world no similar germination of the intelligence and of wisdom shall ever have been heard of. In the meantime, vice will not be able to suppress the good, nor will the resources of those vicious persons, many though they be, cause any loss to the upright.

THE BOOK CONCERNING THE TINCTURE OF THE PHILOSOPHERS

CHAPTER I

I, PHILIPPUS Theophrastus Paracelsus Bombast, say that, by Divine grace, many ways have been sought to the Tincture of the Philosophers, which finally all came to the same scope and end. Hermes Trismegistus, the Egyptian, approached this task in his own method. Orus, the Greek, observed the same process. Hali, the Arabian, remained firm in his order. But Albertus Magnus, the German, followed also a lengthy process. Each one of these advanced in proportion to his own method; nevertheless, they all arrived at one and the same end, at a long life, so much desired by the philosophers, and also at an honourable sustenance and means of preserving that life in this Valley of Misery. Now at this time, I, Theophrastus Paracelsus Bombast, Monarch of the Arcana, am endowed by God with special gifts for this end, that every searcher after this supreme philosophic work may be forced to imitate and to follow me, be he Italian, Pole, Gaul, German, or whatsoever or whosoever he be. Come hither after me, all you philosophers, astronomers, and spagyrists, of however lofty a name ye may be, I will show and open to you, Alchemists and Doctors, who are exalted by me with the most consummate labours, this corporeal regeneration. I will teach you the tincture, the arcanum,* the quintessence, wherein lie hid the foundations of all mysteries and of all works. For every person may and ought to believe in another only in those matters which he has tried by fire. If any one shall have brought forward anything contrary to this method of experimentation in the Spagyric Art or in Medicine, there is no reason for your belief in him, since, experimentally, through the agency of fire, the true is separated from the false. The light of Nature indeed is created in this way, that by means thereof the proof or trial of everything may appear, but only to those who walk in this light. With this light we will teach, by the very best methods of demonstration, that all those who before me have approached this so difficult province with their own fancies and acute speculations have, to their own loss, incurred the danger of their foolishness. On which account, from my standpoint, many rustics have been

---

* The Arcanum of a substance is not the virtue (*virtus*) but the essence (*vis*) and the potency (*potentia*), and is stronger than the virtue; nevertheless, an old error of the doctors conferred the name of virtues upon the potential essences.—*Paramirum*, Lib. IV. Many things are elsewhere set forth concerning the Quintessence, but what is described is really a separation or extraction of the pure from the impure, not a true quintessence, and it is more correctly termed an Arcanum.—*Explicatio Totius Astronomiæ*.

ennobled; but, on the other hand, through the speculative and opinionative art of these many nobles have been changed into clowns, and since they carried golden mountains in their head before they had put their hand to the fire. First of all, then, there must be learnt—digestions, distillations, sublimations, reverberations, extractions, solutions, coagulations, fermentations, fixations, and every instrument which is requisite for this work must be mastered by experience, such as glass vessels, cucurbites, circulators, vessels of Hermes, earthen vessels, baths, blast-furnaces, reverberatories, and instruments of like kind, also marble, coals, and tongs. Thus at length you will be able to profit in Alchemy and in Medicine.

But so long as, relying on fancy and opinion, you cleave to your fictitious books, you are fitted and predestinated for no one of these things.

CHAPTER II

Concerning the Definition of the Subject and Matter of the Tincture of the Philosophers

Before I come, then, to the process of the Tincture, it is needful that I open to you the subject thereof: for, up to the present time, this has always been kept in a specially occult way by the lovers of truth. So, then, the matter of the Tincture (when you understand me in a Spagyrical sense) is a certain thing which, by the art of Vulcan,* passes out of three essences into one essence, or it may remain. But, that I may give it its proper name, according to the use of the ancients, though it is called by many the Red Lion, still it is known by few. This, by the aid of Nature and the skill of the Artist himself, can be transmuted into a White Eagle, so that out of one two are produced; and beyond this the brightness of gold does not shine so much for the Spagyrist as do these two when kept in one. Now, if you do not understand the use of the Cabalists and the old astronomers, you are not born by God for the Spagyric art, or chosen by Nature for the work of Vulcan, or created to open your mouth concerning Alchemical Arts. The matter of the Tincture, then, is a very great pearl and a most precious treasure, and the noblest thing next to the manifestation of the Most High and the consideration of men which can exist upon earth. This is the Lili of Alchemy and of Medicine, which the philosophers have so diligently sought after, but, through the failure of entire knowledge and complete preparation, they have not progressed to the perfect end thereof. By means of their investigations and experiments, only the

---

* The office of Vulcan is the separation of the good from the bad. So the Art of Vulcan, which is Alchemy, is like unto death, by which the eternal and the temporal are divided one from another. So also this art might be called the death of things.—*De Morbis Metallicis*, Lib. I., Tract III., c. 1. Vulcan is an astral and not a corporal fabricator.—*De Caduco Matricis*, Par. VI. The artist working in metals and other minerals transforms them into other colours, and in so doing his operation is like that of the heaven itself. For as the artist excocts by means of Vulcan, or the igneous element, so heaven performs the work of coction through the Sun. The Sun, therefore, is the Vulcan of heaven accomplishing coction in the earth.—*De Icteritiis*. Vulcan is the fabricator and architect of all things, nor is his habitation in heaven only, that is, in the firmament, but equally in all the other elements.—*Lib. Meteorum*, c. 4. Where the three prime principles are wanting, there also the igneous essence is absent. The Igneous Vulcan is nothing else but Sulphur, Sal Nitrum, and Mercury.—*Ibid.*, c. 5.

initial stage of the Tincture has been given to us; but the true foundation, which my colleagues must imitate, has been left for me, so that no one should mingle their shadows with our good intentions. I, by right after my long experiences, correct the Spagyrists, and separate the false or the erroneous from the true, since, by long investigations, I have found reasons why I should be able justly to blame and to change diverse things. If, indeed, I had found out experiments of the ancients better than my own, I should scarcely have taken up such great labours as, for the sake, the utility, and the advantage of all good Alchemists, I have undergone willingly. Since, then, the subject of the Tincture has been sufficiently declared, so that it scarcely could or ought to be exceeded in fidelity between two brothers, I approach its preparation, and after I have laid down the experiences of the first age, I wish to add my own inventions; to which at last the Age of Grace will by-and-by give its adhesion, whichever of the patriarchs, O Sophist, you, in the meantime, shall have made leaders.

CHAPTER III

Concerning the Process of the Ancients for the Tincture of the Philosophers, and a more Compendious Method by Paracelsus

The old Spagyrists putrefied Lili for a philosophical month, and afterwards distilled therefrom the moist spirits, until at length the dry spirits were elevated. They again imbued the *caput mortuum* with moist spirits, and drew them off from it frequently by distillation until the dry spirits were all elevated. Then afterwards they united the moisture that had been drawn off and the dry spirits by means of a pelican, three or four times, until the whole Lili remained dry at the bottom. Although early experience gave this process before fixation, none the less our ancestors often attained a perfect realisation of their wish by this method. They would, however, have had a shorter way of arriving at the treasure of the Red Lion if they had learnt the agreement of Astronomy with Alchemy, as I have demonstrated it in the Apocalypse of Hermes.* But since every day (as Christ says for the consolation of the

---

* The Book of the Revelation of Hermes, interpreted by Theophrastus Paracelsus, concerning the Supreme Secret of the World, seems to have been first brought to light by Benedictus Figulus, and appeared as a *pièce de résistance* in his "Golden and Blessed Casket of Nature's Marvels," of which an English translation has been very recently published. ("A Golden and Blessed Casket of Nature's Marvels." By Benedictus Figulus. Now first done into English from the German original published at Frankfort in the year 1608. London: James Elliott and Co. 8vo. 1893.) Among the many writings which have been fabulously attributed to Hermes, there does not seem to be any record of an *apocalypse*, and it is impossible to say what forged document may have been the subject of interpretation by Paracelsus. As the collection of Figulus is now so readily accessible, it is somewhat superfluous to reproduce the treatise here, but since this translation claims to include everything written by the physician of Hohenheim on the subject of Alchemy and the Universal Medicine, it is appended at this point. It should be premised that Benedictus Figulus complains bitterly of the mutilation and perversion to which the works of Paracelsus were subjected, and the *Revelation of Hermes* seems in many parts to betray another hand, especially in its quotation of authorities who are not countenanced by its reputed author.

Hermes, Plato, Aristotle, and other philosophers, flourishing at different times, who have introduced the Arts, and more especially have explored the secrets of inferior Creation, all these have eagerly sought a means whereby man's body might be preserved from decay and become endued with immortality. To them it was answered that there is nothing which might deliver the mortal body from death; but that there is One Thing which may postpone decay, renew youth, and prolong short human life (as with the patriarchs). For death was laid as a punishment upon our first parents, Adam and Eve, and will never depart from all their decsendants. Therefore, the above philosophers, and many others, have

faithful) has its own peculiar care, the labour for the Spagyrists before my times has been great and heavy; but this, by the help of the Holy Spirit flowing into us, will, in this last age, be lightened and made clear by my theory and practice, for all those who constantly persevere in their work with patience. For I have tested the properties of Nature, its essences and conditions, and I know its conjunction and resolution, which are the highest and greatest gift for a philosopher, and never understood by the sophists up to this time. When, therefore, the earliest age gave the first experience of the Tincture, the Spagyrists made two things out of one simple. But when afterwards, in the Middle Age, this invention had died out, their successors by diligent scrutiny afterwards came upon the two names of this simple, and they named it with one word, namely, Lili, as being the subject of the Tincture. At length the imitators of Nature putrefied this matter at its proper period just like the seed in the earth, since before this corruption nothing could be born from it, nor any arcanum break forth from it. Afterwards they drew off the moist spirits from the matter, until at length, by the violence of the fire, the dry were also equally sublimated, so that, in this way, just as the rustic

---

sought this One Thing with great labour, and have found that that which preserves the human body from corruption, and prolongs life, conducts itself, with respect to other elements, as it were like the Heavens; from which they understood that the Heavens are a substance above the Four Elements. And just as the Heavens, with respect to the other elements, are held to be the fifth substance (for they are indestructible, stable, and suffer no foreign admixture), so also this One Thing (compared to the forces of our body) is an indestructible essence, drying up all the superfluities of our bodies, and has been philosophically called by the above-mentioned name. It is neither hot and dry like fire, nor cold and moist like water, nor warm and moist like air, nor dry and cold like earth. But it is a skilful, perfect equation of all the Elements, a right commingling of natural forces, a most particular union of spiritual virtues, an indissoluble uniting of body and soul. It is the purest and noblest substance of an indestructible body, which cannot be destroyed nor harmed by the Elements, and is produced by Art. With this, Aristotle prepared an apple, prolonging life by its scent, when he, fifteen days before his death, could neither eat nor drink on account of old age. This spiritual Essence, or One Thing, was revealed from above to Adam, and was greatly desired by the Holy Fathers; this also Hermes and Aristotle call the Truth without Lies, the most sure of all things certain, the Secret of all Secrets. It is the Last and the Highest Thing to be sought under the Heavens, a wondrous closing and finish of philosophical work, by which are discovered the dews of Heaven and the fastnesses of Earth. What the mouth of man cannot utter is all found in this spirit. As Morienus says: "He who has this has all things, and wants no other aid. For in it are all temporal happiness, bodily health, and earthly fortune. It is the spirit of the fifth substance, a Fount of all Joys (beneath the rays of the moon), the Supporter of Heaven and Earth, the Mover of Sea and Wind, the Outpourer of Rain, upholding the strength of all things, an excellent spirit above Heavenly and other spirits, giving Health, Joy, Peace, Love; driving away Hatred and Sorrow, bringing in Joy, expelling all Evil, quickly healing all Diseases, destroying Poverty and misery, leading to all good things, preventing all evil words and thoughts, giving man his heart's desire, bringing to the pious earthly honour and long life, but to the wicked who misuse it, Eternal Punishment." This is the Spirit of Truth, which the world cannot comprehend without the interposition of the Holy Ghost, or without the instruction of those who know it. The same is of a mysterious nature, wondrous strength, boundless power. The Saints, from the beginning of the world, have desired to behold its face. By Avicenna this Spirit is named the Soul of the World. For, as the Soul moves all the limbs of the body, so also does this Spirit move all bodies. And as the Soul is in all the limbs of the Body, so also is this Spirit in all elementary created things. It is sought by many and found by few. It is beheld from afar and found near; for it exists in every thing, in every place, and at all times. It has the powers of all creatures; its action is found in all elements, and the qualities of all things are therein, even in the highest perfection. By virtue of this essence did Adam and the Patriarchs preserve their health and live to an extreme age, some of them also flourishing in great riches. When the philosophers had discovered it, with great diligence and labour, they straightway concealed it under a strange tongue, and in parables, lest the same should become known to the unworthy, and the pearls be cast before swine. For if everyone knew it, all work and industry would cease; man would desire nothing but this one thing, people would live wickedly, and the world be ruined, seeing that they would provoke God by reason of their avarice and superfluity. For eye hath not seen, nor ear heard, nor hath the heart of man understood what Heaven hath naturally incorporated with this Spirit. Therefore have I briefly enumerated some of the qualities of this Spirit, to the Honour of God, that the pious may reverently praise Him in His gifts (which gift of God shall afterwards come to them), and I will herewith shew what powers and virtues it possesses in each thing, also its outward appearance, that it may be more readily recognised. In its first state, it appears as an impure earthly body, full of imperfections. It then has an earthly nature, healing all sickness and wounds in the bowels of man, producing good and consuming proud flesh, expelling all stench, and healing generally, inwardly and outwardly. In its second nature, it

does at the proper time of year, they might come to maturity as one after another is wont to ascend and to fall away. Lastly, as after the spring comes summer, they incorporated those fruits and dry spirits, and brought the Magistery of the Tincture to such a point that it came to the harvest, and laid itself out for ripening.

CHAPTER IV

Concerning the Process for the Tincture of the Philosophers, as it is shortened by Paracelsus

The ancient Spagyrists would not have required such lengthened labour and such wearisome repetition if they had learnt and practised their work in my school. They would have obtained their wish just as well, with far less expense and labour. But at this time, when Theophrastus Paracelsus has arrived as the Monarch of Arcana, the opportunity is at hand for finding out those things which were occult to all Spagyrists before me. Wherefore I say, Take only the rose-coloured blood from the Lion and the gluten from the Eagle. When you have mixed these, coagulate them according to the old

---

appears as a watery body, somewhat more beautiful than before, because (although still having its corruptions) its Virtue is greater. It is much nearer the truth, and more effective in works. In this form it cures cold and hot fevers, and is a specific against poisons, which it drives from heart and lungs, healing the same when injured or wounded, purifying the blood, and, taken three times a day, is of great comfort in all diseases. But in its third nature it appears as an aerial body, of an oily nature, almost freed from all imperfections, in which form it does many wondrous works, producing beauty and strength of body, and (a small quantity being taken in the food) preventing melancholy and heating of the gall, increasing the quantity of the blood and seed, so that frequent bleeding becomes necessary. It expands the blood vessels, cures withered limbs, restores strength to the sight, in growing persons removes what is superfluous and makes good defects in the limbs. In its fourth nature it appears in a fiery form (not quite freed from all imperfections, still somewhat watery and not dried enough), wherein it has many virtues, making the old young and reviving those at the point of death. For if to such an one there be given, in wine, a barleycorn's weight of this fire, so that it reach the stomach, it goes to his heart, renewing him at once, driving away all previous moisture and poison, and restoring the natural heat of the liver. Given in small doses to old people, it removes the diseases of age, giving the old young hearts and bodies. Hence it is called the Elixir of Life. In its fifth and last nature, it appears in a glorified and illuminated form, without defects, shining like gold and silver, wherein it possesses all previous powers and virtues in a higher and more wondrous degree. Here its natural works are taken for miracles. When applied to the roots of dead trees they revive, bringing forth leaves and fruit. A lamp, the oil of which is mingled with this spirit, continues to burn for ever without diminution. It converts crystals into the most precious stones of all colours, equal to those from the mines, and does mayn other incredible wonders which may not be revealed to the unworthy. For it heals all dead and living bodies without other medicine. Here Christ is my witness that I lie not, for all heavenly influences are united and combined therein. This essence also reveals all treasures in earth and sea, converts all metallic bodies into gold, and there is nothing like unto it under Heaven. This spirit is the secret, hidden from the beginning, yet granted by God to a few holy men for the revealing of these riches to His Glory—dwelling in fiery form in the air, and leading earth with itself to heaven, while from its body there flow whole rivers of living water. This spirit flies through the midst of the Heavens like a morning mist, leads its burning fire into the water, and has its shining realm in the heavens. And although these writings may be regarded as false by the reader, yet to the initiated they are true and possible, when the hidden sense is properly understood. For God is wonderful in His works, and His wisdom is without end. This spirit in its fiery form is called a Sandaraca, in the aerial a Kybrick, in the watery an Azoth, in the earthly Alcohoph and Aliocosoph. Hence they are deceived by these names who, seeking without instruction, think to find this Spirit of Life in things foreign to our Art. For although this spirit which we seek, on account of its qualities, is called by these names, yet the same is not in these bodies and cannot be in them. For a refined spirit cannot appear except in a body suitable to its nature. And, by however many names it be called, let no one imagine there be different spirits, for, say what one will, there is but one spirit working everywhere and in all things. That is the spirit which, when rising, illumines the Heavens, when setting incorporates the purity of Earth, and when brooding has embraced the Waters. This spirit is named Raphael, the Angel of God, the subtlest and purest, whom the others all obey as their King. This spiritual substance is neither heavenly nor hellish, but an airy, pure, and hearty body, midway between the highest and lowest, without reason, but fruitful in works, and the most select and beautiful of all other heavenly things. This work of God is far too deep for understanding, for it is the last, greatest, and highest secret of Nature. It is the Spirit of God, which in the Beginning filled the earth and brooded over the waters, which the world cannot grasp without the gracious interposition of the Holy Spirit and instruction from those who know it, which also the whole world desires for its virtue, and which cannot be prized enough. For it reaches to the planets, raises the clouds, drives away mists, gives

process, and you will have the Tincture of the Philosophers, which an infinite number have sought after and very few have found. Whether you will or not, sophist, this Magistery is in Nature itself, a wonderful thing of God above Nature, and a most precious treasure in this Valley of Sorrows. If you look at it from without it seems a paltry thing to transmute another into something far more noble than it was before. But you must, nevertheless, allow this, and confess that it is a miracle produced by the Spagyrist, who by the art of his preparation corrupts a visible body which is externally vile, from which he excites another most noble and most precious essence. If you, in like manner, have learnt anything from the light of Aristotle, or from us, or from the rules of Serapio, come forth, and bring that knowledge experimentally to light. Preserve now the right of the Schools, as becomes a lover of honour and a doctor. But if you know nothing and can do nothing, why do you despise me as though I were an irrational Helvetian cow, and inveigh against me as a wandering vagabond ? Art is a second Nature and a universe of its own, as experience witnesses, and demonstrates against you and your idols. Sometimes, therefore, the Alchemist compounds certain simples, which he

---

its light to all things, turns everything into Sun and Moon, bestows all health and abundance of treasure, cleanses the leper, brightens the eyes, banishes sorrow, heals the sick, reveals all hidden treasures, and, generally, cures all diseases. Through this spirit have the philosophers invented the Seven Liberal Arts, and thereby gained their riches. Through the same Moses made the golden vessels in the Ark, and King Solomon did many beautiful works to the honour of God. Therewith Moses built the Tabernacle, Noah the Ark, Solomon the Temple. By this Ezra restored the Law, and Miriam, Moses' sister, was hospitable ; Abraham, Isaac, and Jacob, and other righteous men, have had lifelong abundance and riches ; and all the saints possessing it have therewith praised God. Therefore is its acquisition very hard, more than that of gold and silver. For it is the best of all things, because, of all things mortal that man can desire in this world, nothing can compare with it, and in it alone is truth. Hence it is called the Stone and Spirit of Truth ; in its works is no vanity, its praise cannot be sufficiently expressed. I am unable to speak enough of its virtues, because its good qualities and powers are beyond human thoughts, unutterable by the tongue of man, and in it are found the properties of all things. Yea, there is nothing deeper in Nature. O unfathomable abyss of God's Wisdom, which thus hath united and comprised in the virtue and power of this One Spirit the qualities of all existing bodies ! O unspeakable honour and boundless joy granted to mortal man ! For the destructible things of Nature are restored by virtue of the said Spirit. O mystery of mysteries, most secret of all secret things, and healing and medicine of all things ! Thou last discovery in earthly natures, last best gift to Patriarchs and Sages, greatly desired by the whole world ! Oh, what a wondrous and laudable spirit is purity, in which stand all joy, riches, fruitfulness of life, and art of all arts, a power which to its initiates grants all material joys ! O desirable knowledge, lovely above all things beneath the circle of the Moon, by which Nature is strengthened, and heart and limbs are renewed, blooming youth is preserved, old age driven away, weakness destroyed, beauty in its perfection preserved, and abundance ensured in all things pleasing to men ! O thou spiritual substance, lovely above all things ! O thou wondrous power, strengthening all the world ! O thou invincible virtue, highest of all that is, although despised by the ignorant, yet held by the wise in great praise, honour, and glory, that - proceeding from humours—wakest the dead, expellest diseases, restorest the voice of the dying ! O thou treasure of treasures, mystery of mysteries, called by Avicenna " an unspeakable substance," the purest and most perfect soul of the world, than which there is nothing more costly under Heaven, unfathomable in nature and power, wonderful in virtue and works, having no equal among creatures, possessing the virtues of all bodies under Heaven ! For from it flow the water of life, the oil and honey of eternal healing, and thus hath it nourished them with honey and water from the rock. Therefore, saith Morienus : " He who hath it, the same also hath all things." Blessed art Thou, Lord God of our fathers, in that Thou hast given the prophets this knowledge and understanding, that they have hidden these things (lest they should be discovered by the blind, and those drowned in worldly godlessness) by which the wise and the pious have praised Thee ! For the discoverers of the mystery of this thing to the unworthy are breakers of the seal of Heavenly Revelation, thereby offending God's Majesty, and bringing upon themselves many misfortunes and the punishments of God. Therefore, I beg all Christians, possessing this knowledge, to communicate the same to nobody, except it be to one living in Godliness, of well-proved virtue, and praising God, Who has given such a treasure to man. For many seek, but few find it. Hence the impure and those living in vice are unworthy of it. Therefore is this Art to be shewn to all God-fearing persons, because it cannot be bought with a price. I testify before God that I lie not, although it appear impossible to fools, that no one has hitherto explored Nature so deeply. The Almighty be praised for having created this Art and for revealing it to God-fearing men. Amen. And thus is fulfilled this precious and excellent work, called the revealing of the occult spirit, in which lie hidden the secrets and mysteries of the world. But this spirit is one genius, and divine, wonderful, and lordly power. For it embraces the whole world, and overcomes the Elements and the fifth Substance. To our Trismegistus Spagyrus, Jesus Christ, be praise and glory immortal. Amen.

afterwards corrupts according to his need, and prepares thence another thing. For thus very often out of many things one is made, which effects more than Nature of herself can do, as in Gastaynum it is perfectly well known that Venus is produced from Saturn; in Carinthia, Luna out of Venus; and in Hungary, Sol out of Luna; to pass over in silence for the time being the transmutations of other natural objects, which were well known to the Magi, and more wonderfully than Ovid narrates in his Metamorphoses do they come to the light. That you may rightly understand me, seek your Lion in the East, and your Eagle in the South, for this our work which has been undertaken. You will not find better instruments than Hungary and Istria produce. But if you desire to lead from unity by duality in trinity with equal permutation of each, then you should direct your journey to the South; so in Cyprus shall you gain all your desire, concerning which we must not dilate more profusely than we have done at present. There are still many more of these arcana which exhibit transmutations, though they are known to few. And although these may by the Lord God be made manifest to anyone, still, the rumour of this Art does not on that account at once break forth, but the Almighty gives therewith the understanding how to conceal these and other like arts even to the coming of Elias the Artist, at which time there shall be nothing so occult that it shall not be revealed. You also see with your eyes (though there is no need to speak of these things, which may be taken derisively by some) that in the fire of Sulphur is a great tincture for gems, which, indeed, exalts them to a loftier degree than Nature by herself could do. But this gradation of metals and gems shall be omitted by me in this place, since I have written sufficiently about it in my Secret of Secrets, in my book on the Vexations of Alchemists, and abundantly elsewhere. As I have begun the process of our ancestors with the Tincture of the Philosophers, I will now perfectly conclude it.

CHAPTER V

Concerning the Conclusion of the Process of the Ancients, made by Paracelsus

Lastly, the ancient Spagyrists having placed Lili in a pelican and dried it, fixed it by means of a regulated increase of the fire, continued so long until from blackness, by permutation into all the colours, it became red as blood, and therewith assumed the condition of a salamander. Rightly, indeed, did they proceed with such labour, and in the same way it is right and becoming that everyone should proceed who seeks this pearl. It will be very difficult for me to make this clearer to you unless you shall have learnt in the School of the Alchemists to observe the degrees of the fire, and also to change your vessels. For then at length you will see that soon after your Lili shall have become heated in the Philosophic Egg, it becomes, with wonderful appearances, blacker than the crow; afterwards, in succession of time, whiter than the swan; and at last, passing through a yellow colour, it turns out more red than any blood. Seek, seek, says the first Spagyrist, and you shall find; knock, and it shall be opened unto you. It would be impious and indecorous to put food in the mouth of a perfidious bird. Let her rather fly to it, even as I, with others before me, have been compelled to do. But follow true Art; for this will lead you to its perfect knowledge. It is not possible that anything should here be set down more fully or more clearly than I have before spoken. Let your Pharisaical schools teach you what they will from their unstable and slippery foundation, which reaches not its end or its aim. When at length you shall have been taught as accurately as possible the Alchemistic Art, nothing in the nature of things shall then at length be so difficult which cannot be made manifest to you by the aid of this Art. Nature, indeed, herself does not bring forth anything into the light which is advanced to its highest perfection, as can be seen in this place from the unity, or the union, of our duality. But a man ought by Spagyric preparations to lead it thither where it was ordained by Nature. Let this have been sufficiently said by me, concerning the process of the ancients and my correction of the Tincture of the Philosophers, so far as relates to its preparation.

Moreover, since now we have that treasure of the Egyptians in our hands, it remains that we turn it to our use: and this is offered to us by the Spagyric Magistery in two ways. According to the former mode it can be applied for the renewing of the body; according to the latter it is to be used for the transmutation of metals. Since, then, I, Theophrastus Paracelsus, have tried each of them in different ways, I am willing to put them forward and to describe them according to the signs indeed of the work, and as in experience and proof they appeared to me better and more perfectly.

CHAPTER VI

Concerning the Transmutation of Metals by the Perfection of Medicine

If the Tincture of the Philosophers is to be used for transmutation, a pound of it must be projected on a thousand pounds of melted Sol. Then, at length, will a Medicine have been prepared for transmuting the leprous moisture of the metals. This work is a wonderful one in the light of Nature, namely, that by the Magistery, or the operation of the Spagyrist, a metal, which formerly existed, should perish, and another be produced. This fact has rendered that same Aristotle, with his ill-founded philosophy, fatuous. For truly, when the rustics in Hungary cast iron at the proper season into a certain fountain, commonly called Zifferbrunnen, it is consumed into rust, and when this is liquefied with a blast-fire, it soon exists as pure Venus, and never more returns to iron. Similarly, in the mountain commonly called Kuttenberg, they obtain a lixivium out of marcasites, in which iron is forthwith turned into Venus of a high grade, and more malleable than the other produced by Nature. These things, and more like them, are known to simple men rather than to sophists, namely, those which turn one appearance of a metal into another. And these things, moreover, through the remarkable contempt of the ignorant, and partly, too, on account of the just envy of the artificers, remain almost hidden. But I myself, in Istria, have often brought Venus to more than twenty-four (al. 38) degrees, so that the colour of Sol could not mount higher, consisting of Antimony or or Quartal, which Venus I used in all respects as other kinds.

But though the old artists were very desirous of this arcanum, and sought it with the greatest diligence, nevertheless, very few could bring it by means of a perfect preparation to its end. For the transmutation of an inferior metal into a superior one brings with it many difficulties and obstacles, as the change of Jove into Luna, or Venus into Sol. Perhaps on account of their sins God willed that the Magnalia of Nature should be hidden from many men. For sometimes, when this Tincture has been prepared by artists, and they were not able to reduce their projection to work its effects, it happened that, by their carelessness and bad guardianship, this was eaten up by fowls, whose feathers thereupon fell off, and, as I myself have seen, grew again. In this way transmutation, through its abuse from the carelessness of the artists, came into Medicine and Alchemy. For when they were unable to use the Tincture according to their desire, they converted the same to the renovation of men, as shall be heard more at large in the following chapter.

CHAPTER VII

Concerning the Renovation of Men

Some of the first and primitive philosophers of Egypt have lived by means of this Tincture for a hundred and fifty years. The life of many, too, has been extended and prolonged to several centuries, as is most clearly shewn in different histories, though it seems scarcely credible to any one. For its power is so remarkable that it extends the life of the body beyond what is possible to its congenital nature, and keeps it so firmly in that condition that it lives on in safety from all infirmities. And although, indeed, the body at length comes to old age, nevertheless, it still appears as though it were established in its primal youth.

So, then, the Tincture of the Philosophers is a Universal Medicine, and consumes all diseases, by whatsoever name they are called, just like an invisible fire. The dose is very small, but its effect is most powerful. By means thereof I have cured the leprosy, venereal disease, dropsy, the falling sickness, colic, scab, and similar afflictions ; also lupus, cancer, noli-me-tangere, fistulas, and the whole race of internal diseases, more surely than one could believe. Of this fact Germany, France, Italy, Poland, Bohemia, etc., will afford the most ample evidence.

Now, Sophist, look at Theophrastus Paracelsus. How can your Apollo, Machaon, and Hippocrates stand against me? This is the Catholicum of the

Philosophers, by which all these philosophers have attained long life for resisting diseases, and they have attained this end entirely and most effectually, and so, according to their judgment, they named it The Tincture of the Philosophers. For what can there be in the whole range of medicine greater than such purgation of the body, by means whereof all superfluity is radically removed from it and transmuted? For when the seed is once made sound all else is perfected. What avails the ill-founded purgation of the sophists since it removes nothing as it ought? This, therefore, is the most excellent foundation of a true physician, the regeneration of the nature, and the restoration of youth. After this, the new essence itself drives out all that is opposed to it. To effect this regeneration, the powers and virtues of the Tincture of the Philosophers were miraculously discovered, and up to this time have been used in secret and kept concealed by true Spagyrists.

HERE ENDS THE BOOK CONCERNING THE TINCTURE OF THE PHILOSOPHERS

THE GRADATIONS OF METALS

PREFACE

WE now purpose to speak concerning gradations,* and those of such a kind that a metal dissolved or digested in them can be promoted to the degree of Sol and transmuted. Many persons endeavour to transmute the lesser metals into silver, and others, of a mediocre nature, into gold, with some difference, however, in their conjunction, so that in a cineritium, by transmutation of graduation, the lesser metals may be brought to the perfection of the greater ones—a perfection which answers any suitable tests. We will set down, then, in this place, fourteen gradations. Many others may be found, but these we willingly forget, and have collected those only which are established by experience, and are worth writing about. These we arrange with a triple differentiation. Some are strong waters, others are oils, and the rest liquids. These we arrange in a like order, as is clear from our method of treating them. That is to say, we put, first of all, strong waters, secondly oils, and lastly liquids.

The First Gradation

Take of Vitriol, Alum, and Salt Nitre, two pounds each; of Flos Aeris, Crocus of Mars, and Hæmatitis, a quarter of a pound each; of Cinnabar, a pound and a half; of Antimony,† three-eighths of a pound; of Arsenic, one-eighth of a pound.

Let all be distilled with a very strong fire into strong water, which purify and clarify after the method of such waters, and dissolve therein cemented Luna, or Part with Part, Luna and Venus; then put it in digestion for a month. Afterwards take out the residuum by fulmination, and thus you will

---

* The term gradation is used by Paracelsus in more than one sense; here it is the process by which one substance is developed into another. Care must be taken to distinguish between this and the grades of metals, etc. Thus, in gold there are said to be twenty-four grades; in silver thirty-two grades of softness; in iron forty-six grades of hardness; in lead eighteen degrees of fluxibility; twelve of malleation in copper; in Mercury eighty-three properties or branches. — *Chirurgia Minor*, Lib. III. Preface.

† From the time of Basil Valentine, Antimony played almost as important a part in the operations of Alchemy as it performed in Medicine. It is variously described by Paracelsus. Sometimes the term is made to include all marcasites, cachimiæ, talcs, ogerta, etc.—*De Morbis Metallicis*, Tract III., c. 3. Again, Antimony is a mucilage, or, that you may understand me the better, firnisium.--*Ibid.*, c. 7. It transmutes Saturn into Venus.—*De Aridura*. It receives its body from Mercury, and is the most gross nature of Mercury, after it has been purged out (that is, expelled from the prime principle). It retains all the powers and virtues of Mercury. Of all products coming forth out of the three first principles, there is none which retains the virtue of Mercury more patently than Antimony. It is nothing but Mercury coagulated through the Spirit of Salt and Sulphur. But, at the same time, understand that it is derived from the gross and rough, not from the subtle nature of the said Mercury.—*De Elemento Aquæ*, Tract V., c. 5.

find it transmuted. Let that which is still in the aquafortis be precipitated and fulminated as above, and thus the remainder of the silver can be obtained. Care should be taken that the aforesaid simples be prepared and separated, first of all, in purgation, because any impurity hinders this work of transmutation very seriously.

THE SECOND GRADATION

In this second gradation it is worth while to note carefully another process, it being one which can be adopted with greater gain and subtlety, as follows : Take of Saltpetre and of Cinnabar each one pound. Let them be pounded together, and the water distilled from them; which water preserve. Do the same with an equal weight of Antimony and Arsenic. Mix together the three waters, and add of Salt Nitre, Alum, and Vitriol each one pound. Distil all again, after the manner of Aquafortis. Afterwards pour this on its *Caput Mortuum*, which has been finely powdered. Again distil it to its ultimate spirit, and clarify it, just as any other aquafortis. In the case of all metals which have been dissolved in it, and have remained in digestion until perfect, its powers of operation are incredibly great. In very truth, there are latent in a composition with these ingredients all the forces of those metals which have in themselves a corporeal matter. For three of such distillations tinge so powerfully, by the force of the water, that scarcely any greater or more powerful means of working with strong waters could be found.

THE THIRD GRADATION

The third gradation, which is reckoned as last among the strong waters, is to be understood and compounded as follows :—Take of Cinnabar, Arsenic, and Antimony, each half a pound, of Saltpetre two pounds, and of Sulphur half a pound. Let these be pounded together, mixed, and distilled to strong water with a very powerful fire. Afterwards take two parts of this water, of Common Alum and Alumen Plumosum each a part and a half, of Vitriol one part, of Verdigris and Crocus of Mars each half a part. Let all these be distilled together into a strong water with a very violent fire. At length, for the whole of this water, take two parts of the *Caput Mortuum* ; and of Antimony, Verdigris, Cinnabar, and Sulphur, half a part each. Distil these from their dregs by strongly driving them into a receiver. Afterwards, in this water, when it has been clarified, dissolve half a part with ten parts of flowers of brass and crocus of Mars, and let it be digested therein. You will afterwards find more of the residuum transmuted to Sol than you would deem possible to the Art.

THE FOURTH GRADATION

Let us now speak about gradations made with oils, which do not dissolve after the mode and form of strong waters, but in digestion, and thus accomplish their perfection. The first gradation of this kind is made with Oil of Antimony, in which is latent a wonderful tincture of redness. Let the following be the process adopted:—Take of Antimony one pound, and of sublimated Mercury half a pound. Let them both be distilled together over a powerful fire by means of an alembic, and the redness will ascend as thick as blood. This tinges and graduates all Luna into Sol, and brings the latter when pale to the highest degree of permanent colour.

THE FIFTH GRADATION

The fifth gradation, whereof the oil is reckoned second in order, is made in the manner which follows:—Take of the Oil of the Philosophers one pound, with which mix half a pound respectively of Calcined Alum and of Citrine Colcothar. Distil a second time, and afterwards rectify to purity and a constant colour. Put in Luna, and let it remain in digestion. Reduce what remains, separate it in aquafortis, and, lastly, fulminate by means of Saturn.

THE SIXTH GRADATION

The sixth gradation, third in order among the oils, is made in the following way:—Take of Live Sulphur* two pounds, and of Flax Oil (linseed oil) four pounds. Let these be formed into a compound, and this be distilled into an oil. To this let there be added the same quantity of Live Sulphur, and let it be treated just as it was for the compound. Let it be digested in horse-dung for a month, or if longer, so much the better. After this let there be added one-eighth of a pound of each of the following: Salt Nitre, Vitriol, Alum, Flos Aeris, Crocus of Mars, and Cinnabar. Distil whatever will ascend. Remove the liquids, keeping only the oils. Put these apart in a glass cucurbite, adding the species as above, and the *Caput Mortuum* in powder. Distil again as before. Afterwards pour it off again from the dregs, let it be putrefied a second time for a month, and further distilled. When the colours are gone or separated, keep the red, and rectify it as required. Lastly, let plates of Luna be digested at the proper time, and at length reduced by a process of fulmination.

THE SEVENTH GRADATION

Gradations which are produced by liquids are found in two different forms, namely, the tenacious and the watery. First let us speak concerning the tenacious. Take one pound of Honey, and in it decoct one-quarter of a pound each of Vitriol and of Alum with an eighth of a pound of Iamen. Distil the water from these over a strong fire, and add thereto one-eighth of a pound each of the *Caput Mortuum* from a human cranium, and half a pound of Sulphur. Decoct into the form of a hepar and digest for a month; then distil and rectify with water until pure. Afterwards add one-eighth of a pound each of Sal Ammoniac, Flos Aeris, Crocus of Mars, and Alum; a quarter of a pound of Vitriol, and two ounces each of fixed Antimony and fixed Red Arsenic. Pound

---

* Live Sulphur is that of which fragments or particles will cohere without it being in a dissolved or fluid state.—*De Præparationibus*, Lib. I., Tract 2.

these together, put into water, and let them stand in heat for ten days. Afterwards let the liquid be separated from the dregs. Purify and project into the mixture metallic plates; then let them remain in moderate digestion until perfect. Lastly, let the matter be burnt, separated, and fulminated.

The Eighth Gradation

The gradation by the second liquid is as follows:

Take a sufficient quantity of aquafortis. In one part thereof dissolve Sol, in another part Venus, in another part Mars. Mix these solutions together, and afterwards distil the water from them. Pour this again on its dregs, and once more distil and pour as before, until a thick liquid is produced therefrom. To this add five parts of distilled and prepared Honey. Let all be digested for a month, and afterwards separate the phlegma. Keep the liquid, and in it let projected metallic plates be digested for a month. Lastly, let it be coagulated into a mass, and into one distinct body. Let this be subjected to a process of fulmination and quartation. Fulminate a second time, and thus will be found an excellent transmutation by means of liquid.

The Ninth Gradation

By the third liquid the ninth gradation is made in the following manner:

Take aquafortis, in which dissolve Verdigris, and let both be kept together in horse dung for the space of a month. Now distil the water from the dregs, pour it on again, distil and pour it over several times, until an oil is produced from it. Into that liquid put metallic filings, and in the course of its being digested you will find a transmutation. Although the liquid may be small in quantity, nevertheless it graduates most effectually, and affects the very largest amount of metal in proportion to its own quantity.

The Tenth Gradation

In the following manner the fourth liquid is to be understood:

Take the best Aquafortis, and in it dissolve as much Steel as possible. Let these remain in digestion for a month, and from them will be formed a compound of one colour. Let this compound also be distilled into a liquid, in which metallic filings must remain in digestion until the liquid is incorporated. Then let both be fulminated together—that is to say, the liquid and the metal—by means of Saturn. Then will be found this transmutation, which must be separated and prepared in the usual way.

The Eleventh Gradation

The eleventh gradation is made by the first aqueous liquid according to the formula subjoined:

Take four pounds of the most highly purified Saltpetre, and repurge this from its phlegma by combustion. Add two pounds of Common Salt duly prepared. Mix these together, and distil with an alembic six or nine times, until the Salts altogether pass over through the alembic into the receiver placed ready for the purpose. Then take two pounds of this Water; two ounces each of Flowers of Antimony, Flos Aeris, Flos Martis, and Flower of Sulphur, with two ounces and a half respectively of Sal Ammoniac and of Alum. Mix all these together, and let them remain in digestion for four and twenty days. After this let them be separated in the purest water. Afterwards let Luna and the metals be graduated by digestion, fulminated by Saturn, separated by quartification, and fulminated a second time.

TWELFTH GRADATION

By means of the second aqueous liquid the twelfth gradation is produced in the following manner:

Take three pounds of the most highly corrected *Vinum Ardens;* one pound of the Water of Saltpetre; half a pound of the Water of Common Salt; and three quarters of a pound respectively of Vitriol, Alumen Plumosum, and Alumen Aochi. Let these be combined to form a mixture, and distil this six times from the *Caput Mortuum.* In this water let metals be digested, when they will be fixed and transmuted, as we have said above concerning the others.

THE THIRTEENTH GRADATION

By the third aqueous liquid the thirteenth gradation is produced in the following manner:

Take one pound of Isteris of Blood. Distil it thirteen times from its dregs, and place in it two ounces each of Flos Aeris and of Sulphur. Let them be dissolved in horse-dung for a month. Afterwards put in Calx Lunæ, so that the colour and substance may be consumed. Afterwards let them be coagulated and fulminated in Saturn. Know that in this liquid common Mercury, as well as that of metals, is coagulated according to the conditions of Transmutation.

THE FOURTEENTH GRADATION

The fourth liquid in this place is the Water of Mercury, which is made for the fourteenth gradation as follows below:

Take one pound of Mercury sublimated twenty times with Sal Ammoniac, and one ounce respectively of the Flowers of Venus, Mars, Sulphur, and Antimony. Grind and mix all together, and then let them be resolved into a water. This water requires no other labour whatever. Metals projected into it, digested for a short time, and afterwards fulminated, are graduated in a wonderful manner.

HERE ENDS THE BOOK OF GRADATIONS

THE TREASURE OF TREASURES FOR ALCHEMISTS

By Philippus Theophrastus Bombast, Paracelsus the Great

NATURE begets a mineral in the bowels of the earth. There are two kinds of it, which are found in many districts of Europe. The best which has been offered to me, which also has been found genuine in experimentation, is externally in the figure of the greater world, and is in the eastern part of the sphere of the Sun. The other, in the Southern Star, is now in its first efflorescence. The bowels of the earth thrust this forth through its surface. It is found red in its first coagulation, and in it lie hid all the flowers and colours of the minerals. Much has been written about it by the philosophers, for it is of a cold and moist nature, and agrees with the element of water.

So far as relates to the knowledge of it and experiment with it, all the philosophers before me, though they have aimed at it with their missiles, have gone very wide of the mark. They believed that Mercury and Sulphur were the mother of all metals, never even dreaming of making mention meanwhile of a third; and yet when the water is separated from it by Spagyric Art the truth is plainly revealed, though it was unknown to Galen or to Avicenna. But if, for the sake of our excellent physicians, we had to describe only the name, the composition, the dissolution, and coagulation, as in the beginning of the world Nature proceeds with all growing things, a whole year would scarcely suffice me, and, in order to explain these things, not even the skins of numerous cows would be adequate.

Now, I assert that in this mineral are found three principles, which are Mercury, Sulphur, and the Mineral Water which has served to naturally coagulate it. Spagyric science is able to extract this last from its proper juice when it is not altogether matured, in the middle of the autumn, just like a pear from a tree. The tree potentially contains the pear. If the Celestial Stars and Nature agree, the tree first of all puts forth shoots in the month of March; then it thrusts out buds, and when these open the flower appears, and so on in due order, until in autumn the pear grows ripe. So is it with the minerals. These are born, in like manner, in the bowels of the earth. Let the Alchemists who are seeking the Treasure of Treasures carefully note this. I will shew them the way, its beginning, its middle, and its end. In the following treatise I will describe the proper Water, the proper Sulphur, and the proper Balm thereof. By means of these three the resolution and composition are coagulated into one.

Concerning the Sulphur of Cinnabar

Take mineral Cinnabar and prepare it in the following manner. Cook it with rain water in a stone vessel for three hours. Then purify it carefully, and dissolve it in Aqua Regis, which is composed of equal parts of vitriol, nitre, and sal ammoniac. Another formula is vitriol, saltpetre, alum, and common salt.

Distil this in an alembic. Pour it on again, and separate carefully the pure from the impure thus. Let it putrefy for a month in horse-dung; then separate the elements in the following manner. If it puts forth its sign,* commence the distillation by means of an alembic with a fire of the first degree. The water and the air will ascend; the fire and the earth will remain at the bottom. Afterwards join them again, and gradually treat with the ashes. So the water and the air will again ascend first, and afterwards the element of fire, which expert artists recognise. The earth will remain in the bottom of the vessel. This collect there. It is what many seek after and few find.

This dead earth in the reverberatory you will prepare according to the rules of Art, and afterwards add fire of the first degree for five days and nights. When these have elapsed you must apply the second degree for the same number of days and nights, and proceed according to Art with the material enclosed. At length you will find a volatile salt, like a thin alkali, containing in itself the Astrum of fire and earth.† Mix this with the two elements that have been preserved, the water and the earth. Again place it on the ashes for eight days and eight nights, and you will find that which has been neglected by many Artists. Separate this according to your experience, and according to the rules of the Spagyric Art, and you will have a white earth, from which its colour has been extracted. Join the element of fire and salt to the alkalised earth. Digest in a pelican to extract the essence. Then a new earth will be deposited, which put aside.

Concerning the Red Lion

Afterwards take the lion in the pelican which also is found [at] first, when you see its tincture, that is to say, the element of fire which stands above the water, the air, and the earth. Separate it from its deposit by trituration.

---

* The *Sign* is nothing else than the mark left by an operation. The house constructed by the architect is the sign of his handicraft, whereby his skill and art are determined. Thus the sign is the achievement itself.—*De Colica*.

† The earth also has its Astrum, its course, its order, just as much as the Firmament, but peculiar to the element. So also there is an Astrum in the water, even as in the earth, and in like manner with air and fire. Consequently, the upper Astrum has the Astra of the elements for its medium, and operates through them, by an irresistible attraction. Through this operation of the superior and inferior Astra, all things are fecundated, and led on to their end.—*Explicatio Totius Astronomiæ*. Without the Astra the elements cannot flourish. . . . In the Astrum of the earth all the celestial operations thrive. The Astrum itself is hidden, the bodies are manifest. . . . The motion of the earth is brought about by the Astrum of the earth. . . . There are four Astra in man (corresponding to those of the four elements), for he is the lesser world. - *De Caducis*, Par. II.

Thus you will have the true *aurum potabile*.* Sweeten this with the alcohol of wine poured over it, and then distil in an alembic until you perceive no acidity to remain in the Aqua Regia.

This Oil of the Sun, enclosed in a retort hermetically sealed, you must place for elevation that it may be exalted and doubled in its degree. Then put the vessel, still closely shut, in a cool place. Thus it will not be dissolved, but coagulated. Place it again for elevation and coagulation, and repeat this three times. Thus will be produced the Tincture of the Sun, perfect in its degree. Keep this in its own place.

Concerning the Green Lion

Take the vitriol of Venus,† carefully prepared according to the rules of Spagyric Art, and add thereto the elements of water and air which you have reserved. Resolve, and set to putrefy for a month according to instructions. When the putrefaction is finished, you will behold the sign of the elements. Separate, and you will soon see two colours, namely, white and red. The red is above the white. The red tincture of the vitriol is so powerful that it reddens all white bodies, and whitens all red ones, which is wonderful.

Work upon this tincture by means of a retort, and you will perceive a blackness issue forth. Treat it again by means of the retort, repeating the operation until it comes out whitish. Go on, and do not despair of the work. Rectify until you find the true, clear Green Lion, which you will recognise by its great weight. You will see that it is heavy and large. This is the Tincture, transparent gold. You will see marvellous signs of this Green Lion, such as could be bought by no treasures of the Roman Leo. Happy he who has learnt how to find it and use it for a tincture!

This is the true and genuine Balsam,‡ the Balsam of the Heavenly Stars, suffering no bodies to decay, nor allowing leprosy, gout, or dropsy to take root. It is given in a dose of one grain, if it has been fermented with Sulphur of Gold.

Ah, Charles the German, where is your treasure? Where are your philosophers? Where your doctors? Where are your decocters of woods, who at least purge and relax? Is your heaven reversed? Have your stars wandered out of their course, and are they straying in another orbit, away

---

Aurum Potabile, that is, Potable Gold, Oil of Gold, and Quintessence of Gold, are distinguished thus. *Aurum Potibile* is gold rendered potable by intermixture with other substances, and with liquids. Oil of Gold is an oil extracted from the precious metal without the addition of anything. The Quintessence of Gold is the redness of gold extracted therefrom and separated from the body of the metal.—*De Membris Contractis*, Tract II., c. 2.

† If copper be pounded and resolved without a corrosive, you have Vitriol. From this may be prepared the quintessence, oil, and liquor thereof.—*De Morbis Tartareis.* Cuprine Vitriol is Vitriol cooked with Copper.—*De Morbis Vermium*, Par. 6. Chalcanthum is present in Venus, and Venus can by separation be reduced into Chalcanthum.—*Chirurgia Magna*. Pars. III., Lib. IV.

‡ There is, indeed, diffused through all things a Balsam created by God, without which putrefaction would immediately supervene. Thus, in corpses which are anointed with Balsam we see that corruption is arrested, and thus in the physical body we infer that there is a certain natural and congenital Balsam, in the absence of which the living and complete man would not be safe from putrefaction. Nothing removes this Balsam but death. But this kind differs from what is more commonly called Balsam, in that the one is conservative of the living, and the other of the dead.—*Chirurgia Magna*, Pt. II., Tract II., c. 3. The confection of Balsam requires special knowledge of chemistry, and it was first discovered by the Alchemists.—*Ibid.*, Pt. I., Tract II., c. 4.

from the line of limitation, since your eyes are smitten with blindness, as by a carbuncle, and other things making a show of ornament, beauty, and pomp? If your artists only knew that their prince Galen—they call none like him—was sticking in hell, from whence he has sent letters to me, they would make the sign of the cross upon themselves with a fox's tail. In the same way your Avicenna sits in the vestibule of the infernal portal; and I have disputed with him about his *aurum potabile*, his Tincture of the Philosophers, his Quintessence, and Philosophers' Stone, his Mithridatic, his Theriac, and all the rest. O, you hypocrites, who despise the truths taught you by a true physician, who is himself instructed by Nature, and is a son of God himself! Come, then, and listen, impostors who prevail only by the authority of your high positions! After my death, my disciples will burst forth and drag you to the light, and shall expose your dirty drugs, wherewith up to this time you have compassed the death of princes, and the most invincible magnates of the Christian world. Woe for your necks in the day of judgment! I know that the monarchy will be mine. Mine, too, will be the honour and glory. Not that I praise myself: Nature praises me. Of her I am born; her I follow. She knows me, and I know her. The light which is in her I have beheld in her; outside, too, I have proved the same in the figure of the microcosm, and found it in that universe.

But I must proceed with my design in order to satisfy my disciples to the full extent of their wish. I willingly do this for them, if only skilled in the light of Nature and thoroughly practised in astral matters, they finally become adepts in philosophy, which enables them to know the nature of every kind of water.

Take, then, of this liquid of the minerals which I have described, four parts by weight; of the Earth of red Sol two parts; of Sulphur of Sol one part. Put these together into a pelican, congelate, and dissolve them three times. Thus you will have the Tincture of the Alchemists. We have not here described its weight: but this is given in the book on Transmutations.[*]

So, now, he who has one to a thousand ounces of the *Astrum Solis* shall also tinge his own body of Sol.

If you have the Astrum of Mercury, in the same manner, you will tinge the whole body of common Mercury. If you have the Astrum of Venus you will, in like manner, tinge the whole body of Venus, and change it into the best metal. These facts have all been proved. The same must also be understood as to the Astra of the other planets, as Saturn, Jupiter, Mars, Luna, and the rest. For tinctures are also prepared from these: concerning which we now make no mention in this place, because we have already dwelt at sufficient length upon them in the book on the Nature of Things and in the Archidoxies. So, too, the first entity of metals and terrestrial minerals have been made sufficiently clear for Alchemists to enable them to get the Alchemists' Tincture.

---

[*] It is difficult to identify the treatise to which reference is made here. It does not seem to be the seventh book concerning The Nature of Things, nor the ensuing tract on *Cements*. The general question of natural and artificial weight is discussed in the *Aurora of the Philosophers*. No detached work on Transmutations has come down to us.

This work, the Tincture of the Alchemists, need not be one of nine months; but quickly, and without any delay, you may go on by the Spaygric Art of the Alchemists, and, in the space of forty days, you can fix this alchemical substance, exalt it, putrefy it, ferment it, coagulate it into a stone, and produce the Alchemical Phœnix.* But it should be noted well that the Sulphur of Cinnabar becomes the Flying Eagle, whose wings fly away without wind, and carry the body of the phœnix to the nest of the parent, where it is nourished by the element of fire, and the young ones dig out its eyes: from whence there emerges a whiteness, divided in its sphere, into a sphere and life out of its own heart, by the balsam of its inward parts, according to the property of the cabalists.

---

* Know that the Phœnix is the soul of the Iliaster (that is, the first chaos of the matter of all things). . . . It is also the Iliastic soul in man.—*Liber Azoth*, S. V., *Practica Lineæ Vitæ*.

HERE ENDS THE TREASURE OF THE ALCHEMISTS

CONCERNING THE

TRANSMUTATIONS OF METALS AND OF CEMENTS

By THEOPHRASTUS

JUST as we have given instructions concerning other transmutations, in the same way also we will fulfil our task with reference to cements. We will make mention of six cements, under which, indeed, all the kinds of cements will be comprised, with which we shall deal singly in serial order according to the recipes and modes of operation. The consideration has to be made general in all respects, so that all the cements may be reduced to one mode of fixation and colouring. These two conditions should not be separated, but they should always persist and remain together in one cineration, coloration, and quartation, just as the properties of true gold are conditioned.

This book on cements does not state how inferior metals are to be transmuted into others, as lead into copper, or iron into copper, etc. But this at least it teaches: how metals may be cemented into the chief metal, gold, that is, Sol. For cements with other recipes cannot perfectly fulfil their operation for transmuting to other metals; but in these only there is a complete and rapid work of transmutation into Sol, which masters all the other metals, not, indeed, by quartation, but by colouring and tincture.* And notice should be taken what are the bodies which confer their own concordance as though belonging to the same species. For there are some bodies which are receptive, and others which are not so. Some have first of all to be reduced

---

* From all that has come down to us concerning the labours and investigations of the old philosophers, we see how indefatigable was their search after the best method for preserving and lengthening life. But being themselves devoid of a perfect instruction in the preparation of medicines, they did not hesitate to have recourse to the Alchemists, and thus, by the combined labours of both parties, there at length arose a genuine science of pharmacy, which then, by means of various chemical experiments devoted to medicine, was marvellously propagated and increased. But that which they call the Tincture excelled all. Yet, at the same time, it had fallen into a certain discredit, owing to the gold-makers, who thought it chiefly useful for the transmutation of metals. The philosophers of old having compounded the Tincture, whereby they transmuted the colours of metals and purged away their dross, as might have been expected, next began to think of making use of it for the purpose of Medicine; and seeing that the flowers of the metals were endowed with greater virtues than the metals themselves, they attempted to utilize these in the interests of the physician. Accordingly, whether from the benignant disposition of Heaven or through the fertility of their minds, those Tinctures were discovered and improved, the efficacy of which is borne witness to by ancient manuscripts, which manuscripts have been suppressed by the crowd of pseudo-medici; but we do not hesitate to publish them.—*Chirurgia Magna*, Tract III., c. 1.

to their flowers; with others this is not necessary. In like manner, some species colour according to the red Sol, others according to the clayey Sol; some in flux, some in half cement. In like manner, too, diligent attention should be paid to fire, as being that wherein all cements chiefly lie concealed, and wherein they gain their power of operation. Fire contains within itself the whole of Alchemy by its native power to tinge, graduate, and fix, which is, as it were, born with it and impressed upon it.* Every elaborator of cements, too, must attend carefully to the method of the process; for the method is even of greater moment than the prescription or recipe.

So, then, let us proceed to the series of the six cements, as being those whereby all cements are regulated. Among these the first is the Royal Cement. Paying little heed to the method of the ancients, we will follow experience as our guide, and those prescriptions which experience proves to be of no use we will omit. Thus :—

Take Flower of Brass, Antimony, Brick Tiles, Common Salt, of each half a pound. Having pounded all these very small and mixed them together, let them be imbibed with wine and dried. Repeat this process twenty-four times. With this powder let plates of Luna be cemented, in a moderate degree of fire, for four hours. Then at length take Regulus, plated and crude from the former process, and cement it with the same materials and an equal degree of fire, repeating the operation four or five times. Afterwards, having fulminated it again with cinders or ashes, reduce it once more to Luna. The instruments, such as the melting vessels, etc., must be thoroughly luted and stopped up. And although what we here set down is a somewhat lengthy process, yet you must know we make it so for the reason that experience teaches us there is no good in short processes by fire. Indeed, seeing that the continuance and force of the fire supply the chief fixation for the Royal Cement, it would really be far better if the substances spoken of were left for four days in the same kind of cement.

Note, too, that the flowers of brass should be extracted from copper by means of vitriol. For herein is contained some natural fixation when it is taken thus, and that for many reasons.

So, too, the tiles should be taken from a good house; for the roof has the power of fixing the vapours which exist in Luna, which otherwise all fly away and escape.

---

* Fire tries everything; what is impure it removes, and it brings about the manifestation of pure substances.— *Paramirum*, Lib. I., c. 1. Fire separates the fixed from the fugitive.— *De Morbis Metallicis*, Lib. II., Tract I., c. 1. Whatsoever pertains to separation belongs to the science of Alchemy. It teaches how to extract, coagulate, and separate every substance in its peculiar vessel.— *De Morbis Tartareis*, c. 16. Whatsoever man does the planets do also, but in an alchemistic sense and process. Accordingly, as the Alchemist seeks saltpetre in nitre, mercury in dung, sulphur in fire, so he also seeks the firmament, which is invisible Vulcan. When he has collected these substances and has united them, detonation follows, of that kind which in heaven is called a thunderbolt, but the Alchemist terms it *Bombard*. For he has the power of producing thunder, as in magic, which is the philosophy of Alchemy. There are foolish people who confound it with (what is now understood as) Nigromancy, yet there is a sense in which it is properly such, and in which Nigromancy is its true name from its earliest origin, being derived from the word blackness, because its initiates walk about as black as charcoal burners. They are all Nigromantics who serve Vulcan.— *De Colica*, s. v. *Paralysis*.

Salt corrects and fixes leprous Luna, cleansing it from its blackness. These four details should exist and be put into practice together; but it is the fire that must be specially observed and noticed. To this cement no other metal is applied; and after it is fulminated it discloses and exhibits gold. Therefore this cement should be considered sufficient for one.

Concerning the Second Cement

The cement which we wish to put down in the second place is only for Sol, and with regard to it there are four objects which have to be considered. The first is that Sol is sometimes found in this cement defective in the fourth or middle part, because it is not fixed, or not brought to its degree without its deficiencies, as we set down in our treatise on Gradations. Wherefore it has to be cemented in order that it may be able to retain its own volatile body, which otherwise sometimes flies off in the cement, or in the process of incineration, or else in aqua fortis.

The second object is that a good deal of Sol is found which is defective in colour, and it is necessary one should know how to bring it to its perfection of colouring without diminution of its bulk, and so that the colours may remain in the specimens.

The third is that this white, imperfect colour of Sol, having the appearance of Luna, may be cemented, so that it may retain the colour thence acquired in every specimen.

The fourth concerns the weight in which the Sol is sometimes defective, so that it is esteemed as somewhat common. This must be entirely restored to it, when it proves its higher quality by its colour, and a higher grade of Sol exists, for many reasons. For the weight deceives no true artist, as the probe may, also, for many reasons.

By means of cement gold can be perfected in these four particulars so that afterwards no defect shall be found in it, nor any volatile or unfixed condition. Let the preparation of this cement be as follows:—Antimony and Flower of Brass, of each half-a-pound; of coagulated mercury, one-fourth of a pound; let all be mixed together and imbibed with red oil of antimony until the whole is reddened. Afterwards cement with it gold in very thin plates or grains, liquefied by fire for twenty-four hours, without the heat being allowed to decrease, in a fusing vessel closely fastened. When this time has elapsed, take out the Regulus not acted upon by the crude antimony. Let it be liquefied with the addition of copper-green or borax, and afterwards pour it into a form. So you will have the very best and most abundant cement for rendering Sol free from defects and in its highest degree, fixed and permanent in all cements, incinerations, and quartations.

Concerning the Third Cement

So far we have set down the two fixations or cementations for Luna and Sol, which ought to be adopted when these metals are to be multiplied. But others, too, have to be cemented, and afterwards placed in a colouring cement. This third cement is adapted for perfecting the other metals and rendering them fit in themselves for the tincture of the other cements. For where a metal has not been prepared and smeared over for this tincture it is not able to take it, or only in a very slight degree, and by a dangerous process.

Quicksilver,* which is comprised under this cement, is not among the number of the metals, but only among metallic materials and malleable bodies. The cement is to be made thus : Antimony, one pound; Saltpetre, one pound ; common Salt and Salt of Tartar, half a pound each. Having mixed all these together, put them in a dish, placed layer by layer with plates or filings of the metals. Let them be closely shut up and cemented for twelve hours with a most powerful fire, which had been originally for the first two hours only a gentle one. When this time has elapsed, let all that remains be extracted, that is to say, the loppa (refuse) along with the Regulus. This must be noted, that all cementings of this kind do not exhibit the Regulus, but some of them only the loppas. These should be afterwards treated with Saturn, according to the ordinary method, and Saturn of the same kind burnt in the vessel ; then the metal will be found fixed upon it. And here mark the difference of the separation by means of the jar, the cupella, or the cineritium. The metal enters into the cineritium or the cupella, but in the other case it remains on the jar. Metal of this kind, which remains on the surface of the jar, you will cement a second or a third time, as above, so that it may turn out more fixed and remain on the cineritium. When this has been done, it is fixed for receiving the tincture, which is given it by cementation.

With regard to this cement it should be remarked that two or three metals can be blended together in one mixture and one body, which will be better than before. The following is the method. Take filings of Mars, Venus, and Jupiter, or Saturn. Let them stand in a fire of liquefaction for twelve hours. *Addition.* The cement will be more useful, if besides the above there be taken of Antimony and Salt each one pound; filings of Mars, Venus, and Saturn, half a pound each. Treat them as before mentioned.

The Fourth Cement

The fourth cement is that which is composed of minerals containing within them a perfect metal, and losing it by means of liquefaction. Here it should be noted that metals cannot be better fixed than when they are crude. They vanish altogether in the process of liquefaction. Transmutation of this kind takes place in minerals and metals before liquefaction, so that the metals may be fixed in their own nature, or may be transmuted into some other metal. Therefore we will here comprise two cements under one. The former is

---

* Quicksilver is generated from the Mercurial prime principle. It is not ductile, and is opposed to ductility. Although of all metals it is chiefly assimilated to Mercury, it differs in this, that it has not received ductility from the Archeus, through the weakness occasioned by its small quantity of salt and sulphur. It can, however, by the Spagyric process of addition, become a ductile metal, as is demonstrated by the philosophy of transmutation, which shews that it is capable of conversion into any metal.—*De Elemento Aquæ*, Tract III., c. 7.

for fixing a metal in a mineral without transmutation; the latter for transmuting the metal of Venus into Sol, or some other metal. It must be remembered that there are far more excellent spirits in minerals than in metals themselves. These are they which assist the gradations and the fixations of minerals when they contain in themselves the tincture and the colours of the matter, which properties have not yet been destroyed by fire, as we fully point out in other books. The following is the prescription for fixing a metal in its own mineral :—

Take of the mineral of Mars, well ground, one pound; to which add two pounds respectively of Antimony and Saltpetre. Cover them closely, lute them, and let them be kindled in a fire of liquefaction for twenty-four hours. Then pour them out. Diminish with some reduction and fulminate with Saturn. Thus you will find metal of the same kind, with good colour of its own, which can be tested in many ways and demonstrated naturally. So with other minerals whereof we make no mention here. For the spirits existing crudely in metals take precedence according to their own colours and essences.

The following is the prescription for the transmutation of minerals :— Crocus of the flowers of Mars and flowers of the Crocus of Venus, each one pound. Vitriol and Alum, each half a pound. Prepared common salt, one pound. Of the mineral, two pounds. Let all be liquefied, deprived of humidity, and cemented for twelve hours. Afterwards let them be liquefied and fulminated in Saturn. When this is done there will be found in the vat a transmutation of the cineritium. You can even, for a transmutation of this kind, add a mixture of metals, taking into account, however, the special aptitude existing in them by means of which one can be more easily transmuted than another.

The Fifth Cement

This fifth cement concerns only volatile bodies, as of common Mercury, and metals such as Saturn, Venus, Jupiter, Mars, etc. It must be remarked that the corporal Mercuries from the metals differ from the common Mercury in their tincture, since they demand more tincture from the proved metals than that common one does. So, too, it should be understood that both Mercuries, the corporal and the common, should be first of all coagulated in order that they may be able to resist the cement, and to recover their corporal substance, together with their tincture and colouring, as the best metals should.

The coagulation of Mercury* is as follows :—Take Aquafortis, weakened by a solution of Luna to such a degree that it no longer has any corrosive force nor sharpness for dissolving. Into this put either of the Mercuries before spoken of; let the water be warmed a little and afterwards stirred to a thick mass. Then the Mercury will coagulate and harden into the form of

---

* Mercury is coagulated by Lead, for no metal has greater affinity for Mercury than lead possesses. Coagulation is performed thus : Take ℥ii. of fine lead. Melt it in a vessel of clay ; remove it from the fire, and let it cool somewhat. When it approaches congelation, pour into it the same quantity of living Mercury.—*Archidoxis Magicæ*, Lib. VII.

metal. Take it out of the water; wash until clean, and then cement it with the following:—Borax, two drams and a half; Sal ammoniac, two ounces; Crocus of the Flower of Brass, and Flower of the Crocus of Mars, each six ounces; calcined Vitriol and calcined Alum, each two ounces; Hæmatitis and Bolus Armeni, each two ounces. Let them be well pounded, mixed, and imbibed several times in urine. Afterwards let them be placed, layer by layer, in a fusing vessel, with the junctions closed and luted. You will cement by observing the degrees of fire, gently for one hour, and then with a stronger heat for the next hour. Cement for four hours, and keep in a state of fluxion. Then put it in Saturn and fulminate. So you will have the transmutation of Mercury as we said above.

But when it has been cemented otherwise than in the way now described, it can still be transmuted with the following cement:—Cinnabar and Borax, of each half a dram. Let these be liquefied into one body, which sublimate after the method of Cinnabar, so that it shall still be one body. Then add the following: Calcined Common Salt, Flowers of Brass, Crocus of Mars, Bolus, of each two ounces; of the above-mentioned body, one dram. Let them be placed layer by layer in a crucible; afterwards let them be slowly heated for the first six hours, and for the next six treated with a greater fire, and at last for twelve hours subjected to the most violent heat. This having been done, again sublimate as above by the aforesaid process, and on the fourth or fifth cement you will find the cinnabar fixed, which reduce and fulminate by Saturn. You will then have its transmutation as aforesaid. In this way you can proceed to transmutation with other volatile bodies.

The Sixth Cement

It now remains to be said in what way Part with Part comes to be cemented so that it receives more of the tincture, and receives it sooner than by other like operations of the Artists, because Sol is fixed and graduated by the cement. It should be understood, too, that these should be cemented and both raised to the highest degree, prepared, subtilised, and re-purified; afterwards liquefied at the same time, and made into plates in equal weight, then stratified in a crucible closely shut, with the following powder: Cinnabar, Flower of Brass, Bloodstone, half a dram each; Sal Ammoniac, Calamine, Sulphur, Common Salt, Vitriol, Alum, and Crocus of Mars, two ounces each.

After having been well ground and mixed into one body, let them be exposed to a gentle fire, and afterwards imbibed with urine, and at length use it for the aforesaid stratification. Let them be placed at the fire six hours to liquefy: then renew and liquefy for another six hours. Do the same a third time for twelve hours; a fourth time for twenty-four hours. Lastly, liquefy by a fulmen of Saturn. In this way you have transmutation. If, however, you have selected other metals, such as Venus or Mars, add more of the powder and more heat, that they may be able to mix and be brought to a state of transmutation.

CONCLUSION

In these few words we would conclude our book on cements, believing that we have treated these matters with sufficent clearness. Although many other prescriptions for similar cementations are in vogue, we exclude them from our own enumeration, putting down in this place only those which have been by experiment proved more useful.

HERE ENDS THE TRANSMUTATIONS OF METALS AND CEMENTS

THE AURORA OF THE PHILOSOPHERS

By THEOPHRASTUS PARACELSUS

WHICH HE OTHERWISE CALLS HIS MONARCHIA *

CHAPTER I

Concerning the Origin of the Philosophers' Stone.

ADAM was the first inventor of arts, because he had knowledge of all things as well after the Fall as before.† Thence he predicted the world's destruction by water. From this cause, too, it came about that his successors erected two tables of stone, on which they engraved all natural arts in hieroglyphical characters, in order that their posterity might also become acquainted with this prediction, that so it might be heeded, and provision made in the time of danger. Subsequently, Noah found one of these tables under Mount Araroth, after the Deluge. In this table were described the courses of the upper firmament and of the lower globe, and also of the planets. At length this universal knowledge was divided into several parts, and lessened in its vigour and power. By means of this separation, one man became an astronomer, another a magician, another a cabalist, and a fourth an alchemist. Abraham, that Vulcanic Tubal-cain, a consummate astrologer and arithmetician, carried the Art out of the land of Canaan into Egypt, whereupon the Egyptians rose to so great a height and dignity that this wisdom was derived from them by other nations. The

---

* The work under this title is cited occasionally in other writings of Paracelsus, but is not included in the great folio published at Geneva in 1688. It was first issued at Basle in 1575, and was accompanied with copious annotations in Latin by the editor, Gerard Dorne. This personage was a very persevering collector of the literary remains of Paracelsus, but is not altogether free from the suspicion of having elaborated his original. The Aurora is by some regarded as an instance in point; though no doubt in the main it is a genuine work of the Sage of Hohenheim, yet in some respects it does seem to approximate somewhat closely to previous schools of Alchemy, which can scarcely be regarded as representing the actual standpoint of Paracelsus.

† He who created man the same also created science. What has man in any place without labour? When the mandate went forth: Thou shalt live by the sweat of thy brow, there was, as it were, a new creation. When God uttered His fiat the world was made. Art, however, was not then made, nor was the light of Nature. But when Adam was expelled from Paradise, God created for him the light of Nature when He bade him live by the work of his hands. In like manner, He created for Eve her special light when He said to her: In sorrow shalt thou bring forth children. Thus, and there, were these beings made human and earthy that were before like angelicals.˙ . . . Thus, by the word were creatures made, and by this same word was also made the light which was necessary to man. . . . Hence the interior man followed from the second creation, after the expulsion from Paradise. . . . Before the Fall, that cognition which was requisite to man had not begun to develop in him. He received it from the angel when he was cast out of Paradise. . . . Man was made complete in the order of the body, but not in the order of the arts.—*De Caducis*, Par. III.

patriarch Jacob painted, as it were, the sheep with various colours; and this was done by magic: for in the theology of the Chaldeans, Hebrews, Persians, and Egpytians, they held these arts to be the highest philosophy, to be learnt by their chief nobles and priests. So it was in the time of Moses, when both the priests and also the physicians were chosen from among the Magi—the priests for the judgment of what related to health, especially in the knowledge of leprosy. Moses, likewise, was instructed in the Egyptian schools, at the cost and care of Pharaoh's daughter, so that he excelled in all the wisdom and learning of that people. Thus, too, was it with Daniel, who in his youthful days imbibed the learning of the Chaldeans, so that he became a cabalist. Witness his divine predictions and his exposition of those words, " Mene, Mene, Tecelphares." These words can be understood by the prophetic and cabalistic Art. This cabalistic Art was perfectly familiar to, and in constant use by, Moses and the Prophets. The Prophet Elias foretold many things by his cabalistic numbers. So did the Wise Men of old, by this natural and mystical Art, learn to know God rightly. They abode in His laws, and walked in His statutes with great firmness. It is also evident in the Book of Samuel, that the Berelists did not follow the devil's part, but became, by Divine permission, partakers of visions and veritable apparitions, whereof we shall treat more at large in the Book of Supercelestial Things.* This gift is granted by the Lord God to those priests who walk in the Divine precepts. It was a custom among the Persians never to admit any one as king unless he were a Wise Man, pre-eminent in reality as well as in name. This is clear from the customary name of their kings; for they were called Wise Men. Such were those Wise Men and Persian Magi who came from the East to seek out the Lord Jesus, and are called natural priests. The Egyptians, also, having obtained this magic and philosophy from the Chaldeans and Persians, desired that their priests should learn the same wisdom; and they became so fruitful and successful therein that all the neighbouring countries admired them. For this reason Hermes was so truly named Trismegistus, because he was a king, a priest, a prophet, a magician, and a sophist of natural things. Such another was Zoroaster.

CHAPTER II

Wherein is Declared that the Greeks drew a large part of their Learning from the Egyptians; and how it came from them to us

When a son of Noah possessed the third part of the world after the Flood, this Art broke into Chaldæa and Persia, and thence spread into Egypt. The Art having been found out by the superstitious and idolatrous Greeks, some of them who were wiser than the rest betook themselves to the Chaldeans and

---

* No work precisely corresponding to this title is extant among the writings of Paracelsus. The subjects to which reference is made are discussed in the *Philosophia Sagax*.

Egyptians, so that they might draw the same wisdom from their schools. Since, however, the theological study of the law of Moses did not satisfy them, they trusted to their own peculiar genius, and fell away from the right foundation of those natural secrets and arts. This is evident from their fabulous conceptions, and from their errors respecting the doctrine of Moses. It was the custom of the Egyptians to put forward the traditions of that surpassing wisdom only in enigmatical figures and abstruse histories and terms. This was afterwards followed by Homer with marvellous poetical skill; and Pythagoras was also acquainted with it, seeing that he comprised in his writings many things out of the law of Moses and the Old Testament. In like manner, Hippocrates, Thales of Miletus, Anaxagoras, Democritus, and others, did not scruple to fix their minds on the same subject. And yet none of them were practised in the true Astrology, Geometry, Arithmetic, or Medicine, because their pride prevented this, since they would not admit disciples belonging to other nations than their own. Even when they had got some insight from the Chaldeans and Egyptians, they became more arrogant still than they were before by Nature, and without any diffidence propounded the subject substantially indeed, but mixed with subtle fictions or falsehoods; and then they attempted to elaborate a certain kind of philosophy which descended from them to the Latins. These in their turn, being educated herewith, adorned it with their own doctrines, and by these the philosophy was spread over Europe. Many academies were founded for the propagation of their dogmas and rules, so that the young might be instructed; and this system flourishes with the Germans, and other nations, right down to the present day.

CHAPTER III

What was Taught in the Schools of the Egyptians

The Chaldeans, Persians, and Egyptians had all of them the same knowledge of the secrets of Nature, and also the same religion. It was only the names that differed. The Chaldeans and Persians called their doctrine Sophia and Magic*; and the Egyptians, because of the sacrifice, called their wisdom priestcraft. The magic of the Persians, and the theology of the Egyptians, were both of them taught in the schools of old. Though there were many schools and learned men in Arabia, Africa, and Greece, such as Albumazar, Abenzagel, Geber, Rhasis, and Avicenna among the Arabians; and among the Greeks, Machaon, Podalirius, Pythagoras, Anaxagoras, Democritus, Plato, Aristotle, and Rhodianus; still there were different opinions amongst them as to the wisdom of the Egyptian on points wherein they themselves differed, and whereupon they disagreed with it. For this reason Pythagoras could not be

---

* Before all things it is necessary to have a right understanding of the nature of Celestial Magic. It originates from divine virtue. There is that magic which Moses practised, and there is the maleficent magic of the sorcerers. There are, then, different kinds of Magi. So also there is what is called the Magic of Nature; there is the Celestial Magus; there is the Magus of Faith, that is, one whose faith makes him whole. There is, lastly, the Magus of Perdition.—*Philosophia Sagax*, Lib. II., c. 6.

called a wise man, because the Egyptian priestcraft and wisdom were not perpectly taught, although he received therefrom many mysteries and arcana; and that Anaxagoras had received a great many as well, is clear from his discussions on the subject of Sol and its Stone, which he left behind him after his death. Yet he differed in many respects from the Egyptians. Even they would not be called wise men or Magi; but, following Pythagoras, they assumed the name of philosophy: yet they gathered no more than a few gleams like shadows from the magic of the Persians and the Egyptians. But Moses, Abraham, Solomon, Adam, and the wise men that came from the East to Christ, were true Magi, divine sophists and cabalists. Of this art and wisdom the Greeks knew very little or nothing at all; and therefore we shall leave this philosophical wisdom of the Greeks as being a mere speculation, utterly distinct and separate from other true arts and sciences.

CHAPTER IV

What Magi the Chaldeans, Persians, and Egyptians were

Many persons have endeavoured to investigate and make use of the secret magic of these wise men; but it has not yet been accomplished. Many even of our own age exalt Trithemius, others Bacon and Agrippa, for magic and the cabala*—two things apparently quite distinct—not knowing why they do so. Magic, indeed, is an art and faculty whereby the elementary bodies, their fruits, properties, virtues, and hidden operations are comprehended. But the cabala, by a subtle understanding of the Scriptures, seems to trace out the way to God for men, to shew them how they may act with Him, and prophesy from Him; for the cabala is full of divine mysteries, even as Magic is full of natural secrets. It teaches of and foretells from the nature of things to come as well as of things present, since its operation consists in knowing the inner constitution of all creatures, of celestial as well as terrestrial bodies: what is latent within them; what are their occult virtues; for what they were originally designed, and with what properties they are endowed. These and the like subjects are the bonds wherewith things celestial are bound up with things of the earth, as may sometimes be seen in their operation even with the bodily eyes. Such a conjunction of celestial influences, whereby the heavenly virtues acted upon inferior bodies, was formerly called by the Magi a Gamahea,†

---

* Learn, therefore, Astronomic Magic, which otherwise I call cabalistic. —*De Pestilitate*, Tract I. This art, formerly called cabalistic, was in the beginning named caballa, and afterwards caballia. It is a species of magic. It was also, but falsely, called Gabanala, by one whose knowledge of the subject was profound. It was of an unknown Ethnic origin, and it passed subsequently to the Chaldæans and Hebrews, by both of whom it was corrupted.— *Philosophia Sagax*, Lib. I., s. v. *Probatio in Scientiam Nectromantricam*.

† The object which received the influence and exhibited the sign thereof appears to have been termed Gamaheu, Gamahey, etc. But the name was chiefly given to certain stones on which various and wonderful images and figures of men and animals have been found naturally depicted, being no work of man, but the result of the providence and counsel of God.--*De Imaginibus*, c. 7 and c. 13. It is possible, magically, for a man to project his influence into these stones and some other substances.—*Ibid.*, c. 13. But they also have their own inherent virtue, which is indicated by the shape and the special nature of the impression.- *Ibid.*, c. 7. There was also an artificial Gamaheus invented and prepared by the Magi, and this seems to have been more powerful.—*De Carduo Angelico*.

or the marriage of the celestial powers and properties with elementary bodies. Hence ensued the excellent commixtures of all bodies, celestial and terrestrial, namely, of the sun and planets, likewise vegetables, minerals, and animals.

The devil attempted with his whole force and endeavour to darken this light; nor was he wholly frustrated in his hopes, for he deprived all Greece of it, and, in place thereof, introduced among that people human speculations and simple blasphemies against God and against His Son. Magic, it is true, had its origin in the Divine Ternary and arose from the Trinity of God. For God marked all His creatures with this Ternary and engraved its hieroglyph on them with His own finger. Nothing in the nature of things can be assigned or produced that lacks this magistery of the Divine Ternary, or that does not even ocularly prove it. The creature teaches us to understand and see the Creator Himself, as St. Paul testifies to the Romans. This covenant of the Divine Ternary, diffused throughout the whole substance of things, is indissoluble. By this, also, we have the secrets of all Nature from the four elements. For the Ternary, with the magical Quaternary, produces a perfect Septenary, endowed with many arcana and demonstrated by things which are known. When the Quaternary rests in the Ternary, then arises the Light of the World on the horizon of eternity, and by the assistance of God gives us the whole bond. Here also it refers to the virtues and operations of all creatures, and to their use, since they are stamped and marked with their arcana, signs, characters, and figures, so that there is left in them scarcely the smallest occult point which is not made clear on examination. Then when the Quaternary and the Ternary mount to the Denary is accomplished their retrogression or reduction to unity. Herein is comprised all the occult wisdom of things which God has made plainly manifest to men, both by His word and by the creatures of His hands, so that they may have a true knowledge of them. This shall be made more clear in another place.

CHAPTER V

Concerning the Chief and Supreme Essence of Things

The Magi in their wisdom asserted that all creatures might be brought to one unified substance, which substance they affirm may, by purifications and purgations, attain to so high a degree of subtlety, such divine nature and occult property, as to work wonderful results. For they considered that by returning to the earth, and by a supreme magical separation, a certain perfect substance would come forth, which is at length, by many industrious and prolonged preparations, exalted and raised up above the range of vegetable substances into mineral, above mineral into metallic, and above perfect metallic substances into a perpetual and divine Quintessence,* including in itself the

---

* Man was regarded by Paracelsus as himself in a special manner the true Quintessence. After God had created all the elements, stars, and every other created thing, and had disposed them according to His will, He proceeded, lastly, to the forming of man. He extracted the essence out of the four elements into one mass; He extracted also the

essence of all celestial and terrestrial creatures. The Arabs and Greeks, by the occult characters and hieroglyphic descriptions of the Persians and the Egyptians, attained to secret and abstruse mysteries. When these were obtained and partially understood they saw with their own eyes, in the course of experimenting, many wonderful and strange effects. But since the supercelestial operations lay more deeply hidden than their capacity could penetrate, they did not call this a supercelestial arcanum according to the institution of the Magi, but the arcanum of the Philosophers' Stone according to the counsel and judgment of Pythagoras. Whoever obtained this Stone overshadowed it with various enigmatical figures, deceptive resemblances, comparisons, and fictitious titles, so that its matter might remain occult. Very little or no knowledge of it therefore can be had from them.

CHAPTER VI

Concerning the Different Errors as to its Discovery and Knowledge

The philosophers have prefixed most occult names to this matter of the Stone, grounded on mere similitudes. Arnold, observing this, says in his "Rosary" that the greatest difficulty is to find out the material of this Stone; for they have called it vegetable, animal, and mineral, but not according to the literal sense, which is well known to such wise men as have had experience of divine secrets and the miracles of this same Stone. For example, Raymond Lully's "Lunaria" may be cited. This gives flowers of admirable virtues familiar to the philosophers themselves; but it was not the intention of those philosophers that you should think they meant thereby any projection upon metals, or that any such preparations should be made; but the abstruse mind of the philosophers had another intention. In like manner, they called their matter by the name of Martagon, to which they applied an occult alchemical operation; when, notwithstanding that name, it denotes nothing more than a hidden similitude. Moreover, no small error has arisen in the liquid of vegetables, with which a good many have sought to coagulate Mercury,* and afterwards to convert it with fixatory waters into Luna, since they supposed that he who in this way could coagulate it without the aid of metals would succeed in becoming the chief master. Now, although the liquids of some vegetables do effect this, yet the result is due merely to the resin, fat, and earthy sulphur with which they abound. This attracts to itself the moisture

---

essence of wisdom, art, and reason out of the stars, and this twofold essence He congested into one mass: which mass Scripture calls the slime of the earth. From that mass two bodies were made—the sidereal and the elementary. These, according to the light of Nature, are called the *quintum esse*. The mass was extracted, and therein the firmament and the elements were condensed. What was extracted from the four after this manner constituted a fifth. The Quintessence is the nucleus and the place of the essences and properties of all things in the universal world. All nature came into the hand of God – all potency, all property, all essence of the superior and inferior globe. All these had God joined in His hand, and from these He formed man according to His image.—*Philosophia Sagax*, Lib. I., c. 2.

* All created things proceed from the coagulated, and after coagulation must go on to resolution. From resolution proceed all procreated things.—*De Tartaro* (fragment). All bodies of minerals are coagulated by salt.—*De Naturalibus Aquis*, Lib. III., Tract 2.

of the Mercury which rises with the substance in the process of coagulation, but without any advantage resulting. I am well assured that no thick and external Sulphur in vegetables is adapted for a perfect projection in Alchemy, as some have found out to their cost. Certain persons have, it is true, coagulated Mercury with the white and milky juice of tittinal, on account of the intense heat which exists therein; and they have called that liquid "Lac Virginis"; yet this is a false basis. The same may be asserted concerning the juice of celandine, although it colours just as though it were endowed with gold. Hence people conceived a vain idea. At a certain fixed time they rooted up this vegetable, from which they sought for a soul or quintessence, wherefrom they might make a coagulating and transmuting tincture. But hence arose nothing save a foolish error.

CHAPTER VII

Concerning the Errors of those who seek the Stone in Vegetables

Some alchemists have pressed a juice out of celandine, boiled it to thickness, and put it in the sun, so that it might coagulate into a hard mass, which, being afterwards pounded into a fine black powder, should turn Mercury by projection into Sol. This they also found to be in vain. Others mixed Sal Ammoniac with this powder; others the Colcothar of Vitriol, supposing that they would thus arrive at their desired result. They brought it by their solutions into a yellow water, so that the Sal Ammoniac allowed an entrance of the tincture into the substance of the Mercury. Yet again nothing was accomplished. There are some again who, instead of the abovementioned substances, take the juices of persicaria, bufonaria, dracunculus, the leaves of willow, tithymal, cataputia, flammula, and the like, and shut them up in a glass vessel with Mercury for some days, keeping them in ashes. Thus it comes about that the Mercury is turned into ashes, but deceptively and without any result. These people were misled by the vain rumours of the vulgar, who give it out that he who is able to coagulate Mercury without metals has the entire Magistery, as we have said before. Many, too, have extracted salts, oils, and sulphurs artificially out of vegetables, but quite in vain. Out of such salts, oils, and sulphurs no coagulation of Mercury, or perfect projection, or tincture, can be made. But when the philosophers compare their matter to a certain golden tree of seven boughs, they mean that such matter includes all the seven metals in its sperm, and that in it these lie hidden. On this account they called their matter vegetable, because, as in the case of natural trees, they also in their time produce various flowers. So, too, the matter of the Stone shews most beautiful colours in the production of its flowers. The comparison, also, is apt, because a certain matter rises out of the philosophical earth, as if it were a thicket of branches and sprouts, like a sponge growing on the earth. They say, therefore, that the fruit of their tree tends towards heaven. So, then, they put forth that the whole thing hinged upon natural vegetables, though not as to its matter, because their stone contains within itself a body, soul, and spirit, as vegetables do.

CHAPTER VIII

Concerning those who have sought the Stone in Animals

They have also, by a name based only on resemblances, called this matter Lac Virginis, and the Blessed Blood of Rosy Colour, which, nevertheless, suits only the prophets and sons of God. Hence the sophists* gathered that this philosophical matter was in the blood of animals or of man. Sometimes, too, because they are nourished by vegetables, others have sought it in hairs, in salt of urine, in rebis; others in hens' eggs, in milk, and in the calx of egg shells, with all of which they thought they would be able to fix Mercury. Some have extracted salt out of foetid urine, supposing that to be the matter of the Stone. Some persons, again, have considered the little stones found in rebis to be the matter. Others have macerated the membranes of eggs in a sharp lixivium, with which they also mixed calcined egg shells as white as snow. To these they have attributed the arcanum of fixation for the transmutation of Mercury. Others, comparing the white of the egg to silver and the yolk to gold, have chosen it for their matter, mixing with it common salt, sal ammoniac, and burnt tartar. These they shut up in a glass vessel, and purified in a *Balneum Maris* until the white matter became as red as blood. This, again, they distilled into a most offensive liquid, utterly useless for the purpose they had in view. Others have purified the white and yolk of eggs, from which has been generated a basilisk. This they burnt to a deep red powder, and sought to tinge with it, as they learnt from the treatise of Cardinal Gilbert. Many, again, have macerated the galls of oxen, mixed with common salt, and distilled this into a liquid, with which they moistened the cementary powders, supposing that, by means of this Magistery, they would tinge their metals. This they called by the name of "a part with a part," and thence came—just nothing. Others have attempted to transmute tutia by the addition of dragon's blood and other substances, and also to change copper and electrum into gold. Others, according to the Venetian Art, as they call it, take twenty lizard-like animals, more or less, shut them up in a vessel, and make them mad with hunger, so that they may devour one another until only one of them survives. This one is then fed with filings of copper or of electrum. They suppose that this animal, simply by the digestion of his stomach, will bring about the desired transmutation. Finally, they burn this animal into a red powder, which they thought must be gold; but they were deceived. Others, again, having burned the fishes called truitas (? trouts), have sometimes, upon melting them, found some gold in them; but there is no other reason for it than this: Those fish sometimes in rivers

---

* So acute is the potency of calcined blood, that if it be poured slowly on iron it produces in the first place a whiteness thereon, and then generates rust.—*Scholia in Libros de Tartaro*. In Lib. II., Tract II.

and streams meet with certain small scales and sparks of gold, which they eat. It is seldom, however, that such deceivers are found, and then chiefly in the courts of princes. The matter of the philosophers is not to be sought in animals: this I announce to all. Still, it is evident that the philosophers called their Stone animal, because in their final operations the virtue of this most excellent fiery mystery caused an obscure liquid to exude drop by drop from the matter in their vessels. Hence they predicted that, in the last times, there should come a most pure man upon the earth, by whom the redemption of the world should be brought about; and that this man should send forth bloody drops of a red colour, by means of which he should redeem the world from sin. In the same way, after its own kind, the blood of their Stone freed the leprous metals from their infirmities and contagion. On these grounds, therefore, they supposed they were justified in saying that their Stone was animal. Concerning this mystery Mercurius speaks as follows to King Calid :—

"This mystery it is permitted only to the prophets of God to know. Hence it comes to pass that this Stone is called animal, because in its blood a soul lies hid. It is likewise composed of body, spirit, and soul. For the same reason they called it their microcosm, because it has the likeness of all things in the world, and thence they termed it animal, as Plato named the great world an animal."

CHAPTER IX

Concerning those who have sought the Stone in Minerals

Hereto are added the many ignorant men who suppose the stone to be three-fold, and to be hidden in a triple genus, namely, vegetable, animal, and mineral. Hence it is that they have sought for it in minerals. Now, this is far from the opinion of the philosophers. They affirm that their stone is uniformly vegetable, animal, and mineral. Now, here note that Nature has distributed its mineral sperm into various kinds, as, for instance, into sulphurs, salts, boraxes, nitres, ammoniacs, alums, arsenics, atraments, vitriols, tutias, hæmatites, orpiments, realgars, magnesias, cinnabar, antimony, talc, cachymia, marcasites, etc. In all these Nature has not yet attained to our matter; although in some of the species named it displays itself in a wonderful aspect for the transmutation of imperfect metals that are to be brought to perfection. Truly, long experience and practice with fire shew many and various permutations in the matter of minerals, not only from one colour to another, but from one essence to another, and from imperfection to perfection. And, although Nature has, by means of prepared minerals, reached some perfection, yet philosophers will not have it that the matter of the philosophic stone proceeds out of any of the minerals, although they say that their stone is universal. Hence, then, the sophists take occasion to persecute Mercury himself with various torments, as with sublimations, coagulations, mercurial waters, aquafortis, and the like. All these erroneous ways should be avoided, together with other sophistical preparations of minerals, and the purgations and fixations of spirits and metals. Wherefore all the preparations of the stone, as of Geber, Albertus Magnus, and the rest, are sophistical. Their purgations, cementations, sublimations distillations, rectifications, circulations, putrefactions, conjunctions, solutions, ascensions, coagulations, calcinations, and incinerations are utterly profitless, both in the tripod, in the athanor, in the reverberatory furnace, in the melting furnace, the accidioneum, in dung, ashes, sand, or what not; and also in the cucurbite, the pelican, retort, phial, fixatory, and the rest. The same opinion must be passed on the sublimation of Mercury by mineral spirits, for the white and the red, as by vitriol, saltpetre, alum, crocuses, etc., concerning all which subjects that sophist, John de Rupescissa, romances in his treatise on the White and Red Philosophic Stone. Taken altogether, these are merely deceitful dreams. Avoid also the particular sophistry of Geber; for example, his sevenfold sublimations or mortifications, and also the revivifications of Mercury, with his preparations of salts of urine, or salts made by a sepulchre, all which things are untrustworthy. Some others have endeavoured to fix Mercury with the sulphurs of minerals and metals, but have been greatly deceived. It is true I have seen Mercury by this Art, and by such fixations, brought into a metallic body resembling and counterfeiting good silver in all respects; but when brought to the test it has shewn itself to be false.

CHAPTER X

Concerning those who have sought the Stone and also Particulars in Minerals

Some sophists have tried to squeeze out a fixed oil from Mercury seven times sublimed and as often dissolved by means of aquafortis. In this way they attempt to bring imperfect metals to perfection: but they have been obliged to relinquish their vain endeavour. Some have purged vitriol seven times by calcination, solution, and coagulation, with the addition of two parts of sal ammoniac, and by sublimation, so that it might be resolved into a white water, to which they have added a third part of quicksilver, that it might be coagulated by water. Then afterwards they have sublimated the Mercury several times from the vitriol and sal ammoniac, so that it became a stone. This stone they affirmed, being conceived of the vitriol, to be the Red Sulphur of the philosophers, with which they have, by means of solutions and coagulations, made some progress in attaining the stone; but in projection it has all come to nothing. Others have coagulated Mercury by water of alum into a hard mass like alum itself; and this they have fruitlessly fixed with fixatory waters. The sophists propose to themselves very many ways of fixing Mercury, but to no purpose, for therein nothing perfect or constant can be had. It is therefore in vain to add minerals thereto by sophistical processes, since by all of them he is stirred up to greater malice, is rendered more lively, and rather brought to greater impurity than to any kind of perfection. So, then, the philosophers' matter is not to be sought from thence. Mercury is somewhat imperfect; and to bring it to perfection will be very difficult, nay, impossible for any sophist. There is nothing therein that can be stirred up or compelled to perfection. Some have taken arsenic several times sublimated, and frequently dissolved with oil of tartar and coagulated. This they have pretended to fix, and by it to turn copper into silver. This, however, is merely a sophistical whitening, for arsenic cannot be fixed[*] unless the operator be an Artist, and knows well its tingeing spirit. Truly in this respect all the philosophers have slept, vainly attempting to accomplish anything thereby. Whoever, therefore, is ignorant as to this spirit, cannot have any hopes of fixing it, or of giving it that power which would make it capable of the virtue of transmutation. So, then, I give notice to all that the whitening of which I have just now spoken is grounded on a false basis, and that by it the copper is deceitfully whitened, but not changed.

Now the sophists have mixed this counterfeit Venus with twice its weight of Luna, and sold it to the goldsmiths and mint-masters, until at last they have transmuted themselves into false coiners—not only those who sold, but those who bought it. Some sophists instead of white arsenic take red, and this has turned out false art; because, however it is prepared, it proves to be nothing but whiteness.

Some, again, have gone further and dealt with common sulphur, which, being so yellow, they have boiled in vinegar, lixivium, or sharpest wines, for a day and a night, until it became white. Then afterwards they sublimated it from common salt and the calx of eggs, repeating the process several times; yet, still, though white, it has been always combustible. Nevertheless, with this they have endeavoured to fix Mercury and to turn it into gold; but in vain. From this, however, comes the most excellent and beautiful cinnabar that I have ever seen. This they propose to fix with the oil of sulphur by cementation and fixation. It does, indeed, give something of an appearance, but still falls short of the desired object. Others have reduced common sulphur to the form of a hepar, boiling it in vinegar with the addition of linseed oil, or laterine oil, or olive oil. They then pour it into a marble mortar, and make it into the form of a hepar, which they have first distilled into a citrine oil with a gentle fire. But they have found to their loss that they could not do anything in the way of transmuting Luna to Sol as they supposed they would be able. As there is an infinite number of metals, so also there is much variety in the preparation of them. I shall not make further mention of these in this place, because each

---

[*] One recipe for the fixation of arsenic is as follows:—Take equal parts of arsenic and nitre. Place these in a tigillum, set upon coals, so that they may begin to boil and to evaporate. Continue till ebullition and evaporation cease, and the substances shall have settled to the bottom of the vessel like fat melting in a frying-pan; then, for the space of an hour and a half (the longer the better), set it apart to settle. Subsequently pour the compound upon marble, and it will acquire a gold colour. In a damp place it will assume the consistency of a fatty fluid.—*De Naturalibus Rebus*, c. 9. Again: The fixation of arsenic is performed by salt of urine, after which it is converted by itself into an oil.—*Chirurgia Minor*, Lib. II.

would require a special treatise. Beware also of sophisticated oils of vitriol and antimony. Likewise be on your guard against the oils of the metals, perfect or imperfect, as Sol or Luna; because although the operation of these is most potent in the nature of things, yet the true process is known, even at this day, to very few persons. Abstain also from the sophistical preparations of common mercury, arsenic, sulphur, and the like, by sublimation, descension, fixation by vinegar, saltpetre, tartar, vitriol, sal ammoniac, according to the formulas prescribed in the books of the sophists. Likewise avoid the sophisticated tinctures taken from marcasites and crocus of Mars, and also of that sophistication called by the name of " a part with a part," and of fixed Luna and similar trifles. Although they have some superficial appearance of truth, as the fixation of Luna by little labour and industry, still the progress of the preparation is worthless and weak. Being therefore moved with compassion towards the well meaning operators in this art, I have determined to lay open the whole foundation of philosophy in three separate arcana, namely, in one explained by arsenic, in a second by vitriol, and in a third by antimony; by means of which I will teach the true projection upon Mercury and upon the imperfect metals.

CHAPTER XI

Concerning the true and perfect special arcanum of Arsenic for the White Tincture

Some persons have written that arsenic is compounded of Mercury and Sulphur, others of earth and water; but most writers say that it is of the nature of Sulphur. But, however that may be, its nature is such that it transmutes red copper into white. It may also be brought to such a perfect state of preparation as to be able to tinge. But this is not done in the way pointed out by such evil sophists as Geber in "The Sum of Perfection," Albertus Magnus, Aristotle the chemist in "The Book of the Perfect Magistery," Rhasis and Polydorus; for those writers, however many they be, are either themselves in error, or else they write falsely out of sheer envy, and put forth receipts whilst not ignorant of the truth. Arsenic contains within itself three natural spirits. The first is volatile, combustible, corrosive, and penetrating all metals. This spirit whitens Venus and after some days renders it spongy. But this artifice relates only to those who practise the caustic art. The second spirit is crystalline and sweet. The third is a tingeing spirit separated from the others before mentioned. True philosophers seek for these three natural properties in arsenic with a view to the perfect projection of the wise men.*

But those barbers who practise surgery seek after that sweet and crystalline nature separated from the tingeing spirit for use in the cure of wounds, buboes,

---

* Concerning the kinds of arsenic, it is to be noted that there are those which flow forth from their proper mineral or metal, and are called native arsenics Next there are arsenics out of metals after their kind. Then there are those made by Art through transmutation. White or crystalline arsenic is the best for medicine. Yellow and red arsenic are utilised by chemists for investigating the transmutation of metals, in which arsenic has a special efficacy.—*De Naturalibus Rebus*, c. 9.

carbuncles, anthrax, and other similar ulcers which are not curable save by gentle means. As for that tingeing spirit, however, unless the pure be separated from the impure in it, the fixed from the volatile, and the secret tincture from the combustible, it will not in any way succeed according to your wish for projection on Mercury, Venus, or any other imperfect metal. All philosophers have hidden this arcanum as a most excellent mystery. This tingeing spirit, separated from the other two as above, you must join to the spirit of Luna, and digest them together for the space of thirty-two days, or until they have assumed a new body. After it has, on the fortieth natural day, been kindled into flame by the heat of the sun, the spirit appears in a bright whiteness, and is endued with a perfect tingeing arcanum. Then it is at length fit for projection, namely, one part of it upon sixteen parts of an imperfect body, according to the sharpness of the preparation. From thence appears shining and most excellent Luna, as though it had been dug from the bowels of the earth.

CHAPTER XII

General Instruction concerning the Arcanum of Vitriol and the Red Tincture to be extracted from it [*]

Vitriol is a very noble mineral among the rest, and was held always in highest estimation by philosophers, because the Most High God has adorned it with wonderful gifts. They have veiled its arcanum in enigmatical figures like the following: "Thou shalt go to the inner parts of the earth, and by rectification thou shalt find the occult stone, a true medicine." By the earth they understood the Vitriol itself; and by the inner parts of the earth its sweetness and redness, because in the occult part of the Vitriol lies hid a subtle, noble, and most fragrant juice, and a pure oil. The method of its production is not to be approached by calcination or by distillation. For it must not be deprived on any account of its green colour. If it were, it would at the same time lose its arcanum and its power. Indeed, it should be observed at this point that minerals, and also vegetables and other like things which shew greenness without, contain within themselves an oil red like blood, which is their arcanum. Hence it is clear that the distillations of the druggists are useless, vain, foolish, and of no value, because these people do not know how to extract the bloodlike redness from vegetables. Nature herself is wise, and turns all the waters of vegetables to a lemon colour, and after that into an oil which is very red like blood. The reason why this is so slowly accomplished arises from the too great haste of the ignorant operators who distil it, which causes the greenness to be consumed. They have not learnt to strengthen Nature with their own powers, which is the mode whereby that noble green

---

[*] The arcanum of vitriol is the oil of vitriol. Thus: after the aquosity has been removed in coction from vitriol, the spirit is elicited by the application of greater heat. The vitriol then comes over pure in the form of water. This water is combined with the *caput mortuum* left by the process, and on again separating in a *balneum maris*, the phlegmatic part passes off, and the oil, or the arcanum of vitriol, remains at the bottom of the vessel. – *Ibid.*

colour ought to be rectified into redness of itself. An example of this is white wine digesting itself into a lemon colour; and in process of time the green colour of the grape is of itself turned into the red which underlies the cœrulean. The greenness therefore of the vegetables and minerals being lost by the incapacity of the operators, the essence also and spirit of the oil and of the balsam, which is noblest among arcana, will also perish.

CHAPTER XIII

Special Instruction concerning the Process of Vitriol for the Red Tincture

Vitriol contains within itself many muddy and viscous imperfections. Therefore its greenness* must be often extracted with water, and rectified until it puts off all the impurities of earth. When all these rectifications are finished, take care above all that the matter shall not be exposed to the sun, for this turns its greenness pale, and at the same time absorbs the arcanum. Let it be kept covered up in a warm stove so that no dust may defile it. Afterwards let it be digested in a closed glass vessel for the space of several months, or until different colours and deep redness shew themselves. Still you must not suppose that by this process the redness is sufficiently fixed. It must, in addition, be cleansed from the interior and accidental defilements of the earth, in the following manner:—It must be rectified with acetum until the earthy defilement is altogether removed, and the dregs are taken away. This is now the true and best rectification of its tincture, from which the blessed oil is to be extracted. From this tincture, which is carefully enclosed in a glass vessel, an alembic afterwards placed on it and luted so that no spirit may escape, the spirit of this oil must be extracted by distillation over a mild and slow fire. This oil is much pleasanter and sweeter than any aromatic balsam of the drug-sellers, being entirely free from all acridity.† There will subside in the bottom of the cucurbite some very white earth, shining and glittering like snow. This keep, and protect from all dust. This same earth is altogether separated from its redness.

Thereupon follows the greatest arcanum, that is to say, the Supercelestial Marriage of the Soul, consummately prepared and washed by the blood of the

---

* So long as the viridity or greenness of vitriol subsists therein, it is of a soft quality and substance. But if it be excocted so that it is deprived of its moisture, it is thereby changed into a hard stone from which even fire can be struck. When the moisture is evaporated from vitriol, the sulphur which it contains predominates over the salt, and the vitriol turns red.—*De Pestilitate*, Tract I.

† The diagnosis of vitriol is concerned with it both in Medicine and Alchemy. In Medicine it is a paramount remedy. In Alchemy it has many additional purposes. The Art of Medicine and Alchemy consists in the preparation of vitriol, for it is worthless in its crude state. It is like unto wood, out of which it is possible to carve anything. Three kinds of oil are extracted from vitriol—a red oil, by distillation in a retort after an alchemistic method, and this is the most acid of all substances, and has also a corrosive quality—also a green and a white oil, distilled from crude vitriol by descension.—*De Vitriolo*. Nor let it be regarded as absurd that we assign such great virtues to vitriol, for therein resides, secret and hidden, a certain peculiar golden force, not corporeal but spiritual, which excellent and admirable virtue exists in greater potency and certainty therein than it does in gold. When this golden spirit of vitriol is volatilized and separated from its impurities, so that the essence alone remains, it is like unto potable gold.—*De Morbis Amentium*, *Methodus* II., c. 1.

lamb, with its own splendid, shining, and purified body. This is the true supercelestial marriage by which life is prolonged to the last and predestined day. In this way, then, the soul and spirit of the Vitriol, which are its blood, are joined with its purified body, that they may be for eternity inseparable. Take, therefore, this our foliated earth in a glass phial. Into it pour gradually its own oil. The body will receive and embrace its soul; since the body is affected with extreme desire for the soul, and the soul is most perfectly delighted with the embrace of the body. Place this conjunction in a furnace of arcana, and keep it there for forty days. When these have expired you will have a most absolute oil of wondrous perfection, in which Mercury and any other of the imperfect metals are turned into gold.

Now let us turn our attention to its multiplication. Take the corporal Mercury, in the proportion of two parts; pour it over three parts, equal in weight, of the aforesaid oil, and let them remain together for forty days. By this proportion of weight and this order the multiplication becomes infinite.

CHAPTER XIV

Concerning the Secrets and Arcana of Antimony, for the Red Tincture, with a view to Transmutation

Antimony is the true bath of gold. Philosophers call it the examiner and the stilanx. Poets say that in this bath Vulcan washed Phœbus, and purified him from all dirt and imperfection. It is produced from the purest and noblest Mercury and Sulphur, under the genus of vitriol, in metallic form and brightness. Some philosophers call it the White Lead of the Wise Men, or simply the Lead. Take, therefore, of Antimony, the very best of its kind, as much as you will. Dissolve this in its own aquafortis, and throw it into cold water, adding a little of the crocus of Mars, so that it may sink to the bottom of the vessel as a sediment, for otherwise it does not throw off its dregs. After it has been dissolved in this way it will have acquired supreme beauty. Let it be placed in a glass vessel, closely fastened on all sides with a very thick lute, or else in a stone bocia, and mix with it some calcined tutia, sublimated to the perfect degree of fire. It must be carefully guarded from liquefying, because with too great heat it breaks the glass. From one pound of this Antimony a sublimation is made, perfected for a space of two days. Place this sublimated substance in a phial that it may touch the water with its third part, in a luted vessel, so that the spirit may not escape. Let it be suspended over the tripod of arcana, and let the work be urged on at first with a slow fire equal to the sun's heat at midsummer. Then at length on the tenth day let it be gradually increased. For with too great heat the glass vessels are broken, and sometimes even the furnace goes to pieces. While the vapour is ascending different colours appear. Let the fire be moderated until a red matter is seen. Afterwards dissolve in very sharp Acetum, and throw away the dregs. Let the Acetum be abstracted and let it be again dissolved in common distilled water.

This again must be abstracted, and the sediment distilled with a very strong fire in a glass vessel closely shut. The whole body of the Antimony will ascend as a very red oil, like the colour of a ruby, and will flow into the receiver, drop by drop, with a most fragrant smell and a very sweet taste.* This is the supreme arcanum of the philosophers in Antimony, which they account most highly among the arcana of oils. Then, lastly, let the oil of Sol be made in the following way :—Take of the purest Sol as much as you will, and dissolve it in rectified spirit of wine. Let the spirit be abstracted several times, and an equal number of times let it be dissolved again. Let the last solution be kept with the spirit of wine, and circulated for a month. Afterwards let the volatile gold and the spirit of wine be distilled three or four times by means of an alembic, so that it may flow down into the receiver and be brought to its supreme essence. To half an ounce of this dissolved gold let one ounce of the Oil of Antimony be added. This oil embraces it in the heat of the bath, so that it does not easily let it go, even if the spirit of wine be extracted. In this way you will have the supreme mystery and arcanum of Nature, to which scarcely any equal can be assigned in the nature of things. Let these two oils in combination be shut up together in a phial after the manner described, hung on a tripod for a philosophical month, and warmed with a very gentle fire; although, if the fire be regulated in due proportion this operation is concluded in thirty-one days, and brought to perfection. By this, Mercury and any other imperfect metals acquire the perfection of gold.

CHAPTER XV

Concerning the Projection to be made by the Mystery and Arcanum of Antimony

No precise weight can be assigned in this work of projection, though the tincture itself may be extracted from a certain subject, in a defined proportion, and with fitting appliances. For instance, that Medicine tinges sometimes thirty, forty, occasionally even sixty, eighty, or a hundred parts of the imperfect metal. So, then, the whole business hinges chiefly on the purification of the Medicine and the industry of the operator, and, next, on the greater or lesser cleanliness and purity of the imperfect body taken in hand. For instance, one Venus is more pure than another; and hence it happens that no one fixed weight can be specified in projection. This alone is worth noting, that if the operator happens to have taken too much of the tincture, he can correct this mistake by adding more of the imperfect metal. But if there be too much of the subject, so that the powers of the tincture are weakened, this error is easily remedied by a cineritium, or by cementations, or by ablutions

---

* Antimony can be made into a pap with the water of vitriol, and then purified by sal ammoniac, and in this manner there may be obtained from it a thick purple or reddish liquor. This is oil of antimony, and it has many virtues.—*Chirurgia Magna*, Lib. V. Take three pounds of antimony and as much of sal gemmæ. Distil them together in a retort for three natural days, and so you will have a red oil, which has incredible healing power in cases of otherwise incurable wounds.—*Chirurgia Minor*, Tract II., c. 11.

in crude Antimony. There is nothing at this stage which need delay the operator; only let him put before himself a fact which has been passed over by the philosophers, and by some studiously veiled, namely, that in projections there must be a revivification, that is to say, an animation of imperfect bodies—nay, so to speak, a spiritualisation; concerning which some have said that their metals are no common ones, since they live and have a soul.

Animation is Produced in the Following Way

Take of Venus, wrought into small plates, as much as you will, ten, twenty, or forty pounds. Let these be incrusted with a pulse made of arsenic and calcined tartar, and calcined in their own vessel for twenty-four hours. Then at length let the Venus be pulverised, washed, and thoroughly purified. Let the calcination with ablution be repeated three or four times. In this way it is purged and purified from its thick greenness and from its own impure sulphur. You will have to be on your guard against calcinations made with common sulphur. For whatever is good in the metal is spoilt thereby, and what is bad becomes worse. To ten marks of this purged Venus add one of pure Luna. But in order that the work of the Medicine may be accelerated by projection, and may more easily penetrate the imperfect body, and drive out all portions which are opposed to the nature of Luna, this is accomplished by means of a perfect ferment. For the work is defiled by means of an impure Sulphur, so that a cloud is stretched out over the surface of the transmuted substance, or the metal is mixed with the loppings of the Sulphur and may be cast away therewith. But if a projection of a red stone is to be made, with a view to a red transmutation, it must first fall on gold, afterwards on silver, or on some other metal thoroughly purified, as we have directed above. From thence arises the most perfect gold.

CHAPTER XVI

Concerning the Universal Matter of the Philosophers' Stone

After the mortification of vegetables, they are transmuted, by the concurrence of two minerals, such as Sulphur and Salt, into a mineral nature, so that at length they themselves become perfect minerals. So it is that in the mineral burrows and caves of the earth, vegetables are found which, in the long succession of time, and by the continuous heat of sulphur, put off the vegetable nature and assume that of the mineral. This happens, for the most part, where the appropriate nutriment is taken away from vegetables of this kind, so that they are afterwards compelled to derive their nourishment from the sulphur and salts of the earth, until what was before vegetable passes over into a perfect mineral. From this mineral state, too, sometimes a perfect metallic essence arises, and this happens by the progress of one degree into another.

But let us return to the Philosophers' Stone. The matter of this, as certain writers have mentioned, is above all else difficult to discover and abstruse to understand. The method and most certain rule for finding out this, as well as other subjects—what they embrace or are able to effect—is a careful examination of the root and seed by which they come to our knowledge. For this, before all things else, a consideration of principles is absolutely necessary; and also of the manner in which Nature proceeds from imperfection to the end of perfection. Now, for this consideration it is well to have it thoroughly understood from the first that all things created by Nature consist of three primal elements, namely, natural Mercury, Sulphur, and Salt in combination, so that in some substances they are volatile, in others fixed. Wherever corporal Salt is mixed with spiritual Mercury and animated Sulphur into one body, then Nature begins to work, in those subterranean places which serve for her vessels, by means of a separating fire. By this the thick and impure Sulphur is separated from the pure, the earth is segregated from the Salt, and the clouds from the Mercury, while those purer parts are preserved, which Nature again welds together into a pure geogamic body. This operation is esteemed by the Magi as a mixture and conjunction by the uniting of three constituents, body, soul, and spirit. When this union is completed there results from it a pure Mercury. Now if this, when flowing down through its subterranean passages and veins, meets with a chaotic Sulphur, the Mercury is coagulated by it according to the condition of the Sulphur. It is, however, still volatile, so that scarcely in a hundred years is it transformed into a metal. Hence arose the vulgar idea that Mercury and Sulphur are the matter of the metals, as is certainly reported by miners. It is not, however, common Mercury and common Sulphur which are the matter of the metals, but the Mercury and the Sulphur of the philosophers are incorporated and inborn in perfect metals, and in the forms of them, so that they never fly from the fire, nor are they depraved by the force of the corruption caused by the elements. It is true that by the dissolution of this natural mixture our Mercury is subdued, as all the philosophers say. Under this form of words our Mercury comes to be drawn from perfect bodies and from the forces of the earthly planets. This is what Hermes asserts in the following terms: "The Sun and the Moon are the roots of this Art." The Son of Hamuel says that the Stone of the philosophers is water coagulated, namely, in Sol and Luna. From this it is clearer than the sun that the material of the Stone is nothing else but Sol and Luna. This is confirmed by the fact that like produces like. We know that there are only two Stones, the white and the red. There are also two matters of the Stone, Sol and Luna, formed together in a proper marriage, both natural and artificial. Now, as we see that the man or the woman, without the seed of both, cannot generate, in the same way our man, Sol, and his wife, Luna, cannot conceive, or do anything in the way of generation, without the seed and sperm of both. Hence the philosophers gathered that a third thing was necessary, namely, the animated seed of both, the man and the woman, without which they judged that the whole of their work was fruitless and in vain. Such a sperm is Mercury, which, by the natural conjunction of both bodies, Sol and Luna, receives their nature into itself in union. Then at length, and not before, the work is fit for congress, ingress, and generation, by the masculine and feminine power and virtue. Hence the philosophers have said that this same Mercury is composed of body, spirit, and soul, and that it has assumed the nature and property of all elements. Therefore, with their most powerful genius and intellect, they asserted their Stone to be animal. They even called it their Adam, who carries his own invisible Eve hidden in his body, from that moment in which they were united by the power of the Supreme God, the Maker of all creatures. For this reason it may be said that the Mercury of the Philosophers is none other than their most abstruse, compounded Mercury, and not the common Mercury. So then they have wisely said to the sages that there is in Mercury whatever wise men seek. Almadir, the philosopher, says · "We extract our Mercury from one perfect body and two perfect natural conditions incorporated together, which indeed puts forth externally its perfection, whereby it is able to resist the fire, so that its internal imperfection may be protected by the external perfections." By this passage of the sagacious philosopher is understood the Adamic matter, the limbus of the microcosm,* and the homogeneous, unique matter of the philosophers. The sayings of these men, which we have before mentioned, are simply golden, and ever to be held in the highest esteem, because they contain nothing superfluous or without force. Summarily, then, the matter of the Philosophers' Stone is none other than a fiery and perfect Mercury extracted by Nature and Art; that is, the artificially prepared and true hermaphrodite Adam, and the microcosm. That wisest of the philosophers, Mercurius, making the same statement, called the Stone an orphan. Our Mercury, therefore, is the same which contains in itself all the perfections, force, and virtues of the Sun, which also runs through all the streets and houses of all the planets, and in its own rebirth has acquired the force of things above and things below; to the marriage of which it is to be compared, as is clear from the whiteness and the redness combined in it.

CHAPTER XVII

Concerning the Preparation of the Matter for the Philosophic Stone

What Nature principally requires is that its own philosophic man should be brought into a mercurial substance, so that it may be born into the philo-

---

* Man himself was created from that which is termed limbus. This limbus contained the potency and nature of all creatures. Hence, man himself is called the microcosmus, or world in miniature.—*De Generatione Stultorum*. Man was fashioned out of the limbus, and this limbus is the universal world.—*Paramirum Aliud*, Lib. II., c. 2. The limbus was the first matter of man. . . . Whosoever knows the limbus knows also what man is. Whatsoever the limbus is, that also is man. -*Paramirum Aliud*, Lib. IV. There is a dual limbus, man, the lesser limbus, and that Great Limbus from which he was produced—*De Podagra*, s. v. *de Limbo*. The limbus is the seed out of which all creatures are produced and grow, as the tree comes forth from its own special seed. The limbus has its ground in the word of God.—*Ibid.* The limbus of Adam was heaven and earth, water and air. Therefore, man also remains in the limbus, and contains in himself heaven and earth, air and water, and these things he also himself is.—*Paragranum Alterum*, Tract II.

sophic Stone. Moreover, it should be remarked that those common preparations of Geber, Albertus Magnus, Thomas Aquinas, Rupescissa, Polydorus, and such men, are nothing more than some particular solutions, sublimations, and calcinations, having no reference to our universal substance, which needs only the most secret fire of the philosophers. Let the fire and Azoth therefore suffice for you. From the fact that the philosophers make mention of certain preparations, such as putrefaction, distillation, sublimation, calcination, coagulation, dealbation, rubification, ceration, fixation, and the like, you should understand that in their universal substance, Nature herself fulfils all the operations in the matter spoken of, and not the operator, only in a philosophical vessel, and with a similar fire, but not common fire. The white and the red spring from one root without any intermediary. It is dissolved by itself, it copulates by itself, grows white, grows red, is made crocus-coloured and black by itself, marries itself and conceives in itself. It is therefore to be decocted, to be baked, to be fused; it ascends, and it descends. All these operations are a single operation and produced by the fire alone. Still, some philosophers, nevertheless, have, by a highly graduated essence of wine, dissolved the body of Sol, and rendered it volatile, so that it should ascend through an alembic, thinking that this is the true volatile matter of the philosophers, though it is not so. And although it be no contemptible arcanum to reduce this perfect metallic body into a volatile, spiritual substance, yet they are wrong in their separation of the elements. This process of the monks, such as Lully, Richard of England, Rupescissa, and the rest, is erroneous. By this process they thought that they were going to separate gold after this fashion into a subtle, spiritual, and elementary power, each by itself, and afterwards by circulation and rectification to combine them again in one—but in vain. For although one element may, in a certain sense, be separated from another, yet, nevertheless, every element separated in this way can again be separated into another element, but these elements cannot afterwards by circulation in a pelican, or by distillation, be again brought back into one; but they always remain a certain volatile matter, and aurum potabile, as they themselves call it. The reason why they could not compass their intention is that Nature refuses to be in this way dragged asunder and separated by man's disjunctions, as by earthly glasses and instruments. She alone knows her own operations and the weights of the elements, the separations, rectifications, and copulations of which she brings about without the aid of any operator or manual artifice, provided only the matter be contained in the secret fire and in its proper occult vessel. The separation of the elements, therefore, is impossible by man. It may appear to take place, but it is not true, whatever may be said by Raymond Lully, and of that famous English golden work which he is falsely supposed to have accomplished. Nature herself has within herself the proper separator, who again joins together what he has put asunder, without the aid of man. She knows best the proportion of every element, which man does not know, however misleading writers romance in their frivolous and false recipes about this volatile gold.

This is the opinion of the philosophers, that when they have put their matter into the more secret fire, and when with a moderated philosophical heat it is cherished on every side, beginning to pass into corruption, it grows black. This operation they term putrefaction, and they call the blackness by the name of the Crow's Head. The ascent and descent thereof they term distillation, ascension, and descension. The exsiccation they call coagulation; and the dealbation they call calcination; while because it becomes fluid and soft in the heat they make mention of ceration. When it ceases to ascend and remains liquid at the bottom, they say fixation is present.

In this manner it is the terms of philosophical operations are to be understood, and not otherwise.

CHAPTER XVIII

Concerning Instruments and the Philosophic Vessel

Sham philosophers have misunderstood the occult and secret philosophic vessel, and worse is that which is said by Aristoteles the Alchemist (not the famous Greek Academic Philosopher), giving it out that the matter is to be decocted in a triple vessel. Worst of all is that which is said by another, namely, that the matter in its first separation and first degree requires a metallic vessel; in its second degree of coagulation and dealbation of its earth a glass vessel; and in the third degree, for fixation, an earthen vessel. Nevertheless, hereby the philosophers understand one vessel alone in all the operations up to the perfection of the red stone. Since, then, our matter is our root for the white and the red, necessarily our vessel must be so fashioned that the matter in it may be governed by the heavenly bodies. For invisible celestial influences and the impressions of the stars are in the very first degree necessary for the work. Otherwise it would be impossible for the Oriental, Chaldean, and Egyptian stone to be realised. By this Anaxagoras knew the powers of the whole firmament, and foretold that a great stone would descend from heaven to earth, which actually happened after his death. To the Cabalists our vessel is perfectly well known, because it must be made according to a truly geometrical proportion and measure, and from a definite quadrature of the circle, so that the spirit and the soul of our matter, separated from their body, may be able to raise this vessel with themselves in proportion to the altitude of heaven. If the vessel be wider, narrower, higher, or lower than is fitting, and than the dominating operating spirit and soul desire, the heat of our secret philosophic fire (which is, indeed, very severe), will violently excite the matter and urge it on to excessive operation, so that the vessel is shivered into a thousand pieces, with imminent danger to the body and even the life of the operator. On the other hand, if it be of greater capacity than is required in due proportion for the heat to have effect on the matter, the work will be wasted and thrown away. So, then, our philosophic vessel must be made with the greatest care. What the material of the vessel should be is understood only by those who, in the first solution of our fixed and perfected matter have brought that matter to its own primal quintessence. Enough has been said on this point.

The operator must also very accurately note what, in its first solution, the matter sends forth and rejects from itself.

The method of describing the form of the vessel is difficult. It should be such as Nature requires, and it must be sought out and investigated from every possible source, so that, from the height of the philosophic heaven, elevated above the philosophic earth, it may be able to operate on the fruit of its own earthly body. It should have this form, too, in order that the separation and purification of the elements, when the fire drives one from the other, may be able to be accomplished, and that each may have power to occupy the place to which it adheres; and also that the sun and the other planets may exercise their operations around the elemental earth, while their course in their circuit is neither hindered nor agitated with too swift a motion. In all these particulars which have been mentioned it must have a proper proportion of rotundity and of height.

The instruments for the first purification of mineral bodies are fusing-vessels, bellows, tongs, capels, cupels, tests, cementatory vessels, cineritiums, cucurbites, bocias for aquafortis and aqua regia; and also the appliances which are required for projection at the climax of the work.

CHAPTER XIX

Concerning the Secret Fire of the Philosophers

This is a well-known sententious saying of the philosophers, " Let fire and Azoc suffice thee." Fire alone is the whole work and the entire art. Moreover, they who build their fire and keep their vessel in that heat are in error. In vain some have attempted it with the heat of horse dung. By the coal fire, without a medium, they have sublimated their matter, but they have not dissolved it. Others have got their heat from lamps, asserting that this is the secret fire of the philosophers for making their Stone. Some have placed it in a bath, first of all in heaps of ants' eggs; others in juniper ashes. Some have sought the fire in quicklime, in tartar, vitriol, nitre, etc. Others, again, have sought it in boiling water. Thomas Aquinas speaks falsely of this fire, saying that God and the angels cannot do without this fire, but use it daily. What blasphemy is this! Is it not a manifest lie that God is not able to do without the elemental heat of boiling water? All the heats excited by those means which have been mentioned are utterly useless for our work Take care not to be misled by Arnold de Villa Nova, who has written on the subject of the coal fire, for in this matter he will deceive you.

Almadir says that the invisible rays of our fire of themselves suffice.

Another cites, as an illustration, that the heavenly heat by its reflections tends to the coagulation and perfection of Mercury, just as by its continual motion it tends to the generation of metals. Again, says this same authority, "Make a fire, vaporous, digesting, as for cooking, continuous, but not volatile or boiling, enclosed, shut off from the air, not burning, but altering and penetrating. Now, in truth, I have mentioned every mode of fire and of exciting heat. If you are a true philosopher you will understand." This is what he says.

Salmanazar remarks: "Ours is a corrosive fire, which brings over our vessel an air like a cloud, in which cloud the rays of this fire are hidden. If this dew of chaos and this moisture of the cloud fail, a mistake has been committed." Again, Almadir says, that unless the fire has warmed our sun with its moisture, by the excrement of the mountain, with a moderate ascent, we shall not be partakers either of the Red or the White Stone.

All these matters shew quite openly to us the occult fire of the wise men. Finally, this is the matter of our fire, namely, that it be kindled by the quiet spirit of sensible fire, which drives upwards, as it were, the heated chaos from the opposite quarter, and above our philosophic matter. This heat, glowing above our vessel, must urge it to the motion of a perfect generation, temperately but continuously, without intermission.

CHAPTER XX

Concerning the Ferment of the Philosophers, and the Weight

Philosophers have laboured greatly in the art of ferments and of fermentations, which seems important above all others. With reference thereto some have made a vow to God and to the philosophers that they would never divulge its arcanum by similitudes or by parables.

Nevertheless, Hermes, the father of all philosophers, in the "Book of the Seven Treatises," most clearly discloses the secret of ferments, saying that they consist only of their own paste; and more at length he says that the ferment whitens the confection, hinders combustion, altogether retards the flux of the tincture, consoles bodies, and amplifies unions. He says, also, that this is the key and the end of the work, concluding that the ferment is nothing but paste, as that of the sun is nothing but sun, and that of the moon nothing but moon. Others affirm that the ferment is the soul, and if this be not rightly prepared from the magistery, it effects nothing. Some zealots of this Art seek the Art in common sulphur, arsenic, tutia, auripigment, vitriol, etc., but in vain; since the substance which is sought is the same as that from which it has to be drawn forth. It should be remarked, therefore, that fermentations of this kind do not succeed according to the wishes of the zealots in the way they desire, but, as is clear from what has been said above, simply in the way of natural successes.

But, to come at length to the weight; this must be noted in two ways.

The first is natural, the second artificial. The natural attains its result in the earth by Nature and concordance. Of this, Arnold says: If more or less earth than Nature requires be added, the soul is suffocated, and no result is perceived, nor any fixation. It is the same with the water. If more or less of this be taken it will bring a corresponding loss. A superfluity renders the matter unduly moist, and a deficiency makes it too dry and too hard. If there be over much air present, it is too strongly impressed on the tincture; if there be too little, the body will turn out pallid. In the same way, if the fire be too strong, the matter is burnt up; if it be too slack, it has not the power of drying, nor of dissolving or heating the other elements. In these things elemental heat consists.

Artificial weight is quite occult. It is comprised in the magical art of ponderations. Between the spirit, soul, and body, say the philosophers, weight consists of Sulphur as the director of the work; for the soul strongly desires Sulphur, and necessarily observes it by reason of its weight.

You can understand it thus: Our matter is united to a red fixed Sulphur, to which a third part of the regimen has been entrusted, even to the ultimate degree, so that it may perfect to infinity the operation of the Stone, may remain therewith together with its fire, and may consist of a weight equal to the matter itself, in and through all, without variation of any degree. Therefore, after the matter has been adapted and mixed in its proportionate weight, it should be closely shut up with its seal in the vessel of the philosophers, and committed to the secret fire. In this the Philosophic Sun will rise and surge up, and will illuminate all things that have been looking for his light, expecting it with highest hope.

In these few words we will conclude the arcanum of the Stone, an arcanum which is in no way maimed or defective, for which we give God undying thanks. Now have we opened to you our treasure, which is not to be paid for by the riches of the whole world.

HERE ENDS THE AURORA OF THE PHILOSOPHERS

CONCERNING THE SPIRITS OF THE PLANETS*

PROLOGUE

HAVING first of all invoked the name of the Lord Jesus Christ our Saviour, we will enter upon this work; in which we shall not only teach how to change any inferior metal into better, as iron into copper, copper into silver, and silver into gold, but also to heal all infirmities which to the pretentious and presumptuous physicians seems impossible; and—what is more still—to preserve men to a long, healthy, and perfect age. This Art was bestowed by the Lord our God, the supreme Creator, graven, as if in a book, in the body of the metals from the beginning of Creation to this end, that we might diligently learn from them. When, therefore, any man desires thoroughly and perfectly to become acquainted with this Art from its veritable foundation, it will be necessary that he should learn the same from the Master thereof, that is, from God, who created all things; who also alone knows what nature and properties He has placed in every creature. He, therefore, is able to teach every one certainly and perfectly; and from Him we can be taught absolutely what he means when he says, "Of Me ye shall learn all things." For nothing in Heaven or on earth is found so occult that He who created all things does not see through its properties, and know and perceive all. We will therefore take Him to be our Master, Operator, and Leader into this most veritable Art. Him alone will we imitate, and through Him learn and attain to the knowledge of that Nature which He Himself has, with His own finger, engendered and written on the bodies of these metals. Hence it will come to pass that the Most High Lord God will bless all His creatures in us, and will sanctify all our ways, so that in this work we may be able to bring our beginning to its desired end, and to attain the deepest joy and charity in our hearts.

But if any one shall follow his own mere private opinion, he will not only greatly deceive himself, but also all others who shall cast in their lot with him, and will bring them to great trouble. For man is assuredly born in ignorance, so that he cannot know or understand anything of himself, but only that which he receives from God, and understands from Nature. He who learns

---
* This treatise is not included in the Geneva folio, and, both in style and in the method of treatment, it corresponds closely to the Aurora. The edition made use of for this translation is the Basle 8vo. of 1570. A considerable portion of the work enters into the Paracelsican congeries, entitled *De Transmutationibus Metallorum*, Frankfort, 1681.

nothing from these is like the heathen teachers and philosophers, who follow the subtleties and crafts of their own inventions and opinions. Such teachers are Aristotle, Hippocrates, Avicenna, Galen, and the rest, who based all their arts simply upon their own opinions. Even if, at any time, they learnt anything from Nature, they destroyed it again with their own fantasies, dreams, and inventions, before they came to the final issue. By means of these, then, and their followers, nothing perfect can be discovered.

This it is which has moved and incited us to write a special book concerning Alchemy, basing it not on men, but on Nature herself, and upon those virtues and powers which God, with His own finger, has impressed upon metals. The initiator of this impression was Mercurius Trismegistus. He is not without due cause called the father of all wise men, and of all who followed this Art with love and earnest desire. He teaches and proves that God is the only author, cause, and origin of all creatures in this Art.* But he does not attribute the power and virtue of God to creatures or to visible things, as did the heathen mentioned above, and others like them.

Seeing, then, that all art must be learned from the Trinity, that is, from God the Father, God the Son, Jesus Christ, and God the Holy Ghost, three distinct persons, but one God, we will also divide this our alchemical work into three short treatises. In the first of these we will lay down what it is which the Art itself embraces, and what is the property and nature of every metal. Secondly, by what method a man may work and bring similar powers and forces of metals to a successful issue; and, thirdly, what tinctures are to be produced from the Sun and from the Moon.

---

* All arts which flourish on this earth are divine, all are from God; from no other principle do they originate. The Holy Spirit is the enlarger of the light of Nature. . . . Man of himself can discover nothing. . . . What things soever are found by the enlargement of this light of Nature within us, the same does the devil seek to corrupt, adulterate, and convert into falsehood. Thus are all arts and operations corrupted at this day. Even so is Alchemy debased and given over to lying tongues and depraved professors.—*Paragranum*, Tract IV.

CHAPTER I

Concerning Simple Fire

IN the first place, it is necessary to state clearly what this Art comprises, what is its subject, and what its peculiarities.

First and chiefly, the principal subject of this Art is fire, which always exists in one and the same property and mode of operation, nor can it receive its life from anything else.* It possesses, therefore, a state and power, common to all fires which lie hid in secret, of vivifying, just as the sun is appointed by God, and heats all things in the world, both occult, apparent, and manifest, as the spheres of Mars, Saturn, Venus, Jupiter, Mercury, and the Moon, which can shine only as they borrow their light from the Sun, and are in themselves dead. When, however, they are lighted up, as said above, they live and work according to their special properties. But the sun receives light from no other source than God Himself, Who rules it, so that in the sun God Himself is burning and shining. Just so is it with this Art.† The fire in the furnace may be compared to the sun. It heats the furnace and the vessels, just as the sun heats the vast universe. For as nothing can be produced in the world without the sun, so also in this Art nothing can be produced without this simple fire. No operation can be completed without it. It is the Great Arcanum of Art,‡ embracing all things which are comprised therein, neither can it be comprehended in anything else. It abides by itself, and needs nothing; but all others which stand in need of this can get fruition of it and have life from it, wherefore, first of all, we have undertaken that this shall be made clear.

CHAPTER II

Concerning the Multiplicity of Fire from whence spring the varieties of Metals

Having first written concerning the simple fire which lives and subsists *per se*, it now remains to speak of a manifold spirit or fire which is the cause of variety or diversity of creatures, so that not one can be found exactly like

---

* Fire is not to be regarded as an element, and so there is a distinction between fire and the firmament, which latter is an element. Fire is a matter which cooks and disintegrates, reducing into the ultimate matter, and, in this sense, fire and death are alike. For fire, like death, consumes and devours everything. Therefore, fire cannot be an element, but it can be, and is, a visible and sensible death. The other death is invisible, and is seen by no man.—*Lib. Meteorum*, c. 1.

† The congeries *De Transmutationibus Metallorum*, to which reference has already been made, gives the following variation in the reading at this point: Just so in the Spagyric art is this fire of athanor and the secret fire of the philosophers, which heats the furnace, the sphere of the vessel, and the fire of the matter, just as the sun is seen to operate in the whole world.

‡ All arcana derive from the firmament.—*Fragmenta Modus Pharmacandi*, Lib. II., Tract 1. But that fire which is an element is the firmament, and the stars are the fruits thereof.—*Lib. Meteorum*, c. 1.

another and identical in every part. This may be seen in the case of metals where no one has another exactly like itself. The Sun produces gold; the Moon another and widely different metal, namely, silver; Mars, another, namely, iron; Jupiter, tin; Venus, copper; and Saturn yet another, namely, lead; so that all these are unlike. In the same way does it hold good with men and other creatures, and the cause of this diversity is the manifoldness of fire. For example, the *Venter Equinus* produces one kind of creature through the moderate heat generated by its corruption; the *Balneum Maris* produces another; ashes another; sand, in like manner, another; the flame of fire another; coals another, and so on. This variety of creatures is not produced by the first simple fire, but from the regimen of the elements, which is various, not from the sun, but from the courses of the seven planets. And this is the reason why the universe contains no likeness amongst its individuals. For as the heat is changed every hour and minute, so all other things vary. For this transmutation takes place in the elements, on the bodies whereof it is impressed by this fire. Where there is no great mixture of the elements, Sol is produced; where it is a little more dense, Luna; where still more so, Venus; and thus according to the diversity of mixtures are produced different metals, so that no metal appears in its mineral exactly like another. It should be known, therefore, that this variety of metal is occasioned by the mixture of the elements, because that the spirits of these elements are found to be diverse and without likeness: whereas, if they were born of simple fire they would be so much alike that one could not be distinguished from another. But the manifold fire intervening, variety of form is introduced among creatures. Hence it may be easily gathered why so many and such varied forms of metals are found, and why no one is like another.*

CHAPTER III

Concerning the Spirit or Tincture of Sol

Let us now come to the spirits of the planets, or of the metals. The spirit or tincture of Sol took its beginning from a pure, subtle, and perfect fire,

---

* That fire, then, is manifold which is varied according to the diversity of the subject whereinto it flows, and by means whereof it is afterwards kindled in other subjects, as the fire of ashes, sand, the bath, filings, etc., has a mediated heat flowing from an immediate source into the subject-matter of the instrument, and from hence into the matter underlying the Art. In that manifold fire there is a difference of position. This is for the reason that nothing in the nature of things can be seen which is in all respects like to any other thing, though both come under the same species, nay, though both may be members in the same individual. One metal produces gold from that which generates silver; another brings forth the metal of Saturn, of Venus, or of Mars. Each one of these is varied according to the difference of the place whence it proceeded and was created. No two men, no two members of the same body, no two leaves of the same tree, are found exactly alike: and so of the rest. Dissimilarity proceeds not from the first fire of created things, but from the differing rule over the elements by means of the planets, and not by the sun. Every moment, by this disposition of things, the heat in the elements varies, and at the same time the form of decomposed things from their compounds, though not from the simples. Where the mixture of the elements is not so great, there is generated Sol; where it is a little greater, and less pure, is generated Luna; from that which is still more imperfect, Venus; and so of the rest, according to the mixture of the elements, the mineral of each metal is not like another, nor do the spirits of them in all respects agree one with the other. If they were generated from the simple fire alone, without the intervention of the manifold, no distinction of forms could occur either in metals or in any other created things. Why there are in use no more than seven metals, of which six are solid and the seventh fluid and thin, is explained in adept philosophy but not in Alchemy.—*De Transmutationibus Metallorum*, c. 3. But this statement concerning the seventh fluidic metal seems to be at variance with other teaching of Paracelsus, to which a congeries that has been subject to editing must naturally defer.

for which reason it far surpasses all the other spirits and tinctures of the metals. It remains constantly and fixed in the fire, nor does it fly therefrom, nor is consumed by it, but rather by its agency it becomes clearer, purer, and more beautiful. Nothing either hot or cold can injure it, or any other accident, as they can injure the other spirits or tinctures of metals, and for this reason: that the body which it once assumes it defends from all accidents and diseases, and enables it to sustain the fire without injury. This body has not such power and virtue in itself; but derives it from the spirit alone which is shut up within it. For we know with regard to the body of Mercury that it cannot sustain or endure the fire, but flies from it; but when in Sol it does not fly off but remains fixed and constant, this affords a most certain proof that it receives such a constancy from the spirit or tincture of Sol.* If, therefore, this spirit can be in Mercury, any one can infer that it would have some similar effect in the bodies of men when it is received therein. In our *Chirurgia Magna* we have said concerning the tincture of Sol that it will not only restore and preserve from weaknesses one who uses it, but also conserve him for a long and healthy life.† In like manner, the strength and virtue of other metals may be known from true experience, not from the wisdom of men and of the world, which is foolishness with God, and with His truth; and all who build and rest their hope on that wisdom are miserably deceived.

CHAPTER IV

Concerning the Spirit and Tincture of Luna

After having spoken with sufficient clearness concerning the tincture of Sol, it remains to put forward something about the tincture of Luna, and of the White Tincture which, in like manner, is produced from the perfect spirit, though it be less perfect than the spirit of Sol; but, nevertheless, it excels in purity and subtlety all the other tinctures of the metals which follow it in order This, indeed, is well known to all who handle Luna, even rustics. It does not

---

* It is well understood that the body of Sol is Mercury, which cannot at all stand the fire, but flees from it.—*De Transmutationibus Metallorum*, c. 10.

† In the collection of treatises to which reference is here made, there is the following process for the manufacture of a tincture of gold :— Let the body be first deprived of its metallic and malleable nature; that is to say, let it be corrupted; then let the residue be cleansed with sweet water, and the colour extracted by means of spirit of wine, when the desired tincture will remain at the bottom. To compose the Water of Salt : Take very white salt, but not that which has been whitened artificially; melt it several times; reduce it to an exceedingly subtle powder; mix it with the sap of raphanum. Shake it. Distil, after resolution, with an equal portion of the sap of blood. Then again distil five times. Thin plates of gold which have been purged by antimony are easily reduced to powder in this water. The powder thus prepared must be washed with sweet distilled water until it no longer savours of salt. As the salt does not penetrate into its substance it is easily removed by ablution. To compose the Spirit of Wine : Take one sextarius (about a pint) of generous wine; let it be poured into a circulatory vessel of appropriate size, that is, of such capacity that the wine can be shaken therein. Place it in a Balneum Maris to the depth which the wine occupies, and decoct for ten days. Seal all apertures of the vessels, so that nothing can escape. Then place in a cucurbite, and abstract the spirit by a slow fire. As soon as it has passed away (which you will know by the usual signs), cease to urge the fire, for the residue is a simple sublimate. Pour the spirit of wine upon the above-mentioned powder (which should be like alcohol) to the height of a palm, enclose it in a glass, keep it for a month in a warm bath to digest, when the colour will be separated and commingled with the spirit. A white powder will remain at the bottom. Having separated all these things, melt the powder, and it will be separated into a metallic water. Evaporate the spirits according to art, and the desired spirit will remain at the bottom. Perform its gradation in a retort of the proper size. This is done most conveniently by elevation, which is highly attenuating.—*Chirurgia Magna*, Part II., Tract III., c. 2.

acquire rust, nor is it consumed in the fire like the other metals, all of which Saturn draws with himself when flying from the fire, but not this one.* Hence it may be gathered that this tincture is far more excellent than those set down below, for it preserves in the fire the body it has assumed without any accident or loss. Hence it is quite clear that if this in its own corruptible body by itself produces Mercury, what it will be able to effect when extracted from it into another body. Will not that in the same way protect and defend from accidents and infirmities? Surely if it produces this Mercury in its own body, it will do the same in the bodies of men.† And it not only preserves health, but causes long life, and cures diseases and infirmities, even those which are beyond its own special grade. For the higher, more subtle, and more perfect a medicine is, so much the better and more perfectly it cures. Wherefore those are mere ignorant physicians who waste their skill only on vegetables, as herbs and the like, which are easily corrupted. With these they endeavour to accomplish results which are firm and fixed, but they do this vainly as those who beat the air. But why speak at length about these? They have not learnt better in their universities. If they were compelled to go back to the beginning, learn and study, they would think it a great disgrace. Therefore they remain in their former ignorance.

CHAPTER V

Concerning the Spirit of Venus

We have before made mention of a White Spirit, or colourless Tincture. Now we proceed to speak of a red spirit, which is produced from a thick elemental mixture of the former, to which also it is subject, though, nevertheless, it is more perfect than the spirits and tinctures of the succeeding metals. On this account it remains in the fire more constantly than the rest, so that it is not so soon burnt, nor does it so soon pass away as the other spirits which follow. The air also and the moisture of water are not so injurious to it as to Mars, just as it remains more fixedly and for a longer time in the fire. Venus has this force and property, that is to say, its body has, on account of the spirit which has been infused into it. Since, then, it produces this effect in its own body, that is, in Venus, it accomplishes as much also in man as is by Nature conceded to it. It preserves wounds in such a way that no accident can affect them, nor can the air or the water injure them. It also drives away all such diseases as are under its degree. This spirit further breaks up the bodies of metals so that they lose their malleability.‡ In the bodies of men, too,

---

* Molten lead destroys all the metals, including itself, by means of the fire, except Sol and Luna.—*Congeries Paracelsica*, c. 10.

† Since, then, the spirit of Luna is able to protect from all injury by fire or other accidents the body into which it enters, that is to say, Mercury, and to render it consistent, it is easy to gather from this, if it produces such an effect in the case of an instable and volatile body like Mercury, how much more powerfully it will act when disengaged from its own body and projected into the human body.—*Ibid.*

‡ On the other hand, if it be mixed with certain metals, even among those which are perfect, it tears asunder their bodies, so that they are no longer malleable, or capable of being treated in any way until they are set free from it,—*Ibid.*

when it is taken for a disease to which it is not suitable, it produces inconvenient results.* It is necessary, therefore, that the physician who desires to use these should be experienced, and have a good knowledge of metals. It is far better, then, to use the more perfect spirits, which may be taken without any such fear of danger. Still, since the spirits of Sol and Luna are costly, so that it is not every one who can use them for curative purposes, every one must take according to his means whatever he can get and pay for.† Every one is not of such wealth that he can prepare these medicines, so each is compelled to do what he can. Every one will easily be able to gather from what has been said that metallic medicines far exceed vegetable and animal products in their strength and power of healing. So far we have said enough, and more than enough, concerning the spirit of Venus.‡

CHAPTER VI

Concerning the Spirit of Mars

Speaking of the Spirit of Mars, this comes from a more dense and combustible mixture of the elements than was the case with the others which precede. But Mars is furnished with greater hardness than the other metals, so that it is not melted in the fire as they are. True, it is hurt by the water and the air more than they are, insomuch that it is altogether destroyed by these influences, and it is also burnt in the fire, as experience proves. So, then, its spirit is less perfect than that of any of the above. But in hardness and dryness it exceeds all the metals above or below. For not only does it render the perfect metals, Sol and Luna, proof against the hammer, but even those which rank below itself, as Jupiter, Saturn, and the like.§ Since, then, it produces this effect on metals, this is a sign that it has the same effect on the bodies of men, that is, it produces a struggling; especially when it is taken for a disease to which it is not adapted, it contorts the limbs with great pain. But when it is used and applied for wounds which do not exceed its degree, it is of powerful cleansing qualities. So, then, this spirit is endowed with no less power and potency than are of those above, so far as regards those things for which it was appointed by God and by Nature.

CHAPTER VII

Concerning the Spirit of Jupiter

Concerning the spirit of Jupiter this should be known, that it is derived from the white and pale substance of fire, together with a nature of peculiar

---

* In these cases it produces contraction of the limbs.—*Ibid.*

† It would, however, be safer to use only the spirits of the perfect metals, unless gold and silver are too expensive for a patient's resources, or too difficult in their preparation for the talent and skill of any particular physician. In that case he may be compelled to do what he has learnt to do, that is, to treat such cases with vegetable and animal preparations.—*Ibid.*

‡ Under favourable astrological circumstances, many tinctures can be extracted from Venus.—*De Causis et Origine Luis Gallicæ*, Lib. I., c. 11.

§ Nevertheless, it surpasses any other metals in hardness and dryness, destroying and decomposing them by admixture with them, and this in the case of the perfect no less than of the imperfect metals.—*Congeries Paracelsica*, c. 10,

crepitation and fragility, not malleable like Mars. It, therefore, heats other metals, and renders them capable of being broken with hammers. An example of this may be seen when it is joined with Luna, for it can scarcely be brought to its former malleability, except with the greatest labour.* The same effect it produces in all other metals, with the single exception of Saturn. If it produces this effect in the bodies of metals, it will do the same in human bodies. In these it corrodes the limbs with severe burnings and decay, so that they are completely cut off from their perfect workings, and lose them, so that they are unable to fulfil the necessary requirements of Nature. Nevertheless this spirit has in it the virtue of removing cancer, fistulas, and other similar ulcers, especially those which are of its own nature, and which do not exceed the degree which God and Nature have given to it.

CHAPTER VIII
Concerning the Spirit of Saturn

The spirit of Saturn is concrete and formed from a dry, dark, cold admixture of elements. Hence it results that, amongst all others, it has the least power of remaining and living in the fire. When, however, Sol and Luna have to be proved and purified, Saturn is added to them, and this has the effect of thoroughly purging them. Nevertheless, it is of that nature that it takes away their malleability.† It has the same effect on men, with great pains, as Jupiter and Mars. Being mixed with cold, it cannot act mildly.‡ It has the very greatest powers and virtues, whereby it cures fistulas, cancer, and similar ulcers, which come under its own degree and nature. It drives the same kind of diseases from man as it expels impurities from Luna. But if it does not go out altogether at the same time, it brings more harm than it does good. Consequently, whoever would use it must know what diseases it cures, against what it should be taken, and what effects Nature has assigned to it. If this be well considered it can do no harm.

CHAPTER IX
Concerning the Gross Spirit of Mercury

The spirit of Mercury, which is only subjected to the spirits above, has no determinate or certain form in itself. Hence it happens that it admits every metal, just as wax receives all seals, of whatever form. So this dense elementary spirit may be compared to the other spirits of the metals. For if it receive into itself the spirit of Sol, Sol will be produced from it; if Luna, Luna; and in like manner it does with the other metals. It agrees with them and takes their properties to itself. For this reason, so far as relates to its

---

* By mixture with other metals it corrupts and decomposes them, especially Luna, and only with great labour can it be separated therefrom.—*Ibid.*

† It leaves them broken and decomposed after washing.—*Ibid.*

‡ It distorts the limbs . . . with more severe pains than even tin and iron; but seeing that this spirit is coagulated with a much more intense cold than others, it does not act so violently.—*Ibid.*

body, it is appropriated to the spirits spoken of above, just as a woman to a man. For Sol is the body of Mercury, save only that Sol fixes Mercury and becomes fixed. The common Mercury is inconstant and volatile; nevertheless it is subject to all the abovementioned; and generates again not only the aforesaid metallic spirits and tinctures, but the metal itself by which the beforenamed tinctures arrive at their working. But if moderation be not observed it is impossible to perfect a tincture of this kind. If the fire which ought to vivify this tincture be too fierce, the operation will be fruitless; and so if it be too weak. Therefore it is necessary at this point to know what is the mean in this Art, and what powers and properties it has; also by what means it is to be ruled, and how to tinge the tinctures, or bring them to their perfect operation, so that they may germinate and become apparent. With these few words we would conclude this first tract.*

---

* It is prepared, then, so far as the body is concerned, from the aforesaid spirits, just as his wife is prepared for a husband, not by corporeal admixture, but when the spirit has been educed from its own metal and projected, after preparation, into Mercury, then at length it exhibits its transmutation. – *Ibid.*

THE END OF THE FIRST TREATISE

THE SECOND TREATISE

Concerning the Philosopher's Mercury, and the Medium of Tinctures

IN the first treatise we have written concerning the spirits of the metals, their tinctures, etc., making clear their properties and natures, and what each separate metal generates. In this second we will treat of the medium of tinctures, that is, the Philosophers' Mercury, whereby are made tinctures and fermentations of the metals; in seven chapters, as follows:—

CHAPTER I

From what Tinctures and Leavens are Made

Whoever wishes to have a tincture of the metals, must take Philosophers' Mercury, and project it to its own end; that is, into the quick mercury from whence it proceeded.* Hence will ensue that the Philosophers' Mercury will be dissolved in the quick mercury, and shall receive its strength, so that the Philosophers' Mercury shall kill the quick mercury and render it fixed in the fire like itself. For there is between these two mercuries as much agreement as between a man and his wife. They are both produced from the gross spirits of metals, except that the body of Sol remains fixed in the fire, but the quick mercury is not fixed. The one, however, is appropriated to the other as grain or seed to the earth, which we will illustrate by an example, thus: If anyone has sown barley he will gather barley; if corn, corn, etc. None otherwise is it in this Art. If anyone sows Sol he will gather gold, while from Luna he will collect silver, and so with regard to the other metals. In this way we say here tinctures are produced from the metals, that is, from the Philosophers' Mercury and not from quick mercury. But this produces the seed which it had before conceived.†

---

* Notwithstanding, the tincture of mercury is a supreme secret.—*De Ulcerum Curatione*, c. 10.

† The dead wife of the metal, like an uncultivated field or soil, if it be macerated or revivified by the philosophic plough (the wife remaining fixed and incorrupt during the process), it is united to the aforesaid corporal spirit by the grades of fire, into its own nature and substance, and this with the dead body of the metal. Now, this cannot be done with the crass spirit of mercury. And although the mercury or quicksilver of Sol exists and is fixed, nevertheless the common mercury, not as yet fixed, never attains to resurrection. For the resurrection of the metals is an immortal regeneration, and the medium whereby tinctures of this kind are advanced to their generation. On this account, therefore, it cannot be united to dead bodies so as to bring about their fixation, but only to extracted spirits, as to those corporeal ones above-mentioned, which are subject to the metals just as common mercury is to all metallic spirits. The crass spirit of mercury can no more generate this tincture in its substance than a concubine can bring forth legitimate offspring. In the same way must it be judged concerning the crass spirit of mercury, until the metallic and corporal spirit is produced by means of the natural matter. Without this medium it will be impossible for anything good or perfect to be accomplished in tinctures of this kind. Moreover, if the fire be too intense it cannot generate; if too slack, the same result ensues.—*De Transmutationibus Metallorum*, c. 10.

CHAPTER II

Concerning the Conjunction of the Man with the Woman

In order that the Philosophers' Mercury and the quick mercury may be joined, and this latter united with the fixed, it must of necessity be known how much of it must be taken, since more or less than the proper quantity may hinder or altogether destroy the whole business. For by superfluity the seed is suffocated, so that it cannot live until it is fixed by the Philosophers' Mercury. But by defect, since the body cannot be altogether dissolved, it is also destroyed so that it is able to produce no fruit. Wherefore it should be clearly ascertained how much of the one and the other ought to be taken, if, indeed, the artificer would bring this work to its legitimate end. Let the receipt be as follows, namely: Take one part to two, or three to four, and you will not err, but will arrive at the desired end.

CHAPTER III

Concerning the Form of the Glass Instruments

When the matter has been rightly joined, it is necessary that you should have properly-proportioned glass vessels, neither larger nor smaller than is right. If they are too large, the woman, that is, the phlegm, is dispersed, whence it ensues that the seed cannot be born; where they are too small the germ is suffocated so that it cannot come to fruit, just as when seed is sown under a tree, or among thorns, it cannot germinate, but perishes without fruit. No slight error, therefore, may arise through the vessels; and when once this has occurred it cannot again be remedied in the same operation, nor can it arrive at a satisfactory issue. Wherefore note what follows, namely, that you take three ounces and a half and four pounds; thus, having proceeded rightly, you will save the matter from being dispersed, and prevent the phlegm, or the germination, from being impeded.

CHAPTER IV

Concerning the Properties of Fire

After you have placed the matter in the proper vessels, you will cherish it with natural heat, so that the outside shall not exceed the inside. For if the heat be excessive, no conjunction will take place, because by the intense heat the matter is dispersed and burnt, so that no advantage arises from it. On this account the mid region of the air has been arranged by Nature between heaven and earth; otherwise the sun and the stars would burn up all the creatures on the earth, so that nothing could be produced from it. Take care, therefore, that between the matter and the fire you interpose an airy part of this kind, or a certain distance. In this way the heat will not easily be able in any way to do injury, nor to disperse, and still less to burn. For if the heat be insufficient neither will the spirit rest acting in no way upon its own humidity; so it will be dried or fixed. For the spirits of metals are of themselves dead, and rest, and can effect nothing unless they are vitalised. None otherwise in the great world the seed cast into the earth is dead, and cannot grow of itself unless it be vitalised by the heat of the sun. In the very first place, therefore, is it necessary to build the fire for this work in just proportion, neither too large nor too small; otherwise this work will never be carried on to its desired and perfect end.

CHAPTER V

Concerning the Signs which appear in the Union of Conjunction

When the regimen of the fire is moderated, the matter is by degrees moved to blackness. Afterwards, when the dryness begins to act upon the humidity, various flowers of different colours simultaneously rise in the glass, just as they appear in the tail of the peacock, and such as no one has ever seen before. Sometimes, too, the glass looks as though it were entirely covered with gold. When this is perceived, it is a certain indication that the seed of the man is operating upon the seed of the woman, is ruling it and fixing it. That is, the fixed Mercury acts on the quick, and begins to embrace it. Afterwards, when the humidity has died out before the process of drying, those colours disappear, and the matter at length begins to grow white, and continues to do so until it attains the supreme grade of whiteness. In the very first place, care should be taken not to hasten the matter unduly, according to the opinion of those who think that such a process is in all respects like what is perceived in the growth of corn, or in the production of a human being, the latter process occupying nine months, the former ten or twelve. Sol and Luna do not ripen so soon, or are born so soon, as the child from its mother's womb, or the grain from the womb of the earth. The higher and more perfect anything is and should be in its nature, the longer time is necessary for its production. For it should be known that everything which is born quickly perishes quickly. Both herbs and men afford a proof of this. In proportion as they are quickly produced or born is their life short. It is not so with Sol and Luna; but they have a more perfect nature than men; whence it ensues that they exhibit a long life for men and preserve them from many accidental diseases.

CHAPTER VI

Concerning the Knowledge of the Perfect Tincture

In the preceding chapter we have said how the matter itself is graduated. In this we will make clear by what means it may be recognised when it is perfect. Do this: When the White Stone of Luna stands forth in its whiteness, separate a morsel from it with the forceps, and place it glowing over the fire on a plate of copper. If the Stone emits smoke it is not yet perfect, wherefore it must be left longer in decoction, until it comes to the grade of a perfect Stone. But if it emits no smoke, you may believe it to be perfect. In the same way proceed with the Red Stone of Sol in its due gradation.

CHAPTER VII

Concerning the Augmentation or the Multiplying of Tinctures

When you wish to augment or to multiply the tincture which you have found, join it again with the common mercury. Proceed in all respects as before, and it will tinge a hundredfold more than it did previously. You can repeat this as often as you wish, so as to have as much of the matter as you desire. The longer it remains in the fire, the more highly graduated it becomes, so that one part of it will transmute an infinite number of parts of quick mercury into the best Luna and the most perfect Sol. Thus you have the whole process from the beginning to the end. With these few words we will conclude this second treatise, and will now begin the third.

The End of the Second Treatise

THE THIRD TREATISE

IN the second treatise we have described the method by which the tinctures or fermentations should be produced. In this third we will say how the tinctures of Sol and Luna are made. This we shall make clear at sufficient length, and in what manner Sol, with the other planets, should be produced, namely, with the furnace and fire.

CHAPTER I

Concerning the Building of the Furnace, with the Fire

Mercurius Hermes Trismegistus says that he who perfects this Art creates a new world. For in the same way as God created the heaven and the earth, the furnace with its fire must be constructed and regulated, that is to say, in the following manner: First, let a furnace be built at a height of six palms, with the fingers and thumb extended, but in breadth only one palm; round within and plain, so that the coals may not adhere to it. At the bottom let a little mound be raised, sloping on all sides to the border. Let holes be left open underneath, four fingers in breadth, and to each hole let its own furnace be applied with a copper cauldron, which contains water. Then take the best and most lasting coals, and break them into lumps the size of a walnut. With these fill the long furnace, which must then be closed, so as not to burn out. Afterwards, add coals below, right up to the holes. If the fire is too great, put a stove before it: if too little, let the coals be stirred with an iron rod, that they may meet the air and the heat may be increased. In this way you will be able to regulate the fire, according to the true requirements of its nature, so that it shall not be excessive or defective, but adapted to the movement of the matter. This is compared to the firmament. And there is another firmament in this place, namely, the matter contained in the glass. After these things follows the form of the world. The furnace then is to be placed as the sun in the great world, which affords light, life, and heat to the whole furnace itself, and to all the instruments and other things which it encloses.

CHAPTER II

Concerning the Conjunction of the Man with the Woman

Since we have treated of the furnace in which the tinctures are to be prepared, and of the fire, we now propose to describe more at length how the man and the woman meet and are joined together. This is the manner. Take Philosophers' Mercury, prepared and purified to its supreme degree. Dissolve this with its wife, that is to say, with quick mercury, so that the woman may dissolve the man, and the man may fix the woman. Then, just as the husband loves his wife and she her husband, the Philosophers' Mercury pursues the quick mercury with the most supreme love, and their nature is moved with the greatest affection towards us. So then each Mercury is blended with the other, as the woman with the man, and he with her, so far as the body is concerned, to such an extent that they have no difference, save as regards their powers and properties, seeing the man is fixed, but the woman volatile in the fire. For this reason, the woman is united to the man in such a way that she dissolves the man, and he fixes her and renders her constant in every consideration as a consequence. Conceal both in a glass vessel, thoroughly fastened, so that the woman may not escape or evaporate; otherwise the whole work will be reduced to nothing.

CHAPTER III

Concerning the Copulation of the Man with the Woman, etc

When you have placed the husband and the wife in the matrimonial bed, in order that he may operate upon her and impregnate her, and that the seed of the woman may be coagulated into a mass by the seed of the man, without which she can bring forth no fruit, it is necessary that the man should perform his operation on the woman.

CHAPTER IV

Concerning the Philosophic Coition of the Husband with His Wife

As soon as you see the woman take a black colour, know for a certainty that she has conceived and become pregnant: and when the seed of the man embraces the seed of the woman, this is the first sign and the key of this whole work and Art. Therefore preserve a continuous natural heat, and this blackness will appear and disappear through being consumed, as one worm eats another, and goes on consuming until not one is still left.

CHAPTER V

Concerning the Black Colour

As soon as the blackness appears and is manifest, it may be known that the woman has become impregnated. But when the peacock's tail begins to appear, that is, when many and various colours shall be seen in the glass, it is a sign that the Philosophers' Mercury is acting on the common mercury, and extending its wings until it shall have conquered. When, therefore, the dry acts on the moist these colours appear.

CHAPTER VI

Concerning the Bud appearing in the Glass

When you have seen the different colours, it is necessary that you persevere in the work, by constantly continuing the fire, until the peacock's tail is quite consumed, while the matter of Luna becomes white and glittering as snow, and the vessel attains its degree of perfection. Then at length you may break off a morsel of the regulus, and place it on a heated copper plate. If it remains firm and fixed there, and tinges it, then it is a fermentation brought to the highest perfection of Luna. That King has strength and power, not only for transmuting metals, but also for healing all infirmities. He is a King worthy to be praised, and adorned with many virtues, and so great power, that he transmutes Venus, Mars, Jupiter, Saturn, and Mercury into Luna, which will stand all tests. He also frees the bodies of men from an infinite number of diseases, as fevers, the falling sickness, leprosy, the gallic disease, and many mineral ailments which no herbs or roots, or anything of that kind, can remove. Whoever uses constantly this medicament, prepares for himself a fixed, long, and healthy life.

CHAPTER VII

Concerning the Red Colour

After the King has assumed his perfect whiteness, the fire must be continued perseveringly, until the whiteness takes a yellow tint, this being the colour which succeeds the white; for so long as any heat acts on the white and dry matter, the longer such action lasts, the more is it tinted with yellow and saffron colour, until it arrives at redness, like the colour of a ruby. Then at last the fermentation is prepared for gold, and the oriental King is born, sitting in his seat, and powerful above all the princes of this world.

CHAPTER VIII

Concerning Increase and Multiplication

The multiplying of this fermentation should be noted, which is performed in the following manner. Let it be dissolved in its own moisture, and afterwards subjected to the regimen of fire as before. It will act on its own humidity more quickly than it previously did, and will transmute into its own substance, just as a little leaven seems to transmute into leaven the whole of a large quantity of flour. Wherefore it is an unspeakable treasure on the earth, of which the universe has not the equal, as Augurellus witnesses.

Conclusion

This secret was accounted by the old Fathers who possessed it as among the most occult, lest it should get into the hands of wicked men, who by its aid would be able more abundantly to fulfil their own wickedness and crimes. We, therefore, ask you, whoever have attained to this gift of God, that, imitating these Fathers, you will treat and preserve this divine mystery in the most secret manner possible, for if you tread it under foot, or scatter your pearls before swine, be sure that you will hear pronounced against you the severe sentence of God, the supreme avenger.

But to those who, by the special grace of God, abstain most from all vices, this Art will be more constantly and more fully revealed than to any others. For with a man of this kind more wisdom is found than with a thousand sons of the world, by whom this Art is in no way discovered.

Whoever shall have found this secret and gift of God, let him praise the most high God, the Father and Son, with the Holy Spirit. And from this God alone let him implore grace, by which he may be able to use that gift to God's glory and to the good of his fellow-man. The merciful God grant that this may be so for the sake of Jesus Christ His Son, and our Saviour!

HERE ENDS THE BOOK CONCERNING THE SPIRITS OF THE PLANETS

THE ECONOMY OF MINERALS*

ELSEWHERE CALLED THE GENEALOGY OF MINERALS

PREFACE TO THE READER

ALTHOUGH order seems to demand that we should have treated of the generation of minerals and metals before speaking of their transmutations: still, since theory cannot be more lucidly taught than by its practice, I have thought it best for those who study this art to begin from the very beginning. For, above all else, Alchemy is a subject which is not comprised in mere words, but only in elaborate facts; just as is the case with the rest of those arts, familiarity with which is gained rather by putting them in practice than by any mere demonstrations. It is true that these demonstrations do a very great deal for those who are some way advanced rather than for initiates. For these it is best that from the very first they should have a finger in the pie (as the saying is), and gradually learn from the very mistakes they make. Nobody ever acquired even the easiest art without making such blunders; and certainly no one will be able to follow up Alchemy without making mistakes before he gets at the truth. No one, again, will ever enter the true path so long as he holds back from the goal through fear of making a false step, or fails to correct his own errors by imitating the course of Nature. It will not be so easy to learn if we fail to compare alchemical with natural methods. So, then, it was thought well to let artificial Alchemy precede the natural, so that we may recall those who are venturing forth in this art to the genealogy of minerals, as if to a safe anchorage. It seemed opportune, nay, even necessary, to provide some such anchorage for this purpose in the case of those who are studying Alchemy.

CHAPTER I

CONCERNING THE GENERATION OF MINERALS

When I had most carefully read through the writings of the ancients concerning the generation of minerals, I found that they had not under-

---

* This treatise in the recension here chosen for translation is not found in the Geneva folio, and is translated from another collection of the works of Paracelsus, namely, the Frankfort 8vo. of 1584. A corresponding treatise, entitled *De Mineralibus*, which is included in the Geneva edition, goes over much the same ground, and is, indeed, in parts identical with that given in the text. But at the same time, it has differences sufficiently marked to require that both versions should be provided It will, accordingly, be found in an appendix at the end of this volume, under the title of A Book about Minerals.

stood the ultimate matter thereof, and, in consequence, much less did they understand the primal matter. If the beginning of any matter is to be described, its end should first of all be noted down. I therefore determined first of all to lay before you the ultimate matter of minerals, and from this you will easily understand the primal matter whence they derive their origin. We may bring forward an example from Medicine, where a disease has to be studied from its issue and not from its origin. Of this latter there is no knowledge, because it was secretly introduced, and he who observes it is virtually blind. But the end is visible from the issue towards which we see that disease tending, as though towards a mark set up for it to aim at. Now a thing cannot be better judged than by getting to know for what end it was created by God; otherwise it will often happen that the true use of this creation of God turns to its abuse. Whosoever, therefore, undertakes any work with anything ought thoroughly to understand that with which he works, so that he may accomplish his task in the order prescribed by God, lest on account of his imperfect knowledge or utter ignorance of the matter, things may turn out ill, and the devil's work rather than God's be done, through abuse of the matter and of appliances. For a rough example, take the case of an axe or club in the hand of a man who does not know how to use the one or the other. They become mere instruments of destruction. He alone should handle such tools who knows how to use them, and how, out of the material he has, to construct something that shall be to his neighbour's benefit, and preserve that material for the purpose for which God created it. On this account God wills that everything He has created should be possessed by one who knows how to use it; and every man ought to apply himself to that pursuit whereto he feels in his own conscience called, and not to learn some other fanciful thing suggested by the devil.

Know, then, that the ultimate and also the primal matter* of everything is fire. This is, as it were, the key that locks the chest. It is this which makes manifest whatever is hidden in anything. In this place, then, we understand by the ultimate matter of everything that into which it is dissolved by fire; so that among the three universal things which I have discussed elsewhere in different places, this should be regarded as the first and predominating one. You have an illustration in the case of a metal dissolved in the fire. It at once makes it clear that its first beginning was Mercurial Water, not Sulphur, since its resolution is not accompanied with flame, as would be the case with resins. It is also proved not to be Salt, because the first sign of its resolution is not a crumbling besides liquefaction and flame, as would be the case with earth and stones. Every metal, it is true, contains within itself Sulphur and Salt,

---

* I call the ultimate matter of anything that state in which the substance has reached its highest grade of exaltation and perfection, as, for example, gold, when it has been separated from all superfluities, foreign matter, etc., and remains in its pure virtue, without any admixture, has been educed into its ultimate matter.—*Chirurgia Magna*, Pt. II., Tract II., c. II.   For example, every body made from the first matter is compelled to metamorphose into the ultimate matter. Thus the great ultimate matter has its beginning in the end of the increase of the first matter.—*Ibid.*, Part III., Lib. III.

but Mercury holds the principal place therein. Now, it has seemed good to God to create water an element, and that from it should be every day produced minerals for the use of men. Thus it becomes the mother of those things which are developed in her, as it were in her matrix; that is to say, Mineral Fire, Salt, and Mercury are formed into metals, stones, and all mineral substance, albeit the offspring is quite unlike the mother. In this way the Most High has created all things with their own nature: the birds of the air for one purpose, which is different from that of the fishes in the sea. And so of the rest. Everything is to be committed to His divine will, Who makes everything as it is, and wills that what He makes shall be eternal. As, therefore, water is not like its metallic offspring, nor the son like the mother, in the same way the earth itself is, as it were, wood and not wood, because it comes from that same source. In the same way, stone and iron are produced from water, which, however, becomes such water as never before existed: and the earth, too, becomes something which in itself it is not. So also man must become that which he is not.*

In a word, whatever is to pass into its ultimate matter must become something different from what its origin was—varied and diverse, though from one mother. Thus God willed to be One in all, that is, to be the one primal and ultimate matter of all things. He is such, and so wonderful, an original artificer of all things as never has existed, nor will another ever exist. As, then, you have now heard so far concerning the mother of the minerals, we will in the sequel teach you more fully. The ancients have falsely written that this is the earth; but they have never been able to prove it.

CHAPTER II

Concerning the Ultimate and Primal Matter of Minerals

The first principle with God was the ultimate matter which He Himself made to be the primal, just as a fruit which produces another fruit. It has seed; and this seed ranks as primal matter. Likewise, out of the ultimate matter of minerals the primal element was made, that is, it was made into seed, which seed is the element of water. This resolves it, so that it becomes water. It has been entrusted to it by Nature, or so arranged that it should produce the ultimate matter, and this is in water. Nature, therefore, takes under its own power and separation whatever there is in water; and whatever relates to a metal it puts on one side by itself for each particular metal. So also for gems, stones, the magnet, and other things of that kind, each separately and according to its own kind. For as God has appointed to the wheat its proper time for harvest, and the autumntide for fruits, and to other things like these in their elements, so for the element of water He has

---

* It is needful for man to be born a second time from a virgin, not from a wife, by water and by the spirit. For the spirit vivifies that flesh wherein there is no death possible for ever. The flesh wherein death abides profits nothing, and nothing towards eternal salvation can it confer upon man.—*Philosophia Sagax*, Lib. II., c. 2.

willed that there shall be a proper season of harvest and autumntide; and for all other things, each according to its kind, He has foreordained times for the collection of their fruits. So, then, the element of water is the mother, seed, and root of all minerals; and the Archæus therein is he who disposes everything according to a definite order, so that each comes to its ultimate matter, which at length man receives as a sort of artificial primal matter: that is, where Nature ends, there the Art of man begins, for Nature's ultimate matter is man's primal matter. After such a wonderful method has God created water as the first matter of Nature, so soft and weak a substance, yet from it as a fruit the most solid metal, stones, etc.—the very hardest from the very softest:—and so that from the water fire should issue forth, beyond the grasp of man's intelligence, but not beyond the power of Nature. God has created wonderful offspring from that mother, as appears also in men; if they be looked at even in their mother, each will be found peculiar in his intellect and his properties, not according to his body, but according to his own state of constitution.

CHAPTER III

Concerning the Field, the Roots, and the Trees of Minerals

The Most High created the element of water to be, as it were, a field in which the roots of mineral trees, springing forth from their seeds, should be fixed, and thence the trunk and the branches should be thrust forth over the earth. He separated it, therefore, from the other three, so that neither in the air, nor in the earth, nor in heaven, but placed on the lower globe, it should exist by itself as a free body, to be on the earth and to have its centre there where it was founded, created after such an admirable order that it should bear man upon it like the earth; so that man borne in a ship should speed over the water and get possession of it. What is more marvellous still is that though it surrounds our globe in every direction, the water does not fall down from its own limits, though the part at our antipodes seems to hang downwards, just as our part seems to them, and yet each remains spread out a plane surface on its own sphere, wherever you look at it, as if some pit should be imagined which, descending perpendicularly to the abyss, should find no bottom nor be sustained by the earth. It is even more wonderful than the egg in its shell, provided with all that it requires. The generations of minerals, then, from the element of water are protruded into the earth, just as from the element of earth all fruits are pushed forward into the air, so that nothing but the root remains in the earth. Exactly so, all metals, salt, gems, stones, talc, marcasites, sulphurs, and every similar substance, pass from their mother, the water, to another mother, namely, the earth, in which the operation of their trees is perfected, while their roots are fixed in the water. For as those things which grow from roots in the earth are finished in the air, in like manner, those which derive their origin from the water are altogether completed by Nature in the earth, so that they reach, as the others did, their ultimate matter. The ancients, led astray by this opinion, because they saw that metals were found in the earth, were so little advanced that they did not see their error when, on the subject of minerals, they wrote that out of the earth grew nothing but wood, leaves, flowers, fruits, and herbs, and that everything else was produced from water. No less mythical was the saying of that man who asserted that all things which were produced on the earth had their origin from the air, because they are in the air and are perfected there, though he saw their roots in the earth. Because he did not see the roots of minerals with his bodily eyes he would even feign that they are fixed in the earth. Such is the physical science of the Greeks, deduced only from what is seen, recognising nothing occult by mental experiment. It is just a fiction of lazy men who presume to chatter about natural science from eyesight alone; and who do not experiment so as to observe those occult things which underlie the things which are manifest, the one over against the other.

CHAPTER IV

Concerning the Fruits and the Harvest of Minerals

Just as all the fruits of the earth have their harvest and autumn on the earth and in the air, according to the predestined time in their generation, so the fruits of the water, that is to say, minerals, are gathered at their own time of maturity. When the mineral root first germinates they rise to their own trunk and tree, that is, into the body from which minerals or metals are subsequently produced; just as a nut or a cherry is not immediately produced from the earth, but first of all a tree, from which at length the fruit is generated. In like manner, Nature puts forth a mineral tree, that is, an aqueous body, in the element of water. This tree is produced in the earth so far as it fills the pores thereof, just in the same way as the earth itself fills the air. From this are eventually produced fruits according to the nature and property of its species, at the extremities of its branches, just as occurs in trees which we see on the surface of the earth. We must seek, then, first of all, for the aqueous tree, and by-and-by for its fruits, by a method not inaptly borrowed from agriculture, and in the following manner. Some of the visible trees produce their fruits covered up; for instance, chestnuts under a prickly bark, walnuts under one that is green and bitter, under that a wooden covering, and under this again a bitter membrane, and then at last the kernel. So it happens in minerals, the kernels of which, that is to say, the metals, are separated just like those others by barks. Other trees produce their fruits naked, such as plums, cherries, pears, apples, grapes, etc., where there is no such separation as that just described. So also some aqueous trees produce their gold, silver, corals, and other metals of that kind, free and naked, according to the condition and nature of the water. As we know by the rind what fruit lies concealed within it, and as the spirit is known by its body, just so, in the case of minerals, the spirit of the metal is recognised, though hidden, beneath its corporeal or mineral bark. The spirit of the aqueous element produces the body, of one kind in the mineral, of a different kind in the fruit. Although, then, gold may be in a mineral body, nevertheless that body is of no moment ; it has to be separated from the gold as impure, while the gold itself is pure. There are, therefore, in a mineral two bodies, of which one is the fruit, the pure body of gold, wherewith its spirit is inseparably incorporated. So the fruits are first introduced from the element into the tree, as the spirit into an impure body, and with that at last into the earth, as something noble and pure. The same thing is seen in man, to whom have been given two bodies, one corrupt, but the other incorrupt, which will be eternally united with him, since it is the image of God, and by its possession especially man differs from all other creatures.*

CHAPTER V

Concerning the Death of the Elements, especially of Water

Elements die, as men die, on account of the corruption in them. As water at its death, as it were, consumes and devours its own fruit, so does the earth its own fruits. Whatever is born from it returns to it again, is swallowed up and lost, just as the time past is swallowed up by yesterday's days and nights, the light or darkness of which we shall never see again. It is no weightier to-day than yesterday, not even by a single grain, and will after a thousand years be of the same weight still. As it gives forth, so, in the same degree, it consumes. The death of the water, however, is in its own proper element, in that great terminus and centre of water, the sea, wherein the rivers, and whatever else flows into it, die and are consumed as wood in the fire. Rivers, indeed, are not the element of water, but the fruit of that element, which is the sea ; from this they derive their origin, and in this they receive both their life and their death.

CHAPTER VI

Concerning the Death of the Tree of Minerals

After Nature has planted the mineral root of a tree in the centre of its matrix, whether to produce a metal, a stone, a gem, salt, alum, vitriol, a saline or sweet, cold or hot spring, a coral, or a marcasite, and after it has thrust forth the trunk to the earth, this trunk spreads abroad in different branches,

---

* The flesh and blood which man received from Adam can in no wise enter into the kingdom of God. For nothing can ascend into heaven which did not come forth out of heaven. Now the Adamic flesh is earth. Thus it cannot enter heaven, but is again converted into earth. It is mortal, subject to death, and nothing mortal can enter heaven. There is no fire which can purge it from its stains in such wise as to make it fit for heaven. It admits not of purging or glorifying. At the same time, man cannot enter heaven unless he be true man, clothed upon with flesh and blood. For it is only by flesh and blood that man is distinguished from the angels, for, otherwise, both are of the same essence. Herein man hath more than the angels, in that he is endowed with flesh and blood, and for man was the Son of God born into the world ; for him He died upon the cross, that so man might be redeemed and made eligible for the kingdom of heaven. But when God had thus shewn His love for man, his flesh still excluded him from heaven, whence He gave him another flesh and blood which was built up of the Son, and then this creature, not of the Father, but the Son, enters heaven. For the Adamic flesh is of the Father, and returns whence it came, though had Adam not sinned his body would have remained immortal in Paradise. But Christ, compassionating our calamity, gave us a new body. Of the spirit who

the liquid of whose substance—both of branches and stalk—is formally neither a water, nor an oil, nor a lute, nor a mucilage; in fact, it can only be conceived as wood growing out of the earth, which is, nevertheless, not earth, though sprung therefrom. They are spread in such a manner that one branch is separated from another by an interval of two or three climates and as many regions: sometimes from Germany to Hungary, and even beyond. The branches of the different trees of the same kind are extended over the whole sphere of the earth, just as the veins in the human body are extended into various limbs far apart from each other. But the fruits put forth by the extremities of the twigs, by the nature of the ultimate matter, soon fall to the earth. There is a momentary coagulation of them, and then at length, when all its fruit is shed, this tree dies and is utterly consumed by dryness, its offspring being left in the earth. Afterwards, according to its state of nature, a new tree appears. So, then, the first matter of minerals consists of water; and it comprises only Sulphur, Salt, and Mercury. These minerals are that element's spirit and soul, containing in themselves all minerals, metals, gems, salts, and other things of that kind, like different seeds in a bag. These being poured into water, Nature then directs every seed to its peculiar and final fruit, incessantly disposing them according to their species and genera. These and like things proceed from that true physical science, and those fountains of sound philosophy from which, through meditative contemplation of the works of God, arises the most intimate knowledge of the Supreme Creator and of His virtues. To the minds and mental sight of true philosophers, no less than to their carnal eyes, the clear light appears. To them the occult becomes manifest. But that Greek Satan has sown in the philosophic field of true wisdom, tares and his own false seed, to wit, Aristoteles, Albertus, Avicenna, Rhasis, and that kind of men, enemies of the light of God and of Nature, who have perverted the whole of physical science, since the time when they transmuted the name of Sophia into Philosophy.*

CHAPTER VII

Concerning the Variation of the Primal Matter of Minerals, in proportion to the different species and individuals thereof: also concerning the various colours, etc.

We have before said that the primal matter exists in its mother, just as if in a bag, and that it is composed of three ingredients meeting in one.

---

gives life cometh forth a living flesh, wherein is no death but life. This is the flesh whereof man has need, that he may become a new man, and in this flesh and in that blood, at the last day, shall he arise, and shall possess the kingdom of heaven with Christ. Now, this flesh which has its life from the spirit was first born, without the generation of male seed, from a daughter of Abraham, by promise, and became man by the Holy Ghost. So, also, we who aspire to the kingdom must be born again out of a virgin and faith, incarnated by the Holy Spirit. Thus man must to eternity be flesh and blood; thus is there a dual flesh—that which is Adamic and is nothing, and that of the Holy Spirit which is vivific.—*Philosophia Sagax*, Lib. II., c. 2.

* So high and so lofty is human wisdom that it hath in its power all the stars, the firmament itself, and universal heaven. And as the power thereof pervades all the earth, so also it extends over heaven. The Sun and Moon are its subjects. Even as the hand changes and compels the soil, so also the inner microcosmus compels the zenith to obedience.—*De Peste*, Lib. II., c. 2.

But there are as many varieties of Mercury, Salt, and Sulphur as there are different fruits in minerals. For a different Sulphur is found in lead, iron, gold; in sapphire, and other gems; in stones, marcasites, and salts; likewise a different Salt in metals, salts, etc. So, too, is it with Mercury: one kind exists in gems, another in metals. Besides, in respect of the composition of these, different individuals are found under the same species. Gold is sometimes found, one specimen heavier or more deeply coloured than another: and so of the rest. Moreover, there are as many Sulphurs of gold, Salts, and diversities of Mercury of gold, and of the others, as there are greater and lesser degrees. Nevertheless, all which among them receives particularity from the subject always is comprised under the universality of one and the same Sulphur, Salt, and Mercury, mysteriously comprehended in universal Nature. In this respect Nature may be compared parabolically to a painter, who from some few colours paints an infinite number of pictures, no one exactly like another. The only difference is, that Nature produces living pictures, but the artist only imitates these. He represents the same things to the eye; but they are dead things. Now, all natural colours proceed from the Salt of Nature, in which they exist together with the balsam of things and coagulation. Sulphur exhibits the substance of bodies and their building up; Mercury, their virtues and arcana. God alone assigns life to all, so that from every one should be produced that which He, from all eternity, had predestinated to be thence produced, as He determined and willed that all should be. Whoever, therefore, wishes to understand the bodies of natural things, let him learn from natural Sulphur that which he may first of all have well understood, if he seeks natural colours as the foundation from Salt. But if he wishes to know the virtues of things, he must scrutinise the arcana belonging to the Mercury of that thing whose virtues he wishes to learn. All these matters does that one and the same Nature at once embrace in one, and separate; at the same time distributing, removing, or completely blotting out the colours from such. Consider, I beseech you, this tiny grain of seed, black or brown in colour, out of which grows a vast tree, producing such wonderful greenness in its leaves, such variegated colours in its flowers, and flavours in its fruits of such infinite variety; see this repeated by Nature in all her products, and you will find her so marvellous, so rich, in her mysteries that you will have enough to last you all your life in this book of Nature without referring to paper books. If God, then, shews Himself to our discernment in Nature so powerful and so wise, how much more glorious will He reveal Himself by His Holy Spirit to our mind if we only seek Him? This is the way of safety which leads from below to above. This is to walk in the ways of the Lord, to be occupied in admiring His works, and to carry out His will, so far as is in us, or as it should and can be in us. This has been my Academia, not Athens, Paris, or Toulouse. After I had read many deceitful books of wise men I betook myself to this one alone, from which I learnt all that I write, which also I know to be true. Still, I confess, there are many more things which I do not know, but which will surge up to the surface in God's own time. There is nothing so occult which shall not be revealed when the Almighty wills it so to be.

This, however, I know, that after me will come a disciple of this school, one who does not yet live, but who will disclose many things.

CHAPTER VIII

Concerning the Natural Dispenser of Minerals, and His Ministers

In the manufacture of minerals by men for preparing them and adapting them for use, not one man alone, but many in succession, are required, and each of these has his own special gift and duty. Who is benefitted by a metal being dug from the bowels of the earth, unless it be its separator, preparer, or liquefactor? What is he, again, without the smith? He, too, is of no avail without some buyer, nor the buyer unless there be someone who knows how to adapt those metals for use. Nature does not need all these; but still she needs her own people. Among these is, first of all, Archeus, the dispenser of minerals, who has ministers under him.* He himself, the minister of Nature, has the following: the first, who exhibits the corporeal matter into which the operation falls, namely, the mineral Sulphur, is this or that condition and nature; a second, who fabricates the properties and virtues, and operates on the previously existing matter, say, for instance, Mercury; a third, who, by compaction and coagulation, unites all the single portions together into one body, that is to say, the Salt, which is the confirmer of the work. When all these are brought together into one, and enclosed in an athanor, Archeus decocts them, exactly as the seed in the earth; and not only so, but they are decocted mutually together, one with the other, in the following manner: The Sulphur submits its body to the other two, that they may do with it what they will, and lead to that end whereto is destined that which has to be done. Mercury is added with the properties of its virtues, and this is decocted by the other two. When all the decoctions of this kind are fulfilled, then, at length, the salt begins to operate on the other matters associated with it, and on itself. By first condensing, afterwards congealing, and, lastly, coagulating, it strengthens the work for its autumn and harvest, so that nothing is wanting except a harvester and a smith.

Briefly, then, we have gone through the whole genealogy of minerals. It remains that we specially, but still concisely, hear the force and virtue of each in Alchemy and in Medicine respectively, so far as it is necessary to learn these for the aforesaid faculties. I would admonish my readers to put aside for awhile the mere dreams and opinions of others who romance about these things, until they see that they are only philosophers on paper, not in Nature, who have been taught by men like themselves, and with the same amount of

---

* Archeus is Nature and the dispenser of things.—*Annotationes in Libros duos de Tartaro.* The anatomy of the Archeus is the anatomy of life. - *Fragmenta Anatomiæ.* Archeus is the separator of the elements and of all those things which exist in them, dividing each thing from the rest, and gathering it into its own place.—*De Elemento Aquæ*, Tract II., c. 1.

learning, to think by rote and not by experience, while they shew themselves to others such as they really are. Though they may not care to see, I will still place them so that at least they may perceive the light and nature and life more easily, without being disturbed through the darkness of death. Beginning, then, from the first principles of minerals, which are Salts, we will run through each, that is to say, right up to the very end of the metals.

CHAPTER IX

Concerning the Virtues and Properties of Salts in Alchemy and in Medicine

God, in His goodness and greatness, willed that man should be led by Nature to such a state of necessity as to be unable to live naturally without natural Salt. Hence its necessity in all foods. Salt is the balsam of Nature,* which drives away the corruption of the warm Sulphur with the moist Mercury, out of which two ingredients man is by Nature compacted. Now, since it is necessary that these prime constituents should be nourished with something like themselves, it follows as a matter of course that man must use ardent foods for the sustenance of his internal Sulphur; moist foods for nourishing the Mercury, and salted foods for keeping the Salt in a faculty for building up the body. Its power for conservation is chiefly seen in the fact that it keeps dead flesh for a very long time from decay; hence it is easy to guess that it will still more preserve living flesh. Coming, at length, to its kinds, there are three which are considered specially useful for man's life. The first of these is Marine Salt, the second is Spring Salt, and the third Mineral Salt. Spring Salt is chiefly conducive to health; in the second place, Mineral Salt; and, lastly, Marine Salt. This last and the first are decocted by Art, the other only by Nature. This and the Marine Salt are not comprised under the nature of muria (brine), but that which is decocted is first of all turned thereinto, before it is separated from the water into coagulated salt. There are, therefore, two descriptions of Salt to be put forward by us, one from muria, the other from wholly refined salt. But, first, consideration should be given to that condition which is common to every Salt. Where Salt has not been used with foods there is no correction; and if the stomach receives those foods it is unable to digest them. There is in Salt an expulsive force, acting through the excrement or through the urine, and unless these are kept in their regular course and motion, all the vital faculties are prostrated in their endeavours

---

* White vitriol, red vitriol, cuprine vitriol, rock alum, alumen plumosum, alumen scissum, alumen entali, alumen usnetum, sal gemmæ, rock salt, mountain salt, sea salt, spring salt—all these species originate from the salt of the three prime principles, and are subject to calcination, reverberation, or sublimation. Now, if all these things subsist in a proper proportion or, so to speak, essence, they are called by one universal name, liquor of Nature, or liquor of salt, or balsam of salt. Besides these, there are arsenic, realgar, ogertum, black auripigment (that is, orpiment), antimony, mercury, asphalt. These, in like manner, are subject to calcination, reverberation, distillation, etc., and if they subsist wholly in one essence they are called the balsam of Nature, the liquor of Mercury, or the balsam of Mercury. Finally, there are the various species of sulphur, petroleum, carabe, pitch, etc., which are also subject to the same processes, and if they subsist unseparated in a single essence they are called the tincture of Nature, liquor of sulphur, or balsam of sulphur.—*Fragmenta Medica*, No. 3.

and in their powers of expulsion. The blood is in its own nature salt, and does not receive unsalted nutriment. If it does, through extreme hunger, sometimes receive such, it passes away to decay. In order that such a fault might be avoided, Salt has been appointed as an addition to alimentary foods, so that the natural outlets may not be obstructed, or the members be deprived of their due nutriment. Moreover, there lies hid in Salt a solvent faculty for opening the obstructions which accidentally occur in the pores of the skin, and driving them out by resolving them into urine. The urine is only the salt of the blood, that is, the salt from the natural salt which is associated with the microcosmic salt, and so they both act powerfully for the expulsion of the excrements. Now, this natural conjunction can only be made when tempered with a proper quantity of alimentary salt, otherwise through the stoppage they easily remain and adhere somewhere. Every physician ought to know the power there is in Salt as a medicine, especially when he wishes to purge the natural Salt. Let him more freely prescribe this, especially the kind that comes from gems, which, above all others, has the faculty of attacking and expelling this natural Salt. The operations of these three different kinds of Salt should be carefully watched in practice, a method which opens the eyes far better than any letter or description.

CHAPTER X

Concerning Muria

I just now mentioned two kinds of Salt, Muria and dry Salt. First of all, Muria has the greatest power of drying up all superfluous moisture. It does more in one hour than dry Salt could effect in a month. Although this has been reduced to Muria, it has not the same power as the natural in curing moist gout, dropsy, moist tumours of the tibia, and other tumours as well, in a word, for consuming all unnatural leprous liquids. Its heat should be so tempered that a patient could sit in it as in a bath without injury. The proof of perfection in Muria is that an egg shall swim on its surface when thrown into it. It should be noticed that a bath of this kind is only adapted for stout people. People who are of a spare habit should not use it, as it dries up too much. If after one or two baths the tumours return, it would be best to live for a time in those places where the decoctions of Muriæ and Salt are made.

CHAPTER XI

Concerning Dry Salt

There are various species of dry Salts, such as the common sort used with food, that from gems, stones, and earths, and that which comes through the cones of congelated bodies. Note the common virtue of each. If any one of them be mixed with Sulphur and applied to wounds as a plaster, and then as a lotion, it keeps them from worms, and even if the worms have already been produced, it drives them away and prevents any more from coming. By cleansing alone, and without the use of any medicament, Nature heals wounds, unless any complication prove an obstacle to the free action of the natural balsam. In Salts of this kind is a great remedy for ulcers, scabies, and the like, if they are resolved in baths. The power of Muria is much stronger, and this can be increased by dissolving Salt in it. The same is useful for curing baldness, and other ailments of that kind, especially if these Salts are corrected by addition, or increased in power by the following method : Take equal quantities of dry Salt and Salt of Urine,* as much as you will, let them be calcined together for two hours, and let Muria afterwards be dissolved; or let them be put by themselves in a cold, damp place. They will exhibit artificial Muria very little less strong than the natural in external surgical cases, but much weaker in internal cures. The aforesaid Salts will never be found in any other things, even though the alcali be decocted from them. This Salt is not like those before named, but is called the alcali of natural things or Corporeal Salt, because it is fed by the salts of nutriment in the human body, or by the preceding, even the dry and specially nutritive ones. For Alchemy, the Water of Salt is made from the same kind of Salts calcined into a spirit in a vessel where gold is dissolved into an oil and separated from it so that it remains excellent and potable—Drinkable Gold. Before it arrives at this final condition, as we have heard from jewellers and ironmasters, it is an excellent artifice for gilding silver or iron, and would be a constant treasure if they only knew how to prepare it chemically. It should be remarked, too, concerning pure Salt, congelated by Nature alone either into cones or into the salt of a gem, that this is particularly adapted for the ordinary cementations of silver, and renders the metal malleable without the customary burnings. It does the same with copper by means of a cement reduced to a regulus.

CHAPTER XII

Concerning Salt Nitre

There is also another kind of salt which is called nitre.† It is composed naturally of the natural salt of animals' bodies, and the salt of nutriment in those bodies combined. One salt having thus been formed from two, the superfluity is decocted into urine, and, falling on the earth, is again decocted in due course. The two constituents are more and more closely united, so that from them results one single and perfect salt through the chemical separation brought about by artificial decoction from its earth. It shews itself very clearly in the form of cones or of clods, provided it be thoroughly separated from the superfluous nutrimental Salt not yet digested by the animal decoction

---

* Every urine is a resolved salt.—*De Judicio Urinarum*, Lib. II.   Salt passes into urine.—*De Tartaro*, Lib. I., Tract III., c. 1, *expositio*.

† Nitre forms in the pens and stables where cattle make water. For the earth whereon they make water is afterwards cooked and the salt nitre obtained from it. For all urine is salt.—*De Tartaro*, Lib. I., Tract III., *annotationes in c. 2*.   Nitre is excrement and the dead body of esile and nutrimental matter. And this dead body is that out of which putrefaction grows.—*Fragmenta Medica, De Tararo Nitreo*.   It is an essential spirit and excrement of all salts, possessing a hermaphroditic nature.—*De Pestilitate*, Tract I.

when it is driven off into the urine. In Alchemy its use is very frequent. It would be idle to recount how great was the violence which a first experiment demonstrated therein with disastrous result, when it was compounded with sulphur and formed into blasting powder, whence it has been deservedly called terrestrial lightning. In the same way, from the salt of the liquor of the earth, which is an universal natural balsam, by which all things are built up in their special combinations, returning at length from this by resolution into the earth again—there is produced, as was stated above, a single salt, which afterwards percolating through the pores of the earth is coagulated in the form of cones of ice adhering to the rocks, from which circumstance it changed its name of Nitre into Saltpetre. Neither the one nor the other is particularly useful as an internal medicine, except in the way of reducing too obese bodies; nor is it a very safe remedy, unless the two are mixed with Salt of Copper, or else the three are subjected to a process of extraction, and formed into one body for employment in this special way.

CHAPTER XIII

Concerning the Ill Effects of Nutrimental Salt

All salt used with food which has not been digested by the stomach, which also on being expelled has passed down into the intestine, unless it makes a thorough transit, generates colic and suffering in the bowels which are very difficult to cure. Its corrosive nature causes it sometimes to perforate the intestines, as is shewn by anatomy. If, however, it remains unexpelled in the stomach, eructations and heartburns arise, with many other affections of the stomach. It sometimes happens, too, that the undigested Salt is coagulated in the mesaraic veins, forming a granular deposit, from which proceed many severe diseases which are little understood, and that not only in this particular part of the body but in others also, especially the urinary organs. Enough has been said on the different species of salts, their virtues and their faults. We now pass on to that salt which is more mineral in its character, and is named Vitriol. It excels all others by its utility, both in Alchemy and in Medicine.

CHAPTER XIV

Concerning Vitriol *

Nature produces from the bowels of the earth a certain kind of salt, named Vitriol, possessed of such virtues and powers as can scarcely be described to the full by any. In it are contained perfect cures for the jaundice, gravel, calculus,

---

* An important variation of this and the following chapters on vitriol occurs in the Geneva folio. Concerning the use of vitriolated oil in Alchemy; and in like manner concerning its crude form. By way of saying something about the hidden alchemical powers in Vitriol, I would first of all submit to you, concerning crude vitriol, that each separate kind of crude vitriol makes copper out of iron. It is not the Alchemist who does this, but Nature or Vitriol by the operation of the Alchemist. In the light of Nature it is the subject of no small wonder to observe how any metal, as it were, puts off itself and becomes something else. It is really very much the same as if a woman should be produced from a man. In these matters, however, Nature has her own peculiar privileges conferred upon her by God, for the benefit of

fevers, worms, the falling sickness, and many other diseases which are very difficult to treat, and arising from obstructions, as we shall describe at greater length below. In both faculties, that is to say, in Medicine and in Alchemy, it produces marvellous effects, varying according to the method of its preparation. As from one log of wood different images are carved, so from this body various most excellent medicines are prepared, not only for internal disorders but also for surgical cases, such as ringworm and leprosy. In a word, whatever other remedies are not able to effect against diseases, on account of their own weakness, this it does from the very foundation by removing the cause of the disease. Some of its powers it puts forth in a crude state, others when it is reduced to water, others when it is calcined, others when it is reduced to a green oil, others in the form of a red oil; others, again, it possesses when in the form of a white oil. It assumes new powers with every fresh form of preparation which it receives. It can serve for a fourth part of all the diseases and all the drugs ever thought of. There is no need for the true physician to turn his eyes hither and thither. Like a modest maid, he can keep them fixed on the ground, for there, beneath his feet, he will find more power and wealth in this treasure of Nature than India, Egypt, Barbary, and Greece could bring him.

CHAPTER XV

Concerning the Species of Vitriol and the Tests of it

The species of Vitriol are as varied as the mines or sources from which it is extracted. The tests of its greater or less excellence vary in equal proportion. First, if it tinges an iron plate to the colour of copper, the more deeply it does so, the better it is considered. But this is the highest of all kinds. Secondly, when it is taken internally in a crude form it drives out intestinal worms better than any other medicine, and the more effectually it does this the better it is

---

man. I say this concerning transmutation in order to make you understand how that envious philosopher, Aristotle, in his philosophy, has no sure foundation, but is simply fatuous. I will lay before you, in due course, the recipe itself, so that in all parts of the German nation you may know how to make copper out of iron. From this power of transmutation we can easily gather that many other transmutations are possible, though they are at present unknown to us. It cannot be denied that many arts are still occult, and that these are not revealed by God because we are not worthy of knowing them. Of course the change of iron into copper is not of the same importance as the change of iron into gold. God manifests the lesser, but the greater is kept occult until the time of knowledge and of Elias who is to come. For these arts are not without their Elias. The following is the recipe for transmutation: Take raspings or filings of iron, without any other metal, such as copper, tin, etc., one pound; add quicksilver, half a pound. Put both into an iron pan or pot; pour over them one measure of acetum and a quarter of a pound of vitriol, with one ounce and a half of sal armoniac. Let these be boiled together and constantly well stirred with wood. If the acetum be expended, pour in some more and add fresh vitriol. By this decoction the iron is transmuted into copper. If the copper is made it all passes away to the quicksilver. Having continued the decoction for ten or twelve hours, then separate as much of the quicksilver as is left from the iron, and wash it carefully so that it may be quite clean. Receive the quicksilver in a bag made of soft leather or cotton and squeeze it out. Then you will see an amalgam left. Let that amalgam expend itself, and you will find pure and good copper. Of this copper take half an ounce and the same quantity of silver. Let them pass into a state of flux or liquefaction, and the silver will forthwith ascend to the sixteenth degree. And this is the method of proving that such copper is made from iron. It is not, however, true that the grades are fixed. But whoever can work well with regale will be abundantly rewarded. Everything in this operation depends on skill in working. This is where most operators fail. By the above-mentioned process you can always make copper out of iron. I mention this to confirm the transmutation of one body into another. The nature of vitriol is such that if its colcothar be calcined it is at once, even with slight liquefaction, turned into copper. A remarkable cuprine nature is in it, and there is also an equally remarkable vitriolic nature in copper. If the copper be dissolved in aqua fortis and granulated, all the copper becomes vitriol. There is no more copper left. So, also, copper is made out of vitriol, and no more vitriol

considered to be as a medicine. A third test is when it transmutes iron into copper. The more perfectly and the more rapidly it does this, the better should it be esteemed in both faculties, for there is the greatest affinity between these two metals. Nor is this remarkable when by means of borax water quicksilver is made in like manner from lead. There are other kinds of cachimiæ which convert metals; and besides these there is a fountain in Hungary, or rather a torrent, which derives its origin from Vitriol, nay, its whole substance is Vitriol, and any iron thrown into it is at once consumed and turned to rust, while this rust is immediately reduced to the best and most permanent copper, by means of fire and bellows. A fourth test is when its red colcothar, subjected to a strong fire, exhibits copper of itself. This is comparatively weak in Medicine, but of great excellence in Alchemy. We must not omit to speak of its colours. That which is altogether cœrulean is not so strong in medicine as that which under the same colour has red and yellow spots mixed together. That which is of a pale sky blue colour should be selected before all others for the prepar- of the green and the white oil. That which inclines to a red or dark yellow colour is best for preparing the red oil from it. The last test is when with gall nuts it makes a very black and dark ink. This should be selected in preference to all the others. The species, therefore, are reckoned according to the tests.

CHAPTER XVI

Concerning the Virtues of Vitriol, Crude or Calcined, in Medicine

For the most severe pains in the stomach and discomforts arising from the inordinate use of food or drink, exhibit crude Vitriol to the extent of six cometz or three drops, say, three grains. To weak patients it should be administered in wine or in water, to stronger ones in distilled wine. It purges

---

remains, unless it be reconverted into vitriol by a sufficient quantity of aquafortis. This kind of kinship between vitriol and copper is remarkable. Whatever is of the nature of copper gives good vitriol. Thus verdigris gives good and highly graduated blue vitriol. Although for us to discuss these matters at any length would perhaps be ridiculous, still none can deny that there is latent in vitriol a tincture, which is of much higher excellence than most people imagine. Happy he who understands this matter! Note other facts about the oil of vitriol. If the oil of quicksilver and this oil of vitriol be joined and thus coagulated according to their own special process, a sapphire of marvellous nature and condition is produced. It is not, indeed, the sapphire stone, but like it, with a wonderful tinge, concerning which I have much more to say. Hence it is evident that stupendous secrets lie hidden in Nature and in the different creations of Nature or of God ; and it would be much more to our credit if we looked into these and investigated them, instead of indulging in revelry and debauchery. At present the palm is given to debauchery, until one-third part of mankind or of the population of the world shall be killed, another shall be finished off by disease, and the remaining third only shall be saved and survive. In the present condition of depravity the world cannot last or the arts flourish. It must needs be that the present condition and order of things go to destruction and be altogether eliminated, otherwise no good thing can be compassed. Then at last will flourish the Golden Age : that is, then at last man will use his intelligence and live as a man, not as a brute ; nor will he act the swine, or live in caves and dens of the earth. Since, then, I have so far communicated to you these facts about vitriol, with every good disposition, I now pray you all, that when you see those unlucky and unhappy creatures suffering from critical disease, for the sake of your own conscience, for God's glory, and the love of your neighbour, you will seriously reflect and not despise or lightly esteem the gifts implanted by God in vitriol. Let love constrain you, so that by night and by day you may be occupied herein, and none be found taking his ease, but all ready to do anything for his neighbour's good. Will this not move you lawyers? Listen to what Christ says : " Woe unto you, lawyers ! " This saying is not effete. Nor do you theologians place a stumbling block in the way, you who think so much more of your returns and your salaries than about your sick folk. These are they who pass by on the Jericho road. Be you like the Good Samaritan, and follow the example of his virtue. Then God will so enlarge your gifts that in helping the sick you shall suffer no lack. All that you need shall be given you. You only sell this treasure !

out every failing from the roots, driving it up and down. In arcana it is called Vitriol Grillus, or Grilla. Neither hellebore, nor colocynth, nor digridion purge so strongly or cure so perfectly as this, nor have they the same faculty for driving out worms. For curing the falling sickness, too, the purgation by Vitriol is of all methods the best. These properties accrue to it from its twofold nature, that is to say, its acetosity and its saltness. On this account it is a much nobler medicine than others. Its colcothar, or, as they call it, its red *Caput Mortuum*, should not be taken internally, unless as an adjunct to surgical treatment for putrid ulcers of the first grade of malignity ; but its oil may be taken for those of the second or third grade. Its medicinal virtues are contained in other medical books, as, for instance, in the treatise entitled *De Naturalibus Rebus.* Here we had intended only to treat and to bring to one focus what it does for Alchemy in the way of transmutation. Sometimes medical topics tempt one to stray from one's set purpose. Let us see, then, what Vitriol does in Alchemy beyond the transmutation of iron into copper, as we described above, giving the formula at the outset. Although, then, it is not so difficult a work to transmute iron into gold, God wills that the lesser operations shall be performed first, and that the greater ones should remain occult until the Elias of the Art arrives. All arts have some one person specially their own, as is understood in other arts. Now, take one pound of iron filing, without the admixture of any other metal, and half-a-pound of Mercury. Over these pour one measure of the strongest Acetum, with a quarter of a pound of Vitriol. Throw in an ounce and a half of Sal Armoniac. Boil all together, and stir constantly with a wooden spoon. As the Acetum wastes pour on fresh, and also Vitriol. After twelve hours let the chief part of the Iron which has been transmuted be entirely separated with Mercury from the other part of the Iron which has not been transmuted ; and when the Mercury has been pressed out by a leather, there will remain a paste of amalgam, and when this is reduced by fire it exhibits the purest copper. Half an ounce of this is at once mixed with an equal part of silver, six degrees being held back, though not fixed but ready to be fixed in *regale*, so that therefrom the industrious Artist may have moderate gain for food and clothing. Vitriol is also made from Venus, dissolved by means of aquafortis and granulated. This does not return again to copper. So also from the colcothar of vitriol Venus is made (as we have mentioned above among the tests), which is not brought back to vitriol of itself except by a special water. Verdigris, in like manner, exhibits a Spagyric Vitriol of highest degree. In Vitriol so great and powerful a tincture lurks as an inexperienced person could scarcely believe, though he can who understands its arcana. As often as Oil of Vitriol is mixed with Oil of Mercury, and both are coagulated together, they change to a stone of wonderful tint and condition, very like a sapphire.

Having dealt with the Salts, let us now pass on to Sulphur.

CHAPTER XVII

Concerning the Threefold Sulphur of Minerals

Sulphur should properly be called the resin of the earth, and in it are latent numberless virtues available in both faculties, though its crude form is useful in neither. Its arcanum alone, when cleansed from impurities, operates in a wonderful way, having been washed to that whiteness which is seen in snow, by means of the Isopic art. It has as many different virtues as it has variety of sources: for every metal or mineral contains Sulphur in itself. As we said above, under the similitude of chestnuts and other nuts, that minerals were likewise enclosed in their rinds, and that the chief excellence lay concealed in their nucleus, which is sustained and nourished by the external integuments, so with regard to Sulphurs, it must be understood that it is the interior one which excels the others, and is Spagyrically termed embryonated, on account of its specific origin, as being the Sulphur of gold, stone, etc. The external Sulphur, in which the embryonated lies concealed, is our mineral. There is also a third kind, extracted from the nuclei of minerals or of metals, which cannot have a better name in the art than "animated" and "Spagyric." It is of universal application in both faculties. In order to better comprehension, the first Sulphur, which we have said to be a resin of the earth, as it were, the mother and the father of other sulphurs, we name universal. The second kind is where it assumes a metallic or mineral appearance, but it is now embryonated; the third, which is repurged from these and exists Spagyrically pure from all superfluities, is Animated Sulphur. There are two conditions of this embryonated Sulphur which are worthy of notice. One, passing from the fixed stage, is made volatile; the other is a pure and living fire which destroys with equal facility a log of wood or a disease. The extraction of the embryonated Sulphur is brought about either by sublimation or by descent. But sometimes it is not found mixed naturally with other ingredients, so that being unable through its great subtlety to stand the heat of the fire in preparations of this kind, it has to be extracted from its minerals by means of aquafortis, and afterwards coagulated. This, when set aside according to its true concordance, contains within itself a golden nature, on which account it is to be sought before all others in Alchemy, because it easily admits of fixation, nay, it fixes the gold in cements, and in other metals where it is not yet mature or volatile. But gold is vainly sought therefrom unless it shall have previously existed there by Nature. It contains no silver, but only gold, one containing more than another, as in the embryonate of Venus, of red talc, of gold or iron marcasite, these rarely lack gold. Now, whoever wishes to turn his hand to these things, let him first of all remember and carefully note to separate Sulphur of this kind from gold with the greatest activity, and cleverly withal, so that nothing shall perish with the gold. I could say more than this, but I must be silent. If it were not diametrically to oppose the will of God, it would be the easiest thing possible to make all rich alike by a very few words, and to fulfil the wishes of everybody. But since riches altogether lead aside the poor from the right path, taking away humility and piety, and putting pride and self-sufficiency in their place, together with petulance and incontinence, one would rather hold one's tongue, leaving poverty as a bridle against these faults in those who are at once poor and greedy of wealth. To come to mineral Sulphur. The leader of our Art has directed his disciples to a recognition of this fact, that nothing can be generated from the woman without her husband. They have seen, therefore, that this Art is the father which arranges all things. He has summoned the spirit of transmutation whereby the mineral Sulphur is joined to linseed oil, and thence, by means of decoction a certain form results in the shape of a liver or a lung, and from thence afterwards a twofold liquid, one as white as milk, thick and oily; the other like oil, very red and as thick as blood; but both of such a nature that one will not mix with the other. The white liquid sinks to the bottom, the red floating on the surface. Attempts have been made to go farther, and make a white tincture from the white liquid; but to no purpose. I know that nothing has been done or can be done in this way, because the matter is weak and useless for this Art. But any crystal or beryl placed therein at the proper time, and remaining there for three years at least, is transmuted into a stone very like a jacinth. Likewise a ruby, which has not been sufficiently tinted by Nature, is, in course of time, rendered so clear and bright that it shines by night like a natural carbuncle, and wherever it is placed it can be found at night without a light. The same result follows with a jacinth; and in the sapphire the cœrulean colour is increased beyond the natural hue, with a translucent green tint inserted. It is also a most excellent tincture for other gems, as well as for Luna. If this be placed therein, it grows black, and lays aside the calx of Sol, though it be not fixed until it has arrived at its complete stage of perfection. Enough on this topic. Whoever wishes to work with this tincture, must first learn by means of Alchemy carefully to accomplish its preparation. It is well nigh the most difficult of all alchemical operations so far as preperation is concerned. This oil excels only in tints. In the greater virtues it is not so much to be trusted for acting, because there is a tincture of colour only in it, not of virtue. Some persons have tried also to extract tinctures from the metals. They have failed; but it would not be well to set down here the cause of their failure. This, however, is very certain, whoever has the Tincture of Sol, will be able to bring the body of gold beyond its natural degree, that is to say, from twenty-four to the thirty-six, and beyond, so intensely that it cannot ascend higher, though it still remains constant and fixed in antimony and in every quartation. The Sulphur of Luna, too, exalts its own body to such a degree that Venus, with an equal weight of this Luna, is taken for the Lydian stone. The Sulphur of Venus fixes copper, so that it will stand the test of lightning, but, nevertheless, it does not tinge. With the Sulphur of Saturn [it is transmuted into] the best steel; with the Sulphur of Jupiter, into excellent iron. So, too, tin is fixed with its own

Sulphur, so that it stands lightning, and Saturn is strengthened and fixed by its own [sulphur], so that it no longer affords any ceruse, or minium, or spirit. The Sulphur of Mercury renders its own body malleable, so that it bears the ignition of Venus, but not its ashes. The Sulphur of Sol tinges Luna, but does not fix it. There occur also with the other sulphurs transmutations of things put in them into some other bodies than their own. But this experiment does not turn out as desired. It should be remarked, meanwhile, that Sulphur demands a very expert operator, not a mere boaster or charlatan.

CHAPTER XVIII

Concerning Arsenic used for Alchemy *

It seems right to connect Arsenic generically with Sulphurs rather than with Mercuries, and to treat it immediately after Sulphurs. Some old chemists, or rather sophists, labouring at chemistry, swelling with jaundice, that is, with desire of gold, and a sort of yellow dropsy, when they saw in arsenic the white Tincture of Venus, and the red tincture in the calamine stone, believing, too, that the true arcanum of the stone was contained in these, thought the white and red electrum were silver and gold until they found out the contrary by tests, and learnt that they had been engaged in a vain work. And not content with that they went on perversely in order to arrive at fixation, and persevered until they had neither house nor possession left. They had wrought a transmutation in themselves rather than in the metals! And what wonder? They approached this work without judgment, and possessing no knowledge of minerals and metals, as so many of those who embark in the Art at the present day do. Since the time when the name of electrum given by the ancients passed into oblivion, there has forthwith followed the ruin of those

---

* In this case, also, the Geneva folio offers considerable variations from the text as it stands above. Concerning the Alchemical Virtues in Arsenic.—A certain name was invented and put forth by our ancestors, namely, electrum. Electrum is a metal proceeding from another metal, and unlike the metal from which it descends. For example : Copper turns to white metal. When its redness is removed it is called electrum. In like manner, from copper, by means of cadmia, is made orichalcum, and this is called red electrum. These different kinds of electrum certain alchemical sciolists and artists reckoned as silver, and sometimes took in place of gold ; nor did they understand or believe anything else save that this was silver, and so that silver could be produced from copper. Omitting the name of electrum, they took it for silver or for gold, and did not leave off their investigations so long as a house or a court remained. I point out this in order that error may be avoided, and that due consideration may be given to the questions, What is electrum ? what is gold ? and what is silver ? and that in this way no rash measures may be taken. Now, I will lay before you a certain medicament. Take the metal arsenic, prepared in a metallic way ; cement the same with Venus in the usual manner, and you will find a large quantity of electrum in the copper. No one need incur great expense for this substance, because it costs a good deal to make electrum. So, then, it is better to leave copper as copper in its own form. In no respect is its electrum better, but rather commoner. So by dissolving it in graduated water it leaves a calx. It is not that silver is produced, but electrum ; and it is rendered so subtle that nothing whatever remains, but it vanishes, and because it is not fixed it is consumed. Thus not only in copper, but also in iron, tin, steel, etc., a residuum is left ; but nothing of a fixed character is present, and in this way many are deceived. Eventually matters came to this crisis, that electrum lost its name and was called silver, whereupon there began for the alchemists destruction, exile, misery, and disappointed hopes. There are many recipes of this kind which it is not necessary to recount. They are well known to artists who follow me in this chapter, who also have well weighed their own error in seeking it in vain elsewhere. There is a good deal of seduction for juniors to desert the method of their elders, and when the pupil wishes to be more learned than the master, and no longer remains in the right path, but judges things for himself, and is prepared to abide by his own opinion. All that comes of it is, he labours in vain, thus atoning for his fault and incurring grievous loss. The ancients called this substance electrum, and such is its proper name. The moderns call it silver—its improper name. Our forefathers avoided all loss because they knew what they were about ; the rising generation do not know, and so incur loss. It has been a constant custom in alchemy that

who changed that name into fictitious gold and silver. That has been the destruction of modern chemists. To define Electrum: it is a metal made from some other by Art, and no longer resembling that from which it was made. For example: arsenical metal, prepared according to the form of metallic preparation, cemented with Venus in the accustomed manner, converts the whole copper into white electrum more worthless than its own copper. What need is there to deprave metals at great expense? Would it not be better to leave the copper in its own natural essence, to keep one's money, and devote time and labour to a more useful work? The ancients called Electrum by its proper name; the moderns falsely call it silver. The ancients were not losers, because they knew the Electrum itself; the moderns, because they have no knowledge of Electrum, throw away their faculties, labour, and time. Now, since in Alchemy all mistakes are constantly propped up with some new hope, it was tried to fix Arsenic by means of reverberations for some weeks, and by other devices. Thence it ensued that the Arsenic became red and brittle like coral, but of no use in Alchemy except for Electrum, as was just now said. Then by descent and precipitation they effected nothing more than by their calcinations. Thus it happens that in Alchemy obstinate men are deceived because they do not learn thoroughly from the foundation all the terms of the Art. It is true that Arsenic does, in its own natural condition, contain gold; and that this gold, by the industry of the artist, can sometimes be separated in a cement, or a projection, or otherwise, into silver, copper, or lead by attraction; but it does not therefore follow that this is produced by his operations and his tinctures. It means only that the gold which was there before has been derived by a process of separation, as it generally is, from its ore. It is nearly always found golden, and very seldom lacks gold, as is the case with many

---

investigations shall be made with persistent good hope. Hence operators have tried to fix arsenic, and to transmute it into another essence, on the chance that it may be, or may be rendered, better, and prove of greater efficacy. Hereupon followed the reverberation of arsenic, and its circulation in a reverberatory of reeds for some weeks, or by some similar process. Arsenic has been rendered like crystal, red and beautiful, like red glass for its hardness, light weight, and fragility. There is no place for the virtues of this arsenic in medicine. It regards only electra, as has already been said. Moreover, it has been attempted to deal with this, too, by another method of preparation, namely, by descent. By this method it is rendered red and yellow, and in potency is equivalent to the species already mentioned. Some have precipitated it, and it has approached, or even reached, a red colour; and yet not all the operators in this way have reaped the fruit of their labours or arrived at the result they contemplated, but only at the electric stage of it, which, on account of their ignorance and inexperience, led many artists astray. Wherefore it is necessary that everyone in these things should be farsighted. He who has not full knowledge and comprehension of all names does nothing, and the heads, however full of brains, do not get at the foundation of the matter. One thing is wanting to them for a foundation—to know electrum and other substances when they see them. Then they understand of themselves whether they can progress with electrum or not. Nevertheless it often happens that arsenic is auriferous in its nature, and contains gold in it. Now, if an operator is skilled in separating gold from arsenic, whether by a cement or by some method of projection. or by another process, so that he can reduce that gold to some metal, such as silver, copper, or lead, without doubt he will find it to be gold, and of excellent quality too. To follow this up so that a tincture shall be produced, or it shall issue forth from a tincture, is nothing; but the gold is in the arsenic, and the whole matter lies in depuration, separation, and kindred processes, according as anyone has experience therein. Arsenic, especially, which comes from auriferous districts, or from gold, is rarely without gold. The only point of importance is that the separation shall be properly made. I know nothing more of arsenic and its species beyond what I have put forward; at least, nothing which it is lawful and expedient to make known, whether with reference to medicine or to alchemical operations. Whoever has prudence ought to be sufficiently skilful for this purpose. If he has it not, let him altogether abstain. No faculty can subdue itself; but failure must ensue if due order and a genuine mode of procedure have not been preserved. You should follow the guidance of your own judgment. The man who follows no other guide is not in a state of subservience to any.

other substances. So far, then, have I given concerning Arsenic what I know, or what it is advisable to write. Let everybody first of all diligently examine its name, so that he may understand. Otherwise error is apt to arise easily in both faculties, which is only at length discovered by the result.

CHAPTER XIX
Concerning Quicksilver

Having dealt with salts and sulphurs, we come to Quicksilver. This cannot be properly termed a metal, but rather a metallic water; but it is called a metal for this reason, because, by means of Alchemy, it is brought to a solid substance and into a metallic colour, sometimes being fixed and sometimes not fixed. It can only be known as the chief material of Alchemists, who are able from it to make gold, silver, copper, etc., which will stand the test. So, too, perhaps tin, lead, and even iron. It is of a wonderful nature, inscrutable save after great labour. In a word, it shews itself to be the first material of Alchemists in metallic degrees, and the chief arcanum in medicine. It is a water which wets nothing it touches, an animal without feet, and the heaviest of all metals. It consists of Sulphur, Salt, and Mercury. The first and last matter it discloses in liquefactions of the metals, especially those which liquefy by heat without fire, and in others by flux.

CHAPTER XX
Concerning Cachimiæ and Imperfect Bodies

There is another kind of mineral bodies which is not saline, nor is it a metal, but metallic; such are marcasites,[*] chiseta red and white, perfect and imperfect antimoniacs, arsenicals, auripigments, various talcics, cobleta, granata, gem-like bodies, etc. I say these are metallic bodies since they have chiefly the first metallic matter, and derive their origin from the first three metallic bodies, to which they fly, as it were, and are incorporated with them as metals, for instance, gold, silver, copper, iron, etc. But since together with them there is incorporated a metallic enemy, they can only be separated

---

[*] Marcasites are to be found in all genera, whether you have regard to colour, brilliancy, form, or any other property. For they are nothing else than the superfluity of metals, that is, matter abundant in metals, being something which metals are unable to bear or contain within them, or convert into their own form. First of all, when the salts are separated from Ares (the occult dispenser of Nature), a separation of metals follows. Out of these, firstly, marcasite is produced which is unfit to become a metal, and yet in that matter it so resides that at first out of Ares there grows that matter of the metals. And it is the first matter, consisting of three things, the spirit of salt, the spirit of mercury, and the spirit of sulphur, but in such a manner that these three are one. Of these all metals and minerals consist. These things being so ordered, Archeus (the occult virtue of Nature) institutes the first operation of metals, so as to produce them and distinguish them into their forms and natures. But before he deals with the metals themselves, he ejects the superfluity which abounds in salt, mercury, and sulphur, and purges the three, after which the superfluity emerges along a simple line into its own yliadum (chaos), and is at first divided into two genera, marcasites and cachimiæ Here it is coagulated into a mineral, consisting of salt, sulphur, and mercury. Yellow marcasite obtains its colour from the predominance of sulphur; the white from the predominance of mercury. For sulphur and cachimiæ acquire their colour from salt, for this is derived from the spirit of salt, just as gravity is derived from mercury in all three. But if the separation be properly effected, each of the minerals, that is to say mercury, sulphur, salt, settles in its own place. Of these three all minerals consist.—*De Elemento Aquæ* Tract III., c. 1.

by means of Alchemy when set free from the tyranny of this foe. There are different enemies of this kind which practise robbery against the metals, just as if anyone seeking refuge with a companion should be robbed in his house and killed by the very man whose help he asked. Some of those spoken of consist chiefly of Sulphur, as marcasites, chiseta, cobleta; others in the body of Mercury, as arsenicals, auripigmentals, antimoniacs, etc. Others in Salt, as all belonging to talc. There are two colours of marcasites, the white and the yellow, according to the imperfect metallic Sulphur arranged in them, which also they need for many purposes. An imperfect metal is made from cobleta. This admits of liquefaction, and passes into a state of flux, is of a blacker colour than lead and iron, but of no brightness or metallic glitter; it barely admits of malleation, so that scarcely anything can be made from it. Its ultimate matter has not yet been discovered, nor the process of its separation. There is no doubt it is a promiscuous race from the male and female, as is the case in iron and steel, but these cannot be perfectly welded until some method of separation is discovered. There is another similar body called zinchinum; not that which is commonly so known, but a peculiar kind in which various metals are found to be adulterated, of a liquefiable nature and not malleable. It differs much in colour from the others, of which the last has not yet been found. In its preparation it is almost as wonderful as Mercury itself. It avoids mixture with anything else, and remains a special glass much to be admired among minerals. Metallic grains are found also in torrents, and are called granates, on account of their outer form. They are liquefied and bear the hammer; but still are not capable of being made into any implement. The properties of these bodies cannot be known unless they are revealed by Alchemy. Many contain adulterated metals, such as silver and gold, which flow to them, as they are accustomed to do to copper and to lead. They consist of a certain dense kind of Sulphur. Some granates of another kind are clear as crystal, and there is gold and silver in them.*

CHAPTER XXI

Concerning Metals free by Nature, Perfect and Imperfect; and first concerning Saturn, or Lead

Saturn has obtained a body the blackest and densest of all (though white, yellow, and red inhere therein), Mercury a similar one, and Salt one

---

* As in the generation of marcasites, so in cachimiæ. The superfluity is ejected from the prime principles. Sometimes mercury, sometimes sulphur, sometimes salt, will predominate, and that which predominates forms a mineral. In marcasites sulphur and mercury prevail, as two very light things which first fly away, then coagulate, and become very heavy. After the superfluity more completely departs, there is more salt and less of the other principles, though they are not altogether absent. Thus originate cachimiæ, tabulated and fissile, out of the nature of salt, which in sulphur and such mercury is of this property. It has all colours, white and red, receiving them from sulphur and mercury as one or the other predominates. But cachimia is more fixed and solid than marcasite, by reason of its fixed salt. Colours, also, are fixed in it, so that it may receive no injury from the fire. Thus marcasite is the superfluity abounding in the first matter of metals in Ares, which is separated by Archeus into Yliadum, whence afterwards are generated about thirty forms of marcasite and cachimiæ, all of which are, nevertheless, comprehended under two names. The multiplicity of these genera, which are all derived from one matter, is owing to the unequal manner in which the three prime principles are combined.—*Ibid.*, c. 2.

above all others fusible. By corruption it is easily reduced to its spirit, to white or yellow cerussa, to minium, and lastly, to glass, like the rest. Tin is made up of white fixed Sulphur and fixed Salt but of Mercury not fixed. And because it is fixed in body, not in Mercury, it easily loses its metallic fusion, the spirit passing away by the fire; and when this is absent it is no longer a metal but an evanescent body. Iron and steel are not of the liquefiable Sulphur, Salt, and Mercury, contrary to tin and lead. Iron is coagulated into the hardest metal of all, and it marries itself: that is, two metals are found in one, steel the male, and the female iron. These can be separated one from the other, each for its special use. Gold is generated from the very purest Sulphur, perfectly sublimated by Nature, purged from all its dregs and spurious admixtures, and exalted to such a transparency that no metal can corporeally ascend higher. This Sulphur is one part of the primal element, and if Alchemists could have this as something easily discoverable in its tree and root, they would be able with due cause to rejoice; for it is the true Sulphur of the philosophers out of which gold is made, not that other gold out of which is made iron, copper, etc. This is its universal test. Its Mercury, too, is by Nature perfectly separated from all terrestrial and accidental superfluity, transmuted separately into its mercurial part, and into extreme perspicuity, which Mercury of the Philosphers is the second part of the primal matter of gold, from which gold is generated. Lastly, Salt is the third part of the primal essence of gold, and of the tree from which gold is to be produced, as roses from rose-seeds—gold which is brought to its supreme crystalline brightness, and purified from all the acridity, acerbity, bitterness, darkness, and vitriolic nature of Salt, so that nothing of this kind appertains to it, now that it rejoices in its lucidity and transparency.

When these three meet together in one, the gold is decocted into a mass, not, however, always of one and the same condition or degree. Nature exhibits thirty-two grains of gold, and these in Art become twenty-four grains in the highest grade of perfection. The cause of this is that the gold is nourished in its tree as a cow in its pastures, or an epicurean in his cook-shop and eating-house. Directly one of these leaves his feeding-place he grows lean, and so is it with gold; it is diminished by eight degrees. And as some of these feeding-places are occasionally inferior, it happens that the degrees of the gold are deteriorated or diminished too; so that Nature's sum total of twenty-six is reduced in Art to ten. The accidents, or rather the incidents, of the stars or of the elements sometimes hinder the generation of gold, so that it becomes ruder and less tractable in its nature. But it is especially inequality in the weights of the three primals which has effect. Too great a portion of Salt renders it too pale. With too much Mercury it grows yellow, and with a too plentiful supply of Sulphur it is rendered red. In Nature, just as much as in the work of man, errors occur by means of these hindrances; but these can be removed by means of antimony, cements, and quartations. In Sulphur nothing should be looked for but a body, in Salt confirmation, but in Mercury all virtue, property, essence, and medicine, which do not exist anywhere else as it does therein; but rather as in a dead body from which the spirit has departed, in which, however, we try to keep some of the elementary powers, as, for instance, the remains of the fire of wine in Acetum, though these are corrosive rather than nutritive or strengthening. Natural objects clearly shew that they are compounded of the four elements; but beyond that the matter is occult. They are made up only of the three we have spoken about, which possess a magnet common to them all. This, in the decoction of the preparation, attracts to itself the trinity of essence. The old philosophers called this state *esse*, because the trinity acquires a condition of unity in which the natural motion reposes and settles the degree. But that magnetic virtue should deservedly be called a fourth *esse* (not element) since it attracts the medicine to the Mercury in which it is found. In the ultimate separation, however, the Mercury loses most of its weight. All these matters being thus arranged by Nature, the gold grows up to a tree, spreading forth first from the root by the trunk to its branches and twigs, on which flowers are produced (as we see on the earth), and when these fade the fruit is not always found at the extremities of the twigs, but sometimes a hundred paces farther off in the tree, occasionally in its very midst, or some degrees towards the surface of the earth. It will sometimes happen that nothing but Mercury is produced, when by its superfluity it has suppressed the other ingredients. If, however, the Salts preponderate, their corrosive nature, like so many worms, consumes the flowers of the tree. By the preponderance of Sulphur everything is burnt up, just as on earth by the too great heat of the sun. Copper is produced by the brown Sulphur, red Salt, and yellow Mercury decocted into a metal. This contains within itself its masculine element, that is, the scoria; and if it be again reduced to a metal after separation, it returns to masculine copper, which can no longer be corrupted; and the female will afford no scoriæ at all. On malleation and fusion they differ from each other only as steel and iron, and can be separated in the same way, so that two different metals are thence produced. Silver is composed of white Sulphur, Salt, and Mercury, naturally prepared and fixed to the highest degree of purity and transparency, next after gold, in ashes, not in antimony, or in royal cement, or in quartation. The difference of fixation between gold and silver can easily be learnt by considering that gold is masculine, and has the male virtues very strongly fixed, while silver, as the female, has them weaker. They are of one and the same primal matter, and differ as to colour and fixing in no other way than as the male and the female. The metals, then, are seven in number, exclusive of Mercury, namely gold, silver, tin, lead, iron, steel, and copper. The last contains within itself the male and female, when both are welded for use, and are not separated by Nature, as steel and iron are; so that they are held as one, and since they possess the same malleability and power of being wrought, they are not commonly separated, except when this is done chemically for purposes of the Art. It should be remarked, too, that metals are not always found with their masculine and feminine portions separated by Nature, as is the case with gold, silver, iron, and steel, each by itself. Often the two are found together, as gold and silver in one metal, also steel and iron together, or tin and lead, the one not hindering the other, or being separated one from the other. Sometimes two adulterated metals are found, as gold and silver naturally mixed with others, on account of their subtlety, especially when several of diverse primal nature meet in one body, just as we see on the earth different fruits engrafted on the trunk of one tree.

Conclusion

A fitting treatise on the natural generation of metals was absolutely necessary in order that it might be understood what is meant by the regeneration of metals brought about through Alchemical Art. The opinion of all those who philosophise on this Art is that the Artist in this profession ought in all things exactly to imitate Nature. So, then, it was necessary to say and to understand how Nature works in the innermost parts of the earth, and what instruments she employs. Whoever has not understood in this way will be little likely to get at the knowledge by his own unaided endeavours. Let him who investigates this difficult and abstruse matter be not so much the disciple of Art as of Nature.

Here ends the Economy of Minerals

THE COMPOSITION OF METALS*

IF any one denies that there is great efficacy in the Composition of Metals so far as relates to supernatural affairs, we will answer him, and bring forward so many proofs as shall support our own opinion and force him to subscribe thereto. For if the seven metals were, in just and due order, compounded, mixed together, and united in the fire, you must certainly hold that in one body were conjoined and linked together all the virtues of the seven metals. It has been seen good to call this body electrum. Its efficacy, power, and operations, moreover, shew themselves to be much greater, even supernaturally so, than exist in a latent form grafted by Nature on metals in their rude condition. In those solid and rude metals are only those powers wherewith God and Nature herself have endowed them. Gold, indeed, is the noblest of all, the most precious and primary metal, if we rightly consider it; and we are not prepared to deny that leprosy, in all its forms, can be thereby removed from the human frame. Nor are we unaware that exterior ulcers and wounds are cured by copper and mercury. The other metals, too, have each their own excellences, and these not by any means to be despised; but we will pass over these for the moment, since you will hear of them when we come to treat concerning the Life of the Metals.† But metals cannot be used in medicine without injury, unless they be first comminuted, altered, and, after being deprived of their metallic nature, transmuted into another essence. You can hope for little result from them unless the preparation which Alchemy teaches shall have preceded their administration; that is, if you have not previously reduced them to their arcana, oils, balsams, quintessences, tinctures, calces, salts, crocuses or the like, and then administered them to the patient. Moreover, the supernatural force or effect of the metals, even though it be present in them, will be of no avail unless you first prepare them according to our method in which we will instruct you. But we greatly desire that our electrum should be compounded, since it can afford great and marvellous results in proportion as it is revealed by practice. If we consented to pass

---

* A considerable portion of this tract belongs more properly to the section concerned with Hermetic Medicine, but it is inserted at this point for the further illustration of the subject of electrum, which is somewhat shortly discussed in the foregoing treatise. The work *De Compositione Metallorum* is printed in separate form in the Basle 8vo, but it really constitutes the sixth book of the *Archidoxis Magicæ*, as they are found in the Geneva folio.

† So far as the *Archidoxis Magicæ* are concerned, this promise is not fulfilled. Possibly Paracelsus intended to carry his subject further than the seventh book, which is devoted to the sigils of the planets, and has nothing of a chemical nature. But possibly, also, a reference is intended to the first book *Concerning the Nature of Things*.

over its praises in silence, we should consider that we were doing it an injury: but since its operation and mighty power surpass belief, we deem it necessary to pronounce an eulogium on its virtues and efficacy. We will defer for the moment any mention of the rude and solid metals, since they admit of no comparison with our electrum. If any appliance used for food or drink be made of this material and diligently watched, it will be impossible for any poison or drug to be placed in it, because in our electrum there is so much sympathy towards man through the force, efficacy, and influence of the planets and the stars of Olympus, that for very pity, and as though in difficulty, directly it is taken in hand it betrays the poison by breaking out into a sweat and projecting spots. For this reason our ancestors used to have their drinking-cups, dishes, and other utensils made of the said material. There still remain in our age many necklaces and ornaments, such as rings, bracelets, remarkable coins, seals, figures, bells, shekels, made out of this, which of old were hidden in the earth. When they were dug up nobody, or very few, understood them, and in their ignorance they gilded them over or tinged them with silver. It is just a mark of the ignorance of our age that it cares nothing for such objects as these. But God would not have it that such a mystery of Nature and such a great treasure of His own should be hid any longer, but that what had been hidden by the more than Cimmerian darkness of the sophists should now, after a long season, come to light again. We do not assume to exhaust the virtues of our electrum. The ribald genius of the sophists would be hurt; the crowd of fools would be offended, and would receive what we said with idiotic laughter. Over and over again we have been on our guard against scandalising this impious crowd; so to avoid such a result it will be safest to pass over these matters in silence. Not, however, that we can altogether pass unnoticed certain stupendous effects of our electrum; since they came under our own eyes we shall be able to speak the more freely concerning them, without any suspicion that we are romancing or making up a story. We have seen rings, for instance, which removed all fear of paralysis or spasm from those who wore them on their fingers. These people, too, never suffered from apoplexy or epilepsy. If an epileptic patient put such a ring on the third finger, even though he be so overcome by the violence of the paroxysm as to be prostrated on the ground, he comes to himself and gets up.

Here, too, should be added something which we do not give from the report of others, for the same we have seen with our own eyes and know by experience. If the abovementioned ring be worn on the third finger by a man in whom any ailment is latent and growing, so that it would presently break forth in an eruption, the ring would forthwith give an indication by bursting out in a sweat, and as if seized with a sudden sympathy would put forth spots and become depraved in appearance, as we shall shew more fully in our book entitled "Sympathy."

Lastly, since I would not pass over or omit any word in favour of electrum, it is a preservation and an antidote against evil spirits. There is latent in it an operation and a conjunction of planetary influence which make us the more easily believe that the old Magi in Persia and Chaldæa attempted and accomplished much by its aid. If we sought to enumerate all the cases specifically, we should indeed enter upon a marvellous chronicle. Not, however, to give any occasion of offence or allow persons to make a handle of this, it will suffice to have touched the subject in few words. The Sophists, who are my deadliest enemies, would not hesitate to proclaim me Arch-Necromancer. But I cannot refrain from telling a miracle which I saw in Spain when I was at the house of a certain necromancer. He had a bell weighing, perhaps, two pounds, and by a stroke of this bell he used to summon, and to bring, too, visions of many different spectres and spirits. In the interior of the bell he had engraved certain words and characters, and as soon as the sound and tinkle were heard, spirits appeared in any form he desired. Moreover, the stroke of this bell was so powerful that he produced in the midst many visions of spirits, of men, and even of cattle, whatever he wished, and then drove them away again. I saw many instances of this, but what I particualarly noticed was that when he was going to do anything new, he renewed and changed the characters and the names. I did not, however, get so far as to induce this man to impart to me the secret and mystery of the names and characters. At length I began to speculate more thoroughly about this circumstance; and there came into my mind—ideas which we will pass over in silence here. There was more in that bell than one can put into words; and of this be very sure, that the material of which it was composed was this electrum of ours. You will therefore have no difficulty in believing that Virgil's bell (Nola) was of such a kind as this. At its stroke all the adulterers and adulteresses in the king's palace were so excited and alarmed that suddenly, as if struck with lightning, they rushed over the bridge into the river. Think not this story a mere fable: the thing really happened. Nor be so dense as to hesitate as to whether such properties can exist. For if, as you know to be the case, a visible man can call another visible man to him by a word, and force him to do what he wants—when a mere word, without the aid of arms, can effect so much, much more can it be that an invisible man can do this, since he commands both the visible and the invisible man, not by the aid of a word, but by the direction of his thought. The inferior always obeys the superior, and stands to him in the light of a subject. So, then, you will easily come round to our opinion if you settle it that the interior or invisible man is a kind of constellation or firmament. For he remains latent in the senses and thoughts of the exterior, visible man, and discloses or reveals himself only by imagination. You will concede, therefore, that there are stars in man and that their constellation is so arranged by the Olympian spirit that the man can be led and changed into quite another man. So, then, I say that the same thing occurs with metals, namely, that things may be so constellated by celestial impression as to make the operation and virtue which Nature originally determined, really arise from the good aspect of the higher stars, and thus unfold itself, as is shewn in other books of the Archidoxis Magica.* I will subjoin, if you wish, an illustration. Let any one reduce to an amalgam gold and mercury, making a conjunction of Sol and Mercury, but with a preponderance of Sol. Let him mix and blend them, and soon, with little labour, the two metals will become fixed. With these, if you will, you can make a tincture on *Mercurius vivus*. That, again, can afterwards be increased and augmented with other *Mercurius vivus* under the same constellation. This is, indeed, a great arcanum of Nature. There will be a similar composition and union of gold or silver with mercury without this conjunction. For if gold be placed above mercury, so that the white fume of the mercury touch and penetrate the body of the gold, the gold will be rendered fragile, and will melt with the greatest ease like wax. The process is the same with silver.

This is the Magnesia of the Philosophers, in the finding of which Thomas of Aquinum and Rupescissa and their disciples, though they worked hard, were unsuccessful. And let nobody think it an easy matter so to blend *Mercurius vivus* in the fire with harder metals and those of tardier solution—as silver,

---

* Moreover, it is altogether certain, and experimentally proved, that the mutations of time have singular force and operation, and this is especially the case when certain metals are melted and elaborated together. Further, no one can prove that the metals are devoid of life. Their oils, sulphurs, salts, and quintessences, which are the best reservatives, have enormous power in nourishing and sustaining human life, and herein altogether surpass in strength all other simples, as, indeed, is entirely the case with all our remedies. How, if they were devoid of life, could they awaken in the diseased and half-dead members and bodies of men a fresh and vital strength, and at the very outset restore them ? . . . I therefore boldly assert that metals and stones, equally with roots, herbs, and fruits, have a life of their own, with this distinction, however, inasmuch as metals are prepared and elaborated according to time. The efficacy of time is well-known, but we will speak only of those things which are difficult, and not to be grasped by the senses, but, indeed, are almost contrary to their evidence. Further, even signs, characters, and letters have their virtues and efficacies. Now, if the nature and property of the metal, as also the influence and operation of the heaven and of the sphere of the planets, the signification and formation of the characters, signs, and letters, together with the observation of the times, days, and hours, harmonise and agree, why should not a sign or seal composed after this manner have its own force and operation ? And why, then, should not such and such a medicine, seasonably applied, benefit the head, another the vision, or a third the veins ? And especially in the case of those who dislike to take other remedies into the body. Yet none of these results are possible without the air of the Father of Medicine Himself, Jesus Christ, the one and true Physician. Objectors may say that words or characters have no force, since they are mere signs or figures, and that none at least can compare in efficacy with the cross. But how is it that the serpent in Helvetia, Algovia, or Suavia, understands the Greek phrase *Osy, Osya, Os*, although in none of these countries is Greek so common that venomous reptiles can acquire it ? How is it that, the moment they hear the words, they draw in their tails, stop up their ears, and, contrary to their nature, lie motionless, without doing harm to any man ? . . . By this it is shewn that characters, words, and signs have a recondite and latent force, not in the least opposed to Nature, nor anything to do with superstition. It is found that these words have the same effect when they are written on paper, and not uttered. So, also, let it not be considered incredible that a man should be cured by medicine, even when he does not take it internally, but carries it suspended like a seal from his neck. That even in dead things there is a certain force, I prove by the example of the kingfisher, for if, when it is dead, you remove its skin, and hang it up, you will see that, although it is dry, it will annually cast its old feathers and produce fresh ones of the same colour.—*Archidoxis Magicæ*, Lib. I. For it is certain that in the very signs themselves of the planets, if they are harmonised and carried about in the required manner, according to a favourable hour and time, as regards their course, there reside great force and virtue. For none can deny that the superior stars and influences of heaven have very great weight in transient and mortal affairs. If the superior stars and planets are able to control, rule, and sway according to their will the animal man, although he be made according to the image of God, and be endowed with life and reason, how much more ought they to rule an inferior thing, that is to say, metals, stones, and images, upon which they impress themselves, or which they so occupy, with all their virtue and efficacy, after the manner of an influence, as though they were substantially present, even as they are in the firmament ? It is possible to man himself to bring these into a certain medium, wherein they may effectually operate, whether this medium be a metal, a stone, or an image. But this is most important of all : to know that the seven planets have greater force in nothing than they possess in their proper metals.—*Ibid.*, Lib. VII

copper, gold, iron, and steel--that they may quickly liquefy. Many tinctures and Elyxeria (*sic*) of metals are prepared thus for transmuting metals, as will be more copiously described in other books on Metallic Transmutations.*

The same is the case with common mercury, which with its fume penetrates all other metals, and, as it were, breaks through them, calcines them, and disposes them to its own nature. Metals will coagulate this by their fume. We assert that the most extreme heat resides in Mercury, and that it cannot be coagulated except by extreme cold, which is seen to exhale copiously from metals in the fire. Nothing affects metals in the fire save what is of extreme cold and unable to bear the vehemence of the fire. Such a metal is arsenic, which being liquefied ascends as a spirit from metals while they are in a state of flux.

Moreover, do not lose sight of the fact that Mercury is a metallic spirit, and that every spirit is more powerful than a body. So is it with Mercury in reference to the other metals. Just as it is easy for a spirit to penetrate walls, so it is not difficult for Mercury to penetrate metals.

How many are the wonderful operations and effects of Mercury on the metals! We cannot detail them all. But shall we send you away empty to some other source? We know from experiment that if *Mercurius vivus* be sublimated from some one of the metals which has been several times calcined, and if then the calcinated metal which remains at the bottom be again reduced to its metal, it is melted in the fire as easily as lead, though it were gold, silver, copper, iron, or steel, even if it be only applied to the flame of a candle like so much wax; or as snow and ice melt before the sun. Afterwards by digestion for a certain time it can be changed into Mercury. We have mentioned this

---

* The fourth book of the *Archidoxis Magicæ* is entitled, *Concerning the Transmutation of Metals and their Time*. It is literally as follows:—If you seek to change gold into silver, or any given metal into any other metal, have regard to the following tabulation. Nor is it of small moment so that you may be able to arrive at the end of your purpose more quickly and thoroughly. Scheme of the Transmutation of Metals.—To transmute Sol into Luna, Venus, Mars, Jupiter, Saturn, or Mercury, begin with Luna occupying the sixth grade of Cancer, Taurus, Aries, Pisces, Aquarius, or Virgo, as the case may be, and always in the hour of that planet into which you wish to convert gold or any of the other metals, namely, Luna, Venus, Mars, Jupiter, Saturn, Mercury. To transmute Saturn into Sol, Luna, Mars, Venus, Jupiter, or Mercury, begin with Luna occupying the twentieth grade of Leo, Scorpio, Cancer, Taurus, Pisces, or Virgo, as the case may be, in the hour of Sol, Luna, Mars, Venus, Jupiter, or Mercury, according to the metal into which you would convert Saturn. To transmute Mercury into Sol, Luna, Venus, Mars, Jupiter, or Saturn, begin with the Moon in the first grade of Leo, Virgo, Cancer, Taurus, Pisces, or Aquarius, as the case may be, in the hour of Sol, Luna, Venus, Mars, Jupiter, or Saturn, according to the metal into which you would convert Mercury. To transmute Luna into Sol, Venus, Mars, Jupiter, Saturn, or Mercury, begin with the Moon in the twelfth grade of Leo, Libra, Scorpio, Sagittarius, Aries, or Gemini, as the case may be, and in the hour of Sol, Venus, Mars, Jupiter, Saturn, or Mercury, according to the metal into which you would convert Luna. To transmute Venus into Sol, Luna, Mars, Jupiter, Saturn, or Mercury, begin with the Moon in the ninth grade of Leo, Cancer, Capricorn, Aquarius, Pisces, or Sagittarius, as the case may be, and in the hour of Sol, Luna, Mars, Jupiter, Saturn, or Mercury, according to the metal into which you would convert Venus. To convert Mars into Sol, Luna, Venus, Jupiter, Saturn, or Mercury, begin with the Moon in the eighty-first grade of Leo, Cancer, Taurus, Sagittarius, Scorpio, or Virgo, as the case may be, and in the hour of Sol, Luna, Venus, Jupiter, Saturn, or Mercury, according to the metal into which you would convert Mars. To transmute Jupiter into Sol, Luna, Venus, Mars, Saturn, or Mercury, begin with the Moon in the third grade of Leo, Cancer, Libra, Virgo, Aquarius, or Pisces, as the case may be, and in the hour of Sol, Luna, Venus, Mars, Saturn, or Mercury, according to the metal into which you would convert Jupiter. For example: If you wish to change gold into silver, make a beginning in the hour of the Moon, when the Moon occupies the sixth grade of Cancer. And so, likewise, understand the rest of this scheme for the conversion of metals. For all terrestrial affairs, occupations, and matters of business, are most conveniently and happily executed in harmony with the motions of the heavens and the planets. For all men, by the dispensations of Almighty God, are ruled and led by the power and operation of the firmament, both as to health and disease. So is it necessary before all things to have regard to this operation in the healing art. Simples very frequently push forth their virtues according to a certain rule of time.

fact in our book on the Resuscitation of Natural Things. This is the Mercury of the Philosophers. In this way you will prepare the Mercury of Gold, of Luna, of Venus, of Mars, of Jupiter, and of Saturn. Although in their books Arnold, Aristotle, and other philosophers boast about this, yet I am well assured that it was never prepared or seen by them. It will now be for you to keep this great secret and mystery of Nature, and to take care that it does not fall into the hands of my adversaries; since it would be an indignity for them to get to know it. A pearl or a precious stone will not please a goose, because the goose does not know its price and value. It would infinitely prefer a turnip. We may fitly say the same of the sophists. It is no injustice to conceal secret mysteries from them. Let us not seem to cast pearls before swine or give that which is holy to dogs, since God sternly forbids us so to do.

But let us proceed to the practical work of our electrum, as we promised at the outset. We would have it prepared, compounded, and conjoined according to the revolution of the heaven and the conjunctions of the planets. We will proceed in this way. First, you must diligently observe the conjunction of Saturn and Mercury; and, before this occurs, have ready the appliances you require. These are, fire, a cauldron, lead cut up into minute pieces, and *Mercurius vivus*. Take care that nothing be wanting which the work in hand requires, or for lack of which the action may be hindered or retarded. Then when the conjunction is just going to take place, let the lead be melted in the fire, and be not quite hot when it shall have fused, lest the Mercury which you pour in escape, or, if the heat be too great, pass off in smoke. Let this be done at the very moment of conjunction. Take out suddenly the cauldron with the liquid lead; pour in the Mercury, and afterwards let them both be coagulated.

Then there will be need of attention when the conjunction of Jupiter with Mercury or Saturn is about to take place, so that you may not be ignorant of the time or pass it by. Let everything you will want be ready to hand as I before admonished you. You must take care, before the actual moment of conjunction, to melt in one vessel fine English tin, and in the other lead with Mercury. At the moment of conjunction move the metals from the fire, slackening the heat a little, and pour all into one crucible. When they have coagulated into one body you will have three metals softer and more easily melting over the fire. When they are united let it not escape your notice that in the very first place these are to be dissolved and conjoined. Then notice when there is a conjunction of any of the other four planets—Sol, Luna, Venus, or Mars—with one of the three former, Saturn, Mercury, or Jupiter. Have all instruments and materials ready. Let them be dissolved singly first; then when liquefied pour them into one at the very point of conjunction, and keep them. In a like way proceed with other metals which are to be joined and copulated with the former, until you have reduced and united all the seven according to the due conjunctions of the planets. So will you have prepared our electrum, concerning which enough has now been said.

CONCERNING THE NATURE OF THINGS

BOOK THE FIRST

Concerning the Generation of Natural Things

THE generation of all natural things is twofold*: one which takes place by Nature without Art, the other which is brought about by Art, that is to say, by Alchemy, though, generally, it might be said that all things are generated from the earth by the help of putrefaction. For putrefaction is the highest grade, and the first initiative to generation. But putrefaction originates from a moist heat. For a constant moist heat produces putrefaction and transmutes all natural things from their first form and essence, as well as their force and efficacy, into something else. For as putrefaction in the bowels transmutes and reduces all foods into dung, so, also, without the belly, putrefaction in glass transmutes all things from one form to another, from one essence to another, from one colour to another, from one odour to another, from one virtue to another, from one force to another, from one set of properties to another, and, in a word, from one quality to another. For it is known and proved by daily experience that many good things which are healthful and a medicine, become, after their putrefaction, bad, unwholesome, and mere poison. So, on the other hand, many things are bad, unwholesome, poisonous, and hurtful, which after their putrefaction become good, lose all their evil effect, and make notable medicines. For putrefaction brings forth great effects, as we have a good example in the sacred gospel, where Christ says, "Unless a grain of wheat be cast forth into a field and putrefy, it cannot bear fruit a hundred fold." Hence it may be known that many things are multiplied by putrefaction so that they produce excellent fruit. For putrefaction is the change and death of all things, and the destruction of the first essence of all natural objects, from whence there issues forth for us regeneration and a new birth ten thousand times better than before.

Since, then, putrefaction is the first step and commencement of generation, it is in the highest degree necessary that we should thoroughly

---

* There is another aspect in which generation is also twofold, as, for example, that of wood and other things takes place naturally out of seed. But the worms which destroy wood are the product of a monstrous sperm. Hence there are two generations – natural and monstrous. Every sperm in living things has within it another sperm which is monstrous, and can promote its likeness. There is also a monstrous sperm in all minerals.—*Paragraphorum* Lib. II., Par. IV.

understand this process. But there are many kinds of putrefaction, and one produces its generation better than another, one more quickly than another. We have also said that what is moist and warm constitutes the first grade and the beginning of putrefaction, which procreates all things as a hen procreates her eggs. Wherefore by and in putrefaction everything becomes mucilaginous phlegm and living matter, whatever it eventually turns out to be. You see an example in eggs, wherein is mucilaginous moisture, which by continuous heat putrefies and is quickened into the living chicken, not only by the heat which comes from the hen, but by any similar heat. For by such a degree of heat eggs can be brought to maturity in glass, and by the heat of ashes, so that they become living birds. Any man, too, can bring the egg to maturity under his own arm and procreate the chicken as well as the hen. And here something more is to be noticed. If the living bird be burned to dust and ashes in a sealed cucurbite with the third degree of fire, and then, still shut up, be putrefied with the highest degree of putrefaction in a *venter equinus* so as to become a mucilaginous phlegm, then that phlegm can again be brought to maturity, and so, renovated and restored, can become a living bird, provided the phlegm be once more enclosed in its jar or receptacle. This is to revive the dead by regeneration and clarification, which is indeed a great and profound miracle of Nature. By this process all birds can be killed and again made to live, to be renovated and restored. This is the very greatest and highest miracle and mystery of God, which God has disclosed to mortal man. For you must know that in this way men can be generated without natural father and mother; that is to say, not in the natural way from the woman, but by the art and industry of a skilled Spagyrist a man can be born and grow, as will hereafter be described.

It is also possible to Nature that men should be born from animals, and this result has natural causes, but still it cannot be produced without heresy and impiety. If a man have connection with an animal, and that animal, like a woman, receives the seed of the man with appetite and lust into its womb, and shuts it up there, then the seed necessarily putrefies, and, through the continuous heat of the body, a man, and not an animal, is born from it. For always, whatever seed is sown, such a fruit is produced from it. If this were not so it would be against the light of Nature and contrary to philosophy. Whatever the seed is, such is the herb which springs from it. From the seed of an onion an onion springs up, not a rose, a nut, or a lettuce. So, too, from corn comes corn; from barley, barley; from oats, oats. Thus it is, too, with all other fruits which have seeds and are sown.

In like manner, it is possible, and not contrary to Nature, that from a woman and a man an irrational animal should be born. Neither on this account should the same judgment be passed on a woman as on a man, that is, she should not on this account be deemed heretical, as if she had acted contrary to Nature; but the result must be assigned to imagination. Imagination is very frequently the cause of this: and the imagination of a pregnant woman is so active that in conceiving seed into her body she can transmute the fœtus in different ways: since her interior stars are so strongly directed to the fœtus that they produce impression and influence. Wherefore an infant in the mother's womb is, during its formation, as much in the hand and under the will of the mother as clay in the hand of the potter, who from it forms and makes what he likes and whatever pleases him. So the pregnant mother forms the fruit in her own body according to her imagination, and as her stars are. Thus it often happens that from the seed of a man are begotten cattle or other horrible monsters, as the imagination of the mother was strongly directed towards the embryo.*

But as you have already heard that many and various things are generated and quickened out of putrefaction, so you should know that from different herbs, by a process of putrefaction, animals are produced, as those who have experience of such matters are aware. Here, too, you should learn that such animals as are produced in and by putrefaction do all of them contain some poison and are venomous; but one contains far more and more potent virus than another, and one is in one form, another in another, as you see in the case of serpents, toads, frogs, basilisks, spiders, bees, ants, and many worms, such as canker-worms, in locusts, and other creatures, all of which are produced out of putrefaction. For many monsters are produced amongst animals. There are those monsters, too, which are not produced by putrefaction, but are made by art in the glass, as has been said, since they often appear in very wonderful form and horrible aspect; frequently, for instance, with many heads, many feet, or many tails, and of diverse colours; sometimes worms with fishes' tails or birds' wings, and other unwonted shapes, the like of which one had never before seen. It is not, therefore, only animals which have no parents, or are born from parents unlike themselves, that are called monsters, but those which are produced in other ways. Thus you see with regard to the basilisk, which is a monster above all others, and than which none is to be more dreaded, since a man can be killed by the very sight and appearance of it, for it possesses a poison more virulent than all others, with which nothing else in the world can be compared. This poison, by some unknown means, it carries in

---

* Here, as elsewhere throughout his writings, Paracelsus lays special stress on the power exercised by the imagination.—It is necessary that you should know what can be accomplished by a strong imagination. It is the principle of all magical action.—*De Peste*, Lib. I.   The imagination of man is an expulsive virtue.—*De Peste*, s. v. *Additamenta in* Lib. I.   The imagination dwelling in the brain is the moon of the microcosm.—*De Pestilitate*, Tract II., c. 2, *De Pyromantica Peste.*   All our sufferings, all our vices are nothing else than imagination. . . . And this imagination is such that it penetrates and ascends into the superior heaven, and passes from star to star. This same heaven it overcomes and moderates. . . . Whatsoever there is in us of immoderate and inhuman, all that is an imaginative nature, which can impress itself on heaven, and, this done, heaven has, on the other hand, the power of refunding that impression.—*De Peste, Additamenta in* Lib. I., Prol.   So, also, a strong imagination is the source of both good and evil fortune.—*De Peste*, Lib. II., c. 2.   Any strong appetite, desire, or inclination nourished by the imagination of a pregnant woman can be and is often impressed upon the fœtus. It is also possible for such a woman, by persistently thinking upon a wise and great man, such as Plato or Aristotle; an illustrious soldier, such as Julius Cæsar or Barbarossa; a great musician, like Hoffhammer; or a painter, like Durer; so to work upon the plastic tendencies of her offspring, that it will exhibit similar qualities. But there must be something also in the mother which shall correspond to the special talents which she has imagined.—*De Origine Morborum Invisibilium*, Lib. III. Imagination can distort and deform the fœtus, and in this manner many wonders are produced, when there are no physical peculiarities in the parent.—*Ibid.*

its eyes, and it is a poison that acts on the imagination, not altogether unlike a menstruous woman, who also carries poison in her eyes, in such a way that from her very glance the mirror becomes spotted and stained. So, too, if she looks at a wound or a sore, she affects it in a similar way, and prevents its cure. By her breath, too, as well as by her look, she affects many objects, rendering them corrupted and weak, and also by her touch. You see that if she handles wine during her monthly courses it soon turns and becomes thick. Vinegar which she handles perishes and becomes useless. Generous wine loses its potency. In like manner, amber, civet, musk, and other strongly smelling substances being carried and handled by such a woman lose their odour. Gold, corals, and many gems are deprived of their colour, just as the mirrors are affected in this way. But—to return to my proposal of writing about the basilisk—how it carries its poison in its eye. You must know that it gets that power and that poison from unclean women, as has been said above. For the basilisk is produced and grows from the chief impurity of a woman, namely, from the menstrual blood. So, too, from the blood of the semen; if it be placed in a glass receptacle and allowed to putrefy in horse dung, from that putrefaction a basilisk is produced. But who would be so bold and daring as to wish to produce it, even to take it and at once kill it, unless he had first clothed and protected himself with mirrors? I would persuade no one to do so, and wish to advise every one to be cautious. But, to go on with our treatise about monsters, know that monstous growths amongst animals, which are produced by other methods than propagation from those like themselves, rarely live long, especially near or amongst other animals, since by their engrafted nature, and by the divine arrangement, all monsters are hateful to animals duly begotten from their own likeness. So, too, monstrous human growths seldom live long. The more wonderful and worthy of regard they are, the sooner death comes upon them; so much so that scarcely any one of them exceeds the third day in the presence of human beings, unless it be at once carried into a secret place and segregated from all men. It should be known, forsooth, that God abhors monsters of this kind. They displease Him, and none of them can be saved when they do not bear the likeness of God. One can only conjecture that they are shapen by the Devil, and born for the service of the Devil rather than of God; since from no monster was any good work ever derived, but, on the contrary, evil and sin, and all kinds of diabolical craft. For as the executioner marks his sons when he cuts off their ears, gouges out their eyes, brands their cheeks, cuts off their fingers, hands, or head, so the Devil, too, marks his own sons, through the imagination of the mother, which they derive from her evil desires, lusts, and thoughts in conception. All men, therefore, should be avoided who have more or less than the usual numbers of any member, or have any member duplicated. For that is a presage of the Devil, and a certain sign of hidden wickedness and craft.*

---

*A special treatise on this subject and cognate matters is found elsewhere in the Geneva folio. It is, briefly, as follows. There are many monsters in the sea which are not products of the original creation, but are born from the

But neither must we by any means forget the generation of homunculi. For there is some truth in this thing, although for a long time it was held in a most occult manner and with secrecy, while there was no little doubt and question among some of the old Philosophers, whether it was possible to Nature and Art, that a man should be begotten without the female body and the natural womb. I answer hereto, that this is in no way opposed to Spagyric Art and to Nature, nay, that it is perfectly possible. In order to accomplish it, you must proceed thus. Let the semen of a man putrefy by itself in a sealed cucurbite with the highest putrefaction of the *venter equinus* for forty days, or until it begins at last to live, move, and be agitated, which can easily be seen. After this time it will be in some degree like a human being, but, nevertheless, transparent and without body. If now, after this, it be every day nourished and fed cautiously and prudently with the arcanum of human blood, and kept for forty weeks in the perpetual and equal heat of a *venter equinus*, it becomes, thenceforth a true and living infant, having all the members of a child that is born from a woman, but much smaller. This we call a homunculus; and it should be afterwards educated with the greatest care and zeal, until it grows up and begins to display intelligence. Now, this is one of the greatest secrets which God has revealed to mortal and fallible man. It is a miracle and marvel of God, an arcanum above all arcana, and deserves to be kept secret until the last

---

sperm of fishes of unlike species coming together contrary to the genuine order of Nature. Thus monsters are sometimes found in the sea exhibiting the form of man, which yet have not been generated *ex sodomia* from men, but arise by the conjunction of diverse fishes. . . . Even among men monsters are sometimes found that remind us partly of a human being, and partly of an animal. This is a repellent subject, but requires to be fully explained, that the first birth may be correctly understood. The same also takes place in the sea. There is, for example, the syren, of which the upper parts are those of a woman and the lower those of a fish. This does not form part of the original creation, but is a hybrid offspring from the union of two fishes of the same kind, but of different forms. Other marine animals are also found, which, without corresponding exactly to man, yet resemble him more than any other animal. However, like the rest of the brutes, they lack mind or soul. They have the same relations to man as the ape, and are nothing but the apes of the sea. As often as they unite, marine monsters of this kind are produced. Another such monstrous generation is the monachus or monk-like fish. But there are many genera of fishes, and many modes of generation, which do not always result from the sperm familiar or customary to them, but happen in various other ways. For example, certain monsters are drowned in the sea, and are devoured by the fishes. Now, if a sperm, constituted in exaltation, were to perish by immersion, and, having been consumed by a fish, were again exalted within it, a certain operation would undoubtedly follow from the nature of the fish and the sperm, whence it may be gathered that the majority of marine animals which recall the human form are in this manner produced. Yet, having the nature of a fish, they live in the waters and rejoice therein. The marine dog, the marine spider, and the marine man are of this class. If they are generated in any other way, it must be set down to sodomia. But there may be a third cause, namely, when spermatica of this kind acquire digestion, and by reason of this conjunction a birth takes place. . . . Monsters are likewise generated in the air, from the droppings of the stars from above. For a sperm falls from the stars. The winds also in their courses bring many strange things from other regions to which they are indigenous. The sperm of spiders, toads, and other creatures floating in the air are resolved, and hence other living things are produced. In this way grasshoppers and other monsters are begotten, their generation being of one only and not of two. Such births are more venomous and impure than are other worms. Therefore, houses ought to be scrupulously cleaned, or else so constructed as not to favour the accumulation of much filth. For the air is efficacious against seeds dispersed in this manner. The earth is, however, the most fruitfu matrix of monstrous growths. There the animals both of land and sea congregate. The basilisk is generated from the sperm of a toad and a cock. The sperm of the cock uniting with that of the hen produces an egg. But if the cock emit his sperm without the hen doing likewise, the egg will be imperfect, and something will be generated unnaturally. There is another kind of basilisk, produced by the union, *sodomitice*, of a cock and a toad. After the same manner, lizards unite with geckoes, and the copulation produces a peculiar worm, partaking of the nature of each, and known as a dragon. The asp is another instance of this unnatural generation. . . . From all that has been set down we may learn that whoever lives for his body alone is a basilisk, a dragon, and an asp, not, indeed, generated as yet, but meanwhile moving alive until he dies. You can now understand the abominable manner wherein unnatural monsters are generated. For if a man lives in sperm, his very sperms turn into worms, and remain worms, and in the day of the resurrection shall they be buried in the deepest parts of the earth, over which shall walk those who have risen.—*De Animalibus natis ex Sodomia.*

times, when there shall be nothing hidden, but all things shall be made manifest. And although up to this time it has not been known to men, it was, nevertheless, known to the wood-sprites and nymphs and giants long ago, because they themselves were sprung from this source; since from such homunculi when they come to manhood are produced giants, pigmies, and other marvellous people, who are the instruments of great things, who get great victories over their enemies, and know all secret and hidden matters.* As by Art they acquire their life, by Art acquire their body, flesh, bones and blood, and are born by Art, therefore Art is incorporated in them and born with them, and there is no need for them to learn, but others are compelled to learn from them, since they are sprung from Art and live by it, as a rose or a flower in a garden, and are called the children of the wood-sprites and the nymphs, because in their virtue they are not like men, but like spirits.

Here, too, it would be necessary to speak about the generation of metals, but since we have written sufficiently of these in our book on The Generation of Metals, we will treat the matter very briefly here, and only in a short space point out what we there omitted. Know, then, that all the seven metals are born from a threefold matter, namely, Mercury, Sulphur, and Salt, but with distinct and peculiar colourings. In this way, Hermes truly said that all the seven metals were made and compounded of three substances, and in like manner also tinctures and the Philosophers' Stone. These three substances he names Spirit, Soul, and Body. But he did not point out how this was to be understood, or what he meant by it, though possibly he might also have known the three principles, but he makes no mention of them. I do not therefore say that he was in error, but that he was silent. Now, in order that these three distinct substances may be rightly understood, namely, spirit, soul, and body, it should be known that they signify nothing else than the three principles, Mercury, Sulphur, and Salt, from which all the seven metals are generated. For Mercury is the spirit, Sulphur is the soul, and Salt is the body. The metal between the spirit and the body, concerning which Hermes speaks, is the soul, which indeed is Sulphur. It unites those two contraries, the body and the spirit, and changes them into one essence. But it must not be understood that from any Mercury, and any Sulphur, and any Salt, these seven metals can be generated, or, in like manner, the Tincture or the Philosophers' Stone by the Art and the industry of the Alchemist in the fire; but all these seven metals must be generated in the mountains by the Archeus of the earth.† The

---

* Elsewhere Paracelsus states that giants are born from sylphs, and dwarfs from pigmies. Of these monsters are produced, as, for example, nymphs and syrens. Albeit these are rare, they have appeared with sufficient frequency, and in such a marvellous manner, that there can be no doubt of their existence.—*De Nymphis, Pygmiis, Salamandris, etc.* With regard to the generation of homunculi there is also the following passage:—Porro hoc etiam sciendum est, sodomitas hujusmodi sperma quandoque etiam in os ejaculari. Quod si in stomachum tanquam in matricem recipiatur, ex ipso ibi monstrum, aut homunculus, aut simile aliud generatur, ac inde morbi multi, iique difficiles surgunt, tamdiv sævientes, donec generatum excernatur.—*De Homunculis et Monstris.*

† As a sure and fundamental conclusion to those things which have been advanced, let it be notified to those who desire to be acquainted with the true essence and origin of metals, that our metals are nothing else than the most potent and best part of common stones—the spirit, gluten, grease, butter, oil, and fatness of stones, which, while still combined

alchemist will more easily transmute metals than generate or make them. Nevertheless, live Mercury is the mother of all the seven metals, and deserves to be called the Mother of Metals. For it is an open metal and, as it were, contains in itself all the colours which it renders up from itself in the fire; and so also, in an occult manner, it contains in itself all metals which without fire it does not yield up from itself. But the regeneration and renovation of metals takes place thus: As man can return to the womb of his mother, that is, to the earth from which the first man sprang, and thus can be born again anew at the last day, so also all metals can return to quick mercury, can become Mercury, and be regenerated and clarified by fire, if they remain for forty weeks in perpetual heat, like a child in its mother's womb. Now, they are born, however, not as common metals, but as metals which tinge: for if, as has been said, Luna is regenerated, it will afterwards tinge all metals to Luna. So gold tinges other metals to Sol, and in like manner it must be understood of all other metals. Now, when Hermes said that the soul was the only medium which joins the spirit to the body, he had no inadequate conception of the truth. And since Sulphur is that soul, and, like fire, it hastens on and prepares all things, it can also link together the spirit and the body, incorporate and unite them, so that a most noble body shall be produced. Yet it is not common combustible sulphur which is to be esteemed the soul of metals; but that soul is another combustible and corruptible body. It cannot, therefore, be burnt with any fire, since it is itself entirely fire, and, in truth, it is nothing but the Quintessence of Sulphur, which is extracted by the spirit of wine from Reverberated Sulphur, and is ruby coloured and clear as the ruby itself. This is indeed a mighty and excellent arcanum for transmuting white metals, and for coagulating quick mercury into fixed and approved gold. Hold this in commendation as a treasure for making you rich; and you should be contented with this secret alone in the transmutation of metals. Concerning the generation of minerals and semi-metals, no more need be known than we stated at the beginning concerning the metals, namely, that they are produced, in like manner, from those three principles, Mercury, Sulphur, and Salt, though not, like the metals, from these principles in their perfection, but from the more imperfect and weaker Mercury, Sulphur, and Salt, yet still with their distinct colours.

The generation of gems takes place by, and flows out from, the subtlety of the earth, from the clear and crystalline Mercury, Sulphur, and Salt, also

---

in the stone, are not good, not pure, not clean, and are altogether wanting in perfection. For this reason they are to be sought, found, and known in stones, and thence, also, must be separated and extracted by pounding and liquefaction. When this has been effected they are no longer stones, but prepared and complete metals, agreeing with the celestial stars; which stones, indeed, are secreted from the terrestrial stars. Furthermore, if anyone desire to investigate and to know minerals and metals, he should clearly realise that they are not always to be sought in the common and familiar mineræ, nor in the depths of mountains, because they are very often found more easily, and in greater abundance, upon the surface of the earth than in its bowels. For this reason, any stone that may offer itself to the eye, whether great or small, rock or flint, should be diligently examined as to its property and nature, for very often a small and despised pebble is of greater value than a cow. So, also, there is common dust and sand which are abounding in Sol and Luna.—*Chirurgia Minor, De Contracturis*, Tract II., *conclusio*.

according to their own distinct colours.* The generation of common stones is from the subtlety of water, by the mucilaginous Mercury, Sulphur, and Salt. For all stones are produced by the mucilage of water, as also pebbles and sand are coagulated from the same source into stones.† This is patent to the eyes: for every stone placed in water soon draws the mucilage to itself. If, now, that mucilaginous matter be taken from such stones and coagulated in a cucurbite, a stone will be produced of the same kind as would of itself be produced and coagulated in the water, but after a long period of time.

---

* The generation of gems in Ares occurs after this manner: When the gross genera of stones have been all extracted out of Ares, a certain subtlety remains, more diaphanous in its nature than are other stones, and out of this the Archeus subsequently procreates gems after such a manner that hardness and very great transparency are first prepared. Hence the gems are afterwards developed, each according to its own form and essence. Very great subtlety and artifice are employed over this generation.—*De Elemento Aquæ*, Tract IV., c. 10.

† The body of every kind of stone is sulphur, as that of metals is mercury. The hardness is from salt, and the density from mercury.—*Ibid.*, c. 5.

CONCERNING THE NATURE OF THINGS

BOOK II

Concerning the Growth of Natural Things

IT is clear enough, and well known to everybody, that all natural things grow and mature by warmth and moisture, as is plainly demonstrated by the rain followed up with sunshine. None can deny that the earth is rendered fruitful by the rain, and all must confess that every kind of fruit is ripened by the sun. Since, then, by the Divine institution, this is possible to Nature, who will deny or refuse to believe that man possesses this same power by a prudent and skilful pursuit of the Alchemical Art, so that he shall render the fruitless fruitful, the unripe ripe, and make all increase and grow? The Scripture says that God subjected all created things to man, and handed them over to him as if they were his own property, so that he might use them for his necessity, that he might have dominion over the fishes of the sea, the fowls of the air, and everything on the earth without exception. Wherefore man ought to rejoice because God has illuminated him and endowed him, so that all God's creatures are compelled to obey Him and to be subject to Him, especially all the earth, together with all things which are born, live, and move in it and upon it. Since, then, we see with our eyes, and are taught by daily experience, that the oftener and the more plentifully the rain moistens the earth, and the sun dries it again with its heat and glow, the sooner the fruits of the earth come forth and ripen, while all fruits increase and grow, whatever be the time of year, let none wonder that the alchemist, too, by manifold imbibitions and distillations, can produce the same effect. For what is rain but the imbibition of the earth? What are the heat and glow of the sun other than the sun's process of distillation, which again extracts the humidity? Wherefore I say that it is possible by such co-optation in the middle of winter to produce green herbs, flowers, and fruits, by means of earth and water, from seed and root. Now, if this takes place with herbs and flowers, it will take place in many other similar things too, as, for instance, in all minerals, the imperfect metals whereof can be ripened with mineral water by the industry and art of the skilled alchemist. So, too, can all marchasites, granites, zincs, arsenics, talcs, cachimiæ, bismuths, antimonies, etc., all of which carry with them immature Sol and Luna, be so ripened as to be made equal to the richest veins of gold and silver, only by such co-optation. So, also, the Elixir and Tinctures of metals are matured and perfected.

Since, therefore, humidity and warmth mature all things and make them grow, let none wonder that, after a long time, in the case of a criminal on the gibbet, the beard, hair, and nails grow ; nor let this be taken for a sign of innocence, as the ignorant read it. It is only natural, and proceeds from natural causes. As long as there is moisture in the body, the nails, beard, and hair grow ; and, what is more, in the case of a man buried in the earth itself, nails, beard, and hair grow up to the second year, or up to the time of the man's decay.

It should be known, too, that many substances grow and increase perpetually in size, weight, and virtue, both in water and on land, in each of which they remain good and effective, such, for example, as metals, marchasites, cachymiae, talcs, granites, antimony, bismuths, gems, pearls, corals, all stones and clays. So also it can be brought about that Sol shall grow and increase in weight and in body, if only it be buried in land looking east, and be constantly fertilised with fresh human urine and pigeons' dung.

It is also possible for gold to be so acted upon by the industry and art of the skilled alchemist that it will grow in a cucurbite with many wonderful branches and leaves, which experiment is very pleasant to behold, and full of marvels. The process is as follows : Let gold be calcined by means of aqua regis so that it becomes a chalky lime ; which place in a cucurbite, pouring in good and fresh aqua regis and water of gradation so that it exceeds four fingers across. Extract it again with the third degree of fire until nothing more ascends. Again pour over it distilled water, and once more extract by distillation as before. Do this until you see the Sol rise in the glass and grow in the form of a tree with many branches and leaves. Thus there is produced from Sol a wonderful and beautiful shrub which alchemists call the Golden Herb, or the Philosophers' Tree. The process is the same with the other metals, save that the calcination may be different, and some other aqua fortis may have to be used. This I leave to your experience. If you are practised in Alchemy you will do what is right in these details.

Know also that any flint may be taken out of river water, placed in a cucurbite, and sprinkled with its own running water until the cucurbite is full. This may again be extracted by distillation, as long as a single drop ascends, until the stone be dry. Let the cucurbite be again filled with this water, and once more extracted. Repeat this until the cucurbite is filled with this stone. In this way, by means of Alchemy, in a few days you will see that a very large stone can be made, such as the Archeus of the waters could scarcely make in many years. If you afterwards break the cucurbite on a stone you will have a flint in the shape of the cucurbite, just as though it had been poured into the glass. Though this may be of no profit to you, still it is a very wonderful thing.

CONCERNING THE NATURE OF THINGS

BOOK III

Concerning the Preservation of Natural Things

IN order that a thing may be preserved and defended from injury, it is necessary that first of all its enemy should be known, so that it may be shielded therefrom, and that it may not be hurt or corrupted by it, in its substance, virtue, force, or in any other way suffer loss. A good deal depends upon this, then, that the enemy of all natural things should be recognised; for who can guard himself against loss and adverse chance if he is ignorant of his enemy? Surely, no one. It is therefore necessary that such enemy should be known. There are many enemies; and it is just as necessary to know the bad as the good. Who, in fact, can know the good without a knowledge of the evil? No one. No one who has never been sick knows how great a treasure health is. Who knows what joy is, that was never sad or sorrowful? And who knows rightly about what God is, who knows nothing about the devil? Wherefore since God has made known to us the enemy of our soul, that is, the devil, He also points out to us the enemy of our life, that is, death, which is the enemy of our body, of our health, the enemy of medicine, and of all natural things. He has made known this enemy to us and also how and by what means we must escape him. For as there is no disease against which there has not been created and discovered a medicine which cures and drives it away, so there is always one thing placed over against another—one water over against another, one stone over against another, one mineral over against another, one poison over against another, one metal over against another—and the same in many other matters, all of which it is not necessary to recount here.

But it ought to be known how, and by what means, each several thing is preserved and guarded from loss: that many things, for instance, have to be kept for a long time in the earth. All roots, especially, remain for a long while in the earth fruitful and uncorrupted. In like manner, herbs and flowers and all fruits keep undecayed and green in water. So also many other fruits, and especially apples, can be preserved in water, and protected from every decay, until new apples are produced.

So also flesh and blood, which very soon putrefy and become rancid, can be kept in cold spring water; and not only so, but by the co-optation of renewed and fresh spring water they can be transmuted into a quintessence, and conserved for ever from decay and bad odour without any balsam. And not only does this process preserve flesh and blood, but (so to say) it preserves all other kinds of flesh and blood, and especially the body of man, from all decay and from many diseases which arise from decay, better than the common mumia does.* But in order that blood may be preserved of itself from decay and ill odour, and not as a quintessence ; and in order, also, to protect other blood, as aforesaid, you must use this process : Let the blood be separated from its phlegm, which moves of itself, and is driven to the surface. Draw off this water by a dexterous inclination of the vessel, and add to the blood a sufficient quantity of the water of salt, which we teach you in our Chirurgia Magna how to make.† This water at once mingles with the blood, and so conserves the blood that it never putrefies or grows rancid, but remains fresh and exceedingly red after many years, just as well as on the first day; which, indeed, is a great marvel. But if you do not know how to prepare this water, or have none at hand, pour on a sufficient quantity of the best and most excellent balsam, which produces the same effect. Now this blood is the Balsam of Balsams, and is called the Arcanum of Blood. It is of such great and wonderful virtue as would be incredible were we to mention it. Therefore you will keep this occult, as a great secret in medicine.

In the conservation of metals the first thing to learn is what are their enemies, so that they may be thereby the better kept from loss. The principal enemies of metals, then, are all strong waters ; all aquæ regiæ, all corrosives and salts, shew their hostility in this circumstance, that they mortify all metals, calcine them, corrupt them, and reduce them to nothing. Crude sulphur shews its hostility by its smoke ; for by its smoke it takes away the colour and redness from Venus, and renders it white. From white metals, as Luna, Jupiter, Saturn, and Mars, it takes away their whiteness and reddens them, or induces in them a reddish colour. From gold it takes away the agreeable yellowness and golden tint, renders it black, and makes it as uncomely as possible.

Antimony shews its hostility in this : that it spoils all metals with which it is liquefied in the fire, and with which it is mixed ; it deprives and robs them ; moreover, like the sulphur, it robs metals of their genuine colour and substitutes another.

Quicksilver, on the other hand, exercises a hostile force upon the metals with which it is conjoined, in that it invades and dissolves them so that it makes an amalgam from them. Moreover, its smoke, which we call the soot of Mercury, makes all metals immalleable and fragile ; it calcines them and whitens all red and gold coloured metals. It is the chief enemy of iron and

---

* According to one explanation : Mumia is man himself. Mumia is balsam, which heals wounds.—*Paramirum - De Origine Morborum*, Lib. II., c. 2. The virtues of all herbs are found in this Mumia.—*De Origine Morborum Invisibilium*, Lib. IV. Whoever seeks opoponax will find it in Mumia (that is, in the Mumia which is man), and so also all other creatures whatsoever.—*Ibid.* Now, this is Mumia : If a man be deprived of life, then his flower bursts forth in potencies and natural arcana.—*Ibid.*

† This process will be found in the second footnote on p. 76 of the present volume.

steel, for if common mercury touches a steel rod, or if the rod be anointed with mercurial oil, it can afterwards be broken like glass and cut off. This is indeed a great secret and must be kept strictly occult. In the same way, too, the magnet should be guarded and kept from Mercury, for it exerts hostility on it as on Mars. For every magnet which common mercury touches, or which is anointed with mercurial oil, or only placed in Mercury, never afterwards attracts iron.* Let no one be surprised at this; there is a natural cause for it, seeing that Mercury extracts the spirit of iron which the magnet holds latent in itself. Wherefore also the spirit of iron in the magnet attracts the body of Mars to itself; and this happens not only in the magnet but in all other natural things, so that the foreign spirit which is in an alien body, which is not of its own nature, always attracts a body agreeing with its own nature. This should be known not only of the magnet, but of all natural bodies, such as minerals, stones, herbs, roots, men, and animals.

After this it should be known that metals exercise hostility amongst each other, and mutually hate one another from their inborn nature; as you see in the case of Saturn, which is the principal enemy of Sol, from its congenital nature. It breaks up all the members of gold, renders it deformed, weak, and destroys and corrupts it even to the death, more than it does any other metal. It also hates tin, and is an enemy of all the metals, for it renders them degenerate, unmalleable, hard and unfit, if it be mixed with either of them in fire or flux.

Since, therefore, you have now heard about the enemies of the metals, learn, moreover, about their preservation and conservation, which guard the metals from all loss and corruption, and, in addition, strengthen them in their nature and virtue, while they graduate them more highly in colour. First, then, it ought to be known concerning gold that it cannot be better and more beautifully preserved than in boys' urine, in which has been dissolved sal ammoniac, or in the water of sal ammoniac alone. In these, with time, it acquires such a high grade of colour as cannot be surpassed. Silver cannot be better preserved and conserved than if it be boiled in common water or acetum in which have been dissolved tartar and salt. In this way any old silver, though blackened and stained, is renewed, if it is boiled thus. Of iron and steel the best and most useful conservative and preservative is fresh, not salted, lard from a gelded sow. This protects all iron and steel from rust if they are anointed therewith once every month. In like manner, if iron be liquefied with fixed arsenic, and occasionally reduced to a flux, it can be so renewed and fixed that, like silver, it never rusts. Copper can be conserved and preserved if only it be mixed with sublimated Mercury, or anointed with oil of salt, so

---

* So, also, it is affirmed that if the magnet be steeped in garlic it will be deprived of its attractive virtue.—*De Morbis Amentium*, c. 5. Should anyone make use of a magnet while he is wearing a sapphire, it will effect nothing till the gem be removed. The same quality seems to reside in carabe, coagulate of gum, resin, and therebotin.—*De Peste*, Lib. II., c. 2.

that for the future it gives forth no vitriol or verdigris, nor does it become of a green colour.

Lead cannot be conserved better than in cold water, and in a damp place, such is its nature. But for the conservation of the magnet nothing is better than filings of iron or steel. If the magnet be placed in these, not only does not its force decrease, but it grows more and more every day.

As to the conservation of salts, and all those substances which are of a salt nature, and are comprised under the name of salt, of which there are more than a hundred, it is well to know that they must be kept in a warm and dry place, and guarded well from the air in wooden chests. They must not be placed on glass, stone, or metal. By these they are dissolved and turn into water and amalgam; but this does not occur in wood.

Moreover, you should learn the method of conserving certain waters and liquids by means of pressed herbs, roots, and other fruits and growing things, which easily absorb all mustiness and mould just as if a skin were wrapped around them. Let these waters, or other liquids, be placed in a glass vessel, narrow at the top and wider below. Let the vessel be filled to the top and then some drops of olive oil added, so that all the water or liquid may be covered. The oil will float at the top, and, in this way, will protect the liquid or the water a long time from mustiness or mould. No water or liquid, if it be covered with oil, can ever become mouldy or smell badly. In this way also two waters, two liquids, two wines, can be kept separately in one vessel, so that they shall not mix; and not only two, but three, four, five, or still more, if only oil be between them, for they are separated by the oil as by a wall, which does not suffer them to be conjoined and united. For oil and water are two contraries, and neither can mingle with the other. As the oil does not allow the waters to mix, so, on the other hand, the water prevents the oils from blending.

For the conservation and preservation of cloth and garments from moth, so that they may not eat them or settle in them, nothing is better than mastix, camphor, ambergris, or musk: but the best is civet, which not only preserves from moth, but drives away and puts to flight moths, with other worms, fleas, lice, and bugs.

All timbers can be conserved, as in buildings or bridges, so that they shall never decay, whether they be in water, under water, or out of the water, in the ground, under the ground, or out of the ground, whether exposed to rain or wind, air, snow, or ice, in summer or winter, and moreover, preventing them from decaying or worms breeding in them when felled. The method of conservation in this case is that grand arcanum against all putrefactions, and so remarkable a secret that no other can compare with it. It is none other than the oil of sulphur, the process for making which is as follows:—Let common yellow sulphur be pulverised and placed in a cucurbite. Over it pour as much aquafortis as will cover four fingers across. Abstract this by distillation three or four times, the last time until it is completely dry. Let the sulphur which remains at the bottom, and is of a dark reddish colour, be placed in marble or glass and easily dissolved into an oil. This is a great secret in the conservation of timber so that it may never decay and may be protected from worms. For if sulphur be prepared as aforesaid, and turned into an oil, it afterwards tinges the timber which has been anointed with it so that it can never be obliterated. Many other things, also, can be conserved and preserved from decay in this oil of sulphur, especially ropes and cables in ships and on the masts of ships, in chariots, fishing-nets, birdcatchers' and hunters' snares, and other like things which are being frequently used in water and rain, and would otherwise be liable to decay and break; so also with linen cloths and other similar things.

The conservation of potable things, too, should be noticed, under which we comprise wine, beer, hydromel, vinegar, and milk. If we wish to keep these five unharmed and in their virtue, it is necessary to know their chief enemy. This is none other than unclean women at the time of their monthly courses. They corrupt these things if they handle or have anything to do with them, if they look at them, or breathe on them. The wine is changed and becomes thick, beer and hydromel turn sour, vinegar is weakened and loses its acidity, milk also becomes sour and clotted.

This, therefore, should be well known before anything is said specially about the conservation of one of these things in particular. Moreover, the chief preservative of wine is sulphur and oil of sulphur, by means of which all wine can be preserved for a very long time, so that it neither thickens nor is in any way changed.

The means of conserving beer is by oil of garyophyllon, if a few drops of it are put in, so that one measure has two or three drops. Better still is the oil of benedicta garyophyllata, which preserves beer from acidity. The preservative for hydromel is the oil of sugar, which must be used in the same way as the oil of garyophyllon or the benedicta.

The preservative of vinegar is oil of ginger, and of milk the expressed oil of almonds. These two must be used as described above.

The preservative of cheese is the herb hypericon or perforata, which protects all cheeses from worms. If it be placed against the cheese and touches it, no worm is produced in it, and if some have been already produced, they die and drop out of the cheese.

Honey has no special preservative, only it must be protected from its enemy. Its chief enemy is bread. If ever so small a quantity of bread made from flour be put or fall into it, the whole honey is turned into ants, and perishes entirely.

CONCERNING THE NATURE OF THINGS

BOOK IV

Concerning the Life of Natural Things

NONE can deny that the air gives life to all corporeal and substantial things which are born and generated from the earth. But as to what and of what kind the life of each particular thing is, it should be known that the life of things is none other than a spiritual essence, an invisible and impalpable thing, a spirit and a spiritual thing. On this account there is nothing corporeal but has latent within itself a spirit and life, which, as just now said, is none other than a spiritual thing.* But not only that lives which moves and acts, as men, animals, worms in the earth, birds under the sky, fishes in the sea, but also all corporeal and substantial things. For here we should know that God, at the beginning of the creation of all things, created no body whatever without its own spirit, which spirit it contains after an occult manner within itself. For what is the body without the spirit? Absolutely nothing. So it is that the spirit holds concealed within itself the virtue and power of the thing, and not the body. For in the body is death, and the body is subject to death, and in the body nothing but death must be looked for. For the body can be destroyed and corrupted in various ways, but not the spirit: for it always remains a living spirit, and is bound up with life. It also keeps its own body alive, but in the removal of the body from it, it leaves the body separate and dead, and returns to its own place whence it had come, that is to say, into chaos, and into the air of the higher and lower firmament. Hence it is evident that there are different kinds of spirits, just as there are different kinds of bodies. There are celestial and infernal spirits, human and metallic, the spirits of salts, gems, and marcasites, arsenical spirits, spirits of potables, of roots, of liquids, of flesh, blood, bones, etc. Wherefore you may know that the spirit is in very truth the life and balsam

---

* Life is a veil or covering which encloses three principles—sulphur, salt, and mercury.—*Paramirum*, Lib. I The life of the body is fire.—*De Ente Astrorum*, c. 6. There is a twofold life in man : there is the life of the soul, which proceeds from the nature of God ; but I speak here as a physician, and not as a theologian. There is also a life of the animal kind, which is of air and fire, and the same is domiciled in the body, which is earth and water. So is man dowered with an animal and a sidereal life.—*De Pestilitate*, Tract I. In another sense the life of man is said t be triplex—necrocomic, cagastric, and salnitric. But this has reference to the animal life only —*Liber Azoth*. That which sustains the body is the life, but the life itself is from God, and not from man. This life consists in four things - humours, complexions, natural species, and gifts or virtues. - *De Generatione Hominis*.

of all corporeal things. Now we will go on to its species, and here will describe to you in detail, but as briefly as possible, the life of each natural thing.

The life, then, of all men is none other than a certain astral balsam,* a balsamic impression, a celestial and invisible fire, an included air, and a spirit of salt which tinges. I am unable to name it more clearly, although it could be put forward under many distinctive titles. Since, however, the chief and the best are here pointed out, we will be silent as to the rest and the inferior names.

The life of metals is a latent fatness which they have received from sulphur. This is shewn from their fluxion, because everything which passes into flux in the fire does so on account of its hidden fatness. Unless this were so no metal could be reduced to a fluid state, as we see in the case of iron and steel, which have the least Sulphur and fatness of all the metals, wherefore they are of a drier nature than all the rest of them.

The life of mercury is nothing but inner heat and outer frigidity. That is to say, within it gives heat, but without it causes cold; and in this respect it is aptly to be compared to a garment of skins, which, like mercury, causes both heat and cold. For if a garment of this kind be worn by a man, it warms him and protects him from the cold; but if he wears the hairless part against his naked body, it causes cold, and defends him from excessive heat. So it came about that in very ancient times, and it is even the custom still, that these coats of skin are worn both in summer and in winter, as much against the heat as against the cold; in summer the hairless part is turned within, and the hairy part outside, but in the cold winter season the hairy part is turned within and the hairless part outside. As it is with the garment of skins, so is it with mercury.

The life of sulphur is a combustible, ill-smelling fatness. Whilst it flames and sends forth its evil odour it may be said to live.

The life of all salts is nothing else but a spirit of aqua fortis: for when the water is abstracted from them, that which remains at the bottom is called dead earth.

The life of gems and corals is mere colour, which can be taken from them by spirits of wine. The life of pearls is their brightness, which they lose in their calcination. The life of the magnet is the spirit of iron, which can be extracted and taken away by rectified *vinum ardens* itself, or by spirit of wine.

The life of flints is a mucilaginous matter. The life of marcasites, cachymiæ, talc, cobalt, zinc, granites, zwitter, vismat (rude tin), is a metallic spirit of antimony, which has the power to tinge. Of arsenicals, auripigment, orpiment, realgar, and similar matters, the life is a mineral coagulated poison.

---

* The flesh and blood of man are preserved and sustained by a certain balsam. Now, this balsam is the body of salt. So, therefore, by salt is man preserved as by a balsam.—*De Morbis Tartareis*, c. 21. The balsam of man exists alike in all his members, and is specialised therein – in the blood, in the marrow, in the bones, the arteries, etc.—*Chirurgia Magna*, Lib. V.

The life of wavelike substances, that is to say, of the dung of men and animals, is their strong and fœtid smell. When this is lost they are dead.

The life of aromatic substances, to wit, musk, ambergris, civet, and whatever emits a strong, sweet, and pleasant odour, is nothing but that grateful odour itself. If they lose this they are dead and useless.

The life of sweet things, as sugar, honey, manna, fistula cassiæ, and the like, is a subtle sweetness, with the power to tinge; for if that sweetness be taken away by distillation, or sublimation, the things are dead, fatuous, and no longer of any value.

The life of resins, as caraba, turpentine, and gum, is a mucilaginous, glittering fatness. They all give excellent varnish; when they no longer furnish this, and lose their glitter, they are dead.

The life of herbs, roots, apples, and other fruits of this kind, is nothing else than the liquid of the earth, which they spontaneously lose if they are deprived of water and earth.

The life of wood is a certain resin. Any wood that is deprived of resin is unable longer to flourish.

The life of bones is the liquid of mumia. The life of flesh and blood is none other than the spirit of salt, which preserves them from ill odour and decay, and spontaneously, as the water is separated from them.

But concerning the life of the elements there is this to be known. The life of water is its flowing. When it is coagulated by the cold of the firmament and congealed into ice, then it is dead, and all power of doing harm is taken from it, since no one can any longer be drowned in it.

So, too, the life of fire is air, for the air makes the fire blaze more strongly and with greater impetuosity. Some air proceeds from all fire, sufficient to extinguish a candle or to lift a light feather, as is evident to the eyes. All live fire, therefore, if it be shut up or deprived of the power to send forth its air, must be suffocated.

The air lives of itself, and gives life to all other things. The earth, however, is of itself dead; but its own element is its invisible and occult life.

CONCERNING THE NATURE OF THINGS

BOOK V

Concerning the Death of Natural Things

THE death of all natural things is nothing else but an alteration and removal of their powers and virtues, an overthrow of their potencies for evil or for good, an overwhelming and blotting out of their former nature, and the generation of a new and different nature.* For it should be known that many things which in life were good, and had their own virtues, retain little or none of that virtue when they are dead, but appear altogether fatuous and powerless. So, on the other hand, many things in their life are evil, but in death, or after they have been mortified, they display a manifold power and efficacy, and do much good. We could recount many examples of this, but that is altogether foreign to our purpose. Yet, in order that you may see that I do not write from my mere opinion, however plausible, but from my experience, it is well that I should adduce one example with which I will quiet and silence the sophists who say that nothing can be gained from dead things, nor anything ought to be sought or found in them. The cause of this assertion is that they value at nothing the preparations of the alchemists, by which many great secrets of this kind are discovered. For look at Mercury, live and crude sulphur, and crude antimony; as they are brought from the mines, that is, while they are still living, how small is their virtue, how lightly and tardily do they exercise their influence. Indeed, they bring more evil than good, and are rather a poison than a medicine. But if, by the industry of a skilled alchemist, they are corrupted into their first substance and prudently prepared (that is, if the Mercury be coagulated, precipitated, sublimated, resolved, and turned into oil; the sulphur be sublimated, calcined, reverberated and turned into oil; and, in like manner, Venus be sublimated, calcined, reverberated, and turned into oil), you see what usefulness, what power and virtue, and what rapid efficiency they afford and display, so that none can fully speak or write of it. For their manifold virtues are not to be investigated, nor can any one search them out. Every alchemist, therefore, and every faithful

---

* Death is the mother of tinctures, for tinctures proceed from the mortification of the body, in which the colours are contained, even as in a seed there are green, yellow, black, blue, and purple colours, which are, nevertheless, invisible until the seed has perished in the earth, and till the sun has prepared and produced them, so that what was first hidden from the senses is now revealed to them.—*De Icteritiis*.

physician, ought to seek into these three things during his whole life, and even up to his death should play with them and find his pastime in them. Most assuredly they will nobly compensate him for all his labour, study, and expense.

But let us come to particulars, and specially describe the death and mortification of each natural thing, what its death is, and in what way it is mortified. First of all, then, with regard to the death of man, it should be understood that, beyond a doubt, it is nothing else but the end of his day's work, the taking away his air, the evanescence of his balsam, the extinction of his natural light, and the entire separation of the three substances, body, soul, and spirit, and the return to his mother's womb. For since the natural earthborn man comes from the earth, the earth, too, will be his mother, into which he must return, and therein lose his earthborn natural flesh, so that at the last day he may be regenerated in a new, a heavenly, and purified flesh, as Christ said to Nicodemus when he came to Him by night. For, as we said, these words apply to regeneration.

But the death or mortification of the metals is the removal of their bodily structure, and of the sulphurous fatness which can be removed from them in many ways, as by calcination, reverberation, resolution, cementation, and sublimation. But the calcination of metals is not of a single kind only. For one is produced by salt, one by mercury, one by strong waters, one by the *fuligo mercurii* and quick mercury. Calcination by salt is when the metal is formed into very thin plates, and stratified and cemented with salt. Calcination by sulphur is when the metal is formed into plates, stratified and reverberated with sulphur. Calcination by strong waters is when the metal is granulated, resolved in aqua fortis, and precipitated therein. Calcination by the *fuligo mercurii* is brought about thus: Let the metal be formed into plates; let the mercury be put into an earthen vessel, narrow at the top but broad below, and afterwards set on a moderate coal fire, which should be blown a little until the mercury begins to smoke, and a white cloud issues from the mouth of the vessel. Then let the plated metal be placed on the orifice of the vessel. Thus the common mercury penetrates the metal and renders it as friable as a lump of coal. Calcination by quick mercury is when the metal is cleft into small particles, made into plates, or granulated, and formed into amalgam with mercury. Afterwards let the mercury be pressed out through a skin, and the metal will remain within the skin in the form of lime or sand. But beyond these mortifications of the metals, destructions and whitenings of their life, you must know that there are many other mortifications of the metals. For beyond the fact that all rusting of iron and steel is a death, there are others which are to be esteemed as more important. For instance, it should be known that all vitriol, or even burnt brass, is mortified copper; all precipitated, sublimated, calcined cinnabar is mortified mercury; all white lead, red lead, or yellow lead are mortified lead; all lazurius is mortified silver. So, also, all Sol, from which its tincture, quintessence, resin, crocus, or sulphur has been withdrawn, is dead, because it no longer has the form of gold, but is a white metal like fixed silver.

But now let us go on to lay before you by what means the mortification of the metals is brought about. First of all, it should be known concerning iron that it can be mortified and reduced to a crocus in the following way: Form very thin plates of steel, beat them red hot, and then extinguish them in vinegar made from wine. Keep on doing this until you see the vinegar has become very red. When you have enough of this red vinegar, pour it all out, and distil therefrom the moisture of the vinegar. Coagulate the residuum into a dry powder. This is the most excellent Crocus of Mars. There is, however, another way of making the Crocus of Mars which partly surpasses the former, and is carried out with much less expense and labour, thus: Stratify very thin plates of steel with equal quantities of sulphur and tartar. Afterwards reverberate. This produces the most beautiful crocus, which should be taken from the plates.

In the same way you should be informed that if any plate of iron or steel be smeared over with aqua fortis, it renders also a beautiful crocus. Such is the result, too, with oil of vitriol, water of salt, water of alum, water of sal ammoniac, water of salt nitre, sublimated mercury, all of which mortify iron, and reduce it to a crocus; but none of these methods is to be compared with the two mentioned above; for they can only be used in Alchemy and not in medicine; so use in preference the first two methods, and avoid the rest.

The mortification of copper, to reduce it to vitriol, verdigris, or burnt brass, can also be accomplished in various ways; and there are various processes with this metal, too, but one is better and more useful than another. Wherefore it will be well to make a note of the best and most useful, and to say nothing about the others. The best, easiest, and most reliable method of reducing copper to vitriol is as follows: Let plates of copper be smeared with water of salt or of saltpetre, and hung or exposed in the air until the plates begin to become green. Wash off this greenness with clear spring water, dry the plates with a rag; again smear the plates with water of salt or saltpetre, and again proceed as before, repeating the process until the water becomes quite green, or sends forth much vitriol to the surface. Then remove the water by tilting the vessel, or by drawing it off, and you will have an excellent medicinal vitriol. For Alchemy, there is no more beautiful, noble, or better vitriol than that which is made by aqua fortis, or aqua regis, or water of sal armoniac. Proceed thus: Let plates of copper be smeared with one of the aforesaid waters, and as soon as the greenness has been extracted, and the plates have been dried, let the greenness be taken off with a hare's foot, or by some other means at pleasure, as white lead is scraped off leaden plates. Let them be again smeared as before, until the plates are entirely consumed, and thence is produced a very beautiful vitriol, such as you cannot fail to admire.

Water of saltpetre is made thus: Purify saltpetre, liquefy and pulverise it.

Afterwards dissolve it by itself in a vessel with boiling water. Thus you have water of saltpetre. Water of sal ammoniac is made as follows: Calcine sal ammoniac and resolve it in a case on marble. This is water of sal ammoniac.

In order to make verdigris from copper there are several ways not necessary to recount here. We will therefore describe two only, with a two-fold method of preparation, one for Medicine and the other for Alchemy. The verdigris used in medicine admits of the ensuing process: Take plates of copper, and smear them with the following compound: Take equal quantities of honey and vinegar, with a sufficient quantity of salt to make the three together the consistence of thick paste. Mix thoroughly, and afterwards put in a reverberatory, or in a potter's furnace, for the same time as the potter bakes his vessels, and you will see a black substance adhering to the plates. Do not let this circumstance cause you any anxiety or detain you at all; for if you suspend or expose those plates in the open air, in a few days the substance will turn green, and will become excellent verdigris, which may be called the balsam of copper, and is highly esteemed by all physicians. And this need not cause surprise, because the verdigris first becomes green in the air, and because the air has the power of transmuting a black colour into such a beautiful green. For here it should be known that, as daily experience in alchemy proves, every dead earth or *caput mortuum*, as soon as ever it comes out of the fire into the air, immediately acquires another colour, and loses its own colour which it had assumed in the fire. The changes of these colours are very diversified. According to the material such are the colours produced, though, for the most part, they flow from the blackness of dead earth. You who are skilled in Alchemy see that every dead earth, flux of powder, or of aqua fortis, comes black from the fire, and the more ingredients there are in it the more varied are the colours displayed in the air. Sometimes they only appear red, as vitriol makes them; sometimes only yellow, white, green, cerulean; sometimes mingled, as in the rainbow or the peacock's tail. All these colours display themselves after death, and as a consequence of death. For in the death of all natural things new colours appear, and they are changed from their first colour into another, each according to its own nature and properties. Moreover, we will say about verdigris that which we dedicate to Alchemy. The process of its preparation is as follows: Form very thin plates of copper, which stratify on a large tile with equal portions of sulphur and tartar, pounded and mixed. Reverberate for twenty-four hours with a strong fire, taking care that the copper plates do not melt. Then take them out; break the tile; expose the plates to the air, with the matter which adheres to them, for a few days, and the matter on the plates will be converted into most beautiful verdigris, which in all strong waters, in waters of gradations, in cements and colourings of gold, tinges gold and silver with a deep colour.

But in order that copper may become *æs ustum*, which is also called the crocus of copper, the following process must be adopted: let copper be formed into plates, smeared with salt reduced into a paste with the best vinegar, then put on a large tile, placed in a blast furnace, and for a quarter of an hour burnt with a strong fire, but so that the plates may not melt. Let these plates, while still glowing, be extinguished in vinegar wherein sal ammoniac has been dissolved—half an ounce in a pound of vinegar. Let the plates be again heated, and extinguished as before; but continually scrape off into vinegar the scales which adhere to the plates after they have been extinguished, or else knock them off by beating the plates, or in any way you can. Keep doing this until the plates of copper are nearly consumed. Then let the vinegar be extracted by distillation, or let it evaporate in an open vessel, and let it coagulate into a very hard stone. Thus you will have the crocus of copper used in Alchemy. Many persons commonly make *æs ustum*, or the crocus of Venus, from Venus by the extraction of alcohol (others of *vinum aceti*), like the crocus of Mars; but I much prefer this method.

The mortification of Mercury, in order that it may be sublimated, is brought about by vitriol and salt. When it is mixed with these two and then sublimated it becomes as hard as crystal and as white as snow. In order that Mercury may be reduced to a precipitate,* nothing more need be done than calcine it in the best aqua fortis; then let the graduated aqua fortis be extracted from it five times, more or less, until the precipitate acquires a beautiful red colour. Sweeten this precipitate as much as possible; and finally distil the rectified wine from it seven or nine times, or as often as necessary, until it burns in the fire and does not escape. Then you have the diaphoretic precipitate of Mercury.

Moreover, here should be noted a great secret concerning precipitated Mercury. If, after its colouration, it be sweetened with water of salt of tartar, by distilling it until the water no longer ascends acid, but is altogether sweet, then you will have the precipitate as sweet as sugar or honey. This is the principal arcanum for all wounds and ulcers and the Gallic disease, insomuch that no physician need wish for better; and it, moreover, brightens up despondent alchemists. For it is an augmentation of Sol, it enters into the composition of Sol, and by it gold is rendered constant and good. Although, then, much labour and toil may be required for this precipitate, it compensates for these and returns to you what you have spent. Moreover, you get sufficient gain from it—more than you could compass by the highest artifice of any kind. You ought, therefore, to rejoice over it, and to thank God and me

---

* It is also stated that there is nothing in medicine to compare with precipitated mercury for the cure of icteritia.—*Fragmenta Medica*, s. v. *Annotationes in Lib. de Icteritiis.* The medical preparation of the precipitate of mercury as a healing unguent has been boastfully claimed to their own credit by many persons, though they are all filched from the writings of the ancient artists and Spagyrists. Vigo was not free from the disgrace of this falsehood. Precipitated mercury is certainly an ancient remedy, but has lain hidden for a long time by the perfidy of physicians. All cavernous ulcers (except those of the eating and spreading kind) are completely cured by its use. But experience teaches us that the oil of argent vive, when outwardly applied, has much greater efficacy.—*De Tumoribus, etc., Morbi Gallici*, Lib. X. The bloodlike redness of the precipitate of mercury has caused it to be ignorantly confused with the ruddy powder into which the sweet balsam of mercury is reduced when it is prepared without sublimation or calcination by means of the water of eggs.—*Ibid.* Precipitated mercury of the metals is the reduction of the metals into their first matter, which afterwards is deposited below.—*Chirurgia Magna, De Impostumis in Morbo Gallico*, Lib. II.

for it. But in order that Mercury may be calcined, I have already said that this must be done in sharp aqua fortis, which must be abstracted by distillation, and the precipitation is made. But in order that Mercury may be reduced to cinnabar,* you must first of all mortify it, and liquefy it, with salt and yellow sulphur. Reduce it to a white powder, then put it in a cucurbite; place an aludel above, and sublimate with great fluxion, as is customary. Thus the cinnabar ascends into the aludel and adheres to it, as hard as hæmatite.

The mortification of lead, in order that it may be reduced to white lead, is two-fold, one for Medicine, the other for Alchemy. The preparation of cerussa for Medicine is as follows: Suspend plates of lead in an unglazed vessel over strong vinegar made from wine, the vessel being well closed so that no spirits may escape. Place the vessel in warm ashes, or, in winter, behind the fire. Then, after ten or fourteen days, you will find the very best cerussa adhering to the plates. Scrape this off with a hare's foot, and replace the plate over the vinegar until you have sufficient cerussa. The other preparation of cerussa for Alchemy is like the former, save that a quantity of the best sal ammoniac must be dissolved in the vinegar. In this way you will have a very beautiful cerussa, most subtle for purging tin or lead, or for removing whiteness from copper. But if we wish to make red lead out of the lead, it must first be calcined to ashes, and afterwards burnt laterally in a glazed jar, stirring it continually with an iron wire until it grows red. This minium is at once the best and the most valuable, and should be used in Medicine as well as in Alchemy. The other, which dealers sell in the shops, is of no use. It is made up only of the ashes which remain in the liquefaction of lead ore, and the potters buy it for encrusting vessels. Such minium is useful only for pictures, but neither for Medicine nor for Alchemy.

In order to reduce lead to a yellow colour a process is required not altogether unlike the preparation of minium. Here, too, the lead must be calcined with salt, and reduced to ashes. Afterwards it must be stirred continually with iron in one of the wide dishes used by those who test minerals, over a moderate coal fire, careful watch being kept lest the heat should be too great or the stirring neglected. Otherwise it would melt and produce yellow glass. In this way you will have excellent yellow lead.

The mortification of silver so that lazurium, or some similar substance, may be produced from it, is brought about as follows: Let Luna be made into plates, mixed with Mercury, and suspended in a glazed jar over the best vinegar in which auratæ have been previously boiled. Afterwards dissolve in it sal ammoniac and calcined tartar. In all other particulars proceed as directed in the case of cerussa. Then, after fourteen days, you will have the most precious and beautiful lazurium adhering to the silver plates, which you will wipe off with a hare's foot.

---

* The physicians of Montepessulano and Salerno committed the error of supposing that cinnabar was different from mercury, when it is clear that they are the same.—*De Tumoribus*, etc., *Morbi Gallici*, Lib. I., c. 8. Cinnabar i extracted from Saturn and Mars by means of mercury.—*Ibid.*, Lib. III., c. 7.

We do not deem it necessary here to repeat the method of mortifying gold so that it may be reduced to its arcana, as, for instance, to tincture, quintessence, resin, crocus, vitriol, and sulphur. These preparations are manifold, and for the most part we have already given such secrets in other books, as the extraction of the Tincture of Sol, the Quintessence of Sol, the Mercury of Sol, Sol Potabilis, the resin of Sol, the Crocus of Sol. These have been given in the Archidoxa and elsewhere. But the secrets omitted there we will impart here. These concern the vitriol of Sol* and the sulphur of Sol, which are by no means the least among such secrets, and, indeed, ought to delight every physician. In order to extract vitriol from Sol, proceed thus: Take two or three marks of pure gold, which form into plates and suspend above boys' urine, mixed with grape-berries, in a wide glass cucurbite closely sealed at the top. Bury this in a glowing heap of grape-berries, as they are taken from the wine-press, and let it stand there for a fortnight or three weeks. Then open it, and you will find a most subtle colour, which is vitriol of Sol, adhering to the plates of Sol. Remove this with a hare's foot, as you have been told in the case of the other metals—the crocus of Mars from the plates of iron, the vitriol of Venus and verdigris from the plates of copper, the cerussa from the plates of Saturn, the lazurium from the plates of Luna—all these being comprised under one process, but not with the same preparation. When, therefore, you have enough of this vitriol of Sol, boil it well in distilled rain water, stirring it continually with some sort of spatula. Then the sulphur of gold rises up to the surface like grease, which remove with a spoon. So also proceed with other vitriol. After the sulphur is taken away, evaporate that rain water to perfect dryness, and the vitriol of Sol will remain at the bottom. This you can easily resolve on marble in a damp place. In these two arcana, that is to say, the vitriol of gold and the sulphur of gold, a diaphoretic virtue is latent. However, we will not describe those virtues here, because we have sufficiently indicated them in the book on Metallic Diseases and elsewhere.

The mortification of sulphur consists in taking away its combustible and fœtid fatness, and reducing it to a fixed substance. This is accomplished in the following way: Take common yellow sulphur, reduced to a fine powder, and abstract from it the very acrid aqua fortis by a threefold distillation. Afterwards sweeten the sulphur which remains at the bottom, and is of a black colour, with sweet water, repeating the process of distillation continually until nothing but sweet water proceeds from it and there is no more smell of sulphur. Reverberate this sulphur in a closed reverberatory, as in the case of antimony. Then it will become, at first white, afterwards, yellow; thirdly, red

---

* Artificial acids are from the minerals of metals and cognate substances. But note here that what is usually called vitriolated acid is really vitriolated copper of Venus. For copper is vitriol. If, therefore, the acidity be extracted from copper, then he who uses it digests copper. It is the same with all the other vitriolates of metals. . . . In all metals there are vitriolated acids, except in gold, which does not know vitriol.—*De Morbis Tartareis*, c. 16.

as cinnabar. When you have it in that form you ought to rejoice; for it is the beginning of wealth for you. This reverberated sulphur tinges any silver very deeply so as to turn it into most precious gold, and the human body it tinges into its most perfect condition of health. Of so great virtue is this reverberated and fixed sulphur.

The mortification of all salts, and whatever is of a salt nature, is the removal and distillation of their watery and oleaginous part, and besides of the spirit of salt; for if these are taken away, they are called afterwards dead earth, or *caput mortuum*.

The mortification of gems and corals is that they shall be calcined, sublimated, and dissolved into a liquid, as the crystal. The mortification of pearls is that they be calcined and resolved in sharp vinegar in the form of milk.

The mortification of the magnet is that it be smeared with oil of mercury or touched by common mercury. Afterwards it attracts no iron.

The mortification of flints and stones is calcination.

The mortification of marcasites, cachymiæ, talc, cobalt, zinc, granites, zwitter, vismut, and antimony, is sublimation, that is, their being sublimated with salt and vitriol. Then their life, which is the metallic spirit, ascends with the spirit of salt. Let whatever remains at the bottom of the sublimatory be washed, that the salt may be removed from it, and you will have dead earth wherein is no virtue.

The mortification of arsenicals, auripigment, orpiment, realgar, etc., is when they are made fluid with salt nitre, are turned to oil or liquid on marble, and fixed.

The mortification of undulous things is a coagulation of the air.

The mortification of aromatic substances is the removal of their good odour.

The mortification of sweet things is that they shall be sublimated with corrosives and distilled.

The mortification of carabæ, resins, turpentine, and gum is their being reduced to oil or varnish.

The mortification of herbs, roots, and the like is that their oil and water shall be distilled from them, the liquid squeezed out in a press, and afterwards the alkali extracted.

The mortification of woods is their being turned into charcoal or ashes.

The mortification of bones is their calcination.

The mortification of flesh and blood is the removal of the spirit of salt.

The mortification of water is produced by fire: for the heat of fire dries up and consumes all water. So the mortification of fire is by water; for the water extinguishes the fire and takes away from it its force and effectiveness.

Thus you are sufficiently informed, in few words, how death is latent in all natural things: how they are mortified and reduced to another form and nature, as also what virtues flow from them. Whatever else is necessary to say we will set down in our book concerning the Resuscitation of Natural Things.

CONCERNING THE NATURE OF THINGS

BOOK THE SIXTH

Concerning the Resuscitation of Natural Things

THE resuscitation and reduction of natural things is not the least important in the nature of things, but a profound and great secret, rather divine and angelic than human and natural. I would, however, on this point be understood with the greatest discrimination, and in no other way than according to my fixed opinion, as Nature daily and clearly points out and experience proves; so that I may not be exposed to the lies and misrepresentations of my enemies the quack doctors (by whom I am constantly ill judged), as if I myself pretended to usurp some divine power, or to attribute that same to Nature which she never claims. Therefore, at this point, the most careful observation is necessary, since death is twofold, that is to say, violent or spontaneous. From the one, a thing can be resuscitated but not from the other. Do not, then, believe the sophists when they tell you that a thing once dead or mortified cannot be resuscitated, and when they make light of resuscitation and restoration; for their mistake is great. It is indeed true that whatever perishes by its own natural death, or whatever mortifies by Nature according to its own predestination, God alone can resuscitate, or that it must be done by His divine command. So whatever Nature consumes man cannot restore. But whatever man destroys man can restore, and break again when restored. Beyond this man by his condition has no power, and if any one strove to do more he would be arrogating to himself the power of God, and yet would labour in vain and be confounded, unless God were with him, or he had such faith that he could remove mountains. To such a man this, and still greater things, would be possible, since Scripture says, for Christ Himself has said—"If ye have faith as a grain of mustard-seed, and say to this mountain: Depart and place yourself yonder, it would do so and place itself there; and all things shall be possible, and nothing impossible, to you."

But let us return to our proposition. What is the difference between dying and being mortified, and from which of these conditions is resuscitation possible? The matter is to be understood thus. Whatever dies by its own nature has its end according to predestination, and as the pleasure and dispensation of God arranges. But this, too, happens from different diseases and accidents, and herefrom there is no resuscitation, nor is there any preservative which can be used against predestination and the cognate end of life. But what is mortified can be resuscitated and revivified, as may be proved by many arguments which we will set down at the end of this book. So, then, there is the greatest difference between dying and mortifying, nor should it be thought that these are only two names for one thing. In very deed these differ as widely as possible. Examine the case of a man who has died by a natural and predestined death. What further good or use is there in him? None. Let him be cast to the worms. But the case is not the same with a man who has been slain with a sword or has died some violent death. The whole of his body is useful and good, and can be fashioned into the most valuable mumia. For though the spirit of life has gone forth from such a body, still the balsam remains, in which life is latent, which also, indeed, as a balsam conserves other human bodies. So, too, in the instance of metals you see that when a metal has a tendency to die it begins to be affected with rust, and that which has been so affected is dead; and when the whole of the metal is consumed with rust the whole is dead, and such rust can never be brought back to be a metal, but is mere ashes and no metal. It is dead, and death is in itself: nor has it any longer the balsam of life, but has perished in itself.

The lime and the ashes of metals also are two-fold, and there is the greatest difference between these two. For the one can be revived and brought back to be a metal, but not so the other. One is volatile, the other is fixed. One is dead, the other is mortified. The ash is volatile and cannot be brought back to be a metal, but only to glass or scoriæ. But the lime of metals is fixed and can be brought back again into its own metal. If you would understand the difference and its cause, know that in the ash there is less fatness and more dryness than in the lime, and it is this which gives the fluxion. The lime is fatter and more moist than the ash, and still retains its resin and its fluxion, and more especially does it retain the salt which of its own special nature is capable of flux, and also makes all metals pass into flux, thereby reducing them. Hence it follows with the ashes of metals that they cannot be brought back into metals. The salt must be extracted; then they are perfectly volatile. This is the chief point, and must be very carefully noted, since no little depends upon it. Among sham physicians a vast error is prevalent. In place of Sol Potabilis, the Quintessence of gold, the Tincture of gold, and so on, they have palmed off on men a leprous Calx of Sol, not considering the difference or the evils resulting therefrom. For two notable and necessary facts must here be observed, namely, that either calcined or pulverised Sol, when given to men, is congregated into one mass in the bowels, or passes out *per anum* with the dung, and so is vainly and uselessly taken; or else by the great internal heat of the body it is reduced, so that it incrusts and clogs the bowels, whence ensue many and various diseases, and at last even death.

And as with Sol, so also in the case of other metals, you should take no metallic arcanum or medicament into the body unless it shall have first been rendered volatile, so that it cannot be brought back to its metallic condition. Wherefore the first step and beginning of preparing Aurum Potabile is this; afterwards such a volatile substance can be dissolved by spirit of wine, so that both ascend together, becoming volatile and inseparable. Just as you prepare gold, in the same way you prepare potable Luna, Venus, Mars, Jupiter, Saturn, and Mercury.

But to return to our proposition, and to prove by illustrations and by adequate reasons that mortified things are not dead and compelled to continue in death, but can be brought back and resuscitated and vitalised by man, according to natural guidance and rule. You see this in the case of lions, who are all born dead, and are first vitalised by the horrible noise of their parents, just as a sleeping person is awakened by a shout. So the lions are stirred up ; not that they are sleeping in the same way—for one who sleeps a natural sleep would necessarily wake—but this is not the case with lions. Unless they were stirred up with this noise they would remain dead, and life would never be found in them. Hence it is understood that they acquire their life and are vitalised by that noise. You see the same thing in all animals, except those which are produced from putrefaction, like flies, which, if they are drowned in water so that no life could be discerned in them, and were so left, would continue dead, and never would revive of themselves. But if they are sprinkled with salt and placed in the warm sun, or behind a heated furnace, they recover their former life, and this is their resuscitation. If this were not done they would remain dead. So you see in the case of the serpent. If it be cut in pieces, and these pieces be put in a cucurbite, and putrefied in a *venter equinus*, the whole serpent will revive in the glass in the form of small worms or the spawn of fishes. Now, if these little worms are—as they ought to be—brought out by putrefaction and nourished, more than a hundred serpents will be produced from the one, any single serpent being as big as the original one. This can be accomplished by putrefaction alone. And just as with the serpent, so many animals can be resuscitated, recalled, and restored. By this process, with the aid of nigromancy, Hermes and Virgil endeavoured to renovate and resuscitate themselves after death, and to be born again as infants, but the experiment did not turn out according to their intention and it was unsuccessful.

Let us, however, pass by these examples, and come to the practical method of resuscitation and restoration. It is advisable to begin with metals, because metallic bodies more frequently resemble human bodies. Know, then, that the resuscitation and renovation of metals are twofold : one brings back calcined metals by a process of reduction to their original metallic body ; the other reduces metals to their first matter. The former is a reduction to *argentum vivum*, and such, too, is the latter process. Calcine a metal by means of the *fuligo Mercurii*. Put this calx and a sufficient quantity of the quicksilver into a sublimatory, and let them stand for some time, until the two are coagulated into one amalgam. Then, by means of sublimation, elevate the Mercury from the calx. When elevated, pound it again with the metallic calx, and sublimate as before. Repeat this until the metallic calx liquefies over a candle, like wax or ice, and the thing is then done. Let this metal be placed in digestion for such time as may be required, and the whole will be changed into *Mercurius vivus*, that is, into its first matter. This is called the Philosophers' Mercurius of Metals. Many alchemists have sought it, but few have found it. So is now prepared *Mercurius vivus* from all metals, namely, Mercurius of Gold, Luna, Venus, Mars, Jupiter, Saturn.

The resuscitation or restoration of calcined Mercury is produced by distillation in retorts. For only *Mercurius vivus* ascends into the cold water, and the ashes of Saturn, Venus, or sulphur are left. But the resuscitation and restoration of sublimed Mercury is brought about in hot water. It is necessary, however, that it should first of all be very minutely pounded, so that the boiling water may resolve from it the spirit of salt and of vitriol, which it raises up with itself in the process of sublimation, and the *Mercurius vivus* runs together at the bottom of the water. If, now, such *Mercurius vivus* be sublimated anew with fresh salt and vitriol, and again be resuscitated in boiling water, and if this be repeated seven or nine times, it will be impossible to purify and renovate it more effectually. Preserve this as a great secret in Alchemy and Medicine, and rejoice over it exceedingly; for in this way all the impurity and blackness and poisonous nature are taken away from Mercury. The resuscitation, restoration, and renovation of Mercury cannot be accomplished without sublimation; for unless after calcination it be sublimated it will never be revivified. Sublimate it, therefore, and afterwards reduce it as you would any other sublimated substance.

The resuscitation of cinnabar, lazurium, aurum musicum, or precipitated gold, in order that they may be revived into *Mercurius vivus*, is effected as follows: Take any one of these substances, pound it very fine in a marble mortar, and make it into a paste with white of eggs and smegma. Then make pills, the size of a nut, which place in a strong earthenware cucurbite. At its orifice arrange an iron plate which has several little holes, and let it be fastened with lute. Distil by descent over a strong fire, so that it may fall into cold water, and again you will have *Mercurius vivus*.

The resuscitation and restoration of wood is difficult and arduous; possible, indeed, but not to be accomplished without exceptional skill and industry. The following is the method of its revival: Take wood which has been first of all carbon, then ash, and place it in a cucurbite with the resin, liquid, and oil of its tree, the same weight of each. Let them be mixed and liquefied over a gentle fire. Then there will be produced a mucilaginous matter, and so you will have the three principles together from which all things are born and generated, namely, phlegma, fat, and ash. The phlegma is

Mercurius, the fat is Sulphur, and the ash is Salt. For that which smokes and evaporates over the fire is Mercury; what flames and is burnt is Sulphur; and all ash is Salt. Now, when you you have these three principles together, place them in a *venter equinus*, and putrefy for the time required by each respectively. If afterwards that matter be buried, or poured into a rich soil, you will see it begin to revive, and a tree or a little log will be produced from it, which, indeed, is in its nature much higher than the original one.

This is really wood, and is called resuscitated, renewed, and restored wood. It was from the beginning wood, but mortified, destroyed, and reduced to coals, to ashes—to nothingness; and yet from that nothingness it is made something, and is reborn. Truly in the light of Nature this is a great mystery, that a thing which had altogether lost its form, and had been reduced to nothingness, recovers that form and becomes something from nothing—something which afterwards is much nobler in its virtue and its efficacy than it had been at first.

But, in order that we may speak generally concerning the resuscitation and restoration of natural things, this should be understood as the principal foundation—that to each thing may be again conceded that which had been taken from it and separated in mortification. It is difficult to explain this specifically here; so we will conclude this book, and in the following book make these things more clear with regard to the transmutations of natural things.

CONCERNING THE NATURE OF THINGS

BOOK VII

Concerning the Transmutation of Natural Objects

IF we are to write concerning the transmutation of all natural objects, it is just and necessary that, in the first place and before all else, we should point out what transmutation is; in the second place, what are the successive steps thereto; and, thirdly, by what means, and in what manner, it is brought about. Transmutation, then, takes place when an object loses its own form, and is so changed that it bears no resemblance to its anterior shape, but assumes another guise, another essence, another colour, another virtue, another nature or set of properties: as if a metal becomes glass or stone; if stone or wood becomes coal; if clay becomes stone and slate; hide, glue; rag, paper; and many such things. All these are transmutations of natural objects. After this it is most necessary to know the steps to transmutation, how many there are. There are not more than seven. For although some persons reckon a greater number, there are, of a truth, only seven principal steps; the rest which may be included among the steps are comprised in these seven. They are the following:

CALCINATION, SUBLIMATION, SOLUTION, PUTREFACTION, DISTILLATION, COAGULATION, TINCTURE

If anyone ascends that ladder, he will arrive at so wonderful a place that he will see and experience many secrets in the transmutation of natural objects.

The first step, then, is Calcination,* under which are comprised Reverberation and Cementation. Among these three there is little difference so far as relates to Calcination. Here, therefore, Calcination is the principal step, for by Reverberation and Cementation many corporeal objects are calcined and reduced to ashes, especially metals. What is calcined is not on that account reverberated or cemented. By Calcination all metals, minerals, stones, glasses, and all corporeal objects, become carbon and ashes; and this is done in a naked fire, strong, and exposed to the air. By means of this all

---

* One of the *Fragmenta Medica* contained in the first volume of the Geneva folio, when explaining the process of calcination from the standpoint of Hermetic Medicine, observes that it is eminently necessary for the physician who concerns himself with Alchemy to understand calcination and the virtue which resides therein.

tenacious, soft, and fat earth is hardened into stone; but all stones are reduced to lime, as we see in the kiln of the lime burner and the potter respectively.

Sublimation* is the second step, also very important for the transmutation of many natural objects. Under this are included Exaltation, Elevation, and Fixation†; and it is not altogether unlike Distillation. For, as from all phlegmatic and watery objects, water ascends in distillation, and is separated from its body, so, in the process of Sublimation, in dry substances such as minerals, the spiritual is raised from the corporeal, subtilised, and the pure separated from the impure. For in Sublimation many excellent virtues and wonderful qualities are found in minerals, and many things are fixed and become permanent, so that they remain in the fire in the following way: Let the body which is sublimated be ground again and mixed with its own dregs. Let it be again sublimated as before, and let this be repeated until it sublimates no longer, but all remains in the bottom and is fixed. Thus it will afterwards become a stone and an oil when and as often as you wish. For if, having refrigerated it, you put it in the air, or in a glass vessel, it is there immediately resolved into an oil. If you once more put it in the fire it is again coagulated into a stone, which is of great and wonderful powers. But this consider a great secret and mystery of Nature, and do not disclose it to sophists. Moreover, as in Sublimation many corrosives become sweet by the conjunction of the two natures, so, on the other hand, many sweet substances become sour or bitter; whilst many bitter things are made sweet as sugar. Here it should be remarked, too, that every metal which is brought to a state of Sublimation by means of sal ammoniac may afterwards be dissolved into an oil in the cold, or in the air, and, contrariwise, may be coagulated to a stone in the fire. This is one of the greatest and most complete transmutations in all natural objects, namely, to transmute a metal into a stone.

The third step is Solution, under which term are comprised Dissolution and Resolution. This step frequently follows after Sublimation and Distillation, as, for instance, when you dissolve the matter which remains at the bottom. Solution, however, is twofold: one by cold, another by heat; one out of the fire, the other in the fire. The cold process of Solution dissolves all salts, corrosives, and calcined bodies. Whatever salt and corrosive quality there may be it resolves into an oil, a liquid, or water; and this takes place in a damp and cold chamber, or otherwise in the air only, in marble or glass. For everything that is dissolved in the cold contains the sharp spirit of salt, which it often acquires and assumes in Sublimation or Distillation. And everything which is dissolved in the cold or in the air is again by the heat of fire changed

---

* By sublimation the lower minerals are separated from those elements which are the source of their poverty and baseness, but in addition to this, the process has many other virtues. For example, the sublimation of quicksilver has this operation, that even the air in its vicinity has a recreative effect. For in the air permeated by mercury all the virtues of mercury are present. In like manner, the sublimation of arsenic releases a fervid spirit into the atmosphere which cures quartan fever and other acute diseases.—*De Morbis Metallicis*, Tract III., c. 5.

† Exaltation, conjunction, opposition, and kindred processes are not materially performed, but after a mode which is altogether spiritual.—*Paramirum*, Tract III., c. 6.

into dust or stone. But the Solution of heat dissolves all fat and sulphurous bodies; and whatever the heat of fire dissolves this the cold coagulates into a mass, and whatever the heat coagulates, this the air and the cold again dissolve. This also should be known, that whatever the air or the cold chamber dissolves, is of great dryness, and holds concealed within itself a corrosive fire. So whatever is dissolved in fire, and by its heat, has in itself sweetness and cold, but not fire. Thus, and in no other way, is Solution to be understood.

Putrefaction* is the fourth step, under which are comprised Digestion† and Circulation. Now Putrefaction is a very important step which might deservedly stand first, only that would be contrary to the just order and to the mystery which lies concealed here, and is known to very few. For these steps should follow one another in turn, as has been said, like the links in a chain, or the rounds of a ladder. For if one of the links of the chain were taken away, the chain would be broken and the captive would escape. And so, too, if one of the rounds of the ladder should be removed from the middle and put in the highest or the lowest place, the ladder too would be broken, and many would fall headlong from it and endanger their lives. So understand here that these steps follow one another in a just order; otherwise the whole work of our mystery would be perverted, and all our toil and pains frustrated and rendered void. Putrefaction is of so great efficacy that it blots out the old nature and transmutes everything into another new nature, and bears another new fruit. All living things die in it, all dead things decay, and then all these dead things regain life. Putrefaction takes away the acridity from all corrosive spirits of salt, renders them soft and sweet, transmutes their colours, separates the pure from the impure, and places the pure higher, the impure lower, each by itself.

Distillation is the fifth step to the transmutation of all natural objects. Under it are understood Ascension, Lavation, Imbibition, Cohobation, and Fixation. By Distillation all waters, liquids, and oils are subtilised, the oil is extracted from all fat substances, the water from all liquids, and in all phlegmatic substances the oil and the water are separated.

Moreover, many things in Distillation are fixed by Cohobation, especially if the substances to be fixed contain water within them, as vitriol does. When this is fixed it is called colcotar. Alum, if it is fixed with its own water, is

---

* Putrefaction is the handmaid of separation.—*Modus Pharmacandi*, Tract I. Putrefaction is a new qualitative generation.—*De Modo Pharmacandi*, Tract III. The firmament produces colours, corruptions, and digestions of nutriment, of nature, etc. And putrefaction produces a succession of colours rapidly.—*Ibid.* All putrefaction is essentially and excessively cold.—*De Tartaro*, Lib. II., Tract II., c. 7. Putrefaction is the separation of virtue, and at the same time is almost a conservation.—*De Naturalibus Aquis*, Lib. IV., Tract 2.

† Digestion is putrefaction.—*De Pestilitate*, Tract I. By the process of digestion, what is bad and unprofitable in a substance is separated so that the substance remains in its essence, as it was created. In so far as it has become vitiated, digestion causes it to purge itself, so that it labours to return into its essence.—*De Tartaro*, Lib. II., Tract II., c. 2. Between digestion performed in the earth and the digestion which takes place in the body of man, there is this difference, that the earth separates nothing, in the sense that it does not cast out anything excrementitiously; it digests, putrefies, generates, and augments by the power and ministry of the stars. There is no excremental separation, but there is a separation of seed into salt, sulphur, and mercury. Yet this is not precisely a deprivation of the earth, because the earth contains in itself salt, sulphur, and mercury. The earth, moreover, requires no nutrimental support after the manner of human beings, but the seed is sown in it just as the male seed is sown in the female womb. The earth generates, augments, and multiplies by means of its own indwelling Archeus.—*De Pestilitate*, Tract II.

called Zuccari. This, too, is resolved into a liquid, and if it be putrefied for a month it produces a water as sweet as sugar, which, indeed, is of great power and an excellent arcanum in medicine for extinguishing the microcosmic fire in men of a metallic temperament, as we write more at length in our books on Metallic Diseases.* And just as you have heard of vitriol and alum, so also salt nitre and other watery minerals can be fixed by cohobation.

The process of Cohobation is that a *caput mortuum* is frequently imbibed with its own water, and this is again drawn off by means of Distillation. Moreover, in Distillation many bitter, sharp, and acrid things become very sweet, like honey, sugar, or manna; and, on the other hand, many sweet things, such as honey, sugar, or manna, become sharp, as oil of vitriol or vinegar, or bitter, as gall or gentian, or sharp, as corrosive. Many excrementitious things lose their excessive stench in distillation, since it passes out into the water. Many aromatic things lose their pleasant odour. And just as Sublimation alters things in their quality and nature, so does Distillation.

Coagulation is the sixth step. There is, however, a twofold process of Coagulation, one by cold, another by heat; that is, one of the air, another of the fire. Each of these, again, is twofold, so that there are really four processes of Coagulation, two by cold, and two by fire. The Coagulations by fire are fixed, the others by cold are not fixed. One, indeed, is produced only by common air, or without fire. Another is produced by the upper firmament of winter stars, which coagulate all waters into snows and ice. The Coagulation by fire is produced by the artificial and graduated fire of the alchemist, and is fixed and permanent. For whatever such a fire coagulates, that becomes permanent. Another Coagulation is produced by the Ætnean and mineral fire in mountains, which, indeed, the Archeus of the earth rules and graduates in much the same way as the alchemist; and whatever is coagulated by such a fire is also fixed and constant, though originally its matter was mucilaginous, and it is coagulated by the Archeus of the earth and by the work of Nature into metals, stones, flints, and other bodies. But it should also be known that fire coagulates no water or moisture, but only the liquids and juices of all natural things. For this reason no phlegm can be coagulated, unless it was originally a corporeal matter, whereto, indeed, it can be again restored by the industry of an experienced alchemist. So every mucilaginous matter or spermatic lentor can, by the heat of fire, be coagulated into a body and corporal material, but cannot again be resolved

---

* Medicines are therefore chosen which are free from coagulation, such as alum, in which humidity and coagulation simultaneously exist. If these two be separated one from another, the quality withdraws into one place, and the element, in like manner, into another. Now, the element of alum is most akin to the element of water. For the element of water also consists in its Hyle, as alum after its excoction, and when it has been separated from its coagulates, it passes into its pure and proper element, despoiled, however, of its medicinal arcana. But alum does not suffer this privation. For water alone prevails against the microcosmic fire. Whence the matter stands thus, that the aquosity must be separated from the alum, and must be rectified therein till it is almost like sugar. The dose is one scruple. If the symptoms of the elementary disease again present themselves, they must be again extinguished as before. There are many such arcana, which I leave to the experience of the school of Vulcan, as it is impossible to enumerate them in this place. – *De Morbis Metallicis*, Lib. II., Tract IV., c. 6.

into water. And as you have heard concerning Coagulation, so know also concerning Solution, namely, that no corporeal matter can be resolved into water unless it originally was water, and such is the case with all mineral substances.*

Tincture is the seventh and last step, which concludes the work of our mystery, with reference to transmutation, makes all imperfect things perfect, transmutes them into their noblest essence and highest state of health, and changes them to another colour. *Tincture, therefore, is the noblest matter with which bodies, metallic and human, are tinged, translated into a better and far more noble essence, and into their supreme health and purity.* For a Tincture colours all things according to its own nature and its own colour. But there are many and various Tinctures, and not only for metallic and human bodies, since everything which penetrates another matter, or tinges it with another colour or essence, so that it is no longer like what it was before, may be called a Tincture. So then there are manifold tinctures, that is to say, of metals, minerals, human bodies, waters, liquids, oils, salts, all fat substances— in a word, of all things which, with or without fire, can be brought or reduced to a state of fluxion. For if the tincture is to tinge, it is necessary that the body or material which is to be tinged should be open, and in a state of flux; for unless this were so, the tincture could not operate. For it would be just as though one were to cast saffron, or some other colour, into coagulated water or ice; it would not tint the ice so quickly with its colour as if one were to put it into other water. And, although it might tinge the ice, it would at the same time reduce it into water. Wherefore, metals also, which we wish to tinge, must be liquefied by fire, and freed from their coagulation. And here it should be known that the more hotly they are liquefied the more rapidly the tincture runs through them, just as fermentation penetrates the whole mass and imparts acidity to it, and the better it is covered up, and the warmer the mass is kept, the more perfectly it ferments, and the better bread it gives: for fermentation is a Tincture of the farinaceous mass and of the bread.†

---

* All created things proceed from a coagulate, and afterwards this coagulate must pass into a liquid. From a liquid, then, all procreated things proceed, whether these be liquids or solids possessing a defined shape. Further, the solid can never be so perfectly liquefied as not to strive to return to its solidity. For example: salt, when it is dissolved in water, seeks to revert into its original state. It is the same with all other substances. Moreover, no solid is so completely dissolved but that it will actually return into its original shape, by means of the nature it retains. Understand that any solid proceeds from one of the three principles—sulphur, mercury, and salt - whichsoever it may be. Sulphur is never liquefied so completely as not to leave some solidity adhering to it. This is also the case with salt and mercury. Great attention must be paid to this solidification and dissolution. The one frequently prevails over the other. . . . Understand, therefore, of things in general, that they proceed from three principles; but that from which they proceed is a solid, as, for example, seed, earth, all fruits, and all growing things. Nothing exists which is not a solid. But this is not the solidification of which mention is made here, but is above it and was before it. For fruits were produced from that liquid, and were again solidified. The result is that here a certain kind of generation takes place, and if it be not followed again by a second digestion, as in the digestion which ensued after the first dissolution into fruits, that which finally remains becomes the principle of tartar.—*Aliud Fragmentum de Tartaro.*

† The brutes themselves have an innate knowledge, good and bad. None the less has man, also, his tinctured knowledge, which is bad and good, being tinctured from the stars as regards his earthy nature and condition. In consequence of this nature a most supreme and exhaustive investigation of philosophy is permissible. The right and proper understanding of the animal condition of human nature is contained in an understanding of the tincture of the animal man. Man has two tinctures, one, as regards his inferior being, from the stars, and the other, supernatural, from God.—*De Pestilitate,* Tract I.

It is also to be remarked that some dregs are of a more fixed substance than their liquid, of a sharper also and more penetrating nature, as you see in the case of *vinum ardens*, which is made from the dregs of wine, and in the case of *cerevisia ardens*, which is distilled from the dregs of cerevisia; and just as *vinum ardens* burns, and as sulphur is kindled, so, if from the dregs of *acetum* another *acetum* should be distilled, as *vinum ardens* is commonly distilled, there will be produced thence an *acetum* of so fiery and acrid a nature, that it would consume all metals, stones, and other substances, like aquafortis.

Moreover, tincture must be of a fixed nature, fluxible, and incombustible, so that if a little of it be thrown on an ignited plate of metal it will presently float like wax, and that without any smoke, and will penetrate the metal as oil penetrates paper, or water a sponge, and tinge all metals to white and red, that is, in the case of Luna and Sol.* These are now the Tinctures of the metals, which must first of all be turned to alcohol by the step of Calcination. Afterwards, by the second step of Sublimation, their own easy and gentle flux must be produced; lastly, by the step of Putrefaction and Distillation the Tincture is evolved, fixed, incombustible, and of changeless colour.

But the Tinctures of human bodies—whereby those bodies may be tinged into their supreme state of health, and all diseases may be expelled, that their lost powers and colours may be restored, and they themselves invigorated and renewed—are these : Gold, pearls, antimony, sulphur, vitriol, and the like, the preparations whereof we give in many other books, so it does not seem necessary for us to repeat them here.

But concerning Tinctures nothing more need be written, seeing that every extracted colour may be called a Tincture, which, indeed, tinges with a permanent colour things which do not enter the fire, or keep their colours fixed in the fire. All these things are in the hand and power of the dyer or the painter, who prepares them according to his own pleasure.†

It is especially necessary, too, in this book to know the degrees of fire, which can be graduated and intensified in many ways, and each degree has its own peculiar operation, while no one gives the same result as another, as every skilful alchemist finds from his daily experience and the practice of his art. One is the live flaming fire which reverberates and calcines all bodies.

---

* I call the tincture of gold the colour of the body itself, which, if separated from the body, so that a white body remains, will be a perfect work. For colour and body are two different things, and for this reason admit of separation. that is to say, the pure (the colour) is separated from the impure (the body). Unless this be done, all the labour will turn out useless. When, accordingly, this separation is accomplished, we must immediately hasten to the clarification of the colour, and to the highest grade of exaltation. But the grade to which the tincture can be exalted is five times double, that is, five times into five times twenty-four, for it cannot become more sublimated.—*Chirurgia Magna*, Part II., Tract III., c. 2.

† Tinctures operate approximately as follows : Just as you see fire completely consume firewood and similar bodies, which, as gold, etc., have no figure of man, so we must believe that tinctures operate. Thus, as antimony purges away all the dross of gold, perfects it, and raises it to the highest grade by cementation, in like manner it becomes manifest that the tinctures themselves have obtained a nature similar to cement, inasmuch as they perform operations completely similar to those of the latter and of fire. The ancient artists marvellously wearied themselves at conjoining tinctures with fire, for they anticipated a medicine in their almost sacred conjunction, but all in vain. –*Ibid.*, c. 8.

Another is the fire of the candle and lamp, which fixes all volatile bodies. Another is the coal fire, which cements, colours, and purges metals from their scoriæ, graduates more highly Sol and Luna, takes the whiteness from Venus, and, in a word, renovates all the metals. Another is the fire of an ignited iron plate, on which the tinctures of metals are probed, which also is useful for other purposes. In another way, scobs (*i.e.*, alkali) of iron produces heat, in another way, sand; in another, ashes; in another, the *balneum maris*, by which many distillations, sublimations, and coagulations are produced. In yet another way operates the *balneum roris*, in which take place many solutions of corporeal things. Otherwise, again, acts the *venter equinus*, in which the principal putrefactions and digestions take place, and in another way operates the invisible fire, by which we understand the rays of the sun, which also is shewn by a mirror, or steel plate, or crystal, and displays its operation and effect, concerning which fire the ancients wrote scarcely anything. By this fire, indeed, the three principles in any corporeal substance can be separated on a table. Of so wonderful a virtue is this fire, that by means of it metals are liquefied, and all fat and fluxible things—all combustible things, indeed—can be reduced to carbon and ashes on a table, and without fire.

Since, then, I have placed before you and disclosed the steps of Alchemical Art, and the degrees of alchemical fire, I will, moreover, point out to you, and describe generically, the various transmutations of natural objects. Before all, one should speak of the metals; secondly, of stones; thirdly, of various objects after their kind. The transmutation of metals, then, is the great secret in Nature, and can only be produced with difficulty, on account of the many hindrances and difficulties. Yet it is not not contrary to Nature or the will of God, as many falsely say. But in order to transmute the five lower and baser metals, Venus, Jupiter, Saturn, Mars, and Mercury, into the two perfect metals, Sol and Luna, you must have the Philosophers' Stone. But since we have already, in the seven steps, sufficiently unveiled and described the secrets of the Tinctures, it is not necessary to labour further about this, but rather rest satisfied with what we have written in other books on the Transmutations of Metals.

But there are further transmutations of imperfect and impure metals, as, for instance, of Mars into Venus. This may be effected in different ways: Firstly, if iron filings are heated in water of vitriol; or, secondly, if iron plates are cemented with calcined vitriol; thirdly, if glowing iron plates are extinguished with oil of vitriol. In these three ways iron is transmuted into the best, natural, and heavy copper, which, indeed, flows very well, and has its own weight as well as any native copper. Iron filing can also be reduced and transmuted as if into lead, so that it becomes entirely soft, like native lead, but it does not flow easily. Therefore proceed thus: Take some iron filing, and the same quantity of the best liquefying powder. Mix them; place them on a tigillum in a blast furnace, make a strong fire, not so much as to melt the iron, but let it stand as if in a cement a whole hour. Afterwards increase the fire vigorously, so that the iron may glow and melt. Lastly, let the tigillum cool of itself, and you will find a regulus of lead on the tigillum, as soft and ductile as native lead can be.

But in order to transmute Venus into Saturn proceed thus: First of all, sublimate copper, and reduce it by fixed arsenic to a white substance, as white as Luna. Then granulate. Of this, and of good reduced powder, take the same quantity; first cement, and, lastly pour into the regulus, when you will have the true leaden regulus.

On the other hand, it is very easy to turn lead into copper, nor is any great skill required. This is the process: Calcine plates of lead in vitriol, or stratify with the crocus of Venus, cement, and, lastly, liquefy. Then you will see as much native lead as you please transmuted into good, heavy, and ductile copper.

If, now, such copper, or any other copper, be made into plates and stratified with tutia and calamine, cemented, and if, lastly, it be cast, it is changed into a splendid amber or red colour, like gold.

If you wish to change Saturn into Jupiter, take plates of Saturn, and stratify with sal ammoniac, cement, and, lastly, cast, as above. So all its blackness and darkness are taken away from the lead, and it becomes in whiteness like the best English tin.

As you have now heard in brief a summary of some transmutations of metals, so, moreover, know concerning the transmutations of gems, which, indeed, are various and by no means alike. For you see how great a transmutation of gems lies hid in oil of sulphur. Any crystal can be tinged and transmuted in it, and in course of time graduated with distinct colours so as to become like a grained jacinth or ruby.

Understand in like manner concerning the magnet. It can be transmuted into ten times its power and virtue in the following way: Take a magnet, and heat it in the coals to such a degree that it may be at a high temperature, but still not red hot. Extinguish this immediately in the oil of the crocus of Mars, which is made of the best Carinthian steel, so that it may imbibe as much as it can take. Thus you will make a magnet so powerful that with it you can pull out the nails from a wall, and do other wonderful things which a common magnet could never accomplish.

Moreover, in the transmutations of gems, it must be known that the world is situated in the two grades of tincture and coagulation. For as the white of an egg can be tinged with saffron, and afterwards coagulated into a beautiful yellow amber, with the dye of a pine into black amber, with verdigris into green amber, like the cyanean or Turkish stone, with green juice into the likeness of an emerald, with lazuleum into a cerulean amber like sapphire, with Brazilian wood into a red amber like the grained jacinth or ruby, with a purple colour like amethyst, or with ceruse made to resemble alabaster—so all other liquids, and especially metals and minerals, can be tinged with fixed colours, afterwards coagulated, and transmuted into gems.

Similarly pearls, too, can be made entirely like true ones in appearance so that by means of their brightness and beauty they can scarcely be distinguished from genuine ones. Proceed thus: Purify as much as possible the white of eggs with a sponge. Into this put and mix some fair white talc, or pearl shell, or Mercury coagulated with Jupiter and reduced to alcohol. At the same time pound it in marble very fine, so that it becomes a thick amalgam, which must be dried in the sun or behind a warm furnace until it becomes like cheese or hepar. Lastly, from this mass make as many pearls as you wish, and fix them on hog bristles. Having thus bored them, dry them as you did the amber, and you have prepared them. If they do not shine sufficiently anoint them externally with the white of an egg, and again dry them. Thus they will become most beautiful pearls, like true ones in form though not in virtue.

Almost in the same way corals are made by those who wish to deceive people as with the pearls just spoken of. Proceed thus: Pound cinnabar with white of eggs in a marble mortar for an hour. Afterwards dry it like potter's earth. Then form from thence pilules or small branches, as you will; lastly, dry them thoroughly, and anoint them externally, as you did the pearls, with white of egg. Dry them again, and thus they will become like native coral in appearance, but not in virtue.

It should also be known that the white of eggs by itself can be coagulated into a very fine varnish, into which coagulation Luna or Sol may be put.

There are many other and various transmutations, whereof I will tell you briefly, and by the way, those which I know and have experimented on. First, learn that any wood, if at a particular time it be put in the water of the salt of a gem, is converted into stone in a manner calculated to cause wonder. So, too, stones are transmuted into coals by Ætnean fire, and these are called stone coal.

In the same way glue is made from hides, paper from linen rag, and silk is produced out of linen with a very sharp lixivium made from lime and the ashes of woad. If the downy parts are taken from feathers and dressed with this lixivium, they can be spun and woven like cotton. Any oil or spermatic mucilage can be coagulated into varnish; any liquid into gum. All these are transmutations of natural objects: whereof we have now said enough, and therefore write our finis.

CONCERNING THE NATURE OF THINGS

BOOK THE EIGHTH

Concerning the Separations of Natural Things

IN the creation of the world, the first separation began with the four elements, when the first matter of the world was one chaos. From that chaos God built the Greater World, separated into four distinct elements, Fire, Air, Water, Earth. Fire was the warm part, Air only the cold, Water the moist, and, lastly, Earth was but the dry part of the Greater World.

Now, that you may learn our method in this Eighth Book as briefly as possible, you must know that we do not propose to treat herein concerning the Separation of the Elements in all natural things, since we have fully and perfectly taught concerning these arcana in our Archidoxa on the Separations of the Elements. But here we touch only on the separation of natural things,* where some one thing is singly, and by itself, materially and substantially separated and segregated, when two, three, four, or more have been mingled in one body, and yet only a single matter is touched and seen. And here it frequently happens that corporeal matter of this kind can be known by nobody, nor be designated by an express name, until the process of separation is instituted. Then sometimes from a single matter two, three, four, five, or more, proceed, as by daily experience in Alchemy is made evident. By way of example for you, there is electrum, which by itself is not a metal, but still conceals all the metals in one metal and body. If this, by alchemical art, be anatomised and separated, all the seven metals, and these pure and unmixed, proceed from it, namely, gold, silver, copper, tin, lead, iron, quicksilver, etc.

But in order to understand what separation is, you should know that it is nothing else but the segregation of one thing from another, whether two, three, four, or more have been mixed: I mean the segregation of three principles, as mercury, sulphur, salt, and the extraction of the pure from the impure, or of the pure and noble spirit and quintessence from the dense and

---

* Separation is grounded in heat, as in a faculty of digestion, whence, sometimes in one way, and sometimes in another, the ultimate matter is formed.—*Modus Pharmacandi*, Tract III. The office of the Archeus is the sequestration of the pure from the impure.—*De Morbis Tartareis*, c. 5. For unless there be separation in the greater world, there can be no metal, and unless there be separation in the smaller world, that is, in the microcosmos, which is man, there can be neither health nor disease, but an equable and perpetual disposition of all things.—*Chirurgia Magna*, Part III., Lib. 2.

elemental body ; and the preparation of two, three, four, or more from one : or the dissolution and liberation of things linked and bound together, which are by nature adverse, and perpetually act contrariwise one to the other, and go on doing so until they mutually destroy each other.

There are many and various modes of separation, all of which are not known to us ; but those among the soluble natural elements which have been investigated by us shall here be set down and described according to their species.

The first Separation of which we speak should begin from man, since he is the Microcosm, the lesser world, and for his sake the Macrocosm, the greater world, was founded, that he might be its Separator. But the separation of the Microcosm begins from death.* For in death the two bodies of man separate from each other, that is to say, the Celestial and the Terrestial, the Sacramental and the Elemental. One of these soars on high, like an eagle ; the other sinks down to the earth, like lead.†

The elemental body decays and is consumed. It becomes a putrid corpse, which, being buried in the earth, never again comes forth or appears. But the Sacramental body, that is, the sidereal and celestial body, does not decay, is not buried, occupies no place. This body appears to men, and is seen even after death. Hence we have spectres, visions, and supernatural apparitions. From these the Cabalistic Art was elaborated by the ancient Magi, which is treated of more at length in the books on the Cabala.‡

After this separation has been made, then, by the death of the man, the three substances separate one from the other, that is to say, the body, the soul, and the spirit, each wending its way to its own place, as to the ark from

---

* There are two kinds of death – one from the Yliadus, and one from the Ens. With that which comes from the Yliadus medicine may attempt to do battle ; with that which comes from the Ens it is useless to attempt to cope.—*De Tartaro*, comment. in Lib. II.

† It has, therefore, seemed good to me that man should first of all be described according to his nature and condition, so that it may become more clearly intelligible what is to be sought in the mortal body, that is to say, mere mortality, and what also is to be sought in the sidereal body, forsooth mere mortality. Afterwards we must become acquainted with the soul, which is by no means mortal, but is the eternal man. You must further know that the soul is flesh and blood, and that it consists of flesh and blood, but that there is a twofold flesh, namely, mortal and eternal. The mortal takes its essence from mortal flesh ; the eternal is perfect flesh and blood unto life eternal. Therefore if man considers within himself who and what he is, and what will be his future condition, he will thence readily understand that in this body, incarnate from the Holy Spirit, he shall see God, his Redeemer, and that whatsoever God our Redeemer operates in us, He does through the man of new generation, because that is not of a mortal but an eternal body. Only this body is secure from the devil. The second is from Adam, and is like a seed in water. The other body is suitable for the performance of works Divine, for a mortal body can accomplish nothing of those things which are celestial. It cares only for things earthly and things of the firmament, and it produces men skilled only in natural light. Hence God ordains man to gain a wider experience from that which is naturally formed, to pass from one to the other, and to emulate Nature. For in a new body and a celestial philosophy is life eternal. Death is inherent in natural strength, but life, on the contrary, consists in eternal strength. The instruction of Nature is from the earth, and she knows not God, except that she admires the Creator in man. Nor yet does man recognise God according to Nature or in Nature. But he who is born from on high is acquainted with supernal things. The first of these is Christ. All who are reborn in flesh and blood, conceived and incarnate from the Holy Ghost, do follow Him, and these same have the knowledge of things above. For they are from Him who cometh from on high. Hence there are two instructions, one of the earth earthy, the other from on high, which He imparts who also is from on high, from whom we derive, whose flesh and blood we are, etc. —*Philosophia Sagax*, Lib. II., c. 2.

‡ The sole work on the cabala which has been preserved in the name of Paracelsus, is a short treatise, which forms a detached portion of the book entitled *De Pestilitate*. It is not cabalistical in the sense which properly attaches to that term, nor does it exhibit any special acquaintance with that section of Jewish traditional literature to which it is referred in name. In its general outline it seems to be fairly in harmony with the great body of cabalistical cosmogony.

which it first of all came forth: the body to the earth, as the first matter of the elements; the soul to the first matter of the sacraments; and, lastly, the spirit to the first matter of the aërial chaos.

What has now been said concerning the separation of the Microcosm should also be understood of the greater world, which the mighty ocean has separated into three parts, so that the universal world is thus divided into three portions, Europe, Asia, and Africa. This separation is a sort of prefiguration of the three principles, because they, too, can be separated from every terrestrial and elemental thing. These principles are Mercury, Sulphur, and Salt. Of these three the world is built up and composed.

From this should be known the separation of the metals from their mountains, that is to say, the separation of metals and minerals. By the separation which is instituted in these, many come forth from one matter. You see that from minerals come forth metal, scoriæ, glass, sand, pyrites, marchasite, granite, cobalt, talc, cachimia, zinctum, bismuth, antimony, litharge, sulphur, vitriol, verdigris, chrysocolla, cœruleum or lazulum, auripigment, arsenic, realgar, cinnabar, fireclay, spathus, gyphus, tripolis, red earth, and other like things; and then of each one of these the water, the oils, the resins, the calx or ash, the Mercury, Sulphur, Salt, etc.

Vegetables in their separation give waters, oils, juices, resins, gums, electuaries, powders, ashes, Mercury, Sulphur, Salt, etc.

Animals in their separation give water, blood, flesh, fat, bones, skin, body, hair, Mercury, Sulphur, Salt, etc.

Whoever, therefore, boasts to be a separator of such natural things, needs long experience, and perfect knowledge of all natural objects. Besides this, he must be a skilled and practised alchemist, to know what is or is not

---

and it is briefly as follows. Earth, water, air, and fire have their origin from three things, which, however, are not to be regarded as of prior creation, for they are and have been fire, air, water, earth. The three have all proceeded from one mother. This mother was water. When the whole world was formed the Spirit of God was borne over the waters, for by the word *Fiat* water was first created, and thence all other creatures, animate and inanimate. These three are called, truly, sulphur, mercury, salt. These, therefore, are the true principle, these the true matter, out of which all animals and man himself are formed. Thus for perfect generation in all things there are three things required—spring, summer, and autumn. This is especially the case in man himself. Now, sulphur, mercury, and salt recognize two rulers. Salt has the Moon, and is thereby governed. It is also a subject of water, in which it is dissolved and liquefied. It is of autumn and winter. But the Sun is king and lord of sulphur, which is fervid, igneous, and dissolved in fire. Now, the Sun is the ruler of spring and autumn. But all things are nothing else save sulphur, mercury, and salt, which, further, are the most certain mark of every true physician. Salt is the body of autumn and winter, and sulphur of spring and summer. Salt gives form and colour to all creatures; sulphur gives body, increase, and digestion. These two are father and mother, from which mediating stars all creatures are produced. But mercury needs daily nourishment, and also continual augmentation from sulphur and salt. Know also that God has put much sulphur and salt into earth and water, and every creature, animate and inanimate, in water and earth, have their proper sulphur and salt, whence they receive nourishment and savour. Salt gives savour and form, sulphur odour and the power of putrefaction. The Sun and Moon assiduously labour to generate these three things copiously, and also to mature the same. The Sun and Moon are the parents of all creatures, while sulphur and salt are the seed. The seed is brought by the parents, and the fœtus, which is mercury, is born. The manner of the nativity of everything has its analogies in the great world. When the death of winter has passed, all things that are capable of receiving life are set in motion by the amenity of May, and all creatures are transported with singular delight, even as a pregnant woman who desires to bring forth. Now, every individual being has assigned to it its own May for its conception and birth, its respective autumn, and its peculiar harvest. So are there various springs, summers, and autumns, according to the infinite varieties of creatures. The doctrine of the three prime principles recurs continually in the writings of Paracelsus, and is elsewhere treated at considerable length in the text of this translation. At the same time, the obscurity which involves the subject seems to warrant the citation of passages such as the above, not exactly to cast light upon the question, but to exhibit the primeval mystery of Paracelsican philosophy with all its available variations.

combustible, what is fixed and what volatile, what does or does not pass into flux, and what thing is heavier than another. He must also have investigated in every object its natural colour, odour, acidity, austerity, acridity, bitterness, sweetness, its grade, complexion, and quality.

Moreover, it is necessary to know the grades of separation, that they consist of distillation, resolution, putrefaction, extraction, calcination, reverberation, sublimation, reduction, coagulation, pulverisation, lavation. By distillation, water and oil are separated from all corporeal substances. By resolution, metals are separated from minerals, and one metal from another, salt and fatness from others, and the light is separated from the heavier. By putrefaction, the fat is separated from the lean, the pure from the impure, the decayed from the undecayed. By extraction, the pure is separated from the impure, the spirit and the quintessence from their body, and the pearl from its dense body. By calcination are separated watery moisture, fatness, natural colour, odour, and whatever is otherwise combustible. By reverberation are separated colour, odour, inflammability, all moisture and wateriness, fat, whatever, in a word, there is in the substance which is fluxible or inconstant, and so on. By sublimation are separated from each other the fixed and the volatile, the spiritual and the corporeal, the pure from the impure, the Sulphur from the Salt, the Mercury from the Salt; and the rest. By reduction, the fluxible is separated from the solid, the metal from its mineral ore, one metal from another, metal from ash, the fat from that which is not fat. By coagulation is separated moisture from mere humidity, water from earth. By pulverisation are separated one from the other dust and sand, ashes and lime, the mineral from the animal and vegetable substance. All powders which are of unequal weight are separated by the process of jaculation, just as the chaff from the corn. By washing or ablution, ashes and sand are separated, the mineral from its metal, the heavy from the lighter substance, the vegetable and animal portion from the mineral, Sulphur from Mercury and Salt, Salt from Mercury.

But now, discarding mere theory, let us approach the practical work of separation, and come down to special details. It must be remarked that the separation of metals is rightly the first of all. For this reason, therefore, we will treat of that first.

CONCERNING THE SEPARATION OF METALS FROM THEIR MINERALS.

The separation of metals from their mineral ores can be effected in many ways, for instance, by ebullition or excoction, or by liquefaction with certain liquefying powders, as salt of alkali, litharge, *sal fluxum*, *fel vitri*, ash, sal gemmæ, saltpetre, etc. Put them into a vessel or dish, and let them liquefy in a furnace. Then the metal as a regulus will subside to the bottom of the vessel, but the matter of the mineral will float on the surface and will become ash. You must then work this metallic regulus in a furnace by means of a reverberatory, until all the pure metal is liberated without any dirt or ash. In this way, the metal is thoroughly digested and (so to say) refined or purged from all its dirt and scoria. Mineral ores of this kind will sometimes contain more than one metal, as is very often the case: for example, copper and silver, copper and gold, lead and silver, tin and silver, etc., may be found in one mineral ore; and the sign of this circumstance will be apparent if the metallic regulus, after being dealt with in the reverberatory, be resolved in a small vat after the proper fashion and mode. All the imperfect metals in it are separated, such as copper, iron, tin, lead, and so they pass away in smoke together with the lead (of which there should be added twice as much as of the regulus), and then only fine silver and gold remain in the vat. A similar result is attained, too, if the metallic rex is liquefied and poured upon the lumps. By that method of fusion the intermixed metals are separated. That which is best and weightiest always sinks to the bottom, while the lighter mounts above.

Two or three metals in admixture can also be separated in acrid and strong water, and one can be extracted from the other, and extended and resolved. But if both metals are resolved together, one of them in that resolution, as sand or calx, can be diverberated and depressed with salt according to the usual method, and so separated.

Besides this, metals can also be separated by fluxion according to the following process. Reduce the metals to a state of flux. When this has been done, throw in for every pound of the metal one ounce of the most perfectly sublimated and refined sulphur. It will there be burnt, and in the course of that operation it will attract to itself, on the surface, one metal, the lightest, whilst it will leave the heavier at the bottom. Let them stand in this way until cool. So in the one regulus two metals will be found, not, as before, mixed together, but opposed to each other, and separated by the sulphur as if by a wall, even as oil cuts off two bodies of water, so that that they cannot join and be commingled. In the same way sulphur acts with these metals. Sulphur, therefore, is an arcanum, worthy of the highest esteem.

Volatile and fugitive metals, such as gold and silver, if they are to be separated from their minerals, since they can neither be treated in the fire nor with strong waters, should be amalgamated, separated, and extracted by means of *Mercurius vivus*. Afterwards the *Mercurius vivus* must be abstracted and separated from the calx of the gold or silver by the grade of distillation.

In this way, other metals, too, as gold, silver, copper, iron, tin, lead, and substances prepared from these, as red electrum, white magnesia, aurichalcum, lead ashes, laton, casting brass, part with part, etc., and whatever transmuted metals of this kind there are, must be abstracted and separated from their extraneous substances by means of *Mercurius vivus*. For this is the nature and quality of *Mercurius vivus*, that it is amalgamated with metals and wholly united with them, but more quickly or more slowly with one than with another, according as the metal is more or less akin to its nature.

In this scale the principal is fine gold, then fine silver, the third lead, fourth tin, the fifth copper, and the last iron. So among transmuted metals the first is part with part, then lead ashes, next laton, afterwards casting brass, then red metal, and lastly white. Mercury, for its part, does not take more than one metal with which it is amalgamated. Afterwards, that amalgam must always be vigorously pressed out by means of goat's skin or a cotton rag, of which a strip is to be inserted, by which means nothing but *Mercurius vivus* alone will pass over. The metal which was attracted will remain on the skin or the rag like lime, and you can afterwards reduce it to a metallic body, by liquefying it with salt of alkali, or some other substance. By this device *Mercurius vivus* is separated from all the metals more quickly and conveniently than by the method of distillation. By this process with *Mercurius vivus*, in the hands of a skilled and active alchemist, all metals can be extracted and separated one from another in turn, after their calcination and pulverisation. In the same manner, with very small outlay of labour, tin, too, and lead can be separated from copper, or from copper vessels, from iron and steel covered with tin, and this without any fire or water, solely by the amalgam of *Mercurius vivus*, as we have said. Again, gold and silver leaf, as also every metal after being ground or pounded, and written with pen or pencil on cloth, parchment, paper, leather, wood, stone, or other material, can be resolved with *Mercurius vivus*, but so that afterwards the *Mercurius vivus* can again be separated and segregated from these metals.

The separation of metals in aqua fortis, aqua regis, and similar strong corrosives, is effected in the following manner: Let the metal which is mixed and joined to another be taken and reduced into very thin plates, or most minute portions. Let it be put into a separating vessel, and a sufficient quantity of common aquafortis be poured upon it. Let these stand, and both be macerated until all the metal is resolved into a transparent water. If it be silver, and contains gold in it, all the silver will be resolved into water, while the gold will be calcined and sink down to the bottom in the form of black sand. By this method the two metals, gold and silver, will be separated. But if you wish to separate the silver alone without distillation, and to drive that to the bottom like black sand, and to bring it back to calcination from its state of resolution, then put into that resolution a small copper plate, and thereupon the silver will sink in the water, and occupy the bottom of the glass vessel like snow, while it will begin gradually to consume the copper plate.

The separation of silver and copper by means of common aquafortis is accomplished in the following way: Reduce the copper which contains silver, or the silver which contains copper within itself, into very thin plates, or into grains; put it into a glass vessel, and add as much common aquafortis as necessary. In this way the silver will be calcined, and will go to the bottom in the form of white lime, while the copper will be resolved and converted into transparent water. If this water, together with the resolved copper, be abstracted through a glass funnel from the silver calx into a separate glass vessel, then the resolved copper can be reverberated with common rain or river water, or with hot salt water, so that it will occupy the bottom of the glass vessel like sand.

The separation of hidden gold from any metal is effected by the degree of extraction through aqua regis; for this water does not approach for the purpose of resolving any metal but fine gold alone.

This same aqua regis also separates fine gold from gilded clenodia. If it be smeared over these, it wipes away and sunders the gold.

Moreover, also, two metals mixed together can be separated one from the other with a cement by the degree of reverberation, especially if they are not in a similar degree of fixation, as iron and copper. A metal which has very little fixation, such as tin and lead, is altogether consumed in the cement by the degree of reverberation. The more fixed any metal is the less is it affected or consumed by the cement.

It should be known, too, that fine gold is the most fixed and perfect of all metals, and can be consumed by no cement. Next to this is fine silver. But if gold and silver be mixed together in one body, which is generally called "part with part," or if silver contains gold, or gold silver, in itself—if these mixtures, I say, be cemented and reverberated together, then the gold always remains entire and inviolate, while the silver is consumed by the cement, and is extracted from the fine gold; and so is copper from silver or iron, or tin from copper and iron, or lead from tin; and so on in order with the others.

Concerning the Separation of Minerals

So far we have explained the separation of metals from their earth and matter, and of one metal from another; and have shewn how it was to be done, using the greatest brevity consistent with accuracy, and following the alchemical art and practical experience. Now, next in order, it will be necessary also that we treat of those things out of which metals grow and are generated, such as are the three principles, Mercury, Sulphur, and Salt, and other minerals, among which is found the first essence of metals, that is, the spirit of metals, as is evident in marchasites, granites, cachimiæ, red talc, lazurium, and the like. In these the first essence of gold is found by the degree of sublimation. So, too, in white marcasite, white talc, auripigment, arsenic, litharge, etc., the first essence of silver is found. In cobalt, zinctum, etc., the first essence of iron. In zinctum, vitriol, atramentum, verdigris, etc., the first essence of copper. In zinctum, bismuth, etc., the first essence of tin. In antimony, minium, etc., the first essence of lead. In cinnabar is found the first essence of silver.

Concerning this first essence, it should be known that it is a fugitive spirit, still existing in a volatile state, as a child lies hidden in the womb of its mother. It is sometimes assimilated to a liquid, sometimes to alcohol. Whoever, therefore, is anxious to have the prime essence of any body, and to separate it, needs great experience and knowledge of the Spagyric Art.

If he has not diligently laboured in alchemy it will avail him nothing, and his labour will be in vain. How the first essence is to be separated from all mineral bodies has been sufficiently explained in the books of the Archidoxis, and need not be repeated here. But as to the separation of minerals, it should be remarked that many things of this kind are separated by means of sublimation, as the fixed from the non-fixed, spiritual and volatile bodies from the fixed, and so throughout all the divisions, as is detailed in the case of metals. With all minerals the process is one and the same, through all the degrees, as the Spagyric Art teaches.

Concerning the Separation of Vegetables

The separation of those things which grow out of the earth and are combustible, such as fruits, herbs, flowers, leaves, grasses, roots, woods, etc., is also arranged in many ways. By distillation is separated from them first the phlegma, afterwards the Mercury, after this the oil, fourthly their sulphur, lastly their salt. When all these separations are made according to Spagyric Art, remarkable and excellent medicaments are the result, both for internal and external use.

But when laziness has grown to such an extent among physicians, and all work and every pursuit are turned only to insolence, I do not wonder, indeed, that preparations of this kind are everywhere neglected, and that coals stand at so low a price. If smiths could do without coals for forging and fashioning metals as easily as these physicians do without them in preparing their medicines, there is no doubt that all the coal merchants would have been before now reduced to extreme beggary. In the meantime, I extol and adorn, with the eulogium rightly due to them, the Spagyric physicians. These do not give themselves up to ease and idleness, strutting about with a haughty gait, dressed in silk, with rings ostentatiously displayed on their fingers, or silvered poignards fixed on their loins, and sleek gloves on their hands. But they devote themselves diligently to their labours, sweating whole nights and days over fiery furnaces. These do not kill the time with empty talk, but find their delight in their laboratory. They are clad in leathern garments, and wear a girdle to wipe their hands upon. They put their fingers among the coals, the lute, and the dung, not into gold rings. Like blacksmiths and coal merchants, they are sooty and dirty, and do not look proudly with sleek countenance. In presence of the sick they do not chatter and vaunt their own medicines. They perceive that the work should glorify the workman, not the workman the work, and that fine words go a very little way towards curing sick folks. Passing by all these vanities, therefore, they rejoice to be occupied at the fire and to learn the steps of alchemical knowledge. Of this class are: Distillation, Resolution, Putrefaction, Extraction, Calcination, Reverberation, Sublimation, Fixation, Separation, Reduction, Coagulation, Tincture, and the like.

But how all these separations are made according to Spagyric and

Alchemical Art by the help of distinct degrees has before been said generally, and to repeat the same thing here anew is vain. To go on to specialities and briefly explain the practical method, let it be known that all cannot be separated by one and the same process; that is to say, the water, spirit, liquid, oil, etc., from herbs, flowers, seeds, leaves, roots, trees, fruits, woods, according to the grade of distillation.

Herbs require one process, flowers another, seeds another, leaves another, roots another, trees, stalks, and stems another, fruits another, woods another, etc. And in this grade of distillation the four degrees of fire have to be considered. The first degree of fire is the *Balneum Mariæ*. This is the distillation made in water. The second degree of fire is distillation made in ashes. The third is in sand, the fourth in free fire, as also distillation is generally made by aqua fortis and other violent waters. Herbs, flowers, seeds, and the like, require the first degree of fire. Leaves, fruits, etc., need the second. Roots, branches, and trunks of trees, etc., require the third. Timber and the like require the fourth. Each of these substances must be minutely cut up or pounded before being brought into the still. So much has been said as to the distillation of waters and vegetable substances. As regards the separation and distillation of oils the same process must be followed as we have spoken of in the separation of waters, except that, for the most part, they have to be distilled by descent. They cannot, like waters, ascend in the still; therefore, in this case the process has to be changed. Liquids, however, are not separated like waters and oils, by distillation, but are squeezed out from their corporeal substances under a press. And here it should be known that some oils, in like manner, just as liquids, are squeezed out from their corporeal substances and separated by means of the press for this reason, that they can bear scarcely any combustion or heat of the fire, but acquire therefrom an unpleasant odour. Of this kind are the oils of almonds, nuts, hard eggs, and the like. This also is to be noted, that all oils, if they are prepared or coagulated according to Spagyric and Alchemical Art, pour forth varnish, electuary, gum, or a kind of resin, which might also be called a sulphur; and if the species left in the still were calcined and reduced to ashes, alkali could be extracted and separated from them with simple warm water alone. The ash which is left is called dead earth, nor can anything more be produced or separated from it.

Concerning the Separation of Animals

It is necessary to preface the separation or anatomy of animals by shewing how the blood, flesh, bones, skin, intestines, etc., stand each by itself, and then how each is separated by Spagyric Art. In this part the separations are principally four. The first draws forth from the blood a watery and phlegmatic moisture. For when the blood has been separated in this manner, according to the process handed down in the book on Conservations,[*] an

---

[*] That is, the *Preservations of Natural Things.—De Natura Rerum*, Book III.

excellent Mumia* comes forth, and a specific so potent that any fresh wound can be healed and consolidated in twenty-four hours by a single ligature.

The second separation is that of fat from flesh. This fatness being separated from human flesh, a most excellent balsam is produced, allaying the pains of gout, of contraction, and others of a like nature, if the members affected be anointed with it while warm. It is also useful for convulsed tendons of the hands or feet, if they are daily anointed with it. It further cures the itch, and all kinds of leprosy. This, therefore, is the chief surgical specific, and of the very first efficacy in all accidents and wounds.

The third separation is that of the watery and phlegmatic moisture with fatness extracted from the bones. For if these two are separated from human bones by Spagyric Art, and according to the degree of distillation, and if, moreover, by the method of calcination they are reduced or burnt to a white ash, and if, lastly, these three be again united in the proper way, so that they are like to butyrus, there will be formed a wonderful arcanum and specific, with which you will be able, without pain, to entirely cure any fracture of the bones after binding them up only thrice, provided only that you treat the fracture by setting it according to the rules of surgical science, and then put on the specific in the form of a plaster. The same also thoroughly cures wounds of the skull, or any contusion of the bones, in the shortest possible time.

The fourth and last separation is that of resins and gums from the skin, intestines, and tendons. For the resin is extracted and separated from these by the degree of extraction according to Spagyric Art, and when coagulated in the rays of the sun it comes out as a clear and transparent paste. When this paste has been prepared, extracted, and separated from the human body according to the prescribed method, a most excellent styptic arcanum and specific issues forth, with which a wound or ulcer can be quickly healed and the lips brought together, just as two sheets of paper are stuck together with paste, if only you apply to the wound two or three drops of that resolved substance. This arcanum, too, is of singular efficacy for burns, and falling off or roughness of the nails, if it be spread over them with a feather. In this way the bare flesh will be covered over with a cuticle.

Many other separations also of one thing or another might be recounted here; but since we have made mention of them in other places, it would be in vain idly to repeat them now.

---

* Mumia is that which cures all wounds, that is, sweet mercury. For mercury is extracted both in a sweet and bitter form. The former is adapted to wounds and the latter to ulcers. Mumia is the liquor diffused through the whole body, the limbs, etc., with the strength that is required. It is divided as follows: in flesh, according to the nature of the flesh; in bone, according to the nature of the bone; in the arteries and ligaments, according to their nature; and so also in the marrow, the veins, and the skin. Hence it follows that the mumia of the flesh cures wounds of the flesh, the mumia of the ligaments cures wounds of the ligaments, etc. Thus the body which has sustained an injury carries its own cure with it; the mumia of the aged, however, is deficient in virtue and strength. The corruption of the mumia, which is often occasioned by the mistakes of ignorant physicians, impedes the cure of wounds. . . . The nobler the animal organism is, by so much is the mumia of the organism enhanced in power and efficacy. The medicaments which benefit wounds perform this operation by attracting the mumia to the place where its office is required.—*Chirurgia Minor*, Lib. I., c. 1.

Here it is only necessary to write down that which we have not mentioned elsewhere.

But at last, at the end of all earthly things, will be brought about the final separation, in the third generation, on that great day whereon the Son of God shall come in His majesty and glory, and before Him shall be borne, not swords, chains, diadems, sceptres, and treasures, or other royal jewels, with which princes, kings, and Cæsars bear themselves pompously, but His Cross, and crown of thorns, and nails piercing His hands and feet, and the spear with which His side was wounded, and the reed and sponge on which they stretched out that which they gave Him to drink, and the rods with which He was scourged and beaten. No crowd of horsemen with far sounding drums shall accompany Him; but the four trumpets shall be blown by the angels towards the four parts of the earth, and at their tremendous sound all who are among the living shall be slain, and these together with the buried dead shall immediately rise again.

For a voice shall be heard, "Rise, ye dead, and come to judgment!" Hereupon the Twelve Apostles shall sit down on thrones prepared from the clouds, and shall judge the twelve families of Israel. In that place the Holy Angels shall separate the bad from the good, the cursed from the blessed, the goats from the sheep. Then the cursed shall be thrown down like stones and like lead; but the blessed shall fly like eagles. Then from the tribunal of God shall issue forth a voice to those standing on the left hand, "Go away, ye cursed, into everlasting fire, prepared from eternity for Satan and the devils. For I was hungry and you did not feed Me; I was thirsty and you gave Me no drink; I was sick, and a prisoner, and naked, but you did not visit Me, did not set Me free, did not clothe Me. In a word, you were not touched with pity for Me. Therefore, here you shall meet with no pity!" Contrariwise to those standing on the right side it shall say thus: "Come, ye blessed and elect, into the Kingdom of My Father, which has from the beginning been prepared for you and for all the angels. For I was hungry and you gave Me food; I was thirsty and you gave me to drink; I was a stranger and you received Me; I was naked and you clothed Me; I was sick and you visited Me; I was in prison and you came to Me. So will I receive you also into My Father's house, in which are the many mansions of the saints. You pitied Me; and so I will pity you!"

When all these things are finished and done, all the elementary subjects shall return to the first matter of the elements, and shall be turned about for eternity, yet never consumed. On the contrary, all sacramental creatures shall return to the primal matter of the sacraments, that is, they shall be glorified, and in eternal joy they shall worship God their Creator, from universe to universe, from eternity to eternity. Amen.

CONCERNING THE NATURE OF THINGS

BOOK THE NINTH *

Concerning the Signature of Natural Things

IN this book, our first business, as being about to philosophise, is with the signature of things, as, for instance, to set forth how they are signed, what signator exists, and how many signs are reckoned. Be it known, first of all, then, that signs are threefold. The first things signed man signs; the second Archeus signs; the third the Stars of the Supernaturals. In this way, then, only three signators exist, Man, Archeus, and the Stars. Moreover, it should be remarked that the signs signed by man carry with them perfect knowledge and judgment of occult things, as well as acquaintance with their powers and hidden faculties.

The signs of the stars give prophecies and presages. They point out the force of supernatural things, and put forth true judgments and disclosures in geomancy, chiromancy, hydromancy, pyromancy, necromancy, astronomy, the Berillistic art,† and other astral sciences.

Now, in order that we may explain all the signs as correctly and as briefly as possible, it is above all else necessary that we put forward those whereof man is the signator. When these are understood you will more rightly attain to the others, whether natural or supernatural. For instance, it is known that

---

* Note with reference to the books *De Natura Rerum*. In most editions, seven books only are included under this heading, but the Geneva folio, from which the translation has been made, gives nine as above. In the other cases the treatises on *Separations* and *Signatures* are regarded as independent works. There can be no doubt that the classification adopted by the Geneva folio is correct, for in method and design these treatises are integrally connected with the rest of the *Nature of Things*.

† Among the branches of astronomy there is one which is called Nigromancy. It has gained this name because it is exercised by night rather than by day. This science is everywhere and by all rejected and cursed as diabolical, yet only by those who are ignorant of it. For this science is a natural one, born of the stars. But above all notice the property of beryls. In these are beheld the past, present, and future. Let no one be surprised at this, because the constellation impresses the image and similitude of its influence upon the crystal in the likeness of that concerning which inquiry is made. This must take place by a compulsion of the constellation, as is recorded in magic. As the splendour of the sun flows in upon the crystal, so the constellation pours it from above upon the object. Moreover, all things which exist in Nature are known to the constellations, and when the stars are subject to man, he can bring them to such obedience that they favour his will. It is universally boasted concerning faith that it can accomplish many things. This is, indeed, not far from the truth, for Christ Himself bears witness to it. And since faith is an operative principle it is evidently nothing else but a virtue and an efficacy. For virtue works in a word, and words make the dead alive. In a similar fashion, what else is there in the stars than that by faith in Nature they are conquered? And as by the word of faith the mountain is cast into the sea, know that it is owing to natural faith that by a word the stars are brought down, so that they may perform their operation according to our imagination, for he is wise who rules the stars—he is wise, I say, who can bring their virtues under his rule, for in this manner are constituted visions in glasses, mirrors, waters, and the rest, according to the quality of the power, and of the union made in conception.—*Explicatio Totius Astronomiæ*.

Jews wear a yellow sign on their cloak or on their coat. What is this but a sign by which anybody who meets him may understand that he is a Jew? So, too, the lictor is known by his parti-coloured tunic or armlet. So, too, every magistracy decks its ministers with its own proper colours and adornments.

The mechanic marks his work with its peculiar sign, so that everyone may understand who has produced it. For what purpose does the courier carry the insignia of his master or his city on his garment, except that it may be clear he is a messenger, that he serves one or another, that he comes from one place or another, and so thus procures for himself a safe passage?

So, too, the soldier carries a sign or symbol, black, white, green, blue, or red, that he may be distinguished from the enemy. Hence it is known that one is on the side of Cæsar, or of the kings; that one is an Italian, another a Gaul, etc. These are signs which relate to rank and office; and many more of them might be enumerated. But, nevertheless, since we have proposed to ourselves to describe other signs of natural and supernatural things, we will not overload our book with those signs that are foreign to our purpose.

It is necessary more clearly to explain those signs which man affixes, and which lead to a knowledge, not only of rank, office, or name, but also of discrimination, intelligence, age, dignity, degree, Next in order, with regard to money, it should be remembered that every coin carries its proof and sign by which it may be known how much that coin is worth, to what power it belongs, where it circulates and is passed. Here comes in the German proverb: "Nowhere is money more acceptable than where it is struck."

The same is to be understood of the customary signs which are affixed by jurors and those appointed for the purpose, after due inspection has been previously made. An instance of this is found in the cloths marked with distinguishing signs by which it may be known that on examination they have been found good and genuine. Why is a seal appended to letters except that there may be a certain force which none will dare to violate? The seal is the confirmation of the letter which gives it authority among men and in trials. A receipt without a seal is dead, useless, empty.

In the same manner, by a few letters, names, or words, many things are designated, just as books which, though lettered outside with only one word, in that way signify their contents.

Such, too, is the condition of the vessels and boxes in drug-shops, which are all distinguished by peculiar names or labels affixed to them. If that were not done, who could distinguish one from the other among so many different waters, liquors, syrups, oils, powders, seeds, ointments, and the like? In the same way, too, the alchemist in his laboratory marks with their own proper names and labels, all the waters, liquors, spirits, oils, phlegmata, crocuses, alkalis, powders, and then all the different kinds of these, one by one, so that he can select from among them whatever he wants. Without this safeguard it is impossible to remember each separately.

Thus also rooms and buildings constructed by men can be signed with a number, so that the age of any of them can be at once known by the first glance at the number affixed.

I determined to lay these signs before you in order that when you had mastered these, I might be more readily understood by you in the rest, and that the meaning of each might be plainer and more evident.

Concerning Monstrous Signs in Men

Many men come to the light deformed with monstrous signs. One man has a finger too many, another a finger too few; and the same may be the case with the toes. Another brings with him from the womb a distorted foot, arm, back, or other member; another has a weak or a hunched back. So also there are born hermaphrodites, androgyni, men, that is to say, possessing both pudenda, male as well as female, and sometimes lacking both. Of monstrous signs like this I have noted many, both in males and females, all of which are to be regarded as monstrous signs of secret sins in the parents. Hence has grown up the old proverb: "The more distorted, the more wicked"; and again: "lame limbs, lame works." These are signs of vices, and rarely denote anything good.

Just as the hangman brands his sons with degrading signs, so also bad parents mark their offspring with mischievous supernatural signs that people may be more cautious when they see the example of wicked men who carry tho stigmata in their forehead or cheeks, or in defective ears, fingers, hands, eyes, or tongues.

Each of these signs of infamy designates some particular vice. If there is a stigma burnt into the face of a woman, or if there be a lopping off of the ears, it, for the most part, indicates theft. Loss of fingers tells of cheating gamblers. The loss of a hand indicates violators of peace. That of two fingers points out perjury. The loss of an eye indicates that people engage in sharp and subtle crimes. The cutting off of the tongue designates blasphemers and calumniators. So you can recognise those who are called mamelukes, or deniers of the Christian religion, by a cross burnt into the heel of their feet, because they denied Christ their Redeemer.

But let us dismiss these matters and return to the monstrous signs brought about by wicked parents. It should be known that all monstrous signs are not produced only by the progenitor, but frequently also from the stars of the human mind, which perpetually at all moments, with the Phantasy, Estimation, or Imagination, rise and set just as in the firmament above. Hence, through fear or fright on the part of those who are pregnant, many monsters are born, or children signed with marks of monstrosity in the womb of their mother. The primary cause of these things is alarm, terror, or appetite, by which the imagination is aroused. If the pregnant woman begins to imagine, then her bosom is borne round in its motion just as the superior firmament, each movement rising or setting. For, as in the case of the greater firmament, the stars of the microcosm also move by imagination, until there comes a sort of bounding, in which the stars of the imagination produce an influence and an impression on the pregnant woman, just as though one should impress a seal or stamp a piece of money. Whence those signs and birthmarks derived from the lower stars are called "impressions." About these matters many men have philosophised and tried to form from them a solid judgment, without being able to do so. For these things adhere to, and are impressed on, the fœtus in proportion as the stars of the mother press frequently or with violence on the fœtus, or the desire of the mother is not satisfied. If the mother, for instance, longs for this or that kind of food, and is unable to get it, the stars are, as it were, suffocated in themselves, and perish. That desire abides with the unborn child throughout all its life, so that it is impossible ever to satisfy it. The same reason explains other matters, too, which we must not discuss here at too great length.

Concerning the Astral Signs in the Physiognomy of Man

The signs of physiognomy derive their origin from the higher stars. This science of physiognomy was held in the highest esteem by our ancestors, and among the first by the heathens, Tartars, Turks, and the rest, whose custom it is to sell men and slaves; nor was it altogether lost among Christians. Many errors, however, which had not yet been perceived by anyone, crept in with it when every fool and every clown took upon himself to judge offhand about everything. It is marvellous that these mistakes were not found out from the evil deeds and limited powers of the men themselves.

Now if anyone at this point argues against us, saying, "The signs of physiognomy are from the stars, but no one has the power of compelling or urging on the stars," he does not speak amiss. Yet, this difference must be noted at the outset, that the stars compel one and do not compel another. This ought to be known, who it is that can rule and coerce the stars, and also who is governed by the stars. The wise man can dominate the stars, and is not subject to them. Nay, the stars are subject to the wise man, and are forced to obey him, not he the stars. The stars compel and coerce the animal man, so that where they lead he must follow, just as a thief does the gallows, a robber the wheel, a fisher the fishes, a fowler the birds, and a hunter the wild beasts. What other reason is there for this, save that man does not know or estimate himself or his own powers, or reflect that he is a lesser universe, and has the whole firmament with its powers hidden within himself? Thus man is called animal and unwise and the slave of all earthly things, when, nevertheless, he received from God in Paradise the privilege of ruling over and dominating all other creatures, and not of obeying them. So it was that God created man last, when all other things had been made before him. This right was afterwards lost by the Fall. Yet, the wisdom of man was not made servile, nor did he lose his freedom. It is right, then, that the stars should follow him and obey him, not he the stars. And although he is the son of Saturn, and Saturn is his parent, still he can withdraw himself from him, and so conquer him that he becomes the offspring of the Sun, and can thus subject himself to another planet, and make himself its son. It happens much in the same way to him as to the miner, who for a long time has hired out his labour to the master of the mines, and managed his department righteously at peril of his life. At length he holds this discourse with himself: "Are you going to spend all your life underground and endanger your body, nay, your very existence, by continuous labours? I will seek release from my master, and follow another where my life shall flow pleasantly on, where I shall have plenty of food and drink, where my garments may shine, where no work and much reward shall be given to me, and where I shall not be oppressed by the mountain overhanging me." In this way he can constitute himself lord where otherwise he would remain all his life a slave and mercenary, wasting away with hard labour and scanty food.

Moreover, as you have now perceived that man rules the stars, and can free himself from a malignant planet and subject himself to another better one, from slavery pass by virtue to freedom, and rescue himself from the prison of an evil planet, so also the animal man who is the son of Sol, Jupiter, Venus, or Mercury, can withdraw himself from that benignant planet and subject himself to Saturn or to Mars. This man is like one who, fleeing from a college of religions, and being tired of their soft life, becomes a soldier, or in other respects a man of no esteem, who must afterwards spend all his life in pain and care. Such, too, is the rich man, who, out of mere levity, wastes all his goods unjustly, gambling, feasting, keeping evil company, until at last, when all is gone, he comes to want, and in miserable conflict with discreditable poverty he deservedly rouses laughter and contempt in all, so that you hear even from the boys in the streets: "Look at yonder worthless man, who, when he could have been master, scorned dominion and preferred to be a slave, a beggar, a servant of servants, so that he cannot now even aspire to his dominion." It is to this that a bad star or a bad parent has led him. Had he not been foolish and wicked, he would not have left to the stars so unquestioned a dominion over himself, but he would have struggled against them. And, although of himself he had not known how to fight against his stars, yet he could have turned his mind to the examples of other men, thinking thus within himself: "See how rich this man was; but by foolish and shameful enterprises he involved himself in mere poverty!" Again, "This or that man lived splendidly, and without any great bodily labour; but, though having got good food and ample pay, he was not able to bear his fair fortune. Now he has to live frugally and sordidly. In place of wine he has to drink water, and whilst his daily labour increases his income is diminished." How often must such a man thus address himself: "What have I done? How have I thrown myself headlong down by wasting prodigally the substance I had collected and acquired? Who will restore it to me? If I could only recover what I have lost, quite another mode of life should be begun, and so I would learn wisdom from my own loss, and compensate for my evil deeds by wiser counsel for the future." But it is well to know that nobody grows wise from his own loss. He who is wise has learnt wisdom from another's loss, not from his own. He who has wasted his substance once will waste it again. He who perishes once, perishes again. He who once throws the dice will throw them again. The man who has once thieved and cheated the gallows tries to steal a second time. So he thus thinks within himself: " My undertaking has succeeded once and again, why should it not succeed a third and a fourth time? If God has once restored what had perished, He will restore it a second and a third time. If in my first misery I have not been deserted, I shall not be in my second or my third." All this does the animal man who is the servant and slave of the stars; who is swayed backwards and forwards by the stars like a reed in the waters. This is the reason why he has to spend his life in misery and so to die in dishonour. Who, then, would bear so disgraceful a slavery and not extricate himself from so squalid a prison? For by bringing to bear his own wisdom, and with the help of his star, anyone can free himself. Look at the matter thus: A fowler, relying on his own prudence, and by the assistance of his star conquering another star, has no need to pursue birds, for the birds will follow him, and though their nature rebel they will fly together to unaccustomed places. In like manner, to the fisherman at his ease and relying on his wisdom, the fishes will swim of their own accord, so that he can catch them with his hands. The hunter exerting his wisdom by means of his star so collects the wild beasts that he has no need to pursue them; they pursue him, contrary to the guidance and impulse of Nature. And so also with other living creatures.

In order to grasp these things it must be remembered that stars are of two kinds, terrestrial and celestial. The former belong to folly, the latter to wisdom. And as there are two worlds, the lesser and the larger, and the lesser rules the larger, so also the Star of the Microcosm governs and subdues the celestial star. God did not create the planets and stars with the intention that they should dominate man, but that they, like other creatures, should obey him and serve him. And although the higher stars do give the inclination, and, as it were, sign man and other earthly bodies for the manner of their birth, yet that power and that dominion are nothing, save only a predestined mandate and office, in which there is nothing occult or abstruse remaining, but the inner force and power is put forth through the external signs.

But to return to our proposition concerning the physical signs of men: know that these are twofold, like indeed in outward form, but dissimilar in power and effect. Some are from the upper stars of heaven; others from the lower stars of the microcosm. Every superior star signs according to birth up to mid-age. That signature is predestined, and is not without its own peculiar force. It is attested by a man's nature and condition of life. But whatever the lower star of the microcosm signs from birth has its origin from the father and the mother, as often as the mother affects by her imagination or appetite, her fear or dread, the unborn child in her body with supernatural signs by means of their own close contact. These are called mothers' marks, or uterine marks. We have spoken of these before, so spare ourselves the labour of repetition, since it is our purpose to treat of physiognomical signs alone, among which we understand those signs of men the like whereof neither the father nor the mother have borne in their body. Of this class are black or grey eyes, too small or too large; a long, crooked, or sharp-pointed nose; hollows in the jaws, high cheekbones, a flat or broad nose, small or large ears, a long neck, an oblong face, a mouth large and drawn down; hair thick or fine, abundant or scanty, black, yellow, or red, etc. Of these signs, if one or more appear in a man, be sure that he will not lack the qualities signified thereby. Only you must judge them according to the rules of physiognomy, and have had experience in the art of signature, according to which you can judge a man by outward signs.

Descending, then, to the practical portion of our subject, let us repeat a few of these signs and their signification.

Black eyes not only denote a healthy constitution, but also, for the most part, a constant mind free from doubt and fear, healthy and hearty, truthful and loving virtue.

Grey eyes are the sign of a crafty man, ambiguous and inconsistent. Weak eyes denote good counsels, clever and profound deliberations, and so on. Bright eyes, which turn up, down, and to both sides, denote a false, clever man, who cannot be deceived, faithless, shirking work, desirous of ease, seeking to gain his livelihood in laziness, by gambling, usury, impurity, theft, and the like.

Small eyes, somewhat deeply sunk, indicate weak sight, and often impending blindness in old age. At the same time, they denote brave men, bellicose, crafty, and adroit, factious, capable of enduring misfortune, and whose departure from life is, for the most part, of a tragic character.

Large eyes denote a greedy, voracious man, especially if they project far out of the head.

Eyes which are constantly winking indicate weak sight, a timid and careful man. Eyes which move quickly hither and thither, under the glance of men, indicate an amorous heart, provident, and of quick invention.

Eyes continually cast down show a reverential and modest man.

Red eyes show a bold, brave man.

Glittering eyes, which do not move readily, point out a hero, a high-minded, brave, quick man, formidable to his foes.

Large ears indicate good hearing, retentive memory, attention, diligence, a healthy brain and head.

Depressed ears are a bad sign. For the most part they point out a man who is malicious, fraudulent, and unjust. They indicate bad hearing, treacherous memory, and a man who readily exposes himself to danger.

A long nose curved downwards is a good sign. It denotes a strenuous, provident man, occult and cruel, but still just.

A flat nose indicates a malignant man, false, lustful, untruthful, inconstant.

A pointed nose indicates a changeable person, given to mockery.

A long nose shews a man slow in business, yet of good odour.

Hollow cheeks denote a talkative, contemptuous, contentious person.

An oblong chin, with a long face, shews an irritable man, one who is slow at his work.

A cleft chin shews a faithful man, officious, of abstruse and diversified speech; a man who says one thing and means another; quick at anger, yet repenting of his passion; ingenious and inventive.

A large, wide mouth shews a gluttonous man, insipid, fatuous, shameless, and fearless. A small mouth indicates the contrary.

Lips drawn together, when the upper is larger than the lower, shew an irritable man, pugnacious, courageous; yet for the most part of heavy, unchaste character, like a pig.

Lips larger below shew a dense, stupid, slow person.

Concerning the hair of the head or beard, the signs are not very plain, since experience teaches us that this can be marvellously varied according as it is black, yellow, red, or white, and hoary, or curled. So, too, hair is rendered soft or hard according to people's wish. Hence it is that many persons, who are in other respects well-skilled in physiognomical science, are woefully deceived when they rashly pass judgment from the hair, imputing to the stars what should rather be ascribed to men. Still it cannot be denied that hair firmly fixed on the head shews good health, both of the head and of the whole body. This is why people who buy horses pluck their tails so as to judge of their soundness. So swine are judged by their bristles, fish from their fins and scales, a bird by its feathers, and so on.

If the neck is unusually long, transcending the limits of Nature, it denotes a careful man, prudent and attentive.

Broad shoulders and back shew a man who is strong for carrying and moving things. Muscular arms also shew a man who is strong and robust in beating, thrusting, throwing, and the like.

Hard hands bespeak a laborious, mercenary man; soft hands, the contrary.

A short body and long legs denote a good runner, one who is easily satisfied with food and drink, but generally a man of somewhat short life.

Large and conspicuous veins in a man below mid age signify that he is full of blood and bodily juices; but above middle age they denote a sickly man who is still, however, vivacious.

With reference to manners and gesture, a man cannot be so easily known or judged from these. Experience teaches us that these can be changed every moment, so as to deceive the signator, and lead him to an erroneous judgment.

This is what astronomers hitherto have not observed with sufficient accuracy. The signator's business is not always to look at the manners and actions, but rather at other bodily signs which are fixed, and cannot by any artifice be counterfeited or changed. For if red hair, motion of the forehead and eyebrows, frequent agitation of the mouth, strong and deliberate step, and light spirits, indicate of necessity a generous, active man, or soldier, such as any one could easily shew himself by his own activity, and so stand better when put to the proof, and command higher pay, so, likewise, must judgment be passed on other manners which betoken wisdom, folly, truth, falsehood, fortune, victory, and the rest.

Concerning the Astral Signs of Chiromancy *

Concerning the signs of chiromancy it should be held that they arise from the higher stars of the seven planets, and all of them ought to be learnt and judged from the seven planets. Now, Chiromancy is a science which not only inspects the hands of men, and from their lines and wrinkles makes its judgment, but, moreover, it also considers all herbs, woods, flints, earths, and rivers—in a word, whatever has lines, veins, and wrinkles. But neither is this science free from its errors, which astronomers have alleged against it. For they have assigned the fingers of both hands to the planets and the principal stars, when, notwithstanding, there are on one hand only five fingers but on both hands ten, while the planets are only seven in number. How can these things be made to agree? Now, if there were seven fingers on each hand, then it might be possible to assign a finger to each of the planets. It happens, indeed, very often that a man only has seven fingers on his two hands, the others being lost by some accident. But still the stumps exist, and, moreover, the persons were not born in this way, so this matter has no relevance here. Besides, if it did so happen that a man was born with seven fingers either on one hand or on both, that would be a monstrous birth, not according to Nature, and therefore not to be assigned to the stars. So here, again, no comparison can be instituted. It would have been better, then, that the planets should cast lots and see which two ought to retire. This, however, could not be done, because the planets had neither dice nor lots up in the firmament; so one wonders who took it upon him to allot the planets by name, giving the thumb to Venus, the index finger to Jupiter, the middle

---

* It is a great error to suppose that chiromancy is concerned only with the hands, for it includes the significance of the lines upon the entire body. Nor is it confined to the body of man, for it deals also with the trunks of trees, and with the tracery upon the leaves of trees. Every peculiarity of line, whether in leaves or in human hands, has its special meaning. No man deserves to be called a doctor who is ignorant of chiromancy, because, for example, the presence upon the hand of those lines which are called *linea architecta*, indicate that the person will be likely to die of the colic; but then there are certain leaves which possess corresponding lines, and these leaves are the cure of colic. So also the *linea ancora* is the line of apoplexy, and this line is found in the acorus (*i.e.*, the sweet flag), which is a medicine of apoplexy. . . . Thus by the same sign Nature indicates the existence of the disease and its remedy. But the physician who is ignorant of the sign is ignorant of everything. But as physiognomy is both outward and inward, so there is an internal and external chiromancy, and that which is without is an evidence of that which is within.—*Duo Alii Libri de Podagricis Morbis*, Lib. I. I have frequently indicated that chiromancy is the inventress of arts, if it be cabalistically treated.—*De Peste*, Lib. II., *Præf.*

finger to Saturn, the ring finger or medicus to the Sun, and the little finger to Mercury. Meanwhile, Mars and the Moon were, so to say, banished. Would one be surprised, then, if in righteous indignation Mars bade his sons kill that allotter, or keep up continual strife with him: or who would wonder if the Moon weakened his brain, or took his wits away altogether? And this is the first error which we say has been committed in chiromancy.

The second mistake is this. It often happens that the original natural lines of the hands are changed by injuries or chance accidents, or become larger or smaller, or appear in other places. It is just as if a road were blocked with some obstacle, or covered by a mountain falling on it, or destroyed by an inundation. Men would make another road near it. So with the old lines of the hand. Sometimes when wounds or ulcers have healed, along with the new flesh new lines come into existence, and the old ones are altogether blotted out. In the same way, by hard work lines are obliterated, or those which were there originally enlarged. Then the same thing happens as with trees. If the growing tree puts forth many leaves, a number of them are cut off and the tree is enlarged in size.

And now let us pass on to the practical part of this science of chiromancy, and in a few words disclose our opinion. I would have you know that, so far as relates to hands, I make no change therein, but I acquiesce with the observations and descriptions of the ancients. But in this practical chiromancy I have undertaken to write only of those matters which the ancients have not mentioned, as concerning the chiromancy of herbs, woods, stones, and the like. And first it should be remarked that all herbs, of whatever kind they are, belong to one and the same chiromancy. If their lines are unlike, and appear greater or less in some than in others, this is through their age. We expressly avow that the chiromancy of herbs confers no other advantage beyond enabling us to know the age of any herb or root.

Someone in arguing may urge and assert that no herb as long as it adheres to its root can be more than four or at the most five months old, that is, reckoning from May to autumn, after which time every herb perishes and drops away from its root. To this I answer that a unique virtue exists in the root, which is the first essence and spirit of the herb, from which the herb is born and sustained to its predestined time, and so is exalted right up to the production of the seed. And this is the sign or indication that the virtue goes back again into the root, and thus the herb withers. But as long as that spirit, which is the supreme force of the herb, remains in the root, every year that herb is renewed, unless it happens that the spirit is taken away, and withers along with the herb. Then for that herb there is no renovation. The root is dead, and no longer has life in it. But how that spirit is taken away with the herb from the root, or with the root from the earth, so that its virtue goes back either into the root, or from the root into the earth, must not be discussed in this place. It is Nature's sublime mystery, not to be put forth for the benefit of sophistical physicians, for whom such secrets are not only a mockery but a cause of contempt. What we here omit we will give in the Herbary.*

The younger and less full of years herbs are the more do they excel in their force and their faculties. For just as man is enervated by old age, and fails in his natural powers, so also is it with herbs.

But in order to know what is the chiromancy, and what the age, of herbs and similar bodies, long experience is required, since the number of years is not written upon them but has to be divined solely by chiromancy, as we have said. Now chiromancy supplies, not numbers, not letters, not characters, only lines and veins and wrinkles, as a means of reckoning the age. The older anything is the larger and more visible are the lines exhibited, and the virtue and operation of the thing are less active. For as a disease of one month or one year is more easily cured than one of two, three, four, five months or years, so a herb of one year more quickly cures its disease than one of two or three years. And on this account for old ills young herbs and those which have fewer years should be given, but for recent ailments old herbs and medicines should be administered. For if old be joined to old, the blind leads the blind and both fall into the ditch. This is the reason why many medicines are inoperative. They are in the body and they fill the limbs, but only as mud sticks to the shoes. Hence the diseases are often doubled.

Now here is a matter which, up to this time, has never been thought out by unskilled sophists, while by their ignorance they have lost more patients than they cured. The very first thing you physicians ought to know is that the medicine must always be younger than the disease, in order that it may get the better of it, and be stronger in expelling it. If the medicine be more powerful than the disease, the disease will be expelled, as fire will be extinguished by water. If the disease be more powerful than the medicine, that medicine turns into a poison, and afterwards diseases are redoubled and made more severe. Thus, if the disease be of iron, the medicine must be steel. Steel cannot be conquered by iron. The more powerful conquers, the weaker is subdued.

Although, therefore, it was no part of my original plan to write in this place anything about medicine, still, for the sake of true and genuine physicians, I could not pass by these matters in silence.

Concerning Mineral Signs

Minerals and metals, apart from fire and dry material, show their indications and signs which they have received at once from the Archeus and from the higher stars, each one telling its genus by differences of colour and of earth. The mineral of gold differs from the mineral of silver. So the mineral of silver differs from the mineral of copper. The mineral of copper differs

---

* The *Herbarius Theophrasti*, concerning the virtues of herbs, roots, and seeds, etc., will be found in the second volume of the Geneva folio. It is an incomplete treatise which discusses the virtues of black hellebore, persicaria, common salt, carduus angelicus, corals, and the magnet. The portions of this treatise to which reference is made above, and again upon p. 189, are apparently in the missing fragments.

from the mineral of iron. So also that of iron from that of tin and of lead. And so with the rest. None can deny, then, that by means of chiromancy all minerals and metallic bodies of mines, which lie hid in secret places of the earth, may be known from their external signs. That is the chiromancy of mines, veins, and lodes, by which not only those things which are hidden within are brought forth, but also the exact depth and richness of the mine and yield of metal are made manifest. Now, in this chiromancy three things are necessary to be known, the age, depth, and breadth of the veins, as was said just now in the case of herbs. For the older its veins, the richer and more abundant in metals is the mine. On this subject one would reason that all metals, so long as they remain in their matrix, so long do they continually increase. Whence this, too, is clear, that any growing thing, even when placed outside its matrix, cannot grow less, but is thereupon increased, that is, multiplied, and goes on growing in substance, measure, and weight up to its predestined time. This predestined time is a third part of the destined age of all minerals, vegetables, and animals, which are the three chief genera of all terrestrial things. That which is still in its matrix grows until the matrix itself dies. For there is a predestined period of living and dying, even for the matrix, provided only it be subjected to the external elements. That which is not so subjected has no period, no terminus, other than the elements themselves have, together with which, at the last day, which is the end of those elements, it will perish. Hence it follows that all things which are below the earth are in the least possible degree subjected to the elements. For they feel neither heat nor cold, moisture nor drought, wind nor air, by which they may be destroyed. Bodies so situated, therefore, cannot decay, nor do they gather rust and corruption, nor perish, so long as they remain below the earth in their own chaos. This relates so far to metals and stones, but it applies also to men, many of whom have supported themselves for a hundred years in mountain-caves, as did the giants and the pigmies, concerning each of which I have written a book.*

---

* Men of abnormal height, who, however, are naturally begotten, are distinguished by Paracelsus from another genus of giants who belong to a wholly different order of existence. Concerning the generation of giants and dwarfs, it is to be understood that giants are born of sylphs and dwarfs of pigmies. These beget various monsters, and it should be noted that both giants and dwarfs are possessed of remarkable strength. They are not a *lusus naturæ*, but are the product of a singular counsel and admonition of God. They deserve consideration on account of the great achievements they accomplish. Moreover, being monsters, produced in a singular manner by God, they finish without offspring as to body and blood. Their parents have not the same kind of soul as themselves. They are the offspring of animal men, and hence it follows that they have derived no souls from their parents, although they have performed many great deeds, have studied the truth, and have accomplished many other things, from which the possession of a soul might be argued. God, had he so willed, could have endowed these creatures with souls, as is shewn by the union of man with God, and of the nymphs with man. Whatsoever good deeds they may perform they are not on that account partakers of salvation. While it is impossible to give a clear account of the way in which such monsters originate, it may be compared to the generation of erratic stars and comets in the firmament, and it is actually the result of a *bizarre* conjunction in the firmament of the Microcosm. . . . . . . Pygmies, like other creatures of this kind, that is, like nymphs, sylphs, and salamanders, are not of the generation of Adam, though they bear the likeness of men, but are equally diverse from humanity and from all animals. . . . . . . Pygmies and Ætnæi are regarded as spirits, and not such creatures as they appear. But it should be understood that they are what they seem to be, namely, beings of flesh and blood. At the same time, they are as agile and swift as a spirit. They know all future, present, and past things, which are not present to the eyes themselves, but are hidden. Herein they serve man by revelations, premonitions, etc. They have reason in common with man, save only the soul. They have the knowledge and the reason of spirits, if we except those things which pertain to the nature of God. Endowed with such great

In pursuit of our present purpose, then, I pass on to a very brief practical exposition concerning the chiromancy of mines. The deeper and broader the veins are, the older they may be known to be. When the tracts of the veins are stretched to a very long distance, and then gape, it is a bad sign. For as the courses of the veins gape, so the mines themselves gape, which fact they indicate by their depth. Although sometimes good mines are found with a very deep descent, they for the most part vanish more and more, so that they cannot be worked without great expenditure of toil. But where those veins are increased by other accessory ones, or in any other way are frequently cut off, that is a fortunate sign, indicating that the mines are good not only on the surface, but that they increase in depth and are multiplied, so that they are rendered rich mines, and yield most ample treasure.

It is not altogether beside the subject that many metallurgists praise those mines whose course is straight down, and which verge from east to west. But then reasoning and experience in the mines themselves also teach us that very often veins which stretch from west to east, or from south to north, or, contrariwise, from north to south, abound in metal no less than others. No one vein, then, is to be preferred before another, nor is there any need of further discussion on this point.

Then with regard to those signs which concern the colours of minerals and inner earth, one may dispose of them briefly. When miners come upon clayey soil, from which issues a vein of pure and fresh metal, that is a very good sign, indicating that the metal of which this is a vein is now not far off.

In like manner, if the earth which is dug out lacks metal, indeed, but is fat, and of a white, black, clayey, red, green, or blue colour, then that, too, is a favourable sign of good metal lying hid there. Then the work which has been begun should be briskly carried on, and no pause be made in the digging. Metallurgists especially regard brilliant, glittering, and primary colours, as are green earth or chrysocolla, copper green, lazurium, cinnabar, sandarach, auri-

---

powers, they lead and attract man to make experiments and to believe about Him. Wherefore God hath produced them that man may learn from his acquaintance with them what great things God works in those creatures. Gnomes (*i.e.*, pigmies) are like unto men, but of stunted stature. They are about half the size of man, or a little taller. . . . The devil at times enters into gnomes and ministers unto them. . . . . If the gnomes have once bound themselves to our service, they abide by their bargain, but they require to be served in turn, and those things ought to be given to them which they request. If the pacts into which we enter with them are fulfilled on our part, they remain sure, constant, and faithful in their office, especially in obtaining money. For the gnomes abound in money, which they coin themselves. You must understand this as follows : The spirit has whatsoever it wishes, for if a gnome desires a certain sum of money, he obtains it and has it. In this manner they give money to many men inhabiting the mountains to persuade them to go away again. The lot of man is very hard. To hope or to wish will profit him nothing, and he must work for all he wants ; but the gnomes have whatever they seek without any labour in getting or preparing it. . . . . Concerning their day and night, their sleeping and waking hours, the case is exactly the same with them as with men. Moreover, they have a sun and a firmament no less than we have, that is, the gnomes have the earth which is their chaos. This is to them only as our atmosphere ; it is not as earth to them in our sense. Hence it follows that they see through the earth just as do we through the air, and the sun shines for them through the earth as it does for us through the air. For they have the sun, the moon, and the whole firmament before their eyes, even as have we men. . . . The gnomes dwell in the mountain chaos in which they construct their dwellings. Hence it is that very often arches, caves, and other similar constructions are found in the earth, about a cubit in height, the work of these men, and their habitation. . . . . The gnomes pass through solid rocks or walls like spirits, for all these things are to them chaos, that is, nothing. . . . . The more crass the chaos, the more subtle is the creature, and *vice versa*. The gnomes have a crass chaos and are therefore subtle. — *De Pygmæis et Salamandris*.

pigment, litharge of gold and silver, etc. Nearly every one of these points out some special metal and mineral. Copper green, chrysocolla, and green earth indicate generally copper. So, too, lazurium, or white arsenic, or litharge of silver, mark copper metal. So cinnabar and sandaracha point out sometimes gold, sometimes silver, or the two together in combination. In the same way, auripigment, red sulphur, or litharge of gold, for the most part portend gold. So, too, when chrysocolla with lazurium, or lazurium with chrysocolla and auripigment, are found mixed and combined, excellent and rich minerals are generally indicated. When stones and earths of a ferruginous colour are seen they certainly designate iron mineral.

It should be remarked that it sometimes happens the Archeus of the earth occasionally thrusts forth, and, as it were, eructates from the lower earth some metal or other through a hidden burrow. That is a good sign when it appears. Diggers, therefore, should not relax their labours in face of such a sure and remarkable hope of hidden metal. If, moreover, slight metallic foliage, like talc, adheres to the stones or rocks, it is a sure and a good sign.

Then as to coruscations. These should be carefully and closely watched. They are most certain signs that lodes of some particuliar metal exist, also of their extent, and of that special kind of metal. Here, too, it should be remarked, that metals of this kind have not yet come to perfect maturity, but are still in their first essence. In whichever direction the coruscation extends, in that direction also extends the metallic lode.

Then, too, it must be known that the coruscation is threefold in colour, as, for instance, white, yellow, and red, for example, like white Luna. In this way all the metals which they indicate to us are recognised. A white coruscation points out white metals, such as tin, lead, silver. A red coruscation denotes red metals, like copper and iron. A yellow coruscation reveals golden metals. Add to this that a slight and subtle coruscation constitutes the best sign. It is just as you see in the case of trees; where there are fewer flowers you get better fruit. So, too, small and subtle coruscations indicate subtle and excellent metals, and *vice versa*. In addition to this, it should be known that so long as these effulgences appear, be they great or small, of this colour or of that, the metal is not yet perfect and matured in its ore, but still exists in its first essence, like the man's sperm in the matrix of the woman.

Now let us explain what this coruscation is. It appears sometimes during the night in mines like scintillating fire, just as gunpowder, scattered in a long train and when lighted at one end, exhibits a protracted fire. In the same way, this coruscation, or scintillation, is borne along its own track, sometimes from east to west, or, contrariwise, from west to east, from south to north, or *vice versa*. And so, a straight line drawn from any hour or part of the mountain map towards the nearest hour opposite, divides into two parts the map which is marked off into twenty-four hours or parts.

All these coruscations, whenever they appear, afford most reliable indications of metallic lodes, so that from them may be recognised the metals too as certain gifts of God coming out of the earth. For whatever God has created for the use of men that He has put in man's hands as a property, so that it should not remain hidden. And although He has created it hidden, yet He has added these particular outward signs leading to investigation. Here His marvellous predestination ought to be recognised. Just in the same way, men themselves, if they bury treasure, mark the place by the addition of some sure signs. They bury them at landmarks, or statues, or fountains, or some other object, so that, if need be, they themselves can find them again and dig them up. The old Chaldeans and Greeks, if in time of war they feared siege and exile, buried their treasures, and only marked the place by proposing to themselves a certain fixed day, hour, and minute of the year. They waited until the sun or the moon cast a shadow there, and in that spot they hid or buried their treasures. This art they called Sciomancy or the Art of Shadows. From these studies of shadows many arts arose, and many occult matters were revealed, as, for example, the methods by which all spirits and sidereal bodies might be distinguished. These are the infallible cabalistical signs; and should be carefully watched.

You must take particular care, however, not to let yourselves be beguiled by divinations obtained through uncertain arts. These are vain and misleading; and among the first of them are the divining rods, which have deceived many miners.* If they once point out rightly, they deceive ten or twenty times. In like manner, no confidence should be placed in other deceitful signs of the devil, which appear by night or at unseasonable times, out of the way of Nature, such as are spectres, visions, and the like. Be sure that the devil gives these signs merely from fraud, and with intent to trick you. No temple is ever built where the devil does not have his chapel; no chapel where he has not his altar. Good seed is never sown, but he sows tares along with it. That is the meaning of visions and supernatural apparitions, the same in all, be it in crystals, mirrors, waters, or the like. The ceremonial necromancers have foully abused the commandment of God and the light of Nature itself in this way. Visions, however, are not altogether to be rejected. They have their place, but only when produced by a different method. We are now no longer living in the first but in the second generation. By us Christians then, in our regenerate state, ceremonies and conjurations are no longer to be used, as the ancients used them in the Old Testament, for these people were living in the first generation. These men were foreshadowings for us who were to live under the New Testament. Whatever, therefore, the ancients, under the Old Testament, or the first generation, accomplished by means of ceremonies and conjurations, all these things, we Christians, who belong to the second generation, and live under the New Testament, ought to obtain by prayer, that is, we should seek it in faith by praying, knocking, and asking. In these three primary points consists the whole foundation of magical and cabalistical

---

* Elsewhere Paracelsus says that it is faith which turns and directs the divinatory rod in the hand.—*De Origine Morborum Invisibilium*, Lib. I.

science, by which we can gain all we desire, so that to us as Christians nothing shall be impossible. Having written, however, much about this in the book on Visions,* and other cabalistical institutions, I forbear to repeat it here. See how wonderfully, in His love for us, Christ, the Son of God, works in us, faithful Christians, by means of His angels, and how fraternally He associates with us. We are very angels, and members of Christ, since He is our head, that is, He lives in us, that so we may live in Him, as is handed down in the books on The Lord's Supper.†

But to return to our subject of mineral signs, and especially to the coruscations from metallic veins. Know that as all metals which are still in their first essence exhibit their coruscations, that is, their signs, so also the Tincture of the Philosophers, which transmutes all imperfect metals into good silver or gold (white metals into silver, red into gold), removes all these particular signs, such as coruscations, if it be astrally perfected and prepared. For as soon as ever a little morsel of it is thrown into the fused metal, so that the two meet in the fire, a natural coruscation or brightness arises, just as fine gold or silver flashes in the vat or vessel, which is a sign that this gold or silver is free and purified from all admixture of other metals. But how our Philosophic Tincture is rendered astral is a thing that ought to be learnt. Every metal, so long as it lies hid in its first essence, has its own peculiar stars. Gold has the stars of the sun; silver the stars of the moon; copper

---

* Natural sleep is the rest of the body, which recuperates its wasted energies. Now the day pertains to bodies, night to spirits; bodies work in the day, spirits at night. The sleep of the body is the waking time of the spirit, for the two cannot operate together, being contraries, and mutually incompatible things. Whatsoever is done by the body during sleep is really performed by the spirit. For some speak and give answers in their sleep; some arise and walk therein, but all this is done by the spirit governing the body. Hence it happens that if such a man be called by his name, he wakes up because the spirit in him is terrified by being called by the name of the man, for spirits are no less terrified by the voice of a man than are men by the voice of a spirit. The man in baptism receives a name, but not so the spirit. Therefore the spirit is terrified when the man is called. Hence sleep-walkers should by no means be left alone in their rooms, and this is especially the case with those who are afflicted by the Sagæ, i.e., divinatory spirits, because it is of great importance that such persons should be addressed by name, for thus all nocturnal divining spirits, and all formidable spectres, and all waking visions, are driven away and dispelled. But it should be noted, that all men, promiscuously, who talk in their sleep, are not thus to be invoked or shouted at, because they may be in communion with a spirit whose voice is not heard. For, although the spirit voice may be much clearer than that of humanity, it is not audible commonly by humanity, for the material ear can be, and is, closed by the power of such an intelligence, as is well known to those who divine by nigromancy by means of the spirits of the air, who are intermediate spirits, neither precisely good nor evil. No man holding such a conversation should be disturbed, so long as his accents are cheerful, but if he answers with trembling, fear, and consternation, this is a sign of a bad apparition, and such a person ought to be awakened by shouting. Such conversations are not, however, always conducted with the bodily organs of voice on the part of the sleeper, but also with those of the spirit, in which case there is no audible sound, and this last kind of speech is not only more frequent but of greater importance. It was profoundly investigated by the ancient Magi, who by this means could extract from the spirits of the departed a knowledge of those secrets which they had concealed from the whole world while they lived in the body. In this way they became acquainted with the mysteries of Alchemy, Astronomy, Astrology, Medicine, Theology, etc., namely, by direct communication of their spirits with the spirits of those who had professed these sciences on earth. In order to acquire the arcane method of communication with such intelligences, the first requisite is to implore by faith the mercy of God in the matter; then we must, also with faith, make an image of that man with whom we desire to communicate. On the body of such image the name of the man must be written, and also the question to be asked. Put this image at night under your head and sleep upon it. That man himself will then appear to you spiritually, and will answer your questions, teaching you whatever he can. There is, however, a more certain and better manner. This dispenses with the image, and has recourse only to faith and imagination. No danger attaches to this experiment, but it requires great confidence in the validity of the operation. I have several times had practical evidence of its truth. - *De Philosophia*, Tract V.

† A work of Paracelsus, entitled *De Cœna Domini*, exists in the Harleian collection among the MSS. of the British Museum. It is numbered 508, and is a large volume, very legibly written. No printed copy is known to the present editor.

the stars of Venus; iron the stars of Mars; tin the stars of Jupiter; lead the stars of Saturn; quicksilver the stars of Mercury. As soon, however, as they have come to their perfection, and are coagulated into a fixed metallic body, their stars withdraw from every one of these, and leave their body dead. Hence it follows that all the bodies alike are dead and inefficacious, and that the unconquered star of the metals subdues all of them, converts them into its own nature, and so makes them all astral. For this reason, our gold and silver, which are tinged and prepared with our tincture, are much more noble and more excellent for the composition of medicinal arcana, than that gold itself which Nature generates in mines, and afterwards segregates from other metals. So also corporal *Mercurius*, made astrally from another metal, is much nobler and more fixed than common mercury. In the same way you may judge of other metals. I assert, therefore, that every alchemist who has the star of gold, turns all red metals into gold by tingeing them. So by the star of silver, all white metals are turned into silver; by the star of copper, into copper; by the star of quicksilver, into corporal Mercury; and so with the others. How all these stars are prepared by Spagyric art, it is no part of our present purpose to declare. The explanation belongs to the book on the Transmutation of Metals.

So far as relates to the true signs of these, I would have you know that our red tincture, which contains within itself the stars of gold, is of a substance fixed above all consistency, of most rapid penetration, and deepest redness, its powder recalling the colour of the saffron, and its entire body that of the ruby. Its tincture is fusible as resin, clear as crystal, brittle as glass, but very heavy in weight.

The white tincture, which contains the stars of Luna, is, in the same way, of fixed substance, of changeless increment, of consummate whiteness, fluid as resin, clear as crystal, brittle as glass, in weight like the adamant. The star of copper is of supreme citrine colour, like emerald, fusible as resin, and much heavier than its own metal.

The star of tin is whiteflowing as resin, somewhat dark, and suffused with a claylike colour. The star of iron is of remarkable redness, clear as granatum, fusible as resin, brittle as glass, of fixed substance, and much heavier than its own metal. The star of lead is like cobalt, black, but transparent, fluid as resin, brittle as glass, equal to gold in weight, heavier than other lead. The star of quicksilver is of a white, glittering colour, like snow in a deep frost, very subtle, penetrating, and of corrosive sharpness, clear, like crystal, easily melted as resin, very cold to the touch, but extremely warm within the fire, volatile, moreover, and of a substance which easily flies before fire.

From this description you will know the stars of the metals, and you will understand that for the preparation of either tincture, the red or the white, you must take at first, not the body of gold or of Luna, but the first essence of gold or of Luna. If a mistake is made at the outset, all the subsequent work and labour will be thrown away.

Moreover, this fact applies to metals, that each of them in the fire puts forth some peculiar sign by which it can be recognised. Among these are, sparks, flames, brightness, colours of the fire, smell, taste, etc. For instance, in the reverberation of gold or silver, the genuine sign is a brightness above the vessel or vat. When this appears, it is certain that the lead, and other accessory metals, have disappeared in the fumes, and so the gold and silver are thoroughly purified. Iron, which is completely fused in the furnace, sends forth limpid, clear sparks, which rise to a height. As soon as these appear, unless the iron be at once removed from the fire, it will be burnt up like straw.

In the same way, every earthly body exhibits its own peculiar and distinct signs in the fire, whether it has any Mercury, sulphur, or salt, and of which of these three principles it has most. If it smokes before it bursts into flame it is a sign that it contains more Mercury than sulphur. If, on the other hand, it burns with a flame and blazes forth without any smoke, it is a sign that a good deal of sulphur, and no Mercury, or very little, lies hidden within it. This you see take place with fatty substances, as with fat itself, oil, resin, and the like. But if without any flame nothing goes forth through the fumes, it is a sign that much Mercury and very little sulphur exists therein. This you see take place with herbs, flowers, and the like; and also with other vegetable substances and volatile bodies, such as minerals and metals, as yet in their first essence, and not yet mixed with corporeal sulphur. These send forth only smoke, and no flame.

Minerals and metals which in the fire emit neither fume nor flame—that is, neither smoke nor blaze—shew an equal mixture of Mercury and sulphur, and a fixity and perfection beyond all consistency.

Concerning Certain Particular Signs of Natural and Supernatural Things

We must now, in due course, speak of some peculiar signs, concerning which nothing up to this time has been handed down. In this treatise it will be very necessary that you who boast your skill in the science of signatures, who also wish to be yourselves called signators, should rightly understand what we say. In this place we are not going to speak theoretically, but practically, and we will put forth our opinion comprised in the fewest possible words for your comprehension.

First of all, know that the signatory art teaches how to give true and genuine names to all things. All of these Adam the Protoplast truly and entirely understood. So it was that after the Creation he gave its own proper name to everything, to animals, trees, roots, stones, minerals, metals, waters, and the like, as well as to other fruits of the earth, of the water, of the air, and of the fire. Whatever names he imposed upon these were ratified and confirmed by God. Now these names were based upon a true and intimate foundation, not on mere opinion, and were derived from a predestinated knowledge, that is to say, the signatorial art. Adam is the first signator.

Indeed, it cannot be denied that genuine names flow forth from the Hebrew language, too, and are bestowed upon each thing according to its nature and condition. The names which are given in the Hebrew tongue indicate by their mere bestowal the virtue, power, and property of the very thing to which they belong. So when we say, "This is a pig, a horse, a cow, a bear, a dog, a fox, a sheep, etc.," the name of a pig indicates a foul and impure animal. A horse indicates a strong and patient animal; a cow, a voracious and insatiable one; a bear, a strong, victorious, and untamed animal; a fox, a crafty and cunning animal; a dog, one faithless in its nature; a sheep, one that is placid and useful, hurting no one. Hence it happens that sometimes a man is called a pig on account of his sordid and piggish life; a horse, on account of his endurance, for which he is remarkable beyond all else; a cow, because he is never tired of eating and drinking, and his stomach knows no moderation; a bear, because he is bigger and stronger than other people; a fox, because he is versatile and cunning, accommodating himself to all, and not easily offending anybody; a dog, because he is not faithful to anything beyond his own mouth, and shews himself unaccommodating and faithless to all; or a sheep, because he hurts nobody but himself, and is of more use to anyone else than to himself.

In the same way many herbs and roots have obtained their names. So the euphrasia or *herba ocularis* is thus called because it cures ailing eyes. The sanguinary herb is thus named because it is better than all others to stop bleeding. The scrofulary (*chelidonium minus*) is so called because it cures the piles better than any other herb. And so with many other herbs, of which I could cite a vast number, all of which were named on account of their virtue and faculty, as I have shewn more at length in my Herbary.

Then, again, many herbs and roots got their names, not from any one inborn virtue and faculty, but also from their figure, form, and appearance, as the Morsus Diaboli, Pentaphyllum, Cynoglossum, Ophioglossum, Hippuris, Hepatica, Buglosum, Dentaria, Calcatrippa (*consolida regalis*), Perforata, Satyrion or Orchis, Victorialis, Syderica, Petfoliata, Prunella, Heliotrope, and many others which need not be recounted here, but separately in the Herbary.

The same is true as to the signs of animal matters, because, in like manner, from the blood and its circulation, from the urine and the circulation thereof, all diseases which lie hid in men are recognised. From the liver of a slaughtered animal all its flesh can be judged whether it is fit for food or not. For if the liver be not clear and of a red colour, but livid and yellow, rough and perforated, it is inferred that the animal was sick and that, on this account, its flesh is unwholesome. It is no marvel that the liver indicates this by natural signs. The origin of the blood is in the liver, and hence it flows forth through the veins over the whole body, and is coagulated into flesh. For this reason, from a sickly and ill-affected liver no healthy and fresh blood can be produced, just as from morbid blood no wholesome flesh can be coagulated. But, nevertheless, even without the liver, the flesh, as well as the blood, can be distinguished. If both are sound, they have their true and natural colour, which is purple and bright, with no extraneous colour, such as yellow or livid. These extraneous colours always indicate sickness and disease.

But, moreover, there are other signs which are worthy of our wonder, when, for example, the Archeus is the signator and signifies on the umbilical cord of the fœtus by means of knots, from which it can be told how many children the mother has had or will have.

The same signator signs the horns of the stag with branches by which its age is known. As many branches as the horns have, so many years old is the stag. Since there is an addition of a new branch to the horn every year, the age of the stag can be set down as twenty or thirty years.

So, too, the signator marks the horns of the cow with circles from which it is known how many calves she has borne. Every circle indicates one calf.

The same signator thrusts out the first teeth of the horse so that for the first seven years its age can be certainly known from its teeth. When the horse is first born it has fourteen teeth, of which it sheds two every year, so in seven years all of them fall out. For this reason a horse more than seven years old can only be judged by one who is very skilled and practised.

The same signator marks the beak and talons of a bird with particular signs, so that every practised fowler can judge its age from these.

The same signator marks the tongues of pigs with blisters, by which their impurity can be known. If the tongue is foul, so is the whole body.

The same signator marks the clouds with different colours, whereby the tempests of the sky can be prognosticated.

So also he signs the circle of the moon with distinct colours, each one of which has its own special interpretation. Redness generally indicates coming wind; greenness or blackness, rain. The two mixed, wind with rain. At sea this is a sign which generally portends tempests and storms. Brightness and clear whiteness are a good sign, especially on the ocean. For the most part they presage quiet and serene weather.

So far we have confined our remarks to natural signs. With regard to supernatural signs this is a matter of special science and experience, as Magical Astronomy and the like.*

Now here it is most necessary to have certain knowledge. Hence proceed many arts, such as geomancy, pyromancy, hydromancy, chaomancy, and

---

* Whatsoever Nature generates is formed according to the essence of the virtues, which is to be understood as follows: According to the soul, the property, and the nature of any man, the body is constituted. For this proverb is often quoted—the more distorted the more wicked. Adam was originally created in such a manner that he was without inherent vice of body or soul; but when he distinguished between good and evil, Nature then commenced to mark each person according to his constitution. Adam was well pleasing to God before he knew good and evil; but afterwards, God repented having made man. Man was therefore made subject to the rule of Nature, so that Nature treats him even as a flower of the field, which she marks, and so makes recognisable to all. Man also is marked like a flower of the field, so that one person can be discerned from another, after the same way that flowers and all growing things are distinguished each from each. And since there is nothing hidden in man but must be revealed, this must be made known by three different methods—either by the signs of Nature, or the proper mark, or by the judgment of God. Omitting the two latter, I will speak of the first, that is to say, the signs which are exhibited by Nature. It is

necromancy, each of which has its own particular stars, and these stars sign in a supernatural manner.* The stars of geomancy sign or impress their marks on the terrestrial bodies of the whole world in many and various ways. They change the earth, produce earthquakes and landslips, make hills and valleys, bring forth many new growths, produce gamahei on nude figures and images having remarkable powers and potencies, which they receive from the seven planets, just as the shield or target receives the pellet or the dart from a slinger. But to know how these signs and images of the gamahei may be distinguished one from the other, and what they signify in magic, requires great experience and knowledge of Nature, nor can it be in any way perfectly dealt with here. But this must be noticed, that every stone or gamaheus possesses only the power and properties of one planet, and so can be endowed only by that one planet. And though, indeed, two or more planets may be conjoined in earthly bodies, as in the higher firmament, nevertheless, one is oppressed by the other. For as one house cannot have two masters, but the one thrusts out the other, so is it here also. One remains master; the other becomes a slave. Or as when one is keeping a house another comes upon him, thrusts him out by force, and makes himself master, arranging all things by his will and pleasure, while the other is reduced to slavery, so also one star expels the other, one planet the other, one ascendant the other, one influence another,

---

known to all that if a seed be cast into the earth and concealed therein, the latent nature of that seed, at the proper time, manifests it above the earth, and anyone may see clearly what manner of seed has lain in that place. It is the same with the heart (*cor*) and seed of man : out of that seed Nature produces a body so that anyone can see what kind of heart has been there. And, although there be a great difference between herbs or trees and men, yet art in man sufficiently demonstrates and proves those things. We men in this world explore all things which lie hidden in the mountains by means of traces and external signs. For we investigate the properties of all herbs and stones by their signed sign (*signum signatum*). Similarly, nothing can lie hidden in man which is not outwardly marked on him, for, as the physician has his own knowledge, so, also, the astronomer explores from the signed (*ex signato*). So now there are three things by which the nature of man and of everything that grows is revealed : Chiromancy, which concerns the extremities, as, for example, the hands, the feet, the veins, the lines, and the wrinkles ; Physiognomy, which regards the constitution of the face and the parts belonging to the head ; Proportion, which considers the condition of the whole body. These three should be combined : according to these three every created thing can be recognised : by the physician, that is to say, the remedy ; by the astronomer, that is, the man ; and by the metallurgist, that is, the metal. Such is the condition of the mother which manifests that which is latent in anything. He who is incapable of understanding these three things can be in no sense a natural philosopher, astronomer, or doctor, or know anything of the arcana and mysteries of Nature. The foundation is in this, that all things have seed, and in seed all things are contained, for Nature first fabricates the form, and afterwards she produces and manifests the essence of the thing. *Explicatio Totius Astronomiæ.*

* The *Liber Philosophiæ*, in a treatise *De Arte Presaga*, regards the varieties of sortilege discussed in this book from a totally different standpoint. The four arts of Geomancy, Hydromancy, Pyromancy, and Necromancy are thus noticed : Spirits which are (normally) unable to communicate visibly with men, have by lying arts invaded their imagination, and have raised up therein Geomancy, Pyromancy, Hydromancy, and Necromancy, arts not invented from the light of Nature or of men, but instilled by spirits, who, by their frauds, after they had descried some one or other discoverer suitable for their purposes, then added fitting disciples to these, namely, cultivators and admirers of the said arts. The first discoverers were obsessed by the devil, and sought out through his power and instigation arts of this kind. There are some, indeed, who, hiding the matter, affirm that they have been revealed from God ; but they are deceived, for God is not the author and teacher of inquiries into the future by means of such devices. He in no wise created us that we might devote ourselves to the investigation of what is to come, but ordered rather that, directing His attention to His commandments, we should seek out the knowledge of Himself and His manifest will. It is, therefore, a false pretence that these arts proceed from God when they emanate from spirits alone. It is, indeed, true that the spirits extracted them from God, not from the devil. But we on the earth derive them from spirits, not from God. Now, communication with such spirits is forbidden, though they themselves neglect the mandate. It is equally forbidden to the spirits to teach these arts, but here, also, they pay no attention to the command. And this is the reason why they are silent and tell lies when it is least becoming to do so. Thus, in order that man may act disobediently towards God, and plunge into superstitions, they have devised the four above-mentioned methods for inquiring into the future. Geomancy is the art of points, having sixteen signs and figures, which they have arranged according to their property. To these they added translations, creta (*sic*), form, points, and similar things, and have taught the erection of the whole figure, fixing certain rules by which each figure could be understood, each recognised in its own house, with a sufficient and necessary interpretation.

one impression another, and one element another. As water extinguishes fire, so one planet strikes out the property of the other and brings in its own. And so is it with their signs, which are manifold, and not only characters, as some think, but all those which are found in the entire map of the planets, that is, everything which is cognate with those planets or subject to them.

To make myself more easily understood, let me add an example. To the planet Sol there belong the crown, the sceptre, the throne, all the royal power and majesty, all the domination, all the riches, treasures, ornaments, and paraphernalia of this world.

To the planet Luna are subject all agriculture, navigation, travelling, and travellers, and everything concerned with matters of this kind.

To the planet Mars are subject munitions (as they call them), all breast-plates, cuirasses, spears, and all arms, with everything relating to war.

To the planet Mercury are subjected all literary men, all mechanical instruments, and every requirement of art.

To the planet Jupiter are subject all judgments and laws, the whole Levitical order, all ministers of the church, the decorations of temples, ornaments, and whatever else belongs to this class.

To the planet Venus are subject all things relating to music, musical instruments, amatory exercises, loves, debaucheries, etc.

The method is as follows: They guide the hand and mark the points until a judgment is made concerning the proposed matter. But the spirits know exactly how many points are required to make a figure which will explain the matter. If their direction be right, the figure also is correct and valid. For example, suppose I ask who is standing at the door, and what kind of tunic does he wear? Take the seven colours, to each of which attribute a geomantic sign, and consult that figure. Then, whatever sign falls indicates the colour. Now, if I knew what colour it were, but you did not know, I might so direct your hand, forming certain points in one line that, by obliterating or wiping off, there would remain the colour red, and supposing the tunic itself was red, then you would reply rightly: It is a red tunic. But I knew that before, and directed your hand to those points. The spirits do likewise with all the figures; and, since they know all things, it is easy for them to describe the figures and to guide your hand. Every rhombus is described by guiding the hand. In this manner Geomancy is constituted. Moreover, many superstitions are added thereto by men to augment it, as, for example, that it should be performed when the sky is clear and serene, or in the quiet and silence of night. Also, that you should not operate for your own purposes. Again, that you should say such and such a prayer at the beginning, and commence under good auspices, etc. All these are human superstitions: for, not knowing the foundation on which the art depends, they increase it, but it is as much an art as a superstition. Geomantia, as it was called at first, is so constituted that the ascendant is twofold—natural and of spirits. For the natural has its art, namely, Astronomy. The spirit has its Pyromancy. Accordingly, if a nativity be constituted out of the stars it is astronomically erected. If it be made according to spirits it is Pyromancy. But Pyromancy consists in the spirit being connected with the ascendant, and it leads the infant for example, into whoredom, thefts, lies. And as the art comes forward and succeeds, the spirits suggest to astronomers that if a conjunction of this or that star takes place, say, this or that event will take place, not because Nature herself will accomplish such things, but I myself will see to it, and, being everywhere, will bring about such and such effects; but as no one can trace my actions, they will be imputed to the stars or the elements. Hence it comes to pass that people pay more attention to the stars than to God. This is an astute feat of the devil. It is the spirits who cause the astronomical and other predictions to be fulfilled that the credit of the art may be sustained, so that men may be involved in errors and loss, while, intent on vain fantasies, they forget the true God. Their devices are favoured by their dupes, for in the case of twenty prophecies, if only one be fulfilled, they will never cease from inquiring until the other nineteen lies have been fulfilled also. Meanwhile, they are so deluded by the spirits themselves that they cannot arrive at the true *fundamentum*. For it is the property of spirits to lie. We have finished, then, with the foundation so far as they are concerned. Now one thing is wanting, now another; now the fault lies with the house, now with the exaltation, etc. In this discipline men have laboured for many thousands of years, nor have yet discovered the truth, which, indeed, is impossible to find, as the whole foundation is on falsehood. We now see for what reason astronomy is called Pyromancy when the operation proceeds pyromantically. The same spirits make their way into the third element, that is, water. For Geomancy has been named from the earth, as if it arose from the nature of the earth. Nor without reason, for the earth also has its own heaven or stars; but the spirits who are pyromantically recognised have devised them. Similarly, in the element of water there is a star wherein the pyromantic spirits dwell who have instituted Pyromancy, chiefly in the times of the Greeks, who, being easily led into all manner of delusions, promptly subjected themselves to the spirits. Pyromancy is an art consisting of signs and figures harmonising with the universal figure of the heaven. The process is as follows: Take a basin full of water, which set down, and notice the direction of the wavy movements as the water quiets down. Notice, also, the tremor, the rest,

To the planet Saturn are subjected all those who work in and under the earth, as metallurgists, miners, sextons, well-diggers, with all the tools used by them.

Pyromancy puts forth its signs by the stars of fire; in common fire by sparks, flames, crackling, and so forth; in mines by coruscations; in the firmament by stars, comets, thunder and lightning, nostoch, and the like; among spectres by salamanders, ethnic, and other similar spirits which appear in the form of fire.

Hydromancy gives its signs by the stars of water, by waves, inundations, droughts, discolorations, lorindi, new floods, washing away of territory. In magic and necromancy by nymphs, visions, and supernatural monsters in the waters and the sea.

Chaomancy exhibits its signs by the stars of the air and the wind, by discoloration, the loss and destruction of all tender and subtle things, to which the wind is opposed, by shaking off and stripping flowers, leaves, fronds, stalks. If the stars of chaomancy are excited the Necromicæ fall down from the upper air, and frequently voices and answers are heard. Trees are plucked up from the earth by their roots, and houses are thrown down. Lemurs, Penates, Undines, and Sylvans are seen. So also Tereniobin, Tronosia, and Manna fall upon the trees.

Necromancy puts forth its signs by the stars of death, which we also call Evestra, marking the body of the sick and those about to die with red, livid, and purple spots, which are certain signs of death on the third day from their appearance. They also sign the hands and fingers of men with clay-coloured spots, which are sure signs of something, good or bad, about

---

and the bubbles. These four give four figures, and the figures give twelve. Near the figures, rules and such things are found. Now, the spirit moves the bubbles, originates the shaking, the rest, the calm, according to the necessity of the sign, so that there may result a figure which indicates what is desired. Those, therefore, who have well-disposed spirits, to whom few things are forbidden, make good sorcerers in the art. On the other hand, a bad sorcerer has a mute and mendacious spirit. Among spirits one may be more mute and lying than another. When, therefore, one sorcerer is said to be more certain than another, it does not follow that he has greater skill, for he may possess a more reliable spirit. Now, the spirits delight by means of vexing and deluding men to cause them to hate one another, and this, indeed, is their first object. Were the foundation of this art more closely investigated by men, it would be seen that it was a hoax of the spirits. Yet, even if men arrived at perfection in this art, what solid advantage would it confer on them but a futile prediction and a pretext for wasting time. Suppose I desire to marry, and consult an omen as to the result, even if I get an answer I shall be uncertain of its truth; it is just as likely to speak falsely as truly. But i the prediction be fulfilled, it may be by the devil's arrangement. In any case, how will it help me? If I escape this evil, it will take shape in another way. Consequently, no faith can be placed in these arts. In addition to the methods which have been already mentioned there is Necromancy, which is the art of the air. And although others define differently what is meant by *Necro*, this is genuine—that it is the art of shades, for shades only are in the air, and these things are known by the shades. . . . Some people, at night, see figures in the air, as in heaven sometimes figures appear which have a certain signification. This is Necromancy. Men appear walking in the air, the clash of arms is heard, etc. Wondrous shades are likewise occasionally visible in water. The cause of all these things is, that the spirits display what they wish according to their own pleasure. A part of their deception is to make men fancy that the spirits must be propitiated by prayers, or compelled by force and conjuration to produce prodigies. Now, all these things are sheer superstition. It is also thought that men can compel spirits, through God, to do this or that; but it is highly displeasing to God that we should be occupied with such triflings, and the spirits are rejoicing meanwhile that, in opposition to God, we have become their accomplices. The prayers, conjurations, fasts, and other ceremonies are nothing but a cloak to superstition. The pronunciation of various words is committed to memory, but these are not the real names of the spirits, and they are altogether unimportant. For although each spirit has his own peculiar name, yet they salute one another by different names at different times, and so make game of men. Now, concerning the nature of shades, whatever is seen in a figure or image is to be considered such. He who is favoured by spirits sees many things, but otherwise, little or nothing. Did God permit it, these beings would be always in our midst, enticing us to desert God, and devote our mind to them. But if we carefully regard what they have performed during a given year we shall see that it has been mere trifling, devoid of use and profit, destructive to body and soul, health and property, praise and honour, in a word, disgraceful allurements, frauds, and devices, sprung from the root of lies itself.

immediately to happen. When the stars of necromancy are moved, then the dead give forth miracles and signs, the deceased bleed, dead things are seen, voices are heard from graves, tumults and tremblings arise in the charnel-house, and the dead appear in the form and dress of the living, are seen in visions, mirrors, beryls, stones, and waters under different appearances. Evestrum and Tarames give signs by knocking, striking, pounding, falling, throwing, and so on, where only a disturbance or sound is heard, but nothing seen. All these are sure signs of death, presaging it for him in whose dress the spectres appear, or for some one in the place where they are heard.

Concerning these signs much more could be set down than has so far been said. But since these bring with them bad, hurtful, and dangerous phantasies, imaginations, and superstitions, which may be the cause not only of misfortune, but even of death, we pass them over in silence. We are forbidden to reveal them, since they belong only to the ancient school and to the Divine power. So now we bring this our book to an end.*

HERE END THE NINE BOOKS CONCERNING THE NATURE OF THINGS

---

* In certain editions the following dedication is prefixed to the Nine Books containing the *Nature of Things*.—Theophrastus Paracelsus gives greeting to the honourable and prudent gentleman. John Winckelstein of Friburg, his initimate friend and dearest brother :—It is right, O intimate friend and dearest brother, that I should satisfy your friendly and assiduous prayers and petitions which you have addressed to me in your several letters, and since, in your latest letters of all, you have earnestly and courteously requested that I should at length come to you, if it were consistent with my convenience, it is not meet for me to conceal from you, that this course is, by reason of various hindrances, impossible. But with regard to the second request you have made to me, that I should furnish you with an excellent and clear instruction concerning certain matters. I neither can nor will refuse you, but am compelled to gratify you therein ; for I am well acquainted with your disposition ; moreover, I know that you hear and behold with delight anything that is fresh or marvellous in this art. I know, also, that you have devoted a great portion of your life to the arts, which have formed the chief element of your curriculum. Since, therefore, you have displayed, not only benevolence, but fraternal fidelity towards me, I am rightly powerless to forget either your fidelity or your benefits, but am indeed of necessity grateful, and, in case I should not see you in person again, I must leave a brotherly farewell to you and yours, as a memorial of myself. For herein I shall not only answer and clearly explain those points oncerning which you have consulted me and asked me in brotherly fashion, but will dedicate to you a special treatise on those points, which treatise I shall name *Concerning the Nature of Things*, and shall divide it into nine books. This work satisfies all your requests, and, indeed, more than you have requested of me, although you will greatly wonder at its matter, and will doubt whether things are just as I have described them. But do not so act, nor think that they are mere theories and speculations, whereas they are of practice and proceed from experience. And, in spite of the fact that I have not personally verified them all, notwithstanding, I both possess, have proved, and know these things by experience from and by means of other persons, as also from the light of Nature. But if in certain places you do not rightly understand what I say, and in one or more processes require of me a further explication, write to me secretly, and I will put the matter more clearly before you, and give you a sufficient instruction and understanding, although I do not believe that there will be any need for this, but that you will easily comprehend without it, since I know how richly you have been endowed by God with the arts and with good sense. Moreover, you know myself and my feelings, wherefore you will easily and quickly take my meaning. But, above all, I hope and am confident that you will look upon the present work, and will fittingly regard it as a treasure, will by no means publish it, but exclusively keep it in great secrecy for you and for yours, exactly as a vast hidden treasure, noble gem, and precious thing, which is not to be cast before swine, that is, before sophists, contemners of natural blessings, arts, and secrets, which persons are not worthy to read, much less to have, know, and understand them. And, although this book be very small, containing few and scanty words, yet it is full of many great mysteries, for herein I shall not write from speculation and theory, but practically from the light of Nature and experience itself, nor will I burden you and render it tedious by much speech. Wherefore, dearest friend and most intimate brother, since I have addressed this book out of love to you alone, and to no one else, I request you to keep the book as a precious and secret thing, and not to part with it until your dying day. After death, in similar fashion, command your children and heirs to preserve it also in secrecy. Furthermore, it is my special request that it should remain only in your family, and at no time become so public as to fall into the hands of sophists and mockers, who despise all things which do not agree with them, and cover them with calumny ; who also are pleased only with that which is their own, as is the case with all fools ; who are pleased only with their own trumpets, but not with that of another ; and do hate all wisdom, regarding that as of small account and even as folly, which is greater than theirs, that is to say, what is in their own head, because it does them no good, nor do they know the use of it. One workman cannot use the tools of another, and so in the same way a fool can use no better instrument than his own key, nor is any sound sweeter to his ear than the tinkling of his own bells. Wherefore, dearest friend, be faithfully admonished, as I have entreated you ; do that which I expect of you, so shall you do well and rightly. Farewell, under the care of God.—Given at *Villacus*, in the year 1537.

THE PARACELSIC METHOD OF EXTRACTING MERCURY FROM ALL THE METALS

TO extract Mercury from metallic bodies is nothing else but to resolve them, or to reduce them into their first matter: that is, running Mercury, such, in fact, as it was in the centre of the earth before the generation of the metals, namely, a damp and viscous vapour, containing invisibly within itself natural Mercury and sulphur, the principles of all metals. Such Mercury is of unspeakable power and possesses divine secrets.

The reduction spoken of is made by mercurial water, which was not known to John of Rupescissa, or to others, however they may boast. It must, therefore, be carefully studied and treated with unwearied assiduity. Let the aforesaid mercurial water be thus prepared :—

Take three pounds of Mercury sublimated seven times by Vitriol, Salt-Nitre, and Alum; one pound and a half of Sal ammoniac, clear and white, three times sublimated from salt. Grind these well together, alcoholise them, and sublimate in a sublimatory by means of sand for nine hours. When the mass has cooled, remove the sublimate with a feather, and sublimate with the rest as before. Repeat this operation four times, until it will no longer sublimate, and in the bottom there remains a black mass of fluid like wax. Having cooled this, take it out; grind it again, and imbibe it in a glass dish several times with the prepared water of Sal ammoniac. When it is spontaneously coagulated, imbibe it again and dry it, repeating this process nine or ten times, until it will scarcely coagulate any further. Grind it very small on marble in a damp place, and dissolve it into a beautiful oil, which you must rectify from all its dregs and residuum by distillation in ashes. Carefully preserve this water, for it is by far the chief of all waters. Take eight ounces of it, and put in it plates of the purest gold or silver carefully cleansed, an ounce and a half in weight. Place this in a closed vessel for digestion over hot ashes during a period of eight hours. Then you will see your body at the bottom of the vessel transmuted into a subtle vapour or Mercury. Having made a solution of the whole mercurial water, separate it, by sublimation in an alembic over a slow fire, from its first matter, and keep it carefully in a glass vessel. You will thus have the true Mercury of the body, the use whereof in desperate cases, provided only it be carefully employed, is marvellous and celestial*; and on that account, therefore, not to be revealed to unworthy persons.

---

* For example, the red Mercury of Gold constitutes a good medicament for the cure of wounds and of the plague, that is, if it be reduced to a precipitate to prevent vomiting. This is accomplished by the upward separation of its laxative part. For in every preparation of gold the chief point is to remove superfluity from it. In the plague there is no necessity for purging. Gold, however, is a laxative, a tonic, and an astringent. Take it away; preserve the rest. The medicaments for the plague are divided into those used for the *accidentia* and those adapted for its cure. Understand concerning the cure that the spirits of gold and of gems are the best medicines whereby all plagues, wheresoever located in the body, are most successfully healed. The principal is gold; the second are gems, for gems are tonics and preventives. It should at the same time be remembered that all sores are, as far as possible, to be cured from within. For this reason there is no more excellent medicine—speaking of vulnerary potions - than is internal Mumia. No wound is properly healed from without. Internal Mumia is the perfect curative. Otherwise, there is no more sublime incarnative than gold itself.—*Fragmentum de Peste.*

THE SULPHUR OF THE METALS

THE Sulphur of the metals is an oiliness extracted from the metals themselves, endowed with very many virtues for the health of man.* Another sulphur is drawn from metals before they have undergone the fire, as from the golden and silver marchasites and others, which take rank and excellence according to the nobility of the mineral. So also is it drawn from the mineral of marchasite and cobalt, according to the nature and property of each.

The more common mode of extraction is to take Acetum carefully distilled, which has stood for twenty-four hours on a *Caput Mortuum* made out of distilled Vitriol, Salt, Nitre, and Alum, which also has itself been distilled by means of an alembic. This, I say, you must pour on the pulverised metallic body in a glass vessel so that it shall stand above it by the height of seven fingers. Then place it to digest in horse-dung for nine days. The coloured Acetum distil in the ashes until it comes to a superfluous oil, which you will rectify in a bath, or in the sun. You will then have the very truest Sulphur of the metallic body, which you will rightly use at your discretion.

The extraction can also be made by means of a sharp and thoroughly separated lixivium. But other sulphurs are less suitable for the internal bodily use on account of the alkali of the ashes, out of which we make a clavellated corrosive substance, and also on account of the lime of which such lixivia are composed. The Sulphur thus extracted can be washed with sweet water and precipitated. The subsequent digestion requires a double space of time. The lixivium also ought to be rectified from all earthy deposit by means of sublimation, so that such sulphurs may not be incorporated with it and become corrosive so as to cause injury to sick persons. It is to prevent this that the separation spoken of should be made. So far concerning the crude materials.

But now, these having been fused and depurated, you may draw forth their sulphur. There is no more certain, noble, or better way than by the water of salt or by its oil, prepared in the way I have clearly described in my

---

* The Sulphur of Metals, and, indeed, that Sulphur which can also be extracted from minerals, is said to be of special utility in dropsy, for it is of a drying nature, and is, as it were, a sun, or solar heat, which disperses this rain of the body, and causes it to pass off in vapour.—*De Hydropisi.*

treatise on Alchemy. Such a water extracts from the very foundations and roots their natural liquid out of all metallic bodies, or a sulphur and a crocus most excellent for all medicinal as well as alchemical purposes. It resolves and breaks every metal changing it from its metallic nature into some other, according to the different intention and industry of the operator.

THE CROCUS OF THE METALS, OR THE TINCTURE

THE Crocus of the Metals is of four kinds: of the Sun, of Venus, of Mars, and of Chalybs. The best is that of Chalybs. It is extracted by reverberation or by calcination, reducing the aforesaid bodies to dust. In like manner, filed iron is consumed by rust. The consumption of the rust is made by the imbibition of those things which produce rust, and by a decoction extracting the colour of rust.

Take old Urine poured away from its deposit, several cups of it, in which dissolve three handfuls of ground Salt. When you have strained it, boil it and skim it carefully. In this again dissolve a handful of bruised Vitriol, with two or three ounces of bruised Sal Ammoniac, and then carefully skim again. With this liquid imbibe some filings, and boil until it can be pulverised. The dust thus produced reverberate over a powerful fire, continually stirring it with an iron rod, until it changes from its own colour to another, and at last into the hues of most brilliant violet. From this you can easily, with spirits of wine or distilled acetum, draw off the Tincture, and when it is extracted by separation of the elements you will collect what remains at the bottom of the glass, by means whereof you will be able to produce wondrous effects, both within and without the body.

For making the crocus of Venus, take one or two pounds of copper-rust carefully alcoholised, pour on it plenty of distilled Acetum, and stir it well three times every day. Gently pour off the coloured Acetum, and thoroughly sublimate it in ashes until it is dry. Let this powder be afterwards washed nine times with warm water from all acridity, and then dried. You will then have the prepared Crocus of Venus, or Flower of Brass, from which, if you wish, you can easily extract an oil according to the instructions given in the great work on Surgery, where also its use is explained.

The Crocus of the Sun should be extracted by the water of salt, whereby the metallic nature, or malleability, is destroyed. When the residuum has been washed with warm water, the Crocus can be extracted with spirits of wine; and, this being again separated, the Crocus will remain at the bottom. This is changed into the liquid, or truest quintessence of the Sun, by means of elevation, and sublimating with five different grades of fire. With this you can produce marvellous effects. But there is need not of a merely imaginary, but of an active and skilled, operator.*

---

* The flow of blood from wounds can be stopped by means of the most skilfully reverberated Crocus of Mars.—*Chirurgia Magna*, Tract II., c. 10. Moreover, the Crocus and Flower of Mercury may be successfully made use of for the cure of ulcers.—*Chirurgia Magna*, Pars. III., Lib. V. The Crocus of Iron, if it be reduced by the reverberatory into alcool, is supposed to cure the same ulcers that are successfully treated by the Oil of Iron, provided they have ceased to flow, and have reached their proper maturity.—*De Tumoribus et Pustulis Morbi Gallici*, Lib. X. By artificers and mechanics certain arcana are discovered in the things which they daily use. Thus workers in brass have stopped the flow of blood with burnt brass, and have dried flowing wounds. Workers in iron have used their burnt iron, which is called Crocus of Iron, for wounds. Potters also have made some discoveries with what they call silver or golden litharge. Many are the inventions of the vulgar which have been called experiments; many more, which need not be described here, such as minium, ceruse, and the like, have resulted from the various attempts of the alchemists upon various substances.—*Chirurgia Vulnrum*, c. 9. The Crocus or Flower of Copper, which is usefully applied to the cure of corrosive ulcers, is usually prepared in two ways, one of which is that the greenness is abstracted by means of distilled Botin, and the said Botin is then again extracted. Notwithstanding, the strength of Venus is feeble unless vitriol be added to it. But I regard that as vitriol which is extracted from the body of Venus.—*De Tumor. et Ulcer. Morbi Gallici*, Lib. X.

THE PHILOSOPHY OF THEOPHRASTUS CONCERNING THE GENERATIONS OF THE ELEMENTS*

BOOK THE FIRST

Concerning the Element of Air

TEXT I

IN the beginning, Iliaster, which is nothing, was divided, thus giving and arranging the four elements.† It was even as the seed from which springs the stem. What the seed gives forth it does not receive in the same form into itself again. But this Iliaster again attracts to itself the four elements. Thus, that is dissolved and becomes what it was before the four elements were produced, provided only one year of the world has elapsed. The four elements are the growth produced from the Iliaster. And the seed does not give those very things from which the infant is produced after this year of the world; but the four elements are both mothers and daughters. Of this family nothing is found surviving after death; but its end is the same as its origin; and so whatever is in it perishes at the same time. Although another world follows after, which is the daughter of this one in name, still, it is not so in form, in essence, or the like. For this will not pass away, but will remain like the

---

* The philosophy of Paracelsus concerning the generation of the four elements and concerning the three prime principles, Sulphur, Mercury, and Salt, appears to have been regarded by himself and by his editors as an essential part of his doctrine and practice of alchemy. To include it in the first section of this translation is by no means outside the issues of Hermetic Chemistry. Paracelsus was not the first adept who regarded the process in the accomplishment of the *Magnum Opus* as offering a rigorous analogy with the creation of the greater world. All alchemy insists on it. He who succeeded in accomplishing the Grand Magisterium, the confection of the Philosophers' Stone, became initiated thereby into the secret of the *Mysterium Magnum;* and, on the other hand, an exact comprehension of the true principles which obtained in the universal genesis, was enough to possess anyone with a full and practical illumination concerning the arcanum of philosophy. The cosmological philosophy of Paracelsus is the necessary complement of his alchemy, and whether or not their combined study is likely to throw light upon either, an opportunity must be offered to the student for the comparison of the two. The treatises which have been selected for the purpose are translated from the second volume of the Geneva folio, and the copious notes which have been added are derived from analogous writings which Paracelsus left unfinished, or which, for some other reason, have come down to us in an imperfect state.

† When God determined to form the world and deliberated with His Divine Prudence concerning its nature and the manner of its creation, He divided it into four parts or bodies, which he designed to be the mother of all things, but subject to him whom God intended to create after His own image, even the man Adam. When, therefore, the matter had been deliberated on and decreed by God, the four said bodies were created—that is, heaven, earth, water, air. For, as the Scripture saith, heaven was created first, then earth, and subsequently the two others. Hence you must know that these four bodies, mothers, or matrices, exist that they may produce fruit, and furnish the necessities for man's nourishment. Thus, for example, the earth brings forth its peculiar products, but it is man and not the earth who makes use of them. Similarly, heaven is a body, free by itself, whence fruits proceed simply for the use of man. —*Liber Meteorum*, Pref.

soul, which is indeed made and created but not mortal. Such is also the lot of this world.

TEXT II

Now, it is quite certain that the Eternal Father, who is not only the father of His own Son, but also of all things, mortal and immortal, permanent and transitory, blessed and damned together, created *Domor*, that is, heaven and earth, the firmament and the water, to which He also gave His own Divine will. We will not further discuss this subject here, but the same things can be read in the Paramira.* He formed the natural from the non-natural. From that which had never perceived any nature, He produced another nature, and following that nature He willed that yet another nature should be produced, whilst a year revolves, wherein His majesty Himself carries on the Divine rule, which man now moderates and possesses. Yet these primal natures differ, so that from the earth springs the pear-tree, from the sand the thistle, from the water cachimiæ, from the sky chaos, and from the fire snow. But seeing how wonderful these things are, and how unlike they seem to the first source from which they sprang, we ought to make it a matter of knowledge and of philosophy, that the element of water is not water only, but a mineral as well; that the element of earth is not earth only, but a grape as well, and so with the rest. For that philosophy is vain which gives it out that the earth is an element, indeed, but not a nut, or that fire is an element, but not snow. So, too, those who say that the four elements exist in all and everything, advance mere nonsense.

TEXT III

The earth is an element, and whatever is produced from it. So is the water and all produced therefrom. So then that is an element which produces. And an element is a mother, and there are four of them, air, fire, water, earth. From these four matrices everything in the whole world is produced. And the speech is inconsiderate of those who assert that an element is simply endowed with a complexion, warm, dry, cold, moist, or a compound of these. All these things are in all these four elements. You can understand it thus: the earth is cold and dry, cold and moist, warm and dry, warm and moist. This is how matters stand. Whatever thing which is warm and dry grows out of the earth, grows out of that which in the earth is warm and dry. Whatever is or is produced cold and moist, is produced from that in the earth which is of a similar nature. So also from fire four complexions proceed. Snow, for example, from that in the fire which is cold and dry; and lightning from that in the fire which is warm and dry. It is the same with the other two elements. I would have you then, at this point, before all to be advised not to determine the elements according to their com-

---

* But more completely and copiously in the treatises and fragments of treatises from which the ensuing notes have been rendered.

plexions, but according to their forms, that is, what are the four matrices which they have within them. The earth is material, clayey, conglutinous. Such it is whether it be warm, dry, cold, or moist. The water is humid, sensible, tangible, but not corporeally, not materially. And such is the element, whether it be cold or warm. The fire is a firmament, and is the element of fire, though it be in one place warm, in another cold. The air is a heaven which comprises all things, and is moist, warm, cold, or dry, as shall hereafter be set forth.

TEXT IV

Now, in order to advance towards the established principle with regard to the elements, understand this. The Iliaster was originally distributed into four parts—the air, which is a heaven embracing all things; fire, which is a firmament producing day and night, cold and heat; earth, which affords fruits of all kinds and a solid foundation for our feet; and water from whence are given forth all minerals and half the means of nutriment for living things.

These nutriments are twofold, one found in air and fire, the other in earth and water. The two former nourish us as if spiritually and invisibly; the two latter materially and corporeally. These four elements are divided into two classes. One is constituted of air and fire; the other of earth and water. The air sustains fire, the earth water. Air and fire hold water and earth; while these two hold air and fire. So then all things were created in due order, that the one might support, seek for, and nourish the other. Thus the Iliaster was divided into one *domor*, of which there are two globules, an outer and an inner, each enclosed with two elements.

Beyond is nothing, so far as we know. Within is what we see, and touch, and what the light of nature suggests to us. He who created these things is not among us, but dwells without us. But He who was begotten of Him is amongst us. Still we must not philosophise further concerning the four elements than Nature teaches and points the way for us.

TEXT V

In the beginning the body of the four elements was founded with that form and amplitude in which the heaven lies extended; and it was made corruptible or perishable so far as the air surrounds it.* There was the throne

---

* But now we must understand what is the nature of the body of heaven. Earth, water, air have each their peculiar bodies, but, indeed, all the four bodies of the four elements are made of nothing, that is, they are made only by the Word of God. This nothing, whence is produced something, turns into substance and body, which body of all the four elements is distinguished into three species, so that the creative *fiat* resulted in a triple body. Thus the earth and the other elements are all threefold. At the same time, there is such a distinction between the elements that the four things are not one body. The air is one body, the earth is another, the water a third. So also would be heaven if these four had a like body. But the earth has three bodies, and so also have water, heaven, and air, and yet a piece of wood is one body, a metal another, a stone another, a sponge another. So also the four elements of bodies are distinct and separate, as though someone were to take lead and make of it minium, ceruse, glass, and spirit of Saturn. So then, these three species are distributed into four elements, a peculiar body being assigned to each. To pay more exact attention to these numbers, God Himself chose three, and constituted all things out of three, and separated all three. For the origin of this number is immediately from God, the principle of God and the centre of His Kingdom, from which centre the world was created, but so that it should be something mortal and perishable created by God. To rightly understand this you must know that from that centre the world arose and was made material. On this seat Christ hung from the cross; on this seat sat the prophets; it is the footstool of God. Here, therefore, material and corporeal things are made God, and His work, the centre of His Kingdom, and His throne.

It should be known, then, at the outset, and before the philosophy itself is unfolded, that God has made the centre of His heaven, and even Himself, perishable. For as corporeally He is called the Son, so the world is His house. But although it be thus made and created, still we must believe that it will not perish as it was produced. Of man the heart will endure: of the world the flower will be permanent.

TEXT VI

As to the manner in which God created the world, take the following account. He originally reduced it to one body, while the elements were developing. This body He made up of three ingredients, Mercury, Sulphur, and Salt, so that these three should constitute one body. Of these three are composed all the things which are, or are produced, in the four elements. These three have in themselves the force and the power of all perishable things. In them lie hidden the mineral, day, night, heat, cold, the stone, the fruit, and everything else, even while not yet formed. It is even as with wood which is thrown away and is only wood, yet in it are hidden all forms of animals, of plants, of instruments, which any one who can carve what else would be useless, invents and produces. So the body of Iliaster was a mere trunk, but in it lay hidden all herbs, waters, gems, minerals, stones, and chaos itself, which things the supreme Creator alone carved and fashioned most

---

in the Deity being three. Now, the word also was threefold, and the word is the beginning of heaven and earth and of all creatures. All things are synthesized in three, and there is nothing on earth which consists not of and in three, and is reduced again into that three. On the one hand, then, it is evident that each creature can be distributed into three, each in its place; but, on the other hand, what they dogmatize concerning the four things or elements, to the effect that each thing consists of four elements—that is false; each thing, however, contains in itself one complexion and not more, nor can it have any other element than that which it receives from its mother. For instance, every herb has only one element—that is, of the earth; every stone has one element—that is, of the water. But in addition to this it receives a complexion, frigid and humid, frigid and dry, warm and humid, warm and dry. Yet that is not a whole element, but the element is the matrix, as water or earth. For instance, man is taken from the slime of the earth; but the element is not slime, it is quintessence. Yet it again becomes an element, that is, it returns to the element with the distinction which subsists between an element and flesh. Hence the elements only recur into three, and these three are the prime matter of the elements. However, the fashion of the prime matter of water, earth, air, and heaven is diverse, for the number three constitutes only three species in reality, which three make a perfect body, and these same are found by art in all bodies of Nature. These three are the first matter and have only one name. The first matter is as God; and as in the Deity there are three persons, so here each species is separate by itself as to its office, but the three offices are comprehended under the one name of the first matter. This first matter has been distributed by God among four parts or elements. Whatsoever resides in the first matter of the earth is being separated or has been separated into earth. The case is the same with the other elements. So, everything has been ordained into its predestinated form, earth having been ordained to be earth, with its office, and so of the rest. So all things consist of one body, and yet there are four bodies, and the four elements are all distributed into four bodies, and are formed from one matter which is in itself triple, having been originally formed out of the word. The three first things are three parts, namely, fire, salt, and balsam. All bodies consist of these three—all elements and all fruits thereof. Earth is threefold in its body—fire, salt and balsam—while that which grows from it is similarly distributed into three species. The body of a tree is fire, salt, and balsam, and the things which are generated from balsam are

subtly, having removed and cast away all that was extraneous. First of all He produced and separated the air. This being formed, from the remainder issued forth the other three elements, fire, water, earth. From these He afterwards took away the fire, while the other two remained, and so on in due succession.

TEXT VII

The four fields, therefore, having been in this way set apart and separated, there remained also four storehouses for keeping the four elements, namely, the hot, the cold, the moist, the dry. Each of these was far from being unimportant. First the air was arranged; afterwards the fire; then the earth; and, lastly, the water, in the following way: From the air proceeded chaos, the throne, the chain, the foundation. From the fire, night and day, the sun and the moon. From the earth, trees and herbs, grasses and fruits. From the water, minerals and stones. Of these the succession was so arranged that from the superfluity was continually produced something else. For instance, from the Iliaster of the earth beech wood was extracted and the wood of apples removed. Each was disposed in its own place; nothing being corrupted or intermixed. In water gold was separated from the rest of the metals, and afterwards the others also were removed in turn. In the fire, the cold withdrew from the heat, the light from the darkness. In the air, chaos was set in order for preserving all things, and for separating earth from heaven. These four Iliastri having been created and arranged according to elements, that is, according to the matrices of their fruits, the air was prepared before all else; then afterwards the fire. These two were linked together in union. Afterwards the earth, too, and the water, being separated from the two former, were joined in one. These are now conjoined Iliastri. The air is by itself, and the fire. In like manner, also, the earth and the water.

---

fire, salt, and balsam. It is the same with those fruits which have water for their matrix. It is the same with heaven, of which the fruits are the sun, etc. It is in like manner with snow and rain. The art, therefore, of Nature does not, then, teach us how to extract anything out of fruits except fire, salt, and balsam, which also are so separated from one another by the force of fire that the fire, salt, and balsam become separate. Now, fire is also called sulphur; salt, balm; and liquor, mercury. It is necessary, however, that we should have a clear idea what an element is. Now, man has a large body, containing many substances. But that which is the man himself, namely, soul and spirit, is a small thing. The reason why the body is called man is because the man remains hidden in the body. So also the eye is a considerable part in man, but the force which sees is very small in respect of the eye. In like manner, the earth is called an element, whereas it is a rude body, and its true element is hidden therein, invisibly, like the spirit in man. It is the same with the other elements, which are, indeed, corporal, but are yet spirits according to their nature and substance. So often, then, as you hear that this or that proceeds from an element, understand that it proceeds from the element itself, and not from its body. In man the tongue speaks and does not speak, for the spirit speaks in it, whose intimate permixture and union with the body causes it to be thought that the body does everything. . . . . The odour of the box tree is the spirit of the box tree; what there is else is its body. The soul of musk is in its odour. In corals the colour is the spirit. Thus, all fruits, like their element, have spirit as well as body, and the true fruit is not seen by the eyes. Yet there is a certain difference between the natural and the supernatural spirit, for the first is corporeal and material, subsisting in a corporeal body, but the second is altogether destitute of a body. The body of the natural spirit is clothed by Nature with another body of its own element. But concerning heaven it is to be noted that God has given it the name of firmament. The firmament is the heaven and its whole substance. The three other elements are included in the firmament, as the egg in its shell. By the demonstration of the name which He has given to it God teaches that He has endowed the firmament with power that it may be as a sure shell, wherein all the creatures of Nature are firmly contained. And, just as the yolk remains immovable in its place, whether the egg be put up or down, so is it with heaven. Wherever we dwell, we live at a high level or a low, and can call ourselves dwellers on high or dwellers below. For a circle has neither summit nor base.—*Liber Meteorum*, c. 2.

Thus it was that God made the material centre of His throne, and afterwards sundered it in three primal elements, from which constantly emerges everything that is born. Without these three, nothing in the four Iliastri can grow. But while they grow they are elements, and so, moreover, they lose their name of Iliastri and are called elements.

TEXT VIII

These four elements were sundered into their own places and seats, so that none of them should be mixed. All these were removed, just as a sculptor when making a statue throws away what does not suit the intended image. So there are four elements, but only three primary ones; three in the air, three in the fire, three in the earth, and three in the water. Everywhere there is only a single triad of the primaries, that is, one Mercury in all, one Sulphur in all, one Salt in all. Yet they differ in their properties. Whatever is growing, herb, leaf, grass, or the like, was relegated to the earth. Whatever is mineral withdrew into the water. Whatever is warm, cold, day, night, betook itself to the fire. Whatever is air spread itself out over chaos. And all these three are one, each in itself. It is just as when a stone is divided into four parts, and out of one is made a statue, out of another a pitcher, out of a third some other kind of a vessel, and out of the fourth a milestone; yet all are stones, nay, all one stone, though divided into four portions.

Of these Iliastri there are four, and no more; these being sufficient. So God disposed the world in a quaternary. He was satisfied with this number, though He could have made eight parts. One portion of nutriment He conferred on the air, a second on the fire, a third on the earth, a fourth on the water. Nowhere was there any deficiency.

And now it is further necessary that in the course of our philosophising we should go on to treat of these four under the name of elements, to tell of their possibilities and performances, and to state in what they excel. We will begin with the air, and conclude our philosophy with the water, adding such explanations as the nature of insensible things requires.

TEXT IX

The element of the air was appointed for no other purpose than to be the abode of the other three, each to be conserved, as it were, within its close in the following way.* The air encloses in itself every mortal thing,

---

* The elements and all that exists are built upon the element of air, even as a house upon its foundations. We should philosophise, however, concerning that which sustains the air. This power is situated in the exterior part of the air in which the Triune God dwells, so ruling and sustaining the air that it does not yield, nor is broken. For it is impossible that perishable things should fall into the sphere of the imperishable. Moreover, it cannot fall, because all things tend upwards, nothing downwards, nor is there any bottom or profundity. For the air is so compacted and confirmed in its circle that it can no more be broken or dissolved than the external kingdom can perish till its time arrives, when it will collapse inward towards the centre, the air and stars rushing towards the globe of earth, and then the globe shall by them be so utterly consumed that not a single ash shall remain. For the manner of this destruction shall be such that nothing shall collapse outwardly from the circle, but all inwardly to the centre. And this is the highest secret of philosophy—that the circle rushes to the centre because there is no profundity outside.—*Alius Liber Primus Meteorum, De Elemento Aeris.*

and shuts it off from what is immortal, as a wall divides a city from the fields. It strengthens the world and keeps it together, as a dam does a marsh. And just as there is nothing in an egg to one who looks at it from without, or outside the egg, which agrees with what is inside, so the sky is a shell dividing heaven and earth, just as the egg-shell separates the egg from what is outside it. The air, again, is like a skin in which is stored up a body, the whole world, to wit, and wherein the earth is contained and preserved. The air, then, is this sky, a skin, or egg-shell, or wall, or mound, beyond which nothing can burst through, and within which nothing can break in. Moreover the air is breath, from which all draw their life. This is truly air itself, and puts forth the air which nourishes the four elements, and at the same time sustains the life of man. Without it none could live. Without this no element could advance, no wind could blow, no rain or snow could fall, no sun could shine, no summer could flourish, no water could flow, no earth could sustain. All this force proceeds from the air, and is attracted by the four elements. For as the lungs every moment inhale air, so does the earth, while the water and the fire each do the very same thing. That is a palpable error which lays it down that winds are caused by the air. They burst in upon us like poison, not as a means of life. The first element brings air, but fire gives the winds.

TEXT X

From this same element, too, flows forth a power by which fire is joined to the air, so that it may not fall down. Thus it is like a chain which, without materiality or visibility, holds together and binds. This it does by means of its chaos, which it inserts between the pellicle and the earth. There is also a middle space extending from heaven to earth, in which are balanced the fire, the earth, and the water. And as the chicken is sustained in the egg by its albumen without touching the shell, so chaos sustains the globe and prevents it from tottering. This chaos is invisible, though it appears of a slight green tint. It is an intangible albumen, having the power and property of sustaining, so that the earth shall not fall from its position. As the chick in its albumen, so this globe of earth and water is balanced in the air. As a ship is borne up by the ocean, so is this globe by the air. It is one vast and marvellous albumen which invisibly supports the globe of earth and water. It bears up even the firmament itself, which is placed in it as the seed of the cucumber is placed in its mucilage. And as every morsel of flesh lies in its own liquid, or the generating seed in the sperm, so the stars lie in this albumen, and move therein like a bird in its flight. In no other way are they borne up than in what is clear from the illustrations which are named. There is at least only this difference: that the chaos is unlike the albumen or the sperm, in that it is impalpable and extremely subtle. Otherwise, in all its powers and energies it corresponds exactly to those things which have been enumerated.

TEXT XI

While discussing the powers of this element, it should, moreover, be pointed out that the air and its chaos and the sky exist in a round form which is inherent in them. No one can point out or distinguish what is above or what is below. Let us give an example. If it could be brought about that one should be shut up within an egg, it would be impossible to know which part looked towards the sky and which towards the earth.* The rotundity prevents there being any "up" or "down." So we are prisoned within a shell, and do not know which is up and which is down. Walking over the whole world, we look up to the sky, and everywhere there is height, whilst at the same time everywhere there is depth. The cause lies in the rotundity of the globe and of the sky, and thus it is natural to every mortal body that all things grow in a threefold line, and not only man walks, but also trees, veins of metal, and springs take this course. As God created the circle of the globe and the sky, so he founded also the semicircle, the diameter and the meridian—a threefold line—and other similar ones. For in heaven and earth, in fire and water, are found all lines and all circles. Here, too, are the true Geography, Cosmography, and Geometry. By the elementary geography of the air are conserved the structures of the air, that is, the sun and moon, all the stars, the trees of the earth, and other things, as the minerals of the water and the rest. Here, too, beyond a doubt, is found the true basis of all geometry, where man stands like the straight line looking up to heaven. Of this geometry God alone is the artificer, the mason, the geometrician. From this line nothing falls away or emerges, be it water, fire, earth, tree, man, beast. All things tend towards this aërial geometry, which God made and graved as a mason does the statues on a tower.

TEXT XII

Now, as to the philosophy of the three prime elements, it must be seen how these flourish in the element of air. Mercury, Sulphur, and Salt are so prepared as the element of air that they constitute the air, and make up that element. Originally the sky is nothing but white Sulphur coagulated with the spirit of Salt and clarified by Mercury, and the hardness of this element is in this pellicle and shell thus formed from it. Then, secondly, from the three primal parts it is changed into two—one part being air and the other chaos—in the following way. The Sulphur resolves itself by the spirit of Salt in the liquor of Mercury, which of itself is a liquid distributed from heaven to earth, and is the albumen of the heaven, and the mid space. It is clear, a chaos, subtle, and diaphanous. All density, dryness, and all its subtle nature, are

---

* Air preserves the elements and all creatures, so that they may persist in their course and centre. Land and sea are the centre of a circle of which the air is the circumference. Earth and water constitute one globe, resting on nothing, but free on all sides, being encompassed by the element of air, which is like a vast chaos, which conceals that which is called heaven by the ignorant. Within this chaos all creatures are included and involved. Between the circle of the air and the globe of earth and water which is at the centre, a sustaining operation intervenes, which may be compared to the albumen interposed between the shell and yolk of an egg.—*Ibid.*

resolved, nor is it any longer the same as it was before. Such is the air. The third remnant of the three primals has passed into air, thus: If wood is burnt it passes into smoke. So this passes into air, remains in its air to the end of its elements, and becomes Sulphur, Mercury, and Salt, which are substantially consumed and turned into air, just as the wood which becomes smoke. It is, in fact, nothing but the smoke of the three primal elements of the air. So, then, nothing further arises from the element of air beyond what has been mentioned. Many of the ancients and later writers, nay, even some now living, ascribe wind to the air, making out its cause to be the mobility of the sky. That is all nothing. It never reaches the sky; and the air is by itself, coming forth from its element as smoke from wood. Whoever wishes to understand more clearly about it, and what its motion is, let him read about the properties of fire, where more is set down than can be here comprised.

THE PHILOSOPHY OF
THE GENERATION OF THE ELEMENTS

BOOK THE SECOND

Concerning the Element of Fire

TREATISE I   TEXT I

WE have spoken thus far concerning the element of air, according to the position in which the elements have been arranged. The air is first in position; next to it is the fire. These two constitute and surround the entire globe.* We shall next philosophise as to what concerns the element of fire.

First of all, from the Iliaster were separated the air and the fire. Afterwards these two were sundered the one from the other, so that the air occupied the first place, as we pointed out in the former book. The next place to this the fire occupied. By a process of separation, these two elements, air and fire, were divided. From the air were produced the heavens; from the fire came forth the firmament. As in the air there is only chaos and nothing besides, so, in the element of fire we find nothing but heat and cold, light and darkness. But, whatever withdraws from the globe and from the air, is sustained in the element of fire. It is not, however, called the element of fire because it can only burn, as many have foolishly said. It is not the element of fire which burns, but that which burns and is contrary to it, is congelation. The element of fire is not by its constitution warm and dry; the cold and the moist come from the element of fire. They are quite beside the mark, then, who seek the element of fire in the element of earth or of water. Though these probably produce something of a warm complexion, still that warmth does not constitute the element of fire. This element is not, therefore, called an element because it is fire, but rather because in it the whole firmament subsists. It is an element from which should proceed day, night, brightness, white or red, rain, tempests, winds, and all impressions. It is also the place and portion of the four parts of creatures. Therefore it is called an element. For as the earth gives heat and cold together, though it be the element of

---

* Fire and air constitute the chaos which encircles the globe of earth and water. The two superior elements send down their impressions upon the two inferior. Fire is disposed and digested by God into the stars. – *Ibid.*, c, 7.

earth, so is it to be understood also of fire. Yet there is a difference, because material fire is called an element when it is not really an element. It is not even produced by the element of fire, but it is like elementary fire in that position when it looks towards the sun. So also the water is like the element of fire in a place where it rains. Material fire, which we use, is in the four elements; it is called Tristo, and exists in them thus: The element of water requires the element of fire for its operation. That fire remains in the element of water, and shews itself in steel and in those stones where it exists. So is it with the air, and so with the other elements. Each has its own Tristo within itself, as is demonstrated in the Nature of Things. So, too, the sun can shew its element in wood, can kindle and burn it, because it is of the same nature as that by which the element of fire moistens the earth with rain. As the element of fire moistens the earth, and it is its nature and property to do so, it kindles wood also by a mirror in the sun. The material fire is brought to the globe just as rain to the earth. Both come from one element divided as to their nature. But the fire which is extracted from stones and metals has penetrated thither from the sun by means of its own Ares. As the earth is nourished by the sun, so is the one element by another. Of the three primaries, Salt could not coagulate unless the element of fire were in it. So Mercury could not give a body unless it contained in itself the element of water. So neither is Sulphur without its terrestrial quality. The air is without material or body, impalpable. Therefore, of itself, like the other elements, it cannot give a body; but it works together with it, as the rest do.

TEXT II

Having thus far explained the separation of the two elements, fire and water, it remains to speak of their order, which is as follows:—Originally the distribution of them was made into the sun, the moon, and the other stars. Beyond these there is no element of fire. Whatever virtue they are endowed with beyond this is only trifling. This is more fully shewn in the treatise *De Natura*. Here it is sufficient to know that this element, the firmament, to wit, is nothing but stars. What these produce and send on the earth, as snow, rain, wind, hail, cold, heat, night, day, summer, winter, and the like; all these things come from the element of fire, as an infant from its mother, or an apple from its tree.

This element of fire is placed in the element of air. For as the water and the earth are comprised in one globe, so the fire and the air are mingled in one, neither injuring the body of the other. They move freely in the air, not leaning or propped up on any foundation. As birds fly in the air, so the sun moves in the sky, that is, in the air. For just as it is appointed that man walks on the earth, the bird flies in the air, the fish swims in the water, and the gnome lives within the earth, so has it been arranged concerning the elements, that one lies still, another flies, one is in this mode, another in that, not moving from one seat or place. Every star has its own special orbit, nor does one collide with another. For as no one man walks exactly like another, and yet there is one mode of progress for all, so is it with the stars. And as men by Nature are not precisely alike, neither are the stars; so manifold is their nature and condition. On this topic one need not philosophise more deeply than to say that all these things are arranged and constituted by fate.

TREATISE THE SECOND

Concerning the Sun, Light, Darkness, and Night *

IN the first treatise it was stated that the primal Iliaster was furnished with all the colours, and with brightness and splendour all mingled together. From thence the four elements were secreted. Herein shall be stated in due succession what was added or subjected to the element of fire. In the beginning the first element, that is to say, the air, was extracted from the Iliaster, and afterwards the element of fire. From this a separation was made. First of all the white brightness was drawn out, and therefrom was made a material body, the sun. Therein is all the white brightness of the element of fire, and besides this is no white brightness at all in the whole element. The red transparency was also extracted and transferred to the stars, that is, to the moon and the other stars, which were distributed into many parts. While the white brightness was conglobated into one form, the red brightness was divided into many parts. Hence now follow day and night. For since all the white brightness was coagulated into one globe, it will be day wherever that globe is. Where that globe is not, there is no white brightness, but it is night and darkness; for the red brightness transfers no light to the white brightness. Moreover, it must also be known that in the element of fire two natures exist, a warm and a cold one. Heat is universal in the white brightness, cold in the red. All fire which is warm is in the sun, and not in any element besides. All coldness is in the stars; there is none in the sun. Hence it is clear that summer comes from the sun, winter and cold from the stars. In the sun is an expulsive heat, in the stars an expulsive cold—thus: The sun emits from himself heat to the earth by means of his rays. For just as the wind blows from its cave, or as from the ground a stalk rises above the earth, so heat goes forth from the sun over the globe.

---

* All the clarity which in the element produces night is fiery and twofold—white and red. The white is from mercury and salt, the red from pure sulphur. These two colours inhere in the three principles by reason of the predominant fire in the substance. The same are divided, the red into one part, the white into the other. The first is distributed among all the stars, the second into one only. But if the red, like the white, were compacted and digested into a single star, instead of into so many, the red splendour would be equally great with reference to redness as is the white with reference to whiteness. On the other hand, were the white star distributed after the manner of the red, there would be a faint and perpetual daylight. Such a perfect and condensed splendour would not illuminate the earth, but one weaker and more divided. The universal splendour of the mercury has, however, been concentrated from the three prime principles into one orb or star, which receives its motion according to the will of the Creator. The motion of this star takes place round the globe. When it radiates upon the earth there is day, but elsewhere night reigns, for all the brilliance of day is in it, and without its radiation there is no brilliance upon earth. The red brilliance of the other stars is the light of the fire in red, only in sulphur, where there is no mercury or salt.—*De Meteoris*, c. 32

Heat is the fruit of the sun on the globe, and it has no other fruit. Hence it follows that the sun has two operations, a greater and a lesser heat, in this way. The sun divides his heat in two modes. Hence it is granted to the stars to lose their coldness. The matter stands thus: For us Germans if the sun is supreme his heat is greatest with us. Then the autumn and harvest are at hand. In winter the cold comes on, not because the sun is low and depressed (for it is the same sun which can by his rays shed heat everywhere), but because his harvest is not then imminent as in June. All fruits are then in a state of repose, and have been harvested. But below us, in Ethiopia and other places which verge towards the antarctic pole, the sun is warm while with us he is cold, for this reason: because it is his harvest-time, but with us the fallow season. This fallow season he makes more or less. Everything which has to produce fruit needs rest and sleep; and unless the sun were lying fallow, its heat would be equally intense with us in winter as it is in summer.

In the meantime, while the sun is lying fallow, the harvest and autumn of the cold stars are substituted, so that during the whole year there shall be no sterility, but fruits shall be constantly produced. Now the snow falls, and the north wind blows. Then follow the east and the south winds, which are the attendants of the sun. Thus are produced winter and summer, night and day, and the whole year. In this way is there transition from one autumn to another through the year of the sun and the year of the stars.

Moreover, on this subject it must be remarked that dryness and humidity occur thus: Dryness is in heat, that is, in the sun. There is no other dryness in the whole element of fire save that which the sun has in himself. Moisture is in cold, that is, in the cold stars, which are of red brightness. This is the true state of the case. Humidity cannot coexist with heat. Heat consumes all moisture and brings back dryness. Coldness never coexists with dryness. What is cold is dissolved if heat coagulates itself. Thus the element of fire is divided into two. In one is dryness, and this is in the sun; in the other is humidity, and this is in the cold. If coldness sometimes seems to be dry, the dryness is only as when one sweats in the sun, where that moisture is quite foreign. So is that coldness foreign too. It is true, indeed, that a humid body on the earth can be dried by the stars, though not on account of their dry nature, but on account of their cold nature, whereby they are able to coagulate so that a thing seems dried up. Thus must their nature be understood as frozen water. Such, too, is the method of the sun for rendering moist. By its heat it melts wax, so that it liquefies, as does tallow. But what has this to do with the matter? Nothing. These things are only given as illustrations. There is dryness, too, in the stars, for instance, snow, hoar-frost, sleet, hail, lightning and the like, as metals and stones coming thence. But what is the dryness of snow, which does not last? In what respect is a metal dry which returns to its original matter? And so of the sun. Where is his humidity? It does not last. What is it if, indeed, he moistens fat? No sooner has he withdrawn than it is dry again. Afterwards it is no longer moistened. It is the same, too, with fire. It dries wood so that it never afterwards grows damp again, that is, so long as it is ashes. But what does the star do? It wets Salpalla so that it never again returns to dryness, but always remains moist. The stars moisten the rain, which always remains moist, and is never again dried. Wherever it is poured out, wherever it breaks forth, it is always moist, always wet. So, that which is dry remains in heat; what is wet, in cold; dryness never grows wet, and moisture never grows dry. In like manner, lime remains lime; glass, glass; wine, wine, etc.

But in order that the element of fire may be more thoroughly understood, we will, in the first place, describe the sun, the account of which is as follows: All heat is drawn together and rounded out into Magdalion. The whole white light is therein. Thus, then, white light and heat make up Magdalion, composed of ignited white Sulphur, congested into one body of noblest Mercury, pre-eminent over all the other elements, and coagulated by the most subtle spirit of salt. Out of these three the sun exists, so dry and so warm that there is place for no humidity, but it would all be consumed. In this way, both the daily rain and whatever water is poured out by the three other elements is consumed by the force of the sun, lest a too copious supply of water should cause inundation. So, then, the sun is the death of aqueous nature, both of the sea and of the Rhine, Danube, Nile, and Tiber. They are consumed by the heat of the sun so that they do not increase in volume. Death exists in all things for this very purpose, that they may not increase too much but may keep within bounds. So man has his own form of death, which is invisible. So dryness has its death, namely, water. So, too, the waters have their death, fire; and it is not true to say, that what the fire consumes reappears elsewhere. It perishes entirely in its own form.

But the spirit remains, and this the sun consumes. It is the veritable death, consuming and taking away the other three elements—alike with man on the earth and with the bear in the cave.

But now to philosophise more about the sun. It regulates its course by Divine providence, which decides when and how all things should exist. By this it is arranged that the sun going round the globe rounds out its circle for the sake of this autumn and harvest of the sun. In this course is nothing but day and night, summer and winter, light and darkness; and the darkness which falls upon some lands is intercepted from others in due succession. From this impetus and motion no wind is aroused; but the sun moves and proceeds just as a ball is driven along the surface of the earth, without any wind arising, or as a ship in the sea, which does not of itself generate any wind. So neither does the sun produce wind. It does not grow warm by its motion; for, although the globe should roll on for a hundred years, it would not of itself grow warm. If it be warm, it must have been warm before. So the sun going upon its course is a globe, and may be compared to birds in its mode of motion. It even diminishes heat, that is to say, if it be fallow. But its brightness remains always and under all circumstances. For Magdalion is fixed, and will remain from the first point of time to the latest in one shape and appearance and one proportion of light, of Sulphur, Salt, and Mercury. These have only one year of their fixation, which will endure from the first Iliaster to the last Iliaster, wherein the world will be renewed. I say there is one year, the year of the sun. In like manner all the stars have fixity. That is the year of fire, or the stellar year, yielding place to the time of the year, as if to its own daughter.

But, now, in due course, we must speak of the other stars in which exist coldness and red brightness, as, for instance, in the moon, planets, and the rest. In this red brightness is a different kind of rest from that in the sun. For the moon has no fallow season, but simply dies and departs. The seed only is left there, from which the new moon is born. And the generation is of such a nature that it gains its power of increase from the sun. Whatever grows does so by force of the sun's heat, and without that heat nothing grows at all. When, therefore, the Creator made the moon after such a manner as that she should wane and wax, He did it for this purpose, that the moon, like seed, should be united with the sun, and should thence acquire her power of increase. Thus it is that she increases and comes to fulness, and then afterwards wanes. For whatever increases, the same also decreases. As man by disease wastes away and dies, so the decrease of the moon is her sickness even to death, wherein she passes away, leaving only her seed behind. The moon is, in fact, the phœnix of the firmament, from which, when it dies, a new one constantly issues forth. So, in like manner, there are other stars, and they are made up of the redness of Sulphur, Mercury, and Salt. And there is a cold of Sulphur, Mercury, and Salt, too, which has its origin in that virtue from which the sun, too, received its own. Thus it is that the moon has such strong influence over the earth on account of her coldness and her humidity. She is superior to all the other stars in this element of the coldness of fire. The other stars, too, are composed of these three primal elements; but, still they are divided into many parts. For the cold in the element of fire is divided into a thousand essences and natures. Thus, in some stars are produced winds circling over the entire globe; in others, snow, rain, and the like, have their origin. In truth, so manifold are they, that manifold natures and virtues flow down from them to the earth; and this could not otherwise be the case if there were but one Magdalion, like the sun, possessing only a single nature, heat. Therefore, in the stars there are many cold natures. Now, cold produces many more forms of effluence than heat. A warm man is a healthy man. A cold man is exposed to more misfortunes than twenty warm ones. Since, therefore, cold has a nature which is contrary and opposed to the sun, the element of fire is divided into many stars, so that each virtue should exist by itself without the impediment of another nature. From these come forth warm winds, warm showers, warm tempests, and the rest, coruscations, dragons, lanceæ, and the like. Yet, all these are cold fire, without ardour. On the other hand, what is warm and burns has its origin from accident, as the special chapters demonstrate. An entire section follows on the properties of the stars, as to the necessities they produce, and giving what is necessary for a description of their natures.

Concerning Winds *

Through the course of the globe, there are scattered the windy stars which continually bring round their autumn and harvest. They surround Zedoch in a circle and at the same time embrace the globe above and below. As, therefore, the firmament goes round the globe in its rotundity, and the round globe lies therein, so the stars consist in the circle of Zedoch, and the globes touch Zedoch in the midst. Two winds, therefore, proceed to the two sides, and separate above and below, that is, one part to the arctic and the other to the antarctic pole. These stars are actual stars of the winds, because they blow upon us annually, and have their own year, which is the year of the winds. The other stars of the winds blow above and below us, not according to the year, but sometimes they blow and sometimes they do not, and infringe upon one region only, wherever that may be. The true stars of the winds blow each according to its year continually, above and below, across the whole globe, and are without hail, without lightning, without frost, without coruscation.

---

* Since the meteorological principles have now been abundantly explained and recognised, the next thing is to impart some information concerning meteoric things generated, or their generations. But we will first write of the rise or generation of the winds, proceeding from their predestined circles. There are four parts of the orb and circle of the winds; one looks to the east, another to the west, a third to the south, a fourth to the north. The manner of the circles is as follows. As in the middle of the firmament there are placed two elements, earth and water, and the element of air stands between the element of heaven and the lower globe—as, I say, the earth is placed in the middle, and the heaven surrounds it completely, so there proceeds or advances a circle transversely on a level in the middle of heaven, earth, and water, similarly surrounding. . . . In the same way you will further note that heaven goes round the world with a certain circle. In this circle stand the mother of the winds, and the places whence arise the predestinated winds. If these are about to emerge from that circle, they blow upon the globe through the element of air. But while they are arriving at the rotundity of the terrestrial globe and dash upon it, it is possible that the winds may be either stirred up below the globe and impelled towards those who live below us, or may be driven above the globe to the dwellers on higher ; or again may be divided and driven in either direction through the heights and the depths of the globe. Thus the winds are impelled through the air beyond land and sea, and persist until they are worn out by reason of the distance, the way, or the violent motion, etc. Each of these four parts has a nature peculiar and proper to itself, for the oriental part is warm and dry, not being so on account of the sun, or because it occupies the east, but because such a nature is derived from the three prime principles. Therefore, also, in the true south-east wind and its satellites no other nature and operation are perceived than warm and dry. On the other hand, the west wind, by the setting of the sun, is cold and humid, not because it rises from the west (for the complexion of the east and the west is one and the same), but because the matter of the winds has been created cold and humid in the west. From the north blow winds of a cold and dry nature, which they also impart to those regions, not that the winds are so affected by the regions, but the regions receive that nature from the winds. The south wind is warm and humid, not because much water is accumulated there, or that moist and humid places abound there, but because such is the peculiar nature of this wind, and it is imparted to the region that it occupies. For this is to be observed, that the winds acquire no property from without, but are tempered from themselves, and are not affected by their regions. The generations, therefore, of the winds are circular, from their proper nature. They are produced from their stars, and the stars are their mothers. Stars of this kind are innumerable in the four quarters From these all the winds proceed. For although winds are also stirred up by the stars of rain or hail, yet they are not enumerated with the circle or the four cardinals. And since we have already spoken of the place and dispersion of the winds, because they flow from the farthest heaven across sea and land, it must now be added that those stars have the power of generating winds, and disposing of them according to their nature and quality. As a tree puts forth its fruit out of its internal nature, which consists in wood and marrow, so also the same is to be understood of the stars. But the seed of the winds is the first matter of the three principles, salt, sulphur, and liquor. These three are the mothers from which are born those fœtuses which we call winds. In the northern quarters they are of a cold and dry nature ; in the south, warm and humid ; in the east, warm and dry ; in the west, cold and humid. For as is the nature of the three principles, so are their fruits. Moreover, you must know that the winds arise from their stars by

There are very many which surround the whole Zedoch, like the Galaxy, and over against the Galaxy is Deneas. Concerning the elementary nature of these stars it may be said that they are all humid. Antiquity has given to these four names which we retain, though not with the ancient interpretation. All those stars which are situated at the north throughout the entire Zedoch are called Boreas. Censeturis is dry and cold, yet not altogether dry. It is cold and congelated, that is, its humidity is coagulated, whence it appears dry. Zephyrus comprises the western stars, all being humid and cold, but not congelated. So it is that by comparison with these Boreas is accounted dry, on account of its congelation. The other stars in Zedoch, Eurus and Auster, are altogether cold. As soon, however, as the winds issue forth from their stars, they become warm by the sun whose beams they pass through, and thus they are held to be warm, which they are not by their own nature. Eurus is accounted dry, but is not so. The sun consumes the humidity which it possesses until it comes to that moisture contained within it, which the sun cannot take away. Auster is called humid, and is so because the sun does not take away so much of its humidity as in the case of Eurus. That is prevented by the sea, which supplies to the sun sufficient moisture for its consumption. So Auster with its humidity bursts forth on us throughout the lower and the upper part of the globe.

---

rule of time and season. For they retain the nature of the three principles. The variations of their strength are in proportion to the distance they have travelled. . . . Boreas is affected by the summer but not by the winter stars, and if it be impregnated with sulphur it produces sulphureous maladies ; if with salt, it dries up and cracks the skin. The south wind at its proper season, namely, in spring, is most healthful. These risings of the winds we are able to prove by a terrestrial example. Water boiling in a jar emits a wind ; so do all boiling substances, whether dry or humid. Moist coction, as of water, produces a moist wind ; the dry coction which is known to the alchemists occasions a dry wind. There is no other generation of the winds than when the three principles are set in motion and driven to their work by Vulcan. This action produces wind, and imposes its own nature thereon, whether warm, cold, humid, or dry. We must understand that God has constituted a generation of the firmament of such a nature that the three principles should generate and produce all things in their places to which they were ordained by God, and should by their operations tend towards the centre of the earth. Above all things, therefore, it is necessary that the three principles should be rightly recognised. These three principles are all of an igneous nature till they arrive at their operation, that is, at their ultimate matter. Sulphur is a fire which burns ; salt nitre burns also ; and it is in like manner with mercury. Now, fire cooks wind, and in the generation of winds the stars are vials and cucurbites, containing in themselves meteoric sulphur, mercury, and salt, which operate in these phials by means of our ethereal Vulcan. From these ethereal operations ethereal works are produced, such as the winds. . . . From earthly examples we understand the operations of the firmament, not, indeed, according to one grade, for as the heaven is higher than the earth, so also is it stronger ; and as the heaven has more of clarity than has earth of grossness, so much more sublimely graded and intense is its operation. That which is unseen by the eye is judged analogically by things which the eye beholds. But you must know that the hour and time of the generation of those winds must be fundamentally understood by astronomy and all its branches. If the winds blow, they advance to places suited for them. Much concordance produces strong wind. To frequently concord and generate is frequently to excite winds. Many species and a strong Vulcan generate mighty and violent winds, which root up trees and demolish houses. For the wind is, according to its own nature, as corporeal, and substantial as stone or any matter hurled down from a great height. And although a stone is one body and wind is another, yet the latter is capable of great bodily destruction, for therein are invisible as well as visible corporeities created by God, diverse in appearance but equal in virtue. Concerning the origin of winds it is then to be concluded that they are generated in windy stars above, and by the operation of Vulcan they are matured at their proper time, when they dash forth into the centre of the globe, transforming all obstacles into their own nature and property. As Boreas coagulates, so the south wind dissolves, the east preserves, the west putrefies. They perform their operation according to their implanted nature. But if they blow at those times when their innate malice is removed from them and modified, they effect nothing of importance. Among other things, the wind exercises great force upon the waters of the sea, stirring up tempests, and so penetrating through everything that it enters through the depths of the sea into the earth itself, whence it again issues through mountains, caverns, etc. In this way tremblings of the earth are generated, although this is not the sole cause of such occurrences. Wind has the power of penetrating all stones, all metals, and all things without exception.—*Liber Meteorum*, c. 5.

As to how wind proceeds from the stars, this must be held to be the method. As the sun pours its heat on the world, so in these stars there is no other nature and property except to produce winds, which are decocted from Sulphur, Mercury, and Salt, and issue forth according to Adech. Their wind is daily, hourly, blowing gently and peacefully over the whole world. So the respective winds must be learnt in the course of our exposition as to the other windy stars.

Concerning the Temperate Stars
Zedoch

The following is the theory concerning the stars in the firmament.* Every star has in it a certain amount of frigidity. This causes winds. Cold is the parent of all winds. But the nature of cold is that some cold produces winds with rain, some with snow, some with hail and the like. The truth is that all winds, intermittent and temporary, proceed not from Zedoch, but are collected from all quarters out of particular stars. The mode in which all winds are generated is as follows :— By means of frigidity the stars periodically beget their own vacuum, which is manifold in

---

* The philosophy concerning the stars in the firmament and, generally, concerning the constitution of heaven, is discussed elsewhere at greater length by Paracelsus, and in connection with the four elements, as follows: Of the elements it has been said that they are four. Man has need of these. But they are divided into four complexions, which are by no means as the ancients have imagined them, as, for example, that the earth is cold and dry. This is without foundation; certainly in some places it is cold and dry, but in others it is cold and wet, while in yet others it is warm and dry. Nevertheless, it is an element, that is, the mother of these things. It is called an element, inasmuch as it is the mother of these things, not on account of the complexion. The case is the same with water. This is specified to be cold and humid. It ought certainly to be humid, but not equally cold. At the same time, that humidity is often dry and warm, by reason of the virtue. The body itself, in its corporal nature, is humid. The earth is dry, so that its fruit can be conceived in it and come forth from it. So, also, the heaven is not of one complexion, but of many complexions. It is not fire, but is understood as fire, because it proceeds therefrom. The fire thereof is at times a water, at others a fire, now cold, now warm, etc. We must consider, therefore, that the elements are only matrices, nor are they restricted to one complexion. For, as the offspring is, so is that which generated and produced it. Thus, a flammula proceeds from a flammula, and solatrum from solatrum, Accordingly, hearken concerning these things. It pertains to the earth to bear and sustain man and his dwelling-place, as also rocks, stones, sands, and all growing things. Hence it is clear that the earth is necessarily compact and solid, so as to be capable of bearing them. Consequently it is hard, and requires to be cut and ploughed. It is, in like manner, equally necessary for water to be moist, so that fishes may move and swim through it, which can by no means take place on land. The same ought also to produce salt and stones. Now, all these things must be humid in their first matter, and must pass from humidity into a coagulate. But that which is born from the earth has seed, that is, a dry body, such as are seeds, roods, trees, etc. For all these things are dry and compacted from the first matter. But in the first matter of water there is no compact body, the whole being liquid. The matrix hereof is from the element of water. In this element grow those ultimate matters, the principle of which is liquid and humid. The third body is the air. This element has need of another kind of body, which must not be humid like water, or solid like earth. Out of this element whatsoever things are born have their ingress in the body of the air, just as the fish in water. Man has been surrounded by an aeriform vehicle that he may walk in it, as the fish moves in water. Thus the air also sustains all trees and whatsoever grows. It is necessary, therefore, that the air should be a chaos; not earth, not water, but something perspicuous, diaphanous, impalpable, invisible, so that the palpable and the visible may be insphered (literally, *touched*) by it, and may be seen through it as through glass. Furthermore, the heaven is a body of this kind, not humid as the water, not perspicuous as air, nor solid as the earth, but one of another essence, so that heaven is not earthy, but is yet compact in its essence, not, however, with terrene compactness. It is also tenuous and permeable as water, but still is not water; it is likewise limpid and perspicuous without being air. It is most comparable to smoke, which is diverse from other bodies in respect of corporality and substance—that is, it is not like stone or wood, earth, water, or air, but is a body without mixture or affinity with others. It is in like manner with heaven, and the bodies which are born therefrom are at once bodies and not bodies, compact and not compact, permeable like water, yet not water, perspicuous and impalpable as air, yet not air. Such a body is the sun, such is the moon, and such are the other stars. Heaven is without complexion and the element of fire, and the matrix out of which fire is generated and grows. For as fire has a certain corporality, so have heaven and the stars, which take their nature and substance from heaven. Consider, therefore, that such corporality is derived from heaven, the peculiar quality thereof, and the very element of fire; and whatsoever fire is about to do, the same is performed by heaven, whence fire proceeds. · But we must make inquiry concerning the colour of heaven.

character. But as to the winds, the following is the received theory. The stars have their own emunctories, by means of which they excrete those things produced in them to which the emunctories refer. The duration of the wind is as long as that required for the purpose of emptying. The stars Zedoch perform this process of emptying every day, and raise up winds in the world for moderating the heat of the sun and dispersing the cold in the frigid portions of the earth. They mitigate both heat and cold, and are the most perfect moderators of summer and of winter alike. When these winds do not obviate such a result chaos is frozen just like water. The reason why water is frozen is that the winds of Zedoch do not penetrate it. They penetrate chaos, and therefore do not allow it to be frozen. There is no other use of winds except to mitigate each season of the year, and to moderate their excesses, which might otherwise do damage.

The common nature of all the stars comprised in the sky and the firmament is that, every day, nay, every hour and moment, they exude. For the stars attract to themselves the heat of the sun, just as the fruits of the earth absorb the same. The solar heat causes the stars to be resolved from their

---

That of earth tends towards black. Whatsoever is of another colour therein belongs to minera. So water has its own colour alike through all things. Its colour, however, has no name, for it is neither white, nor grey, nor blue, nor green, and yet it can be called all these. Earth, too, is really neither black nor purple, and yet up to a certain point it corresponds with both. The case is the same with air, which is pure and pellucid in chaos, and yet is neither white, nor blue, nor citrine, etc., while it is still partially assimilated to these. So also heaven has its special colour - like blue, like red, like green, and yet none of these colours is present therein otherwise than apparently. For elementary bodies are so formed as to have no perfect colour by which they may be named. But the things which are produced from them have their distinct, determinate colours, and to these names can be given. Thus, many colours are produced from the elements, and they are therefore composed of many, even of that number which they produce from themselves. From the earth proceed blue, red, black, etc., while from water all colours come forth, and so also from air and heaven. Accordingly, colours are collected from many into one, heaped together and mixed, and such mixture produces no express, determinate, and definite colour. Give heed to an example taken from heaven and its fruits. For ye see that everything which grows from the earth has its palpable foot and root, as are trees and herbs, etc. But the stars are the fruits of heaven, yet they do not put forth their roots in heaven, for they stand immovable below the heaven, without any support or attachment. Earth and heaven are opposite in this respect—one yields its fruits with roots, the other without; one tends upwards, the other downwards, and as fishes rest upon nothing, and, without feet, swim about in the water, so in like manner, stars swim about in the heaven, that is, in the body of heaven, preserving that order which God has prescribed them, some moving at a higer, some at a lower, level, at different distances apart, and with a quicker or slower motion. The details of this question must be referred to astronomers, but this, at least, should be remembered, that heaven is a body which, like water, is capable of sustaining a swimming thing, yet it is not water, but dry, while that which floats in it is also dry. It is not strictly swimming, but has analogy therewith ; it is not going or running, since it is not effected by hands or feet ; it is the miraculous work of God, and an element which contains and includes all the rest, and drives them in a round or a circle. The stars were born from heaven, and stand therein as if they flew like a bird through the air, according to the order and circle, even as God has destined and formed them to motion. Having been once formed, they henceforth remain for ever the same. The trees and fruits of the earth fall and are re-born. The stars can perish once only, namely, at the end of the world. Whatever else is formed in the elements is eaten away by mould, moth, and death. It is only the stars of the celestial heaven which remain in immunity, and yet their fruits rise and fall, as rain, snow, etc. But they have a unique and special colour, which is fiery. Thus, earth chiefly displays greenness, though it has also other colours. The sun is peculiar in colour, and if the same be igneous, it is not after the manner of wood, but of an element. . . . . We must repeat concerning fire that it has been enumerated as one of the elements, but with manifest absurdity. The earth, indeed, exhibits itself as an element, water in like manner, and so also air. But consider the fourth element. This cannot be fire, for it confers nothing elemental and no fruits upon man, nor does it possess any affinity with man, or *vice versa*, but it has an altogether fatal power, whereby the soul is separated from the body. It is, therefore, necessary that heaven should be regarded as the fourth element, for this is akin to man, nor can man dispense therewith, whereas he can dispense with fire and can live without it. The possibility of his dispensing with fire shews that it is not an element, but such rather is that heaven which brings forth day and night, summer and winter, increasing all fruits, and helping the other elements. The Scripture states that God created the heaven and the earth first. In heaven are the other elements, and even as the jar is made ready before the wine is pressed out, so the element of heaven is in reality the first element, which we have here named for the fourth.—*Liber Meteorum*, c. 1.

frigidity. This resolution is one and the same with that of a cold stone, which exudes on account of the vapour which it has acquired from Mercury, Sulphur, and Salt. That vapour exists in all elementated bodies. For as man, by natural exercise and the process of excretion, purges the phlegm from his nostrils, so do the stars also and all the elements undergo these excretions. This vapour flows down every day from the stars, and falls on the earth. During the day it is consumed by the sun. But by night it glides down to the earth before the sun rises, and is called dew. Through the winter, or during a cold autumn, it is frozen, and becomes hoar-frost. This is nothing else than the exudation of the stars in the whole firmament, which thus falls drop by drop. For as boiling water evaporates upwards towards the sky, or sends its dew on high, so the stars send their exudation downwards.

Concerning Nebulæ

Nebula is nothing else but a vapour of this kind, differing from the former only in this, that while not yet quite matured, it is excreted by certain stars. When it falls to the earth like hoar-frost, it rests on the earth and the water, and is like smoke. It cannot be completely resolved into hoar-frost or dew. A certain part of the vapour passes into dew, the rest into nebula. Nebula is imperfect dew which has not yet fully matured. If it is thin it falls to the earth and vanishes. But if it is dense, but not yet prepared, it descends to a higher region of its own, where it is consumed by the sun. If, however, it be mixed with rain-clouds, then rain is produced from it, but of a more subtle kind than other rain. Very often nebulæ of this nature descend and produce a spell of rainy weather. For if the stars are rainy they cannot be resolved into dew, but only into nebula. But if sometimes they bring clear weather, the cause is that the nebula, being more subtle in its preparation, disappears on the surface of the globe.*

---

* Earth is black, gross, rough, clayey, impure, dirty, and nothing could be cruder. Water is more subtle, pure, and clear, so that the eye can penetrate far into its depths. The air is completely pellucid and intangible, so perfectly purified that nothing foreign can be seen in it. Heaven is, however, by far superior to the air, but, though it is the clearest of all the elements, it is yet a body, which is proved by the fact that its fruits are bodies, such as rain, snow, hail, the thunderbolt, etc., for a body can only be generated from a body. But inasmuch as the heaven is more subtle than the earth, so are its fruits in comparison, and not only in subtlety but in operation. We have said enough of the heaven, but there remains something to be imparted concerning the stars and their risings. The stars bear the same relation to the sky as do trees to earth. But whereas trees have their roots in the earth, the stars are without foundation in heaven. The reason is this: trees do not need to be removed from the place where they are planted, but the stars must describe their orbit, for which reason they are separated from the heaven, while at the same time they are in the heaven. At the same time, they do not remove from their own mansions any more than the tree from its garden. Now, so often as there is a new genus among trees, there is likewise a new genus among stars. The same must be understood of herbs and all things that grow on the earth. Growing things correspond exactly to the number of influences and stars. Every genus corresponds to its like. But as some trees produce pears and others apples, so some stars yield rain, others snow, hail, etc., and in this fashion is generated whatsoever falls from heaven. The qualities which are specialised on earth exist more strongly in the heaven, because that element is superior to earthly things. And as the magnet attracts towards itself, so also the stars attract in the heaven. Accordingly, as certain natures on earth are dry and others humid, so throughout the whole firmament some stars are drier than others. Concerning the operations of the stars, they are produced out of congenital properties, and they arise from the three prime principles. That meteorology is false which makes absurd statements about the heat of the sun, of its motion, or other modes of generation, made by attraction from the earth. There is no star which attracts rain, and then again pours it down. The operation of rain proceeds from a nature congenital thereto. Even summer and winter are produced from the stars, the sun being supreme among those of the calorific kind, which arise at the beginning of summer, and

CONCERNING METALS, MINERALS AND STONES FROM THE UPPER REGIONS

TEXT I

Concerning Metals *

THE metals which come from the upper regions derive their origin from the seven planets. But these planets are manifold. There are many suns, many moons, many Marses, Mercuries, Jupiters, and Saturns. They are only called seven because they produce seven metals, and one kind of metal is ascribed to each planet. Those are not planets which the astronomers point out; and they are in error when they assign these to the metals: nor are they unanimous among themselves in what they do say. From these seven kinds of planets proceed the seven metals, and they are the same in the first three, just as in the element of water. The only difference is that in the first three they are volatile, not fixed, in their species. In this way the metals which are found do not stand the test of the lower metals. Neither, again, do the lower metals stand the test of the superior ones. There is not one and

---

are strengthened by their own heat till they reach the supreme grade, when again they gradually fail. Then the winter stars rise in their turn, display their own nature, afterwards die out, and are succeeded by another summer. The varying cold of winter and the varying heat of summer are occasioned by mutations in the potency of the respective stars. The moon is chief among the stars of winter, and is furnished with no small escort. Were the summer stars to fail, there would be no summer, for the sun, whether high or low, dispenses an even heat. Unless, therefore, the summer stars were to arrive, perpetual winter would prevail. The summer stars, however, derive their increment from the sun. So, also, we must not assign a diverse origin to day and night. The day arises from the light of the sun, but the night from the light of the moon. The departure of the sun by no means causes night. It is the peculiar nature of certain stars to produce darkness, which is so gross that unless the moon interfered with her presence, nothing whatever would be visible. Such a course, therefore, has God imposed upon the stars, that, going round the whole firmament, they retain their order and continual progress. For lest they should cease, or have a general holiday, God has ordained that when some are absent, others are present to fulfil their operations. So the nocturnal stars take the place of the receding sun. The bodies of the three prime principles are the cause of those bodies whence day and night proceed. The sun is a perspicuous and diaphanous salt, clarified and extracted from these principles, being purified from all obscurity. Its brilliancy has been extracted from the mass of the first matter of heaven. And whereas that is a white brilliancy which has been digested into the sun, so has a red into the moon and stars. The transparency and perspicuity of the white were extracted in sulphur, salt, and liquor, to make the sun thereof. Afterwards the brilliance of the red was put into a body of sulphur. Thus salt is the body of the sun, sulphur that of the moon, while liquor is the body of darkness.—*Ibid.*, c. 3.

* Metallic natures also subsist in the element of fire, for as in heaven there are stones, so also there are metals, but differentiated beyond all recognition from those of earth. Fiery thunderbolts, with their corruscations, are only metals, harder than all iron or steel, fluxible as copper, mixed with colours, and formed like a thunderbolt. Their fall is solely owing to some miraculous conjunction of elements, which produces them in bodily form. Many marvellous matters are carried up into the heaven and fall down to us. If it were possible for the stars of mercury, salt, and sulphur to be joined in a like copulation, several impressions of this kind would fall hourly. But the disposition of things is not favourable herein, except in the case of the thunderbolt.—*De Meteoris*, Lib. II.

the same ductility, or fluxibility, or hardness in the one as in the other. Neither are they uniform in colour; there is a distinct difference in them. So, again, there is a volatile nature of this kind in the element of fire, which is the metallic operation and nature of all the seven stars, which also falls down from them to the earth at the same time, just like rain and similar effluxes. Many such metals lie under their own stars, some in Asia, a few in Africa, and fewer still in Europe. These stars do not reach our earth, so that these metals are not found amongst us. All those grains, however, which are among the seven metals, and are rough in external appearance, come down from the stars, and not from the element of water. And all the metals which are coagulated without fire, and are rounded in shape like pulse, of whatever kind they are, have come down from the seven stars, whether they lie above them or not; and the earth strikes against them just as rivers do. But where they are found is neither their source nor their root, but they come forth just like kidneys. Their origin is in the stars, and all have come down from thence. For there, in the element of fire, is no rudeness or density to mix itself up. It purges itself according to its own stars, and coagulates of itself purely and entirely. These metals, just like those in the element of water, exist in commixture with Sulphur, Salt, and Mercury, save that the igneous metals have not a watery fixation, just as the aqueous metals have not a firmamental fixation.

TEXT II

When, then, the three primals have completed their effect in the metallic star—as when, in the star of the sun, a composition has been formed of the Mercury of the sun with the solar Sulphur and Salt, then they are digested into a perfect metal, by Adech, who shapes therefrom the form of his own gift. Then at length the star throws off its efflux, warm and liquefied, as if from some furnace. This is shaken in falling, is coagulated in the cold, and lights upon the globe. In the same way, also, the star of the moon makes a composition of Mercury, Sulphur, and Salt. When these are brought to their effect (just as in the case of the sun) it casts them forth. The same thing takes place with Saturn, Mars, Venus, Mercury, and Jupiter. It must be remarked, however, that out of the seven kinds of the seven stars, each one embraces the three primals of one metal; not as in the element of water, where in one Ares the seven are latent. The names of the seven metals, therefore, bear reference to the seven metals not of the earth but of the stars. In the same way, too, many liquids fall down from the stars, being not yet in a state of coagulation. If the earth be moistened with these, a brightness rises thence like cachimiæ, talcs, and sometimes marcasites, though it does not fully and perfectly arise from any of these, nor perfectly bears reference to the same. Hence it will be inferred that the superior metals excel those of the lower earth by many degrees, in goodness, in purity, and in nature, and so in all respects deserve greater praise.

TEXT III

Concerning Stones from Above[*]

In the same way there are also other stars which cast forth from themselves gems, granates, and other forms of stones. For Sulphur, Salt, and Mercury in the element of fire possess a powerful force for generating gems. There are many stars which consist of ruby Sulphur, many of sapphire Salt, and many which are powerful in emerald Mercury. There are also stars which contain the primals of copper, vitriol, salt, or alum. Hence, many of this kind appear rainy. If these are prepared they manifest themselves. From these stars are generated sapphires of lazurium. There, Salt is the body, solidly coagulated with pure Sulphur and with the spirit of Mercury. In the emerald Mercury is the body, having the nature but not the body of copper. It has its colour but not its body from copper. In this way, all the colours of gems which proceed from fire are found in proportion to the nature and condition of the three primals which are found united in the ratio of colours in the metals. For instance, in copper there is redness. But these three primals, if they have not a metallic body, become green. So, from silver, if the metallic body be wanting, lazurium is produced; from iron, a red body; from lead, the same; from Jupiter, a clay-coloured one mixed with white; from gold, a purple body; from mercury, one that is saffron-coloured. In like manner, also, if only the Salt predominates, it produces various colours, such as are conspicuous in certain stones, purple or blue, either lightly or deeply impressed. Equally, too, that which comes only from Mercury is marked by many colours, saffron, red, etc. That which is from Sulphur has for its prevailing colours, white, red, saffron, black, cœrulean, and so on. These stones are very rare, and those which are of a metallic nature are exceedingly precious. Thus, the emerald is a copper stone; the carbuncle or jasper is a golden stone; the ruby and chalcedony are iron stones; the sapphire lazurius is a silver stone; the white sapphire is a stone of Jupiter; the jacinth is a mercurial stone. After this manner, then, stones are generated in their own stars, which closely adjoin the planets, and then are ejected, just as metals are ejected, and so are found in the ftiest parts of the earth, according to the ratio of their generation.

---

[*] In the height of the firmament stand the three principles from which impressions arise. These are so high and so lofty that we cannot behold their form, and yet they have a form. We see, however, the green which is their colour. Hence it is gathered that in the element of fire generations of stone also take place. But where stones are generated they fall. Although this be considered wonderful, rare, and unheard of, it more frequently happens in the sea than with us. The generations of these stones take place as follows. If the principles of thunderbolts are present, any number of thunderbolts may be generated, for with every peal there is a stone. The matter of such stones exists first of all in an aerial condition, and is afterwards coagulated into an earthy one, so that the air can retain them no longer, and they ultimately fall to the earth. Furthermore, the matter of these stones may collect into one place in the absence of any tempest, but it will remain aerial until it comes in contact with a contrary nature, when it will at once begin to coagulate and to fall, even as a cloud is precipitated downward in the form of rain.—*Ibid*,

TEXT IV

Concerning Crystals and Beryls

Of crystals and beryls it should be known that they are generated from the snowy stars, which produce snow, in the following manner: In the snowy stars, the power of congelation is so strong that sometimes they are of a double nature; that is, one and the same star contains within it both snow and congelation, and so becomes twofold. Now, a star of this nature, which has gained at the same time the power of congealing and also of producing snow, easily generates the crystal, the citrine, and the beryl. For, if snow falls, and frost accompanies it, and, moreover, a place be given to him on the globe where Boreas predominates, while the sun or the solar nature does not prevail strongly, then the water which is in combination with the snow is coagulated into a stone. Now, if this water is caught by an intense frost midway, while the snow is falling, stones are formed from it before they fall on the globe. Thus, large or small granules are found in proportion as the frost has caught the snow in falling. But, if this seizing has not been so sudden, the frost collects and drives together all the water contained in the snow, which, however, is not itself snow, into one centre towards the bottom of the earth, and when it is massed there, coagulates it into ice. This, however, does not again liquefy like other frozen bodies, nor is it dissolved, and that because it is derived from snow-water. Other waters, it is true, which are frozen, are partly snowy, but the snow is dissolved with them. Here, this should not take place, but the water is extracted from the snow. The fact that the snow remains, happens only through the snowy star, wherein, also, the power of congelation subsists, so, that, wherever they meet in one place on the earth, the snow is not liquefied, but goes on to the end of the intention or operation. In snows of this kind are produced stones, such as crystals and the like, pure and dark together, for this reason, because S.S. of Mercury and Salt have clarified and purified themselves. Very often, too, crystals, beryls, and citrines of this sort, are found in places which are not snowy. The reason of this is, that they have been coagulated in the higher regions and have fallen down in that form. They are nothing but coagulated snow-water. But their shape and species and angularity are bestowed upon that in proportion as the Salt in them exists in a subtle or a dense state.

THE PHILOSOPHY OF THE GENERATION OF THE ELEMENTS

BOOK THE THIRD

Concerning the Element of Earth

TEXT I

Concerning the Earth, Per Se.

TO philosophise concerning the element of earth, its matter was first made on the following principle: Its three primals were separated, as if out of the great Iliaster, from the two primal elements into another form and nature, so that in the beginning not only the element of earth, but the element of water was segregated, and these were afterwards joined together into one globe, which is the centre of the exterior elements. From these two elements, first the earth was completed, afterwards the water. But concerning the earth, it should be known that all the force and nature which lay hid in the Great Iliaster for nourishing not only man, but cattle, by means of food and other necessaries, were collected into the element of earth, and consisted of all trees, herbs, and other growths. But they were so divided from the other three elements that this virtue exists in the element of earth alone, and not in any other element. Therefore this Iliaster is peculiar to the element of earth so as to afford aliment. For this cause the earth is called, and is, an element, because therein consist all the force and power of nourishing things which are due to living beings.

TEXT II

These three—Sulphur, Salt, and Mercury—are the earth, taken out of the great Iliaster, out of that nature which is the element of earth. For there the element and the three others were one Iliaster, in which the four elements existed. They were, however, divided one from the other, and the Iliaster was divided. Nevermore, then, can the four elements from henceforth be joined or stand together, but each subsists separately by itself in its own place. Those, therefore, labour in vain who endeavour to separate the four elements, or to seek besides these a fifth essence.

From these three primals, disjoined from the other elements, was the matter of the earth produced, in such form as it now is and is seen. And as the air was made heaven, the fire, the firmament, the water, the sea, etc., so did separation bring it about that this element should pass into matter and end in a globular form, and that in it should be included all the virtues of trees, herbs, fungi, so that from it should be procreated in the world all those genera which had been silently sown and had lain hid within it.

TEXT III

In this element of earth was hidden the seed of wood, of roots, of herbs, of fungi, and also the force whereby the stem rises, and is formed and planted according to the will and pleasure of its cultivator. The seed is here invisibly proceeding from the nature of the element, which alone is that seed, as the abode and seat of the same, in which it is elaborated and prepared. But originally that force is separated into its own genus, so that the two do not remain joined in one, but each genus exists solely and separately, one in wood, a second in the herb, a third in fungus. Each of these, again, passes separately, this into cedar, that into anthos, this, again, into balsam, and that into botin.* Of herbs, too, one passes into meligia, another into a lily with thorns—and so with the rest. But in order that this seed may be rightly understood according to its distribution, it should be remarked that in the separation of the great Aniadus the nature of trees was collected into one place, botin into a second, and ebony into a third. So, too, with others. Equally, too, the great Aniadus so disposed of herbs that into one portion of earth was cast grass, into a second trefoil, and into a third lavendula. For so to each land is given its own herb, and its own tree. We should pay attention to what has been the distribution made by the Aniadus.

TEXT IV

As to why the Aniadus thus fell among trees so that in one soil should be produced the orange, in a second the plum, in a third the fig, and in a fourth acorns, the cause may be supposed to have been that the fig and the orange require their soil to be of a peculiar kind which should be favourable to their increase, just as they also require an appropriate climate. If now the

---

* In the botin, the pine, and the fir, there exist two kinds of sulphur—one passes away into coagulation, the other is separated therefrom, and is not coagulated. From the sulphur which is susceptible of coagulation, the wood of the trees is prepared, and the same abounds in salt. It is owing to this sulphur that wood burns, and it goes on burning so long as there is sulphur in it. Whatsoever remains is salt, and this is in the form of ashes. And that truly is salt which the sulphur in trees coagulates into wood, whence glass is made. For salt is fluid. And this glass is the ultimate matter of any salt of wood whatsoever. But the other sulphur which is not susceptible of coagulation gives terebinth, resin of the fir and pine, which inheres chiefly in the wood, and by reason of its subtlety penetrates through the pores outside the bark, either by liquefaction or by a natural resolution. The sulphur which is in botin is more subtle than the sulphurs of the fir or pine, while that of the pine is more subtle than that of the fir. But all three are of one generation, proceeding from the Aniadus, which is united through Mercury. The bark is nothing else but sulphur coagulated after the manner of resin, and it is educed into this form by the Aniadus. For it is a hard congealed sulphur. And as there is no outside in any body without hardness, so is the bark formed from the hardest parts of the sulphur which exists in a growing thing. The branches, the shoots, etc., as also the fruits, proceed, in like manner, from the Aniadus, and derive their special form and character therefrom. This is to be understood also concerning other trees.—*De Elemento Terræ*, Tract II., Tex I.

soil be unsuitable and the climate ill-adapted, the one fruit or the other cannot emerge, but its seed of necessity perishes and never bursts forth. For though it be present there and lie in the earth, it is, nevertheless, dried up by the climate and oppressed by the unfavourable constitution of the soil, which is varied by the variety of the climate, not by its own nature. For the soil is everywhere one; but variety and change accrue to it from the climate, which either encourages or impedes the growths themselves. The sun burns up the genus of lilies, or some other genus; but this rarely happens, for the seed is ready to hand, which Nature produces from the tree or flower. This material seed is the cause why the sun cannot burn up the whole genus of this or that flower or tree, but allows it to come to a condition of vigour: unless perchance it happens that the force of the sun is less than suffices for fertility. Thus in the work of planting, herbs and trees are produced which, on account of the aforesaid defect in the soil, would not otherwise be forthcoming.

TEXT V

But we must proceed with our philosophy of the earth. The fruits proceeding from the element of earth are twofold. The earth either produces them of itself or by means of seed. In this way all growths are produced by the element out of the soil in two ways; that is to say, either from the proper seed of the soil, or from seed entrusted to the earth. The proper seed is when the earth puts forth a herb which springs from itself. Seed that is sown is foreign and not proper. Here the gifts of herbs are twofold. Neither spelt, nor wheat, nor lily, nor pear-tree, nor anything of this kind, grows spontaneously out of the earth, but all have to be sown. Here the philosophy of this treatise is deep, to find out whence come those seeds which do not issue from the earth itself. If neither spelt nor wheat be sown, none of these things will be produced. But herbage and grass do grow. Herbage and grass, therefore, are growths of the earth itself, not like apple trees and cherry trees. So there remains another philosophy by which we learn whence are produced spelt, whence apple trees and pear trees. You must know that the seeds of all these growths are propagated from Paradise, sown outside it, then planted and cultivated far and wide. These fruits of Paradise come to be understood in the same way as we understand that Christ was God and yet a mortal man.

TEXT VI

As to the method whereby the seed passes into its shoot, it must be known that the seed takes from the earth nothing more than its increment and formative power. The other is from Paradise, and is taught in the Paramirum.* But as to how much of an element is taken from the earth,

---

* Every seed is threefold; that is, the seed is one, but three substances exist and grow therein. But even as the seed appears one, so are these three to be understood as one only. Every individual thing is united in its seed, and not divided, but the same is a conjunction of unity. An illustration may be taken from trees, which have their bark, their wood, and their roots, which are distinct in themselves, and yet co-exist in a single seed.—*Paramirum*, Lib. II., c. 1.

that may be understood from the fact that in the beginning the three primals of the earth mix with the seed, so that it tends towards the end destined for it, and becomes that which it is before. For the seed is that which is of itself, but not yet manifested. Out of this proceeds first the root; from this, afterwards, the stalk. From the root and stalk issue forth the branches. From these three burst out the leaves. After this appear the flowers and fruits. This shoot or growth is formed by the great Aniadus, and is like a man. It has its skin, which is the bark. It has its head and hair, which are the root. It has its figure, its signs, its mind, its sense in the stalk, the lesion whereof is followed by death. Its leaves and flowers and fruit are for ornament, as in man hearing, vision, and the power of speech. Gums are its excrement, and the parasite is its disease. Philosophise as we will about its growth, this is nothing more than its Aniadic nature, which arranges all forms and directs them into their essence for which they were created. Its death and passing away are the period of its years. A pear-tree will stand for ten or twenty years. After that time it dies. Thus a shoot or a tree growing in the earth dies according to the time appointed for its death. Its decay is the element of fire. That is, fire destroys wood, leaves, grass. Whatever is left in the field decaying and passing into rottenness is consumed by the sun and the movement of the galaxy, so that it is no more left on the earth than as though it had never grown there, as happens to wood in the fire. Thus are growing things consumed and eaten away so that no relic remains, but all are removed like dust. The very remnants are so dispersed by a strong wind that not a fragment survives and remains at the expiration of a year.

TEXT VII

Since, then, trees, herbs, corn, and vegetables are produced out of the earth, the power of this element should before all else be learnt: because some growing things are food and aliment, as vegetables and fruits; others are drink, as grapes and berberis; others purge the body, like turbith, hellebore, and colocynth; others strengthen it, as cinnamon, carraways, mace; others have their virtue in the root, as parsnip and gentian; others in the leaves, as pot-herbs and cabbage; others in the flowers, as ox-tongue; others in the fruits, as apples, pears, etc.; others in the seeds, as pepper, nuts, and the like. Now, it is worth while to know how all these things take place. It is the Aniadus of the Earth who thus distributes them. The nutrimental virtues he arranges in three parts, the seed, the roots, the extremities. Thus the apple is a fruit on the tree because the Aniadus thrusts it forth, and shapes the fruit into the form of an apple, or a pear, or a fig, etc. In the nucleus is a species of seed, as in wine there is a species of drink. So, then, the Aniadus, before man, operates the first preparation, and man directs the second for his own convenience. After these, whatever is of a laxative nature degenerates into another growth, as into the mountain brook-willow, the rhabarbarus, or hermodactylos. Whatever is of a sweet nature passes into sugar, fœnogræcum, liquorice root and flowers. Hence it is that pears and figs derive their sweetness, and bees their honey. Bitterness turns to amarissima, warmth to pepper and grains of Paradise, coolness into nenuphar and camphor. For as in the element of fire everything by itself is divided from another, so also the virtue of the element of earth is divided to its own growth. And yet it often happens that two or three natures link in a single substance. So in cassia there are heat, sweetness, and a laxative nature. In mace there are odour, goodness, and strengthening power. Such is the case with many others, and yet one does not on that account destroy the other. In the same way the power of the element of the earth either makes for health, as in the tare, in persica and gamandria: or it is of a consolidating nature, as in the comfrey and the red artemisia: or in the odour, as in the lily of the valley and narcissus: or in its stench, as in the dane-wort. These are all either produced from the Aniadus, or distributed for the use of those who live on the earth. In this way the mighty gifts are learnt, just as the virtues of the elements which have flowed down from the great Iliaster.

---

* As out of the element of earth trees pass off into wood, so in the same element there is a certain sulphur which can be separated and passes off into food. Of this kind are vegetables and cereals. Dry and humid sulphurs are united, being the three principles duplicated according to nature and essence. One of these is for use, the other is not. Thus the avena is sulphur, but it is not edible. The seed, however, is edible. The non-esile sulphur is first of all developed into stalk, etc., and subsequently the esile sulphur is collected into the grains of the cereal.—*De Elemento Terræ*, Tract III., Text I.

THE PHILOSOPHY OF THE GENERATION OF THE ELEMENTS

BOOK THE FOURTH

Concerning the Element of Water with its Fruits

TEXT I

CONCERNING the element of water, the first things to be considered are: What is its origin, into what divisions it is broken up, and what the element is *per se*. The element of water is a seed from Yle, bringing forth stalks and fruits, that is, water, and its fruits, such as stones and metals of various kinds. Concerning the seed of the element of water, it must be laid down that it is latent in its workshop, just as seed lies in the soil. From this workshop proceed the stalk and its branches and fruits, in this way. Out of this seed is produced the stalk, breaking out of the soil into the light, whilst it remains lying in the earth. For, as the element of earth bears its fruit in the body of chaos, so, in like manner, the earth is a body, which sustains growing things such as trees and fruits from the tree of the element of water. There is no element but requires a body by which it may be sustained. Chaos bears impressions. The element of fire sustains the fruits of the earth, the earth bears up the fruits of water, the water those of the air. Thus, the fruits of each element are borne by some other element. Now, as from the seed of the element issues its tree, so its tree is a flowing stream, distributed throughout the whole earth. All things are one tree, with one origin, one root, from one stalk. And the streams of the whole globe are the branches of this one stalk. All the humour of the whole globe is Abrissach, which falls down from the branches of this tree, and pervades all the pores of the globe with its distillation. For, as the fragments from the fir-trees fall down from above to the earth, so these branches from the water fall down into the hollows of the earth. In this way takes place the generation of the element of water. All the water and all its fruits come forth from the element of water; but they are not the element itself. The element itself is never seen by any, and yet, nevertheless, there is an element of water. From it emanates nothing but water. It is called an element on account of the water and its fruits, not on account of its own complexion and quality, just as is the case with the other elements.

TEXT II

But concerning its course and goal, as also its seats and termini, the truth is, that the tree has its exit and end of itself, rises and falls, is produced and perishes. Thus, all water that flows forth from it is new, not old, and was never before seen. For, as the element of water lies in the middle of the globe, so, the branches run out from the root in its circuit on all sides towards the plains and towards the light. From this root very many branches are born. One branch is the Rhine, another the Danube, another the Nile, etc. So, there are also smaller branches, all born out of that root which rises from the seed, whence proceeds the element of water. And all the stalks belong to one tree, which is born of the root along a triple line in the circle of the outer firmament of the two elements, fire and air. So, then, the tree is distributed by this triple line over the universal globe, tending towards the light. So the stalk and its branches grow out from the centre of the globe until they reach the two external elements where the line ends. It does not go on to its own body, or Yliadum. For, unless the Yliadum were so placed in that position, every tree would spring right up to the sky, extending itself further outside the earth than from above, where it is fixed in the earth. So, neither do the fruits of the element of the earth grow farther than to the prescribed limit of the Yliadum, which is the lower chaos of the earth, not occupying more of the earth than the height to which growing things rise. Chaos, therefore, is twofold. That which is above is the chaos in which fire is sustained; and, unless the Yliadum were opposed, the element of earth would extend its fruits to the mid heaven. So, too, the element of water. The course and progress of the stalk of the tree is, that it goes on to its Yliadum above the plain of the earth, where its height ends. But how far it extends since it lies in its Yliadum, this must be sought from philosophy, because all the branches reach their Yliadum in the sea, where they all meet. For, as there is one root, so is it compelled to reach one summit or canopy, which is the sea. The sea itself is of itself neither the stalk nor the tree, but, as it were, the canopy of the stalk, which is not first or proximately born from the root, but composed of the branches. Why it is salt, is on account of its position and because salt waters flow together into it, as will hereafter be shewn where we speak about the subject of salts. The cause of its ebb and flow is that all the fruits (or the humours) flow down by night, but by day they swell to a height, that is, clissus. And this clissus in water is the same as in other fruits, increasing and decreasing, going and returning.

TEXT III

Now, since it is well to know all these things, so their death, that is, their consumption, should be understood. Nothing is free from this consumption. It should be understood, then, that everything, when it comes to its Yliadum, is subject to putrefaction and is consumed. Putridity is a kind of consumption, and the passing away of that thing to which it appertains, so that it is consumed just as if it had never existed. This is the operation of its nature. As Nature produces things, so does she again remove them. As the thing proceeds from nothing, so it returns to nothingness again. Hence it is clear that the element of water itself is subject to putrefaction or corruption. If it comes to its canopy, that is, to the sea, it grows putrid and is consumed of itself, no extraneous agency being accessory thereto, but through its own nature and arrangement. As the fire consumes and extinguishes itself, so, in like manner, does the water. This is the way, then, in which the tree of the element of water and its branches are distributed. What fruits lie hid in it remain to be seen, as also concerning their nature and the generation of the outgrowths. The nature and property of this element is that some of its fruits it bears within itself, others it casts out, and some it altogether throws away. It must also be separately learnt concerning this in how many modes of nature and essence its effluents and streams arise. But in order that all things born of water may be understood in its death, it should be realised that the branches, but not the fruits, pass away to their canopies. Concerning the death of fruits it should be said that they all flow into Drachum. In that hour they are consumed, as lastly it should be understood and held on the subject of Drachum.

TEXT IV

By way of simplifyng any study concerning the origin of fruits, we will consider that the following are the fruits of the element of water :*—Salts, minerals, gems, and stones. There are, therefore, four kinds of growths out of the seed of the element of water, in this way. Sweet water is the stalk. Afterwards its nature is manifold in the matrix. One matrix is of salt, one of minerals, one of gems, and, lastly, one of stones.† Each of these, again, is divided in a different way. For instance, there are three fruits of salt—salt, vitriol, and alum. And each of these has many genera ranged under it. There are many kinds of salt, many of vitriol, many of alum.‡ Some are metals, some marcasites, some cachimiæ. But even these, again, singly, admit of more kinds. There are seven metals, nine marcasites, twelve

---

* The fruits of water are born from the seed of Ares. Archeus, who is the separator of the elements and of all things which lie in them, divides one thing from the other, and collocates it into its place. In the seed of the element of water Archeus removes everything, and ordains it into its Nedeon, for the Yliadum of the earth, separates the germs of salt from all other natures, and in like manner the germs of sweet water and things which are of an acid quality. When he has divided these things and educed them into Nedeon, the operation of Nedeon goes on into Yliadum, together with its maturation to which it is ordained.—*De Elemento Aquæ*, s. v. *De Generibus Salium*, c. 1.

† Metals, minerals, and stones, while they are all generated out of water, do yet owe their development and perfection to the element of earth. There is a twofold corruption of these substances—one which results from a too prolonged connection with the foreign element, and the proper corruption which takes place in their own element, even as the fruit at last passes into putrefaction on its own tree.—*De Naturalibus Aquis*, Lib. III.

‡ For example, the origin of vitriol, as also of alum, is as follows. For as salt is extracted solely according to its own essence, so also are separated vitriol and alum. But the form which is manifested in salt, even as in vitriol and alum, is known from this, that all the fruits of the element of water are minerals, and share the nature of metals. But from all those things which arise out of salts, none is more akin to mineral virtue than vitriol, because the salts are minerals, and all minerals lie hidden in one mass and Ares. But vitriol is the ultimate in the separation of minerals. It is followed as closely as possible by the separation of metals, of which Venus is the first. Hence vitriol adheres to the nature of Venus. It is partly salt and partly mineral. So in every vitriol there is copper, and by reason of this

cachimiæ. So in turn every metal by itself is manifold: as fixed gold and not fixed, fixed silver and not fixed, and Venus is both copper and zinc. Such also is the case with the others. So there is a vast variety of marcasites and cachimiæ. As to their origin and progress, their autumn and the rest—as, for example, their harvest and ingathering—suffice it to say that all the fruits proceeding from the element of water are divided into their branches and trees. So salt has its own mode of egress, together with sweet waters, even to the boundary of its Yliadum. The same is the case with the rest. But with regard to their division and separation, all such fruits consist of one root, out of which each nature is separately born according to its condition. So from one seed is born one tree, and in this the wood, the bark, the fruit, the leaves are all separate, yet all are but one tree. So also from one root innumerable fruits are produced, but each fruit passes to its own Yliadum and triple line, as the founder has arranged. If, therefore, the distribution proceeds in this way, from Yle into its own stalk, and fruit is produced after its kind, then different things are found proceeding from the element of water —on one stalk salt, on a second a mineral, on a third something else. As, therefore, in the earth every seed produces its own fruit, so the seed of water is the seed of numberless things springing forth from it. Now, if these are brought to their Yliadum, and await their autumn-tide, then at length the autumn and harvest come for the fruit of every branch, which fruit is in itself of this autumn-tide and this generation.

---

metallic affinity vitriolic salt is of venereal nature. Copper, in like manner, is combined with vitriol. Indeed, its generation instructs us that it is wholly vitriol. At the same time vitriol in itself remains a salt, and derives its body from the liquor of the metals. For this reason it acquires a certain fiery quality and brilliancy. Alumen, on the other hand, by no means has affinity with metals, but is a free salt, consisting solely of acetosity, and having a body which is devoid of earthy quality, unlike vitriol, which arises solely from a permixture of metallic bodies. Hence it exhibits a similitude with marcasites and cachimiæ, which come forth in the first generation of metals. The medium which unifies and conglutinates copper with vitriol is a phlegma.—*De Elemento Aquæ*, s. v. *De Generibus Salium*.

HERE ENDS THE PHILOSOPHY OF THE GENERATION OF ELEMENTS

APPENDICES

APPENDIX I

[In the Geneva folio of 1658, which is by far the largest, as it is also the best, collected edition of the works of Paracelsus, there are many treatises included which conspicuously overlap each other; and further, there are many treatises, independent in themselves, which are devoted to precisely the same subjects. For example, the *Philosophia Sagax* occupies, and at equal length, a similar ground to the *Explicatio Totius Astronomiæ*, and the latter is substantially identical with another astronomical interpretation included in this translation. It is much after the same manner that the *Economy of Minerals* corresponds to the *Liber Mineralium*, but, having regard to the metallurgical importance which, from the Hermetic standpoint, attaches to both these works, it has been thought well to include in an appendix the treatise which here follows.]

A BOOK ABOUT MINERALS

SINCE I have considered well beforehand, and come to the resolution of writing about minerals in general, all that relates to minerals, and everything bearing on the generation and nature of minerals, I would have you know before all else, that not a few persons have the priority of myself in publishing on the origin of minerals. When I read their works, I found that they were involved in many errors. As far as one can judge from their writings, they have never fully understood what the ultimate matter was. Now, if the ultimate matter be not understood, what, pray, will happen to the first matter? Whoever can describe the beginning will probably be certain about the end and ultimate. What is a theologian who is ignorant of the end? What is an astronomer who is full of boasting, indeed, but without experience of light? Since, then, these authors are detected as in a state of hallucination about the end, that is, the ultimate matter, how will they be more worthy of credit about the beginning? I repudiate their writings and their letters; this is not the foundation. But, in order that you may have proof positive in a short space as to my possessing much greater dexterity for writing about this matter than those my predecessors had, I will first of all explain to you the ultimate matter of minerals, so that you may plainly know on what basis I treat this subject, and hence may more rightly understand what is the beginning. It is necessary that a physician should first be familiar with the disease with which he has to deal; when he knows this, the method of treatment will spontaneously unfold itself. But to know a disease is the end, not the beginning. The art resides in the departure, not in the entrance. The entrance is dark and dubious; the issue is evident. In this knowledge lies hid. I point out this, therefore, as the foundation, namely, that every matter must be thoroughly known at its commencement, so that it may also be more exactly understood for what purpose the matter has been framed. Now, if man ought to lay out before himself the works of God, and rightly use them, it is necessary that they should not be hidden from him; otherwise he will be sure to abuse them. What good is an axe to a person who is ignorant of its purpose? Let him hand it over to one who knows all about it. In the same way, whatever God has created ought to be in the hands of a man who knows how he ought to employ it. Men should know and learn these things, not mere trifles and phantoms conjured up by the devil.

But when I propose to write about the origin of minerals, I shall do this not of myself, but from my experience, and by means of him from whom I myself received it. What I said in my first paragraph, I here repeat, namely, that the last must be known before the first, and from the last the first should be understood. I make this clear from the example of Christ, who was not understood until He sent the Holy Spirit, who, at His coming, revealed all things. By Him we understand Christ, though He came after Christ. So, from the same ultimate, that is, by the Holy Spirit, we now understand both the Father and the Son.

Now this fits in exactly with the philosophy of minerals, because the ultimate matter is made up of those things which teach the beginning of their mother, or of their birth. From them this birth must be understood. Already in other philosophic paragraphs I have named these three substances, Sulphur, Salt, and Mercury, as being the principle of all those things which spring from four matrices, that is, the four elements. In the generation of minerals it is necessary to explain that iron, steel, lead, emerald, sapphire, flint, duelech, etc., are nothing else than Sulphur, Salt, and Mercury. Everything produced by Nature is frail and corruptible, and it can be ascertained by Art from what it has issued forth. And here is a proof from Nature, since those three substances just spoken of are in the air, no less than in other things, such as fire, balsam, mercury, etc. If, by the aid of Art you resolve steel, gold, pearls, or corals, you will still find Sulphur, Salt, and Mercury. When these are extracted by Art, nothing more of that mineral remains, but all is dissolved. Seeing, then, that the dissolution of substances reveals particularly what they are, and what is in them, you can gather that those things are three, namely, Sulphur, Salt, and Mercury. These three are the body, and everywhere there is one body and three substances. Concerning these three substances I will now begin my teaching, by which you may know that in the ultimate matter there are three substances, neither more nor fewer, and out of these three all minerals have been formed. Furthermore, how God created Nature shall also be stated. On this basis nothing shall be found lacking.

In the beginning it pleased God to make one element—water—whereinto He infused the power of generating minerals, so that they might forthwith grow, and thus adapt themselves to human needs. Water, I say, He destined for this office, that it should be the Matrix of the Metals, by means of these three substances spoken of—namely, Fire, Salt, and Mercury. In this arrangement so much foresight and discrimination were observed that from the one element of water were produced metals, gems, stones, and all minerals. And though the fruit be unlike its parent, so God willed that each should be produced according to its own nature. One is a bird of the air, another a fish in the water. And just as these differ one from the other, so do the natures of other created things. All these depend on the power of God, who willed that His good pleasure should be fulfilled in them.

Now, it should first of all be realised that the element of water is the mother of all minerals, though water itself is utterly unlike these. So also is the earth related to wood, though earth is not wood. Nevertheless, wood comes from it. In the same way, stone, iron, etc., are from water. Water becomes that which of itself it is not. It becomes earth, which it is not. So is it necessary for man also to become that which he is not. Whatever is destined to pass into its ultimate matter must necessarily differ from its beginning. The beginning is of no avail.

Now, in water is the primal matter, namely, the three first substances, Fire, Salt, and Mercury. These have certain different natures in them, as will hereafter be pointed out. They have metals, they have gems, they have stones, they have flints, and many things of this kind. One is a metal, another a stone, another a flint. So in the sky, too, one is snow, another thunder, another the rainbow, another lightning. In like manner on earth, too, one thing is wood, another a herb, another a flower, and another a fungus. Such an artificer has God shewn Himself, the Master of all things, whose works no one is able to rival. He alone is in all things. He is the primal matter of all : He is the ultimate matter. He *is* all things. Then, when we come in due succession to explain minerals, we will, in the ensuing discourse, speak before all else concerning the properties of the matrix, that is, the element of water. The things whereof I write were supposed by the ancients to spring from the earth. Their meaning was good ; but the position was incapable of proof. In this point they were defective, as also in the materials for establishing that proof.

The principle, then, was first of all with God, that is, the ultimate matter. He reduced this ultimate matter into primal matter. It is just in the same way as the fruit, which is to produce other fruit, has seed. The seed is in the primal matter. So in the case of minerals, the ultimate matter is reduced to the primal, as in the case of seed. The seed here is the element of water. God determined that there should be water. Then He conferred upon it, besides this nature, that it should produce the ultimate matter, which is in water. This water He subjects to special preparation. That which is metallic He separates into metals and arranges each metal separately by itself. That which belongs to gems He also digested into its own nature. That which is stony in like manner. The same is the case with marcasites and other species.

Moreover, if God created time—harvest for the corn and autumn for the fruits—He also appointed its own special autumn for the element of water, so that there might be a certain harvest and definite autumn for all things. So, too, the water is an element, is the matrix, the seed, the root of all minerals. The Archeus is he who in Nature disposes and arranges all things therein, so that everything may be reduced to the ultimate matter of its nature. From Nature man takes these things and reduces them to their ultimate matter. That is, where Nature ends man begins. The ultimate matter of Nature is the primal matter of man. So, then, by an admirable design, God has appointed that the primal matter of Nature should be water, which is soft, gentle, and potable. Yet its offspring or fruit is hard, as metals or stones, than which nothing is harder. The very hardest, therefore, derives its origin from the very softest—the fire from the water—in a way beyond the capacity of man to grasp. But when the element of water becomes the matrix of minerals, this is not beyond the capacity of Nature. God has produced a wonderful offspring from that mother. You judge a man by his mother. Every one has his own special feelings and properties, not according to his bodily organization, but according to his nature. Thus all metals according to their body are water, but according to their special properties they are metals, stones, or marcasites. In no other way can reason grasp that these things are diverse in substance and in body.

Thus, then, God created the element of water, that it might be the element of all metals and stones; and He separated it from the other three elements into a peculiar body which was not in the air, in the earth, in the sky, but was something special, different from these. This he placed on the lower globe so that it might be above the earth and occupy the cavity in the earth where it lies. He founded it with such wonderful ingenuity that together with the earth it should carry men, who might walk and move upon it. And the first thing which moves our wonder in this respect is that it surrounds and encircles the globe and yet does not fall away from its appointed station; so that the part lying under us is turned upwards just as we are, and in the same way hangs suspended downwards. Then our wonder is increased, seeing that the bed or pit of this genuine element, at its centre of greatest depth, is quite bottomless, so that the water receives no support from the earth on which it lies; but it stands freely and firmly in itself like an egg, nor does anything fall away from the shell; and this is a clear miracle of God.

Now, in this element are the generations of all metals and stones, which exhibit themselves under multifarious natures and forms. Moreover, as you see, all fruits grow out of the earth into the air, and none of them remain in the earth, but go out of it and separate themselves from it, so, growing out of the water, there go forth metals, salts, gems, stones, talcs, marcasites, sulphurs, etc.—all proceeding from the matrix of this element into another matrix, that is, into the earth, where the water completes its operation, but the root of minerals is in the water, as the root of trees and herbs is in the earth. But they are brought to perfection above the earth, and pass on to their ultimate matter, which is entirely in the air.

In like manner is completed on the earth that which grows in the water. So, then, when the root is in the water the growth takes place on the earth, and hence the doctrine of those writers is clearly erroneous who advance the the opinion that minerals grow out of the earth, and that all these minerals, how many soever they be, recognise the earth as their mother. This idea is worth nothing. Indeed, nothing grows from the earth save leaves, grasses, woods, herbs, and the like. Everything else is from the water. Otherwise, by the same method of reasoning, it might be said of the growing things of the earth that they grow in the air since they live in the air; but this is clearly fallacious. Their roots are found in the earth, and hence we learn that their origin is in the earth, but their perfecting in the air. In the same way, that which originates in the water acquires its perfection in the earth. The growth of minerals follows the same course, convincing us that they are aqueous, and proceed from the water, existing in the water as the primal matter of those same minerals, just as all fruits of the earth are generated in the earth, and after the predestined period they burst forth into harvest, or autumn, and generate that which is in them. When a root of this kind is born, it first rises into its own special tree, that is, its body, from which the particular mineral, metal, or other growth, should be produced in the earth. In like manner, also, the nut or the cherry does not spring straightway out of the earth, but first of all the tree is produced, and afterwards the fruit; so, also, in the water Nature first puts forth a tree, which is the aqueous body, and this afterwards grows out into the earth; that is, it occupies the pores of the earth, just as the tree fills the air. When this tree is now put forth into the earth, the fruits are forthwith born, congenital with the tree, according to their nature and condition. Here the metal grows in its own special kind, there some sort of salt is produced, there again some genus of sulphur breaks forth, and elsewhere some sort of gem is protruded. And, just in the same way as many cherries or pears are found on one tree, so similar fruits of the water are found at the extremities, and, as it were, on the shoots of the trees appertaining to the element of water. Again, like as some trees put forth many fruits, and others only few, so, in this case too, there is a similar property, nature, and condition. Trees of this kind, therefore, should first be sought, and afterwards their fruits. Thus, the rustic who pursues his culture in the element of water will be taught and instructed, as the husbandman who plies his craft in the soil is taught how he should pursue his husbandry and where fruits must be found.

Careful attention, too, should be given in this method of generation, so that the illustration from the earth may hold good—in this way : There are some trees which bear their fruit, not nakedly, but under mixed conditions. The chestnut, the nut, and other similar growths, have a bark, thorny in appearance, and inside another, while, lastly, a thin skin encloses the kernel. So, in like manner, there are metals, also, and minerals lying hid in flesh and skin, such as are the ore of iron, the ore of silver, and so on. These have to be removed in order that, after separation, the desired fruit may be extracted. On the other hand, there is another kind which puts forth its fruits nakedly, as cherries, plums, grapes. From these nothing is thrown away, but all is useful and good. So in the aqueous fountain are found pure and naked silver, gold, coral, carabe, and the like. These are all so arranged by Nature that there may be different sorts of trees and of barks, in which the mineral lies, which also depend upon the variety and division of water, climate, and geographical position. That which lies hid within has to be extracted from the bark or shell, just as in the case of fruits. And yet further, as you see in the kernel a body and the kernel itself, so be well assured that, similarly, in the element itself there is a body and a spirit, so that the body has first to be sought for, and then the spirit in the body. Now, it is the spirit that makes the body, and so it makes also the mineral (or the nutriment). The mineral has one body, the fruit another. That is the same as saying that, although there may be gold in a body, and the body is worthless, because impure, and it must be separated by the goldsmith, so gold has a body which is not impure. There are two bodies. In the second is incorporated the fruit of the mineral, which need not be separated from that gold. So then the fruits are first developed out of the element into a tree, afterwards into a body, and within the first shell that which is precious and good. Just as man is a twofold body, a dense body which is worthless, and within this another body which is good, so is it with all growths. Whatever God has created He perfects its corporality by a similar process. He has made man in one way, a tree in another, and a stone in another. But He made man more carefully, because He would that man should be created in His own likeness, so that eternity, in which other created things have no share, might reside in man.

The same judgment is to be passed concerning the death of elements, because water has its own death no less than other things. Indeed, water is its own death, eating into, strangling, and consuming its own growth. We have proof of this in the earth. That which grows from it returns to it and perishes, so that no part of it any longer survives. So yesterday perishes and no man will ever see it again, and it is in like manner with the night past. In like manner also pass away all things born of the earth, which return to the earth, and are consumed by it, and yet it is not heavier by half an ounce then it was yesterday, nor is it heavier to-day than it was a thousand years ago.

Its weight remains one and the same. God has gifted His elements with this peculiarity, that they should give fruits and consume their superfluities, but whither those superfluities have gone no man knows, any more than he knows whither yesterday has gone. In like manner, the element of water is its own death, inasmuch as it consumes and mortifies its own fruits. That death is in the great centre and terminus of water, the open sea, into which all water flows. Whatever passes hereinto dies and decays, passing away even as wood is consumed in the fire. And as, year by year, new fruits emerge from the earth, while the old ones perish, so, every day new minerals are begotten, be they metals, marcasites, gems, stones, salts, or springs. These all come forth girt about with death, as an infant who brings along with it death bound up with life. By the same method of reasoning, metals, too, bring with their own beginning their own death too, and they die in the terminus of the water, that is, in the open sea. The Rhine, the Danube, the Elbe, and other rivers are not the element itself; they are its fruits. The element is in the open sea. It is that out of which all grow and into which all must perforce return, and thus they acquire death whence life is allotted to them This death will be more fully described hereafter in distinct paragraphs, when it is pointed out separately how each mineral comes into being and dies.

Now, with regard to the tree of the element of water, mark this. When Nature is about to put forth any growth into the world—be it gold, silver, copper; be it gem, emerald, sapphire, granate; be it a spring, sweet or brackish, warm or cold; be it coral or marcasite—she then raises up, from the element of water, a tree on the earth, so that its root is fixed in the centre of the sea (or of the matrix). That tree sends forth its seed into the earth, and spreads forth its branches. Know, therefore, that its stock has the form of a liquid, which is not water, oil, bitumen, or mucilage. It has the appearance of wood produced from the earth, but still it is not wood, nor seed (or stock) and yet it is of the earth, and each has its own body. That liquid is the stock, and its branches are that same liquid, just as a tree is wood, and its branches are like in kind. So, then, the mineral tree is formed into a body of this kind, and afterwards divided into its ramifications, so that one branch very often extends from another into a second or third, running out and separately extending itself to a space of twenty, forty, or sixty miles. One branch turns to the German Alps, another to Lungia, another to the Valley of Joachim, and another to Transylvania. Such is its distribution throughout the whole world. In this way innumerable trees are interwoven, wherever the earth extends. As trees grow forth in this fashion, one after another on all sides, their extremities extend to the uttermost parts of the earth. Sometimes they crop up to the surface of plains under the open sky; sometimes they remain in the earth according to the nature and condition which is special to each tree. Hence it follows that at the extremities of the branches the nature of the element of water pours forth its fruits on the earth. As soon as ever these fruits drop on the earth they are at once coagulated, and there is produced from every such tree just what should be produced in proper kind and quantity. When its fruit has been completely shed, that tree withers and dies within itself. It perishes like all other things, and itself passes on to the consummation where all things find their end; while, lastly, according to its nature, a new growth emerges thence.

From this you may learn that the primal matters of all minerals are put together in water, and that this primal matter is neither more nor less than Sulphur, Salt, and Mercury, which are now made the soul, spirit, and true essence of the element. These three substances contain within them all metals, salts, gems, and the like. And when, at the predestined period, it is about to beget those fruits which it cannot help producing, then each genus and species gives birth to that which is like itself. Thus, if any person had different seeds, as many as ever the world produces, mixed together in a bag, and if he were to cast these forth, or to sow them in a garden, Nature, being equal to the occasion, would by-and-by allot to each its own fruit, bringing every separate seed to its own vigour and perfection without injury to the others. Exactly the same is it with the element of water, as though this were the bag filled with seeds of all kinds to be sown. Here, too, every genus and species is brought to its own nature and perfection. God, according to His marvellous plan, has gifted the four elements with these miracles of creation. These are the elements from which issue forth fruits destined for the service of man. Every different kind has been created by God. By such investigations as these the mighty works of God are explored and understood.

Surely, therefore, that philosophy is worthy of all praise which puts forward only the works of God for our consideration. Every man is bound to learn all he can about these, so that he may know what, and how much, his Creator has done for his sake.

True, the enemy has intruded and sown his tares in this philosophy. Such as this are Aristotle, Albertus, and Avicenna, with their accomplices, who are mere tares of the field. That enemy bursting in has devastated everything and begotten other noxious philosophers whose system is destitute of all knowledge of Nature, and is without any foundation at all. Lacking all light of experience, such philosophy violates in the most disgraceful way the light of Nature. Its professors are the busy-bodies who, mixing themselves up with all good things, exhibit themselves to the devil as sons of perdition.

So far, you have heard that the primal matter is conjoined in the matrix as in a bag, being compounded of three parts. As many as are the fruits, so many are the different kinds of Sulphur, Salt, and Mercury. There is one kind of Sulphur in gold, another in silver, another in iron, another in lead, tin, and so on. So, there is one kind in the sapphire, another in the emerald, another in the ruby, crysolite, amethyst, magnet, etc. Furthermore, there is a different kind in stones, flint, salts, fountains, and the rest. And there are not only so many Sulphurs, but so many Salts. There is one Salt in metals, another in gems, another in stones, another in salts, another in vitriol, another in alum. Such, too, is the case with Mercury. There is one kind in metals, another in gems, and so on as before. Yet these things are still only three. One essence is Sulphur, one Salt, one Mercury. Add to this, that all these are still more specially divided. Gold is not one but manifold, as also a pear, an apple, is not one but manifold. There are, therefore, just as many Sulphurs of gold, Salts of gold, Mercuries of gold. The same remark applies to metals and gems. As many sapphires as there are, some more valuable, others more common, so many Sulphurs of sapphire, Salts of sapphire, and Mercuries of sapphire are there. The same is true of turquoise and all other gems. All these things Nature holds, as it were, as in one closed hand, from which she puts forth every separate kind, the best and noblest that she has. Thus, she contributes metals to one genus, and divides that genus into other and various species, all comprising metals. In this way the three primals are to be understood, namely, that they embrace as many created species as grow; and yet they are only composed of one Sulphur, one Salt, and one Mercury. As a painter with one colour depicts numberless figures and forms, no one of which is like another, so Nature is like that painter. In this alone they differ; Nature produces these things with life, while the painter produces only dead ones. Nature's productions are substantial; the painter's are mere shadows.

Then again, the reasoning about colours leads to a similar conclusion. On that head, notice this brief information, that all colours proceed from Salt. Salt gives colour, gives balsam and coagulation. Sulphur gives body, substance, and build. Mercury gives virtues, power, and arcana. So these three ought to be combined, nor can one exist without the other. God gives life to those whom He has predestined to derive it from these as it has seemed good to Him. Now Nature herself extracts the colours from the Salt, giving to each species that colour which is suitable. The body which is appropriate to each it takes from Sulphur. Thus, too, the necessary virtues are derived from Mercury. So, then, whoever wishes to learn the bodies of all things must before all else make himself acquainted with Sulphur. Again, he who desires to know colours must seek his knowledge from Salt. He who wishes to learn virtues let him scrutinise the secrets of Mercury. So he will have laid the foundation for examining the mysteries of every growing thing as Nature has infused these mysteries into each separate species. But you should know that Nature has mixed up such bodies, colours, virtues, one with the other; yet with a little effort it is possible for any one who will, and to whom God gives the power, again to separate them, to form, colour, and endow them. You see and know how it wakens our wonder when from a dusky black seed emerges a tree adorned with its bright and joyous colours, with leaves, fruits, and flowers. This mystery of Nature, as it exists in flowers, is so sublime and great that no one can fully investigate it. God is very much to be admired in His works, and from the contemplation of these one ought not to withdraw by night or day, but constantly to take delight in the study of them. This is in the truest sense to walk in the ways of God.

Moreover, it will be in consonance with my subject, and of practical use as well, if I advise you in one course of the order observed in this book about minerals. This order is different to that which has been pursued by others. First the metals will be treated, and these are not of one kind but distributed according to their own essences and also according to the uses which they supply for men. Some of these are fragile, others durable, and in proportion they are subservient to human convenience. So, also, some gems are useful to man not in their metallic form, but in order that they may be worn, or minister to human health. Such as these are the sapphire, the magnet, the cornelian, etc. These are created in a special form, so that a man may be able easily to carry them about with him. Then, again, there is another kind of stones which man does not use as he uses a metal or a gem, but which he employs for building houses or other receptacles necessary for human life. Further still, another genus is composed of Salts, of more than one species, which are neither metals, nor gems, nor stones, which also are useful for purposes which are subserved neither by metals, gems, nor stones. Moreover, a special order has been assigned to springs, some of which do good to the internal organs of the body, others help it externally. Some are warm and others cold, some acid whilst others are sweet. There are so many different species that one could not exhaustively define them. There are also different kinds of marcasites, two, for example, coloured like gold and silver. But there are very many species in which God has held several things in reserve, which also are put in man's hand that he may seek what he will, and extract from them whatever God has conferred upon them. There are also things that belong to a different genus; talc, of which there are four sorts, red, white, black, and clay-coloured. This genus comprises neither metals, gems, stones, salts, springs, nor marcasites, but something special and by itself. It gives also sand, with a supply of silver. Of this more need not be said than that it is useful for buildings and for making cements. There is also another genus given to us, namely, sulphureous minerals, of which there are two, the clay-coloured and the black; and there are also carabæ.

There are more of this nature, and especially one genus which is allied to no other, in which the health of men is to be found, and it can also be applied to external uses. Besides this there is another genus not like the above-mentioned, namely, corals. Of these the red and the white are well known. Other colours are also found, and forms such as are described in the paragraphs devoted to the subject. Moreover, after these there remains another genus, beyond what is natural, which, by the will of Nature, becomes an instrument of various forms and properties, as the eagle-stone and the buccinæ, cockles, patellæ, etc. The origin of these from the element of the water, you can find in my succeeding paragraph. From the element of water, too, many kinds of fruits are produced; and though I shall only describe those which are known to me, I have found out much more, because the lower globe and the higher sphere, in all their parts, above, below, and on every side, are crammed with such as have been mentioned. I should, therefore, be fully competent to write about these. But still it is true that many are hidden in the world about which I know nothing. Yet neither do others know them. It is, indeed, true that many and various things are about to be revealed by God, concerning which none of us has hitherto even dreamed. For it is true that nothing is so occult that it shall not at length be made manifest. Some one will come after me whose great gift does not yet exist, and he will manifest this.

You should know, however, that there are three parts in this Art, to which the perfections of minerals are compared. These three artifices in the nature of the element are congenital with the three primals. For as man has his gifts in the arts, by which he excels, so also Art affords to them in the matter of the three primals. And it should next be understood that no man can bring to perfection any thing or any work by himself, without some one to help him. No one is superior to another save that man alone who knows how to conjoin what should be conjoined. Iron ore, for example, is ready to hand. But what can it do of itself? Nothing, unless there be added one who will fuse and prepare it. Secondly, this is nothing without a smith to forge it. This, again, is of no practical use unless there be someone to buy it and to apply it to its purposes. Such is the condition of all things. The same thing likewise occurs in Nature, where it is not one thing only which makes a mineral. Others must be added, analogous to the fuser, buyer, seller, and user. If Nature does not supply this work, she deputes it to man, as the primal matter whose duty it is to supply what is lacking. Nature, nevertheless, has need of a dispenser, who will arrange and set in order what ought to be joined together, so that what should be done may find accomplishment. One is ordained by God for this conjunction, and that is the Archeus of Nature. He afterwards requires his operatives to co-operate with him, to fashion the thing, and bring it into that condition for which it is appointed. Hence it follows that three things must be taken which reduce every mineral to its appointed end. These are Sulphur, Salt, and Mercury. Those three perfect all things. First of all there is need of a body in which the fabrication shall be begun. This is Sulphur. Then there is necessary a property or virtue. This is Mercury. Lastly, there is required compaction, congelation, unification. This is Salt. Thus at last the thing is brought about as it should be. But it is not every Sulphur which is a body for gold, nor every Mercury for its virtue, nor every Salt for its unification; but just as there are many blacksmiths, one doing this thing, another that, so also here. God, therefore, has appointed that the Archeus should set in order those things which are to be conjoined, just as a baker, cooking bread, joins together what has to be joined, or a vine-dresser seeks out and joins what has to be joined for the purpose of cultivating his vineyard. Everything is appointed to its own purpose, and everything finds out what is necessary for its own special purpose. Now, if the Archeus has his lead ore, and it be necessary to form a tree in gold, iron, jacinth, granate, duelech, marble, sand, cachimia, or what not, then he takes and combines the three simples, Sulphur, Salt, and Mercury, which are of this nature, and do serve his purpose. Afterwards he casts them into his Athanor, where they are decocted, as seed in the earth. They are decocted again in such a way that Sulphur may add its body, in which the operation consists. They prepare it according to their judgment for that which it ought to be or to become. Next, out of the other two Mercury is decocted for its properties, so that those may be present which ought so to be. When these decoctions have been made, there follows, lastly, conservation, which is brought about by means of Salt. In this way all is coagulated; that is, the Salt first unifies, next congeals, and lastly, coagulates. Now it is strengthened, so that already the autumn is ready and he is at hand who is to beat out the metal. Let this brief account suffice for every generation of metals, namely, in what way they are conjoined. Concerning each one separately, how it is to be dealt with, instruction shall be given in the particular chapter. And this teaching, indeed, concerning minerals is necessary in order that everything may be more rightly and plainly understood, and that you may not be led away by the deceits of the old writers and their followers. They are puffed up with vast self-esteem, and are only approved by those like them, who are as unskilful as themselves, but do not take their ease quite so much, hoping that they may search into and gather these things by more exact study.

II

Concerning the generation of metals, you may be assured that there is a great number and vast variety of them. A metal is that which fire can subdue, and out of which the artisan can make some instrument. Of this class are gold, silver, iron, copper, lead, tin. These are called metals by every one. But there are also, besides these, certain metals which are not reckoned as metals, either in the writings and philosophy of the ancients nor by the common people, and yet they are metals. To these belong zinc and cobalt (which are subdued and forged by force of fire), as also certain granates (accustomed to be so called) of which there are many kinds, themselves also metals. But many more are those which up to this time are not as yet known to me, as are many different sorts of marcasites, bismuths, and other cachimiæ, which produce metals, but of kinds not yet known. Only the principal ones are known, which are more ready and convenient for use, such as gold, silver, iron, copper, tin, lead. The rest are pretty completely neglected, and nobody cares about their properties—neither the smith nor the ironworker, the tinman, brazier, or goldsmith. Nevertheless, these metals are for other operators, not yet born. No one is competent to learn save in one way and by a single art. The assertion that quicksilver is a metal has no truth in it. It belongs to another class of minerals; not being a metal, a stone, a marcasite, or a sapphire, etc. It is a peculiar growth of Nature, gifted with its own body like the rest, and provided with its properties. The custom is passing away, too, of arranging seven metals for the seven planets. From this it arose that, not having full knowledge of metals, people reckoned quicksilver as one of them. According to their comparison of things, gold is Sol, silver is Luna, copper is Venus, lead Saturn, and tin Jupiter. But come, arrange these things. If you join Venus and copper you will soon see how they square and agree with one another. Join and compare lead with Saturn, and notice what happens. Compare tin and Jupiter, and see what fruit will arise. Such philosophy is nothing but rubbish and confusion. Not the slightest vestige of any foundation or light appears in it. Such remarks are merely barbarous, and not philosophy at all. Of the same kind is the assertion that quicksilver is Mercury. Compare the complexion, nature, working, quality, properties, and various virtues and essences, and see how they square one with another and agree. They are quite incongruous. One has not the least likeness to the other. It is true that the Philosophy of Plants has arranged seven herbs according to the seven planets; but these are the mere dreams of physicians, with no stability or power of proof in them. According to them, mercurialis is Mercury, heliotrope Sol, and lunaria Luna. But do you think - you " Fathers "—that you can fly away to the sky and have the power of comparing earth with heaven without any astronomy or philosophy, when you cannot even get a glimpse of what lies hid in so common a growth as the heliotrope? This distribution, therefore, should be admitted by nobody, but ought to be relegated to those who do not judge according to the light of Nature, but by their own long stoles. The chapter on metals teaches you that those metals are six in number, so far as they are known to me, and I have given them above. To these are added a few others —some three or four – which are known to me, and the number and species whereof shall be given in due course. I think it very likely that a large number still remain. For by provings of the metals, many proofs present themselves which are metallic, that is, they are reckoned according to the nature of the six metals, though they do not altogether agree thereto; so that I should augur from this that a great number of metals still remain. Every mineral can be thoroughly known and discriminated if subjected to a sufficient examination.

With regard to the generation of Gold, the true opinion is that it is Sulphur sublimated to the highest degree by Nature, and purged from all dregs, blackness, and filth whatever, so transparent and lustrous (if one may say so) as no other of the metals can be, with a higher and more exalted body. Sulphur, one of the three primals, is the first matter of gold. If Alchemists could find and obtain this Sulphur, such as it is in the auriferous tree at its roots in the mountains, it would certainly be the cause of effusive joy on their part. This is the Sulphur of the Philosophers, from which gold is produced, not that other Sulphur from which come iron, copper, etc. This is a little bit of their universality. Moreover, Mercury, separated to the highest degree, according to metallic nature, and free from all earthly and accidental admixtures, is changed into a mercurial body with consummate clearness. This is the Mercury of the Philosophers which generates gold, and is the second part of the primal matter. The third part of the primal matter of gold, or of the tree from which gold ought to grow, as a rose from a rose-seed, is salt, crystallized to the highest degree, and so highly separated and purified from all its acridity, bitterness, acetosity, aluminous, and vitriolic character, that it no longer has anything of the kind appertaining to it, but is carefully illuminated in itself to the very supreme point, and advanced to the highest transparency of the beryl. These three ingredients in conjunction are gold, which is decocted in the way of which we have already spoken.

Moreover, the genus of gold is not single, but manifold. Its grade is not one only, but Nature of herself gives thirty-two degrees to the finest gold. In our Art, twenty-four degrees are found for establishing the best gold. The cause of this is that gold in its tree is like a cow in the pastures, or like Epicurus in the kitchen. As soon as he has gone out all vigour and animation become fallen and diminished. So is it with gold : because if it be reduced so as to be the first matter of man, then, as if gone out of its kitchen, it at once loses eight out of the thirty-two degrees to which allusion has been made. But there are diversities in the kitchens, too, some being better and others worse. Accordingly as the gold falls into this one or the other, so it is either increased or diminished in degrees from twenty-six degrees as a maximum down to ten degrees as a minimum. The grades below this are too pale and not recognisable. For it is the nature of gold to be either light or dense. This happens from some impediment which occurs from the stars or other elements which aid in the decoction. As one man is more dense or more subtle than another, so neither does gold always attain its complete grade, principally for this reason that too much body, or Salt, or Mercury, has been added, from which fault and error are sure to arise. Too much Salt causes too great paleness. Too much Mercury makes the gold too much the colour of corn. Too much Sulphur confers excessive redness. And it must be remembered, too, that sometimes the weights are unequally divided. Nature sometimes errs as well as men. If this happens, the grade is unequal. It reaches a point from twelve to twenty-four. But if the superfluous weight be removed (as it can be by Art), say, by antimony, by quarta, as it is called, by regal cement, or by other means, the irrelevant weights are removed and the twenty-four degrees remain. Let not the Alchemist, then, attempt rashly to graduate gold, which is done in this way. For the weight in excess is unfit to assume its degree and to be reduced to a just standard. But what is not good of its kind cannot be exalted. Yet it may be that gold which is too pallid in its decoction may be graduated. But a principal item of knowledge with regard to this is that it does not lose its body in regale, antimony, and quarta. Indeed, it persistently retains both its colour and its weight. This is a property of good gold.

Gold becomes white by Sulphur in the manner already detailed. But the other two, Mercury and Salt, are white, and of a golden nature. These so tinge a sulphurous body that it loses its redness and grows white. Sulphur takes the tint of other colours. For though the whole be red, or white, or clay-coloured, its colour is changed by the tincture which is composed of Mercury and Salt. When, therefore, the body is Sulphur, the tincture of Alchemy can easily change its colour. It is necessary, however, in this case, that the other tincture, the Alchemical to wit, should tinge the Mercury and Salt from whiteness to redness. In this way gold assumes the colour which it ought to have. And it should be realised that there are complexions in gold and in other metals, just as there are in man himself.

Another fact which should be accepted is that the white complexion also is changed by corporal transmutation. So also is redness. These two colours separately inhere in redness. Yellowness inheres in whiteness; and these are subject to the primary colours. This transmutation can be effected by means of Alchemy, but under the condition that it shall be directed to the complexions, and that it shall first of all be tested in man, so that one shall be made of a melancholy or a sanguine temperament, just as cattle may be made black or white, and that by a tincture. Nature, indeed, in her mineral working, acts exactly as she does with man in his generation. In the same way man also ought to act in the generation of Nature, as being superior to Nature in this respect, if only Nature has gifted him with the astral mysteries of the arts. This method of treatment, however, I now relegate to astronomy.

Attention also must be paid to the fact that at this juncture Nature takes the lead in matters of the kind described. In Sulphur there is nothing save a body, in Salt nothing, only in Mercury. Sulphur and Salt are so far available that the one gives the body in which is gold, the other adds strength. In what relates to the nature, force, and virtue, all this is due to Mercury. Whatever property there is in Sulphur belongs to all alike. There is nothing in it except body where Mercury is not present. So in Salt. But know that Salt is a balsam, and conserves Mercury so that its virtues and properties shall not putrefy or decay. Thus, this virtue is incorporated with gold, and if it be separated after coagulation in Salt it cannot be detected by Art, as neither can the properties of Sulphur be discovered. But all these are readily found in Mercury. So when Art separates, it deserts the body, nor takes any heed of its medicine. In like manner, it deserts Salt, together with its medicine. And although the body has some influence as a body, and Salt as Salt, still, these medicines must not be sought therein, but only in Mercury, which contains all things. For this is the *rationale* of creation, that in all the outgrowths from the four elements of Nature, not only are those things present which are of themselves seen and understood, but these also contain within them the magnet which, in decoction and preparation, attracts to itself the essences of the three primals, that is, the Quintessence, as the ancients term it, though they ought rather to call it the quart-essence. For the mineral consists of three; and besides these there is the magnet, which is a medicine. The magnet has attracted this and it is found in Mercury. But Mercury itself, too, in its ultimate separation, loses much of its weight.

When Nature is thus prepared and lead to such increase, at first the gold becomes a tree after its kind. This spreads itself, and afterwards are generated the branches. The flower follows; then the fruit. The flower in the earth, like that in any other tree, is at the extremity. And as the flower is at the extremity, and the nucleus too, while yet immature, so there is the same method observable in the generation of gold and of all metals. When the flower falls the fruit is born in its place. This, it is true, does not always burst forth where the flower had stood, but this is the nature of the auriferous tree, that the fruit flourishes sometimes at the distance of several hundred ells in the interior of the tree itself, some straightway in the open air, and others midway between the two. There is thus some difference amongst auriferous trees, the natures of which vary one from the other. Hence they are found distributed in different ways, just as their own peculiar mode of growth is assigned by God to other trees.

Besides, with regard to gold, this fact also deserves to be well weighed, namely, that it is sometimes overloaded with impediments, so that occasionally nothing takes place except a generation of Mercury. If this takes place, it leads one astray. If corrosive salts fall on the flowers, they are eaten away, just as the actual flowers on trees are eaten by worms. The gold, too, is chilled by Mercury or burnt by Salts. There are many mishaps of this kind. The earth, and the firmament, and the air may destroy it. Unless these are fruitful they bring forth no good. As trees are burnt up by a blazing sun, so here also it takes place in the water. The light of philosophy teaches us all these matters, and they are abundantly established by experience. The minerals of gold, therefore, and others, are forced to submit to hindrances of this kind. There is nothing in existence which is not occasionally shaken with its tempests. But there are other impediments which are wont to effect the degree. Of this class are cachimiæ, resins, and other marcasites, which insinuate themselves into the workings, and send forth their tinctures. All these are rejected in the Art.

Concerning Silver

Silver is generated from white Sulphur, Salt, and Mercury, which, being most subtly prepared and rendered transparent, have been restored to a fixed nature, that is, they are fixed from their special nature nearest to gold in a fire of ashes, but not with antimony, regale, and quarta. Here is the difference in fixation between gold and silver, in this respect, that gold is male and possesses masculine virtues, while silver is female and is possessed of feminine virtues. Herein lies the difference between the fixation of gold and of silver. Since gold is male it can bear more fixation, but silver less. Thus the matter of silver is comprised in its primals, as is the case with a woman. Gold and silver, indeed, are of one and the same primal matter; but the same distinction supervenes as exists between a man and a woman.

Concerning Jove

Of the generation of Jove it should be known that it is produced from fixed white Sulphur, fixed Salt, and from Mercury that is not fixed; and for this reason, because Jupiter is fixed according to body, but not in the substance of Mercury. It loses all its fusion and malleability. Afterwards it ceases to be a metal; for the metallic spirit is separated therefrom by Art. As soon as ever this has been done, it is nothing else but white Sulphur, and Salt, and dried Mercury.

Concerning Saturn

Saturn is born from a black, sulphurous, and dense body beyond all other metals. On account of its density it consists of the thickest Mercury and the most fluid Salt, so that there is received into Saturn the most fluid body of Sulphur, Salt, and Mercury. These same, moreover, are the three most dense natures of all the metals. If this metal be dissolved and ceases to be lead, it becomes ceruse, spirit of Saturn, lead ochre, and finally glass. It consists of three colours, the lemon colour it gets from Sulphur, and the white from Mercury. It gets its spirit from Salt, and from all together its vitreous nature, just as all the metals have.

Concerning Iron and Steel

On the other hand, iron is generated from the least fluid Sulphur, Salt, and Mercury, being the very opposite of tin and lead. It is coagulated into a hard metal, and copulated in itself. For two metals are joined together in one, iron and steel. Iron is feminine and steel masculine. This conjugation resembles that of gold and silver, that is to say, the male and female grow together. They can, therefore, be in their turn separated, the female to her sex, the male to his. The female can be applied to her uses, and the male to his in like manner.

Concerning Venus

Copper is generated from purple Sulphur, red Salt, and yellow Mercury. If these three colours be mixed with one another, copper is produced. Now, copper contains within itself its own female element, that is, its scoriæ. If these are separated by Art, and the body reduced, it comes out male. The nature of each constituent is such that the male does not suffer itself to be again destroyed, and the female no longer emits scoriæ. They differ from one another in fluxibility and malleability, as iron and steel differ. If that separation be made, and each consigned to its own nature, two metals are produced, differing altogether in essence, species, and properties.

Note

Such and so many in number are the metals, as I have reckoned them up, namely, gold, silver, tin, lead, iron, steel, female copper, and male copper. Thus they are eight in number. But if—as cannot be the case—iron and steel, and male and female copper respectively, are reckoned each as one metal, there would be only six, and the arrangement would be inconvenient. There are seven well-defined and publicly known metals: gold, silver, tin, lead, iron, steel, and copper, the last being reckoned as one metal, since the male and female are wrought together and not separated, as they ought to be.

Of Mixed Metals

You perceive, from what has been already said, that the male is not always solitary without a consort, but often they co-exist, as in the cases of gold and silver, iron and steel, which grow together in one working, from which each retains its own special nature, but still they are mixed so that one does not impede the other, nor are they of their own accord separated one from the other. Such, too, is often the case with tin and lead. But where they are thus joined no good result ensues from them. They do not square into one body; but it is better that each should be separated into its own body.

Concerning Spurious Metals

Metals can be adulterated. Only gold and silver mix with the other metals, for the reason that they are the most subtle. Only, therefore, when such a primal matter is present, does each grow up together by itself. It may easily be that six or seven different fruits shall be grafted together on the same tree; and there is the same marvellous kind of implantation here in Nature.

Concerning Zinc

Moreover, there is a certain metal, not commonly known, called zinc. It is of peculiar nature and origin. Many metals are adulterated in it. The metal of itself is fluid, because it is generated from three fluid primals. It does not admit of hammering, only of fusion. Its colours are different from other colours, so that it resembles no other metals in the condition of growth. Such, I say, is this metal that its ultimate matter, to me at least, is not yet fully known. It does not admit of admixture; nor does it allow the fabrications of other metals. It stands alone by itself.

Concerning Cobalt

Moreover, another metal is produced from cobalt. It is fluid like zinc, with a peculiar black colour, beyond that of lead and iron, possessing no brightness or metallic sparkle. It is capable of being wrought, and is malleable, but not to such an extent as to fit it for practical use. The ultimate matter of this substance has not as yet been discovered, nor its method of preparation. There is little doubt that the male and female elements are joined in its constitution, as in the case of iron and steel. They are not capable of being wrought, but remain such as they are, until Art shall discover the process for separating them.

Concerning Granates

Besides these, there is another peculiar metal which is found in streams and marshes, in the form of a seed like a large or small bean. It is founded and wrought by itself, but not so as to fit it for making instruments. It is of no practical use, nor is it known what properties it comprises. Unless Alchemy shall disclose its nature, it is not likely to be made clear at all. It allows many mixtures of silver and gold, which penetrate it as they do copper or lead. It is produced from citron-coloured Sulphur.

Note.—Concerning Gems

There are other transparent granates in the form of crystal, wherein are latent both silver and gold.

Concerning Quicksilver

There is, moreover, a certain genus which is neither hammered nor founded; and it is a mineral water of metals. As water is to other substances, so is this with reference to metals. So far it should be a metal as Alchemy reduces it to malleability and capacity of being wrought. Commonly it has no consistence, but sometimes it has. The right opinion about it is that it is the primal matter of the Alchemists, who know how to get from it silver, gold, copper, etc., as the event proves. Possibly also tin, lead, and iron can be made from it. Its nature is manifold and marvellous, and can only be studied with great toil and constant application. This, at all events, is clear, that it is the primal matter of the Alchemists in generating metals, and, moreover, a remarkable medicine. It is produced from Sulphur, Mercury, and Salt, with this remarkable nature that it is a fluid, but does not moisten, and runs about, though it has no feet. It is the heaviest of all the metals.

Note

So far, then, all the metals have been thus described, up to the point that they are known to me, according to their substance and origin, following that guide, and based upon that foundation, which is supplied by the ultimate matter. By means of this the first three are found out, what is their species, and whence they are derived. Indeed, the generation of the others cannot be explained in any way save by experience, which is finally proved by the primal matter in Vulcan. In this way none can err.

Concerning Cachimiæ, that is, the Three Imperfect Bodies

Attention should be paid to a certain genus of minerals which is, indeed, of a metallic nature, but is not a metal. The things which belong to this genus possess peculiar qualities, of which I shall give several instances. For example, all marchasites, which are multifold, red and white, as also pyrites, which are also multifold, white and red, and of another genus than marchasites. There are, moreover, the genera of antimony, which are many, perfect and imperfect; next the varieties of arsenicalia. To these also pertain talcs, auripigments, and many cachimiæ of this kind, which differ with the regions in which they are found. Concerning these we must set down that they are to a certain extent metallic, in that they have a proximate metallic first matter, and descend from the first three metallic principles. Metals such as gold, silver, copper, lead, etc., are incorporated with them. But because they incorporate also a metallic foe, nothing can be extracted from them without alchemy; but these same foes are of great capacity. These are generated in the following order: Marchasites, pyrites, antimonies, cobalts, talcs, auripigments, sulphurs, arsenicalia. I am acquainted with all of these.

General Recapitulation concerning Generation

This chapter and text is entitled Concerning the Three Imperfect Bodies for this reason, that it is concerned with a metallic growth which bears the same relation to metals as tumourous fleshly excrescences bear to natural flesh, as the fungus bears to the herb, or the ape to the man. Of these things some are in the body of sulphur, as marcasites, pyrites, cobalts; others are in the body of mercury, as antimony, arsenicalia, and auripigment; yet others are in salt, as talc.

Of the Generation of Marcasites

Marcasite is of two colours, citrine and white, metallic and brilliant. It is generated from imperfect metallic sulphur, which is destined to become marcasite by a natural necessity.

At the conclusion of the Book about Minerals *there follows in the Geneva folio a brief fragment which is concerned with the three prime principles in their connection with man. It is entitled an*

Autograph Schedule by Paracelsus

There are, then, in human beings only seven planets; four of which are bodies *per se*, not forming part of anything else. There are also other minerals, those of the three primals to wit, which come from Sulphur, Mercury, and Salt, and are specially called mineral, because they are either themselves minerals or form parts of minerals. There are two minerals, and several parts, which enter partially into their composition. Gold, for instance, bears with it three parts, Salt, Sulphur, and Mercury; and all species comprised under minerals are made up of these three parts. Every planet has a perfect Yliadus. The other parts have not the same, as, for instance, sal gemmæ, forming a species, not a part; a marcasite is a species, cachimiæ is a species. But spirits have species in them, as the salt of a gem has Arsenic, fixed Sulphur, and liquid Mercury. The Yliadus, however, differs from the former Yliadus, because the former has his substance and mineral perfect. Minerals have such species; not a manifest body as planets have. Wherefore the Yliadus is to be understood in a twofold sense, one referring to the body, and one to the spirits. The corporal Yliadus is partaker with the spirits of the Yliadus; but the spiritual is not partaker with the former.

APPENDIX II

[The alchemical importance which attaches to a proper conception of the four so-called elementary substances is explained in a note appended to the Philosophy of Paracelsus Concerning the Generation of Elements. The origin, nature, and operation of the three prime principles are, however, of no less moment. As these principles are evidently to be distinguished from salt, sulphur, and mercury of the vulgar kind, it is requisite to accentuate the distinction by contrasting at some length the references to the principles which are contained in the text of the present volume with the knowledge exhibited by Paracelsus on the subject of ordinary salt, sulphur, and mercury. The treatise concerning the first of these substances, which has been here selected for translation, is derived from a collection entitled *De Naturalibus Rebus*, which will be found in the second volume of the Geneva folio.]

CONCERNING SALT AND SUBSTANCES COMPREHENDED UNDER SALT

GOD has driven and reduced man to such a pitch of necessity and want that he is unable in any way to live without salt, but has most urgent need thereof for his food and eatables. This is man's need and condition of compulsion. The causes of this compulsion I will briefly explain.

Man consists of three things: sulphur, mercury, and salt. Of these consists also whatever anywhere exists, and of neither more nor fewer constituents. These are the body of every single thing, whether endowed with sense or deprived thereof. Now, since man is divided into species, he is therefore subject to decay, nor can he escape it except in so far as God has endowed him with a congenital balsam which also itself consists of three ingredients. This is salt, preserving man from decay; where salt is deficient, there that part which is without salt decays. For as the flesh of cattle which is salted is made free from decay, so also salt naturally infused into us by God preserves our body from putrefaction. Let that theory stand, then, that man consists of three bodies, and that one of these is salt, as the conservative element which prevents the body born with it from decaying. As, therefore, all created things, all substances, consist of these three, it is necessary that they should be sustained and conserved by their nutriments each according to its kind. Hence, also, it is necessary that all growths of the earth should gather their nutriment from those three things of which they consist. If they do not, it is inevitable that these first creations perish and die in their three species. These nutriments are earth and rain, that is, liquid. Herein there are threefold nutriments. In sulphur is its own sulphur, in mercury its own mercury, and in salt its own salt. Nature contains all these things in one. So from this liquid, which is the nutriment of natural things, natural salt is decocted.

Hence by parity of reasoning it is clear that man himself also must be nourished in the same way: that is to say, that his sulphur must receive nutrimental sulphur, mercury its nutrimental mercury, and the congenital salt its nutrimental salt, whereby, from these three, man may be sustained and conserved in his species. Whatever burns is sulphur, whatever is humid is mercury, and that which is the balsam of these two is salt. Hereupon depends the diversity of human aliments. Man has need of ardent foods for the sustentation of his sulphur; he wants moist foods for keeping up his supply of mercury, and eats salt to cherish his nature of salt. If this order be violated, that species in the body perishes, whichever species is neglected; and when one part perishes the rest perish with it. This order must be kept in due series. The Academics know nothing of this philosophy, a fact not be wondered at, since in other matters they neither know nor can do anything.

Now, all the world over, there are ardent foods such as flesh-meat, fish, bread, etc. So there are humid foods, as springs, flowing streams, seas. In like manner, there is salt everywhere. These things are distributed over the whole world, so that everywhere the supply of them is ready to hand.

Now, with regard to the nature of man, the following should be accepted. The reason man desires food is on account of his sulphur. Why he needs drink, whether it be water or wine, is on account of the mercury; and the reason of his desiring salt is on account of his salt in himself. These facts are little known, but nevertheless nature does crave for these things. And this is not the case with men only; but animals, too, become fatter, stronger, more useful, and more healthy with salt than without it. If the due quantity of salt be not supplied, some defect arises in one of the two species, so that the animal decays and dies. Its nature is no longer supported by those necessary aliments which it requires. The condition of man is similar. Without nutriments of this kind he cannot live. The appetite of the nature with which he is born requires some satisfaction proportioned to his need. It is reported, indeed, that in certain newly-discovered islands men prepare no food cooked with salt, nor supply such food to their animals, but it is quite certain that their own nature and that of their cattle needs the salt water of the sea, and that they have cooked their food mixed with this. Nature never rests at ease, but constantly catches at and seeks for that which its necessity and use require, and thus compels cattle, not to mention man, to lick salted things.

For ourselves, custom and necessity alike prescribe that we eat salt in our food. Such an ordinance is natural and prudent. In this way three nutriments meet; that is to say, salt and food in one, and with these a third, namely, drink. By these nature is nourished and sustained.

I have said of salt that it is the natural balsam of the living body. That is, so long as the body lives, so long the aforesaid salt is its balsam against putridity. By this balsam the whole body of man, as well as that of other creatures, is kept and conserved. But if there accrue to man any decay or—if I may so term it—any cadaverousness, as in the disease called Persian fire, the reason is that. Now, if everything in creation is to be dissolved, it is clear that even the very balsam itself contains the elements of dissolution, and when once this dissolution begins, its strength and power increase. If the balsam is dissolved or corrupted (and the various modes in which this may take place are given in my Theory of Medicine), then forthwith corruption and decay begin, according to the mode in which the salt has been corrupted. If the salt has not undergone corruption, then neither the external nor the internal body of man decays. Hence we must conclude that salt is like a balsam in man; and that the natural salt which man eats is his food and aliment. I have discussed the subject of salt at some length, for the sake of securing fuller intelligence of the matter. Putting aside, therefore, the idea of a natural balsam, I would point out, moreover, concerning the salt in food, how it is an aliment, and with what gifts it is endowed by God, both for preserving the health of men and for warding off many diseases. But since nothing is so good as not to have some evil combined with it, it remains for us to recount the evil there is in salt, so that in this way the good and evil may be conjoined, and the one separated from the other. The nature and condition of salt are very remarkable. If salt can preserve the dead body or corpse, much more will it preserve the live flesh. If by its power and efficacy salt preserves the dead body from worms, much more the living body, and for this reason, that it is not only an aliment, but a necessary food and a medicine useful for old and young alike. Salt must be supplied to all.

But there are three kinds of salt. There is sea salt, which is salt of itself, not salted by others. As wine differs from water, so the sea in its nature differs from other waters. Other waters are sweet; this is salt. Secondly, there are some springs which are sweet yet salt at the same time. These have a special nature, insomuch as they have that nature not in common with the sea, but of themselves contain a different kind of salt. Thirdly, there are also mineral salts, with the appearance of a stone, of a different kind from other metals or minerals. The best salt is from springs. Next comes that from minerals. The harder it is the better. Then there is sea salt. And as salt is divided into many kinds, so also is it sundered into many and various properties distinct from one another. As to the way in which salt is prepared, there is no need to discuss that subject here, since it is clear enough. Neither is this the place to describe how it grows. That topic belongs rather to the Book on the Generation of Minerals. My intention is to enlarge upon the virtues and vices of salt. In this case there is no need to speak of sea-salt. Whatever is written about white salt applies also to sea salt. Of rock salt not decocted again it is not treated here so much as of salt which has been so decocted. All salt, which is prepared either from water, or out of a saline and mineral, preserves the common order and virtue of salt; for the strongest foundation is in liquid. Sea-salt and rock-salt do not become liquid. But salt which is decocted passes into a liquid before it is separated from the water into coagulated salt. The description of salt, then, is twofold. One is that of salt from liquid; the other of salt which is entire and definite.

It should be known at the outset that this is the nature of every salt in its kind; it is a corrective of foods. When salt is defective food is not corrected. For example : if the stomach takes food which has no salt, its decoction is languid, and its assimilation imperfect. From salt proceeds an expulsive force in the excrement and the urine. If these two functions do not proceed regularly, and the expulsions are not genuine, everything is wasted. Moreover, if the food is not properly salted, it is certain that those liquids in man which take nothing unsalted cannot be fed. The blood becomes disorganised. Where salt is not incorporated or united with the food it is not attracted by the blood. Whatever is sluggishly and faintly attracted occasions decay in the blood. Now, in order to avoid this, and for the sake of those particular members, foods should be salted, so that they may not be deprived of their due nutriment. Moreover, there is a solvent power in salt. If any obstructions of the pores or other accidents arise, salt takes away or removes these, so that they pass away in the urine. Urine is the salt of the blood; that is, it is the salt of natural salt. Natural salt is united with nutrimental salt, and that conjunction causes the excrements to be expelled. If, however, salt is not supplied in due mode and sufficient quantity, a natural conjunction cannot be effected. Now, let every physician know that, since natural salt is wont to issue forth or be expelled by means of salt, the use of salt should be so much the more frequent. It is a great advantage if the salt called sal gemmæ is used, as being much more available than all other salts for expelling the natural salt. It is peculiarly the duty of physicians, therefore, not to neglect the three species of salt and the operations of each, but diligently to use them.

I have said above that the description of salt is twofold, one as a liquid the other as a solid or dry substance. Concerning the liquid, note this fact, that all salt dries up every description of humour that proceeds from the body. Nevertheless, the liquid itself in one hour has more effect than the dried salt would have in a whole month; so much more of a drying nature is there against superfluous humours than in dry salt. Even if dry salt be reduced it is not of equal excellence, as you will learn in its addition and correction. It is accordingly of great importance that the liquid of salt should be correctly described. If the liquid be prepared of such a consistency that it will bear up

Concerning Salt and Substances comprehended under Salt

and sustain a vessel or an egg when thrown into it, its virtue is as follows: whatever diseases are produced from humours, infesting the natural humours, these are purged when the liquid is exhibited. Of this class are moist gout, dropsy, humid tumours, and legs swollen by the influx of humours. To speak summarily, whatever leprous humour not existing naturally it touches, it consumes. It produces such effect in this way: the liquid itself is like a warm bath or hot springs. If it be so refrigerated that the patient can sit in it, he should wash in it as is customary in hot springs, and the like. This, however, should be done on the advice of a prudent physician, as to how long and to what extent the treatment should be continued. Thus those humours are absorbed, the feet cease to swell and are reduced to their natural condition. A sound and firm nature consists in a dry body, not a fat, adipose, and humid one. A dry and muscular body is the best and healthiest. Whatever bodies are not so constituted, but are fat, humid, and flaccid, should all be washed in that bath; thus they will be dried and become healthy. But if it happens that after a bath of such kind in progress of time the superfluous humours again invade the body after an interval, care should be taken that the patient spend his life and dwell near salt springs. A long life is better than a short one, and the pleasures of this world must not be considered. What diseases are of a kind to need this treatment you must learn from physicians.

But now, turning to dry salt, it should be known at the beginning that there are several different kinds, as common table salt, clear salt, sal gemmæ, rock salt, earth salt, and sal stiriatus. Whatever be the case with these, it should be known that any kind of salt put into water and used for washing wounds, preserves them from putrefaction and from worms, and so effectually removes any worms which may have been produced, that none are ever generated again. If wounds are kept pure and clean, they are healed by the operation of Nature herself, even if they are very severe, provided only they have not assumed a poisonous aspect, in which case, for the most part, not even a balsam does any good. So also in virulent ulcers salt is a singular remedy. Besides this, if salt be put into a bath, and a patient washes therein, he is freed from all sorts of scab. In this respect the liquid is more powerful, for it is a potent cure of scab and itch. And here, too, should be noticed the possibility of correction by which dry salt may be to a certain extent reduced to this form.

Salt is useful in many other cases than we have so far recapitulated in external diseases of the body. So many virtues be hid in the use of salt. In conclusion, it should be remarked that in process of time the liquid removes and cures baldness and mange.

Correction and Addition on the subject of a Second Time Correcting and Reducing Dry Salt

The following is a recipe for correcting and reducing back again dry salt: Take common salt and the salt of urine in equal quantities. Let them be calcined according to the rules of Alchemy for two hours. Afterwards let them be resolved in a cell in the usual manner. Thus you will have the reduced liquid. This is of such powerful virtue that in surgical cases it differs little from the true liquid of salt. For internal disarrangements of the body it is much slower in operation. In applying and administering it you will observe the method first mentioned. It should be known, also, that no addition is advisable, since the virtues peculiar to salt are found in no other substance. The less salt there is in other things the fewer similar virtues can be found; and therefore every accessory preparation is useless. If alkalis be decocted these are not a genus of salt, that is, they are not salt, but alkali. There is a difference between salt and alkali in that alkali is natural salt in bodies derived from the three species. But salt is nutrimental, feeding and nourishing even alkali. Therefore, no addition can be made, or any other correction, save only that the salt should be kept by itself without any addition, as was said on the subject of calcination. The same is true concerning the water of salt, which is distilled into a spirit from the calcined substance. This spirit resolves gold into an oil. But if it be again extracted and carefully prepared, potable gold of the most excellent character will be the result. But if without such extraction the gold be resolved, then it is a most subtle object of art for goldsmiths in gilding, and a constant and priceless treasure to other artificers for the same purpose. But, nevertheless, they must be skilled in Alchemy for the work of preparation.

Concerning clear salt, sal stiriatus, and the salt of gems, the fact is that these are most of all adapted to Alchemy, so that silver can be cemented in them after the common mode. In these salts, any Luna, that is, silver, becomes very malleable, and without the aid of fire is wrought almost as easily as lead is. It is also the best purifier of copper if it be reduced to a cement.

Besides the conditions of salt already mentioned, one other property remains. It is this. In whatever place the urine of men or animals is deposited, there salt nitre is afterward produced. The urine being collected and prepared so as to form another salt, is called salt nitre. Now, salt nitre is salt formed from the natural corporeal salt and the salt of food. If these two are joined in man they expel from him what is superfluous by means of the urine, which is nothing else than natural, corporal, and nutrimental salt meeting with other humours. Now, if the urine be excreted into nitre, and stand for some time, then the spirit of salt meeting together in its operation, prepares one salt out of two, and that, indeed, of a peculiar kind. This the Alchemists afterwards extract from the nitre, clarify by alchemical art, and separate that which is not salt from the salt which has been produced. That they clarify again, and then the salt nitre manifests its conditions. In the preparation, however, a separation of the salt may be brought about, so that the true and genuine salt may again be extracted from a certain part, and the rest mixed with the salt of the nitre. Now, the reason why the genuine salt can be again extracted by decoction is, that this salt is not digested in man or in the animal, but is passed out in a crude state, so that it can be detected as such. But that which has been digested is mixed, and, as one may say, incorporated with the corporeal salt so that afterwards it cannot be separated, but passes into the form of salt nitre. No salt in the universe is like this one. Alchemy found it lying hid in nitre, reduced it to the form of a coagulated salt, and then evolved the latent virtue from it, only for purposes of Alchemy and the manual art. They tried to distil sulphur and salt nitre together, but this could not be accomplished on account of the violent chemical action produced. Having accomplished this afterwards by the addition of carbon, the Alchemists discovered gunpowder, and gradually so augmented this by new inventions that now it breaks through walls like a thunder-bolt. Hence it is with good reason called terrestrial lightning. By means of this salt many of the arcana in Alchemy are brought about which need not be described here. We have not yet got at the true foundation or any good end. It is best, therefore, not to write on this subject at all, so that no one may be led astray.

But, so far as relates to the art of Vulcan, it cannot be denied that great secrets be hid there. This subject relates in no way to the health of men, but purely to igneous preparations, which demand a chapter to themselves. The nature of man is indeed wonderful, since, from the body of man or brute, simply from its excrements, and by an internal motion, such a generation is contrived that when it proceeds from living beings it is so violent against life that nothing more destructive can be imagined. It destroys man's life with such swiftness that no defence is sufficiently strong against it. But these matters must be referred to metaphysical science in the Paramirum.

In the beginning of this chapter I said that Nature had incorporated salt in the liquid of the earth. From this salt all growing things have proceeded, and it is the balsam of salt which I have mentioned. It should be known, too, that from this salt another salt is found also in the earth, and like salt nitre. For Nature having pores, cavities, and cataracts in the earth, deposits in them stalactites and long dependent growths with the form and appearance of salt. If these are taken and prepared by the art of salt, they put forth two kinds of salt, table salt and salt nitre. It is called saltpetre, because it adheres to rocks, from which circumstance the name originates. Salt nitre and saltpetre, however, are distinguished by a certain difference. In the probation of salt the nature of each can be easily discriminated. A certain difference, too, can be observed in the species and powers of salt, so far as they relate to health and other matters. At the same time, I do not think it advisable that the salt which is formed from the salt nitre and saltpetre for food should be given man to eat, unless you wish to make him lean and dried up. Otherwise, it is very useful for gunpowder. It acquires another spirit, a different nature and condition.

Now, one must speak of the losses and injuries of salt, for it is well to write of the evil as well as the good. Let this be understood concerning salt, that if it be not digested it is driven from the stomach through the intestines, and in its transit causes so severe a colic and bowel complaint that it can scarcely be cured even by the most careful treatment. It acquires such a strong corrosive force that it seems as though it wished to eat away all the intestines. It has been often discovered by anatomy that a separated salt of this kind has produced perforation of the bowels.

Besides this, if it remains in the stomach it causes craving, heat of stomach, and other ailments, all of which arise from crude salt adhering to the orifice of the stomach. In the case of these patients the physician must take great care to observe whether that salt has proceeded from salted, smoked, or dried foods. Salt is not added in equal portions to every kind of food; and this circumstance should be diligently considered by the physician.

It also happens sometimes that this salt enters the mesenteric veins, and is there granulated and constipated, whence arise many unusual diseases, not only local but extending over the whole body. The same may also occur in those parts to which the urine penetrates on its passage to the emunctories. All this we leave to be weighed by the prudent physician.

Now, therefore, we will conclude as to the matter of salt in its kind. We thought it should be specially described, as it is a German growth. Many more things could be said of it here, but they are not all relevant, and many of them would be injurious, so that I have been unwilling to discuss them. What seemed to me useful I have done my best to impart as the result of my experience.

APPENDIX III

[The treatise which follows constitutes the seventh chapter *De Naturalibus Rebus*, and may be compared with *The Economy of Minerals*, c. 17. It is an addition of considerable importance to the Hermetic Chemistry of Paracelsus.]

CONCERNING SULPHUR

GOD created the resin of the earth and endowed it with many unspeakable qualities, not only for remedying diseases, but also for alchemical operations. Other virtues also are conspicuous in sulphur, which is a resin of the earth. It will be suitable, then, not only to discuss the medical virtues of sulphur, but also to treat of its alchemistical and other uses. Much has been written and published on the subject of sulphur, but no one has ever yet reached the source of its true power. Many authors have undertaken to describe everything, but they understood nothing. They piled up heaps of matter, but deduced nothing from the source as a good writer should do. They did not understand the subject themselves; and though ambition led them to keep on compiling books, those books were without spirit or life, in fact, a mere dead letter.

I, as an experienced man, will lay before you what I have learnt about sulphur, and what is comprised in it as regards medicine, alchemy, and in other respects. Unless God Himself interposes and hinders, the operations of sulphur are stupendous, so that the natural light in man cannot sufficiently admire them. If God does not hinder, any defect is in the artificers, who handle their sulphur so that the result does not correspond to its innate virtue. When every simpleton is made a doctor and every trifler poses as an alchemist, this fact accounts for science not being brought out into open light. And the foundation is that so many arcana and powers of both faculties are contained in sulphur, that they cannot be thoroughly investigated by any— because, I repeat, such excellent virtues are latent therein, they are deservedly the subject of universal wonder. After long experience gained in both faculties, these powers of sulphur were discovered and understood by me, and I realised that scarcely any exist which are superior to them, or which can even be compared to them in medicine and in Alchemy. Sulphur confutes Aristotle when he says that the species of things cannot be transmuted.

Sulphur transmutes them; and if Aristotle were alive at the present day, he would be completely put to the blush and made ashamed of himself.

One who practises as a physician or an alchemist does not use Sulphur as it exists *per se*, but rather as it is separated into its arcanum, and so cleansed from its impurity that it becomes in its virtue whiter than snow. This is accomplished by Ysopus, that is, the art of separating, which was anciently called the Ysopaic art in Alchemy and in all kinds of sequestration. Even when crude, it is remarkable for common use and for all external purposes. But, in order to be quite accurate in explaining Sulphur, I will differentiate it first according to its nature. It is not produced from one matrix, but from many. Hence it has diverse modes of operation, and comprises many natures, differing one from the other. These I will detail separately, so that no physician may make any mistake, and so that it may be clearly known what is its use in medicine, and also how far it is serviceable for Alchemy. When these points are established I will go on to specify its daily uses. So, then, when we shall have explained accurately and in due order its use in medicine, in Alchemy, and in other respects, all its operations will be understood by everybody, so that they will be able to handle it without danger of error.

Concerning the Kinds of Sulphur

As often as you get new metal, so often you get sulphur; because no metal is without sulphur. Every metallic body consists of three things, sulphur, salt, and mercury. In the perfection or generation of metals, however, the superfluous sulphur is removed. You see a nut generated, not simply *per se*, but with a skin and a shell, and you know that these are superfluous save for the embryonic conservation of the kernel, as is explained in the treatise concerning generation. I adduce this illustration to shew that there are as many kinds of sulphur as of metals, each bearing relation to the nature of its own metal. And this is true not only of metals, but of stones. There are as many kinds of sulphur as of stones. All bodies having their own substance are made up of the three constituents just mentioned. On this account they have an embryonated nature. Hence arise different names of sulphur, for example, the embryonic sulphur of gold, silver, sapphire, marble, etc. The sulphur is distinguished by the name of the embryo, which arises from the generation of a single product, be it metal or stone. Nor do I speak of metals and stones only, but also of all the different corporalities, such as vitriol, alum, marcasite, bismuth, antimony, etc. Each of these comprises an embryo, which takes its name according to the speciality of its own generation. For instance, the embryonate sulphur of Mars is different from the embryonate sulphur of vitriol or of jaspis. The same holds good concerning growing bodies of the earth, as woods, herbs, and the like, each of which contains in itself a sulphur of this kind.

One thing should here be mentioned. It sometimes happens that embryonal sulphur of this kind produces metals of fair quality, gems pure and bright, and other matters of like nature, because in that generation whereby they are produced, something is united therewith which is, as it were, a certain spirit of that body. And not only the spirit, but with that same also a corporality, but a subtle and ephemeral one, which cannot sustain any fire. Apart from the Vulcanic operation it is produced in those metals whence it arises, in gold from gold, in lead from lead. Similar preparations are also sometimes made in the sulphureous embryos of gems, by which are separated mutually from one another dead sulphur, of a weak character, and a precious stone, latent in it, all which things have been discovered and investigated by art. But this stone was like that from which it was produced, granate from granate, hyacinthus from hyacinthus. Relegating these things, however, to the alchemical process, we will here point out only what experience has taught and confirmed in the science of finding out secrets of this kind. Let the alchemist, therefore, in investigations of this nature, give his attention to finding out the embryo, lest by chance he light upon something else. Let so much be said, then, concerning one kind of sulphur, as to its origin. Besides this there is another generation of sulphur, *per se*. This I will now describe, and will set forth its virtues in medicine, Alchemy, and other arts.

Sulphur, then, has still another generation, and one peculiar to itself, without any embryonic nature and condition, so that it is a thing growing by itself, like a beech or an oak, separated from other substances by its own special genus. This is called mineral sulphur. This sulphur is a mineral *per se*. And as the Vulcanic art teaches how to separate minerals so that the true body may be taken away from the false,—as silver or iron from its ore,— so also in mineral sulphur there is a body which is extracted, as tin from its zwitter.

That body is mineral sulphur. Of this sulphur there are many different kinds, no one exactly like another. Thus you see in all those things in which Nature abounds for us, that the genus is distributed not into one but many species. There is not only one lead, one copper, one gold. So, there is not only one sulphur; since one sort is of a higher, another of a lower grade, or they have more or less of transparency and clearness. For this reason medical properties also should be sought therein. And this difference should be especially kept in view by alchemists, so that the particular species which is sought may forthwith be found. From this, it is sufficiently clear what are the different kinds and conditions of sulphur, and how they are to be recognised. But beyond these, I should wish you to know of another kind which is a special secret, as follows :—In alchemical separation, gold is dissolved from its corporality, as also silver, every metal, and gems, from all which the sulphur withdraws, is prepared, and extracted. Of this kind are the sulphur of gold, the sulphur of jaspis, the sulphur of vitriol, etc. And in truth, various secrets are here used ; but this sulphur is so excellent an arcanum that nothing like it can be put forward, nor, indeed, ought to be in this place since this matter relates to Vulcan.. So, thus far, we have put forward a triple sulphur. Of these three, I will point out how they are useful to the physician, the alchemist, and the soap maker respectively.

Concerning Embryonated Sulphur

Concerning embryonated sulphur it should be known that it has different virtues according to that from which it is derived, that is, from its generators. Let us use an illustration to explain our meaning. A nut, *per se*, is simply the kernel. But the kernel contains in itself an integument which corresponds to the nut. As in foods, the kernel differs from its integument; so do their virtues differ. Over against this, again, a dry shell is produced, which is of a nature altogether different from the nut. As the bodies differ, so do their properties. Over this, finally, grows a green rind or bark, where the same diversity is once more observed. The chestnut, for example, has these two coatings. And as the chestnut differs from the bark when masticated in the mouth, so do the properties differ. I say this in order that you may understand how embryonated sulphur is also a similar impurity from its embryo and differs from its true products by as wide an interval as its form, essence, substance, and corporality differ. The virtue of the nut is not to be looked for in the shell; so neither is it in embryonated sulphur that one must seek the virtue of gold, silver, tin, copper, emerald, or jacinth; but another virtue must be selected for medicine. Many virtues are hidden in these sulphurs, each differing from the other. This, also, which now we are going to say should be noted before all else, namely, that with all these sulphurs the spirit of arsenic blends, more subtly in one than in another. As is that which is generated, so also is that arsenic sometimes like realgar, sometimes like auripigment, sometimes like crystalline, etc. I adduce these facts in order that you physicians may understand that you ought to be naturalists—not sophists—so that you may know natural substances, and discover what is this arsenic in embryonated sulphur, so that you may not treat men as though you were robbers. They only know the sulphur of the hucksters' shops. They would not even know that if they had not heard it talked about. Yet, all these things ought to be known thoroughly from Nature herself, if we would not lend ourselves to robbery, but would have a good conscience towards God. You Academicians think nothing of this, content with one thing—if money flows into your pockets. You care nothing for God, the Creator of yourselves and of all Nature.

Moreover, note this with reference to the embryonated sulphur of the metals. It can be clearly seen how it firmly conserves and restores its own particular member. For all the seven members require minerals only, and no other remedies for their ailments. Thus the sulphur of gold is beneficial to the heart, the sulphur of silver to the brain, the sulphur of copper to the kidneys, the sulphur of lead to the spleen, the sulphur of iron to the gall, the sulphur of tin to the liver, the sulphur of quicksilver to the lungs. But all these avail in one disease only, as in the suffocation of these members, if there be a flux of humours in them which threatens such suffocation. Although among the ancient and rival physicians no recipes are found against suffocations of this kind, still they one and all decline learning how to prepare these embryonate sulphurs, and to administer them to their patients when necessity requires. I write here, therefore, concerning this one sole virtue, because no medicine has been found for suffocations, which is able to do what these metallic embryonated sulphurs do. As to their other virtues, these will be dealt with under the head of mineral sulphur. They suit all operations; but the metallic are stronger than the mineral sulphurs, and must be used with greater caution.

Moreover, there are also the sulphurs of gems in which precious stones lie as a chestnut within its thorny bark. The constitution of eagle-stones is well known. In the same way, also, all gems are by Nature enclosed in something which is their embryonate. In that embryonate, sulphur lies hid. When this is extracted you have no less virtue than in the stone itself, not, indeed, for wearing, but for using in place of medicine. So it is well known that in the sapphire is concealed the virtue of removing anthrax, and reducing it to an eschara above all other corrosives, and yet without any corrosion. Of the same nature is its sulphur, if, indeed, it be extracted from the body and be used as a plaster. Laid on thus, it produces the same effects. And this is the case not only with anthrax, but also with cancer and Persian fire, especially at the beginning, if it breaks forth with an abscess. Care must be taken, therefore, that from those gems which we Germans have we extract the virtues which are applicable to these special uses. If you have these virtues in gems you will have them also in sulphur, with the same mode of operation. They are not, it is true, equally strong in the sulphur, but still they are there. The application, separation, and gradation cause it to accomplish the same result. The correction and gradation alone tend thereto, otherwise none of these results could be brought about. As in the beginning, I took an illustration from the difference of the shell and the kernel in a nut, so is it to be understood here. But if the kernel of the nut be corrupted or dissolved, so that it is no longer useful for food, the nut still has the same properties as its shell. Let us take a further illustration. Suppose the kernel is burst, and an alkali formed from it, then, in like manner, the shell becomes an alkali too, and both tinge with a black colour those substances which were not previously black. When, therefore, I say that the virtue of the embryonate is like that of the generated, I would be understood thus, if the generated be dissolved and reduced to a Vulcanian preparation. The same must be understood of all the embryonates of gems and the rest.

But with regard to embryonated sulphurs in cachimiæ, such as marchasites, antimony, talc, etc., it should be known that if they are extracted from their bodies and from the matters adhering to them, they produce a similar clear and bright sulphur. In proportion as the degrees hereof are graduated in the operation, the operation itself and the virtue answer to that degree. As this is extracted, so are all other embryonates, of which there is more to be said in their alchemical operation, which cannot properly be recounted here. But the virtue is this, that it rivals those which are generated if it be corrupted in the preparation. Secondly, they are specially useful in all phlegmatic cases, especially in phthisis, peripneumonia, empyemata, and every kind of cough. Whatever can be naturally supplied in any way, that this sulphur brings to its condition. I have no greater desire or longing than that the state of the world to-day, among its princes, kings, and magnates, may be the same as it was in the time and age of the Magi. Then the virtues in all things would so shine forth that all men would admire God, being such a profound artificer as He is, since He has hidden so many miracles in Nature, in order that man may trace them out. The Magi passed away, and the drunkards rushed into their place, and now nothing remains but whoremongers, mockers, robbers, and thieves. One ought to grieve from the heart that there is to-day no Magus flourishing among princes, but all things on every side have degenerated into mere trifling and ineptitude, while wolves sit at our councils, and have the mastery, who by their exactions and their usuries make more than enough gain for themselves and their lords. This fate awaited the Science of the Secrets of Nature, that after the passing away of the Magi, or of Magic, all the sciences also perished together by the same fate; and in their place arose scribes with long garments, and rapacious wolves, who, swaying all rights by their mere nod, threw all things into a state of terrorism. What shall I say? The arts have perished, and in their place a den of robbers has been substituted.

Next in order, concerning the embryonated sulphur in vitriol and its cognates, which are species of vitriol. Know this, that they all produce a wonderful sulphur when animated bodies are separated from their embryonates, as from salt, from the sal gemmæ, from different species of alum, from vitriols, etc. Here I will lay down a general rule for you, namely, that all sulphurs formed from vitriolated salts are stupefactive, narcotic, anodyne, and sleep-producing, with this special property, however, that here the somniferous condition is so placid and gentle that it is free from all harm, and does not act as an opiate, as is the case with henbane, pepper, mandragora, etc., but safely, quietly, effectually, yet without evil consequences. Such a sleep-producing stupefactive, therefore, decocted, prepared, and corrected by Nature herself, is worthy of the highest praise. Physicians are agreed that soporifics of this kind produce many wonderful effects. In opiates, on the contrary, there is so much poison that, except in the form of a quintessence, they cannot be used; and the more confidence should be placed in this present soporific, since we know that there are many diseases which are not curable without anodynes, and of which the whole remedy has been placed by God in these anodynes. This is the reason why I write the more carefully about this sulphur. How it is found and prepared is described in the alchemical process. Here, however, concerning this same sulphur, it may be mentioned that of all the productions of vitriol it is the best known extract, because it is fixed of itself. Then, too, it has a certain amount of sweetness in it, so that poultry will eat it. It sends them to sleep for some time, but they wake up by-and-bye without feeling any evil effects from it. Concerning this sulphur there cannot be two opinions; in all diseases curable by anodynes, without any ill effect, it lulls all passions, soothes all pains, reduces all fevers, and prevents the severe symptoms of every disease. This ought to be the first remedy and preventive in all ailments, being followed up by the quintessence as a tonic. What other means can raise physicians to a higher position, beyond all Apollos, Machaons, Hippocrates, and Polydores? And this is called the philosophers' sulphur, because all philosophers aim at these results—to prolong life for many centuries, to make men live in health and resist disease, and they have found this faculty in its highest degree in this sulphur. That is why they have given it this name. Give your utmost attention that you may learn how to graduate, separate, and purify it.

Besides this there is another kind of embryonated sulphur in wood. This sulphur is only fire, which none can kindle save in wood, which also perishes with the wood. This sulphur exists in all substances which are wooden, or which in burning can be reduced to ashes. It is vegetable, not fixed, and available only for those substances which have to be prepared by fire. Everyone knows that this sulphur indicates the virtue of other sulphurs in that way. As it is itself fire, consuming all things, so every sulphur is an invisible fire consuming diseases. As fire consumes wood visibly, so does the other invisibly. For this reason the element of fire is a great arcanum in all diseases. Whatever physician has not this element of fire in its arcanity— if I may coin that word—cannot boast that he is a true and tried physician. He is a mere tyro, and pilferer of people's purses. One may now say, then, that sulphur is the element of fire. But if you contend that sulphur is fire in its medicinal effect, you must take care that it be reduced to its proper volatility, so that it may vanish like flame, that is, it shall be so subtilised that it will leave its own body, and its own body is separated from it, because it is not an element of fire. The sulphur being reduced to subtlety and volatility, then at length the consuming body must be consumed, that, namely, which is not fixed by Nature. So diseases are not fixed; but the body is fixed against the element; and the element of fire, at least, is opposed to that which is not fixed against it, that is, it is opposed to diseases. Now, if sham physicians had acted thus, if this our philosophy had found a place and acquired development in the schools of medicine, while the triflers and mountebanks, with their blind eyes, were banished, there is no knowing what position might have been reached, while these people would have avoided any number of homicides of which they have been guilty by their rashness. In the meantime, since they have no consciences, what can one do but let them pose as sham physicians? But whoever wishes to be a true physician must hunt out the virtues of the elements in natural things. There he will find, not only truth, but how to cure his patients. There are, then, two kinds of embryonated sulphur, one fixed, but made volatile, the other pure fire. That is to say, one is living fire, the other insensible fire. Each, however, the sensible as well as the insensible, has a like consummation, the one in wood, the other in diseases.

Concerning Mineral Sulphur

The following is a brief dissertation on mineral sulphur. Of the mode of separation from its scoria it is not necessary here to speak. This is treated of in the book on "The Generation of Minerals." It is well, however, to know something of its virtues. It must not be used in its crude form for medicinal purposes, but has to be separated from its fæces. In this way it is a remarkable medicine, if it be raised in the second or third degree from aloes and myrrh. It is an excellent preservative in the plague, in pleurisy, in all abscesses and putridities of the body. Taken in the morning it prevents the pestilence for that day, or pleurisy, or abscesses, especially if it be prepared according to the following prescription. Rec. Of purified sulphur as above described, ʒx.; of Roman myrrh, ʒi. and a half; best aloes (*aloe epaticus*), ʒi.; oriental saffron, ʒss. (half). Mix and make into a powder. Moreover, if it be elevated several times from vitriol (the oftener the better), it then takes into itself the essence and virtues of vitriol. In this way it is a preservative in all fevers, and a curative in every kind of cough, whether recent or of long standing. It is also a preservative against the falling sickness, and a curative in childhood. If it be taken daily it preserves the health, and prevents anything untoward from happening. In business and commerce it is a corrective of wine, so that it remains sound and uncorrupted, and is wholesome for those who drink it. It must not, however, be used in a crude state. It is so powerful a preservative for wine that it leaves nothing impure in the wine, but drives it all out. If wine is treated herewith it does not produce gravel or calculus, apoplexy, abscesses of any kind, fluxions, coughs, fevers, etc. Nothing can be found like it, or of equal efficacy with it, when it is prepared. It is not without reason, therefore, that I here sound its praises. If one had time, a very few pages of this our writing would suffice to establish this point in discussion with the academic doctors. Pearls are not to be cast before swine; and these would rather see people sicken and die than yield a jot of their opinion, although they are not able to be of the slightest use to the sick. But to return to mineral sulphur: observe once more that it must not be used in a crude state, but prepared. The more carefully it is prepared, the better it turns out; at length it throws off all its dregs and poisonous character, and everything in it that is useless retires from it; what remains is a pearl of price and the most desirable of medicines.

Crude sulphur has the property of bleaching red colours with its fumes. It turns red roses into white ones. If it be used medicinally in an elevated state it produces whiteness, but only externally. Moreover, it should be observed that there are several kinds of sulphur, differing in colour. There is, for instance, the yellow, the yellowish, that which is red in a greater or less degree, purple, black, white, ash-coloured; but of these colours none is any use except the yellow.

The more yellow sulphur is, and the more it inclines to gold colour, the better and more wholesome it is. The others contain a good deal of arsenic, realgar, etc., and so are avoided in medicine. But so far as concerns alchemy, these others are better on account of the ingress which they have through such spirits of realgar.

Moreover, it is worth mentioning that this sulphur removes skin diseases and other external affections of the body. In these cases the coloured sulphurs are better than the yellow, on account of their subtle arsenical spirits. If these sulphurs are sublimated with vitriol, alum, sal gemmæ, sal plumosum, etc., several times, they become so subtilised that they completely eradicate skin disease and ring-worm. This treasure is so precious because it removes externally those blemishes which have an internal origin. As the magnet attracts iron to itself, so that it moves from its position and does not remain where it was, so here are magnetic powers which cannot be altogether explained. A single experiment in the Vulcanic art opens up these marvels of Nature.

God has supplied medicine in sufficient quantity. The blindness lies in the fact that no one attempts their preparation, so that the useless may be separated from what is useful. They think it suffices if, like apothecaries, they jumble a lot of things together and say " Fiat unguentum." This has been so far esteemed learning: and the world has returned to such a condition that medicine is mere trifling, and not, as it once was, an art or a science. It is not the artists in medicine, but the mere sophists, who have the pre-eminence. Yet, if medicine were handled by artists, a far more healthy system would be set on foot. Note, then, with regard to sulphur, that when it is granulated it is a most useful medicine for man, not, indeed, taken internally, but exhibited externally, even in the form of fumes. In this way, as we have said, it preserves and conserves, with the addition of some grains of juniper, rosemary, etc.

Concerning Metallic Sulphur: that is, Sulphurs prepared from the Entire Metals

Alchemy has devised certain arts and modes whereby metals are drawn out of their bodies, so that they are no longer metals but a certain destroyed matter which has lost its former condition. On this subject it should be remembered that every metal is made up of three constituents, salt, sulphur, and mercury. Since these three, then, are the primal material of the metals, it follows from hence that these three can be destroyed and dissolved and so subjected to art, that they can be reduced to another essence and transmuted. This destruction having been made, the three primals can be still further separated by art, so that the sulphur remains solitary and by itself, as does the salt, and the mercury respectively. We will speak here of the sulphur, leaving the other two on one side. Sulphur is separated from other metals in this very way. Whatever forces I have assigned to sulphur generally, these also exist in the metallic sulphurs; and the more so because the metal has acquired a special nature from that which makes it a metal. Of these virtues some are conferred on sulphur, so that the metallic is more excellent and more noble than other sulphurs. And the physician ought to know that all the virtues of sulphur are present in this kind of sulphur, graduated to their very highest degree (if I may so say), and endowed with the condition of the metal. Hence, sulphur acquires from gold the virtues of gold, from silver the virtues of silver, from iron the virtues of iron. Whatever iron does, whatever the crocus of Mars does, whatever the topaz of iron does, all these same things the sulphur of iron does. In like manner is it with the sulphur of castrum, of lead, and of other metals. Every physician, therefore, should get possession of these sulphurs. The dose of them is small, but the effect is marked. These should convince the physician that God has set a remedy over against every disease. It this be true, the physician should be produced by magic, whereby he may understand all the secrets of Nature. Thus it will be made clear that Nature has such resources as to heal even the lepers. The physician who is unacquainted with magic is a mere tyro, and will remain such so long as he lives. It is a difficult matter to have understood medicine, and to have visited its innermost shrines, at all events for those who are unacquainted with the Cabbala and with magic.

Concerning the Alchemical Virtues of Sulphur: and First Concerning Embryonated Sulphur

The extraction of embryonated sulphur is brought about by sublimation, and sometimes by descent, if the sulphur be properly ripened and there be a plentiful supply, without the admixture of other bodies. Sometimes, if it be too subtle, it will not admit of sublimation or descension, but must be extracted with strong waters, so that by means of other bodies it may be reduced to water and then coagulated again from the water. There are many kinds of these strong waters, which we will not recount here, but they should be of such a kind as not to take away or change the power of the sulphur. For if they be extracted by art, according to their own concordance, they will not, indeed, be golden, but in alchemy they will be very convenient sulphurs for other preparations. They admit of fixation, and so produce in cements a volatile subtle gold in metals, in such a way that they bear separation in strong waters and put forth their gold. Otherwise, from this sulphur nothing can be hoped for in alchemy, unless it be extracted, according to its concordance, from those things in which it is latent, and afterwards be fixed. If, as is often the case, it contains gold, that is discovered by fulmination. It is likewise so fixed for retaining all volatile gold that it cannot otherwise be restrained, nor is it taken in separation on account of its tenuity of subtle corporality. Many processes have, indeed, been tried for making a tincture out of sulphur. These have not succeeded, because there is no tincture in it. It is, therefore, labour in vain. Unless gold were contained therein, nothing can be sought there, nor ought it to be attempted that gold should be produced in other bodies. There is none of a silver character, only golden, and one kind more so than another. As far as concerns antimony, red talc, gold, marcasite, etc., they are rarely deficient in gold. Whoever wishes to treat this, let him take care to separate the sulphur so subtly that nothing shall depart from the gold. And unless God opposes (for He does not wish all to be rich, and Himself knows the reason why goats have not longer tails), much could be here imparted in few words. But since riches lead the poor man astray, and take away his modesty and humility, adding haughtiness and pride in their place, therefore it is better to be silent and let these people remain poor.

Concerning Mineral Sulphur

Next in order I will impart to you some marvels, though I am aware that this discourse concerning the wonderful use of sulphur in alchemy will be unacceptable to many. It is known to all that the spirit of the sciences does not take holiday, but works constantly and unremittingly, that it may hunt out and discover those facts in the secret things of Nature which God has hidden. With this spirit there goes together for the most part another bad and false spirit, not only in this art, but in others too, even those which regard the soul. But concerning this false spirit I keep silence. The devil, indeed, mixes himself up in all matters, but I make no remark on his deceits. For the sake of this mineral sulphur the alchemical art has made many attempts to form something from it which shall be more than sulphur. Now this itself is a miracle—to make something more out of a thing than the thing of itself is. This, however, God has allowed to be done by art. Now, since this would be the very potency of art, the great Master of the art Himself has, by superintending the art, made experiment as to what can be formed from sulphur, and how; something which is not in the sulphur itself, but, however, can be obtained from it. The woman by herself cannot beget children, but she begets them with the man. If this begetting is to be accomplished it must be by means of two. Here art is the man and the father who brings all things to perfection. But now that stage of the operation has been reached when the spirit of transmutation has given its prescription for making the liver or lung from the oil of flax and sulphur. The distillation of this liver or lung is manifold. But it is found out by operating that this liver is given by milk, which differs in no respect from common milk, but is thick and fat. It also gives a red oil, like blood. This milk and that blood have not confounded their colour and essence in the process of distillation, but these have remained distinct and separate, the white subsiding to the bottom and the red ascending to the top. Art, it is true, has urgently sought to form silver out of the white or the milk and gold out of the red. But I am certain that this has never been able to be done, either by the ancients or by those of more recent times. I say, therefore, that the milk is dead, and nothing is contained in it.

But concerning this red oil, which gives the liver-mark. Any crystal or beryl which has been previously well polished, if it be placed in this oil for some time, namely, for three years, becomes a jacinth. If there be placed in it a ruby which is not highly graduated, in a space of nine years it becomes so clear and bright that it shines in the darkness like a burning coal, and can be seen everywhere. This has been proved experimentally. Alchemists, indeed, have tried to make a carbuncle of it by placing a jacinth of good quality for some time in the oil. But my experience says that this cannot be done. And this colouring does not take place only in the ways that have been mentioned; but the same oil tinges a sapphire also a blue colour, mixed with green. In the same way it colours other gems. Over glass and similar substances it has no power. But it so exalts gems that they attain their highest degree of excellence, a higher one, indeed, than that to which they could be exalted by Nature. Concerning other gradations and colourings of gems nothing more has been heard or written than that the red blood of sulphur colours and tints them. And here observe that all silver, if it be placed therein and left for a due time, by-and-bye grows black, and deposits a calx of gold, which until the proper season is not fixed, but is a volatile and immature substance. If, however, it reaches its proper limit, by its own despatch it hastens on other things, about which I must not say more here. So, then, remark concerning sulphur, that if it be duly graduated, the more subtle, beautiful, high, and quick in operation it is, the higher and greater will be the result. In this way metals and stones are formed. Let him who is about to make the attempt not think, but be sure, that he can do it. For this is the most perilous work of all in alchemy, needing for its accomplishment great experience and continual practice. It should not depend on mere hearsay, but on manifold practice. Of the virtues themselves and how they are graduated, I cannot say anything. I speak of the colourings only, that they should be exalted to the highest degree. But that this should take place in colours I do not think possible. This is a tincture not of virtue, but of mere colour.

Concerning the Use of Sulphur of the Metals in Alchemy

I have several times in this chapter mentioned sulphur prepared from the decomposed metals, and added something as to their use in medicine. So far as relates to alchemy, I would have you know that many have tried to extract from it a tincture with which they should change things from one tint to another. This has not been successful, for a reason not to be mentioned here. But whoever has the sulphur of gold can by means thereof graduate other gold from 24 to 36 grains or more, so that gold cannot mount any higher, whilst it abides and remains in antimony and quartarium. But the sulphur of silver, too, so exalts silver in its whiteness, that if copper and silver are mixed in equal proportions, they cannot be discriminated by the needle or the Lydian stone, but both seem to be equally pure and choice silver. In the same way, by the sulphur of copper, the metal copper can be brought to such a state that it is proof against lightning, even though it be not graduated, and retains its own colour. From the sulphur of Mars is made the best and most excellent steel. From the sulphur of Jove, the best tin, which will bear the lightning. From the sulphur of Saturn is made fixed Saturn, which gives neither white lead, nor minium, nor any other spirit. The sulphur of quicksilver reduces quicksilver to such a point that it can be wrought with the hammer, and bears the fire as well as copper does. The ashen fire, however, it does not bear. This is the power exercised by the sulphur of the metals over its own special metals. This is *per se*. If the sulphur of gold is applied to silver it colours it, but has no power of fixation; and this is always the case with the transmutation of sulphur into some other metal.

So far, then, you have learnt how many kinds there are, and what are the nature, properties, and essence of sulphur. Whoever wanted to say all that can be said about sulphur would consume a great deal of paper. The subject demands a careful workman, a ready and skilled artist, one who does not shout or traffic in trifles, who does not deal with his art by mouth and tongue only, but puts it to the test of work itself. Miracles will abound for such an one. He who knows nothing about sulphur is a man of no worth, unskilled both in philosophy and medicine, and conscious of none of Nature's secrets.

APPENDIX IV

THE MERCURIES OF THE METALS

IN the year 1582 an octavo edition of the *Archidoxorum Libri Decem* was published in Latin at Basle, and included several other treatises of great importance, some of which are absent from the Geneva folio. Among these there is one upon the Mercuries of the Metals, which fills a somewhat curious lacuna in the writings of Paracelsus, as there is no other extant work attributed to him which treats individually of Mercury, while concerning Salt and Sulphur there is an abundance of material which not a little embarrasses selection. It is entirely devoted to experiments, and it will be consequently of the more value to practical students of early chemistry.

A LITTLE BOOK CONCERNING THE MERCURIES OF THE METALS, BY THE GREAT THEOPHRASTUS PARACELSUS, MOST EXCELLENT PHILOSOPHER AND DOCTOR OF BOTH FACULTIES

Extract aquafortis out of 4 lbs. of salt nitre, with 3 lbs. of green vitriol, ʒii. of alum, and ʒi. of sal ammoniac. After it has subsided with a little copper, dissolve in this water ʒi. of crude sal ammoniac, which has previously been slightly pounded. Let there be hence produced aqua regis through V. In this water dissolve ʒii. of gold, which has previously been well and most exactly purged by antimony. After the dissolution has taken place let the calx subside; effect separation by drawing off the aquafortis; and then reduce the calx by washing to a sweet condition. For this purpose wash six or seven times with sweet water until no sharpness of the aquafortis any longer remains. Subsequently dry the calx over a slow fire, weigh it, and you will find that a third part of the weight has been extracted. Thereunto add an equal proportion of very finely pounded sulphur, a double quantity of vitriol, and white calcined tartar to the weight of all the aforesaid things. Pound all of them very finely, place in a glass vessel, and pour upon the top exceedingly strong vinegar, together with salt water, so that aqueous matter may swim upon the top to the height of two fingers, more or less. Seal the vessel effectually, and place it in a cupel, or alchemistic furnace, for thirty days. The furnace must not be of sufficient heat to burn the finger when placed therein. At the expiration of the time specified break the glass, when the matter will be in the form of washed silver, or calx of silver which is friable into small grains. Mercury, meanwhile, is not visible. Therefore place the said matter in a mortar, and pound with a wooden pestle, for Mercury is compelled by pounding. Let this process continue until Mercury shall become complected, and a live matter, or body, shall have been produced. Nevertheless, it is not so quickly produced or composed as Mercury of Saturn. Next cleanse the remaining matter with fresh and clear water; dry it perfectly; and you will have Mercury of the Sun, when the gold will be no longer fixed but voluble, and can be sent through the corium, whereby any impurity which may chance to remain is separated.

Mercury of the Moon

Let silver be reduced to thin plates, in such a way that it may be easily removed, and at the same time well purified. Sprinkle one of such plates with strong vinegar, and set aside in a humid place for a short space of time, until it becomes completely blue. Then dissolve with aquafortis separated by the separation of solution, and after it subsides, and the aquafortis has been affused but not sweetened by washing, and dried gradually, pour vinegar again upon the calx, and then separate until the whole becomes completely blue.

Then take ʒii. of mountain or mineral cinnabar ground to a very fine powder, and afterwards ʒi. respectively of calx of silver, cinnabar, alumen, sulphur, and vitriol. When ground subtly place all these in a jar, including the calx of silver, which ought to sink to the bottom. Furthermore, cover the surface of the matter, or compound, at the top of the jar, with welding sand, such as the workers in iron are accustomed to use. Afterwards place this jar, mouth downward, on the top of another jar, which must be filled with pure water, and hidden in the earth by descent. About the upper jar kindle a slow fire, and increase it more and more until the whole of the said upper jar shall become white with heat. Let it cool a little, and the Mercury of the Moon will be found in the lower jar. Let the jar remain for two hours, more or less, at a white heat, and thus out of ʒii. of the Moon is produced ʒi. and a half of Mercury, which is altogether like crude Mercury. This is again pressed through the corium, so that the pure may be separated from the impure.

Mercury out of Venus

Take copper reduced to thin plates and purged to the utmost of all dross. Divide it into small particles, and confect with salt on a tigillum, layer upon layer. Seal the upper orifice of the tigillum, so that nothing may evaporate. Place the said tigillum on the hottest part of a brick furnace for nine days. Then take out the copper, when it will be of red colour approaching blackness. Pound it with salt in a mortar as soon as it has been removed from the tigillum. Macerate the powder in strong wine, and let there be added to each ʒv. of subtly ground arsenic, ʒi. and a half of copper. Leave each of these together for the space of fifteen days. Let the measure of wine be sufficient to swim over the powder to the height of two straws. When removed there will remain an excellent, brilliant, whitish calx. Wash this in fresh spring water.

Take of the Calx, ℥ii.
of Sulphur, ℥ii.
of Gluten of Sulphur, ℥ii.
of Vitriol ⎫
of Arsenic ⎬ each ℥ss.
of Alum ⎭

Mix each of these, when very finely pounded, with half a measure of the best vinegar. Let them all be distilled through the alembic until no further water can be extracted. Then add fire, remove the water, and there will collect on the side of the top of the alembic a white powder. This is the Mercury of Venus. Sprinkle this upon hot water, and it will flow together. It is sufficient if the cucurbite be at a white heat. From one pound of Venus ℥ii. and a half of Mercury are obtained; such Mercury is altogether thin and subtle, and is so soluble that it escapes in boiling water. Wherefore the said water is only tepid.

Mercury out of Mars

Reduce Mars into coarse filings, but avoid chalybs, wherefore filings *de calcaribus* are the best. Take thereof ten pounds, and sprinkle well with salt water; leave for ten days or longer—the longer the better. Afterwards wash Mars in such a fashion as to avoid separating the turbidity. At length the water becomes clear, for the turbidity sinks to the bottom in the form of a red viscosity. Separate the water gradually by straining; keep the matter; dry it so that no excremental or gross part may remain. Take of this viscid matter ℥v., of pounded sulphur ℥xxx., compound delicately to the form of fine flour. Place it in the tigillum. Seal up securely, so that nothing may escape, and let the tigillum glow for an hour. Then let it cool, break it, and a grey powder will be found. Add thereto ℥i. of spume of glass, ℥ss. of sal ammoniac, and ℥v. of vitriol. Place on a smooth stone in a humid spot, and the water will flow out. But leave it for ten days; crush it in the hands, and you will have live Mercury, which is Mercury of Mars. Out of ten pounds of Mars one drachm and a half will be obtained. It is black and dull in colour.

Mercury of Jupiter

Jupiter is calcined in the following fashion: Take filings corresponding in grossness to the back of a knife. Place in good distilled vinegar for twelve hours. Dry, and there will adhere a whitish cuticle. Remove this carefully with the hare's foot and set it apart. Again moisten, and again dry the filings; separate a similar cuticle a second time, and repeat this process till there is enough of the white calx. This take, and subject to all the processes to which the Calx of Lead is subjected, but avoiding the addition of Succinum, or white vitriol. Put green copper rust in place thereof, and the work will be accomplished. Jupiter does not yield so much Mercury as Saturn, for one pound of the metal produces less than ℥vi.

Mercury of Saturn

Take Villarensian Lead, or any other in which there is no silver, otherwise it must be purged in the following manner: If it has been calcined, let the calx boil for two whole hours in a lixivium composed of willow ashes, in which have been first dissolved one ounce of alumen and eight ounces of salt. In coction it is purged of all sulphur and other viscous matter. Calcine this lead in the following manner with salt: Melt the lead, pour it into a wooden receptacle, mix it well with common salt, and it will be reduced to a powder like sand. Cleanse the salt away ten or twelve times, till no saltness remains in the lead. Dry the calx gradually by continual agitation. When it has been dried, produce water as follows: Take of white vitriol, otherwise called succinum, five ounces, and one measure of vinegar, to six pounds of calx of lead. Dissolve the vitriol in vinegar. Sprinkle the calx of lead with this water, or perfectly saturate, or so place the calx in water that it protrudes above it. Leave it for thirty-six hours, so that it becomes an ashen-coloured powder. Then take a light marble vessel, the larger the better. Put it obliquely in a humid place, or in a wine cellar, and in front of it so place a wooden receptacle that it will receive whatever shall flow out. Calx of lead may be dissolved with three measures thereof. Again take this water and add to it a small quantity of fresh matter, which will concrete in the form of flour at the bottom. Place it in a similar marble, put a copper operculum over it, and make a small charcoal fire at the top of the operculum. When the said matter receives the heat Mercury comes forth; the fire is preserved in good order and grade until no more of the calx of lead remains. Therefore, Mercury of Saturn which flows into that wooden receptacle should be well washed and purified, so that if perchance crude Saturn flows down at the same time, as often happens, the same may be separated. From ten pounds of Saturn are made eight pounds, and often eight pounds and a half, of Mercury. Note.—Let not the fire in the operculum be too great or fierce, for otherwise much crude matter of Mercury will flow down at the same time. To the said marble apply a copper operculum corresponding to the size of the marble, which operculum should be at its sides and ends of the height of a spithmia, and should be shaped like a frying-pan. Let the front part be open for the passage of the Mercury. Then take greyish powder made from lead, together with succinous matter, and add to it the following water:

> Take of Alum, ℥i.
> of Salt Nitre, ℥ii. and ss.
> of Mountain Verdigris, ℥ss.
> of Rock Salt, ℥ii.

Pound these substances minutely, saturate with stale wine, then distil, and there will proceed water of yellowish colour (golden, or crocus). To this water add semi-vitrified calx of lead, and the calx will sink at the bottom. Afterwards gradually pour off the water. Set the same apart, because it never turns putrid, and a centenarius of fire should therefore be maintained in an equal and moderate grade. If the crude Mercury flow forth at the same time, it remains after passing through the corium, and must be cooked out from the rest, for another confection, and thus thou hast Mercury of Saturn by the most simple way.

APPENDIX V

DE TRANSMUTATIONIBUS METALLORUM

IN the year 1581 a *Congeries Paracelsicæ Chemiæ de Transmutationibus Metallorum* appeared in octavo at Frankfort. In the notes to the *Aurora of the Philosophers, Concerning the Spirits of the Planets*, and elsewhere, some references have been already made to this work, which antedates by nearly a century the Geneva edition of the writings of Paracelsus. As its title indicates, it attempted to collect and digest into a single methodical treatise the whole substance of alchemy, as taught and practised by Paracelsus. While in many respects the digest was passably well done, and affords a tolerably representative notion of the opinions and experiments of Theophrastus, it is perhaps needless to say that, as it was included in the compass of a small volume, it is really very meagre. There is, however, one point in which it may be of value to the student. The *Congeries* is, in all probability, an adaptation of autograph manuscripts, and where its readings, which is by no means invariably the case, can be distinguished from editorial interpolations and extensions, they may be useful in so far as they vary from the readings of the Geneva folio and some other less carefully supervised editions. Perhaps, after all, the value, such as it is, of this point, is likely to be appreciated only by that very small circle of readers who believe that in ancient practical alchemy there are chemical secrets hidden which are unknown to the chemistry of to-day. For these the importance of a perfect text of the old alchemical processes, whether in the case of Paracelsus or in that of any other recognized master, is no doubt very high. In the present instance, the difficulty of distinguishing between the text and its editor, in so far as there are substantial variations, makes it needless to tabulate the readings, and the purpose of this appendix is of a far less pretentious character. There are a few paragraphs in the *Congeries* which it has not been found possible to identify in the collected editions of Paracelsus, or at least they offer very conspicuous differences, and these it is desirable to cite. The first has regard to the erection of the philosophical furnace, which *The Aurora of the Philosophers* affirms it is difficult to describe, at least as regards its form, while the specific direction contained in the third treatise, *Concerning the Spirits of the Planets*, only partially corresponds to what is stated in the following excerpt, which constitutes the fourth chapter of the *Congeries Paracelsicæ Chemiæ*. The editorial argument which follows is also worth inclusion, as it is concerned with a matter which, in more than one instance, must have struck the reader of the present translation, namely, that it is not altogether easy, in every case, to harmonize Paracelsus with himself.

Concerning the Visible and Local Instruments : and first of all concerning the Spagyric Uterus

Before we come to the matter, we must describe in order all the instruments, both actual and local, which are required in this art. We have said that the first actual instrument is the fire. The first local instrument is the furnace, designated by the ancients under the alchemical name of athanor. This takes the place of the uterus in spagyric generation.

Hermes Trismegistus, though he was not the inventor of this art, no less than Paracelsus in spagyric medicine, deserves to be called its restorer.

He asserts that this spagyric work, in which human philosophy reaches its extreme point, originated in the meditative contemplation of the greater world, intimating that the spagyric athanor ought to be constructed in exact imitation of the heaven and earth. In order to exercise the ingenious it will not be amiss to examine this comparison, and I think I shall be able thereby to profit my readers.

No philosopher will deny that the sun generates a sun like itself; but it is not every one who will acknowledge that this foetus exists in the centre: least of all will those disciples of the philosophers who have no other knowledge of the Actnæan fire than that which comes from the fleshly eye, just like rustics in this respect. This terrene sun of the lower, or elementary, machinery is kindled by the fire of the higher sun. Just in the same way the centre of our matter is kindled by the centre of our world, or athanor, which is a fire, discharging after a manner the function of the natural sun.

Who does not see—I ask you, my brethren—that the form of the whole created universe has the similitude of a furnace, or, to speak more respectfully, the form of that which contains the matrix of a womb—the elements, that is to say, in which the seeds of the sun and the moon, cast down by the stars in their different influxes, are decayed, concocted, and finally digested for the generation of all things? These things are transparently clear, I will not say to philosophers, but even to boys, wherefore we will not insist upon them further.

Let us come, then, to the construction of our athanor.

First let a furnace be built seven spans in height, and let the rounded interior be the height of one span, the lower part a little broader than the upper part inside, and let it be polished, so that the coals, when put in, may not stick through the roughness of the surface, but may be able easily to fall down through the grating while they are being burnt. To equalise this let there be two or three holes, with which two or three lateral or uterine furnaces—or, if you like, a single one—shall correspond: the breadth of the mouths should be four fingers. To every furnace let a brazen vessel be fitted, and these are to be filled with water. Let the others be shut up; as the egg in the hen, so is the glass in its uterus for the work of the magistery. Then, when you are going to work, and all has been carefully prepared, having broken coals into pieces the size of walnuts, or a little larger, fill up the turrets with these and kindle the fire at the door beneath; but let the top be kept shut, so that the coals at the top or in the middle may not be kindled, thus stirring up a heat that shall destroy the whole work, and burn everything together. When the heat shall appear to exceed what is proper, it can be controlled by applying a small brick or tile to the mouth; on the other hand, if it be too slack, let the coals be stirred up with an iron rod beneath the grating. The fire can be still more readily controlled by registers (which are called governors). Experience teaches the uses of those things which are necessary in preparations before you have arrived at this stage. The fire, then, having been regulated to a just proportion, as Nature teaches in all things, the heat will cause a fermentation, and by-and-bye this will affect the matter lying hid in the egg.

Henceforwards, just as the sun in the great universe shines, illuminates, and gives life to the rest of the stars and to the elements, so the spagyric fire, illuminating its athanor, with all the instruments, and warming the sea-bath, acts just like a hen which hatches its animated egg.

But I hear a giant roaring like a lion against the furnace, and seeking Paracelsus to devour him. "See," says he, "how he contradicts himself! Just now he told us, and that with considerable severity, not to build a fire with coals; now he is teaching the use of coals for this art of his!" You have touched the matter, no doubt, but only in the same way as you have judged that the other writings of Paracelsus are contradictory. Open the other eye, my one-eyed friend, or you will act the part of a blind man passing a judgment about colours. Can you not see what is the meaning of this particle—simple and without middle meaning—added to the interdiction of coals? Do you not see how Paracelsus, though dead, answers you and his other calumniators in his living works, saying—"You, who adjudge me to err, yourselves err, even when judged by yourselves. Have I not often admonished you, and those like you, envious people that you are, in almost all my works, not to pass over even a little word which you have not thoroughly appropriated, lest the same thing happens to you, giants, fighting with my pigmy homunculus, as formerly happened to Goliath fighting with the boy David? Take care, I say, lest you collide with this stone of ours as the great mystery, and sink down with it to the abyss whereto you would consign me!" Thus seems Paracelsus to thunder forth in his tomb. We must not, my brothers, speak unfairly of the dead, even of those whose deserts were small. Let them all remain at rest, and all await their deeds. It is easy to carp, but to avoid judgment to-day is difficult; at the last it will be impossible. What are you which I am not, or what am I which you are not? Again, what has happened to both of us save that which may occur to another, namely, to err? We are all men; and error happens to men more than it does to brutes without reason, who are stirred solely by the promptings of Nature. I confess that I err in very many things, and you err. It is yours to confess it, and mine to admonish, not to judge. Be it your duty, as it is mine, not enviously to disclose those things of your brothers which have not been duly done, before that, with a certain amount of modesty, you have admonished him according to the discipline of true philosophy, otherwise neither you, nor he, nor I, if we act in a contrary manner, are worthy of the name of philosophy. But this philosophic discipline (alas, that it should be so!) is impugned even by the most learned: and so much has the dogma of heathen philosophers prevailed, being at the present day very celebrated among these people, that they do not take a comprehensive glance at what is without and within; they display nothing beyond a mere ambition for honour and renown. This is the chief end of their study and toil. Hence it has come about that everybody tries to get credit for himself by tripping up or blackening the character of somebody else. But these wretched people do not consider that no great ill can be done which will not incur a greater punishment still, at least if we are all foolish together, and nobody even approaches wisdom. Many are wise, a few very wise, who still preserve no medium. Let us look to it, then, lest what we parade as wisdom may, even in this our day and generation, be turned by the good God into open folly, and that through our own efforts. If, then, we have erred at all, and our own conscience tells us that we have done so, by which error on our part it seems only too likely that we may mislead others, let us in the presence of God and men retract, and that without reserve. The wisest of men are ever ready to acknowledge their common error, but stubborn men and fools are not ready to do this. Every made course seems the straight road for them, and *vice versâ*.

The following passage appears as a sequel to the *Treasure of Treasures*, which, in a somewhat modified form, occupies the thirteenth chapter of the *Congeries*. It may be entitled

The Phœnix of the Philosophers

The exposition of the cabalists has, under the name of the Phœnix, that it is the Flying Eagle, whose feathers fly without the wind, and bear the body of the phœnix to its nest, in which is nourished the element of fire. Its young peck out their mother's eyes with their beak, and there is produced a whiteness in its separated sphere. In this consists the life of its heart and the balsam of its intestines. According to the Cabalists this refers to the sulphur of cinnabar, to which Paracelsus alludes. Very lately, when electrum was being treated, I referred the reader to cinnabar, and not without cause, since it has the greatest affinity therewith. What is cinnabar but a composition or mixture of two minerals, sulphur and quicksilver? What, too, is electrum but a mixture of two or more, whether minerals or metals? The sulphur of Sol, therefore, joined artificially with philosophic Mercury of Luna,— why should not this be electrum, why not cinnabar? Whether each is made by Nature or by chemistry, the component parts do not differ.

The last citation which it will be necessary to make is the fifteenth chapter of the *Congeries*. It is an exceedingly concise abridgment of the fifth book of the *Archidoxies* as regards the section on the Stone of the Philosophers, and it is inserted at this point as an illustration of the method of the editor. It is called

A Very Brief Process for Attaining the Stone

I neither am nor wish to be a teacher or a follower of that Stone which is taught in different ways by very many. Leaving, therefore, this process for its attainment, I have proposed to describe in very few words that which has been discovered by me through practice and experience. This, no less than the other, affects the bodies of men, though it is not prepared by the same process. Take, then, mercury, otherwise the element of mercury, and separate the pure from the impure. Afterwards let it be reverberated even to whiteness, and sublimate this with sal ammoniac until it is resolved. Let it be calcined and dissolved again, and digested in a pelican for one month, being afterwards coagulated into a body. This is no longer burnt, or in any way consumed, but remains in the same condition. The bodies penetrated by it are permanent in the cineritium, so that they cannot be reduced to nothing or be altered; but it takes away, as we have often said, all superfluous qualities both from sensible and insensible things.

Although I have here set down a very brief way and process, it requires long labour, and one that is involved in many intricate circumstances; demanding, at the same time, an operator who is unassailable by fatigue, and in the highest degree diligent and expert.

APPENDIX VI

THE VATICAN MANUSCRIPT OF PARACELSUS

IN the nineteenth chapter of his *Rituel de la Haute Magic*, Eliphas Levi observes that "amongst the rare and precious books which contain the mysteries of the Great Arcanum, there must be placed in the first rank the *Chemical Pathway*, or *Manual* of Paracelsus, which contains all the mysteries of demonstrative physics and of the most secret cabala. This manuscript work, unique and priceless, exists in the library of the Vatican. A transcription of it was made by Sendivogius, and this was made use of by Baron Tschoudy for the compilation of the Hermetic Catechism contained in his work entitled *The Burning Star*. This catechism, which we point out to instructed cabalists, as a substitute for the incomparable treatise of Paracelsus, contains all the veritable principles of the great work, after so satisfactory and explicit a manner that a person must be absolutely wanting in that quality of intelligence which is requisite for ocultism if they fail to attain the absolute truth when they have studied it." The manuscript to which reference is made in this interesting citation, is still an unedited treasure, although, as will be seen in the next appendix, there has been at least one *Manual* attributed to Paracelsus, which has been in print for four centuries. In the absence of the Vatican treatise, the student who desires to make acquaintance with a work of Paracelsus which adepts in the Art of Alchemy seem to prefer before all published writings of the same author, must make shift with the Hermetic Catechism, as suggested by Eliphas Levi. He will find it an exceedingly, succinct, and simple presentation of the fundamental alchemical theories. Though it may not initiate the reader, whatever the quality of his intelligence, into the mystery of the Great Arcanum, it is, in its way, very lucid and direct. Whether this merit belongs to Paracelsus or his interpreter, is an unprofitable subject of speculation in the absence of the original text, which few persons have had the opportunity or disposition to to consult. The work of Baron Tschoudy was published in two volumes at Hamburg, in 1785, and later on there was another edition at Paris. Having regard to its period, it is a sensible, though somewhat romantic, attempt to trace back Free Masonry to its historical origin, while, over and above this, it constitutes a valuable hand-book of the analogies which subsist between that system and Hermetic science, more especially Alchemy. The catechism itself, which is the most important section of the *Burning Star*, teems with analogies of this kind, which, of course, are the creation of the editor, and are suppressed in the translation which follows, in part because they exceed the intention of the present work, and in part for other reasons.

A SHORT CATECHISM OF ALCHEMY
FOUNDED ON THE MANUAL OF PARACELSUS PRESERVED IN THE VATICAN LIBRARY

Q. What is the chief study of a Philosopher?
A. It is the investigation of the operations of Nature.
Q. What is the end of Nature?
A. God, Who is also its beginning.
Q. Whence are all things derived?
A. From one and indivisible Nature.
Q. Into how many regions is Nature separated?
A. Into four palmary regions.
Q. Which are they?
A. The dry, the moist, the warm, and the cold, which are the four elementary qualities, whence all things originate.
Q. How is Nature differentiated?
A. Into male and female.
Q. To what may we compare Nature?
A. To Mercury.
Q. Give a concise definition of Nature.
A. It is not visible, though it operates visibly; for it is simply a volatile spirit, fulfilling its office in bodies, and animated by the universal spirit—the divine breath, the central and universal fire, which vivifies all things that exist.
Q. What should be the qualities possessed by the examiners of Nature?
A. They should be like unto Nature herself. That is to say, they should be truthful, simple, patient, and persevering.
Q. What matters should subsequently engross their attention?
A. The philosophers should most carefully ascertain whether their designs are in harmony with Nature, and of a possible and attainable kind; if they would accomplish by their own power anything that is usually performed by the power of Nature, they must imitate her in every detail.
Q. What method must be followed in order to produce something which shall be developed to a superior degree than Nature herself develops it.
A. The manner of its improvement must be studied, and this is invariably operated by means of a like nature. For example, if it be desired to develop the intrinsic virtue of a given metal beyond its natural condition, the chemist must avail himself of the metallic nature itself, and must be able to discriminate between its male and female differentiations.

Q. Where does the metallic nature store her seeds?
A. In the four elements.

Q. With what materials can the philosopher alone accomplish anything?
A. With the germ of the given matter; this is its elixir or quintessence, more precious by far, and more useful, to the artist, than is Nature herself. Before the philosopher has extracted the seed, or germ, Nature, in his behalf, will be ready to perform her duty.

Q. What is the germ or seed, of any substance?
A. It is the most subtle and perfect decoction and digestion of the substance itself; or, rather, it is the Balm of Sulphur, which is identical with the Radical Moisture of Metals.

Q. By what is this seed, or germ, engendered?
A. By the four elements, subject to the will of the Supreme Being, and through the direct intervention of the imagination of Nature.

Q. After what manner do the four elements operate?
A. By means of an incessant and uniform motion, each one, according to its quality, depositing its seed in the centre of the earth, where it is subjected to action and digested, and is subsequently expelled in an outward direction by the laws of movement.

Q. What do the philosophers understand by the centre of the earth?
A. A certain void place where nothing may repose, and the existence of which is assumed.

Q. Where, then, do the four elements expel and deposit their seeds?
A. In the ex-centre, or in the margin and circumference of the centre, which, after it has appropriated a portion, casts out the surplus into the region of excrement, scoriæ, fire, and formless chaos.

Q. Illustrate this teaching by an example.
A. Take any level table, and set in its centre a vase filled with water; surround the vase with several things of various colours, especially salt, taking care that a proper distance intervenes between them all. Then pour out the water from the vase, and it will flow in streams here and there; one will encounter a substance of a red colour, and will assume a tinge of red; another will pass over the salt, and will contract a saline flavour; for it is certain that water does not modify the places which it traverses, but the diverse characteristics of places change the nature of water. In the same way the seed which is deposited by the four elements at the centre of the earth is subject to a variety of modifications in the places through which it passes, so that every existing substance is produced in the likeness of its channel, and when a seed on its arrival at a certain point encounters pure earth and pure water, a pure substance results, but the contrary in an opposite case.

Q. After what manner do the elements procreate this seed?
A. In order to the complete elucidation of this point, it must be observed that there are two gross and heavy elements and two that are volatile in character. Two, in like manner, are dry and two humid, one out of the four

being actually excessively dry, and the other excessively moist. They are also masculine and feminine. Now, each of them has a marked tendency to reproduce its own species within its own sphere. Moreover, they are never in repose, but are perpetually interacting, and each of them separates, of and by itself, the most subtle portion thereof. Their general place of meeting is in the centre, even the centre of the *Archeus*, that servant of Nature, where coming to mix their several seeds, they agitate and finally expel them to the exterior.

Q. What is the true and the first matter of all metals?

A. The first matter, properly so called, is dual in its essence, or is in itself of a twofold nature; one, nevertheless, cannot create a metal without the concurrence of the other. The first and the palmary essence is an aerial humidity, blended with a warm air, in the form of a fatty water, which adheres to all substances indiscriminately, whether they are pure or impure.

Q. How has this humidity been named by Philosophers?

A. Mercury.

Q. By what is it governed?

A. By the rays of the Sun and Moon.

Q. What is the second matter?

A. The warmth of the earth—otherwise, that dry heat which is termed Sulphur by the Philosophers.

Q. Can the entire material body be converted into seed?

A. Its eight-hundredth part only—that, namely, which is secreted in the centre of the body in question, and may, for example, be seen in a grain of wheat.

Q. Of what use is the bulk of the matter as regards its seed?

A. It is useful as a safeguard against excessive heat, cold, moisture, or aridity, and, in general, all hurtful inclemency, against which it acts as an envelope.

Q. Would those artists who pretend to reduce the whole matter of any body into seed derive any advantage from the process, supposing it were possible to perform it?

A. None; on the contrary, their labour would be wholly unproductive, because nothing that is good can be accomplished by a deviation from natural methods.

Q. What, therefore, should be done?

A. The matter must be effectively separated from its impurities, for there is no metal, how pure soever, which is entirely free from imperfections, though their extent varies. Now all superfluities, cortices, and scoriæ must be peeled off and purged out from the matter in order to discover its seed.

Q. What should receive the most careful attention of the Philosopher?

A. Assuredly, the end of Nature, and this is by no means to be looked for in the vulgar metals, because, these having issued already from the hands of the fashioner, it is no longer to be found therein.

Q. For what precise reason?

A. Because the vulgar metals, and chiefly gold, are absolutely dead, while ours, on the contrary, are absolutely living, and possess a soul.

Q. What is the life of metals?

A. It is no other substance than fire, when they are as yet imbedded in the mines.

Q. What is their death?

A. Their life and death are in reality one principle, for they die, as they live, by fire, but their death is from a fire of fusion.

Q. After what manner are metals conceived in the womb of the earth?

A. When the four elements have developed their power or virtue in the centre of the earth, and have deposited their seed, the *Archeus* of Nature, in the course of a distillatory process, sublimes them superficially by the warmth and energy of the perpetual movement.

Q. Into what does the wind resolve itself when it is distilled through the pores of the earth?

A. It resolves itself into water, whence all things spring; in this state it is merely a humid vapour, out of which there is subsequently evolved the principiated principle of all substances, which also serves as the first matter of the Philosophers.

Q. What then is this principiated principle, which is made use of as the first matter by the Children of Knowledge in the philosophic achievement?

A. It is this identical matter, which, the moment it is conceived, receives a permanent and unchangeable form.

Q. Are Saturn, Jupiter, Mars, Venus, the Sun, the Moon, etc., separately endowed with individual seed?

A. One is common to them all; their differences are to be accounted for by the locality from which they are derived, not to speak of the fact that Nature completes her work with far greater rapidity in the procreation of silver than in that of gold, and so of the other metals, each in its own proportion.

Q. How is gold formed in the bowels of the earth?

A. When this vapour, of which we have spoken, is sublimed in the centre of the earth, and when it has passed through warm and pure places, where a certain sulphureous grease adheres to the channels, then this vapour, which the Philosophers have denominated their Mercury, becomes adapted and joined to this grease, which it sublimes with itself; from such amalgamation there is produced a certain unctuousness, which, abandoning the vaporous form, assumes that of grease, and is sublimised in other places, which have been cleansed by this preceding vapour, and the earth whereof has consequently been rendered more subtle, pure, and humid; it fills the pores of this earth, is joined thereto, and gold is produced as a result.

Q. How is Saturn engendered?

A. It occurs when the said unctuosity, or grease, passes through places which are totally impure and cold.

Q. How is Venus brought forth?

A. She is produced in localities where the earth itself is pure, but is mingled with impure sulphur.

Q. What power does the vapour, which we have recently mentioned, possess in the centre of the earth?

A. By its continual progress it has the power of perpetually rarefying whatsoever is crude and impure, and of successively attracting to itself all that is pure around it.

Q. What is the seed of the first matter of all things?

A. The first matter of things, that is to say, the matter of principiating principles is begotten by Nature, without the assistance of any other seed; in other words, Nature receives the matter from the elements, whence it subsequently brings forth the seed.

Q. What, absolutely speaking, is therefore the seed of things?

A. The seed in a body is no other thing than a congealed air, or a humid vapour, which is useless except it be dissolved by a warm vapour.

Q. How is the generation of seed comprised in the metallic kingdom?

A. By the artifice of *Archeus* the four elements, in the first generation of Nature, distil a ponderous vapour of water into the centre of the earth; this is the seed of metals, and it is called Mercury, not on account of its essence, but because of its fluidity, and the facility with which it will adhere to each and every thing.

Q. Why is this vapour compared to sulphur?

A. Because of its internal heat.

Q. From what species of Mercury are we to conclude that the metals are composed?

A. The reference is exclusively to the Mercury of the Philosophers, and in no sense to the common or vulgar substance, which cannot become a seed, seeing that, like other metals, it already contains its own seed.

Q. What, therefore, must actually be accepted as the subject of our matter?

A. The seed alone, otherwise the fixed grain, and not the whole body, which is differentiated into Sulphur, or living male, and into Mercury, or living female.

Q. What operation must be afterwards performed?

A. They must be joined together, so that they may form a germ, after which they will proceed to the procreation of a fruit which is conformed to their nature.

Q. What is the part of the artist in this operation?

A. The artist must do nothing but separate that which is subtle from that which is gross.

Q. To what, therefore, is the whole philosophic combination reduced?

A. The development of one into two, and the reduction of two into one, and nothing further.

Q. Whither must we turn for the seed and life of metals and minerals?

A. The seed of minerals is properly the water which exists in the centre and the heart of the minerals.

Q. How does Nature operate by the help of Art?

A. Every seed, whatsoever its kind, is useless, unless by Nature or Art it is placed in a suitable matrix, where it receives its life by the coction of the germ, and by the congelation of the pure particle, or fixed grain.

Q. How is the seed subsequently nourished and preserved?

A. By the warmth of its body.

Q. What is therefore performed by the artist in the mineral kingdom?

A. He finishes what cannot be finished by Nature on account of the crudity of the air, which has permeated the pores of all bodies by its violence, but on the surface and not in the bowels of the earth.

Q. What correspondence have the metals among themselves?

A. It is necessary for a proper comprehension of the nature of this correspondence to consider the position of the planets, and to pay attention to Saturn, which is the highest of all, and then is succeeded by Jupiter, next by Mars, the Sun, Venus, Mercury, and, lastly, by the Moon. It must be observed that the influential virtues of the planets do not ascend but descend, and experience teaches us that Mars can be easily converted into Venus, not Venus into Mars, which is of a lower sphere. So, also, Jupiter can be easily transmuted into Mercury, because Jupiter is superior to Mercury, the one being second after the firmament, the other second above the earth, and Saturn is highest of all, while the Moon is lowest. The Sun enters into all, but it is never ameliorated by its inferiors. It is clear that there is a large correspondence between Saturn and the Moon, in the middle of which is the Sun; but to all these changes the Philosopher should strive to administer the Sun.

Q. When the Philosophers speak of gold and silver, from which they extract their matter, are we to suppose that they refer to the vulgar gold and silver?

A. By no means; vulgar silver and gold are dead, while those of the Philosophers are full of life.

Q. What is the object of research among the Philosophers?

A. Proficiency in the art of perfecting what Nature has left imperfect in the mineral kingdom, and the attainment of the treasure of the Philosophical Stone.

Q. What is this Stone?

A. The Stone is nothing else than the radical humidity of the elements, perfectly purified and educed into a sovereign fixation, which causes it to perform such great things for health, life being resident exclusively in the humid radical.

Q. In what does the secret of accomplishing this admirable work consist?

A. It consists in knowing how to educe from potentiality into activity the innate warmth, or the fire of Nature, which is enclosed in the centre of the radical humidity.

Q. What are the precautions which must be made use of to guard against failure in the work?

A. Great pains must be taken to eliminate excrements from the matter, and to conserve nothing but the kernel, which contains all the virtue of the compound.

Q. Why does this medicine heal every species of disease?

A. It is not on account of the variety of its qualities, but simply because it powerfully fortifies the natural warmth, which it gently stimulates, while other physics irritate it by too violent an action.

Q. How can you demonstrate to me the truth of the art in the matter of the tincture?

A. Firstly, its truth is founded on the fact that the physical powder, being composed of the same substance as the metals, namely, quicksilver, has the faculty of combining with these in fusion, one nature easily embracing another which is like itself. Secondly, seeing that the imperfection of the base metals is owing to the crudeness of their quicksilver, and to that alone, the physical powder, which is a ripe and decocted quicksilver, and, in itself a pure fire, can easily communicate to them its own maturity, and can transmute them into its nature, after it has attracted their crude humidity, that is to say, their quicksilver, which is the sole substance that transmutes them, the rest being nothing but scoriæ and excrements, which are rejected in projection.

Q. What road should the Philosopher follow that he may attain to the knowledge and execution of the physical work?

A. That precisely which was followed by the Great Architect of the Universe in the creation of the world, by observing how the chaos was evolved.

Q. What was the matter of the chaos?

A. It could be nothing else than a humid vapour, because water alone enters into all created substances, which all finish in a strange term, this term being a proper subject for the impression of all forms.

Q. Give me an example to illustrate what you have just stated.

A. An example may be found in the special productions of composite substances, the seeds of which invariably begin by resolving themselves into a certain humour, which is the chaos of the particular matter, whence issues, by a kind of irradiation, the complete form of the plant. Moreover, it should be observed that Holy Scripture makes no mention of anything except water as the material subject whereupon the Spirit of God brooded, nor of anything except light as the universal form of things.

Q. What profit may the Philosopher derive from these considerations, and what should he especially remark in the method of creation which was pursued by the Supreme Being?

A. In the first place he should observe the matter out of which the world was made; he will see that out of this confused mass, the Sovereign Artist began by extracting light, that this light in the same moment dissolved the darkness which covered the face of the earth, and that it served as the universal form of the matter. He will then easily perceive that in the generation of all composite substances, a species of irradiation takes place, and a separation of light and darkness, wherein Nature is an undeviating copyist of her Creator. The Philosopher will equally understand after what manner, by the action of this light, the empyrean, or firmament which divides the superior and inferior waters, was subsequently produced; how the sky was studded with luminous bodies; and how the necessity for the moon arose, which was owing to the space intervening between the things above and the things below; for the moon is an intermediate torch between the superior and the inferior worlds, receiving the celestial influences and communicating them to the earth. Finally he will understand how the Creator, in the gathering of the waters, produced dry land.

Q. How many heavens can you enumerate?

A. Properly there is one only, which is the firmament that divides the waters from the waters. Nevertheless, three are admitted, of which the first is the space that is above the clouds. In this heaven the waters are rarefied, and fall upon the fixed stars, and it is also in this space that the planets and wandering stars perform their revolutions. The second heaven is the firmament of the fixed stars, while the third is the abode of the super-celestial waters.

Q. Why is the rarefaction of the waters confined to the first heaven?

A. Because it is in the nature of rarefied substances to ascend, and because God, in His eternal laws, has assigned its proper sphere to everything.

Q. Why does each celestial body invariably revolve about an axis?

A. It is by reason of the primeval impetus which it received, and by virtue of the same law which will cause any heavy substance suspended from a thread to turn with the same velocity, if the power which impels its motion be always equal.

Q. Why do the superior waters never descend?

A. Because of their extreme rarefaction. It is for this reason that a skilled chemist can derive more profit from the study of rarefaction than from any other science whatsoever.

Q. What is the matter of the firmament?

A. It is properly air, which is more suitable than water as a medium of light.

Q. After the separation of the waters from the dry earth, what was performed by the Creator to originate generation?

A. He created a certain light which was destined for this office; He placed it in the central fire, and moderated this fire by the humidity of water and by the coldness of earth, so as to keep a check upon its energy and adapt it to His design.

Q. What is the action of this central fire?

A. It continually operates upon the nearest humid matter, which it exalts into vapour; now this vapour is the mercury of Nature and the first matter of the three kingdoms.

Q. How is the sulphur of Nature subsequently formed?

A. By the interaction of the central fire and the mercurial vapour.

Q. How is the salt of the sea produced?

A. By the action of the same fire upon aqueous humidity, when the aerial humidity, which is contained therein, has been exhaled.

Q. What should be done by a truly wise Philosopher when he has once mastered the foundation and the order in the procedure of the Great Architect of the Universe in the construction of all that exists in Nature?

A. He should, as far as may be possible, become a faithful copyist of his Creator. In the physical chaos he should make his chaos such as the original actually was; he should separate the light from the darkness: he should form his firmament for the separation of the waters which are above from the waters which are below, and should successively accomplish, point by point, the entire sequence of the creative act.

Q. With what is this grand and sublime operation performed?

A. With one single corpuscle, or minute body, which, so to speak, contains nothing but *fæces*, filth, and abominations, but whence a certain tenebrous and mercurial humidity is extracted, which contains in itself all that is required by the Philosopher, because, as a fact, he is in search of nothing but the true Mercury.

Q. What kind of mercury, therefore, must he make use of in performing the work?

A. Of a mercury which, as such, is not found on the earth, but is extracted from bodies, yet not from vulgar mercury, as it has been falsely said.

Q. Why is the latter unfitted to the needs of our work?

A. Because the wise artist must take notice that vulgar mercury has an insufficient quantity of sulphur, and he should consequently operate upon a body created by Nature, in which Nature herself has united the sulphur and mercury that it is the work of the artist to separate.

Q. What must he subsequently do?

A. He must purify them and join them anew together.

Q. How do you denominate the body of which we have been speaking?

A. The RUDE STONE, or Chaos, or Iliaste, or Hyle—that confused mass which is known but universally despised.

Q. As you have told me that Mercury is the one thing which the Philosopher must absolutely understand, will you give me a circumstantial description of it, so as to avoid misconception?

A. In respect of its nature, our Mercury is dual—fixed and volatile; in regard to its motion, it is also dual, for it has a motion of ascent and of descent; by that of descent, it is the influence of plants, by which it stimulates the drooping fire of Nature, and this is its first office previous to congelation. By its ascensional movement, it rises, seeking to be purified, and as this is after congelation, it is considered to be the radical moisture of substances, which, beneath its vile scoriæ, still preserves the nobility of its first origin.

Q. How many species of moisture do you suppose to be in each composite thing?

A. There are three—the Elementary, which is properly the vase of the other elements; the Radical, which, accurately speaking, is the oil, or balm, in which the entire virtue of the subject is resident—lastly, the Alimentary, the true natural dissolvent, which draws up the drooping internal fire, causing corruption and blackness by its humidity, and fostering and sustaining the subject.

Q. How many species of Mercury are there known to the Philosophers?

A. The Mercury of the Philosophers may be regarded under four aspects; the first is entitled the Mercury of bodies, which is actually their concealed seed; the second is the Mercury of Nature, which is the Bath or Vase of the Philosophers, otherwise the humid radical; to the third has been applied the designation, Mercury of the Philosophers, because it is found in their laboratory and in their minera. It is the sphere of Saturn; it is the Diana of the Wise; it is the true salt of metals, after the acquisition of which the true philosophic work may be truly said to have begun. In its fourth aspect, it is called Common Mercury, which yet is not that of the Vulgar, but rather is properly the true air of the Philosophers, the true middle substance of water, the true secret and concealed fire, called also common fire, because it is common to all mineræ, for it is the substance of metals, and thence do they derive their quantity and quality.

Q. How many operations are comprised in our work?

A. There is one only, which may be resolved into sublimation, and sublimation, according to Geber, is nothing other than the elevation of the dry matter by the mediation of fire, with adherence to its own vase.

Q. What precaution should be taken in reading the Hermetic Philosophers?

A. Great care, above all, must be observed upon this point, lest what they say upon the subject should be interpreted literally and in accordance with the mere sound of the words: For the letter killeth, but the spirit giveth life.

Q. What books should be read in order to have an acquaintance with our science?

A. Among the ancients, all the works of Hermes should especially be studied; in the next place, a certain book, entitled *The Passage of the Red Sea*, and another, *The Entrance into the Promised Land*. Paracelsus also should be read before all among elder writers, and, among other treatises, his *Chemical Pathway*, or the *Manual* of Paracelsus, which contains all the mysteries of demonstrative physics and the most arcane Kabbalah. This rare and unique manuscript work exists only in the Vatican Library, but Sendivogius had the good fortune to take a copy of it, which has helped in the illumination of the sages of our order. Secondly, Raymond Lully must be read, and his *Vade Mecum* above all, his dialogue called the *Tree of Life*, his testament, and his codicil. There must, however, be a certain precaution exercised in respect to the two last, because, like those of Geber, and also of Arnold de Villanova, they abound in false recipes and futile fictions, which seem to have been inserted with the object of more effectually disguising the truth from the ignorant. In the third place, the *Turba Philosophorum*, which is a collection of ancient authors, contains much that is materially good, though there is much also which is valueless. Among mediæval writers Zachary, Trevisan, Roger Bacon, and a certain anonymous author, whose book is entitled *The Philosophers*, should be held especially high in the estimation of the student. Among moderns the most worthy to be prized are John Fabricius, François de Nation, and Jean D'Espagnet, who wrote *Physics Restored*, though, to say the truth, he has imported some false precepts and fallacious opinions into his treatise.

Q. When may the Philosopher venture to undertake the work?

A. When he is, theoretically, able to extract, by means of a crude spirit, a digested spirit out of a body in dissolution, which digested spirit he must again rejoin to the vital oil.

Q. Explain me this theory in a clearer manner.

A. It may be demonstrated more completely in the actual process; the great experiment may be undertaken when the Philosopher, by the medium of a vegetable menstruum, united to a mineral menstruum, is qualified to dissolve a third essential menstruum, with which menstruums united he must wash the earth, and then exalt it into a celestial quintessence, to compose the sulphureous thunderbolt, which instantaneously penetrates substances and destroys their excrements.

Q. Have those persons a proper acquaintance with Nature who pretend to make use of vulgar gold for seed, and of vulgar mercury for the dissolvent, or the earth in which it should be sown?

A. Assuredly not, because neither the one nor the other possesses the external agent—gold, because it has been deprived of it by decoction, and mercury because it has never had it.

Q. In seeking this auriferous seed elsewhere than in gold itself, is there no danger of producing a species of monster, since one appears to be departing from Nature?

A. It is undoubtedly true that in gold is contained the auriferous seed, and that in a more perfect condition than it is found in any other body; but this does not force us to make use of vulgar gold, for such a seed is equally found in each of the other metals, and is nothing else but that fixed grain which Nature has infused in the first congelation of mercury, all metals having one origin and a common substance, as will be ultimately unveiled to those who become worthy of receiving it by application and assiduous study.

Q. What follows from this doctrine?

A. It follows that, although the seed is more perfect in gold, it may be extracted much more easily from another body than from gold itself, other bodies being more open, that is to say, less digested, and less restricted in their humidity.

Q. Give me an example taken from Nature.

A. Vulgar gold may be likened to a fruit which, having come to a perfect maturity, has been cut off from its tree, and though it contains a most perfect and well-digested seed, notwithstanding, should anyone set it in the ground, with a view to its multiplication, much time, trouble, and attention will be consumed in the development of its vegetative capabilities. On the other hand, if a cutting, or a root, be taken from the same tree, and similarly planted, in a short time, and with no trouble, it will spring up and produce much fruit.

Q. Is it necessary that an amateur of this science should understand the formation of metals in the bowels of the earth if he wishes to complete his work?

A. So indispensable is such a knowledge that should anyone fail, before all other studies, to apply himself to its attainment, and to imitate Nature point by point therein, he will never succeed in accomplishing anything but what is worthless.

Q. How, then, does Nature deposit metals in the bowels of the earth, and of what does she compose them?

A. Nature manufactures them all out of sulphur and mercury, and forms them by their double vapour.

Q. What do you mean by this double vapour, and how can metals be formed thereby?

A. In order to a complete understanding of this question, it must first be stated that mercurial vapour is united to sulphureous vapour in a cavernous place which contains a saline water, which serves as their matrix. Thus is formed, firstly, the Vitriol of Nature; secondly, by the commotion of the elements, there is developed out of this Vitriol of Nature a new vapour, which is neither mercurial nor sulphureous, yet is allied to both these natures, and this, passing through places to which the grease of sulphur adheres, is joined therewith, and out of their union a glutinous substance is produced, otherwise, a formless mass, which is permeated by the vapour that fills these cavernous places. By this vapour, acting through the sulphur it contains, are produced the perfect metals, provided that the vapour and the locality are pure. If the locality and the vapour are impure, imperfect metals result. The terms perfection and imperfection have reference to various degrees of concoction.

Q. What is contained in this vapour?

A. A spirit of light and a spirit of fire, of the nature of the celestial bodies, which properly should be considered as the form of the universe.

Q. What does this vapour represent?

A. This vapour, thus impregnated by the universal spirit, represents, in a fairly complete way, the original Chaos, which contained all that was required for the original creation, that is, universal matter and universal form.

Q. And one cannot, notwithstanding, make use of vulgar mercury in the process?

A. No, because vulgar mercury, as already made plain, is devoid of external agent.

Q. Whence comes it that common mercury is without its external agent?

A. Because in the exaltation of the double vapour, the commotion has been so great and searching, that the spirit, or agent, has evaporated, as occurs, with very close similarity, in the fusion of metals. The result is that the unique mercurial part is deprived of its masculine or sulphureous agent, and consequently can never be transmuted into gold by Nature.

Q. How many species of gold are distinguished by the Philosophers?

A. Three sorts :—Astral Gold, Elementary Gold, and Vulgar Gold.

Q. What is astral gold?

A. Astral Gold has its centre in the sun, which communicates it by its rays to all inferior beings. It is an igneous substance, which receives a continual emanation of solar corpuscles that penetrate all things sentient, vegetable, and mineral.

Q. What do you refer to under the term Elementary Gold?

A. This is the most pure and fixed portion of the elements, and of all that is composed of them. All sublunary beings included in the three kingdoms contain in their inmost centre a precious grain of this elementary gold.

Q. Give me some description of Vulgar Gold?

A. It is the most beautiful metal of our acquaintance, the best that Nature can produce, as perfect as it is unalterable in itself.

Q. Of what species of gold is the Stone of the Philosophers?

A. It is of the second species, as being the most pure portion of all the metallic elements after its purification, when it is termed living philosophical gold. A perfect equilibrium and equality of the four elements enter into the Physical Stone, and four things are indispensable for the accomplishment of the work, namely, composition, allocation, mixture, and union, which, once performed according to the rules of art, will beget the lawful Son of the Sun, and the Phœnix which eternally rises out of its own ashes.

Q. What is actually the living gold of the Philosophers?

A. It is exclusively the fire of Mercury, or that igneous virtue, contained in the radical moisture, to which it has already communicated the fixity and the nature of the sulphur, whence it has emanated, the mercurial character of the whole substance of philosophical sulphur permitting it to be alternatively termed mercury.

Q. What other name is also given by the Philosophers to their living gold?

A. They also term it their living sulphur, and their true fire; they recognize its existence in all bodies, and there is nothing that can subsist without it.

Q. Where must we look for our living gold, our living sulphur, and our true fire?

A. In the house of Mercury.

Q. By what is this fire nourished?

A. By the air.

Q. Give me a comparative illustration of the power of this fire?

A. To exemplify the attraction of this interior fire, there is no better comparison than that which is derived from the thunderbolt, which originally is simply a dry, terrestrial exhalation, united to a humid vapour. By exaltation, and by assuming the igneous nature, it acts on the humidity which is inherent to it; this it attracts to itself, transmutes it into its own nature, and then rapidly precipitates itself to the earth, where it is attracted by a fixed nature which is like unto its own.

Q. What should be done by the Philosopher after he has extracted his Mercury?

A. He should develop it from potentiality into activity.

Q. Cannot Nature perform this of herself?

A. No; because she stops short after the first sublimation, and out of the matter which is thus disposed do the metals engender.

Q. What do the Philosophers understand by their gold and silver?

A. The Philosophers apply to their Sulphur the name of Gold, and to their Mercury the name of Silver.

Q. Whence are they derived?

A. I have already stated that they are derived from a homogeneous body wherein they are found in great abundance, whence also Philosophers know how to extract both by an admirable, and entirely philosophical, process.

Q. When this operation has been duly performed, to what other point of the practice must they next apply themselves?

A. To the confection of the philosophical amalgam, which must be done with great care, but can only be accomplished after the preparation and sublimation of the Mercury.

Q. When should your matter be combined with the living gold?

A. During the period of amalgamation only, that is to say, Sulphur is introduced into it by means of the amalgamation, and thenceforth there is one substance; the process is shortened by the addition of Sulphur, while the tincture at the same time is augmented.

Q. What is contained in the centre of the radical moisture?

A. It contains and conceals Sulphur, which is covered with a hard rind.

Q. What must be done to apply it to the Great Work?

A. It must be drawn out of its bonds with consummate skill, and by the method of putrefaction.

Q. Does Nature, in her work in the mines, possess a menstruum which is adapted to the dissolution and liberation of this sulphur?

A. No; because there is no local movement. Could Nature, unassisted, dissolve, putrefy, and purify the metallic body, she would herself provide us with the Physical Stone, which is Sulphur exalted and increased in virtue.

Q. Can you elucidate this doctrine by an example?

A. By an enlargement of the previous comparison of a fruit, or a seed, which, in the first place, is put into the earth for its solution, and afterwards for its multiplication. Now, the Philosopher, who is in a position to discern what is good seed, extracts it from its centre, consigns it to its proper earth, when it has been well cured and prepared, and therein he rarefies it in such a manner that its prolific virtue is increased and indefinitely multiplied.

Q. In what does the whole secret of the seed consist?

A. In the true knowledge of its proper earth.

Q. What do you understand by the seed in the work of the Philosophers?

A. I understand the interior heat, or the specific spirit, which is enclosed in the humid radical, which, in other words, is the middle substance of living silver, the proper sperm of metals, which contains its own seed.

Q. How do you set free the sulphur from its bonds?

A. By putrefaction.

Q. What is the earth of minerals?

A. It is their proper menstruum.

Q. What pains must be taken by the Philosopher to extract that part which he requires?

A. He must take great pains to eliminate the fetid vapours and impure sulphurs, after which the seed must be injected.

Q. By what indication may the Artist be assured that he is in the right road at the beginning of his work?

A. When he finds that the dissolvent and the thing dissolved are converted into one form and one matter at the period of dissolution.

Q. How many solutions do you count in the Philosophic Work?

A. There are three. The first solution is that which reduces the crude and metallic body into its elements of sulphur and of living silver; the second is that of the physical body, and the third is the solution of the mineral earth.

Q. How is the metallic body reduced by the first solution into mercury, and then into sulphur?

A. By the secret artificial fire, which is the Burning Star.

Q. How is this operation performed?

A. By extracting from the subject, in the first place, the mercury or vapour of the elements, and, after purification, by using it to liberate the sulphur from its bonds, by corruption, of which blackness is the indication.

Q. How is the second solution performed?

A. When the physical body is resolved into the two substances previously mentioned, and has acquired the celestial nature.

Q. What is the name which is applied by Philosophers to the Matter during this period?

A. It is called their Physical Chaos, and it is, in fact, the true First Matter, a name which can hardly be applied before the conjunction of the male—which is sulphur—with the female—which is silver.

Q. To what does the third solution refer?

A. It is the humectation of the mineral earth, and it is closely bound up with multiplication.

Q. What fire must be made use of in our work?

A. That fire which is used by Nature.

Q. What is the potency of this fire?

A. It dissolves everything that is in the world, because it is the principle of all dissolution and corruption.

Q. Why is it also termed Mercury?

A. Because it is in its nature aërial, and a most subtle vapour, which partakes at the same time of sulphur, whence it has contracted some contamination.

Q. Where is this fire concealed?

A. It is concealed in the subject of art.

Q. Who is it that is familiar with, and can produce, this fire?

A. It is known to the wise, who can both produce it and purify it.

Q. What is the essential potency and characteristic of this fire?

A. It is excessively dry, and is continually in motion; it seeks only to disintegrate and to educe things from potentiality into actuality; it is that, in a word, which coming upon solid places in mines, circulates in a vaporous form upon the matter, and dissolves it.

Q. How may this fire be most easily distinguished?

A. By the sulphureous excrements in which it is enveloped, and by the saline environment with which it is clothed.

Q. What must be added to this fire so as to accentuate its capacity for incineration in the feminine species?

A. On account of its extreme dryness it requires to be moistened.

Q. How many philosophical fires do you enumerate?

A. There are in all three—the natural, the unnatural, and the contra-natural.

Q. Explain to me these three species of fires.

A. The natural fire is the masculine fire, or the chief agent; the unnatural is the feminine, which is the dissolvent of Nature, nourishing a white smoke, and assuming that form. This smoke is quickly dissipated, unless much care be exercised, and it is almost incombustible, though by philosophical sublimation it becomes corporeal and resplendent. The contra-natural fire is that which disintegrates compounds, and has the power to unbind what has been bound very closely by Nature.

Q. Where is our matter to be found?

A. It is to be found everywhere, but it must specially be sought in metallic nature, where it is more easily available than elsewhere.

Q. What kind must be preferred before all others?

A. The most mature, the most appropriate, and the easiest; but care, before all things, must be taken that the metallic essence shall be present, not only potentially but in actuality, and that there is, moreover, a metallic splendour.

Q. Is everything contained in this subject?

A. Yes; but Nature, at the same time, must be assisted, so that the work may be perfected and hastened, and this by the means which are familiar to the higher grades of experiment.

Q. Is this subject exceedingly precious?

A. It is vile, and originally is without native elegance; should anyone say that it is saleable, it is the species to which they refer, but, fundamentally, it is not saleable, because it is useful in our work alone.

Q. What does our Matter contain?

A. It contains Salt, Sulphur, and Mercury.

Q. What operation is it most important to be able to perform?

A. The successive extraction of the Salt, Sulphur, and Mercury.

Q. How is that done?

A. By sole and perfect sublimation.

Q. What is in the first place extracted?

A. Mercury in the form of a white smoke.

Q. What follows?

A. Igneous water, or Sulphur.

Q. What then?

A. Dissolution with purified salt, in the first place volatilising that which is fixed, and afterwards fixing that which is volatile into a precious earth, which is the Vase of the Philosophers, and is wholly perfect.

Q. When must the Philosopher begin his enterprise?

A. At the moment of daybreak, for his energy must never be relaxed.

Q. When may he take his rest?

A. When the work has come to its perfection.

Q. At what hour is the end of the work?

A. High noon, that is to say, the moment when the Sun is in its fullest power, and the Son of the Day-Star in its most brilliant splendour.

Q. What is the pass-word of Magnesia?

A. You know whether I can or should answer:—*I reserve my speech.*

Q. Give me the greeting of the Philosophers.

A. Begin; I will reply to you.

Q. Are you an apprentice Philosopher?

A. My friends, and the wise, know me.

Q. What is the age of a Philosopher?

A. From the moment of his researches to that of his discoveries, the Philosopher does not age.

APPENDIX VII

[The manuscript of Paracelsus which is preserved in the Vatican Library is not the only treatise which is attributed to him under the title of *Manual*. The octavo volume, which has already supplied the material for the fourth appendix, contains two extensive collections of processes, the one devoted to chemistry and the other to medicine, which are respectively described as the *Primum* and the *Secundum Manuale*. The latter is wholly outside the scope of this translation, but the first, which here follows, would have assuredly deserved a position of palmary importance in its proper section if there were not grave reason to doubt its genuine character. The preface has already stated that there are no satisfactory rules for distinguishing between the authentic and forged writings which pass under the name of Paracelsus. The early date of the Basle octavo might be regarded as in favour of its contents; it contains the *Archidoxies*, which are themselves indisputable, and it will be seen that the *Primum Manuale* claims to have been printed direct from an autograph manuscript. At the same time it does not correspond in any traceable manner with what is known of the Vatican treatise, and its "demonstrative physics" would appear to belong rather to the most suspicious section of alchemical literature than to serious experimental records. While this, of course, is an individual opinion, it is based upon a somewhat wide acquaintance with the great masters of alchemy, and on the evidence of other writings contained in the present volume which are less open to question. But whatever its actual value, it would by no means be right to exclude it because it is of doubtful authenticity, or because it is not in correspondence with what is known concerning a manuscript to which few have an opportunity of access. It has been, therefore, reserved to an appendix, where it may be accepted for what it is worth. If it be really a work of Paracelsus, the veils of the great mystery have been folded very thickly, and are not of an inviting texture.]

A MANUAL OF PARACELSUS THE GREAT,

That most excellent Philosopher and Doctor of both kinds of Medicine;

THAT IS,

A THESAURUS OF SPECIAL ALCHEMICAL EXPERIMENTS UNDER THE AUTOGRAPH OF THE AUTHOR HIMSELF—PARACELSUS

THE WORK ON MERCURY FOR LUNA AND SOL, WHICH I HAVE DONE WITH MY OWN HANDS

TAKE of calcined tartar 2lb., and of quicklime 1lb. Mix together. Place, in a vessel well luted, in a potter's furnace, that it may be calcined and rendered white. Dissolve that matter in the following lixivium. Let it stand until the calx in the bottom shall become tartar turned into water. Then distil by a filter; afterwards take the lixivium and place on it 1lb. of egg shell with 1lb. of quicklime. Make all boil together, so that it may become a stone. Put this to be again calcined as before, and also place it again in the lixivium; dissolve and distil it by the filter. Treat it again with egg shell and quicklime, as above, and repeat this three or four times; lastly, take the lixivium distilled by the filter, and make it boil until it is congealed. Let that tartar be calcined by itself for 15 or 16 hours; then dissolve it into an oil, and thus you will have oil of tartar. Then take the Mercury, sublimate it with quicklime and egg shell and calcined sulphur three or four times. Imbibe it nine times with the aforesaid oil, and sublimate it. What remains at the bottom, preserve in a glass vessel. What ascends, imbibe another nine times, sublimate, and preserve what remains. Imbibe yet again, and continue doing so until nothing ascends. Then take the mercury, pound it well, and imbibe it yet once again. Then dissolve it into an oil in a cold cellar. When all is dissolved put in 6 oz., 1 oz. of silver foil. Place it in horse-dung, that it may be dissolved into water, and, when dissolved, coagulate it into Luna.

INTO SOL

In the place of Luna take Sol, and in the imbibition of the Mercury add crocus of Mars, that it may become red.

The Lixivium in Made Thus

℞   Quicklime
    Wood ashes
    Egg shell

of each a sufficient quantity; and boil to the thickness you know to be sufficient.

The Work of Sulphur, According to our Operation

℞   Sulphur 1lb.
    Crocus of Mars 1lb.
    Colcothar ½lb.

Place in a glazed vessel, and boil with the aforesaid lixivium until it be well reddened. Then distil, calcine the remains, repeat operation, and do the same twelve times, or even more, till one part of it becomes red and the other remains white. Let that whiteness be distilled by itself so long as it does not burn. Then mix with oil of tartar, and let it fix the mercury. It will congeal it if it be boiled therein, and will be fixed by sublimation and putrefaction, so that one part tinges a hundred parts. Take the red part of that oil, place it on silver foil, and let it stand for a week. Afterwards purify by ashes, and you will have good and excellent Sol. Or boil mercury in the same and a similar result follows. Hamelius first made a lixivium, and says that out of ten pounds there is scarcely one of tartar. Let it be very acid.

Concerning Mercury

Take mercury, and pound it with egg-shells. Boil it with oil of tartar, and afterwards sublimate it. Repeat this fifty times. Then take that mercury, and imbibe it with oil of Luna. Having done this, add a little of the oil of tartar and sublimated mercury. Mix all together and put in a flask, well luted, on a gentle fire for 24 hours, and then for four hours on a fierce fire. You will then find a stone which will perform wonders. Dissolve this in water, and the water tinges to Luna. Dissolve in it as much Luna as possible, and afterwards coagulate.

To Reduce the Dust of Metals into Ashes

Take of the dust of the metal 1 quartal; borax, tartar, 1 quartal each; imbibe with oil of tartar, and afterwards dissolve.

Concerning Stelliones or Spotted Lizards

If stelliones are distilled by descent, they yield an oil, of which it is said that it should fix mercury and convert it into Sol.

Another

We believe that the ashes of a stellio should convert Luna into Sol, if they be projected on Luna.

Another

If you pour into their stomach, by means of a reed, a quantity of *Mercurius vivus*, put it into a luted vessel, and burn it, you will find fixed mercury, which is Luna. Take particular notice of these stelliones.

Concerning Lizards

As we have prescribed in the case of stelliones, so do we also write concerning common lizards. The virtue of these animals should be carefully noted.

To Cut Iron with Iron

Take a leek, a radish, and some earth-worms. Distil the water from them, in which dissolve a small knife as long as may be necessary.

An Opinion on the Fixation of Spirits

We have an opinion that the fixation of mercury can in no way be better effected than in the following: Take the white of eggs and purify it with vitriol, alum, salt of nitre, and colcothar. Distil it again and again. Then distil with calcined tartar and the ashes of eggs from three to six times. Then imbibe the mercury from the colcothar and calcined tartar with the aforesaid oil until it remains on red-hot iron. Place that mercury in a glass vessel and dissolve it in water. That water dissolves Luna. When it has been dissolved, put it again to digest until it is altogether turned to water. Coagulate this with water and project this stone, whereupon it is dissolved and converted into Luna. It will last for ever. But if you desire to turn it to a red colour, take gold instead of Luna, and it is coloured red with water of the crocus of Mars. Investigate concerning the calx of Luna and fixed arsenic.

The Preparation of Tuthia for Perpetual Redness

Take some green tuthia and pound it. Mix it with salt, and place it on a fire, in a well luted crucible, for a day and a night. Afterwards open crucible and sweeten the compound. Repeat this process; then scatter Luna upon it, and it will be reddened for ever.

For Fixing all Spirits

Take of the salt of tartar, vitriol, and saltpetre 1 lb. each, of common salt, wood ashes, oak ashes, vine-wood, and aminon 5 lb. or 6 lb. each, with 1 lb. of calcined tartar, quicklime, and alum. Pour acetum over this, and let it stand for three days. Constantly shake it, and afterwards set it to boil for an hour. Then let it be luted and leave it to stand for fifteen days, it being shaken five times in each day. Then distil it by means of a filter, and keep it. You can pour in fresh acetum upon it until the virtue is thoroughly extracted. Then place spirit in that water, congeal and dissolve it, and it will be congealed.

The Fixation of Spirits

Take of prepared common salt, sweet water, prepared salt of alcali, sal ammoniac, oil of urine, and tartar, 1 lb. each, and of honey 5 lb. Place these in a glass vessel in horse-dung for a period of eight days. Afterwards take them out and let them boil gradually in succession. Then you will find a clear white stone. Project this in acetum, and it is turned into water. With this water imbibe the spirits and the calcined bodies; thus mercury is converted into Luna.

The Fixation of Spirits

Imbibe the spirit with oil of tartar, and then the oil is distilled or extracted from it. Keep doing this until it will stand a somewhat strong fire. Take the sublimated spirit, etc., and pound it on marble with oil of tartar. Place it in an alembic upon the sublimating furnace, and distil the oil from it until it is perfectly dried. Do this until nothing is evaporated from the coal. Afterwards put it in a phial and plunge it in dung for ten days, until it is consumed into water. Afterwards put it in a hot furnace with a clear fire until it is congealed.

Note, that you can fix the spirit with oil of sulphur, or of sal ammoniac, and with water of white vitriol. But for conversion into Sol it must be treated with water of tartar.

How Spirits should be Dissolved

Put into water eggshell, salt of alcali, and sal ammoniac. This water is poured on the warm or fused Spirit and it becomes a powder. Afterwards dissolve and congeal it. Dissolve nine times and it will be fixed.

The Fixation of Salts and of Spirits so that they will remain in the fire and melt like lead

Dissolve salt in acetum and place therein mercury at pleasure, sulphur, and arsenic, and they will melt together.

Concerning Lixivia

Make a lixivium as strong as possible, pour oil on it, and let it stand for two days. Shake it frequently, and it will be converted into milk. Having done this, try whether it will stand the fire. If it does, well and good; but if not, put it on again until it does stand, and it will be fixed. Do this for seven days, imbibing and drying it over a slow fire. It will then flow like lead.

Concerning the Virtue of Oil of Tartar

Distil the oil from crude tartar. It avails in all diseases of the joints and of the spleen, whilst it also absorbs ulcers.

The Fixation and Tincture of Spirits

Take of calcined tartar one part, of water of the body *quantum suff*. Let it boil, and note that it is well imbibed with oil of tartar. Distil until it is dried, and repeat this as long as the ingredients remain in the fire. Afterwards take the spirit, pound it, and place it within a glass vessel in horse dung. Let it remain for ten days, desiccate over fire the water thence produced, and thus it will be fixed. Note that you can fix spirits with oil of sulphur or with water of sal ammoniac, or with water of white vitriol. Distemper the water of tartar in the sun with crocus of Mars, as above, etc.

To Dissolve Tartar in one hour without a Furnace

Take marble and place upon it some tartar. Let it stand in water as long as possible, but so that the water shall not touch the tartar, and it will soon be dissolved.

Note Concerning Minerals

Note concerning alcali, which has been made from the strongest lixivium. If a mineral be imbibed therewith in a glass vessel, it will be fixed.

Another Method of Combustion, which holds good for all Minerals

Take Mineral, 4 lbs.
    Calcined Tartar, 1 quartal.
    Flour, ½ lb.
    Bruised Glass, 1 lb.
    Good lute and oil, *quant. suff.*

Make a ball the size of your fist. Dry it and burn it for ten hours in a closed vessel or a hollow globe. Having done this, break and wash it, imbibe it with litharge and anatron in Saturn, with glass or sand strewn over it, for two or three hours. Afterwards it must be fulminated; but note this, that it should be imbibed with the sand until no smoke ascends, and then fulminated.

Note on the Above

Every mineral should be first evaporated in Saturn before it is placed in the ashes.

For Sulphureous and Antimoniac Minerals

Distil the sulphureous or antimoniac minerals by descent for 6 or 8 hours. Then the sulphur is distilled by descent, and the mineral holds Luna or Sol, which is purged by lead, as you well know how.

But if a lixivium is put into the lower vessel, then the sulphur is converted into an oil.*

To Separate Luna from Venus in Money

Take equal parts of arsenic and saltpetre. Dissolve by successive degrees over a slow fire. Then the money in a state of flux is put in, piece by piece, and left in a state of fusion for about a quarter of an hour. Pour in a regulus, and the Luna is separated from the Venus.

The Coagulation of Mercury

Fill an egg-shaped crucible with mercury, fasten the opening with a lute, then place it in an open vessel. Pour on lead, and put it to cool. Thus you will have it coagulated.

To Redden Sulphur

Let it be distilled by descent, and afterwards let water be placed in the lower vessel. The sulphur will adhere, and will be coloured red.

My own Recipe

Take Sublimated Mercury, lb. ij.
    Sublimated Sal Ammoniac, lb. i.
    Crude Tartar, lb. ij.
    Calcined Tartar, lb. iij.

Mix together and place in a glass vessel in cold water, so that it may be dissolved. When dissolved put in it 6 oz. of calx of Luna. Let this dissolve,

---

* From this point the recipes are given partly in Latin and partly in German, the greater part being often in the latter language.

and when you have done so, coagulate. Dissolve and coagulate a second time.

In Alchemy

Take Oil of Antimony, iiij. parts
    Auripigment, i. part
    Crocus of Mars, i. part
    Salt of Nitre, i. part
    Reddened Flos Aeris, i. part

Let them be imbibed with oil of antimony and dried, until the whole of the oil is imbibed. Take of these powders ʒiij., and of Luna ʒs. Let them remain continuously for 12 hours in a state of fusion. You will find in the separation of the strong waters, lx. d. of good gold, or more ☿ by the side of ☿ white or yellow arsenic, ☿ by the fusion of sulphur . . . . lily. Take of this oil half an ounce, of fixed sulphur and sal ammoniac ʒi., and of burnt salt ʒij.

A Cement of Part with the Part

Take ʒiiij. each of
    Bolus Armenus
    Fixed Salt of Nitre
    Fused and prepared Common Salt
    Red Vitriol.
ʒi. each of
    Sal Ammoniac
    Flos Aeris.
ʒi. each of
    Burnt Brass
    Red Calamine Stone.

Let the powders be imbibed with urine, and the *pars cum parte* of Sol and Luna be cemented for twelve hours. Then let it stand in a state of flux for six hours. Let this process be repeated by cementing three times, and you will have Sol of 24 degrees according to every test.

Given by the Grace of God

Take Oxide of Vitriol ⎫
    Flos Aeris   ⎬ ¼ part each.
    Sal Ammoniac ⎭
    Ematite i. oz.

Boil in a glass vessel and dry to a powder. Place in dung or a cellarium, that it may be dissolved to water. Coagulate this water to a powder. Then take i. part of gold, ii. or iii. of silver. Dissolve and project ½ oz. of this over viii. oz. projected in a state of flux. The more it is burnt, the better it will be.

A Very Good Cement in Sol

Take calcined Venus, vitriol calcined to a gray colour, and use a common salt. Let Sol be cemented in it 12 or 6 or 8, and it will then be graduated to the highest point.

Fixed Oil

Take equal parts of salt, of nitre, and of quick lime. Burn it well for one hour, then dissolve and filter it, and it will be coagulated. Do the same with oil of tartar. Take equal parts of both. Dissolve in dung or in a bath.

Solution of Gold by Marcasites

Take Antimony x. parts.
Common Sulphur }
Crude Tartar } iij. parts each.
Common Salt fused ij.

Let them flow together into a black mass. Put them at the same time to be melted with the marcasites. Pour in the regulus, and purify in a cinder fire.

The Projection of Luna

Take of golden marcasite and of stibium equal parts. Pour them together into a fusorium 27. From the fusorium it is extinguished in strong alcali. Project in succession, and you will find much Sol in the separation of the strong waters.

The Fixation of Antimony

Take Antimony i. lb.
Saltpetre ¼lb.

Put them to melt together in a tigillum, and the mixture will become fixed. It no longer consumes gold or silver, nor is evaporated by fire, and is of a red colour.

Water giving Weight to Sol and Luna

[Also if you wish to give weight to a silver or gold cup.]

A Great Secret

Take Calx of egg-shells ʒij.
Sal Ammoniac 2½ ozs.

Let these be mixed by impastation. Leave the mixture in a damp place that it may be dissolved. Distil by a filter. Then dip gold or silver therein, and they will acquire three times their original weight.

A Good Mode of Whitening, which will stand the Fire, and the Test of Lead

Take Sublimated Mercury ʒiv.
Sublimated Arsenic }
Calcined Luna } each ʒij.
Sal Ammoniac ʒviij.

Sublimate three or four times, and afterwards add the clear part of eggs, cooked *quant. suff.* Pound the whole together until it be dissolved Afterwards distil in an alembic and congeal. If it be then dissolved in a bath, it will have more strength, and let there be placed therein one part in 100 of purified copper, tin, or mercury.

For Minerals not Fixed

Take the unfixed mineral
Common Salt
Flour

in equal parts. Let them be pounded together and moistened, so as to become a thick mass. Let it be dried by the fire, afterwards pounded, washed, and fulminated.

Method of Calcining Mercury and the Preparation of Mercury with Marcasites

Take Salt of Nitre
Calcined Alum

1 lb. each. Take aquafortis, neither more nor less in quantity. Take the marcasite in powder, and place it in a phial with three ounces of Mercury. Pour upon it the aquafortis aforesaid, and let the phial be closed with sealing-wax. When the mercury ceases to move and to leap about, then wash it well, dry it, with it take equal parts of litharge and calcined tartar. Mix all, reduce to a fluid together, afterwards by ash fire.

From the golden marcasite is made Sol, and from the silver, Luna.

A Test for the said Work over Marcasite

Take marcasite, and place it in a tigillum with Mercury over the coals. Thus the Mercury becomes hard and red, and the marcasite fluid. Then it is strong.

Another Method for the above

Place it in lead previously washed. If the lead renders it hard as Luna, and divisible when struck, it is good. Lord Leonhardus Sems (gave) the above.

Note

Take of Cinnabar i. lb.
of Arsenical Sulphur } equal quantities, viz.,
of Calcinated Tartar } i. quartal.
of Alcohol of Soot } each ½ lb.
of Twice-prepared Salt }
of Saltpetre to the weight of all.

Let them be mixed, pounded, and imbibed several times with water of eggs or albumen of tartar. Let them stand in flux for three hours. Then kindle. After this liquefy the mixture for four hours in a very strong fire. Proceed to wash and purge by cineritium, and you will have the Treasure of the World.

Of Zinc

Take aquafortis in which Luna is dissolved; afterwards imbibe with zinc; next by means of the cineritium. Also imbibe and fulminate the dross in the glass.

For Minerals which do not readily Melt

Extract alcali out of the *caput mortuum*, with which alcali every mineral is forcibly dissolved.

Concerning Minerals out of Arsenic

Take, pound well, and lute into slime by *vicediam*. Afterwards burn in the fire till the smoke ceases. Then it will be sublimated.

Another Method

Take equal quantities of slime and of minera. Make a small pellet about the size of a bean. Let it be imbibed successively into Saturn, and afterwards fulminated.

Another Method for Augmenting Minerals

Take of minera and of slime equal quantities. Let them be mingled, placed in a jar, and well luted. Burn three days and nights. After this it will be fulminated and augmented.

For Sulphurous Antimoniacal Minerals

Take of minera i. lb., of saltpetre $\frac{1}{2}$ lb. Burn for an hour. Afterwards wash. Reduce to lead with litharge, and afterwards by means of the cineritium.

Note

Take of filings of Venus twelve parts; of laminated Jupiter one part. Next make the following powder :—

Take xxxij. lb. of Sulphur Sand
iij. lb. of Caput Mortuum
i. lb. of Sulphur

Pound them with the said filings. Make two layers of this powder. Distil in a wind furnace. Let it become a black powder. Next smelt it with half a part fz. Pour the regulus. Join to it three ounces of Luna in a cineritium of light. Then you will have your Luna and $7\frac{1}{2}$ oz. in cineritium.

Reduction of the Minera

Take of minera i lb., of minium lb. i. Cause them to stand in flux half an hour. Then infrigidate. Combine in equal quantities with litharge. Imbibe in Saturn and fulminate. You will then find one dram of Luna and more than half an ounce of pure gold. Dross makes up the larger proportion.

Digestion of the Moon

Take of Alum v. lb.
of Saltpetre } each iiij. lb.
of Vitriol }
of Verdigris iij. oz.
of Cinnabar ij. oz.

Make aquafortis.

Afterwards take of Cinnabar }
crude Mercury } xv. lb. each
reddened Vitriol }

Let the aquafortis above-mentioned be poured over these recipes. Distil and at length purge. Also take iij. lb. of this water. Let there be dissolved therein two marcs of Luna. Afterwards let it stand on the cinders thirty days. You will then find iii. oz. of good gold.

Water Fixing Mercury—A True Recipe

Take of Sal Ammoniac ⎫
         Alcali   ⎬ equal quantities.
         Saltpetre ⎭

Imbibe them with burnt wine and distil. This water fixes Mercury.

Digestion of Luna

Dissolve four ounces of lead, and as it melts a little, inject four ounces of quicksilver. Pulverise, imbibe with oil of tartar into a paste. Pulverise again. Then put the whole aforesaid amalgam in the tigillum. Place above viij. oz. of fixed sulphur. Dissolve thoroughly, mix, afterwards pour out and pulverise. Imbibe thrice with oil of tartar. Dissolve as before. Then take one ounce of purged Luna. Sprinkle gradually over it one ounce of the said powder. You will then find the weight of two grains of good gold.

Separation of Gold from the Cup

Take of white calcined Tartar i. oz.
Sal Ammoniac ij. oz.

Let an oil be produced in the cellarium. Then boil the roots of pyrethrum in vinegar for a long time. Take of oil and water equal quantities. Put them into the cup. *Note*, of sublimated sulphur see elsewhere.

For the Cineritium

Take of Cinder three parts
    of Brick pounded and strained, two parts

Also take again three parts of cinders, with one part of melted salt boiled with one part 27 oz. of small nails. It produces the equivalent of one talent.

To render Iron unusually Hard

Distil water from radish and maw worms. Extinguish twice or thrice, and gems can be cut with it.

To make Light without Fire

Take cantharides, putrefy it in dung, and distil it. Put this water into a hollow crystal, and there will be sufficient light to read by.

For the White and Red

Take Vitriol Rom. lb. j.
Saltpetre lb. vj.
Cinnabar oz. iij.

Distil as aquafortis. But collect the second water, which is saffron-coloured; divide it into two parts, and in the first part dissolve i. oz. of sublimated mercury, in the other part i. oz. of silver filings. Let it dissolve and close it carefully. Having done this, let it distil over a very slow fire by means of an alembic until only a third part of it remains.

Afterwards put it into a good vessel of glass, and place this in cold, damp earth for fifteen days. Then the aforesaid matter falls to the bottom like a small, clear, and crystalline stone. Next, by means of a filter extract the water, so that the stones may remain, and place these, in a well-closed glass vessel, thoroughly luted, for five days in dung, and let them be dissolved into water.

Take this water, put it in another vessel, and place the vessel in some other earth with a moderate light. Make a very slow fire underneath, scarcely more than the light of a lamp, for the space of three days, and that water will be hardened to the consistence of a stone. Take that stone once more and pulverise it on cleansed marble. Then place it in a glass vessel in dung for 30 days. Let it be again dissolved into water and hardened once more by fire in the manner aforesaid. Then once more dissolve it under dung as before, and keep on dissolving and congealing it until it is dissolved in one day. Then you will be better able to congeal and dissolve it. And it will be a water which congeals mercury into true Luna at every trial. One part of Luna is deposited for every thousand parts of mercury.

AND IF YOU WISH TO MULTIPLY

Take the aforesaid aqua fortis in such quantity that you can dissolve in it 40 oz. of Luna in plates. Put it in a vessel by itself, and the same quantity of sublimated mercury by itself in another vessel. Leave it until it is dissolved in the water mentioned above. Into the water named before put one ounce of the elixir already described, which has stood for 12 days, while being made. Put all these ingredients, in a well-closed glass vessel, under the earth, in a cold, damp spot, for nine days; then all will descend to the bottom of the vessel in the form of a crystalline stone. Extract the water by means of an alembic in the method described above, and place the stones in warm dung for nine days as before. They will then be dissolved into water. Then know that this is the water of Luna and Mercurius, and that one ounce of the medicine should be viij. And the saffron medicine which falls will be one on a thousand. For Sol it is made thus :—

Take Vitriol Rom. lb. i.
Saltpetre lb. ½.
Cinnabar lb. ½

Distil, as before, a second prepared water. In one part dissolve quick mercury, as much as you will. Then draw off the water by an alembic, and the calcined mercury will remain at the bottom. Burn this in the fire, as you know how, and it will be like blood. It may be dissolved perhaps in aqua fortis and make this red. Having done this, take viij. oz. of the red water itself and i. oz. of the aforesaid medicine. What you have placed therein will descend to the bottom, in the shape of a crystalline stone. Then draw off the water by means of a filter, as before, and the stones will remain at the bottom. Then take the aforesaid stones, put them in a well-closed glass vessel, under warm dung for nine days, and it will be dissolved into a water. This medicine gives one part for every one thousand parts of quick mercury.

Another Good Method

Make an amalgam of best mercury, carefully washed, iii. parts, and of good Luna i. part. Then wash very carefully with salt and acetum, until all blackness disappears. Pound the amalgam on a stone, and dry it. Steep it in urine after wrapping in linen rag. Take an alembic placed on ashes in a furnace, and make a moderate fire of coals, so that you can touch the top of the urine with the hand. Keep on thus decocting and pounding until it is thoroughly black, for this is a good sign. Then wash it in the water of common salt, thoroughly pure, until no blackness appears. Then once more pound, decoct, wash, and dry on linen rag until all blackness has disappeared. Put it then in a glass vessel for sublimation, when it ascends to the sides of the glass, and the Luna with it. Leave it to cool, and again mix all together; pound and decoct twelve times in succession. Then place it in a vesica with a long neck, together with sublimated sal ammoniac. Pound it with the said amalgam, and let the sal ammoniac be dissolved in warm water, or, better still, in aqua vitæ. Afterwards let the vesica be luted and dissolve in warm dung, changing the dung every week so that the matter shall be thoroughly dissolved, and it then becomes an elixir. Afterwards put it in urine to evaporate, and when the water is evaporated increase the fire, so that all the sal ammoniac may ascend, and the medicine may remain fixed in the bottom. Then remove the sal ammoniac, and congeal the medicine with a slow fire, as if it were Sol, for several days, and the thing is done.

If you wish to multiply this elixir, place one part with 100 parts of quick mercury, purified and heated in a crucible. Melt it completely over a slow fire and it becomes friable Luna. If you put one part of this to 100 parts of quick mercury it will become purest Sol.

Red Oil which fixes Luna and Sol, and from which the Carbuncle is made

Take Crocus Arabicus
Calcanthus
Arabian Verdigris
Litharge
Red Calcined Tin, iij. oz. each
Quick Sulphur
Citrine Arsenic
Red Sublimated Calx, lbs. ij. each
Prepared Sal Ammoniac
Saltpetre, distilled and rectified with distilled
Red Animal Oil, lb. j. each.

Let these ingredients be covered with wax, and, when they have been thus treated twice, put the mixture to dissolve, until a pure red water is produced. Then let it be distilled in an alembic; and with this water imbibe the quicksilver aforesaid. Then dissolve and distil it as before, and waxen with it the sulphur and the arsenic. Leave it to dissolve and to distil until it be clear and red, and coagulates of its own accord in the vessel of coagulation. When it has been coagulated, waxen it as above. Reduce it over a slow fire with the addition of animal oil until it will pour out like wax. Project one ounce of it on 100 parts of Luna or Jupiter. If there be mercury or anything else therein, tinge it, and it will be coagulated, producing gold better than that of Nature. Hereby also carbuncles can be made out of crystals.

The Gradation of Luna

Take Sal Ammoniac, oz. i.
Alumen Jameni, oz. i.
Flos Æris, oz. ij.
Vitriol Rom., oz. ½
Saltpetre, oz. ½
Tuchia, oz. i.

Make an aqua fortis. Then pulverise thoroughly the dregs. Dissolve i. part of Sol and i. part of Luna, and project ½ a dram thereupon. Afterwards granulate *per sturbam* in the aforesaid strong water, and thus, beyond any doubt, you will have gold that will answer any test up to xxiiij. degrees.

Perpetual Water

Take the calx of eggs. Dissolve it with the white of eggs for three weeks. This water placed on a brass plate heated to redness will acquire perpetual whiteness.

Another Mode

Dissolve the calx of eggs with alumen jamenum, sugar, and common salt. Distil once in an alembic, and it serves for Luna.

Precious Water.

Take Salt of Nitre } vj. ozs. each
Sal Ammoniac }
Honey, cooked and skimmed, v. oz.
Boy's Urine, xv. oz.

Let these be mixed; place them in a furnace for two days. Then congeal for one day and dissolve; and it will be perfect water, which congeals mercury that has been purified and heated. It will also transmute brass into good Luna.

The Strong Water called the Oil of the Philosophers

Take one part each of Vitriol Rom., and salt of nitre. Pulverise them and mix with charcoal made from linden wood. Moisten with acetum; then pound, dissolve, and distil, when it will become an ardent water.

For Sol

Dissolve cinnabar with water of sal ammoniac, vitriol, and salt of nitre until it is perfectly fixed and becomes red, which will be after three or four dissolutions. Put this for sublimation with the same quantity of sal ammoniac. Repeat the sublimations until it be poured forth. Then dissolve in acetum, having previously dissolved plumose alum in the bath. Continue until it be as fluid as wax, and part of it tinges prepared Luna.

Corporal Mercuries

Take arsenic and dissolve it in aqua fortis. Then take Luna, and dissolve that. Join them together, and strain them as you know how. Sublimate the arsenic from the body, and it will become a sort of mass. Put it into a damp place, and it will be dissolved into mercury.

Corporal Mercury

Take any body you will and two parts of arsenic. Pulverise, and then sublimate them. Place in hot water and it will come forth as mercury.

Corporal Mercury from the Metals

Take of any metal one part and of purified white arsenic another part. Make thin plates, place them one over the other in layers, and let them be regulated first of all in a slow fire, successively made stronger until it begins to smoke. Then pour quickly into cold water, and you will find mercury.

Mercury from Jupiter and Saturn

Put layer for layer of quicklime and metal, and place them in a sublimatory. Set on the fire, and sublimate as you know how. The mercury ascends, adhering to the alembic, and is quickened.

Corporal Mercury from Saturn

Dissolve Saturn and put salt upon it. Shake it until it turns to a powder. Then wash it with warm water and dry it. Afterwards put it in a glass vessel and pour upon it the white of eggs and water of sal ammoniac. Lute it, put it in horse dung for two days, then take it out, and you will have what you require.

The Mercury of Saturn according to the Experiment of Hamelius

Take thin plates of Saturn, placed layer by layer with common salt in a vessel. Bury it for eight days in the ground. Dilute with common water, and then a part of the mercury will be found at once. Repeat the process with the rest.

Mercury of all the Metals

Take some sal ammoniac, flos æris, vitriol, and plates of any kind of metal. Put them in layers, and sublimate them. Place the sublimated portion in acetum, and you will find *Mercurius vivus*. Do this until all the plates are turned into mercury.

Mercury from all Bodies

Take the calx of any metal and place it in acetum, in which there are two parts of the calx of the body and one part of sal ammoniac. Place these in dung for seven days, and the body becomes mercury. If an amalgam is made from it, together with Sol and Luna, by degrees natural mercury is fixed with it.

Corporal Mercury

Take Saturn, and dissolve it in a tigillum. When it is dissolved, scatter over it sal ammoniac, and stir it with a spatula until it becomes a powder. Afterwards project the powder in boiling water, and the salt will be dissolved in the water. Then desiccate pulverised lead in the bottom, and put it with white of eggs into a glass vessel for 62 days. Afterwards remove the water, and you will find the mercury in a fluid state at the bottom.

Another Method

Take a sufficient quantity of sulphur and linseed oil. Boil the mixture until it comes to the form of a vapour. Put in it plates of Saturn, and it is converted into Mercury in three days.

Corporal Mercury from Luna

Take plates of fulminated Luna; dissolve in aquafortis. Extract the aquafortis by means of an alembic, and wash the calx with fresh water. Mix with it sal ammoniac, alcali, and oil of tartar. Blend all together on marble and pound thoroughly for three or four hours. Then the body gains a soul. Collect this soul carefully, and the four elements.

This is the Foundation about which all the Philosophers have Written

If you cannot pound them so long on marble, then after you have done so for two days, and the soul does not shew any desire for the body, place all together in a glass vessel, and put it in dung for four weeks. Afterwards set the glass vessel in a capella and abstract the water until all moisture has disappeared. Strengthen the fire, and the blessed water ascends. Collect this carefully, and if any remains in the alembic, remove it with a feather. If the body is not totally abstracted, add more oil of tartar, put it in a glass vessel with other simples, as before, pound it well, and set it in dung for eight days. Thus the whole will be extracted.

Corporal Mercury from Saturn

Take Saturn, put it in a patella, and when it is slightly dissolved, stir it well with an iron spatula. Afterwards let it be divided into minute parts and salted with salt. When this is done let it be again dissolved. Pour warm water over it, and it attracts the water to itself. Then take calx of marble, put it in a closed glass vessel, well luted; set it in horse dung, let it stand for a month, renewing the dung each week, and it is transmuted into Mercury.

To Convert White Sol into Aloth

Take oil of vitriol by descent, calcined oil of tartar, and sal ammoniac in equal parts. Mix them, adding calx of Luna, and place in a phial. Set this in horse dung or in a bath of Mary for four weeks, that it may putrefy. Open the vessel and put equal weights of pounded sal ammoniac and salt of alcali, with oil of vitriol. Putrefy as before for four weeks, and the body will be entirely converted into Aloth.

Corporal Mercury

Take vitriol, saltpetre, and alum. Make aquafortis as you know. Then take *caput mortuum* and extract its salt with common water. Reverberate the *caput mortuum* a second time, again extracting the salt. Repeat this process, the oftener the better. Afterwards take the earth that remains. Pour over it aquafortis. Leave it to stand for one day. Then force it through the alembic in that aquafortis. Dissolve the moon. Abstract the phlegm through the alembic in the bath. Afterwards take good sal alcali, reverberate it, and dissolve in oil. Then take good oil of tartar, and afterwards dissolved sal ammoniac, which has been six times sublimated. Take equal quantities of these three oils. Pound away the calx therewith, put in a cupel. Pour the oils over to the height of a finger. Place in horse dung fifteen days. Then set over cinders. Drive at first gently, afterwards violently. Then there ascends a white powder, which proves from a marc to two ounces of Luna. This makes living with brandy or sunst.

Receipt of the Bishop of Strasburg

Take of filings of Luna two ounces. Dissolve in aquafortis. Then abstract thence into the third part of the water through the alembic. Put in a cold place, where white pebbles collect; take them and weigh them. Then take so much of salt of tartar, and half as much of sublimated sal ammoniac, and putrefy them in dung, whereupon the pebbles will become mercury. Dry it by means of fustian (?) cloth. If it does not become Mercury, sublimate it, whereupon you will have Mercury out of Luna.

Freising

Take of aquafortis one part out of two parts of vitriol, and one part of alumen. In it dissolve Luna, then excoct the calx. Then take of the burnt salt of tartar two parts, common salt one part, salt alcali, salt of urine one part each.

Pound these salts together with the calx of the moon, and let the quantity of the salts of the moon be as much as of all the salts. Dissolve it all in water; next put therein wheat meal and a little brandy. Let it become dry as glass (?). Then sublimate it. Thus Luna Mercurius is produced.

Note

Take leaves of Luna beaten extremely thin, a finger's length and the breadth of a creutzer, put it into a mortar (?) and reverberate it nine consecutive days, purify it from the sulphur which then comes forth, and reverberate it five or six times, or until Luna yields no more sulphur. This takes place when the leaves become black.

Impaste this sulphur with quintessence of wine, and there will be quicksilver. Reverberate it with wood of oak or birch.

The Original has

Dissolve tartar in aquafortis, then let it evaporate and be coagulated. Then dissolve it out of a stone into an oil. Pour it upon sulphur of the moon. Stir it in a vessel with a hard wooden spoon, and you have Mercury of the body.

Mercury of the Body

Take of sal ammoniac one part, of Mercury sublimated one part, of calx of lime one part. Pound the sal ammoniac and mercury well together; let them melt gently. Put in the Luna, speedily take it out again (and place) on a dish with water, and stir it. Then dry it with a towel or leather. You will then have Mercury of the Moon, true and good.

Another Method

Take saltpetre which has been well purified. Pound it well and put it into a cucurbit. Set it among warm ashes, whereupon the saltpetre will boil like water. Then take the calcined Luna and put it into water. Let it stand thus, care being taken lest the ashes be too hot. Then mercury of silver will be produced. Strain through a cloth and collect the mercury.

Corporal Mercury

Vitriol also, if it be distilled alone into water and a quantity of filings be placed therein, will in the course of time become Mercury.

Another Method

With Mercury sublimate seven times, pound calx or cinder of every metal; add flour and water. It will descend, for it will turn into living Mercury. Reduce to a cinder after solution and repercussion, cleansing it thoroughly of saltness and spirits of aquafortis.

Take filings of Luna. Afterwards take Mercury, which you must wash with salt and vinegar. Cook Mercury with vinegar and salt. Take three parts of this and of filings of Luna one. Make an amalgam and pound well for two hours on a smooth stone. Then let the Mercury evaporate. Afterwards remove the vinegar and salt with warm water. Calcine that calx of Luna twenty hours, and there will be a woolly Luna. After which calcination put in the distilled vinegar. Then make the remaining calx, working as above. Then take the extracted Luna, pound it with some tartar, and it will be live Mercury.

Note

Dissolve filings of Luna in aquafortis, draw the third part through the alembic and put it in a cold place, where crystals collect. Take it and weigh it; add the same quantity of sulphur and half that quantity of sal ammoniac, etc., as see above, in the Bishop's art.

Another

Take the calx of any metal, place it in vinegar, wherein let there be distilled two parts of sulphur, of sal ammoniac one part, and of Mercury sublimate one part. Let it be placed in dung or in the bath eight or ten days, then let it be distilled among the cinders, and Mercury will ascend. With this there is an amalgam of Luna or of Sol, and you will tinge all bodies. Take of sublimated Mercury, which see, pound it with steel filings, out of sal ammoniac, put it into a wet place, and let it be dissolved into oil without water.

Praxis

Take of Mercury of the sun one part, and of crude Mercury. Put it into a phial and seal. Give it, from the first, four or five gentle fires, increasing the heat gradually until you perceive the red powder. It tinges Mercury and all bodies. Treat the white similarly. Take of Mercury of Luna two parts, of crude Mercury one part. Possibly the red process should have been treated in the same manner.

Note

Take an equal quantity of salt and sulphur, and of water of arsenic. Take as much as you like of calx of Luna, well sweetened. Put it in a cupel, pour water over to the height of three fingers. Let it stay three weeks in dung or in the bath. Let it be well luted. Then place in sand. First distil the water slowly, then drive away the spirits. Afterwards strengthen, whereupon the Mercury ascends; let it become cold and you have it potent.

Another Method

Make aquafortis of equal quantities of saltpetre, of vitriol, and alumen. Of this water take one part, in which dissolve 4 c. of sal ammoniac and 12 of sublimated mercury. Leave it to putrefy fourteen days. Distil slowly through the alembic in the cinders; then take calx of the Moon, well sweetened, pour this extracted aquafortis over, so that it may stand three inches high; let it remain three or four days in a tepid heat, and it will be changed into Mercury.

Mercury of Saturn

Take of soap one part, of living calx one part. Pound them together, distil like an aqua fortis. Then take of ceruse two parts, and of oil of soap one part. Impaste well, let it stand one month in the bath, afterwards pour warm water over it, move with the finger, and you will have the mercury of the bodies.

Mercury of the Bodies

Take one pound of sal ammoniac, purified with ♁ calcined and prepared. Then place two pounds of rock (*petra*) in the middle of that sal ammoniac sublimated, below, above, and around, to the thickness of one finger; let it be compressed and placed in a glass vessel, and then set amid the sand in a jar, which has the lid well luted, and give for four days the fire of a candle, at first with one light, afterwards with two, thirdly with three, and, finally, with four lights. There will then be a black matter. Subsequently, press through the bag.

Note

A tincture must be made after the manner in which oil is extracted from vitriol. Let it come over a slow fire. Then take a quantity of sublimed mercury, and prepare it over a similar fire, till it shall have become white. Take one part of the filings of the body of Luna, with two parts of the Eagle and of sal ammoniac. Mix them well in a crucible, which seal, and set over a good coal fire. When the contents are completely melted remove them, let them cool, and then press through a cloth. Then take mercury; let it remain in the cloth; expel thoroughly by means of Saturn, take a portion of mercurial water and a portion of running Mercury. On account of the low temperature, take care to cover the body produced. Its formation must take place in the water. Afterwards effect its digestion, sealing well, until a powder results. Dissolve this in the cold and coagulate it in the warmth. Perform this twice at least, and you will have a powder which is potent for tinging.

Again

Take vitriol, and pound it thoroughly. Place it in a glass vase, well covered so as to prevent exhalation. Set it above the furnace and keep up a slow fire below for a whole day, and you will find it turned into water. Then take filings of Mars, well washed in sweet water until all the vitriol depart; after which pound heavily in the mortar. Sprinkle the said filings with water of vitriol, then leave to dissolve for two or three days, whereupon there will result a calx of those filings, but at the bottom of the vessel you will discover Mercury.

Mercury of the Body

Make aquafortis out
- Of Vitriol i. lb.
- Of Salt Nitre i. lb.
- Of Calcinated Alumen ½ lb.

And if the alembic becomes red, take the red spirits separately; in it dissolve the body of the sun and moon, so far as they are capable of dissolution. Of the same take one part, and four parts of extracted tartar, with quintessence of wine as aforesaid. Or, put the hot tartar into the quintessence to the extent that the quintessence will receive it. Of the same quintessence, together with the extracted tartar, take, as said above, four parts. Then pour the aquafortis, with the body, into the quintessence, for the first time by drops, and set it in the bath for eight days. Then boil it clear, that no dross may remain with it. Extract this once, and again pour on (the aquafortis), until it becomes a transparent oil. Then it is genuine. Next take the earth previously stirred, and extract its phlegm into the sand. Then give a subliming fire, whereupon the body will arise. Cleanse it and make it living in vinegar or boiling wine. Place in the same Mercury the aforesaid oil separately until it is all brought over; then it will tinge.

Mercury of the Body

Take of Sublimated Mercury four parts,
Sal Ammoniac two parts,
Mercury of Venus, or preferably Salts of Urine, two parts,
Calx of Luna pounded with the salts.

Let them remain in putrefaction for eight days.

Another Method

Take of calx of Luna pounded with the same quantity of sal ammoniac. Sublimate three or four times. Then wash the sal ammoniac from it. Drive the calx through a retort with a small fire, and you will have mercury.

Augment

Take of corporal mercury four parts, and of fine Sol or Luna one part. Make an amalgam in a copper vessel, with an ordinary bath. Then leave it to digest eight days. When it is fixed, augment it with common mercury again, about three times. Afterwards augment it with crude mercury (but perhaps mercury sublimated and rectified is the best). Do this continually, and in eight days it will also be fixed. Or, dissolve Mercury in aquafortis, then let it drive over, and you will again vivify it with that augment.

To make Fixed Augment

Dissolve Luna in aquafortis, permit half of the water to evaporate; if crystals collect, place them in a glass cup in a cellar, and an oil will be produced. Imbibe it with a small augment. Place for eight days on hot cinders with little heat, the glass being open. Afterwards subject it to a powerful fire and pulverise well.

Reduction

Take of Minium one part,
      Ceruse (otherwise litharge) one part,
      The Fixed Augment one part.

Pound the three things well together. Put into a well-luted crucible, and smelt thoroughly for an hour. Then cool and clear away the regulus.

Mercury of the Bodies

Take equal quantities of tartar and of vinegar. Put them in a phial, with long neck, into oil of vitriol. Inject calx of any metal. Stir, and leave to stand eight days. Next inject sal alkali and sal ammoniac in the same quantity as in the previous recipes. Place on dung thirty days, and you will find at the bottom calx converted into Quicksilver.

Note

Take of Mercury sublimate half a pound.
      of Sal ammoniac an equal quantity.

Pound well and put them together in a glass. Let it be well luted. Put it *in carellen*, in warm sand. Kindle a mild fire beneath. Thus it will become one (solid) mass. Remove this; pound it small; and put in a damp compartment to dissolve. Afterwards take of filings of Luna one marc. Put it in water. This also becomes water. Coagulate on soft ashes, and reduce to powder, of which inject one part upon ten of Mercury, of purged Jupiter, or of crude Mercury.

Mercury of Jupiter

Liquefy Jupiter. Then inject the same quantity of Mercury, and thus make an amalgam. Next let them be well pulverised and thoroughly incorporated with water of sal ammoniac, tartar, salt of urine, and the same quantity of common salt. Place in a flask, well sealed up, and set in dung for twenty days, when it will be converted into Mercury.

Augment on Mercury of Saturn

Take of Mercury of Saturn ʒi.
of Sol ʒs.

The amalgam, if kept at a moderate temperature for eight days, changes into a brown powder, which becomes the finest gold. Add to it half-an-ounce of common Mercury. Again let it stand for eight days, when again it becomes a brown or a red powder, and so on with common Mercury. The case is the same for Luna, and for Mercury of Jupiter with Luna. The whole process must occupy eight days.

Note

Dissolve Luna in any quantity of aquafortis, and in vinegar dissolve calcined tartar. Then pour the two solutions properly together, in such a manner that none may run over. Afterwards pour it into a phial, which must be well luted at the top. Set it in horse dung to putrefy for fourteen days. Afterwards put in an (earthen) pan till it becomes inodorous. Precipitate to the bottom whatsoever adheres to the upper part of the pan, and let it cook thus until it becomes as thick as a pottage. Then let it become cold, and stir it under the tartar. If mercury collects, remove it. Then take a measure of water and gradually wash off the tartar, whereupon all the mercury will collect.

Corporal Mercury

Take of Luna out of sweetened aquafortis ʒi.
of Sal ammoniac ʒs.

Mix. Pour over them oil of tartar, which must stand over them to the height of two fingers. Put into a glass well luted, and leave to putrefy four weeks. Extract the humidity from it. Sublimate the remainder. Whatsoever ascends is to be put in warm water, vinegar, or oil of tartar, and thus you will have Mercury.

Oil of Tartar is Produced as Follows

Take of calcined tartar. Pour upon it the quintessence. Allow it to stand for twenty-four hours. Pour out again until at last no more oil remains in the fæces.

Mercury of the Body

Take sal alkali, and pour upon it pure urine of youths, so that it may be dissolved therein. Distil through the filter, and coagulate. Afterwards take some sal ammoniac and twice that quantity of Mercury sublimate. Pound them together, and place them over a glass slab to dissolve. Afterwards take the water thus dissolved, and pour it into a glass. Place the glass in the bath of Mary, that the aquosity may be consumed and vanish. You must then test it by means of the blade of a knife, as you know. Afterwards place therein leaves of Sol, Luna, Venus, Jupiter, or Saturn, for the space of an ordinary day. It will then be converted into Mercury.

Mercury of Saturn

Take of Sal ammoniac half-an-ounce.
of Calcined Saturn two ounces.
of Sal manipulum } one ounce each.
of Calcined alumen }

Mix all together. Put into a glass with a narrow neck, and underneath, at the bottom, put a quantity of genuine mercury. Afterwards place the matter on the top. Close the glass securely. Set in horse dung for four weeks. Then it will become Mercury.

Another Method

Take of calcined Saturn, smear with sap of henbane, dry it, and smear it until it resembles a paste. Then set over a slow fire until the humidity departs. Subsequently, increase the fire, and Quicksilver will come forth.

Salt of Urine for Mercury of the Body

Take of the urine of a man who continually drinks wine, and distil it through the bath. Completely dry the fæces. Then you will have the salt of urine. Then take two ounces of the water of life, four times rectified. In this dissolve half-an-ounce of salt of urine and half-an-ounce of calcined Luna. Pound the calx subtly with burnt salt, the more the better. Cleanse that calx with hot water. Then you must put it into the aforesaid water of life. Let it putrefy fourteen days in dung. Afterwards distil the water of life from the calx. Pound with a little oil of tartar, and it will become Quicksilver.

Mercury of the Moon

Dissolve Luna in aquafortis. Pour in sal ammoniac. Thus Luna is sent down to the bottom. Take one part of this calx. Take an equal part of Mercury sublimate and sal ammoniac. Place in hot cinders, and in three days there will be Mercury of the Moon.

To Convert Metals into Mercury

Take Mercury sublimated seven times, and add to it the same quantity of the purest flour of wheat. Pound them together and saturate the matter with a little pure water. Place in a vessel and subject them to a slow fire, that the moisture may evaporate. This having been done, put the matter into a circular furnace (retort) with the neck of the glass downward. Drive it by descent, and the Mercury will descend. And that Mercury being heated devours all metals, until they are reduced to Mercury.

Mercury out of Bodies

Take of Salt of Tartar ʒii.
Sal Ammoniac ʒi.
Calcined Saturn ʒi.
Luna or Sol ʒi.

Mix all these together and pour over them good vinegar, and let it be distilled. However, let it swim on the top (the height of) one palm or thereabouts. Seal the vessel hermetically. Set in a warm place for a month. Afterwards place in hot cinders, and thence distil the vinegar. Next make a strong fire, whereupon the Mercury will ascend. Collect it, and use it as you will.

Mercury of the Moon

Dissolve Luna in aquafortis. Extract the moisture, even to the spirits. Pour fresh aquafortis upon it. Do this thrice. Then let it dry. Put it with the same weight of four salts in a cellar to dissolve until it will not melt any more. Then put in a glass. Leave it to putrefy for four weeks. Then strain it off, like an aquafortis. Whatever remains behind dissolve again on the stone, putrefy, and strain as before; reduce the residue, and then take the following salts—

> Oil of Salt Alkali
> Oil of Tartar
> Oil of Common Salt
> Oil of Sal Ammoniac

in equal proportions.

Pour all together. Then it is prepared as above.

Mercury of the Body

Let an amalgam of any body be cooked in very strong vinegar and fixed by sal ammoniac for fourteen days, when it will become Mercury; or let it be decocted in water of eggs and sal ammoniac, and it will become Mercury in one day.

Another Method

Take Luna or Sol, calcined or otherwise, and dissolve in aquafortis made out of one part of Mercury, one part of saltpetre, and half a part of ♀. Then cause the water in the bath to evaporate, whereupon Luna or Sol will remain at the bottom like an oil. However, to dissolve the Sun add to the aquafortis one part of sal ammoniac. Next add to the said Sol or Luna, thus dissolved, tartar and distilled water of life. When it has been seven times imbibed and dried at the same time, dilute it, and let the water of life be distilled, so that it may float on the top to the height of three fingers. Afterwards leave to putrefy for eight days in dung or in a bath. Then the water being evaporated over the fire, let an alembic be placed above and set on a good subliming fire. The living and running Mercury will ascend into the receptacle.

Augment in Luna by Count William in Sager

To make Sol out of Luna

Take one ounce of δ, of ♈, and of sal ammoniac. Make a powder out of them. Put into a cucurbite, close it up with a small cloth. Set it in warm sand, so that it (the sand) may melt. Add a part of the calx of Luna. Stir to and fro, whereupon there will settle a liquid matter at the bottom. Cleanse the same from its impurity with warm water. Then take the δ, wash it clean, and although a little silver may still be present, make no mistake in the work. Thus out of every metal you may make ☽. Take of ☽ one marc. Put it in a glass. Pour upon it three ounces of the hereafter to be described aquafortis. Put it in the bath of Mary for six days and nights. Then extract the water and preserve the matter at the bottom. Afterwards take two parts of sulphur and three parts [the original leaves a blank], and impaste over a fire, as cinnabar is made. Pound this very small, then boil it in oil of tartar, according to the process described above, until all the sulphur be excocted hard, and do not burn. Of cinnabar add one marc to the above �, which is to remain in the water. Pound it together. Put it into a glass, but pour six parts of the said aquafortis upon it. Let it stand, however, in the bath of Mary, or in hot sand, for seven days, to digest. Then excoct the water and pound the matter to powder. Fix the powder by the fixation to be described hereafter. Take thin plates of Saturn and place in alternate layers with the powder. Take vitriol, saltpetre, and verdigris in equal quantities. Perlute well. Set it to glow in a fire of calcination, that is, of circulation, for sixteen days, that it may always glow gently. Then reduce it and refine it, whereupon you will have and find a great augment in the Luna, and there will be much Sol therein. To pour on the aquafortis, do as follows:—

> Take of Saltpetre
> of Alumen and Sal Ammoniac } each one pound.
> of Alumen Plumosum, half a pound.

Make aquafortis and extract its moisture.

The Oil of Tartar

Take of oil of tartar two pounds, and of oil of vitriol, extracted by descent, or of oil of calcinated vitriol, two pounds. In this excoct the cinnabar, as explained above.

Mercury of the Body

Take Luna and dissolve in aquafortis. Then dissolve salt in ordinary water, so that it may become thoroughly salt. Pour such a quantity of this salt water as to look like milk upon the aquafortis. Leave for a day and a night; thereupon the sweet moon descends and will be dried. Take ℥viij. of boiling wine. Inject ℥iiij. of calcined tartar and two of sal ammoniac. Distil through the alembic, and the crystals will pass over into the wine; dissolve about eight parts of sal alcali; pour this solution over the ☽, which must go over. Leave for eight days to putrefy. Take it out and extract the moisture. Afterwards lute the glass, and put it in ashes. Give it a vehement fire. Then the Luna will ascend in the form of a powder. Clear this out of the glass, and put it in oil of tartar. Thus in a single night, without fail, there will be produced Quicksilver.

Mercury of Sol or Luna

Take tartar, dissolve, filter, and coagulate again. Place by the fire, that the aquosity may be perfectly removed. Afterwards imbibe with the quintessence four times. Let it stand twenty-four hours. After this pour away the fifth essence again and add another. Repeat the process four times. Then cause the quintessence to evaporate in the fire, and into this oil place calx of the Sun or Moon. Then the Quicksilver, in twelve hours, will be produced, as Maulperger has told us.

Otherwise

Make an aquafortis out of two pounds of vitriol, two pounds of salnitre, and one pound of alum. Take one pound of this aquafortis. Dissolve therein two ounces of sal ammoniac. When this has been effected, next take twelve parts of sublimated mercury. Then dissolve it in aquafortis. After this put the aquafortis into a cucurbite, which must be well sealed up. Leave it to putrefy for fourteen days. Next distil it, as you know. Place the calx of the Sun or Moon in the water. Then you must imbibe the ☿) three or four times in the oil of tartar, and thoroughly dry it again. Afterwards place it in the said water. Leave it to digest for several days, and the calx will be Quicsilver.

Mercury of the Body

Dissolve Luna in aquafortis, sweeten with sweet water, next place the calx, when washed in sal ammoniac (fixed), and suffer to flow in the glass. Stir vigorously with a skewer until it becomes somewhat black. Next place it in hot water, and let the salt be dissolved, whereupon the calx will remain at the bottom. Then distil the water, and afterwards the calx and sal ammoniac. Next imbibe the calx in oil of tartar, dry it, and imbibe again. Repeat the process thrice. Afterwards pour over it the oil of tartar, that it may float on the top to the height of two fingers. Let it stand for a natural day. Then pour out the black oil of tartar, and pour another above. Do as before, and repeat the process until the oil of tartar becomes clear. Pour out the oil, and place the calx of Luna in a glass with a long neck. Pour over it equal quantities of sal ammoniac, oil of tartar, and vinegar. Leave them to putrefy for fifteen days. Afterwards place the alembic above. Distil the vinegar from the matter; then sublime the sal ammoniac. Thus there will remain at the bottom tartar, with salt of the Moon. Then take the matter, wash it with vinegar until the blackness no longer appears. Dry the matter and cover it. Place it in layers in the tigillum, with leaves of pure silver, until the tigillum be filled. Next set the box on a jar wherein is water, as explained below. Also, when you have found the extracted Mercury, imbibe it with vinegar and salt, and wash the same extracted Mercury even as common Mercury.

And Notice

Make an amalgam with extracted Mercury, by the addition of Mercury, sublimated and revivified. Let it stand by a slow fire over the cinders, and you will see Mercury ascending. Make it descend by turning the fixatory until it is fixed and remains with the extracted Mercury at the bottom. Then add another sublimated and revivified Mercury. Fix it, and again add fresh sublimated Mercury, and so multiply infinitely.

Also

Invariably place a little dissolved Luna between the extracted Mercury and revivified Mercury.

MERCURY OF THE BODY

Take of Vitriol and
of Saltpetre equal quantities;
of Calcined alum half a pound.

Make aqua fortis. Dissolve in it filings of Luna, as much as you like. Inject a little salt, and Luna is precipitated. Dry the calx. Place in a cucurbite. If there be five ounces, add two ounces of sal ammoniac and one of calcined tartar. Pour upon these strong vinegar. Let it stand on the top more than the height of two fingers. Place it in horse-dung for four weeks. Afterwards distil as aqua fortis. Then vinegar will ascend first, and afterwards Mercury of the Moon. Collect it with the hare's foot, and you will have Mercury. Take as much hereof as you wish, and put it into the egg of the Philosophers. Close perfectly. Place in a cupel of the wood of the ash. Subject it to a slow fire until a black powder emerges. Afterwards increase the fire until a white powder follows. Add to it half the quantity you require of the corporal Mercury, and the third part of reverberated calx of the Sun. Digest it until it becomes a red powder. Then you will obtain what you desire.

EXTRACTION OF MERCURY OF THE MOON

Take of calx of Luna one mark; of oil of Tartar and sal ammoniac two drams each (or six drams). Mix in a well-closed glass. Put the glass into cold water. Then the calx of Luna becomes solid like a cheese. Next let it stand a day and a night. Then leave it in horse-dung for three weeks. Afterwards take it out, and place it in the bath of Mary for fourteen days. Next set it for three days on cinders, that the water may evaporate, and the matter be completely dried. Then take the matter from the glass. Pour fiercely boiling water over it. Pound it about some time. Thus it will become living Mercury, and there will be scarce four parts out of one marc. That which remains reduce again.

OTHERWISE

Take Luna dissolved in aqua fortis. Then dissolve tartar into vinegar in the same quantity as Luna. Pour the two solutions together by drops, lest it should crackle. Gently extract the moisture. Then extract from it a strong Mercury of the Moon. Thereupon, a greyish powder attaches itself thereto. Take it, and rub with oil of tartar in your fingers, and it becomes living Mercury.

OTHERWISE

Take dissolved Luna, and dissolve tartar in the quintessence of wine. Let the quintessence be four times as much as the aqua fortis. Unite these solutions and there is a ready union, without any commotion. They combine like an adhesive tincture. Take them out, dry, and sublimate. Then the living Mercury comes forth and is produced.

MERCURY

Take Mercury seven times sublimated and revivified, as you know, and place in a warm stove-bath. Supply the same with leaves of Luna to devour.

A Manual of Paracelsus the Great

When they have thus been arranged in the stove-bath, you will perceive that the said Luna has been totally transformed into powder, which is the medicine over the Mercury 3. [? to the third grade.] Then having thus collocated the said Mercury, you are to nourish it with common purged Mercury, so that it may digest well in its hot bath. Thus, also, common purged Mercury is converted into a powder which is a Mercury over other Mercury 3. You may cause it to revert into a body, as you know. Also, you must know that the above mentioned Mercury, if placed in dung, will for a time be converted into oil. Congelate and waxen this with incombustible oil, and its virtue will be infinitely augmented.

Make the Attempt, and you will see Marvels

The process of congealing it without medicine consists in filling a strong vessel to the top therewith, the head of the vessel being closed with salt, lime, and yolk of egg. Let it be permitted to dry, and underneath let a fire be kindled from morning till night. Afterwards examine it, and should you find it fluxible, kindle a fire underneath it for another day. Then extract it, and you will have the same stone, which melts like lead, and is white as silver, nor does it differ therefrom, except that it melts quickly. Melt it again and project it into dissolved salt, until it hardens and becomes silver.

Mercury out of Luna

Dissolve Luna in aquafortis. Then entirely distil the water from it. Dissolve in this water the same quantity of sal ammoniac, and afterwards an equal amount of Mercury sublimate. Let it be distilled through an alembic. This is the qualified water. Place the same over calx of Sol or Luna, etc., wherein there must be dissolved sal ammoniac and oil of tartar. Mingle these together in a long-necked vessel. Let the same stand in dung or the bath, and it will be turned into Mercury.

Mercury of Jupiter

Take of Mercury subl. ℥ij.
Mercury crude, ℥j.
Jupiter ℥ij.

Pound together for five or six hours. The crude will then be converted into water, and Jupiter into Mercury. Preserve all these.

Take of Luna j. part,
Common Mercury ⎫
Body of Jupiter   ⎬ iiij. parts each.
Mercury subl.   ⎭

Make an augment, as you know; place in the glass vessel. Apply at first a slow fire. Afterwards increase it.

Oil of Arcanum

Take some honey with juniper and celandine. Distil thrice (ten times) oil of flax with sulphur, also distilled thrice (water caudi magnæ mirandæ.

Distil thrice oil of yolks of eggs, with the calx of eggs. Let lb. j. each of them be mixed together, and let these species be added:

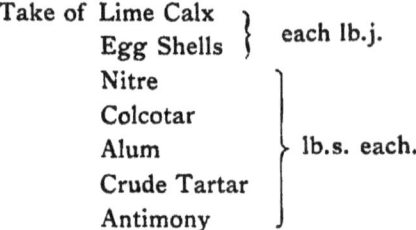

Distil and renew the prescription thrice. Afterwards take of this oil lb.j., of oil of tartar and antimony, with boiling wine extracted and precipitated to the bottom, lb.ij.; of oil of antimony, sublimated, distilled, and red, xiiij parts of a lb.(j. quartale); of red species of aquafortis lb j. Let all these be mixed together into one vessel of good glass. Afterwards take j. quartal of sal ammoniac, of salt petre, of fused salt, all thrice prepared by the calx of cementation, with continual dissolution in red vinegar, and by distillation through a filter, and coagulation with crocus of Mars congealed and five times fused. Of alcali of corrected tartar, and of alcali of corrected soot lb.s Of red arsenic sublimated, dissolved, and congealed, lb.j.; and of the aforesaid oil as much as there is. Let all these be mingled together and dissolved into a glass, and let it become a red and very thick oil, gilding all things, and everywhere making marvellous ingress and tincture of the Sun. But this is not yet perfected. In order to strengthen it, dry into it by the distillation of the alembic these spirits:

Take of Antimony, j. quart.
Colcothar
Salpetre } lb. j. each.
Calcined Alum

Do this thrice; next remove the moisture by means of the bath, and an oil will remain, which it will be lawful to call the arcanum of Christ.

WATER OF MERCURY—A VERY GREAT ARCANUM

Take of Mercury from salt of tartar, as often as it does not ascend, this takes place the seventh time; likewise, take sublimated arsenic and sublimated sal ammoniac, lb.j. of each. Let them be imbibed frequently with oil of the salt of alcali of tartar. Afterwards dissolve over marble into water. Then take as much as there is of this water, and of sublimated sal ammoniac as before. Of Mercury and arsenic take lb.j. Again dissolve into water. In this water dissolve ij. ounces of Luna and one quartal of alcali of soot and of best prepared salt. Mix them together, and coagulate them, by means of an exceedingly gentle fire, into a stone. Imbibe this with water of eggs. Correct and fix very many times. Dissolve again, and coagulate. Imbibe again, and do this eight times or more, when you will have the miraculous stone of tincture. Also let it be imbibed to the red with the oil of the arcanum, so that it may become red; for this is an arcanum not known to all, because it coagulates and fixes Mercury into genuine Luna.

NOTABLE ELIXIR

Take lb.s. of dragon's blood and lb.j. of most white *sal peregrinum*. Dissolve it seven times in water of pomegranates. Let the calx be imbibed, and then frequently dry and desiccate. Afterwards dissolve sal ammoniac in water of atrament. With this water pound the calx. Dissolve for three days. Afterwards congeal in ashes to the elixir. One part changes thousands of prepared Saturn into best Sol, which will be better than the mineral.

ELIXIR AT THE WHITE

Take of fixed sublimated Mercury j. part, and of white sublimated arsenic half a part. Imbibe both with water of eggs, and desiccate five times. Pound as many times, and as often again desiccate, when it will ultimately be converted into a white crystal plate. Take of the same one part. Take one part of this recipe and project over xxx. parts of Venus, or iron burnt through arsenic, and reduced to a solid substance, when it will become silver, perfect under any test.

ANOTHER ELIXIR

Take of Fixed Mercury sublimate lb.j.

Fixed arsenic } lb.s. each.
Sal ammoniac }

Imbibe all these with water of eggs, afterwards place in a glass, and on the top as much of the water as floats above. Close up the glass vessel with wax. Place it in warm horse-dung for fifteen days. Afterwards take it out, and you will find the whole dissolved into water. Take this water and distil through the alembic. Next put the water thus purified in a small vessel. Then devitreate and place over cool ashes; leave it there till it is converted into a plate, which plate does not fear the force of fire. It is upright and deep, tinging and permanent; j. part changes 100 parts of every body into the purest Luna.

NOTE

Water of eggs distilled seven times, and sulphur imbibed therewith over the stone make it fixed and fluxible; thereby Mercury will be congealed.

ELIXIR FOR LUNA

Take some calx of eggs, calcined tartar, and alumen Iamenum. Dissolve them in boys' urine. Then take that powder and dissolve at the bottom of a crucible. Over that powder set Mercury sublimate, so as to completely close it. This having been effected, perfectly close the glass crucible with another crucible by luting it so that the smoke may not escape. Then place by the fire for one hour, and it becomes beautiful Mercury of Luna, which then undergoes increment.

CONCERNING LUNA AND VENUS

Place them in layers and layers with sal ammoniac and laminated Venus. Also lute the tigillum thoroughly, and place on the furnace for three hours.

Afterwards wash the plates with water, and distil by descent; that is, let them be granulated *per scobam*, if not sufficiently whitened. Repeat the process as before; after, add two parts and a half of Luna, and it will stand any test. There is a very great secret *in particularibus*, in which every one may recover their outlay.

NOTABLE ELIXIR

Take of Mercury ij. ounces.

Jewish stone
Sal ammoniac
Common salt
Antimony
} one half ounce.

Pound each separately and mix. Set in layers, first the powders, and the Mercury on the top. Let the tigillum be in such a manner luted and placed to digest that the ☿ does not escape. Arrange coals above and beneath. Afterwards take and place on the cineritium; then purge, and you will have perfect Luna.

REMOVAL OF COPPER

Take of Oil of tartar
Arsenic
} part j.

Place over fire in the cinders in a glass vessel, so that they may become one mass. Pound it, and dissolve it over marble into water. If there be lb.j. of this water, inject j. of white and blind arsenic. Congeal in a glass vessel with a slow fire, of which elixir j. part whitens vj. of copper, and as much of Luna, whence you will rejoice.

NOTE CONCERNING SULPHUR

Take very strong lixivium in any quantity. Distil through a filter. Place in devitreated matter with gallow-stone. Add as much sulphur as you like, and it will be a thin pottage. Leave it to stand for two days. Afterwards cause it to boil for two days, when it will become blood. Distil through the alembic, and you will find at the bottom the sacred divinity. It converts Venus into Sol. Mix it with natural Sol. N.B.

MALLEABLE MERCURY

Cause sulphur to boil in oil; then pour in Mercury. Immediately take it out, and you will find a mass which a hammer will flatten. It does not fear the fire, and you will be able to mingle it with Sol and Luna, with a third part—[lacuna].

FIXATION OF MERCURY

Take equal quantities of sal alcali, ammoniac, and nitre. Let them be imbibed well with boiling wine and water distilled through the alembic. This compound fixes Mercury.

NOTE CONCERNING A PYXIS

Construct a pyxis of iron. Inject Mercury with the sap of gladiolus. It will then audibly groan. When it ceases to cry out, put j. lb. of Mercury into the tigillum, j. lb. of pure tin, and half an ounce of pure Saturn ; let them be dissolved together. This silver stands all tests.

ELIXIR MAKING AN INCREDIBLE QUANTITY OF GOLD

Take ℥iij. of new Saturn and ℥j. of pure Sol. Melt them together. Place on the cineritium in a cask, with Saturn, from morning till evening, until you have expended iij. lbs. of living sulphur, and you will see the same. Out of this project two parts of aqua fortis, and you will have the matter from gold of the usual colour and frangible. This is the medicine, and is called the elixir for the sun. Place j. part over x. of Luna, and it will be ☉. Should you place the first part over two of Luna, there will be *aurum florenorum* transcending credibility.

WATER OF MERCURY

Place iij. parts of Mercury subl. and iiij. parts of sal ammoniac in a luted glass vessel. Let it become a mass and dissolve this in oil. Take that water and close well ; also add j. part of filings of Luna. Dissolve the whole together in water. Then take that water and congeal. One part changes at least lxvj. parts of Jupiter or crude Mercury, which will be the best Luna, standing every test.

TRUE ELIXIR

Take lb.ij. of purged Jupiter and lb.j. of purged Mercury. Dissolve Jupiter, and put in lb.s. of Mercury and arsenic sublimated. Afterwards pound them with lb.s. of sal ammoniac. Place in cucurbit with the addition of the strongest vinegar. Keep that which is distilled. Afterwards increase the fire until it be sublimated. Also pour vinegar over them again, and proceed as above seven times, or until nothing more be sublimated. Then leave to decalcine. Dissolve in vinegar and distil. That which remains at the bottom pound and dissolve into water over the stone, and coagulate in due fashion. j. part changes xxx. parts of Venus, which passes through every test.

FOR LUNA

Take of Arsenic } lb.j. each.
Tartar

Living calx } quart.j. each.
Prepared salt

Let them be pulverised and placed in a luted vessel over a slow fire. Afterwards break and collect the powder. Next dissolve j. lb. each of Luna and Venus. Also project ℥s. of this powder, and it will be good Luna.

TINCTURE FOR LUNA

Take of Salt thrice sublimated
Mercury six times sublimated
Calcined Luna } lb.j.
Water of Sal ammoniac
compounded and rubified

Imbibe altogether in a glass vessel. Afterwards place in cucurbit and distil at first with a slow fire, next with a moderate one, for the space of three days.

Then if it shall have cooled, extract, and when it has thus been distilled, add another pound of water of sal ammoniac, again dissolving as before, restoring to it its water, which had been distilled from the fæces, so that it may be thrice imbibed or distilled, or till iij.lb. of sal ammoniac be consumed. Afterwards you will find at the bottom of the vessel a crystal plate which tinges and is stable and permanent. One part of it tinges 1,000 of Venus into Luna.

To remove Venus

Take iiij. parts of oil of tartar, and j. part of white arsenic. Imbibe them repeatedly until the oil has been consumed. Afterwards dissolve vj. parts of purged Venus with glass and of elixir part j. Make Luna. Let there be added afterwards iij parts of Luna.

Luna out of Mercury

Take of Living Mercury
    Salt fixed by calx and dissolved eight times
    Alcali of soot
    Crude tartar
} part j. each.

Pound, mingle, and burn. Mix after combustion. Place in layers, and let them melt for four hours. Then Mercury will yield Luna, and it will be perfect Luna.

For Luna

Take a tigillum well luted, and at the bottom place sulphur. Also suspend Mercury in a hempen bag above; next let it be everywhere luted. Afterwards set on the fire for one day, or until moisture ceases to appear. The smoke of sulphur then passes into Mercury; next repeat the process and make ij. (*sic*). Afterwards take of this Mercury one half ounce, and project over lb.j. of purged Venus, when it will be natural Luna.

Elixir for Sol

Take Vitriol
    Crocus of Mars
    Flower of Copper
} j. part each.
Prepared sal ammoniac.
Prepared hæmatite.

Let them be pulverised and mixed together. Dissolve in a tigillum until the aquosity is consumed, and it becomes a powder. Afterwards set in a glass vessel on horse dung for several days, or in a damp cellar. Let water be produced. Congeal it with a slow fire. Afterwards take j. part of Sol and ij. or iij. of Luna. Melt, and over xvj. parts of this Luna project one part of Elixir. Thus, the more it is burnt the better. It will be perfect and most beautiful gold.

Cement

Take of Antimony, lb.j.
    Salpetre, lb.ij.
    Calcined tartar, lb.j.

Melt together, and inject lb.j. of pounded ✳ (otherwise tin), and immediately let it be melted. Dissolve again, as above, and melt, until it is very red, one part over two parts of Mercury in flux (of Venus perhaps).

OIL OF ANTIMONY

Antimony converted into oil in a very strong lixivium out of clavellated cinders fixes spirits.

FIXATION OF MERCURY INTO RED

Take equal quantities of salt tartar and sal nitre. Make a strong lixivium. Inject sublimated Mercury. Make it boil, and when it has dried up, pour lixivium in again until it is rubified. Waxen it with calx of Luna.

TO PERMANENTLY RECTIFY MERCURY

Take saltpetre, sal alkali, cinnabar, alum, flower of copper, and sulphur. Let it be imbibed with water of life. Afterwards let it become aquafortis. This water dissolves Mercury, and all bodies with Mercury will remain in the form of a crocus colour; if it has been calcined half a day, it will be redder than cinnabar; it does not diminish in weight, and is dissolved into extremely red water. Afterwards coagulate and, when coagulated, reduce with saltpetre; it thus becomes gold. If further coagulated and dissolved, one part tinges parts (number omitted) of Luna permanently.

PROJECTION OF LUNA

Take of Cadmia
Salpetre } ʒj. each.
Calcined alum

Melt in luted tigillum one hour, and you will find ʒj. of Sol.

NOTE

Oil of iron colours citrine, the oil of chalybs red, and that of lead red.

BEST BORAX

Take Alum
Calcined tartar } equal quantities.
Sublimated salmiax

Let them boil in water together; afterwards let them be strained through a tightened bag. Then let them boil in alcali. You will subsequently have Borax, which dissolves all metals.

PERPETUAL AUGMENT

Take j.lb. of Luna, cemented and purged by means of salt, and iiij.lb. of Mercury purged with salt. Make an amalgam, place in a phial over cinders in a cupel on a slow fire (and avoid closing) for one day, that the vapour may evaporate. Afterwards place over the amalgam as much as there is of fixed saltpetre with yolk of eggs; imbibe thrice or more times over the fire; cook until it is dried up. Imbibe again as before until it be fixed. Afterwards lute it well, and suffer it to stand in a slow fire for eight days. Take out the matter, when you will find it white and hard as crystal. Pound this well; add half the weight of purged gold; place again in glass vessel, as above, and once more superpose the same quantity of fixed saltpetre; permit the moisture to evaporate, as before, and set it to stand for six days. So the process goes on for ever.

REDUCTION TO THE SAME

Take the matter and imbibe the same with oil of albumen of eggs, for the white stage, but for the red with yolk of eggs. Add the same quantity of borax to the tigillum. Melt, afterwards cement.

NOTE TO THE SAME

Cinnabar distilled by descent is the best Mercury.

ANOTHER

To reduce it with raw albumen of eggs, pound and let it become a hard paste. Cause this to melt with borax, as above.

WATER OF NITRE

This retains Mercury.

FIXATION

Take White Tartar
White Arsenic } in equal quantities.
Fused Salt

Boil well in vinegar; add the same quantity of pounded Venetian glass. Let it become a powder. Take one *part* of it, and of the amalgam two ounces. Place in a luted tigillum, and let it melt for one hour, afterwards proceed by cineritium.

PERMANENT ELIXIR FOR THE WHITE BY CINERITIUM

Take of Sal Ammoniac
Sublimated Mercury } $\frac{1}{2}$ oz. each.
Live Calx, 1 oz.

Mix together, place in a glass vessel, and permit evaporation. Afterwards lute and increase the fire, so that it is kindled, and you will find Mercury over the calx with sal ammoniac; also pour over warm water. Next, cause evaporation; again pour on warm water, and it will be possible for the tartar to dissolve. The calx will then arise from the Mercury, and the Mercury will remain at the bottom like snow. Perform this operation twice, and let the sal ammoniac and Mercury be fixed. Afterwards take half an ounce of calx of Luna out of aquafortis, and half an ounce of this powder. Dissolve each of these by itself in aquafortis, and afterwards abstract the aquafortis. Place the matter in a glass pitcher. Pour over a strong alkaline lixivium and coagulate, and so again, etc. Do this four or five times. Afterwards put the matter in a glass vessel, and again pour over it aquafortis and abstract. Do this three times or more, and it will be a hard stone. Also afterwards add Luna; then it preserves the white copper in the lead over the capella. Also, over Mars take half an ounce of white powder of copper. Let it be projected in flux. Also that powder fixes Mercury which has been coagulated without metal.

NOTE

Take Saltpetre
    Alum
    Sal Ammoniac
} in equal quantities.

Make a powder and a pottage, so to speak, with water, and *unge clinodisa*, which are mixed with gold and silver, and it will be a gold colour.

PROJECTION OF LUNA

Take of Burnt Alum
    Sal Ammoniac
    Vitriol
} eight ounces.

of Red Jaspis, ʒiiij.

Cause them to melt together and become a powder. Take ʒs. of this over eight ounces of Luna, and you will have much gold in eight ounces of Luna.

FIXATION OF CINNABAR

Take Cinnabar
    Litharge
    Antimony
} ʒj. each.

Place in an iron pan to boil with strong vinegar. Arrange in layers with Luna. Let a tigillum be luted. Place it by a slow fire for two hours, and the cinnabar will be fixed into gold; but finally make a fire in a wind furnace.

PRODUCTION

Take Good Gold
    Granulated Venus
    Red Sulphur, sublimated by Crocus of Mars
} j. part each.

Melt the gold and copper. Project one part of sulphur, and when it is consumed in the fusibilum, take the regulus and add again the same quantity of Venus. Once more pour on three parts of sulphur. Melt again. Take the regulus and add copper, as before. Do this thirty times, and it will be perfect gold, and of the best colour.

NOTE

I have written as many praises of this powder as I could.

ALSO

Grade together with Sol white and red melted together.

PARS CUM PARTE OF MASTER THOMAS

Take equal quantities of Sol and Luna. Make plates of them.

Take Hæmatite
    Sal Ammoniac
} ʒj. each.

Evaporated Vitriol
    Saltpetre
} ʒiij. each.

Bolus Armenus
    Alum
} ʒs.

Verdigris, one-sixteenth *part*.
Tutia, 16 *parts*.

Make a most subtle powder. Afterwards take lb. j. of vitriol. Distil it. Imbibe the powders in a thrice-devitreated jar. Afterwards take powders and make a layer in the tigillum as thick as a coulter, after having saturated the Luna in the aforesaid water. Thus arrange layer over layer. Perlute the tigillum and set in a circular fire for three hours. Make this cement thrice, and you will have the Sun in all your operations. Then take this gold with three parts of Luna well weighed. Place in aquafortis, and it will be converted in the following manner.

Aquafortis, wherein Luna, When placed, becomes Gold

Take Vitriol ⎫
Saltpetre   ⎬ lb.s.
Alum ʒj.

Precipitate aquafortis with crude Mercury. It afterwards converts Luna into Sol.

Albatio Bambergensis

Take any quantity of Jupiter, and the same of living Mercury. Make an amalgam. Then take sulphur if for the red stage, or arsenic if for the white, and sal ammoniac, all in equal quantities. Pound thoroughly. Place in a well-luted vessel. Sublime for half a day with a slow fire, afterwards with a fiercer fire for a whole day, next with an exceedingly strong one for two days. Then take what is sublimated, and set apart. Let that which has not been sublimated undergo a further process. Then sublime the same from salt well prepared, twice or thrice, until it becomes as snow. Take two parts of these species and one part of filings of Luna. Arrange in layers in a glass vessel, which must afterwards be luted. Burn with a fierce fire, so that they may be melted together. Afterwards pound, dissolve, and coagulate, at least three times, and it will be an Elixir of which one part tinges eight parts of purged Venus into best Luna.

Arsenic Sublimed becomes as Luna

Sublimate it from calx of eggs, and it becomes just like silver.

Note

Sublime Mercury sublimated from Saturn; it makes lead like Luna.

Fixation of Luna

Imbibe cinnabar with oil of tartar, afterwards cement Luna, melt it with borax, and immediately *in coloritio*. [This process seems to be unfinished.]

Sublimation of Mercury

Take aquafortis, in which dissolve as much Mercury as you please. Then add the same quantity of common salt. Abstract aquafortis through the alembic. Then increase the fire, so that the Mercury may be sublimed; an exceedingly fierce fire must then be employed, arrange in layers with that Mercury and plates of Jupiter on a humid place, when the Mercury will become water.

Fixed Luna

Let Luna be cemented for six days in crocus of Mars. Let it not be further affected by aquafortis, and it will take the tincture.

Fixation of Mercury out of R

Take Mercury par. vj.

Borax ʒiiij.

Impaste properly together until the Mercury is completely invisible. Pound and sublimate until ascension altogether ceases, repeating the process till this takes place; then cool. Break the sublimatory. Fulminate that which is found at the bottom thereof. You will then have the best Luna; also in the first eight ounces of this Luna you will have $2\frac{1}{2}$ ounces of best Sol.

Note of Master Albertus

Prepare salt by frequent dissolution in fire, melting and congealing. Repeat this until it melts like wax, fifteen times employing a filter, etc. Also afterwards take iiij. parts of borax, and pound it with well-calcined alum. Pound it again, even seven times. Then take the said salt and white borax in equal quantities. Put it in a glass vessel moderately warm. Then a tincture is produced from this. One part of it is projected over 30 of Mercury, and the same becomes Luna, commencing to flow immediately with the tincture. Blow strongly so that the Mercury may make its way through. Melt, and you will have Luna. *Notice carefully.*

Fixation of Luna

Dissolve eight ounces of Luna in aquafortis. Afterwards congeal. Add to this calx the same quantity of Mercury sublimated. Sublime 15 times, as frequently imbibing and drying, until all that is there is imbibed. Afterwards reduce the dross into a body with oil of tartar, and you will have Luna to the weight of the gold, and it stands in cement.

Note for Luna

Take a globule of earth and fill it with Mercury. Put it into Saturn for three or four hours. Afterwards Saturn attracts Mercury to itself. Over this Saturn and Mercury project fixed borax with saltpetre into the tigillum. Next fulminate in a furnace of wind. This will test the Luna.

Water of Mercury

Take equal parts of tartar and mercury, sublimate thrice, and pound on marble. Then dissolve. The metal is dissolved in that water in one hour, but the matter is to be distilled in water. Afterwards take of pure gold and pure Luna equal parts. Dissolve in the aforesaid water. Then take one part each of borax and camphor. Reverberate one-half of these, distil the water from them and keep it. Next take one drop of this water and project it in the water wherein the body has been dissolved. Place it on a slow fire for seven days, and it will become milk. Replace this, and make in turn two or three drops. Repeat this process until it no longer grows white, but another red liquid remains. Coagulate this; then take one part of the white powder and project it over thirty parts of Venus, when it is rendered white, and answers every test. Finally, take the red powder, project it over the above-mentioned white, and the whole will become red.

Secret Note

Take filings of Venus and put them in wax for three days and nights. Take one pound thereof, four of vitriol, and five of sulphur. Mix; lute between two bricks in a carefully watched furnace, and note the result.

In the Distillation of Vitriol for Fixation

Place camphor in a glass vessel containing oil, and it will become fixed.

Note on Pars cum Parte

Pour *pars cum parte* several times into blessed oil. It gains in grade and acquires great softness.

Rubification

Take vitriol and pound it small on marble. Imbibe it with wine to the consistence of a paste. Then take two stone dishes, one somewhat larger than the other. Place therein imbibed vitriol, and lute with a material composed of dung and gravel. Put it in a furnace for the space of a natural day, and the thing is done.

The Purification of Tin

Project three or four times over oil of tartar.

Note

Oil of tartar is that which is made for common use; but if a plate of copper be frequently placed in it, it becomes white.

For Broken Coins

Take sal ammoniac, place it in the crack and the lesion disappears over a coal fire.

Sal Ammoniac is made thus

Take twenty parts of urine passed by a wine-drinker. Skim and let it cool. Add one pound of sublimated dung and two pounds of salt. Mix these and let it stand for three days; then boil until coagulated.

For Reddening Crystal

Take olive oil and quicklime in any quantity, and shake them well together. Then take two parts of salt alkali and one part of common salt. Mix these so that the oil shall float on the top to the depth of one finger. Distil over a slow fire and let it cool. Set a light to it, and, if it burns, the process is not complete. Repeat until it no longer kindles, and then it is made ready for rubifying the crystal.

Fixation of Mercury

Sublimate with quicklime until complete.

For Gilding

Distil the yolk of eggs. It is converted into a red oil which gilds money and lasts for ever.

Sal Ammoniac

Take one pound of sal ammoniac, pour on it acetum and wine, distil by means of a filter, and add ten pounds of sublimated dung. Set it to boil, when it will be desiccated. Pour urine on it again until a twentieth part of it is consumed. Then add pure water and dissolve. Take the clear water and lay aside the dregs. Boil and dry, either in the sun or in shade.

Sal Borax

Take calcined tartar. Pour on it warm water. Pass this through a straining-bag until the upper part is clarified. Then take common salt dissolved in water, add one part of this, and afterwards boil in an iron dish until it is thickened. Then place in a vessel and desiccate until it becomes friable. Thus you will have borax, which place in glass.

Another Method

Take one-third part of crude tartar. Sift thoroughly, and add six parts of prepared common salt. Boil for a day, until it is converted into water. Set to cool, and distil by means of a filter, afterwards let it boil until hardened, and you will have the very best borax.

Sal Borax for Goldsmiths

Take one part each of starch mastic and sulphur. Of this mixture take two parts, pulverise and boil until thickened. Set it in a glass vessel to putrefy for four weeks, and you will have borax.

Sal Borax

Take Calcined Tartar, lb. $\frac{1}{2}$.
Quick Lime, lb. ij.
Wood Ashes, lb. j.
Crude Tartar, lb. $1\frac{1}{2}$.
Prepared Common Salt, lb. vj.

Boil all together, distil in a filter, and coagulate. Thus is obtained excellent borax.

Lazurium

Take one part of sal ammoniac, two parts of sulphur, and two of mercury. Mix, and proceed as directed above with cinnabar, until a purple smoke ascends; then leave off.

Another and Better Method

Take lapis lazuli. Heat and pound it. Then take two and a half parts of pitch, goat suet, and oil of laurel. Dissolve these together.

Another Method

Take sal ammoniac and pulverize it. Amalgamate this with four parts of mercury, and place it in a glass vessel closely luted. Sublimate for one hour. Then gradually increase it to the smoke of citron wood; afterwards diminish. Finally, you will discover lazurium at the bottom.

Another

Take sulphur and mercury. Amalgamate as above, and sublimate.

Flos Aeris

Take filings of Venus, urine, and sal ammoniac. Mix them together, desiccate, and imbibe a second time.

Cinnabrium

Take sulphur, dissolve one part thereof and two parts of mercury. Cool them, place them in a glazed vessel, and sublimate. First of all a purple smoke ascends, and afterwards a red one. Then cease.

To make Marble

Take quicklime and extract the lixivium with wine. With this is imbibed calcined and pounded flint. Forthwith it is susceptible of colour.

Corals are made thus:

Take of minium one part and of cinnabar half a part, but of quicklime and lime of flint five parts each, with a sufficient quantity of the above-mentioned lixivium and the white of eggs. Add as much salt as you please, and finally boil in linseed oil.

Factitious Corals

Take of good gypsum two parts, pure lime half a part, minium and cinnabar half a part each, white of egg quant. suff. Form and dry.

Pearls from Chalk

Put chalk in the fire until it is friable. Mix with white of eggs and shape. Then harden, and afterwards moisten with spittle. Whiten with silver tablets and again harden with fine powder, either in the sun or over a coal fire, as you please.

To make Pearls better than Natural Ones

Take mother of pearl and pulverize it very small, afterwards adding fine flour. Mix and temper this with Maydew. Shape according to pleasure. Then give them to hungry pigeons to eat. Wash their dung, and you will discover very fine pearls. But notice that the pigeons should be kept for three days without anything to eat.

Another Mode

Take mother of pearl, boil and wash it well. Then take the same quantity of crystal, pound it small and mix with the white of eggs. Shape and dry. Afterwards boil thoroughly in linseed oil, and wash with white wine. Afterwards dry in the sun, or over a fire.

Note for Luna

Take of arsenic and of sublimated mercury one part each, and sublimate them by themselves. Add an equal weight of quick lime. Imbibe with water of fixed sal ammoniac, then cover with wax and place on the fire. Sublimate thrice, and keep in a closed vessel. Afterwards desiccate it, and project one part over twenty-five of purified Venus. Thus Luna will be produced.

Ruby

Take 4lb. of strongest acetum, not distilled, and in it put one ☐ of atramentum. Distil this after the manner of aquafortis, and at last with a very strong fire, so that the spirit may ascend. Take four parts of this acetum, and place it in a glass vessel, into which put one part of filing of Mars. Stir it with a rod, and the mixture begins to boil without fire. Set it apart to stand and you will see a pellicle form and float on the surface. Collect this and put it on one side. Then stir the mixture again for an hour, and once more collect the pellicle. Do this as often as necessary, and put this matter, which is yellow, like gold, into a firmly luted crucible, to be calcined for twelve hours. Then take it again, pound, and dissolve it in fresh acetum as before. Collect once more, as directed, the pellicle floating on the surface; again dissolve and calcine as above, and repeat this process three times. Afterwards dissolve in aquafortis, which dissolves gold, and reduce the water three times. Then it becomes a medicament which tinges Luna to Sol. One part thereof fall upon two hundred parts of Luna.

Aqua Ardens: or Water of Mercury

Take half a pound of sal ammoniac, a sufficient quantity of tartar and live sulphur, with one pound of common salt, and a quart of good white wine. Place all these in a well-luted vessel and submit to the fire, perhaps somewhat severely. Then keep the water well away from the wind. This purifies metals and converts mercury into pure Luna.

Water of Sulphur

Take $\frac{1}{5}$ lb. of sulphur and $4\frac{1}{8}$ lb. of saltpetre. Place these in a well-luted vessel, and put in the ashes for twenty-four hours. Remove and pulverise. Then mix one-fourth of this with 3 oz. of pure water; whereupon the water is turned to a red colour, and money can be coloured therewith into the semblance of most beautiful Sol.

Aurum Musicum

Take of tin and of sulphur each one-third part, amalgamate, and pulverise. Afterwards wash, first with lixivium and then with pure water, until no dirt comes off. Then dry. Next take two pounds of Mercurius vivus and the same quantity of sal ammoniac. Mortify the mercurius with acetum, and at the same time wash it, as before. Then place it in a well-luted phial, so that the glass vessel be half full. Heat it in sand for four hours until you see a golden smoke ascend.

Wonders of Antimony

Take antimony and purify it with calcined tartar. Afterwards make aquafortis. Dissolve it in the water, and congelate, either in the aquafortis or by itself. Then it will be an oil incapable of mixture—oil or stone—in the proportion of one part on three parts of Saturn; and it will be silver answering perfectly to every test.

Wonderful Aqua Ardens

Take old red wine; and put in a glazed vessel one part of auripigment, half of quick sulphur, and a fourth of quicklime. Mix together; afterwards distil by means of a rose alembic and the thing is done.

The Virtues of this Water.

Whoever places his finger in it is burnt as by a light. In like manner, a rag placed in it burns like a candle, and is not extinguished by water.

Augmentation

Make an amalgam of mercury and Luna. Then fill a vessel with quicklime. Take equal parts of salt of alkali and of litharge. Put layer on layer and let it stand for a day and a night. Afterwards let it dissolve.

On Red Venus

Take equal parts of crushed beans and of crude tartar, together with a quantity of tutia exceeding one of those parts. Mix all well together and place layer upon layer with plates of Venus which have been laid all night in acetum. Pour it over the layer, and let it shape wherever it can be poured. You will see the result.

Calcination of Jove and Saturn to White

Make layer upon layer with quicklime and the above-mentioned metals. Cement by night, then take again, dissolve, and pour into a lixivium of quicklime, acetum, and vine-ashes. It will then be as Luna.

To Reduce Calcined Bodies to their Original Matter

Take one-fifth of the metal and two pounds of borax of tartar. Pour together, and place in an iron vessel. It will be as Luna, and can easily become Sol and Luna if first imbibed with oil of tartar.

Calcination of Sol and Luna

Take filing of Sol and place it in acetum for nine days. Then put it in water of sal ammoniac. Imbibe well; then desiccate, and continue the process as long as you please.

The Devil in Alchemy

Dissolve, fix, coagulate, and reiterate.

Water of Salt of Alcali

Take alkali, sal ammoniac, and egg-shell. Pound together with good acetum, and dissolve at the same time.

Purification of Sulphur

Take pounded sulphur, pour on it acetum and wine, let it boil for a day, and skim it. Then pour in urine and boil for two hours. Do this until it ceases to froth. Whatever body there is does not float in the urine, but the urine is clear and sufficient for the purpose.

The Sublimation and Fixation of Sulphur, so that it becomes White

Take of sulphur as much as you will, and pound it. Pour acetum upon it. Heat it until all fatness is removed, and then lay it aside. Secondly, heat it as before in boys' urine, and draw off the fatness. Next, pound it with prepared alum, place it in a sublimatory and sublimate for three hours. Then the white sulphur ascends like snow, and flows down upon the coals just as a snow storm.

Oil of Vitriol

Take as much vitriol as you will, and distil it by descent. It renders a bright green oil, and is called milk of mercury. But it must have a large fire; it is like in its nature to balsam.

Another Method

Distil vitriol with an alembic over a very strong fire for three days and nights. Let it be imbibed thrice with its dregs, and be distilled from them. Afterwards distil five or six times, and thus a great arcanum is produced.

Correction of this Process

After the third distillation mix with it a half portion of vinum ardens, and distil as aforesaid.

Another very Expeditious Mode

Take vitriol and distil it vigorously by descent, afterwards by an alembic, and then by a balneum Mariæ. Finally, do this twice or thrice by means of a retort, and the method is subtle. Some say they have seen the oil of vitriol distilled until a whiteness like milk supervened.

Preparation of Common Salt

Make layer on layer with quicklime. Let it be cemented well, distil by a filtre, and coagulate.

Another

Take urine, quicklime, and salt; reduce to water and boil. Then put them in a vessel, project to water, and coagulate. Place again in the vessel, do as before, and it is ignited at length whether it suffice or not.

Fusion of a Marchasite

Take three-tenths of the Marchasite, ½ of Saturn, iij.⅗ of Venus, and iiij. ♊ of Scoria of Iron. At the same time it is poured upon iiij. of Luna, ♉ of Saturn, and II of Venus.

Alkali is made thus:

Take wood ashes, quicklime, and the ash of beans. Extract the lixivium and coagulate.

The Sublimation of Sal Ammoniac

Take of the salt itself and of pulverised Mars, equal parts, and sublimate.

Crocus of Mars

Take aquafortis, not too strong. Place it on filings of Mars; let it stand for some days. Then heat it over a coal fire and it is made red.

Another Method

Take Antimony, filings of Mars, and crude tartar. Dissolve them together, and a good crocus is produced.

Sal Borax of the Philosophers in which all Metals and Glass are Fused

Take lac tauri purified by a filter, alkali, borax, gem-salt, and goat's blood, equal parts of each. Mix with water, put in a phial, and desiccate. Then Venus or any other substance, whether metal or glass, is fused therein.

Saltpetre

Take quicklime and warm water. Stir for six days and distil by means of a filter. Place it in ☉ until it is consumed and you will have saltpetre.

The Solution of Sol from Silver

Take one part of calcined tartar with two parts of sal ammoniac. Place them on marble and then in a glass vessel. Then take the root of pellitory, pound it with acetum and strain it off. Mix this in the above mentioned water, and put the water in a gilded cup. The Sol is dissolved from the silver and is again reduced with borax.

The Softening of Iron and all Metals

Take of alum, sal ammoniac, and tartar, equal parts. Put in good acetum over a fire and extinguish the metal.

Solution without Antimony

Take Venus which has gold in it, and sprinkle on it the following powder: put it over a fire and let it melt. Then pour into a fusibulum and you will find gold or Luna. Purify over ashes.

Item

Take one part of saltpetre with three parts of sulphur and do with copper as above.

Water of Gradation

Take two parts of vitriol, one part of alum, and half a part of antimony. Distil.

Fixation of Mercury

Take a vessel, in which put quicklime *quant. suff.*, and in the middle of this lime place coagulated mercury. Heat for five hours, and it will be fixed.

Ready Method of Coagulation

Take the pounded root and portions of the herb hermodactyl. Heat mercury, pour it over, and it will be coagulated.

The Fixation of Arsenic

Take of tartar, quicklime, glass, and arsenic equal parts. Imbibe with oil of tartar, and sublimate in a vessel after the usual method. When the whiteness ascends, leave off.

Gilding

Stamped money is entirely gilded in the juice of the aurearia and remains for some time.

Cement

Take Vitriol, j. part.
Sal Ammoniac,
Verdigris,   } ij. parts each.
Alum,
Saltpetre, ½ part.

Pound in an iron mortar. Mix over coals and a black powder will result. It gives a great smoke; and when the powder is desiccated, make layer on layer with Sol. It is graduated to twenty-four degrees.

Fixation of Saltpetre

Take the purest saltpetre. Distil by means of an alembic and a portion of the saltpetre will remain at the bottom. Take one part of this and two parts of quicklime. Dissolve on stone, coagulate, and the thing is done.

Perfect Fixation of Luna

Take of this saltpetre two parts and of Luna one part. Let them melt. When the saltpetre becomes like glass, increase the fire until the saltpetre is consumed. Add crocus of Mars and let it be consumed as before. Afterwards pour it through antimony as is necessary, and let it be fulminated. Then it will be Luna fixed and white.

The Coagulation of Mercury, producing Venus

Take equal parts of Mercury and Venus as much as you like. Boil them in water for three hours and stir continually Afterwards take the Mercury, put it into a linen rag, let it stand for a night, and it will be coagulated as Venus.

Another Coagulation of Mercury

Take an egg-shaped crucible, and fill it with Mercury. Lute it, place it in a patella, pour lead over it, and let it cool. Then take it out, and you have coagulated Mercury.

Purification of Metals

Take two parts of antimony, and two parts each of vitriol and saltpetre. Pound well a moderate weight of filings from the metal, and cement over a slow fire for about an hour. Then let it be heated for 15 hours, and afterwards fulminated. It will be diminished one-third part.

Cement from above

Take equal parts of saltpetre, sal ammoniac, verdigris, common salt, and alum. Pound them together and imbibe them three times with urine. Afterwards with a portion of this make layer on layer for six hours, in the end with a strong fire. Then you will have a golden regulus, but it will not stand in the cineritium.

Oil which tinges Luna into Sol

Take antimony ℥. j., with lb.$\frac{1}{2}$ of sublimated Mercury. From this proceeds a red oil which has the property of gilding.

Oil tinging Brass into Gold of 24 degrees

Take the very strongest lixivium and distil by means of a rose alembic. Then place it in a glass vessel and add saltpetre, sulphur, and crocus of Mars j. ℨ. or more, etc., so that the lixivium may float on the surface to the depth of one finger. Let it stand until it sinks to the bottom. This is done over a coal fire and it will become red like blood. Afterwards take it from the fire and let it cool. Place it again in an alembic and distil it until whiteness supervenes. What remains at the bottom is a tincture, and is a most wonderful production, possessing the virtue of natural gold.

Perfect Fixation of Luna

Take cinnabar, gem-salt, and common salt. Pound them well with plates of Luna. Lute them layer on layer, place on a fire of cement for six hours, then purify by cineritium and repeat this process thrice. Afterwards place in cement for 18 hours and you have fixed Luna.

Take of this Luna three parts, of pure Sol one part, and dissolve them together. Its colour is not diminished in aqua fortis, but remains there.

Glorious Oil of Sol

Take very strong lixivium and distil it in an alembic with sulphur, colcothar, and crocus of Mars, equal parts of each. Dissolve in the lixivium, and afterwards place it on the fire, making it boil until it grows red. Next distil it twelve times, or oftener, until whiteness comes forth. What remains in the bottom is a fixed oil and tinges everything to Sol.

St. Thomas Aquinas

I have very often sublimated Mercury until it became fixed. I afterwards dissolved it in water into its primal matter. In this water I placed calx of Luna and fixed arsenic. This I now dissolved in horsedung and coagulated, whereupon it became a stone of tincture.

Water of Mercury

Take of fixed Mercury $3\frac{1}{2}$. 5., with the same quantity of saltpetre. Pulverise them together. Then place them in a linen cloth in horsedung, with glass below. Above all this put a cloth as a covering, setting the horsedung on all sides above and below. Let it remain for two weeks, and then pound it until quite white. It will be sufficient for Luna.

Water of Sal Ammoniac

Take sal ammoniac, the same quantity of egg shell, and a little acetum. Pound these together. Dissolve them afterwards on marble and you will have water of sal ammoniac.

Lac Virginis

Take pulverised litharge, together with a sufficient quantity of acetum. Let them boil well. Afterwards distil with a white filter. Pour it on again, and distil until the water grows clear. Then take some sal ammoniac and anatron. Treat these in the same way, and afterwards mix them together. Then is produced lac virginis.

Purgation of Venus after the Greek Method

Dissolve Venus and project on it some purified sulphur or arsenic, until it no longer emits any smoke.

For Recovering Luna in Antimony

Take of Antimony j. ▫, of Tartar ij. ▫. Pound together with Sulphur IIj. Pulverise, and melt.

Aqua Fortis

Take equal parts of vitriol, sulphur, and alum. Distil first with a gentle and at last a very strong fire. If you wish to have it stronger, substitute calcined alum for the alum, and citrine colcothar for the vitriol. Then you will have very strong aquafortis. Aquafortis also made from vitriol alone has a wonderful odour as pleasant as musk. This aquafortis dissolves all metals except Sol. If it be desired to dissolve Sol, let there be added to the distilled aquafortis a little sal ammoniac, or common salt. Let these be dissolved together, and then it converts all metals into water.

The Operation Boni Thematis

Take of sublimated fixed Mercury two parts, of calx Lunæ one part. Pound these with water of sal ammoniac, and desiccate seven times; then dissolve. After this distil the water by a balneum Mariæ and place this water in a glass vessel. Set it in the ashes of a sublimatory furnace until the water is entirely consumed, when the tincture will remain. Take one part to a hundred parts of purged Venus, or of Mars for Luna.

For Sol

Take Calx Solis and Crocus of Mars in place of the water of sal ammoniac.

The Operation Bonæ Rapacis

Take ☿ vivus, warm it, and in human blood 7. Thereupon it will be hardened. Boil in white of eggs, and you will have excellent Luna.

For the Same

Take Mercurius vivus and put it in menstrual blood, with an equal quantity of juice from cornflowers and a little euphorbium. Let it stand for four days, and it will coagulate for working.

Calx Peregrinorum, or Marine Calx

Take the bones of large fishes, or signum peregrinorum, or cockle shells. Reduce to a calx, and when it has acquired whiteness you will have the Calx Peregrinorum.

Papua Tincture

Take of vinum ardens *quantum suff.*, and of pounded antimony as much as you will. Wash in the usual way. Take the more subtle portion, and project on dissolved Luna. It becomes in some part Sol, as I myself have seen.

The Work of the Noble Canon in Alchemy

We have seen when we cemented pars cum parte, and the golden regulus was placed in the cement, that the result was, as it were, the best gold. But it did not remain in the cineritium as pars cum parte. Nevertheless it did remain in aquafortis. Thus we made a cement on fixed Luna, and placed that Luna in aquafortis, when it deposited for us a large residuum. When, however, this was fulminated, it grew white again as Luna. Once more we placed it in aquafortis, and not the least thing remained for us in that aquafortis. We believe that the matter in the cementations is either nothing at all, or is not cemented sufficiently, or its realgar is not fixed enough.

Calcination of all Gems

Take any quantity you like of each gem. Pulverise it and mix with sulphur. Set fire to it; thus it burns, and you have the calx of that gem. If it be necessary, wash it, and the powder will be white.

How Bones may be Cut

Take wood ashes and quicklime in equal parts, *quant. suff.* Boil the bones herein until they are softened.

Softening of Metals and Ivory

Take the strongest lixivium of alkali and place in it the metal for fourteen days, when it will be softened. Take it out again, cool it in water, and it will once more become hard. Place ivory in the same way, having previously added the strongest acetum.

To Gild Metals

Smear the metal with varnish, and then place upon it a plate of Sol.

To Whiten Venus

Take aquafortis and dissolve Luna in it. Then with pounded tartar and common salt make a pulp and desiccate it. Thus it can be used with acetum. Some add sal ammoniac, smear the Venus, and heat it, continuing the process until the result is satisfactory.

Light Shining without Fire

Take the eye of a goat, put it in water, and place a mirror above it.

Another

Take some lixivium made out of the best quicklime. Place therein alum and camphor. Put it in a glass vessel with live Mercury and set a mirror above.

Water of Common Salt and Water of Saltpetre

Take some of this with tile and distil it. This water is said to have a marvellous power of fixation. If Luna is melted in it and common salt, it can be distilled with honey.

White

Take alcalum, with juice of white onions. Steep Mars therein, and it becomes Luna.

Method of Quartation

Take one part of Sol, and let two or three parts of Luna be made into plates and put into aquafortis. This is the most consummate and excellent test of gold.

Colouring

Take a little flos æris and sal ammoniac. Make them into a paste with acetum. It is one method of testing and colouring gold.

How a Cloth cannot be Burnt in the Fire

Moisten the cloth with salt water. Let it dry of itself. Prepare it carefully with white of egg. Afterwards desiccate it, and the effect is produced.

When Glass is Destroyed by Fire

Take a half portion of minium, quicklime, and flour. Mix with white of eggs. Apply a cloth moistened with it, and place it for a short time on the fire.

Lute

Take ten parts of well-prepared lute, three parts of cows' hair, five parts of horse-dung, three parts of goat's blood, three parts each of quicklime and common salt, six parts of iron filings, and of white of egg and gypsum a sufficient quantity. Thus is made a lute.

Another for Luting Broken Glass

Take equal parts of calcined flint, quicklime, common salt, and white of eggs. Mix these together, and then take a cloth, place it therein, and smear the glass or fracture. Let it harden; then smear it in water with linseed oil, and a better lute could scarcely be found.

A Very Strong Lute, Which is Proof Against Fire

This is made out of bullock's blood, quicklime, and salt. It is indestructible in fire.

An Excellent Method of Luting Glass

Take sufficient quantities of Venetian glass, finely pounded, and oil of tartar. Make a pulp; lute the fracture and place it before the fire to melt.

To Make Minium

Take of Saturn as much as you please, dissolve it to ashes and it becomes citron-coloured. Afterwards pound it, place it in a vessel over a moderate fire, and it will be coloured red.

The Process of Sulphur

Take of this powder 5. j. Pour over it linseed oil, place it in an iron pan, and let it boil. It gives a red froth and grows thick. Pour it out and it becomes a red substance as thick as hepar. Put the particles into which it is divided into an iron pan with laterine oil and boil it thoroughly for two hours. Afterwards place it in a glass vessel on ashes for three days. Then the sulphur is converted into an oil. Take the glass vessel again and put it in cold water for three days and three nights. Then distil it, first of all over a slow fire, but increasing the heat until it is sufficient. Thoroughly calcine the dregs, which are called the caput mortuum. Imbibe with the first water. Then distil for seven hours. Do this again until the redness of the oil is changed to white, which will take place in three hours. Finally, take again the aforesaid oil; distil it by itself for seven hours, and the process is complete.

Then take a plate of Venus and dip it in the said oil. If it is transmuted into Luna, well and good. If not, distil again until this takes place with the said calcined fæces.

Or:

Imbibe the fæces frequently with the said oil. Let it flow over a copper plate and it becomes white. Afterwards take one part of the fæces, and five parts of Mercury. Place these together in a crucible well luted and set it in the fire. Then cool, and you will find a not very hard substance. Take the crucible and repeat the process with the aforesaid fæces. Proceed as above, and you will find an elixir of which one part is projected on 1000 of purified Mercury, and there is produced perfect Luna. Of this take one part and fuse it with ten parts of calcined and white Venus. Then will be attained a perfection with which none can find fault. Of this elixir take j. 3. and of Mercury 4. j. Let them boil, and good Luna is produced.

If, however, you wish to transmute the above-mentioned Elixir into Sol: Then take the element of fire, place it in a large glass vessel, and desiccate in the sun. After this imbibe with pure water. Then take some of this powder on a silver plate, and it will melt like wax. If not, imbibe it again, until it suffices. Next take one part with two parts of live Mercury. Lute them together, place them for half-a-day over a fire, and do as I before directed. Then take and add of the fire spoken of above the same quantity as the Mercury, and proceed as above. Once more it becomes citron-coloured and as hard as Sol. Of this Elixir take one part to 1000 parts of the Luna which has been made from the said Mercury. Then it becomes gold better than that of Nature, and by separating the said Luna 5. ij. with j. 4.* Mix and then it has admirable power, so that people say this is the most consummate Elixir of all Alchemy.

Operation for the Preceding Work

Take prepared purified sulphur and put it into a vessel so that the vessel may be four parts empty. Place over it a cloth folded double, underneath which is live Mercury. Lute it closely, put it on a fire for three days, when it will harden and become as Luna. If this does not take place, treat it with fresh Mercury, then add one part of that Luna with five of purified Venus, when the whole will become good and pure Luna.

For Sol Boni Thematis

Take   Mercury, iij. 5.

Crude Atramentum, j. 5.

Pour over these

Salt Water, ij. 5.

Jamen Alum, j. 5.

Mix these ingredients together, and place in the sun until the mercury dies and is converted into water. Put in a box thoroughly luted with lead, arsenic, and sulphur, and let it stand in a steady heat for one day. Open the box and you will find the Mercury coagulated. Pound this together with the above-mentioned water made of atramentum and alum. After this desiccate, place it once more in the box, and proceed as before for three hours.

Once again for the fourth time imbibe with the aforesaid water, and place in dung for putrefaction during five days. Then desiccate in ashes, and take of this tincture one part to a hundred parts of Venus. Take one part of the

---

* The translation is literal, though the sense is not intelligible.

aforesaid elixir and of calx peregrinorum. Pound these two together and imbibe with water of mercury; desiccate, imbibe frequently, and it will coagulate. Finally take one part of this to two hundred parts of Venus, and you will have the best Sol that can be found.

Mode of further increasing this augmentation Boni Thematis

Take white of eggs with the same quantity of quicklime. Pound with orpiment water, imbibe, and coagulate. Add water of sal ammoniac, and putrefy in dung for five days. Strain through a cloth and desiccate. This elixir tinges Venus into Sol.

Item

Take some of this elixir, egg shell, and peahen's eggs calcined. Once more imbibe with orpiment water for ten hours. Desiccate, pound with water of prepared salt, and putrefy for one day. Desiccate again, and one part tinges a hundred parts of Venus to Sol.

A Tincture most Effectual for the White and Red

Take of calcined sulphur, white and fixed, two parts, of fixed sublimated mercury ten parts. Take also some water of sal ammoniac and imbibe frequently. Then put it in a glass vessel into horse-dung for fifteen days. Place a little on heated iron and it will melt like wax. Coagulate with a slow fire. Next take one part of this elixir to a hundred parts of Venus purified for Sol. Imbibe with water of sal ammoniac, adding continually a little crocus of Mars until the powder grows red, whereupon putrefy it in dung. Then melt in iron. Take one part to one hundred parts of purified Mars, and it becomes gold better than that of Nature.

Operation for Sol

Take Live Mercury, viij. parts.
Sublimated and Fixed (? Mercury), iiij. parts.
Calcined Luna, ij. parts
White Arsenic, j. part.

Pulverise, put in a silver box, lute well, place in a vessel, and pour over it Saturn for five days in one solution. Then take out the box and you will find a white powder. Pound this with water of sal ammoniac, desiccate, and putrefy for fifteen days until it melts like wax over red hot iron. Take 3.j. of it to one mark of purified Venus. Pour together and it will be true Luna.

For Sol

Then redden the sulphur with crocus of Mars. Instead of the calx Lunæ take calx of Sol, and place it in the box. The substance must afterwards be more carefully strained and putrefied for a longer time.

White

Take Luna, 3 ℥.
Jove, ij. ℥.
Saturn, iij. ℥.

Fuse these until they are thoroughly melted. Let them cool a little; then project 3s., or somewhat more, of warm Mercury, and you will have the white.

Fixation of Mercury

Take a silver phial of any capacity you like, and place in it Mercury until it is a third or half full. Close it up with bread, salt, and white of egg. Then place it among burning coals. Heat it, and then immediately plunge it into hot water. Continue to do this for an hour, when it will give a sound like the hissing of a goose. Then it is sufficient. Take it out, and it will be silver.

Augmentation of Luna

Make an amalgam of Jove and Mercury, mix it with pounded salt, and wash it until no blackness appears. Take this amalgam, put it in a sublimatory, and sublimate the Mercury from the Jove. Then take of the Mercury thus sublimated four loth. Dissolve in aquafortis 21 (*sic*) crocus of Luna, 2 16 (*sic*) loth. of Venus, and dissolve each separately by itself. When all are dissolved mix the whole in one glass vessel, distil the water from the dregs, pour over it fresh aquafortis as before, re-distil the water from the dregs, and afterwards wash with fresh water. Next, dissolve sal ammoniac in acetum, pour this on the dregs, and let it stand throughout the night. Then distil it once more from the dregs by means of an alembic, and reduce the fæces with sal alkali and common salt. Purify the body you have in the cineritium, and you will find six loth. of most excellent Luna.

Water for Gilding

Make aquafortis out of one part of vitriol, one part of saltpetre, and four parts of alum. Into the water so made put four loth. of sal ammoniac. Re-distil. In the same water dissolve an amalgam of Mercury and Sol, as the goldsmiths are accustomed to form amalgams. Let this stand for eight or fifteen days. Let the water boil down to one-third of its volume. When you wish to gild dip a pencil into that water, stir it briskly, and paint over whatever you please. Then let it dry, and afterwards burn it, as the goldsmiths do.

To Blot Out Writing

Take Roman Vitriol, j. 5.
Usifur, 34.
Jamen Alum, ½lb.

Distil a strong water from these. At first it is white; and this you must collect by itself. When you wish to erase any writing, moisten a cloth with this water, touch the letters, and they are obliterated.

Cement

Take Reddened Vitriol ⎫
Verdigris ⎪
Burnt Brass ⎬ Each one part.
Sal Ammoniac ⎭

Alum, to make 1lb. weight of all.

Make layer on layer for eight hours in a closed vessel, and afterwards dip it in urine.

To Make Precious Stones

Take very white silex, calcined and pulverised, one part, and three parts of minium. Place within a crucible in a brisk fire. Then let it cool of itself, and you will have a precious stone. It is coloured as an emerald by ashes of Venus.

Cement as Pars Cum Parte

Take of Bloodstone } j.lb. j₃. each.
of Bolus Armenus }
of Vitriol, a quarter, j. 2.

Pulverise. Then take half a part of the Sun or the Moon. Make plates. Then arrange in layers, as you know, and remove by smoke. Take from the fire, when you will have the best Hungarian Sun. It has been tested by me.

White

Take of Venus, lb.j.
of Luna, lb.s.

Melt. Afterwards sprinkle lb.ij. of salt ammoniac and lb.ij. of pure salt. Project upon it j. quartal. Let it stand for one hour; repeat until eight hours have passed and it is made. Out of that Luna you can make anything you desire.

Beautiful Mercury brought over from Mercury

Take Luna and Mercury equal parts. Dissolve in aquafortis: then abstract the water so that it may remain as a thick pottage. Dissolve this pottage again for eight days, and it will be converted into water. Abstract as before and again resolve four times, when you will ultimately have water which persists through all tests, and makes out of Mercury a Luna which remains everywhere. A drop or 1½ to 2 oz. of Mercury heated to evaporation.

Note the Sulphurous Work

Boil sulphur well in vinegar or urine. Wash it well. Afterwards dissolve it over the fire and project as much as possible over the fire. Place it in a luted instrument and burn gently in a slow fire for 30 days, when you will have at the bottom a Mercury which is not very red. Dissolve in an open glass vessel into oil. This oil tinges in a marvellous manner.

Good and Proved Lazurium

Take of Live Mercury, any quantity.
of Sulphur, a third part.
of Sal Ammoniac, one part.

Mix. Burn like cinnabar, and when you see a purple smoke, take out and mollify the lazurium with boiling water.

Water which makes Mars fluid and also boils in Air

Take Camphor,
Salt of Glass,
Vitriol,
Boiling Wine,

Distil, as you know, and keep well.

Water which makes Luna into Sol

Take red vinegar, sublimated, and live calx. Boil. Then put in that vinegar sal ammoniac and vitriol, dissolve and distil through the alembic. Extinguish plates of Luna therein and it will be converted within and without into the Sun.

For Luna

Extinguish Mercury twelve times in human blood, and it will be hardened. Afterwards boil it in the yolk of eggs for one hour, when it will become good Luna.

Firm Tincture

Take lb.j. of sal alkali, the same quantity of calx of eggs, two parts of clavellated cinder, also four parts of the dew of heaven. Decoct all these to the third part. Afterwards thrice distil through the alembic. Perform this diligently and you will have very strong water, with which Mercury and all bodies of metals are dissolved.

Take of the above water, lb.j.
of Foliated Luna, ʒj.

Place in cinders for three days, and the Luna will be converted into water.

Place Mercury, sublimated and well pulverized, in a phial among the ashes. Pour over some of the said water made from Luna, and it will be congealed. Continue this process, imbibing and desiccating until half part of the water is consumed lb.s. Afterwards place powder from Mercury into the fixatory. Digest slowly. On the third day augment the fire, making it exceedingly fierce, and you will find an everlasting tincture, one part of which falls over 100 parts of crude Mercury, when it will be good Luna standing every weighing and hammering, and lasting for ever.

Concerning the Oil of Sulphur

Take of Oil, one-fourth.
of Sulphur, lb.ij.

Make a *hepar*. Boil it in lixivium so that the oil may be abstracted; afterwards that which remains at the bottom must be distilled through the retort, *secundario per lateres*. Make sufficient oil.

Pars Cum Parte

Take Antimony prepared in Oil of Tartar, ʒj.
Prepared Salt of Nitre,
Prepared Common Salt,   ʒs. each.
Plumose Alum,

Let these ingredients be well mixed twice or thrice and imbibed in oil of tartar, whence will be formed a powder, which place layer by layer with Luna in a cementing fire for six hours; and, when this has been done, let it be sublimated. Then take one ounce of this and half an ounce of pure Sol. Let these be pounded together and formed into plates; and then make layer on layer with the following powder:

Take Sal Ammoniac,
      Saltpetre, } j.ʒ each.
      Prepared Common Salt,

      Verdigris,
      Alum,
      Hæmatite, } $1\frac{1}{2}$ ʒ each.
      Vitriol,

Pound these together and let them be imbibed with oleum laudis ten times or more, when they form a powder. Place them layer by layer in the fire for twelve hours. Afterwards take the regulus, sublimate, add three parts of Venus, and place in the following water of gradation :—

Take Vitriol,
      Saltpetre, } j.lb. each.
      Alum,

      Plumose Alum, } iiij.ʒ each.
      Calcined Alum,

      Cinnabar, } $\frac{1}{2}$ lb. each.
      Sulphur,

      Verdigris, j.lb.
      Antimony, vj.ʒ

Distil twice, that is, once from the caput mortuum and from the residuum which is found at the bottom. Fulminate; and you will have Sol so good as to answer every test.

FIXATION OF MERCURY

Take oil of tartar boiled in best lixivium, and distil through filter. Next boil till it attains to an oily consistency, and place in a good glass.

Take of the said Lixivium, one pocale.
      of Salmiax, five times sublimated, one pound.
      of Mercury, sublimated seven times, lb.j.
      of very strong Water of the Fount, half a pocale.

Let them be mixed together in a Venetian vase. Allow them to stand for a day, so that the boiling may cease. Inject upon the oil ʒiij. of Luna in horse dung for a month, and the whole will be converted into oil. This oil tinges all things into Luna, can be coagulated into a stone, and is the water of Mercury.

PROJECTION.

Of Common Prepared Salt,
    Saltpetre, } ʒj.
    Sal Alkali,

    Prepared Sal Ammoniac,
    Albumen, } ʒs.
    Vitriol,

    Verdigris, ʒiij.
    Crocus of Mars, ʒj.
    Cinnabar, ʒs.
    Prepared Antimony, ʒij.

Make a powder which is to be imbibed with oil of antimony and tartar (and urine) of Mercury ten times. Project over them ʒij. of Luna and ʒiij. Dissolve for six hours. Afterwards let it be fulminated and placed in aquafortis, or cemented as you know.

Preparation of Salts

Make it with quicklime, and by means of a lixivium, as you know. Let it be twice or thrice imbibed with oil of tartar, but the other in oil of tartar.

Concerning Talc

Let it be cemented a whole day with common salt. Afterwards let the talc be collected from the salt and most subtly pounded. Let it be put in a bag. Let there be poured over it a very strong lixivium. Let it be poured over again, until the calx is dissolved. Then it falls in the lixivium to the bottom. Dry previously very well. Afterwards let it be dissolved into oil. It coagulates Mercury into Luna, and similarly Jupiter.

Digestion of Luna

Take Saltpetre,
Vitriol, } lb.j. each.
Cinnabar, ʒiiij.

Make a strong water. Let this water be divided. In one part let there be Luna, in the other cinnabar. Let them be dissolved by the addition of sal ammoniac and then joined together. Let them be digested for fourteen days and the matter distilled. Then let it be reduced, and you will have a double quantity of Luna. Each loto of the salt will have a loto and a half.

Concerning Talc

Let it be cemented with common salt a whole day in aquafortis. Afterwards the talc must be collected, most subtly pounded, and put in a bag. Let it be melted over a very strong lye, and let the process be repeated until the talc is dissolved. It then falls in the lye to the bottom. Dry previously well, then dissolve into an oil. It coagulates Mercury into Luna, and Jupiter likewise.

Separation of Sol and Luna

Take Antimony, ʒviij.
Filings of Mars, ʒvj.
Crude Tartar, ʒiiij.
Common Salt prepared in fluxion, ʒiiij.

Pound all together and fuse in a tigillum. Then there will be a black substance, which also grind to a powder. Take equal parts of this and of Luna. Fuse the Luna until it appears bright and clear, that is, with the powder. Place in cupel with lead, and the antimony entirely evaporates. Afterwards purge by means of a cineritium.

Concerning Scoria of Luna

Take scorias and powders in equal quantities. Melt until you have cocted the silver. Let this process be thoroughly carried out, then refine.

The Fixation of Cinnabar in One Day

Take of beechen ashes viij parts, of quick lime j. part. Make a lixivium, in which dissolve of salt nitre, salt of vitriol, and verdigris, each j. part; of flower of alum and calcined tartar, ½ part each. Dissolve these; boil the cinnabar therein for a whole day, and it will be fixed.

On Antimony

Imbibe the mineral in Saturn. Afterwards scatter sand upon it. Then the sand attracts the antimony to itself. When it becomes scoria, remove it with a spatula and sprinkle sand again, as before, until it will no longer evaporate. Then fulminate.

Concerning the Solution of Magnesia

Take one part thereof, and of sublimated Mercury ij. parts. Grind, mix together, and distil by an alembic. Then a thick and fat water, like linseed oil, is distilled. This tinges Mercury in itself. Then it tinges all bodies projected into it.

Fixation of Antimony

Take Salt of Alkali, ij. oz.
Salt of Nitre, j. oz.
Antimony, lb. j.

Melt together, then let them stand to cool, and the antimony will be fixed.

Oil of Antimony and Mercury, fixing Spirits and of itself dissolving Bodies

Take Mercury, iv. lbs.
Antimony, j. lb.

Dissolve as you know how. This oil dissolves metals.

Oil of Borax

Put borax into a glass vessel and dissolve it. Let it be pulverised, hardened, and placed inside another glass vessel, in a balneum Mariæ. It is converted into an oil which fixes all spirits.

Oil of Gold

Take Sulphur, j. part.
Quick Lime, viij. parts.

Let these be decocted in water and the decoction becomes red. Distil by an alembic, and there remains at the bottom the redness of sulphur. This is called the oil of gold.

Water by which all Spirits are Fixed

Take sal ammoniac, Jamenus, and vitriol. Distil by alembic. Then take any spirit, dissolve it in water, abstract it, imbibe it thrice, and it will be fixed.

Another

Take Antimony,
Crocus of Sulphur, } j. part each.
Sal Ammoniac, iiij. parts.

Imbibe with strong acetum; at last dry it, and then mix with water of fixed sal ammoniac. Abstract, and distil.

Fixation of Sulphur

Grind it with salt of tartar or crude tartar, together with common salt and salt of nitre in equal parts. Dissolve with the water of common salt and congeal. Put it in a vessel, lute, and set it in the fire. Let it get red-hot, afterwards dissolve in fresh water, distil with a filter; congeal a second and third time, and the sulphur will become capable of being melted like wax.

Sublimation to Sol with Sulphur

Take Live Sulphur
    Roman Vitriol   } equal parts.
    Verdigris

Imbibe with the water of common salt. Sublimate thrice. Or, take equal parts of Sulphur, Honey, and Alkali. Let them boil for one day. This mixture makes Sol out of Luna.

Water of Antimony (Sulphur) for Sol

Grind it, and take thereof iij. parts, with one part of sal ammoniac. Place it within a glass vessel, well luted, on ashes for one day. Then pound it in hot water, and it will become like blood. If it were boiled in a lixivium perhaps it would be better.

Fixation of Sulphur

Take sulphur and honey. Imbibe and dry over a slow fire. Then let it boil well in a strong lixivium. Wash the substance until the water appears clear. Renew this process seven times, and you will find the sulphur white like crystal. Afterwards take prepared common salt and the same quantity of sal ammoniac and sulphur. Let these be well ground together twice. Then dissolve with white of egg, and congeal. Distil by means of a filter, and cool. Repeat this three times, and j. part changes xxx. parts of warmed mercury into permanent Luna.

Fixed Oil of Sulphur

Take it, let it boil in alkali for one day, and be sublimated over a slow fire. Having done this, moisten it with acetum four times, abstract by means of a filter, and again moisten three times. It is afterwards abstracted by means of an alembic, and will then be fixed. Let it be dissolved into an oil on marble. Then take the body of the sulphur, dissolve it with oil, and congeal. One part tinges three parts to Luna, and that Luna has many of the properties of gold.

The Whitening and Fixation of Cinobrium

Take it, together with calcined alum and prepared common salt. Pound together with vinum ardens, dry, and sublimate. Then it is whitened and fixed.

FIXATION OF SPIRITS

Take quicklime, salt of alkali, and oil mixed therewith. Distil, and imbibe the spirits with test water, repeating the process until it melts on the plate.

WATER OF MERCURY FIXING ALL SPIRITS

Take Mercury, j. marc.

Sal Ammoniac, ij. marc.

Rub the two together into a glass vessel over a slow fire, and it will become a hard mass. Pound this and it will become a powder, which dissolve in water. Then take j. marc. of pure Luna or Sol carefully made into plates. Put this in the aforesaid water; and this water fixes all spirits.

TO FUSE BONES

Take any quantity of bones and burn them into lime. Having done this, carefully pound it. Take of this lb.iiij., of quicklime ½lb. Mix them together in the powder. Afterwards dissolve some bitumen in a moderate quantity of wine, until the whole of it is melted away. Then place the bones therein, and stir briskly into a thick pulp. Afterwards pour into a mould made of paper. First, however, let it be smeared with oil; set it to cool, and it will be hardened like ivory. You can, in course of fusion, give it any colour with minium, flos æris, or any other tints you like.

TO MAKE A MOULD FOR CASTING ALL KINDS OF IMAGES

Pound tiles very fine, and boil them in strong lixivium so that they form an exceedingly thin paste. Dry this and strain it carefully. Afterwards make a water from the white of eggs. Then let it be pressed on a machine, and it will dry from the top.

TO COLOUR GLASS

Take tartar, wood ashes, and quicklime. Make an alkali from them. Take thereof j. part, dissolve in iij. parts of colour in water (*sic*). Coagulate, and again dissolve with the colour. Do this three or four times until the stone shall be thoroughly coloured. Afterwards melt glass with the aforesaid stone. It will be coloured by the stone, and thus you can fuse crystal.

A METHOD OF COLOURING WHEREBY SOFTENED CRYSTAL CAN BE TINTED AND HARDENED LIKE A PRECIOUS STONE

Take alkali made from tartar, j. part: sublimated salmiax, ½ part: colour ij. parts. Pound well together, dissolve in water, and again pour the colour thereupon, repeating the process until the colour is sufficiently deep. Then it will be coagulated into a stone. Pound this very fine, and mix it with a crystal that has been previously softened with oil of tartar. The red colour is made with cinnabar, and becomes equally red therewith; the citron-colour is made from saffron, and the green with sap-green.

How every Stone can be transmuted into a clear one, though it be itself opaque, and whereby you can tinge a Body, both Crystal and all similar Bodies. Example

Take of ematite stone, very finely pounded, ij. parts. Let it boil fiercely in a lixivium (previously prepared from tartar, wood ashes, and quicklime) for half a day, so that it may be alkalised. Then sprinkle over it iij. parts each of sublimated sal ammoniac and also crude sal ammoniac. Dry over a very slow fire. Afterwards dissolve in water in a damp cellar, and what remains on the stone imbibe again with that water, repeating the operation until the whole is turned into oil. Then put it into a glass vessel and sublimate from the oil itself the sal ammoniac which is not fixed. When you see it has ascended and is no longer in the oil, then pound the stone, and boil it still further in the lixivium for six or eight hours; and then again, as before, dissolve with the sal ammoniac, both sublimated and crude, then by the sublimation of the lixivium and the solution of the sal ammoniac it is changed into a thick oil. Then those spirits which are not fixed are separated from it, as the sal ammoniac by sublimation, in the way before specified. The lixivium will not be separated; but add water, which has been distilled by a filter; then the oil will remain at the bottom, and the lixivium will be raised up with the water. Thus it will be separated, and you will have the oil of the stone alone. It is better, however, not to separate the alkali, but to let it remain in the oil.

Description of the Adept's Fire

Take Vitriol, }
     Alum, } j. lb. each.
     Saltpetre, }
     Calcined Alum, } $\frac{1}{2}$ lb. each.
     Calcined Vitriol, }

Distil in aquafortis for 30 hours. Afterwards take lb.ij. of the following strong water:—

     Calcined Alum, } j. lb. each.
     Calcined Vitriol, }
     Saltpetre, $1\frac{1}{2}$ lb.

Distil as above and renew thrice. Having done this, dissolve in the following strong water:

     Sal Ammoniac, ij. quarts.
     Salmiax, j. quart.
     Sublimated Mercury, j.
     Calcined Tartar and its Alkali (*no quantity given*).
     Sublimated Arsenic, j. quart.

After each has been separately dissolved, let it be distilled by a bath and poured over twice. Afterwards let it be hardened a little and dissolve in a moist bath into a water In this water dissolve iij. parts of Luna, or as much as will dissolve. Coagulate into a hard stone and let this boil in a very strong lixivium into most potent water until it is alkalised. Dissolve and coagulate as long as you please.

For Luna.

Take Sal Ammoniac,
Vitriol,
Rock Alum,
Salt of Alkali or Tartar,
} equal parts of each.

Sublimate in a glass vessel. One part tinges six parts of purified brass. Then add Luna.

The Purification of Brass is effected thus:
With Acetum, Salt, and Tartar.

Oil of Sulphur.

Take three pounds of Hepar Sulphur. Boil it in a lixivium of soap; afterwards add

Rubified Vitriol, lb.ij.
Calcined Alum, lb.j.
Glass, lb j½.
Crocus of Mars, j. quart.
Verdigris, j. quart.

Mix well into one mass. Distil by an alembic, having previously putrefied for three days; and with a very strong fire, so that the spirit may be energetically expelled. Set aside the white part. Pour that which is red upon the caput mortuum again. Re-distil; and keep doing this until no whiteness comes forth. Then thoroughly reverberate the caput mortuum. Distil it; and repeat this process until it no longer burns in the fire. Putrefy from the white oil. These tinge in a wonderful manner.

Mode in which Sol is produced with Pars cum Parte

Take Antimony, ij. parts.
Common Salt, iij. parts.
Vitriol,
Alum,
} j. part.

Grind to a powder. Take ij. parts of this, and j. part of filings of Luna. Mix together, and place in a tigillum on a slow fire for two hours. Then place it in a very strong fire for another two hours, and melt it in glass. Afterwards imbibe in Saturn and purify by means of a cineritium. Next convert it into plates, and put them layer on layer with this powder. Then take

Common Salt, fused,
Salt of Nitre,
Alum,
Vitriol,
} ½ lb. of each.

Ematite,
Flos Aeris,
Calaminaris,
Tutia,
Cinnabrium,
Minium,
Burnt Brass,
} ½ oz. of each.

Dry by the fire. Make into a powder. Imbibe three or four times with urine. Place in a luted tigillum and set in a graduated fire for six hours. On the last day (*sic*) put in the regulus, having purified it by a cineritium. Wash it again and cement it, until it is proof against aquafortis. Then take of that Luna (*sic*) iij. parts, of pure Sol j. part, of copper ij. parts. Dissolve together and cement for ten hours. Let the regulus be dissolved into iiij. parts, and placed in the aquafortis to be hereafter described. Then you will have a residuum of Sol perfect according to every test.

THE AQUA FORTIS

Take Vitriol,
Saltpetre, } j.lb. each.
Alum,
Antimony, 4 ss.
Cinabrium, ℥ 4 vij.
Verdigris, ℥ ss.

Distil all these into a water.

FOR BEAUTY OF FACE

Take Oil of Tartar and Dragaganth. Make an ointment and besmear the face therewith once. Then remove it.

NOTE CONCERNING CEMENTS

Take laminated Luna and let it be cemented with species of cements twice or thrice. Afterwards add one part of purged Venus. Dissolve in water of gradation, and you will have gold if it remains its time, namely, until it acquires a red colour. Note that *pars cum parte* is made as above.

NOTE

Place verdigris in ashes so that it may grow white. Then extinguish in vinegar, when it will become red. Afterwards pound, wash thoroughly, and desiccate again. Tutty is made as above, and becomes red. Also let colcothar be imbibed with vinegar as above. Also take ℥s. of cinnabar and ℥j. of vitriol as above. Next let the vitriol be imbibed with vinegar after the manner of a pottage. Divide it into two parts. Let one part be placed at the bottom of the tigillum. Arrange cinnabar above in layers, well luted. Place by the fire, so that it may not glow, for two hours. Afterwards pound well. Also take ℥ij. of crocus of Mars. Wash in the manner of ceruse. Take the more subtle portion. Desiccate and imbibe with vinegar. Thus dry twice that which floats above.

Also take of Crocus of Mars,
of Tutia, } ℨj. each.
of Verdigris,
of Vitriol, prepared as above, ℥j.

Mix together, pounding well.

Fixed Venus

Take of Filings of Iron.
of Antimony.
of Venus.

Place in a tigillum well luted. Let it stand twelve hours in flux. Afterwards infrigidate and fulminate the king with the same quantity of Saturn. You will then find 12 lotones of fixed Venus of j.lb. It makes no more scoria, nor is further destroyed by R. You can make it red or white.

To the White with Metallic Arsenic

Let it be fixed by means of imbibition of the oil of eggs and of tartar until the tincture be made.

Crocus of Mars

Take any quantity of Mars, and the same of saltpetre. Make it burn and become red. Let it also become sulphur, etc. Also take ʒj.s. of sal ammoniac. Pound. Also take ʒij. of glass, prepared and pounded as above. Mix and imbibe with vinegar twice. Also take iiij.oz. of filings of Mars and two ounces of red vitriol. Make crocus with vinegar, as you have seen. Let the oil of Mars be distilled. Imbibe sal ammoniac as before until it reddens. Take of antimony h. j., well pounded, and of tartar h. j. Mix well. Next dry. Shut up in a jar, so that it may not grow white on the coals. Leave for two hours. Afterwards pound it and pour *vinum ardens* over it, so that it may be inebriated. Place it in a phial for a night and a day. Next distil. Pour over it again, and distil by a slow fire.

To Soften Glass

Take Lybisticum. Press out the sap. Cause the glass to boil in this.

This is the Method of Making Luna

Take j.lb. of Mercury. Heat it, and pour over it the following water: Take common salt j.lb., saltpetre j. quart. Grind them, and then bake j. quart of lime. Mix well in an iron dish over a slow fire until it froths, for one hour. Remove from the fire and cool. Mix all together, and dip it into oil twice. Fulminate this Mercury in a cineritium, and you will rejoice for ever.

Correction of Oil of Tartar, for Beauty and for Luna

Make layer on layer with tartar and lime. Burn well. Then filter for two hours. Lastly dissolve, and you will have corrected oil of tartar.

Note

Take equal parts of sulphur and Mercury. Form a paste like amalgam. Then mix with salt. Let them remain in gentle fluxion for half an hour or thereabouts. Then burn; afterwards wash. The Mercury which you find there grind with a salve of Aza, wax, *vinum ardens*, etc., until it is burnt. Finally reduce in a cineritium with borax, and you will have Luna without any doubt.

Water of Fixation

Rectify the white of eggs with their own shell four times. Take of this lb.j., of well purified Sol j. ʒ, and sal ammoniac j. ʒ. Boil them well with urine. Dissolve these together and distil by means of an alembic. Imbibe therewith sulphur and arsenic, sublimate until fixed, and you will have an elixir.

Note

Take Cinnabar, lb.j.
Sulphur,
Arsenic, } lb.iiij.
Calcined Tartar,
Alkali of Soot, lb.ß.
Salt, nine times prepared, lb.ß.
Salt Nitre, the same weight as all the rest together.

Mix, pound, and moisten several times with the water of eggs or albumen of tartar. Let them remain in a state of fusion for three hours. Afterwards kindle and warm, when it readily dissolves the saltnitre; one part of sublimated Mercury and two parts of sulphur must be stirred continually with a stick. It then speedily loses its smell. The salts are prepaped by frequently evaporating the acetum or urine. Afterwards melt them so as to mix for four hours over a very strong fire. Then wash and purge over ashes; and you will have the Treasure of the World.

Reduction

Take Goldsmith's Borax, part j.
Assafœtida,
Sarcocolla,
Oxicroceus, } parts ij.
Wax,
Galbanum,

Dissolve the gums in vinum ardens, mix with the borax, as above, and it is burnt at the same time. If it does not melt, add more borax until it melts.

Oil which Fixes and Tinges

Take Linseed Oil,
Honey, } oz.viij.
Yolk of Eggs, oz.vj.
Eggshell and Quicklime, each j. quart.
Colcothar, Saltpetre, and
Calcined Alum, } each ij. quarts.
Antimony and Tartar, each lb.ß.
Juniper Wood, 4.

Pound, mix, and distil. Having done this drive out the spirits of nitre, alum, colcothar, and antimony until the water or oil becomes red and thick. Then warm the tartar and antimony, and pour in the oil that the spirit may thus be more reddened. Add iiij. parts of aquafortis, so that the oil may thereby be more fixed. It would be well, too, if this oil were previously rectified from the calx of the eggs and its own fixed spirits by renovating and distilling six times. Lastly, take of this oil j.lb. and ten parts of fixed salt. Let it be fixed by ten cementations, solutions, and fusions, and as often by nitre. At the same time take one quart. each of alkali fuliginis, Jamen alum, and sulphur of tartar. If the sulphur has been imbibed with nitre it renders it red and more fixed, and gives it ingress. Take of this sulphur j. part; of coagulated Mercury ij. parts. Cement for four hours, and it becomes gold more perfect than that of Nature. In like manner all spirits are fixed by this method. They perform wonders, do the like, and ingress and tincture are added to all species. It turns to Sol the calx of Luna when placed upon it, and equally fixed copper in the same manner. It fixes calcined or sublimated Mercury, both into a body and into a tincture, that is, an elixir. This oil can be coagulated over a very slow fire, and you will then have a stone, the virtue of which is to turn into pure Sol, Luna, fixed Venus, and all metals prepared for it. It fixes also cinnabar into Sol.

CAPUT MORTUUM FOR THE SUBJECT OF THIS SUBLIMATION

Take Arsenic,
Sulphur,   } j.lb. each.
Crude Tartar,
Salt, fused and prepared, lb.ij.
Saltpetre to the weight of all.

Dissolve over a slow fire with lb.j. of crude antimony, until two hours are completed. Then kindle, and let them melt for an hour. Let this caput mortuum be imbibed with the said oil. No better caput has been discovered.

READY METHOD FOR COAGULATING MERCURY

Take it when made very warm, steep frequently in warm oil, and it will be hardened.

ARSENICUS MATELLINUS

Take Quicklime,
     Common Salt, } j. p.
     Calcined Tartar, ij. p.

Mix with the clear part of eggs. Make into pills, and distil by descent.

WHITE CINABRIUM

Take equal parts of Alum,
     Calcined Tartar,
     Common Salt,
     Cinabrium.

Sublimate all these. It will become four times as white. It also becomes white in a strong capitellum if you boil it for a night. Jupiter steeped in oil of tartar 7, or more, becomes good, etc.

Note. For making Ice easily Fusible in Fire, and such also as is not Dissolved by Water.

Take aquafortis made from saltpetre and alum, together with oil of tartar, j.lb. each. Pour all together; then put into them a little vinum ardens, and the whole will coagulate into an ice by the power of fire.

Oil which we use for Projection on Melted Luna.

Take Purest Gold, j. loth.
Sal Ammoniac, iij. loth.
Oil of Antimony, x. loth.
Aquafortis, *quant. suff.*

Dissolve, as you know how. Having done this, let it putrefy for seven days, and distil by a bath, continually renewing the whole until the aforesaid materials are converted into a thick oil. Of this take x. loth., with purest gold ij. loth. Dissolve at the same time iiij. loth. each of sublimated Mercury and fixed sublimated sal ammoniac. Proceed as above with fresh aqua fortis, and all these ingredients are converted into a thick oil. In this oil dissolve as much gold as possible. The oil will then transmute into gold all metals on which it is projected, and if it be coagulated into a stone it tinges it beyond measure. By this method you can proceed to silver. Mercury will be fixed with white of eggs, and in the same manner sal ammoniac.

A Wonderful Cement by one Stone over Luna

Take Crocus of Mars, j loth.
Hæmatite, ij. loth.
Verdigris, v. loth.

Pulverise well and boil in a very strong lixivium for ten hours, so that it may be alkalised. Having done so, take of this alkali ½lb., of sal ammoniac iiij. loth., and of sublimated salmiax j. loth. Mix together and dissolve on marble into a watery oil. Pour the oil thus made on the dregs, and keep doing this until the whole matter is turned into water. This being done, coagulate it into a stone, which sublimate in order that the sal ammoniac may evaporate, and the matter alone may remain. Then once more boil the stone in very strong red lixivium. Again add the aforesaid weight, dissolve into oil, and again coagulate the oil, sublimate the sal ammoniac from it, and do this over and over again until the matter of the species flows from them on the marble. Take this water, coagulate it afresh, dissolve it again and put it in j. quartal of powder or j. loth. of water of ducat-gold. When it is dissolved coagulate it. This stone is effectual in cements, and effectual also in projections. It also tinges crystal.

Cement by which half of Luna becomes Sol

Take Hematite, v. loth.
Flos Aeris, ij. loth.
Crocus of Mars, iiij. loth.
Sal Ammoniac, iij. loth.
Nitre, ij. loth.

Boil and desiccate. Afterwards imbibe nine times with oil of antimony; then with equal parts of this powder and of Luna make layer upon layer. Having done this lay aside the Luna and fulminate it. Then it is separated by means of successive applications of water and purged by a cineritium.

Take Antimony, lb.vj.
Verdigris, lb.j.
Calcined Vitriol, lb.ij.
Calcined Alum, lb.j.
Saltpetre, lb.iij.
Sublimated Mercury, lb.½.

Distil, and there issues forth a strong red water which tinges everything into Sol.

Cement whereby four parts of Luna become perfect Gold

Take Luna, iiij. loth.
Venus, j. loth.

Make thin plates and form layer on layer with the following powder. Take: ij. loth. each of red vitriol, calcined alum, and saltpetre; j. loth. each of verdigris, hæmatite, and tutia, with iij. loth. of sal ammoniac. Dry by evaporation over a slow fire, and imbibe several times with white of eggs. Make into a powder, and cement the Luna for seven hours in a graduated fire. Afterwards put it into a cineritium with the following strong water.

Take Saltpetre, ⎫
     Alum,   ⎬ lb.j.
Calcined Alum, j. quart.
Cinnabar, ⎫
Sulphur, ⎬ lb.½.
Verdigris and Calaminaris, each j. quart.
White of eggs boiled in calx of eggs, lb.iij.

Mix, and make into a strong water. Distil again lb.ij. of the aforesaid fæces with lb.½ of calcined alum and one quartal of plumose alum. After a gradation of twenty-four hours reduce with borax, and there will result the most perfect gold.

Zalusia

Dissolve calcined alum, etc., as also calx of alum. Mix with white sugar and camphor, imbibe with quintessence, and burn. Then the camphor is burnt out and consumed, while the alum is transmuted into oil. Place it in aqua vitæ.

Oil of Vitriol for all Weakness

Make vitriol up to redness. Dip that calx in common acetum, as far as it can be dissolved. Pour out that which is dissolved and keep it. That which is not dissolved dissolve again with fresh acetum, as above, until the whole is dissolved. Then let the acetum evaporate to dryness. Calcine the moisture from it anew. Then place it for distillation and you will have an oil. Over this pour the rectified quintessence and place in a bath for putrefying. The superfluous part remains at the bottom. Pour this off clear, and you will have the quintessence by means of the alembic. The oil, of a deep red colour, remains at the bottom.

SALT OF TARTAR

Calcine tartar to whiteness. Dissolve in cooked urine. Filter, coagulate, and you will have salt of tartar. Take it soon; let it dissolve in oil. Dissolve the other part in white vinegar and coagulate into a salt.

ANTIMONY

Take of Antimony ij. p.

of Fused Salt, j. p.

Let them be melted until the water ceases to redden and the antimony becomes white. Put the froth in a glass. Coagulate it. Next place it in a long cap-pot, extract it with an iron scraper, and repeat the process until it becomes white to a powder and then red again. Then take one marc of Jupiter. Sprinkle thereon a quint of this powder when the metal is in flux. Then refine on the test.

CROCUS OF MARS

Dissolve in aquafortis. Then distil the aquafortis from thence. Take as much sal ammoniac as crocus, and sublimate four times from the crocus. Afterwards put the whole substance in acetum to dissolve in a slow heat for two days. Distil the solution by a filter, and then cause the acetum to evaporate. Afterwards strengthen the fire. The sal ammoniac retires, and the crocus remains in a fluid state at the bottom.

WATER OF MERCURY

Take crude Mercury j.lb. Put it in a cucurbit to distil. Give it a slow fire, and a single drop will come forth. Pour this back; distil again, and two drops will come forth. Pour these back again, and continue this process until the whole is converted into water. This water penetrates and dissolves bodies.

NOTE

Take crude tartar and pound it well. Then distil through an alembic. Pour its water again upon the fæces, and distil again until a water or oil collects; then put the same into another receptacle. Afterward rectify it. Take live calx in the same quantity as red oil. Pound it thoroughly. Then extract the water through the alembic until no more fæces remain at the bottom. Next put half as much sulphur into the oil. Putrefy for ten days. Afterwards again distil. Then make an amalgam out of one part of Luna and five parts of Mercury. Heat and extinguish in oil six times. Then Mercury is fixed, and there are many arcana in it.

NOTE

The calx of the body is cast into a fluxion of saltpetre over the fire. Dissolve in a cold chamber. It falls on the bodies.

Water of Luna

Make an aquafortis out of one pound of saltpetre and two of vitriol. Dissolve therein half an ounce of Luna; then dissolve a common salt in warm water. Pour into it aquafortis with Luna, when it will be precipitated. Then extract the water from it, and dry it. Add to the dissolved half ounce one ounce of sugar-candy. Pour upon it again a fresh supply of aquafortis. Extract the moisture therefrom in a bath. Afterwards put it in a sand cupel and extract the water of the moon through the alembic. If it does not ascend more than once, pour the ascended water thereupon until all shall have passed through.

The Fixation of Arsenic

Take two parts of alum and one part of saltpetre. Make a water by means of an alembic, and put into the water a part of sublimated arsenic. Distil to a fixed water.

Fixation of Mercury

Take of Fixed Arsenic, of Sal Ammoniac, two parts each.
of Fixed Sulphur, one part.

Melt them together. Then take one marc of Mercury. Warm it. Next put it into the melted matter therein. Wait an hour. Then refine it in Saturn on the test. You lose nothing, and you will have more than two parts of the Sun. You may then also take the Mercury which is coagulated with the smoke of Saturn.

True Albatio

Take of Sublimated Zaibach, x. parts.
White Sublimated Kybrick, iij. parts.
Sal Ammoniac, iij. parts.

Imbibe frequently with water of sal ammoniac, and dry until they are white and roasted. Again imbibe and roast. At length take water of the eagle, that is, sal ammoniac, double the amount of the powders. Put it to dissolve under moderately warm dung for three weeks. Then take it out of the dung and congelate it into white powders. Of these project one part on 100 parts of purged Venus, and the whole will become silver. These are the truest experiments of many philosophers who have worked by Zaibach, Kybrick, and the eagle. For these are three great spirits. Thus prepared, they tinge.

To Fix Sulphur

Take as much aquafortis as you wish. Inject ʒj. of live sulphur and the same quantity of pulverised alum. Dissolve in water, when the sulphur will become red as blood and fixed. This water dissolves all bodies.

Note

Mercury is called honey, calcined Luna, the assistance. Water of Mercury is called water. Thus Plato.

Fixing Salt

Take Quick Lime, ij. parts.
Soft Smegma, j. part
Wood Ashes, ½ part.

Also some ashes from the dregs of wine. Reduce all these to powder so that they may become a strong lixivium. Strain through a filter; coagulate, and you will have fixing salt, concerning which see below also.

Fixing Oil

Take of olive oil three parts, of quicklime two parts, and of sal ammoniac one part. Mix all together and distil the oil from them. Do this three times, always renewing the dregs.

Soft Soap

Take twelve scutellæ of water, in which place one scutella of wood ashes. Boil until dissolved. Then add half a scutella and boil until the water is reduced one-third. Remove from the fire and distil by a filter. Then to two parts of the water thus strained off add a third part of oil 9. Do this by evaporating over a fire.

Coagulation of Mercury

Make an amalgam from one part of Luna and five parts of purified Mercury. Place it in a glass vessel with a narrow neck which is well smeared below with the lute of wisdom. Place under the amalgam in the glass one layer of salt previously prepared, and also above. Afterwards pour over it the oil previously prepared to the height of three fingers, and let it boil over a slow fire for seven days.

Note

In art we find two co-operators by whose means our working is more easily fulfilled. The one is the destroyer, that is to say, sal ammoniac. But sulphur and arsenic reduced to an oil promote the work. The property of sulphur is to coagulate Mercury. But the property of arsenic is only to inspire and vivify the stone, if it be prepared in due manner. Hence, whoever omits to prepare the oil of Mercury should take the oil of arsenic in its place. But when a philosopher speaks of joining in matrimony the body and the spirit by sal ammoniac, he speaks of the oil extracted from Mercury.

Solution of Bodies

Dissolve honey over a fire, and pour over it tiles heated and pounded so as to imbibe the moisture. Then distil by an alembic, and a saffron-coloured water will be produced, because otherwise the honey cannot be distilled. Into this water let the spirits of aquafortis enter, that is to say, equal parts of vitriol and saltpetre. Then dissolve Luna and Sol in that water, and get ready also an ounce of prepared common salt. Then this converted material can be eaten or drunk without injury, because it is aurum potabile.

Nota Bene
That you may know the Recipe for Solution

Take a clean cloth, like ticken or fustian, as much as an ell in length; pour on it a pint of good wine, and with the wine mix about a half pint of good brandy. Tear the fustian or the cloth at first into fragments. Let it remain thus three days. Afterwards take it out and dry it on a board. When it has dried take one tatter after another on a small stick. Burn it as if to make tinder. Put the burnt tatters one by one into a brazen basin. Place in a cellar, when the whole will speedily become an oil. Then amalgamate Luna or Sol with three parts of Mercury. Let the Mercury evaporate again. Place a calx upon the slab. Pound it with brandy. Then take twice the quantity of the above oil, and twice the quantity of the body of Mercury. Temper it thoroughly on a table made of Saturn. Set it subsequently in a moist place, when you will soon find oil of the body, which use as you know how.

Luna Fixed by Me

Take saltpetre, antimony, and arsenic, and imbibe them well with oil of tartar, so that they may be thoroughly mixed. Dry them somewhat. Afterwards melt them in a tigillum, first over a slow fire and then over a stronger one. Pound the whole together into powders. Take of these one-fifth to one loth., or more if you like, and you have, etc. (*sic*).

Sol is the first which is not altered by fire, nay, it is improved by fire, and cannot (otherwise) become a great or perfect elixir, which is of eternal duration, the rectifier and lightener of all bodies, and is joined with Mars and Saturn. But it cannot be joined with Mars except Mars be filed, but with Saturn it is joined as it is.*

A Noble Work

Take Saturn and melt it. Before it hardens project an equal amount of Mercury. Wash this amalgam thoroughly with water and salt, afterwards with pure water. Grind on a stone, and afterwards add as much sal ammoniacum as there is mercury. Place it in a damp, warm place. When it is entirely dissolved take sublimated arsenic. Imbibe it by pounding it thoroughly on the stone with the aforesaid water. Set it to dissolve, but remember that there should be something there to lighten it. Let it stand until it is quite cold. This preparation serves as a means of lightening when it is projected on one hundred parts of Venus and on two hundred of Saturn. But note that you always ought to place in it salt of alkali, so that the spirit may penetrate better and enter through the whole body.

Fixation and Rubification of Mercury

Take bloodstone, mix it with sal ammoniac and sublimate it. Then the sal ammoniac ascends red, but the stone will remain at the bottom black. Then, having once sublimated the Mercury, grind it with that sal ammoniac, and sublimate it. The mercury will remain at the bottom fused and red.

---

* The text at this point, and in many places throughout the latter part of the collection, seems exceedingly corrupt.

Note

Calcine the cinnabar well with vitriol and salt, when it all goes into the metal.

Note

With regard to what will not readily enter into lead, nothing having been taken out, boil the matters in strong alkali. Add thereto as much as you like of vitriol, and it enters in.

Water of Mercury

Take of mercury sublimated and of antimony ʒiiij. each. Break each in pieces separately. Put together in a well-luted retort. Set in a sand; apply a gentle fire. Refine over. Should it attach itself in the tube, ease it with a coal. Increase the fire until it no longer comes over. When it is cold put the matter upon a stone. Pound it small. Should Mercury still remain, remove it. Then put again into a retort. Refine more effectually than previously, even until all the mercury has become water. Thus you may also make oil of antimony. Take the matter in a glass. Pound small. Pound the moisture. Do this continually and distil until a red oil comes over.

Fixation

Take of Mercury Sublimate, part j.

Sulphur, part ij.

Break up together. Put in a cucurbit. Set it in sand. Let the smell go out, at first weak, afterwards stronger. Should the sulphur not lose its smell, stir with a piece of wood; it will then lose its smell soon.

Proof

Set it upon a plate. Should it not smell it is fixed; if it does, take equal quantities of sulphur and mercury. Pound together, as at first, until it is fixed.

Reduce as Follows

Take one part both of litharge and of Mercury, and a small quantity of sulphur. Put the Mercury into the crucible and thoroughly lute. Apply at first a small fire, then a stronger, until it melts. Afterwards let it cool. Wash the Luna. Refine in Saturn. Then separate in aquafortis.

White

Take the white of forty eggs, break up well and boil. Then distil the water through a cloth. Afterwards take lb.j. of Mercury sublimate, and the same quantity of arsenic. Impaste frequently with the aforesaid water, and dry sufficently. Afterwards (another copy has) let lb.j. of sal ammoniac be added twice. Pour the water over the pounded powder to the height of a finger or two. Set in horse dung in a closed glass for fifteen days. Take out on the sixteenth day, and you will find an aqueous mass. Coagulate this in cinders. Of this j. part tinges one hundred of the bodies, and especially of purged Venus, to which if there be added three or four parts of Luna, it will become perfect Luna. If you add a ferment the Luna will be still more perfect.

Oil of Mercury and of the Sun for Gilding

Take of Sal Ammoniac, ½ oz.
Mercury Sublimate, j. quint,
which are to be well pounded together. Put into a hard boiled egg. Then remove the yolk, and make under or through the floor a small hole with a barrel of a pen. Put the egg on a small glass to eliminate the humour. Do the same with calx of the Sun. Take one quarter of the Sun. Paint it with Mercury. Put therein salt pounded small. Let the Mercury lose its smell. Boil the salt with hot water from the calx. Then take twice as much sal ammoniac as calx. Put this also in an egg with a small hole. Set it as before to dissolve into an oil, but should the sal ammoniac not melt readily, moisten it with brandy.

Oil of Vitriol

Pour aquafortis on carefully-calcined vitriol. Let it stand in putrefaction for fourteen days. Afterwards let the phlegma of the aquafortis be removed, and the oil remains at the bottom.

To Extract the Quintessence of Luna, Saturn, or Jove

Put tartar and sal ammoniac into acetum in a well-closed glass vessel. Put in this the ashes of Saturn, Jove, or Luna. Seal thoroughly and set in warm dung for eight days. Afterwards distil, and the acetum comes first, next the quintessence of the calx, after the manner of quicksilver.

Aquafortis on Venus, to make Luna

Take of Salt,
of Hungarian White Vitriol,
of Alum,
of Arsenic,
} j. lb. each.

Pound small according to the proper manner. Make aquafortis. Dissolve therein as much as you like of Venus, and you will have silver to the quantity of one-half.

Elixir

Take equal parts of Sol, Luna, Mars, and Venus. Melt them and make plates of them, which suspend over acetum. Scrape off the green which will be formed there. Do this until the plates are entirely consumed and turned into green. A portion of this green on 10 parts of Luna transmutes it into the appearance of Sol. If you wash this green with warm water, and afterwards mix with it some water of sal ammoniac, letting it stand for seven days, and dissolve, its effect is doubled If, too, in place of sal ammonia you added aqua fœtida, that is, Mercury, and then operated, as with the water of sal ammoniac, you would certify your work ; nor can you make any mistake on account of the metallic substances being combined in a philosophic way. Note this well.

Another

Rubify vitriol, pound it, and dissolve it in acetum. Then sublimate sal ammoniac over a slow fire, and add what is sublimated to the aforesaid red water in equal weight. Give it a slow fire cautiously for three days, and it will be coagulated and fixed. If it be not fixed, then repeat, and it will become a deeper red. Fix it until it is completely fixed, which you will prove on a plate of copper. Put three parts of this powder to eight parts of Sol, and a sort of fermentation will be produced, one part of which tinges thirty parts of Luna into Sol; and that Sol will change sixty parts of Luna into Sol in my opinion. Add antimony as above; it will then be much better.

One Part of Tincture on Ten Parts of Luna in Sol

Take of Verdigris,
of Mercury, } j.lb. each.
of Vitriol,

Mix thoroughly together. Distil a water out of them, from which Mercury proceeds into the alembic. Take the alembic from it. Subject the other in a glass to a good fire, so that it may become red. Take it out. Make it into a powder. Pour its own water back into it again, and put on the fire. Let the water lose the smell thereof, so that it may become red as scarlet. Pulverise. Melt a ducat in a crucible. Incorporate the powder thoroughly with it, or melt it upon it. Take the gold and throw it upon ten ducats of silver, and melt. Thus it arrives at the twenty-fourth grade. I believe that if a ducat had been first resolved in water of Mercury, and again coagulated with the former powders and resolved, and that three times, it ought to tinge the powders.

Water of Mercury

Mercury thrice sublimated from tartar is thereby turned into water in a cucurbit in hot ashes.

Particular

Take red vitriol, put it while still warm in strong acetum, and dissolve. Add crocus of Mars and leave it thus for eight days. Then take verdigris, grind it very fine, and imbibe it with the former liquid. Next take sulphur, and kill Mercury just as if you wished to make cinabrium, or else take cinabrium itself and sal ammoniac equal in quantity to the above-mentioned ingredients. Grind all together, and place on porphyry to be dissolved. Coagulate the solution. Take one part of that powder to ten parts of Luna, and it will become Sol. Add the Spirit of Venus, and it will perhaps be so much the better.

Aquafortis Purging all Metals and Fixing Sublimated Spirits

Take vitriol, alum, sal ammoniac, and oil of tartar. If you put a metal over warm ashes it is soon melted. It fixes sublimated spirits and purges them from all superfluity.

A Marvellous Fact about Mercury

Take salt of tartar and sulphur in equal parts, and sublimate both together. They will then ascend together into the glass. Take some of that sulphur and sublimated salt, ground small, and place them in a sublimatory on crude Mercury. Sublimate them together; then you will find the sulphur and the salt of tartar above in the glass sublimatory, while the Mercury remains at the bottom, and nobody can ever more revivify it. Take two parts of white sulphur, and also dissolve it. At length mix them together, and congelate. Take also j. ℥ of these powders to 200 (or else 100) ℥ of Mercury made warm, and caused to melt. Then you will find Luna.

Cement

Take one part of common ♃, as it is in itself, two parts of tile dust, M. Make regal cement.

Water fixing all Spirits and dissolving all Metals

Take alum, verdigris, and orpiment, j. oz. each: sal ammoniac and vitriol ij. oz. each: cinnabar ¼ oz. Make a strong water.

The Fixation of Orpiment

Pound it and distemper it with crepine oil. Pound again, imbibe, desiccate, and pound over a stone. Preserve for use. Then put in the fire and melt. If you put heated Mercury therein it will become coagulated and hard. Then take it, place it in a glass, and close securely. Set it over a slow fire for seven days (otherwise seven hours), and it becomes good gold.

How the Spirits of Water are produced: or concerning the Aquification of all Spirits and all Bodies

Take of Calx of the Shells of Eggs, of Sal Ammoniac, } j. pound each.

(I should take fixed sal ammoniac.)

Put into a copper vessel. Cover well with the lid of Venus. Subject to a fire of coals. This smelts the matter. Then pour over the stone. Pound most minutely. Set to dissolve over the stone. Take j. ounce of this water. Pour over whatever spirit you will which flows in the crucible. It is then converted into a powder. Dissolve the same in water. Do this nine times. Then the waters are fixed. Take as much as there is of this water. Pour it over the Sun melting in the crucible. It is thereupon converted into powder, which then dissolves in water. The case is the same with Luna, Mercury, and other bodies. You have all waters fixed to work with them.

Water of Vitriol: very good for the Red

Take vitriol, verdigris, sal ammoniac, sulphur, and, if you will, antimony also. Grind these together and distil with an alembic, until the whole of the water shall pass away, for this helps greatly to redness, and is marvellous for the above (*sic*) work of one day.

Elixir

When sublimated fixed arsenic, sublimated fixed Mercury, and calx of Luna are covered over with water of sal ammoniac and dissolved with it, then is formed an elixir for every purified metal, and it tinges 100. This is the secret of the Greeks.

The Work of One Day for the Red.

Take v. parts of * and grind in a brazen mortar with j. part of sulphur. Place this in a vessel enclosed for four hours in a baker's oven. Afterwards pound and imbibe with water of atramentum and of sal ammoniac. Of this project j. part on xxx. parts of Luna, and thus it will be coloured. If you melt this Luna with Sol it will be very good.

Another for Saturn

Take   Pure Live Sulphur, j. part.
         Crude Mercury, ij. parts.
         Prepared Vitriol, ij. ʒ.

First of all join the sulphur with the vitriol, pulverised in acetum over a slow fire so that it may be resolved. Next add Mercury, by well incorporating it with them, and stirring it with a rod. Then it will be black amalgam. Take this from the fire and grind it to an impalpable powder. Place in a glass vessel well luted with a good seal. Give it a moderate heat for two weeks. Then increase the fire for one week. Break the glass and pound it, giving it continuous heat, until you have a bright red colour. Then cool and pound, placing it in another great and strong vessel, well luted in the midst, and closed with its hermetic seal. Give it a very strong fire for one week, until you see the substance melted like wax or oil. Then refrigerate and take out that most precious substance. If it does not turn to oil as is here described, dissolve it with rectified aqua vitæ, and coagulate three or four times until the oil remains in the fire. Then melt viij.lbs. of Saturn and viij.lbs. of Mercury. Make an amalgam, as you know how, by fusion, and project j.lb. of this medicine leaving it to melt for a quarter of an hour by mixing it with wood. Afterwards cool the black substance; one part thereof tinges the viij.lbs. of Saturn and the viij.lbs. of Mercury into a medicine. Melt again viij.lbs. of Saturn and viij lbs. of Mercury. Take of this medicament one part, do again as above, and you will find Sol.

Augmentation for Sol

Take Saturn. Melt, and pour over it an equal quantity of Mercury. Then take the same quantity of white arsenic. Mix it with Saturn and Mercury, and let it become a powder. Then take three parts of Luna. Melt in *. Inject the powder gradually, and little by little, until it completely enters. Stir it well with a stick. Then purge in a cineritium.

A Secret for the Solution of Sol or Luna

Pound thoroughly calx of Luna or Sol with sal ammoniac. Place it in a glass vessel with the mouth open over hot ashes, and dry with a moderate heat until there shall be seen a white mass in which is no plated Luna. Next place in dung or a bath of Maria for nine or twelve natural days. Thus it is dissolved into water, which putrefy for the proper time. Afterwards congeal and dissolve it; and thus you will have the Stone.

Or Better

Ferment the water thus produced with filings of Luna, or with little lumps of Luna or Sol, with an equal quantity of Mercury dissolved and then re-converted to Stone, in the way you understand. Then dissolve the Luna in the aforesaid water; congeal it frequently according to your lofty intelligence, until, if you shall have operated rightly, the perfect Philosophers' Stone is produced. Also take care that the volatile part does not exceed that of the fixed body. This is the First Secret Way.

Water of Mercury

Take of Mercury, of Sugar Candy, equal quantities.

Pound them together. Leave the whole to putrefy for ten days. Distil it through the alembic twice and you have sublimated arsenic.

Sublimated Arsenic

Take of Crude Arsenic, of Soap, j.lb.

Pound the arsenic well. Mix these together, and so sublimate once or twice. This arsenic dissolves well in aquafortis and is good to fix.

Item

Take equal parts of crude sulphur and mastich.

OR,

Take equal parts of sulphur and white sugar. Distil, and there will come a water or an oil. Take thrice distilled aqua vitæ, and place in it two parts of saltpetre. Distil to a water. It dissolves all bodies and renders Mercury fixed in one hour.

Dissolving Water

Dissolve glass gall in an aquafortis. The same water dissolves all spirits, arsenic, sulphur, cinnabar, Mercury, etc.

Note

Calcine Luna with orpiment. Sprinkle the same powder upon Mercury sublimate. Thus it becomes fine silver.

Note

Take equal parts of orpiment and vitriol. Melt together so that a red powder is produced, of which project some on Luna in a state of flux, and you will find fine gold.

Extraction of the Quintessence from all Metals

Take acetum distilled from wine and sal ammoniac fixed in it. It extracts the quintessence from all metals and is a secret.

Calcination and Solution of Sol

Take laudanum well ground (*sic*), stratify it with Sol perfectly divided into leaves, and burn it. Then the Sol is reduced to a calx, which is to be forthwith dissolved in the quintessence of wine. Let it be distilled by ashes and it will go over through the alembic. If you wish to have pure Sol by itself alone, then put it into water, and distil gently; it will pass over as a quintessence, and remain in the bottom as gold of a citron colour. If you wish it for aurum potabile, use it with the quintessence of wine and rectify; but if for a tincture, you can make these two fixed as you know how; or otherwise, if you join dissolved gold together with its own quintessence to rectified oil of vitriol. Then there is produced a tincture which rejuvenates human bodies, and will transmute all metallic bodies into perfect gold. With this Mercury dissolve white sugar in quintessence of wine. Place at once in it the calx of Sol burnt by laudanum and it is wholly dissolved. These solutions avail for all purposes.

Oil of Tartar which dissolves itself in Heat

Take calcined tartar, or, as I think better, salt of tartar. Pour over it rectified water of life. Let it stand 24 hours in the bath. Extract the water from it, and add fresh aqua vitæ continually. Let it putrefy for 24 hours, as before, and extract it. Do this until it dissolves itself into an oil. Then take crude Mercury (and possibly prepared with Saturn). Inject two drops of that oil and you have etc. This recipe is the best medicine for an old wound.

Augment

Take of Filings of Luna, ½ oz.
of Crude purged Mercury, 1½ oz.

Amalgamate. Add a third part of sublimated Mercury. Place in a jar or glass. Pour on oil of tartar so as to stand higher than two fingers. Shut up; place over a fire, and apply a slow heat for six days, until the matter be hardened. Take out. Pound on a stone. Place again in the vessel. Add more oil as before. Set again over a slow fire of five or six coals. Rule the fire for three days and it will be fixed. Then test. Refine a small quantity in Saturn, but what you do not reduce augment with Mercury, and work as before.

Oil of Tartar is made as follows

Dissolve calcined tartar in good wine vinegar. Distil through the filter. Cause the vinegar to evaporate. Dissolve again, doing so ten times. At length permit the salt to melt by itself into an oil, which use for the above digestion.

Oil of Luna

Take of Verdigris, ℥ij. each.
of Sulphur,

Pound, place in a glass cucurbit. Make therefrom a water like an aquafortis, with a gentle fire at first, latterly with a stronger. Next take ℥ij. of Luna dissolved in common aquafortis. Pour the former water thereupon. Extract the water together with the aquafortis through the alembic even unto the olitet, and the matter becomes brown in a glass; thus the Luna in the Keldt (?) becomes hard, and melts in the warmth like an oil or wax. To the same Luna add viij. oz. of purged ☿. Place in a cucurbit of glass, lute well.

Set in sand, at first with a gentle fire, for eight days until the matter unifies itself; afterwards increase the grade of heat. Although the matter does not ascend, yet it becomes fixed, and half a *part* of it remains.

FOR THE ASHES OF LUNA

Put Luna in a crucible to melt, and when it is in a state of fluxion project on it salt of alkali, thus purging it. Then make filings from it, and calcine them in an open reverberatory apparatus for six days. Thus the salt is extracted, which coagulate and again dissolve. One part thereof coagulates and fixes 40 parts of Mercury, and that Mercury, being coagulated, tinges 50 parts of purified Venus into real gold.

OIL OF SULPHUR

Take Sulphur,
Tartar,
Glassgall, } lb.j. each.

Pound together. Melt. Immediately it liquefies extract it with an iron spoon, and put into a strong lye. Thus it dissolves, and the lye becomes red.

TO FIX SALTPETRE

Take lb.j. of saltpetre in a vitrified vessel. Melt slowly. Sprinkle into flux thereupon one ounce of feathery alum (*alumen plumosum*). Thus it is fixed, and soon becomes an oil. The *alumen plumosum* must previously be pounded small.

COAGULATION OF MERCURY

Take arum in May. Pound it with a wooden hammer. Then distil a water from it. Pour it thereupon, even to the fourth time. Pound the fæces continually, and let the water be consumed. Then make *seltele die druchne* (?). Put Mercury in a crucible. When the bubbling commences, throw the powder thereupon. Thus the Mercury coagulates itself into Luna.

OIL OF MERCURY

Dissolve Mercury in aquafortis. Allow it to boil. Protect it from the fire and stir for three hours in an open glass. Then take the Mercury out. Dry. Pound upon a stone. Put it into a moist cellar. Thus it becomes an oil or water.

KING

Take *alumen de pluma*. Place vinegar over it and distil into water. This water clarifies new and immature pearls.

NOTE. CEMENT

Make the king out of naphtha and marcasite, that is, antimony made white; the latter becomes black as amber, the other white as Luna. Cement therewith Sol and Luna with pounded glass and salt for twenty-four hours, etc.

Fixed Luna, Discovered by Me

Take Saltpetre,
Antimony,
Sublimated Arsenic.

Imbibe thoroughly with oil of tartar, mixing intimately. Dry slightly. Afterwards melt in a crucible, first with a slow fire, then with a strong one. Grind the whole of this to powder. Take j. quint over ʒj., or more, if you like, and you attain the result.

Note

Take equal parts of common salt rendered fluid like wax by a candle, and sulphur. Place in a strong glass phial or in a crucible, setting fire above and below it. Then it will be incorporated. One part of this suffices for 60 parts of Mercury.

Or,

If it be incorporated with sublimated Mercury, or several times with sublimated arsenic, and then be itself sublimated, it works wonders.

The Extraction of Mercury

Take aquafortis, to which join common water, lest it be too strong, and sal ammoniac. If you evaporate this water you have oil of antimony, with which you can test anything.

Note

When you wish to Dissolve Mercury Sublimate and White Marcasite (perhaps Bismuth)

Put the sal ammoniac into aquafortis, and they will dissolve; otherwise they will not. Aquafortis is made of vitriol, saltpetre, alumen, cinnabar, and verdigris.

Note

Take oil of tartar and oil ♋ over amalgam of Luna and Mercury in equal quantities. Mercury will then be fixed as Hans Rormeyer has stated.

The Secret Philosophical Water Coagulating Mercury and Fixing Luna

Make aquafortis out of vitriol and saltpetre in equal quantities. Take away the first water. Take lb.j. of the other. Add viij. parts of verdigris and viij. parts of vitriol beaten small. Seal up well, and put in a bath for nine days. Afterwards distil through a filter. This water dissolves all bodies and coagulates Mercury. It is the secret of the philosophers. Place (? upon) plates the Moon.

Augment of the Moon

Take j. marc of Mercury. Impaste with ʒj. of sulphur and one ounce of arsenic. Then pound with them four ounces of iron filings. Make cinnabar with it all. Take up j. marc. of Luna and oz. iij. of cinnabar. Should the Luna have remained in the test, throw it upon it and it augments itself.

NOTE

Take lb.j. of sugar candy. Put it in a tin can. Pour therein four parts of good white wine. Put this all in a kettle with water. Close the cans (*sic*) well with the upper cowl. Then boil so that at least one can may be well boiled. Then pour again another can of wine upon the remaining matter. Do this in the same fashion five times and you will have the oil. Lay the species therein as you will. Take one ounce of oil or of any species of aromatic. Pour a wine into it again, as above, and let it boil as before. Then you will find a black matter at the bottom of it; extract the oil by means of a filter, and reserve for use as you know. Lay the species according to your pleasure in the oil. Boil it and you will have an artificial balsam. Take a drop or two thereof. Imbibe with wine and it tinges itself.

OF THE WATER WHICH DISSOLVES SOL OR LUNA

Take calx of the Sun or Moon. Pound it. Cleanse with water of salt alkali. Place in a dissolving vessel, and when it has been dissolved you will have perennial water. This water fixes spirits and coagulates Mercury. (Perhaps alkali of the Philosophers would be better.)

NOTE

Take a common aquafortis. Dissolve Luna therein. Place impasted Mercury with sulphur, as you know. Extract it eight times, and prove in the cineritium. If you add oil of soap thereto, it will then be made.

OIL OF SOAP

Shave the soap small. Add to it beaten tile. Distil through retort. Prick the leaves (*sic*). It fixes all spirits.

SALPETRA LUPI

Take equal parts of vitriol, saltpetre, and alum. Make aquafortis and dissolve in it as much Mercury as possible. Then extract the water by an alembic, and finally give a great fire of sublimation, so that the spirits may go out of the water. Having cooled the vessel, let what was sublimated and what remained at the bottom be again dissolved in the same water, which extract once more by an alembic. Do this five times, always giving at the end a fire of sublimation, and at the fifth time it does not ascend. It remains at the bottom as red as blood. Keep this.

Next take sal ammoniac sublimated once by iron filings, j.lb.; and of the above-mentioned Mercury an equal weight. Sublimate five times. Then put it on marble to be dissolved, and let it become a red oil, which also preserve.

Take Sol which has been calcined and five times sublimated, with an equal weight of the above-mentioned sal ammoniac. Put this to dissolve; and you will have a red liquid, which you must retain. Then take oil of Mercury j.℥, liquor solis j.℥. Mix and coagulate. Take one part of this medicine, project it on twenty-five parts of calcined Luna, and you have Sol always remaining at twenty-four degrees.

Note

Take of Calx of Luna, \
    of Saltpetre,   } each part j. \
    of Salt Alkali,

Pound the three substances well upon a stone. Put them in a crucible. Set on a gentle fire until all melts. Then pour out of the crucible on a stone. Next, pound it to a small pwder. Leave the matter with the stone in a damp cellar. When it dissolves in water, take one part of this and one part of water of Mercury. Put both waters together in one glass. Leave in a moderate temperature to coagulate. One part of the powder tinges xij. parts of revivified Mercury.

Fixation of Venus

Make a sharp alkali of *puch* ashes and quicklime.

Take of Arsenic, lb.j. (*alias* qq). \
    of Saltpetre, lb.j. \
    of Calcined Tartar, ʒiiij. \
    of Mercury Sublimate,   } ʒj. \
    of Orpiment,

Make powders. Pour four parts of the lye therein. Let it boil as dry as a powder. Melt Venus. Add an equal weight of the powders in flux thereupon. It will then be fixed Venus.

Antimony

Take some that has been subtly ground. Distil therefrom an ordinary aquafortis, until it becomes white. Next pour the quintessence thereupon. Distil it until it becomes fixed. This proves it.

Take the antimony and calcine well. If it does not diminish in weight it is right. Next, pound it small and pour the quintessence thereupon, when it extracts to itself a certain substance. This you must pour into the same glass vessel; it tinges Mercury into Luna, and it is the fourth part of gold.

Water of Mercury

Take Mercury sublimate, pound it small; imbibe a few times, say, five or six, with oil of vitriol. Continue to pound it dry seven times. Put it in a thoroughly luted glass. Pour upon it six times the weight of oil of vitriol, together with the moisture. Place it in a distilling stove. Distil it as an aquafortis, at first gently, then stronger, and strongest last of all. Then the third part of the sublimated Mercury distils away, and the other stands up beautiful and brilliant like a small oriental pearl in a distillatory; it is such beautiful sublimated Mercury as can never before have been seen. Then proceed continuously as long as you please. Next, gradually extract the phlegm in the bath. Then you will have water of Mercury, which is above all waters. Ferment this with soul of the Sun. Coagulate and dissolve, fix and coagulate, and it will tinge if God has so willed it.

Augment of the Sun and Moon, good and short

Take of Vitriol,
of Verdigris,
of Alum,
of Saltpetre,
} ʒs. each.

of Caluerey, ʒj.

Pound all together and place the powders in the crucible. Put the Sun and Moon over them. Place the powders again on the top. Set among ignited coals for two hours and pour together. Then it is right.

The above-named Fixation

Take Calcined Tartar, lb.j.

Saltpetre, lb.j.

with one measure of acetum distilled by the filter, and extract the salt with the said acid, with which imbibe the amalgam, until the acetum with the salt shall harden. After every imbibition let the acetum evaporate, and so the amalgam will be fixed. Also, if you will, imbibe arsenic, and even sulphur, with oil of tartar three or four times, and add to the former. Thus beyond a doubt will be obtained the fixation of the amalgamated Mercury. Lastly, melt all, and at length purify.

The Hermetic Bird

Take two pounds of strong lixivium, and put therein iiij.oz. of sal ammoniac, with iij.oz. of sulphur. (If the sal ammoniac and sulphur be fixed, and you add salt of alkali, it will be an oil coagulating Mercury.) Set to boil on a slow fire, constantly stirring it until all is turned into water red as blood. Distil by 3 and keep enclosed in glass for future use.

Oil of Luna Dissolving Sol

Take of aquafortis from vitriol j.lb., arsenic j.lb., and stibium j.lb. Purify these by dissolving a little Luna therein. Then take water of alum and pour on that Luna in the glass. It will form a substance white and resembling cheese. Remove this carefully with a spoon. Then you will have water of Luna, to which add aqua vitæ, and distil the moisture thence in a bath. Thus you will find the oil of Luna, wherein you will be able to dissolve gold. Convert into an oil, coagulate, and you will attain your desire.

A Good Aquafortis

Take vinegar and cinnabar in sufficient quantities. Dissolve sublimated Mercury. Coagulate, and you will find *alumen chatinum*, which is made out of glass. Take the same quantity of arsenic out of iron to sublime.

Attinkar of Venus

Take one part of Luna, filed and calcined by means of salt into subtle powders; also one part of Mercury, purged by vinegar and salt. Make an amalgam in vinegar, and afterwards pour into hot water. Then take it and press it delicately out. Next take exactly double the quantity of both. Put them together in an egg with a long neck Set on hot ashes over a gentle fire. Let the humidity come out first. Then thoroughly lute the neck, and the neck shall emerge (*sic*) on the side through the cupel. In order to pound it sublimate it away; let it stand continually for four hours, and stir the egg. Do this for nine days. Then remove and take exactly the same quantity of arsenic as there was of Luna at first. Let the moisture previously escape. Then seal up, and treat in all respects as before for nine days. Then take sal ammoniac sublimated by itself in the same quantity as the matter. Remove the sal ammoniac. Dissolve that which is at the bottom in the white of eggs, as you know, while that which will not dissolve is to be dried. Take again the same weight of sal ammoniac. Pound together. Sublimate the arsenic again therefrom, and dissolve it once more in the small trough. Continue this until it is all dissolved. Then put it all together into the egg. Lute it if the moisture has departed. Treat it as before for nine days, stirring it about. Thus it all becomes fixed. Ultimately it melts in the glass like an enamel, of which one part out of ten parts of purged Venus will be perfect Luna.

To Augment the Tincture

If you have a quint, put it in a crucible. Stir it continually with a small piece of wood. Afterwards, when it is hot, set it on a ½oz. of Mercury. Leave it thus to stand in the fire until you can no longer perceive any smoke. When it is again ready, treat in this manner successively, according to the amount of the weight of the tincture.

When a Thing contains Gold or anything else and will not Separate

In this case cast a lump of arsenic twice or thrice upon it; it then speedily departs and becomes fair.

For Saturn

Take it in filings dissolved with acetum. Distil it by the tongue and again by the alembic. What remains in the bottom dissolve again in acetum, filter and distil by an alembic. Keep on until the Saturn remains at the bottom as a fused oil. Possibly the oil of Jove could be produced in the same way.

Flowers of Bodies

Take the dregs of wine. Place them in a vessel to the thickness of five fingers. Let there be in the middle of the vessel a circulus, and on the circulus place a wooden cross. On this cross put plates, and let the aperture of the vessel be smeared with the lute of wisdom. Place in a hot furnace for seven days. On the eighth day take it out and put aside the flower, washing it with a brush, or simply in water. Let the water subside, strain it through a cloth, and dry the flower. Then operate as before with the plates.

NOTE

Make an extract of lemons in a clean vessel. Put gold filings therein. If water is produced during the night, this is the gold for leprosy, and it keeps a man young.

ALSO

Crush green sloes in a pan.

AMALGAMATION

Take equal quantities of orpiment and vitriol. Smelt together so as to produce a powder. Take one part and throw it upon Luna when it melts, and you will find Luna.

ANOTHER

Take bismuth in any quantity as calcined tartar and pitch (?). Melt and found three or four times. Then add to the bismuth this wilderness. Take one part thereof and one part of Luna. Purge by Saturn and you will have Luna.

Also the *pars cum parte* * will be tincture over Luna. But if the flower of Luna be added to Mercury sublimate or arsenic, and be prepared according to art, it will become a tincture over Venus. Each of them tinges one hundred parts.

NOTE

Liquefy the body Zidar in a large crucible and take it out with an iron spoon, projecting it upon a flat stone. There will be plates which you can select according to your pleasure for the above-mentioned operation.

TO EXTRACT THE SOUL

Take Luna and let it melt. Whilst in flux project within it some talc, which attracts the soul of Luna. Then extract the soul in thrice distilled acetum. The Luna will be fixed if it be frequently melted with some more talc. Note when the soul is extracted. Then take earth and extract from it the combustible oil. Pour on the earth distilled phlegma and extract from it its salt. Then take the spirit in which is the soul, plant it by degrees in the earth, always with an eighth part of water. Imbibe the earth and continually repeat this, always drying it with a gentle heat night and day, until it has drunk up as much as it can. Afterwards distil (otherwise sublimate) as you know how, and as you have the intention.

SOL AND LUNA

Take filings of Sol or Luna, and place them in a glass vessel. Pour thereon a sufficient quantity of undistilled acetum, and let it stand four or five days to putrefy. After this pour away the acetum and wash the filings well with common salt so that it may be pure. Dry on a stone, and then imbibe with water of sal ammoniac. Grind, dry in the sun, and keep doing this as often in the day as you can until the substance shall become black. Before this many colours supervene, to which you need pay no heed. When the substance is after some difficulty coagulated on the stone in the sun, after it has come to the point that it can be dissolved with * in the sun, it suffices. Now put the stone with the substance into a cellar, and it is resolved entirely into oil in the course of the night. Set this oil in dung for five days to putrefy. Then distil with an alembic in a warm and moist place.

Having done this, take a clean sponge, dip it in water of sal ammoniac, which changes over the oil. Thus extract the sal ammoniac. After this is extracted, press the sponge in the hand, and having entirely drawn off the sal ammoniac water from the oil, pour on it common tepid water, and you will remove thus all saltness with the sponge, until you perceive nothing saline remaining and the whole of the oil is sweet.

Precipitation

Take of Calcined Alum, } part j.
of Saltpetre,

Melt them. Two ounces thereof fall upon one marc of Luna. The Luna does not in anywise weaken aquafortis.

Tincture

Take thin plates of Sol in any quantity you like, place them in oil and boil in one glass vessel. Put in it sublimated wine, kindle it, let it burn, and it will become calx of gold. Do this thrice or oftener. Take the calx, grind it well on a stone with an equal quantity of fixed sal ammoniac; place in it a modicum of strong acetum. Having ground these ingredients, put them in a glass vessel carefully closed with the lute of wisdom. Set this in horse dung for eight days and a red water will be produced. Take sublimated Mercury, place it in water and grind on a stone. It becomes a red powder. Place it in dung for eight days and it will become a water. Congelate this and then one part tinges one hundred parts of Luna.

Tested Augment of Luna

Take of Well-calcined Alum, lb.j.
of Saltpetre, lb.ij.

Make aquafortis and purge. Take of that 4 oz. In this dissolve ʒij. of Luna. Take again of the aforesaid water ʒij. Add thereto a little sal ammoniac, and in that water dissolve ij. ounces of metallic or crystalline arsenic. Again take two ounces, with the addition of sal ammoniac, and dissolve ij. ounces of sublimated Mercury. Preserve these waters specially, each by itself. Then dissolve in the said aquafortis lb.j. of purged Venus, which solution takes place, say, in twenty-four lotones. These being dissolved, conjoin all the waters in a grand cucurbit. Pour a fourth part of water thereon. Let it stand three hours to mix. Afterwards distil the water from it by means of an alembic, the fæces will then remain moistened at the bottom. Dry these by means of a slowly burning fire. Then take saltpetre and melted salt. Mix them under the powder in a crucible. Set it in a wind furnace. Melt together. Pour it into a receiver or crucible smeared with honey or suet. Crush and refine upon the test; you will then have twelve ounces of Luna.

Cement Regal

Take of Brick Dust, two parts.
of Salt, one part.

Moisten with vinegar; grade therein.

Physical Water

Take sal ammoniac, which has been thrice sublimated, distilled, and coagulated, and then once more let it be resolved and distilled. Let the process be repeated thrice. If it has been dissolved in aqua vitæ, distilled, coagulated, and finally resolved, it dissolves all calcined and burnt bodies, as also all calcined and sublimated bodies, with a marvellous solution, in a crucible on the fire, within the space of a single hour, and this by the help of God and by virtue of Him.

Item

Let us compound our Philosophic water by the help of God and His virtue. Take of sal ammoniac, dissolved and thrice distilled, vj. drams, and of rectified oil vj. drams. Mix together and imbibe six of the aforesaid plates by degrees on porphyry, or on a glass table until it has thoroughly imbibed, and when all is absorbed, put it in a glass vessel under horse-dung for three days. On the fourth day project this on three pounds of the aqua vitæ already described, and put in a venter equinus for fifteen days. On the sixteenth day you will find the plate dissolved into a white water like milk. Change the dung every fourth day, complete this water, and this is the philosophic oil, a water penetrating and quieting, lighting candles and illuminating the house, and whereby all philosophers are sustained. Herewith every volatile substance is restrained from escaping, such as Mercury, Sulphur, and Arsenic.

The White

Take equal quantities of Mercurialis and Saccharum. Mix and let stand for nine days. After this extract the oil according to the fashion of lay people. That is perfect, if perfectly produced. If it be placed in a glass globe it shines by night, and makes a beautiful colour on the face of a man or a woman. If anyone drinks of this oil every morning, though he be 100 years old, he shall have the complexion of a young man, all his limbs will be light, and he will not be able to be out of spirits. Taking this oil for 40 days on an empty stomach perfectly cures epilepsy of 40 years' standing. So, too, asthma. The white strengthens man's nerves beyond all other kinds. Mixed with castor oil and applied to the bone it removes all contraction of the limbs, and cures paralysis of the nerves. For these reasons we put it forward as precious. One drop of this white put in the eye every morning clears the eyes and keeps them healthy.

Item

Take white (? cadmia), gold leaf, oriental pearls, and rhabarbarum in equal parts. Grind together. This is the best medicine for leprosy, which it entirely cures if it be taken each morning on an empty stomach. It produces a good colour, but care must be taken to drink little and only good wine. A man should use this medicine until his complexion is good, and he himself will be perfectly healthy, please God.

Otherwise

Take equal parts of Mercurialis and Meter. Mix so that the Mercury shall be previously heated. Then allow to stand for nine days. Afterwards extract the oil according to the mode of the lay people. This oil is perfect, etc.

A Light which always Burns

Take any quantity of sugar, and an equal amount of Mercury. Mix thoroughly and leave it to stand for 13 or 14 days. Afterwards distil as an oil. Then take a small linen cloth and press it through. If it were put in a glass it would shine like a light.

Fixation of Mercury

Take alum and dregs of wine. Dissolve in urine, and distil by a filter. Then resolve sulphur in it by boiling. Place Mercury in it and let it boil with a gentle fire, continually stirring it, so that it may escape the fire.

NOTE.—*Certain formulæ and quantities which occur in the* MANUAL OF PARACELSUS *are either peculiar to the treatise or have long fallen into disuse, and it is difficult to identify their meaning.*

HERE ENDS THE FIRST VOLUME OF THE HERMETIC AND ALCHEMICAL WRITINGS OF PARACELSUS.

THE
HERMETIC AND ALCHEMICAL
WRITINGS

OF

AUREOLUS PHILIPPUS THEOPHRASTUS BOMBAST,

OF HOHENHEIM, CALLED

PARACELSUS THE GREAT.

NOW FOR THE FIRST TIME FAITHFULLY TRANSLATED INTO ENGLISH.

EDITED WITH A BIOGRAPHICAL PREFACE, ELUCIDATORY NOTES, A COPIOUS HERMETIC
VOCABULARY, AND INDEX,

BY ARTHUR EDWARD WAITE.

IN TWO VOLUMES.

VOL. II.

HERMETIC MEDICINE AND HERMETIC PHILOSOPHY.

London:
JAMES ELLIOTT AND CO.,
TEMPLE CHAMBERS, FALCON COURT, FLEET STREET, E.C.
1894.

TABLE OF CONTENTS

VOLUME II

PART II

HERMETIC MEDICINE

                                                                         PAGE

THE ARCHIDOXIES OF THEOPHRASTUS PARACELSUS.

BOOK THE FIRST: concerning the Mystery of the Microcosm   3

BOOK THE SECOND: concerning the Separation of the Elements   10

BOOK THE THIRD: on the Separation of the Elements from Metals   15

    Concerning the Separation of the Elements out of Marchasites. Concerning the Separation of the Elements from Stones. Concerning the Separation of the Elements from Oleaginous Substances. Concerning the Separation of the Elements in Corporeal Resins. Concerning the Separation of the Elements from Herbs. Concerning the Separation of the Elements from Fleshly Substances. The Separation of the Elements from Fishes. Concerning the Separation of the Elements from Watery Substances. Concerning the Separation of the Elements out of Water. Concerning the Separation of the Elements from Glasses and those Substances which are of the Nature of Glass. Concerning the Separation of the Elements in Fixed Substances. Concerning the Separation of Fire. Concerning the Separation of Air. Concerning the Separation of Water. Concerning the Separation of the Earth.

BOOK THE FOURTH: concerning the Quintessence.   22

    Concerning the Extraction of the Quintessence from Metals. Concerning the Extraction of the Quintessence from Marchasites. Concerning the Extraction of the Quintessence from Salts. Concerning the Extraction of the Quintessence from Stones, from Gems, and from Pearls. Concerning the Extraction of the Quintessence from Burning Things. Concerning the Extraction of the Quintessence out of Growing Things. Concerning the Extraction of the Quintessence from Spices. Concerning the Extraction of the Quintessence from Eatables and Drinkables.

BOOK THE FIFTH: concerning Arcana   37

    Concerning the Arcanum of the Primal Matter. Concerning the Arcanum of the Philosophers' Stone. Concerning the Mercurius Vitæ. Concerning the Arcanum of the Tincture.

BOOK THE SIXTH: concerning Magisteries   48

    Concerning the Extraction of the Magistery from Metals. The Extraction of the Magistery out of Pearls, Corals, and Gems. Concerning the Extraction of the Magistery out of Marchasites. Concerning the Extraction of the Magistery from Fatty Substances. The Extraction of the Magistery from Growing Things. Concerning the Extraction of the Magistery out of Blood.

vi   *The Hermetic and Alchemical Writings of Paracelsus*

	PAGE

BOOK THE SEVENTH: concerning Specifics — 59

  Concerning the Odoriferous Specific. Concerning the Anodyne Specific. Concerning the Diaphoretic Specific. Concerning the Purgative Specific. Concerning the Attractive Specific. Concerning the Styptic Specific. Concerning the Corrosive Specifics. Concerning the Specific of the Matrix.

BOOK THE EIGHTH: concerning Elixirs — 69

  On Preservation and Conservation by Elixirs. Concerning the First Elixir, that is, of Balsam. Concerning the Elixir of Salt, by the force of which the body is conserved. Concerning the Third Elixir, namely, of Sweetness. Concerning the Fourth Elixir, which is that of Quintessences. The Fifth Elixir is that of Subtlety. The Sixth Elixir, which is that of Propriety.

BOOK THE NINTH: concerning External Diseases — 77

  A Remedy for Wounds. A Remedy for Ulcer. A Remedy against Spots.

BOOK THE TENTH: the Key of Theophrastus Paracelsus Bombast von Hohenheim, from a German Manuscript Codex of Great Antiquity. Comprised in a Preface and Ten Separate Chapters — 81

  The Preface. Chapter I.: concerning the Separation of the Elements. Chapter II.: concerning the Quintessence. Chapter III.: concerning Magisteries. The Preparation of Circulated Salt. Chapter IV.: concerning First Entities; and primarily concerning the Extraction of the Quintessence or First Entity of Common Mercury. Chapter V.: concerning Arcana. Chapter VI.: concerning the Arcanum of the Stone, or of the Heaven of the Metals. Chapter VII.: concerning the Arcanum of the Mercurius Vitæ. Chapter VIII.: concerning the Great Composition, being the chief of our secrets in medicine. Chapter IX.: concerning the Corporal Balsam, or Mercury of the Sun. Chapter X.: concerning the Composition of the Spiritual Balsam, and of the Balsam of the Coagulated Body.

THE MANUAL OR TREATISE CONCERNING THE MEDICINAL PHILOSOPHIC STONE — 94

  Preface to the Reader. The Preparation of the Matter of the Stone. The Remainder of the Preparation. The Use of the Stone.

A BOOK CONCERNING LONG LIFE — 108

THE BOOK CONCERNING RENOVATION AND RESTORATION — 124

  The First Entity of Minerals. The First Entity of Gems. The First Entity of Herbs. The First Entity of Liquids.

A LITTLE BOOK CONCERNING THE QUINTESSENCE — 137

  Concerning the Quintessence out of Oil, also of the Salt of the same. Concerning the Oil or Quintessence of Silver and its Salt. Oil of Mars. Oil of Saturn. Oil and Salt of Jupiter. Of the Quintessence of Antimony. Concerning Oil and Salt out of Marchasite. Oil out of Common Salt. Oil and Salt of Coral, also of Crystal. Oil and Salt of Pearls. Of the Essence and Salt of Things growing on the Earth. The Method of separating Oil and Salt from Oil of Olives. Red Water out of Oil of Olives. Manufacture of Salt from the Red Water. Method of Extracting the Oil and Salt from Pepper. Oil out of Gums. Preparation of Colocynth.

ALCHEMY THE THIRD COLUMN OF MEDICINE — 148

THE LABYRINTHUS MEDICORUM concerning the Book of Alchemy, without which no one can become a Physician — 165

Table of Contents vii

	PAGE

CONCERNING THE DEGREES AND COMPOSITIONS OF RECIPES AND OF NATURAL THINGS IN ALCHEMY.

    THE PREFACE of Theophrastus Eremite of Hohenheim to those desirous of the Medical Art   169

    BOOK THE FIRST   172

    BOOK THE SECOND   178

    BOOK THE THIRD   185

    BOOK THE FOURTH   191

    BOOK THE FIFTH   196

CONCERNING PREPARATIONS IN ALCHEMICAL MEDICINE.

    TREATISE I.: concerning Antimony and Marcasite of Silver, White and Red Cachimia, Fluidic and Solid Talc, Thutia, Calamine, and Litharge   199

    TREATISE II.: concerning Bloodstone, Arsenic, Sulphur, Saxifrage, and Orpiment   208

    TREATISE III.: concerning Gems, transparent and otherwise; Corals, the Magnet, Crystal, Rubies, Garnets, Sapphires, Emeralds, Hyacinths, etc.   214

    TREATISE IV.: concerning Salts, including Sal Gemmæ, Sal Entali, Sal Peregrinorum, Aluminous Salt, Sal Alkali, Sal Nitri, Sal Anatron, Sal Terræ, and Salt from Vitriol   219

    TREATISE V.: concerning Metals, namely, Gold, Silver, Tin, Copper, Iron, Lead, Mercury   225

THE ALCHEMICAL PROCESS AND PREPARATION OF THE SPIRIT OF VITRIOL, by which the Four Diseases are cured, namely, Epilepsy, Dropsy, Small Pox, and Gout. To abolish those errors which are usually committed by Philosophers, Artists, and Physicians   231

        The Process. An Addendum on Vitriol. Of the Oil of Red Vitriol. The White and Green Oil of Vitriol.

THE ALCHEMIST OF NATURE, being the Spagyric Doctrine concerning the Entity of Poison   237

PART II

HERMETIC PHILOSOPHY

THE PHILOSOPHY ADDRESSED TO THE ATHENIANS.

    BOOK THE FIRST   249

viii *The Hermetic and Alchemical Writings of Paracelsus*

	PAGE
BOOK THE SECOND	263
BOOK THE THIRD	277

HERMETIC ASTRONOMY.

PREFACE TO THE INTERPRETATION OF THE STARS	282
THE INTERPRETATION OF THE STARS	284
THE END OF THE BIRTH AND THE CONSIDERATION OF THE STARS	289

    Concerning Man and the Matter out of which man was made. Astrology. Magic. Divination. Nigromancy. Signature. Uncertain Arts. Manual Art. Proof in Astrological Science. Proof in the Science of Magic. Proof in the Science of Divination. Proofs in Nigromancy. Proof in the Science of Signature. Proof in Uncertain Arts. The End of the Proof in Uncertain Arts. Proof in Manual Mathematical Science. Concerning the Knowledge of Stars. Another Schedule. Schedule concerning the Proof of Magic.

APPENDICES

APPENDIX I.: concerning the Three Prime Essences	317
APPENDIX II.: a Book concerning Long Life.	
BOOK THE FIRST	323
BOOK THE SECOND	332
BOOK THE THIRD	339
BOOK THE FOURTH	344
APPENDIX III.: a Short Lexicon of Alchemy, explaining the chief Terms used by Paracelsus and other Hermetic Philosophers	348
INDEX	387

PART II.

HERMETIC MEDICINE.

THE ARCHIDOXIES OF THEOPHRASTUS PARACELSUS*

BOOK I

Concerning the Mystery of the Microcosm †

IF, my dearest sons, we consider the misery by which we are detained in a gross and gloomy dwelling, exposed to hunger and to many and various accidents, from all sources, by which we are overwhelmed and surrounded, we see that we could scarcely flourish, or even live, so long as we followed the medicine prescribed by the ancients. For we were continually hedged in by calamities and bitter conditions, and were bound with terrible chains. Every day things became worse with us, as with others who were weighed in the same balance, whom also the ancients have not so far been able to help or to heal by means of their books. We do not in this place advance the different causes of this misfortune. This only we say, that many teachers by following the ancient methods have acquired for themselves much wealth, credit, and renown, though they did not deserve it, but got together such great resources by simple lies. From which consideration we have wished to elaborate and write this memorial work of ours, that we might arrive at a more complete and happier method of practice, since there are presented to us those mysteries of Nature which are too wonderful to be ever thoroughly investigated.

---

* The ten books of the *Archidoxies* stand in the same relation to Hermetic Medicine as the nine books *Concerning the Nature of Things* stand to Hermetic Chemistry and the science of metallic transmutation. They appear to have been reckoned among the most important works of Paracelsus, and the editions are exceedingly numerous. That which has been selected for translation is derived from the Geneva folio.

† The Microcosmos itself is to be understood thus, namely, as consisting of the four elements, and it is these invisibly. It is formed after the image of Him who created all things, and yet it has remained a creature. Therefore, it is partially one with the earth, because like the earth it has need of the other elements, heaven, air, and fire.—*De Hydropisi*. This, therefore, is the condition of the Microcosmos, or smaller world. It contains in its body all the minerals of the world. Consequently the body acquires its own medicine from the world. Hence it is clear that all minerals are useful to man, if any one of these be joined to its corresponding mineral in the body of the Microcosm. He who lacks this knowledge is by no means a philosopher or physician, for if a physician affirms that a certain marcasite is useful for this or that, it is first of all needful that he should know what is the marcasite of the world and what is the marcasite of the human body.—*Paramirum*, Lib. IV. *De Matrice.* There is a vast variety of things contained in the body of the Microcosm which elude the observation of the senses, though God, the Creator, has willed them to exist in that structure. There are, for example, more than a thousand species of trees, stones, minerals, manna, and metals. Even as the incredible magnitude of the Sun appears small to us by reason of its distance, so other things may be made to look small which are placed nearer at hand, and thus we come to understand that He who filled the shell of the heaven with so many and such great bodies, was able to include as many and as great wonders in the body of the Microcosm. Accordingly, know that the mysteries of the Microcosm are to be mystically understood--that is, we are no to measure local things according to proportion or substance, but must all the more arduously enquire in what matter we are to expect the effect of any agent.—*De Causis et Origine Luis Gallicæ*, Lib. V., c. 10.

Wherefore we have come to consider how that art can be reconciled with the mysteries of Nature, in opposition to those who, so far, have not been able to arrive at the art at all.

The strength of this mystery of Nature is hindered by the bodily structure, just as if one were bound in a prison with chains and fetters. From this the mind is free. For in its operation this mystery is like fire in green wood, which seeks to burn, but cannot on account of the moisture.

Since, then, hindrance arises from this source, one had to see how to get free from it. For such freedom being secured, this art of separation can only be compared to the art of the apothecaries, as light is compared to darkness. And this we say not in mere arrogance, but on account of the great frauds practised by apothecaries and physicians. Wherefore, not undeservedly, we call them darkness, or caves of robbers and impostors, since in them many persons are treated for gain by ignorant men; persons who, if they were not rich, would at once be pronounced healthy, since the practitioners know that there is no remedy or help for these people in their consultations.

This, then, is worthy to be called an art, which teaches the mysteries of Nature; which, by means of the quintessence, can cure a contraction and bring about health in the space of four days, whereas otherwise death would be the result. A wound, too, can be healed in twenty-four hours, which would scarcely yield to bodily treatment in as many days. Let us, therefore, readily approach by experiment this separation of the mysteries of Nature from the hindrances of the body.

First, then, we have to consider what is of all things most useful to man and most excellent. It is to learn the mysteries of Nature, by which we can discover what God is and what man is, and what avails a knowledge of heavenly eternity and earthly weakness. Hence arises a knowledge of theology, of justice, of truth, since the mysteries of Nature are the only true life of man, and those things are to be imitated which can be known and obtained from God as the Eternal Good. For although many things are gained in medicine, and many more in the mysteries of Nature, nevertheless after this life the Eternal Mystery remains, and what it is we have no foundation for asserting, save that which has been revealed to us by Christ. And hence arises the ignorant stupidity of theologians, who try to interpret the mysteries of God, whereof they know not the least jot; and what it is not possible for man to formulate, namely, the will of Him who gave the mystery. But that word of His they twist to their own pride and avarice; from whence arise misleading statements, which every day increase more and more. Hence it comes that we lightly value, nay, think nothing at all of that reason which is not evidently founded on the mysteries. In like manner the jurists have sanctioned laws according to their own opinions, which shall secure themselves against loss, though the safety of the State be imperilled.

Seeing, then, that in these faculties so many practices have come into vogue which are contrary to equity, let us dismiss the same to their proper time.

Nor do we care much for the vain talk of those who say more about God than He has revealed to them, and pretend to understand Him so thoroughly as if they had been in his counsels; in the meantime abusing us and depreciating the mysteries of Nature and of philosophy, about all of which they are utterly ignorant. The dishonest cry of these men is their principal knowledge, whereby they give themselves out to be those on whom our faith depends, and without whom heaven and earth would perish.

O consummate madness and imposture on the part of human creatures, in place whereof it would be more just that they should esteem themselves to be nought but unprofitable servants! Yet we, by custom imitating them, easily learn, together with them, to bend the word of our Teacher and Creator to our own pride. But since this word is not exactly known to us, can only be apprehended by faith, and is founded on no human reason, however specious, let us rather cast off this yoke, and investigate the mysteries of Nature, the end whereof approves the foundation of truth; and not only let us investigate these, but also the mysteries which teach us to fulfil the highest charity. And that is the treasure of the chief good which in this writing of our Archidoxies we understand in a material way.

From the aforesaid foundation we have drawn our medicine by experiment, wherein it is made clear to the eye that things are so. Then, coming to its practice, we divide this our book of Archidoxies into ten, as a sort of aid to the memory, so that we may not forget these matters, and at the same time may speak of them so far openly that we may be understood by our disciples, but not by the common people, for whom we do not wish these matters to be made too clear. We do not care to open our mind and thoughts and heart to those deaf ears, just as we do not wish to disclose them to impious men; but we shall endeavour to shut off our secrets from them by a strong wall and a key. And if by chance this our labour shall not be sufficiently safeguarded from those idiots who are enemies of all true arts, we shall forbear writing the tenth book, concerning the uses of those which precede it, so that we may not give the children's meat to the dogs. Nevertheless, the other nine will be sufficiently understood by our own disciples.

And, to speak more plainly of these matters, it must be known that in this treatise on the microcosm are proved and demonstrated all those points which it contains, which also embrace medicine, as well as those matters which are interconnected therewith. The subject of the microcosm is bound up with medicine and ruled by it, following it none otherwise than a bridled horse follows him who leads it, or a mad dog bound with chains. In this way I understand that medicine attracts Nature and everything that has life. Herein three things meet us, which shew by what forces they are filled and produced. Firstly, in what way the five senses are assisted by the mysteries of Nature, though those senses do not proceed from Nature, nor spring naturally as a herb from its seed, since there is no material which produces them. Secondly, the mobility of the body is to be considered: whence it proceeds, by what power it is moved and exercised, and in what manner it is ministered to. Thirdly, there must be a knowledge of all the forces in the body, and what forces apply to each member, and are transmuted according to the same nature as the particular limb, when originally they are identical in Nature.

First, then, we will speak of these senses: sight, hearing, touch, taste, and smell. The following example teaches us. The eyes have a material substance, of which they are composed, as it is handed down in the composition of the body. So of the other senses. But vision itself does not proceed from the same source as the eye; nor the hearing from sound, or from the same source as the ears; nor touch from flesh, nor taste from the tongue, nor smell from the nostrils, any more than reason proceeds from the brain; but these are the bodily instruments, or rather the envelopes in which the senses are born. For it must not be understood that these senses depend solely on the favour of God, in the sense that they do not belong to the nature of man, but are infused solely by the grace of God above and beyond all Nature, to the end that, if one were born blind, the mighty works of God might be made known to us. We must not think so in this case. For the abovementioned senses have each their own body, imperceptible, impalpable, just as the root of the body, on the other hand, exists in a tangible form. For man is made up of two portions, that is to say, of a material and a spiritual body. Matter gives the body, the blood, the flesh; but spirit gives hearing, sight, feeling, touch, and taste. When, therefore, a man is born deaf, this happens from a defect of the domicile in which hearing should be quartered. For the spiritual body does not complete its work in a situation which is badly disposed.

Herein, then, are recognised the mighty things of God, that there are two bodies, an eternal and a corporeal, enclosed in one, as is made clear in the Generation of Men.* Medicine acts upon the house by purging it, so that the spiritual body may be able to perfect its actions therein, like civet in a pure and uncontaminated casket.

Coming next to the power of motion in the body, let us inquire whence it is produced and has its origin, that is, how the body unites itself to the medicine so that the faculty of motion is increased. The matter is thus to be understood. Everything that lives has its own motion from Nature. This is

---

* The treatise *De Generatione Hominis*, to which reference is here made, appears to be fragmentary in character. It regards the generation of all growing things as twofold—one where the nature and the semen are contained in a single essence, the other where the essence of the nature exists without the semen. In man it recognises the existence of four complexions, and it distinguishes between the outward and the interior man. Men, as regards their mortal part, are nothing but mere cattle. There is, however, an internal man wherein there is no nature of the animal, and to consider him in this his true nature must be counted among the highest branches of philosophy. There is an immortal as well as a corruptible body of man, and it is in this, by the infusion of God's power, that reason, discernment, wisdom, doctrine, art, and generally whatsoever is above mortality, do alone inhere. Man, therefore, endowed with wisdom and subtlety, can emerge from his external body. All wisdom and intelligence which man enjoys is eternal with this body, and man the interior can live after another manner than can man the outer. This internal man is illustrated and clothed upon with truth for ever.

sufficiently proved of itself so far as natural motion is concerned. But the motion of which we think may be described as that which springs from the will, as, for example, in lifting the arm one may ask how this is done, when I do not see any instrument by which I influence it; but that takes place which I desire to take place. So one must judge with leaping, walking, running, and other matters which occur in opposition to, or outside of, natural motion. They have their origin in this, that intention, a powerful mistress, exists above my notions in the following manner. The intention or imagination kindles the vegetative faculty as the fire kindles wood—as we describe more particularly in our treatise on the Imagination.* Nowhere is it more powerful to fulfil its operations than in its own body where it exists and lives. So, in every body, nothing is more easily kindled than the vegetative soul, because it runs and walks by itself and is disposed for this very purpose. For, even as a hidden or buried fire blazes forth so soon as it is exposed and catches the air, so my mind is intent upon seeing something. I cannot with my hands direct my eyes whither I will; but my imagination turns them whithersoever it is my pleasure to look. So, too, as to my motion must it be judged. If I desire to advance and arbitrarily propose this to myself, at once my body is directed to one or the other place fixed on by myself. And the more this is impressed on me by my imagination and thought, the more quickly I run. In this way, Imagination is the motive-power of my running. None otherwise does medicine purify those bodies in which there is a spiritual element, whence it happens that their motion is more easily perfected.

Thirdly, it must be understood that in the body a distribution is made over all the members of everything which is presented to it, either without or within. In this distribution a change takes place by which things are modified, so that one part subserves the constitution of the heart, another accommodates itself to the nature of the brain; and of the rest in like manner. For the body attracts to itself in two ways, from within and from without. Within, it attracts whatever is taken through the mouth. Externally, it attracts air,

---

* *De Virtute Imaginativa* is another treatise which has survived only in a mutilated state. It further insists on the division of man into two bodies, the one visible and the other invisible, but is devoted to the consideration of the second only. The imagination is the mouth of the body which is not visible. It is also the sun of man which acts within its own sphere after the manner of the celestial luminary. It irradiates the earth, which is man, just as the material sun shines upon the material world. As the one operates corporeally, so works the other, after a parallel manner, spiritually. And as the sun sends its force on a spot which it shines upon, so also the imagination, like a star, bursts upon the thing which it affects. Nor are all things posited in heat and cold only, but in every operation. As the sun works corporeally and effects this or that, so also the imagination, by giving fire and fuel, effects all things which the sun effects, not that it has need of instruments, but that it makes those things with which it burns. Consider the matter as follows: He who wishes to burn anything needs flint, fire, fuel, brimstone, a candle, etc., and so he obtains fire; but if the sun seeks to burn, it requires none of these things, doing all things together and at once, no one beholding its steel. Such also is the imagination. It tinges and paints its own surface, but no one sees its pencil, ceruse, or pigments; all things take place with it at once, just as fire from the sun bursts forth without any corporeal instrument. Let no one, therefore, be surprised that from the imagination corporeal works should proceed, since similar results are manifest with other things. The whole heaven, indeed, is nothing else but an imagination. Heaven works in man, stirs up pests, fevers, and other things, but it does not produce these by corporeal instruments, but after the same manner that the sun burns. The sun, indeed, is of one power only, the moon of one power only, and every separate star is of one power only. Man, however, is altogether a star. Even as he imagines himself to be, such he is, and he is that also which he imagines. If he imagines fire, there results fire; if war, there ensues war; and so on in like manner. This is the whole reason why the imagination is in itself a complete sun.

earth, water, and fire. Thus, then, the subject is to be arranged and defined. Those matters which are received from within need not be described. They are known by the foundation of our nature, what they are which are distributed, and we shall speak subsequently as to their division. But, externally, one must understand whatever is necessary to itself the body attracts from the four elements. Unless this were done the internal nutriment would not suffice to sustain the life of man. For instance, moisture, not existing constitutionally in the body, is extracted by the body itself from water, whence it happens that if one stands or sits in water, it is not necessary that he satisfies his thirst from without. It does not, indeed, take place in the same way that heat is extinguished by water, like fire; but the internal heat attracts to itself the moisture from without, and imbibes it just as though it were from within. Hence it happens that in the Alps cattle are able to remain the whole summer without drinking; the air is drink for them, or supplies its place: and the same should be judged with regard to man.

The nature of man, too, may be sustained in the absence of food, if the feet are planted in the earth. Thus we have seen a man who lived six months without food and was sustained only by this method: he wore a clod of earth on his stomach, and, when it got dry, took a new and fresh one. He declared that during the whole of that time he never felt hungry. The cause of this we shew in the treatise on the Appetite of Nature.*

So, in the matter of medicine, we have seen a man sustain himself for many years by the quintessence of gold, taking each day scarcely half a scruple of it. In the same way, there have been many others who for so long as twenty years ate nothing, as I remember to have seen in our times. This was by some attributed to the piety and goodness of the persons themselves, or even to God, which idea we would be the last to impugn or to criticise. But this, nevertheless, is an operation of Nature; insomuch that sorrow and mental despondency take away hunger and thirst to such an extent that the body can sustain itself for many years by its own power of attraction. So, then, food and drink are not thus arranged that it is absolutely necessary we should eat bread or meats, or drink wine or water, but we are able to sustain our life on air and on clods of earth; and whatever is appointed for food, we should believe is so appointed that we should taste and try it, as we shall shew more at length in the "Monarchy of God."† Let us, however, concede this point—that on account of our labours and such things, it cannot be that we do without temporal and bodily food, and that for many causes. Wherefore food was ordained for this purpose, just as medicine was against diseases.

We will make a distinction of things entering into the body after the following fashion, that they are distributed through every part of it none otherwise than as if ardent wine be poured into water. The water acquires

---

* There is no treatise extant on the *Appetite of Nature*.

† There is also no treatise extant under the title of *The Monarchy of God*. *The Aurora of the Philosophers* is the *Monarchia of Paracelsus*, but, as will be seen, it is wholly chemical.

the odour of the wine because the wine is distributed through its whole volume; and, in the same way, when ink is poured into wine the whole of the wine is thereby blackened.

So, too, in the human body, the vital moisture immediately diffuses whatever is received, and more quickly than in the examples we have cited.

But under what form the substance received becomes transmuted depends entirely on the nature of the members that receive it, just as if bread be conveyed into a man it becomes the flesh of a man, if into a fish the flesh of a fish, and so on. In the same way, it must be understood that the substances received are transmuted by the natural power of the members, and are appropriated according to the nature of the parts which take them up. A like judgment must be passed upon medicines, namely, that they are transmuted into members according to the properties of those members. For the limbs gain their own force and virtue from the substance of medicines peculiar to themselves, according to the good or bad dispersion of them, and according as the medicine itself was subtle or gross. This is the case with the quintessence; its transmutation will be stronger and more effectual. But if it be thick it remains the same, just as a picture acquires its tint, its beauty, or its deformity from its colours, and if these be more vivid it will be the same. Wherefore, in order that we may have experience of like matters to fall back upon in those things which happen to us, and that we may lay them up in our memory, so as to have them ready in case of need, we will write these nine books, keeping the tenth shut up in our own brain on account of the thankless idiots. Nevertheless, to our own disciples, these things shall be made sufficiently clear.

And let no one wonder at the school of our learning. Though it be contrary to the courses and methods of the ancients, still it is firmly based on experience, which is mistress of all things, and by which all arts should be proved.

THE END OF THE PROLOGUE AND OF THE FIRST BOOK OF THE ARCHIDOXIES ON THE MICROCOSM, OUT OF THE THEOPHRASTIA

THE SECOND BOOK OF THE ARCHIDOXIES

From the Theophrastia of Paracelsus Theophrastus

Concerning the Separations of the Elements [*]

BEFORE we approach a description of the separations of the elements, we shall explain this separation, for the greater and clearer understanding thereof, seeing that certain matters written about the generation of things are not altogether consonant with the separations of the elements. For every matter is more readily brought to its appropriate end, where mature and intellectual consideration is given beforehand as to what its end may be. Thence the practice becomes clearer. We say, then, that the four elements exist together in all things, and out of them arises to every one of these things its predestined condition. In this way you may understand how it comes about that these four elements, differing so widely from one another, are able to agree and coexist without mutual destruction. Whereas the mixture of the elements is united and strengthened by predestination, it results that no weight is taken account of in them, but the power of one is greater than of another; by which, indeed, it is understood that in the digest and ferment of the predestination, that which is strongest will preponderate and conquer and subdue the other elements. In this way, the remaining three elements cannot attain their perfection, but stand related to that perfect element as

---

[*] The light of Nature teaches that God has separated and divided everything, so that it can exist by itself. Thus are separated light and darkness without any mutual damage, as day and night prove. Moreover, He has also separated the metals, each into its property. Thus, gold has its vein, iron has its minera, silver has its brilliancy. Lastly, every metal has its proper domicile. Moreover, He has separated from one another the marcasites and the genera of salts; in the same manner also summer and winter, elements, herbs, fruits, and every growing thing, so that we hence see how God has created various species from a single Iliaster, while the species of his workshop surpass all the number of the sands. He has so adorned heaven and earth that what they contain can never be sufficiently recognised and considered.—*De Balneis Piperinis*, c. 1. The philosophy of separation in preparing specifics is thus developed in treating of the cure of ulcers. You must also understand concerning separation that there is nothing so noxious but that it has its peculiar use, like the spider, which in addition to its venom possesses a marvellous power for curing all chronic fevers. Good and bad are equally necessary in the constitution of our arcanum, for both the venom and the antidote, the sweet and the bitter, are in the body. Further, it has been proved that all colours and all savours exist in the body. Inasmuch as there exist three colours in Saturn—yellow, white, and red; in Mars also three—purple, red, and black; notwithstanding neither Saturn nor Mars are a colour. Thus we must judge it to be the case with the colours of Salts by means of separation; and just as the redness of Saturn tinges with a red, and the yellow with a yellow colour, by virtue of separation, similarly, alumen works as alumen and salt as salt in ulcers. Concerning the wonders of separation it should be remembered what happens concerning vinegar. Who would judge by his senses that vinegar was present in uncorrupted wine? Nevertheless, no one can doubt that it is there. Similarly, in Venus there is vitriol. Indeed, Venus herself is vitriol, and by separation can be reduced into vitriol. Yet no one would rightly say that Venus was vitriol. Wherefore we must judge concerning separation that it is of the form but not of the species, as in the conversion of Venus into a salt.—*Chirurgia Magna*, Part III., Lib. IV.

the light matter in wood. Wherefore they are not to be called elements, since they are not all perfect, but only one. When, therefore, we speak of the four elements, which finally exist in all things, we are not to understand definitely so that there are four perfect elements therein, but that there is only one such finished element, the rest remaining imperfect through the potency of that excelling element. Hence it happens that they are able to meet and coexist, because in three of them there is no perfection. On which account, too, no corruption can prevail by reason of their contrarieties. Moreover, that an element is predominant in one kind arises from the fact that it is hereto predestinated. Hence no corruption or confusion can accrue, as we lay it down in our treatise on Generations.*

Since, then, there is only one element specially present in everything, it avails not to seek the four elements in things, seeing that three of these elements are not in a state of perfection. In a word, we must understand that the four elements are in all things, but not actually four complications. The matter stands thus: A substance contains water, and then it is nenufar. Besides this element there exists in it no earth, air, or fire. There is no appearance of heat or dryness in it. It has no peculiar operation; but the predestination thereof is water; and the element of water is the only one under which is no dryness nor heat, according to its congenital nature. Though matters be thus, however, yet notwithstanding the three elements are involved in it, still the things have not their origin in those three elements which are not produced in a perfect state, nor have they beginning or aid from them, but from that predestinated element which is united to and impressed upon that particular kind.

Now, although this is at variance with the vulgar philosophy, namely, that one predestinated element has of its own nature the other three elements cohering with it, still it is credible that the element and the substance differ the one from the other. It may be understood thus: the substance is not from that element which gives thereto its special tinge and elementary form; nor, again, are these elements from the substance, but they agree at the same time uniformly, as the body and soul agree. But now every body, as for example of some growing thing, has its own conformation; so has the element. Although the element itself is not visible in the body of the growing thing, or tangible, or demonstrable, because the element is stronger, by reason of its subtlety, and subdues the other elements of the growing thing. They are all, however, in the body, but imperceptibly, just as when water is mixed with vinegar the water becomes like the vinegar; and although the vinegar shall have changed the whole essence of the water, still the complexion of the water remains unchanged, nor does it on that account become

---

* Both from the text and the notes of this translation the reader will see that there are several treatises by Paracelsus on Generations of various kinds, not excepting some fragments upon the generation of fools, who appear, in the days of Paracelsus, to have abounded in the high places of science and religion. The reference above may be either to the *Generation of the Elements* or to the first book *Concerning the Nature of Things*.

vinegar, but remains water as it was before. And although it does not display the properties of water, yet it does not follow that it does not still possess those properties.

In these propositions we wish to make it clear in what way the separations of the elements are to be brought about, concerning which two methods of practice need to be understood. One is that with which agrees the separation of the predestined element, and this we shall elucidate in our treatise on the Quintessence. The other is that to which belong the four substantial elements in growing things. By this it is understood that the Quintessence exists as a predestined element, and that it cannot be separated from itself, but only from the three elements, as follows in the treatise on the Quintessence. But when we speak of the separations of the four elements, we understand those four which are essentially in the body. Hence have arisen various errors, because the four elements have been sought in the predestined element, and also in addition the Quintessence, which cannot in any way come to pass.

Moreover, it must be known when the elements of bodies are to be separated, that one exists as fire, another as water, a third is like air, and a fourth like earth, according to their complications, because elements sometimes appear with their own forms, and at other times with complications, for example, water as water, air as air, earth as earth, and fire as fire. These matters must be subtly understood, and this can be done by means of a similitude, if they are taken for the union of the elements not visibly or in act, or according to the element of fire, but as the environment is warm and dry like fire. In this way its own nature, essence, and condition is assigned to each element, without any breach of propriety. For it is not supposed, because any particular herb is especially warm, as the nettle, that on this account it contains within itself more fire, but rather it is supposed that its own Quintessence is warmer than the Quintessence of the chamomile, which, indeed, has less heat. But the elements of a body receive less or more from their own substance, as, for instance, wood contains within itself more fire than herbs do; and, in like manner, stones have in them more dryness and earth than resins have. Note, also, in like manner, that the bulk and quantity of degree in a Quintessence arises from the predestined element. And the intensity of degree in bodily elements springs out of the appearance of the substance which is unlike.

But we must come to the practice of separating bodily elements from all other things, and this is twofold. One, indeed, teaches us to draw out the three elements from the pure elements as from burning fire, from invisible air, from true earth, and in like manner from natural water, which have not an origin similar to the preceding ones. Another method of practice is with those things out of which these four elements exist, as we have said above, with such difference, however, that this exhibits more of the element of fire, of water, of earth, or of air, with a likeness to the form of the essential elements. When they have been separated in this way they can never be further dissolved, for instance, it is impossible that they should be corrupted beyond their complexions.

It must also be considered that the elements are found by separation to be formally like the essential elements. For air appears like air; and the same cannot in any way be enclosed, as some falsely think, for this reason, because at once, in the moment of separation, it levitates itself, and sometimes bursts forth as wind, and ascends sometimes with the water, sometimes with the earth, and at other times with the fire. And, indeed, this levitation or elevation in the air is very wonderful. Just as if the air were to be separated from the essential element of water, it would be done by boiling. When this begins to take place the air is soon separated from the water, carries off with itself the very light substance of the water, and in proportion as the water is diminished, so the air itself decreases according to its proportion and quantity.

And it must be remarked here that no one of the elements can be conceived or had without air, though of the rest one can be had without another. We do not, therefore, undertake the task of separating the air, since it is in the other three elements, just as life is in the body. For when it is separated from the body all things perish, as we clearly shew in the following practical treatise on separations. Four methods must be considered at this point; one, indeed, in watery bodies, that is, in herbs, which have more water than any of the other elements. A second is in fiery bodies, such as woods, resins, oils, roots, and the like, which contain within themselves more of the fiery substance than of others. A third is to be understood of earthy bodies, which are stones, clays, and earths. The fourth is airy, and this is in all the other three, as we have mentioned. In like manner, also, concerning the pure elements, there are just so many ways to be considered, in the same manner as has been said above concerning the four preceding.

Hence it is easy to learn what the elements are, and how they are to be separated. And among these the separations of the metals first meet us, wherein there are peculiar predestined virtues which are wanting in the other elements. For, although all the elements are alike in form, in heat, in cold, in moisture, and in dryness, still, the dryness or damp, or heat or cold, is not the same in one as in another. In some it is appropriative, but in others specific; and this in various ways, as in each kind they are produced peculiarly and essentially, since no kind of the elements is precisely like another in its properties.

So, also, it must be laid down with regard to the separation of marchasites, which differ from other substances both in the practice and in their elementary nature. For every kind is disposed in a particular separation, and must be dealt with in a special way. Stones and gems must demonstrate their elements afterwards, since they appear in no way similar to the others.

Then, too, salts exist in a peculiar and most excellent nature, with more abundant properties than appear in other substances. There is also a different essence in herbs, which in no way agree with minerals, nor can they be alike so far as relates to their nature. Moreover, the property of woods, of fruits, of barks, and the like, is peculiar; so, too, of flesh, of drinks, and all comestibles, and of things not good or pure, but bad and impure, which have to be separated into their elements.

Of that separation concerning which we think, two methods are found. One consists of the separation of any element confined by itself in a peculiar vessel, without the corruption of its own forces, the air excepted.

There is another method of the separation of the pure from the impure from among the four elements, namely, in the following way. After the elements have been separated, that is to say, one from the other, they have still a dense substance; and, on this account, there follows another similar separation of the already separated elements. Now, we purpose to make clear the practical method in all these cases. For in the first place it must be known that the quintessence of things is to be separated and extracted in this way, because, indeed, the elements of bodies in the nature of a quintessence are not subdued but are left with them. So it is able to tinge these elements more strongly or more lightly. Hence it comes to be understood that the forces in the elements do not perish when the predestined element, that is, the quintessence, is extracted; for this itself is elemental and separable so far as relates to its elementary form, but not as to diverse natures, as is clear from the treatise on the quintessence.

By separations of this kind all elemental infirmities can be cured by one simple method, namely, if the one set of predestinations oppose the other, as we have laid it down in the treatise on predestinations. In these words we have sufficiently unfolded the initial stage of separations. Wherefore we can now speed on to the practice of them; and here there is a tenfold variety: one of metals; a second of marchasites; a third of stones; a fourth of oleaginous matters; a fifth of resins; a sixth of herbs; a seventh of flesh; an eighth of juices; a ninth of vitreous substances; and a tenth of fixed things. For these separations of the elements three methods are adopted: one by distillations; a second by calcinations; and the third by sublimations. In these are comprised all the exercises, as the application of the hands to the fire, the labour, and other necessary things which will be specified in the following pages.

THE END OF THE SECOND BOOK OF THE ARCHIDOXIES AND OF THE FIRST PART CONCERNING THE SEPARATION OF THE ELEMENTS

THE THIRD BOOK OF THE ARCHIDOXIES

From the Theophrastia of Paracelsus the Great

On the Separations of the Elements from Metals

FOR the separation of the elements from metals there is need of the best instruments, of labour, of diligence, together with experience of the art and adaptation of the hands to this work.

Take salt nitre, vitriol, and alum, in equal parts, which you will distil into aqua fortis. Pour this water again on its fæces, and distil it again in glass. In this aqua fortis clarify silver, and afterwards dissolve in it sal ammoniac. Having done this, take a metal reduced into thin plates in the same way, that is, in the same water. Afterwards separate it by the balneum Mariæ, pour it on again, and repeat this until there be found at the bottom an oil, namely, from the Sun, or gold, of a light red colour; of the Moon, a light blue; of Mars, red and very dark; of Mercury, white; of Saturn, livid and leaden; of Venus, bright green; and of Jupiter, yellow.

All metals are not thus reduced to an oil, except those which have been previously prepared. For instance, Mercury must be sublimated; Saturn calcined; Venus florified; Iron must be reduced to a crocus; Jupiter must be reverberated; but the Sun and Moon easily yield themselves.

After that the metals have been in this way reduced to a liquid substance, and have disposed themselves to a disunion of their elements—which cannot be done in a metallic nature, seeing that everything must be previously prepared for the use to which it is adapted—afterwards add to one part of this oil two parts of fresh aqua fortis, and when it is enclosed in glass of the best quality, set it in horse-dung for a month. After that, distil it entirely with a slow fire, that the matter may be condensed at the bottom. And if the aqua fortis which ascends be distilled by a bath in this manner, you will find two elements together. But the same elements will not be left by all metals alike. For from gold there remain in the bath earth and water; but air is in all the other three, and the element of fire remains at the bottom, because the substance and the tangibility of gold have been coagulated by the fire; therefore, the substance will agree in its substantiality. From the Moon there will remain at the bottom the element of water, and in the bath the elements of earth and fire. For from the cold and the moisture is produced the substance and corporality of the Moon, which is, indeed, of a fixed nature, and cannot be elevated. From Mercury there remains fire at the bottom, and earth and water are elevated upwards. From Venus there also remains fire, and both, that is to say, earth and water, remain in the bath. From Saturn there remains the element of earth at the bottom, while fire and water are held in the bath. From Jupiter air remains at the bottom, while fire, water, and earth are elevated therefrom.

So it must be noticed that in the case of Jupiter the air supplies a body, and in the case of no other metal. And of this, although some part ascends together with it and remains mixed inseparably with the other three elements, still it is not corporeal air, but adheres to, and concurs with, the others, and is inseparable from them.

And now, it must be remarked that the residuum, that is, the corporeal element which remained at the bottom, must be reduced into an oil by means of the bath with fresh aqua fortis. So, this element will be perfected, and you will keep it for one part. The rest you will separate by means of a bath in this way. Place them in sand, and press them gently. Then, first of all, the water will be elevated, and will escape; afterwards the fire, for it is known by the colour when these two remain. But, if the elements of earth and water should have remained, the water will ascend first, and afterwards the earth. But if it should be earth and fire, the earth is elevated first and the fire afterwards. If water, fire, and earth be together, the water will first ascend, then afterwards the fire, and last of all the earth. These elements can be so kept in their respective glasses, each according to its own nature; as, for example, from the Sun, the warm and the dry, without any other property; in like manner the cold and the moist, and the cold and the dry. So, also, must it be understood of the others. It must not be forgotten that the corrosive nature of the aquafortis must be extracted as we have handed down in our book on the Quintessence.

Concerning the Separations of the Elements out of Marchasites

Having previously set down the separations of the elements out of metals, it remains that we come to those which can be produced out of marchasites, and that we shew what they are.

Take of marchasite, in whatever form you please, whether bismuth or talc, or cobalt, granite, or any other, one pound; of salt nitre the same. Beat them very small, and, burning them together, distil by means of an alembic without a cucurbite, and keep whatever liquid ascends. But that which remains at the bottom, let this, when ground down, be resolved into a water with aqua fortis. Hereupon pour the water previously collected, and distil it into an oil, as before directed in the case of metals. By the same process, too, you shall separate the elements. So, the golden marchasite is to be understood as gold, the silver as the Moon, bismuth as lead, zinc as copper, talc as Jupiter, cobalt as iron. Let these directions suffice for the separation of marchasites in every kind.

Concerning the Separation of the Elements from Stones

The separation of the elements of stones or gems comes to be understood in the following way. Take a stone well ground, to which add twice the quantity of live sulphur, and, when all is well mixed, put it in a luted pot into an Athanor for four hours, so that the sulphur shall be entirely consumed. Afterwards, let what remains be washed from the dregs and the sulphur, and dried. Let the stony calx be also put into aqua fortis and proceeded with as we have already laid down concerning the metals. Stones, too, are compared with metals; as, for instance, clear gems which are not white or tawny are compared with gold. White, cœrulean, or grey, with silver or the Moon; and afterwards the commoner stones with the other metals, as alabaster with Saturn, marble with iron, flint with Jupiter; but dulech with Mercury.

Concerning the Separation of the Elements from Oleaginous Substances

All oils, woods, roots, seeds, fruits, and similar things, which have a combustible nature, and one fit for burning, are considered oleaginous; and the separation of them is twofold, namely, that of the oleaginous substances and that of the pure oils.

The Separation of the Oleaginous Substances is as follows:

Take such a body, pounded, ground, or reduced to fragments in whatever way you can, wrap it up in linen, fasten it, and place it in horse-dung until it shall be entirely putrefied, which happens sooner in one case than in another. When it is putrefied let it be placed in a cucurbite, and on it let there be poured so much common hot water as may exceed four fingers broad; then let there be distilled in sand all that can ascend. For all the elements ascend except the earth itself, which you will know by the colours; nevertheless, let the hot water first ascend, afterwards the air, next the water, lastly the fire, and the earth will remain at the bottom. Of the pure oils, however, it must be understood that these do not require putrefactions, but they must be distilled alone and without additions. Afterwards their elements, as it has been said above of others, must be separated, and these are discerned by their own colours. None otherwise with resins of liquid substance must it be done, such as pitch, resin, turpentine, gum, and the like. But the corporeal resins which exist, such as sulphur, must be prepared in the following manner.

Concerning the Separation of the Elements in Corporeal Resins

Take sulphur very minutely ground; let this be cooked to hepatic sulphur in a double quantity of linseed oil; let it be shut up in a vessel and place it to putrefy in horse-dung for a space of four weeks. Afterwards let it be distilled in an alembic slowly over a naked fire. The air and the water first ascend, with different and pale colours. Then, the heat being increased, the fire ascends and the earth remains at the bottom. The colours appear pure; the air yellow; the water like thick milk, so much so that it can scarcely be distinguished from milk; the fire like a burning ruby, with transparency and with all the fiery signs; but the earth is altogether black and burnt. The four elements having been thus separated, every one is perfect in its own elemental complexion, and without any admixture, as has been said above.

Concerning the Separation of the Elements from Herbs

So, too, in herbs, the element of water is principally contained when they are cold; but if they are airy, then that element predominates. In like manner must it be understood of fire. The separation of those elements is as follows:

Take sage, and bruise its leaves. After this place it for putrefaction, as aforesaid. Then you will distil it by means of a *venter equinus*, and the element of fire will ascend first so long as the colours are unchanged and the thickness of the water. Afterwards the earth will succeed, and some part of it will remain at the bottom, which part, indeed, is fixed. Distil this water in the sun six days, and afterwards place it in a bath. Then the element of the water will ascend first. It is very minute, and is distinguished by the taste. After the colour changes the element of fire ascends, until the taste, too, is altered. Then at last a part of the earth is elevated, yet it is a very small portion, which, being mixed with air, is found at the bottom. In the same way it must be understood of airy and watery herbs, of which the air ascends first, afterwards the water, and lastly follows the fire, according to the process laid down concerning sage.

Concerning the Separation of the Elements from Fleshly Substances

The separation of the elements from fleshly bodies, and from those which live with blood, comes to be understood thus, because the predominating element in them is more copious and is generally found last of all; as, for instance, water in fishes, fire in worms, and air in edible flesh are the principal elements, as we describe in our treatise on the generation of animals.*

The Separation of the Elements from Fishes is as follows:

Putrefy the fishes perfectly. Then distil by means of the *venter equinus*, and a good deal of water ascends. You will renew this putrefaction and distillation, and increase it until no more water rises. Afterwards distil what remains in sand; then at length the fire ascends in the form of oil, but the

---

* This treatise must also be included in the long catalogue of the missing works of Paracelsus. But many references to the generations of animals will be found in the present translation.

earth remains at the bottom. Thus the whole substance of the fishes is separated into its elements. One need take no account of the fats and the marrows, but must suppose that everything is separated by the putrefaction, and divided into its elements. In the same way is it to be understood of worms, except that there comes forth from them not only water but more fire, unless they are aquatic worms, as serpents; in the distillation of which more wonderful things occur than it is possible to say. Of edible animals, too, it must be understood in the same way, of such, that is, as respectively disclose their elements by separation.

Concerning the Separation of the Elements from Watery Substances

For the separation into their elements of juicy and watery bodies, and of those which have the form of wateriness, as urine, dung, water, and the like, note the following process:—

Take urine and thoroughly distil it. Water, air, and earth will ascend together, but the fire remains at the bottom. Afterwards mix all together and distil again four times after this manner; and at the fourth distillation the water will ascend first, then the air and the fire, but the earth remains at the bottom. Then take the air and the fire in a separate vessel, which put in a cold place, and there will be congealed certain icicles, which are the element of fire. Although this congelation will take place in the course of distillation, still it will do so more readily in the cold.

Concerning the Separation of the Elements out of Water

Make the water boil by means of a dung-heap, and the earth itself sinks to the bottom. Putrefy at the proper time that which ascends, and let it afterwards be distilled by a bath. Then the water will ascend first, and afterwards the fire. Dung, vitriol, tartar, and similar juices, as alum, salts, and other substances of that kind, are to be distilled in ashes, with such an amount of heat and for so long a time, until they cease to rise, and the water and air have ascended, while the earth has remained at the bottom. Afterwards, by means of the heat, the fire will ascend. And in this place it is to be remarked that although the four elements have been separated, there remain still in the earth four occult elements, as though fixed. From vitriol remains a *caput mortuum*, which sublimate with sal ammoniac, and there will issue forth an oil, in which are water and fire, and the earth itself remains in substance. Separate those things which have ascended, and again there will ascend water, while the fire will remain at the bottom. So also must it be understood of tartar and of salts. And although there are in existence additional separations of liquids, yet we shall discuss them more amply when speaking of Transmutations. It must, however, be remembered that there are more elements in a corrosive earth than in ashes. Therefore, the separation must be made by sublimation, as we shall shew.

Concerning the Separation of the Elements from Glasses and those Substances which are of the Nature of Glass

As we lay down above concerning the resolutions of marchasites, so is it to be in like manner understood in this place concerning the glasses: namely, the principal consideration is that they are calcined with sulphur as stones are, then washed away with saltpetre and aqua fortis, and furthermore, as we have before made clear. Their elements also are recognised by the colours in the distillations, not as they shew themselves to the eye; and this is the sum and substance of what we have to say concerning them.

Concerning the Separation of the Elements in Fixed Substances

The separation of the elements in fixed substances is brought about by sublimation, as we teach concerning salts and liquids; with this difference, however, that these are to be calcined with salt nitre, and afterwards to be sublimated. And although there are many other things which are not set down in this place, nevertheless it is to be understood that the separations of all substances should be made in the ten ways already described. Furthermore, concerning the separation of the four elements, it is to be remarked that each of them can be again separated; for example, fire as fire, air as air, water as water, earth as earth, as follows hereafter concerning the respective separations of them.

Concerning the Separation of Fire

It should be known that from the element of fire, the four elements may be separated in this way. When the fire is burning most violently, or ascends, take it in a receptacle or vessel perfectly closed, and place it in horse-dung for a month. Then you will find in that one element four elements, which, when you have opened the vessel, put into a receiver. Thus the vapour or air will mount into the vessel that receives it. Afterwards, distil that which remains by means of a bath, and the water will thus ascend. Next, by means of ashes, the fire will ascend and the earth will remain at the bottom. What is the force of these elements, and why are they described in this place, we will make more clear in other books.

Concerning the Separation of Air

Having received the element of air in a perfect glass vessel, and hermetically sealed it, you must expose and direct it to the sun for the whole of the summer. By circulating, the air is turned into moisture, which increases daily more and more. This quantity you will separate after the following manner. Let it putrefy in horse-dung for four weeks, and afterwards distil it by the bath, like fire. Concerning its potency more is said in another place.

Concerning the Separation of Water

Having filled a glass brimful, leaving no space empty, seal the vessel hermetically, and place it in a warm sun for a month, so that it may receive a daily and equal heat, and would boil, but it cannot on account of the vessel being full. When the time has elapsed, putrefy it for four weeks. Then open it, and distil it by means of an alembic with four necks. In this way the three elements are separated, and in the bottom will remain the earth of that water. The nature of this is said to possess much virtue in many cases.

Concerning the Separation of the Earth

The same process is to be observed with the earth as with the water, save only in the distillation. For this is like that which takes place with fire and is accomplished in the same way. This separation of the elements we have inserted at this point for several causes, because it is very useful, not only in philosophy, but also in medicine. Concerning the separations of the elements, we have thus far written with sufficient fulness. Though much more might be added, it does not appear to be by any means necessary.

Now, we will make clear the separation of the pure from the impure, according to the purpose of our design. This, indeed, is done in the same way as we teach with regard to Arcana and Aurum Potabile; so it need not be put forward here, though that process from its origin is not altogether identical with that which is laid down concerning Arcana and Magisteries. Nevertheless, I do finally assume the same way by the separation of the elements; since in this place those elements are separated after each one of them has been purged from the impurities existing therein, so that no deformity or impediment may arise from them, as might otherwise easily happen.

The End of the Third Book of the Archidoxies concerning the Separations of the Elements

THE FOURTH BOOK OF THE ARCHIDOXIES

From the Theophrastia of Paracelsus the Great

Concerning the Quintessence *

WE have before made mention of the quintessence which is in all things. Already, at the beginning of this treatise, it must be understood what this is. The quintessence, then, is a certain matter extracted from all things which Nature has produced, and from everything which has life corporeally in itself, a matter most subtly purged of all impurities and mortality, and separated from all the elements. From this it is evident that the quintessence is, so to say, a nature, a force, a virtue, and a medicine, once, indeed, shut up within things, but now free from any domicile and from all outward incorporation. The same is also the colour, the life, the properties of things. It is a spirit like the spirit of life, but with this difference, that the life-spirit of a thing is permanent, but that of man is mortal. Whence it may be inferred that the quintessence cannot be extracted from the flesh or the blood of man: for this reason, that the spirit of life, which is also the spirit of virtues, dies, and life exists in the soul, not in the material substance.

---

* *The Correction of the Quintessence.* The books which have been written by so many previous authors concerning the Quintessence, such as those of Arnoldus de Villa Nova and of Johannes de Rupescissa, whence afterwards, under a pretentious title, was composed the *Cælum Philosophorum* [not to be identified with the work of Paracelsus which occupies the first place in this translation], contain nothing of any value. The mere fact that these writings embody a singular and new praxis abundantly demonstrates that their authors have misunderstood the essential nature of diseases, seeing that they have, as it were, devised one form of all diseases, regarding which they have, moreover, invented many marvellous things, adorning their conceptions with monstrous titles, all mere boasting, wherein there is no mention of philosophy, medicine, or astronomy. They are all a mere deluge of absurdities and lies. There can be no doubt that originally most admirable discoveries have been transferred from the chemical art to that of medicine, but the same have been since adulterated by sophistry. For certain people, when they have investigated a chemical preparation, wish to vary it immediately in hundreds of different ways, and thus the truth is foolishly confounded with lies. Now, it should be observed that the severest rebuke which can be given to such impostures is that of paying no attention to promises and to proud titles, and of believing only to that extent which is warranted by good sense and experience. Remedies which require more knowledge of a practical kind than has ever been possessed by monastic pseudo-chemists may, in careless hands, give rise to the most malignant diseases. It may be observed, for example, that the preparations of Mercury which are used against Luis Gallica, and, indeed, all remedies adopted in the cure of this disease, cannot be properly prepared without great skill in chemistry. Pretenders in pharmacy will vainly vaunt such decoctions unless they can compose vitriolated salts, alums, and things similar, by purely chemical artifice, seeing that there are recondite secrets in the remedies which conduce to the cure of Luis Gallica. Yet there is no need to write a new correction; it is enough to adhere to the legitimate mode of preparation, giving no faith to the hollow pretences of alchemists. These men certainly promise more than they are ever able to perform; but this is common to alchemy and not a few other callings, namely, that their professors boast of the harvest before they have finished the sowing.—*Chirurgia Magna, De Imposturis in Morbo Gallico*, Lib. II., c. 13.

For the same reason, animals, too, because they lose their life-spirit, and on that account are altogether mortal, also exhibit no quintessence. For the quintessence is the spirit of a thing, which, indeed, cannot be extracted from things endowed with sensation as it can from those not so endowed. Balm has in itself the spirit of life, which exists as its virtue, as a force, and as a medicine ; and although it be separated from its root, nevertheless, the life and virtues are still in it, and for this cause, that its predestination was fixed. Wherefore the quintessence can be extracted from this, and can be preserved, with its life, without corruption, as being something eternal according to its predestination. If we could in this manner extract the life of our heart without corruption, we should be able to live without doubt, and without the perception of death and diseases. But this cannot be the case ; so from this circumstance death must be looked for by us.

When, therefore, the quintessence of things exists as a virtue, the first thing we have to say is in what form this virtue and medicine are in things, after the following manner. Wine contains in itself a great quintessence, by which it has wonderful effects, as is clear. Gall infused into water renders the whole bitter, though the gall is exceeded a hundredfold in quantity by the water. So the very smallest quantity of saffron tinges a vast body of water, and yet the whole of it is not saffron. Thus, in like manner, must it be laid down with regard to the quintessence, that its quantity is small in wood, in herbs, in stones, and other similar things, lurking there like a guest. The rest is pure natural body, of which we have written in our book on the Separations of the Elements. Nor must it be supposed that the quintessence exists as a fifth element beyond the other four, itself being an element. It is possible that someone may think this essence would be temperate, not cold, not warm, not moist or dry, but this is not the nature of its existence. For there is nothing which exists in this temperature by which it is alienated altogether from the other elements ; but all quintessences have a nature corresponding with the elements. The quintessence of gold corresponds with fire; that of the Moon with water; that of Saturn with earth ; and that of Mercury with air.

Now the fact that the quintessence cures all diseases does not arise from temperature, but from an innate property, namely, its great cleanliness and purity, by which, after a wonderful manner, it alters the body into its own purity, and entirely changes it. For as a spot or film, by which it was formerly blinded, is removed from the eye, so also the quintessence purifies the life for man. All natures are not necessarily of one and the same essence one with the other. Nor do those which are fiery on that account manifest the same operations by reason of their complexion ; as, for instance, if anyone should think that the quintessence of Anacardus should have a like and identical operation with the quintessence of gold, because both are fiery, he would be greatly misled, since the predestination and the disposition make a difference of properties. For as every animal contains within itself the life-spirit, yet the same virtue does not exist in each, simply because they all consist of flesh and blood, but one differs from another, as in taste or in virtue, so is it with the quintessence, which does not acquire its virtue from the elements by a simple intellectual process, but from a property existing in the elements, as we lay down in our book on the Generation of Things. Thence it happens that some quintessences are styptic, others narcotic, others attractive, others again somniferous, bitter, sweet, sharp, stupefactive, and some able to renew the body to youth, others to preserve it in health, some purgative, others causing constipation, and so on. Their virtues are innumerable, and although they are not exhausted here, yet they ought to be thoroughly known by physicians.

When, therefore, the quintessence is separated from that which is not the quintessence, as the soul from its body, and itself is taken into the body, what infirmity is able to withstand this so noble, pure, and powerful nature, or to take away our life, save death, which being predestined separates our soul and body, as we teach in our treatise on Life and Death? In this place it should be equally remarked that each disease requires its own special quintessence, though we tell of some which are adapted to all diseases. How this comes about shall be explained in its proper place.

Furthermore, we bear witness that the quintessence of gold exists in very small quantity, and what remains is a leprous body wherein is no sweetness or sourness, and no virtue or power remains save a mixture of the four elements. And this secret ought by no means to escape us, namely, that the elements of themselves, without the quintessence, cannot resist any disease, but are able to effect only this and nothing more, that is, to produce heat or cold without any force: so that if a disease be hot, it is expelled by cold, but not by that kind of cold which is destitute of force, or things made frigid with snow; since, though, these are sufficiently cold, yet there is no quintessence contained in them by the power whereof the disease might be expelled. Wherefore the body of gold is powerless of itself; but its quintessence alone, existing in that body, and also in its elements, supplies the forces hidden therein. So, likewise, in all other things, it is their quintessence alone that cures, heals, and tinges the whole body, just as salt is the best seasoner of any food. The quintessence, therefore, is that which gives colour, whatever it be, and virtue; and gold when it has lost its colour, at the same time lacks its quintessence. None otherwise must it be understood of metals from which when the colour is taken, they are deprived of their special nature.

In like manner is it with stones and gems; as, for instance, the quintessence of corals is a certain fatness with a red tint, while the body of them is white. The quintessence of the emerald, too, is a green juice, and the body thereof is also white. None otherwise must we judge of all other stones, namely, that they lose their nature, essence, and qualities when they lose their colours, as we particularly shew in our book on the extractions of them.

The same, also, should be understood of herbs, plants, and other growing products. So, too, of flesh and blood, from which no quintessence can be extracted, for reasons already laid down. But, nevertheless, a certain resemblance to a quintessence can be extracted from them by us in the following manner. A morsel of flesh still retains life in itself, because the flesh is yet supplied with all its nature and force. Wherefore there is life in it, which, however, is not true life, but still the life is preserved until putrefaction sets in. So this difference must be noted, whereby dried herbs and the like are to be looked upon in the same way as flesh. That green spirit which is their life has gone from them. Dead things, therefore, can be taken for a dead quintessence, as flesh is able to put forth all its powers from itself, though specially separated as to one part from the body. So, too, with blood and with dried herbs. These, indeed, though they are not living quintessences, none the less demonstrate how a dead quintessence displays some virtues. But metals and stones have in them a perpetual life and essence, and they do not die, but so long as they are metals or stones, so long does their life last. Therefore, they also exhibit perfect quintessences which, in like manner, can be extracted from them.

And now we must see by what method the quintessence is to be extracted. There are many ways indeed: some by additions, as the spirit of wine; others by balsamites; some by separations of the elements, and many other processes which we do not here particularise.

But by whatsoever method it takes place, the quintessence should not be extracted by the mixture or the addition of incongruous matters; but the element of the quintessence must be extracted from a separated body, and, in like manner, by that separated body which is extracted. Different methods are found by which the quintessence may be extracted, for instance, by sublimation, by calcination, by strong waters, by corrosives, by sweet things, by sour, and so on, in whatever way it may be possible. And here this is to be taken care of, that everything which shall have been mixed with the quintessence by the necessity of extraction, must again be drawn off from it, so that the quintessence may remain alone, unpolluted, and unmixed with any other things. For it cannot be that the quintessence shall be extracted from metals, more particularly from gold, which cannot be overcome by itself, without exhibiting some appropriate corrosive, which can afterwards again be separated from it. In this way, salt, which was water, is again extracted from water, so that this water is free from salt. And here this consideration comes in, that it is not every corrosive which is adapted for this purpose, because they cannot all of them be separated; for if vitriol or alum be mixed with the water, neither of these can be separated from it afterwards without loss or corruption, so that they leave behind them a sharp residuum, for this reason, because each of them is watery, and thus two similar things meet, which ought not to occur in this process. Care must therefore be taken that a watery body be not taken for a watery, or an oily body for an oily, or a resinous body for a resinous one; but a contrary ought in every case to separate the quintessence and to extract it, as waters extract the quintessence of oily bodies (which is explained in the case of the metals), and oily substances the quintessence of watery bodies, as we may learn concerning the quintessences of herbs. The corrosives, therefore, after the separation and extraction of the quintessence, must be again separated, and this will be easily done. For oil and water are readily separated, but not so oil from oil; nor, in like manner, can water be separated from water without admixture; and if this is left it may cause great damage to the quintessence. For the quintessence ought to be clear and spotless, and collected without the admixture of anything, that it may possess a uniform substance by means of which it can penetrate the whole body. In truth its subtlety and force cannot be probed to the foundation, no more than its origin whence it first proceeded can be fully known. It has many grades: one against fevers, as in the case of opiates; another against dropsy, as the essence of tartar; one against apoplexy, as that of gold; one against epilepsy, as that of vitriol. The number of these is infinite, and incapable of being proved by experiment. Wherefore the greatest care and diligence should be shewn that to every disease its true enemy may be assigned. In this way Nature will afford help beyond belief, as will be made more clear in what follows. We cannot speak of the grades in the same way as grades are applied to simples in medicine, for this reason that there is no possible comparison between the grades of a quintessence and the grades of simples, nor ought the comparison to be attempted; but when such gradation is made it is found that the excellence and virtue of one is greater than of another, but not the complexion. For it must not be set down that the quintessence of Anthos is hotter than the quintessence of lavender, or the quintessence of Venus is drier than that of the Moon; but the grade of anything should be determined by its great and more excellent virtues, namely, after this manner. The quintessence of antimony cures leprosy, and the quintessence of corals cures spasms and contortions. In order to learn, then, which of these occupies the better and higher grade, there can be no other judgment than this, namely, that the quintessence of antimony is higher and more potent, inasmuch as leprosy is a more severe disease than colic and its belongings. According to the property, therefore, which it has against different diseases its grades are considered. This is, moreover, the case in one and the same disease; for one essence is more powerful than another for curing leprosy. The quintessence of juniper expels it, and the quintessence of amber, the quintessence of antimony, and the quintessence of gold. Now, although there are these four essences in all which cure leprosy, they do it with a difference, since with regard to the cure thereof they do not occupy the same grade. For the essence of juniper drives away this disease by the extreme purgation and purification which it introduces into the blood, and so consumes the poison that it is not so perceptible. Hence it is reckoned in the first grade of that cure. The quintessence of amber also takes away the poison and more. It purifies the lungs, the heart, and the members subject to leprosy; wherefore, the second grade is assigned to it. The quintessence of antimony, beyond both the virtues already spoken of, also clears the skin, and sharpens and renovates the whole body in a wonderful manner; so, then, it holds the third rank. But the quintessence of the Sun by itself fulfils each of these tasks, and then takes away from the roots all the symptoms of leprosy, and renovates the body as honey and wax are purged and purified by their honeycomb. For this cause it occupies the fourth rank.

In this way the grades of the quintessence can be learnt, and the one distinguished from the other, that is to say, which of them is higher or more excellent than another. Even simples should be known by their properties. For whatever be the property in a simple form, such is the property of their quintessence, not more sluggish, but much stronger and more excellent.

But now, moreover, let us learn the differences of quintessences, for some are of great service to the liver by resisting all its diseases, some to the head, others to the reins, some to the lungs, and some to the spleen, and so on. So, too, some operate only on the blood, others on the phlegm, others in melancholy only, others only in cholera, while some others have effect only on the humours, some on the life-spirit, some on the nutritive spirit; some operate on the bones, some others on the flesh, some on the marrow, others on the cartilages, some on the arteries; and there are others which have effect only against certain diseases and against no others, as against paralysis, the falling sickness, contractions, fluxes, dropsy, and so on. Some, also, are found to be narcotics, others anodynes, some soporific, some attractive, purgative, cleansing, flesh-making, strengthening, regenerative, and some stupefying, and the rest.

Some, too, are found which renovate and restore, that is, they transmute the body, the blood, and the flesh. Some are for preserving the continuance of life, some for retaining and preserving youth; some by means of transmutation, others by quickening. Moreover, this must be understood, that some have a specific form, others an appropriated, others an influential, and others a natural form. In a word, more of their virtues exist than we are able to describe, and their effects in medicine are most wonderful and inscrutable. In different ways it happens that some quintessences render a man who is a hundred years old like one in his twentieth year, and this by their own strength and potency. But what man is able to trace the origin of so great a mystery, or to ascertain from whence the first materials naturally take their rise? It belongs to our Supreme Creator to make these things so, or to forbear. For who shall teach us, that we may know by what powers the quintessence of antimony throws off the old hair and makes new hair grow; why the quintessence of balm destroys the teeth, eradicates the nails of the hands and the feet, and restores new ones; the quintessence of rebis strips off and renews the skin; and the quintessence of celandine changes the body and renovates it for the better, as colours renovate a picture?

There are far more matters than these, which here we omit, and reserve for making clear in their proper chapters.

How, then, at length could it be that we should relinquish that noble philosophy and medicine, when Nature affords us such wonderful experiments in and from them? Of these all the other faculties are destitute, and so take their position nakedly in mere cavil. Why should not this fact delight us that the quintessence of the carline thistle takes force away from one and affords it to another who uses it; also that the quintessence of gold turns inside out the whole leper, washes him as an intestine is washed in the shambles, and likewise polishes the skin and the scabs and makes a new skin, loosens the organs of the voice, takes away the whole leprous complexion, and makes him as if he had recently been born of his mother.

Wherefore we will turn our mind to the making of such quintessences, pointing out the way for the extraction or composition of them—one for the metals, another for marchasites, another for salts, another for stones and gems, another for burning things, another for growing things, another for spices, another for eatable and drinkable things, all of which, with their belongings, we will endeavour to make clear in a proper series. But it must be noted, in the practice with quintessences, a good knowledge of theory and of natural science is required; theoretically, that is, of the properties of things with regard to natural diseases. One must not be ignorant that there is a difference between a quintessence, aurum potabile, arcana, magisteries, and other things of that kind; such as this, for example, that a quintessence cannot be again reduced to its own body, but aurum potabile is easily transmuted again to its metallic body; so that far more noble virtues exist in a quintessence than in the other things.*

---

* The distinction which subsists between the quintessence of gold and potable gold is illustrated by the following citation which, though it occurs in an independent treatise, offers only some small variations from a passage in the *Chirurgia Magna* which has been already given in the note on page 76 of the first volume:—We have already said that it is scarcely possible for contractions to be cured in any other way than by medicines existing in the supreme grades, such as is potable gold and the like, concerning which we have treated in the larger grades, whence great care must be taken concerning them, as is stated in the Book about the Quintessence. It is called potable gold so often as it is reduced, together with other spirits and liquors, into a substance which can be drunk. The oil of gold is a golden oil made out of the substance only, saving addition. It is called quintessence of gold when a reddish tincture is extracted therefrom and separated from its body. A virtue, or at least an active force, exists in the tincture. The dose of potable gold is ℈ i., when needed. The dose of golden oil should not exceed ten grains of barley in weight. That of the quintessence should not exceed three similar grains in good water of life, or some other water of equal subtlety. It should be taken morning, noon, and evening, according to the requirements of the medicine, and without the addition of corrosives or corruptives, which can neither alter its nature nor be mixed therewith.—*Description of Potable Gold.* Take of potable gold, pulverized and dissolved in salt, ℥ i., with a sufficient quantity of distilled vinegar. Perform successive separations upon the whole by means of distillation till nothing of the acquired savour shall remain. Then take water of life ℥ v., pour them into a pelican, and digest together for a month, when you will have perfect potable gold, the practice of which you must learn from our book on the quintessence. Though it has not been described in glowing colours there is no equal medicine found in this age.—*Description of the Water of Life.* Take 10lbs. of *vinum ardens*; of roses, balm, rosemary, anthos, cheirus, both species of hellebore, marjoram, ana m. j.; of cinnamon, mace, nutmegs, garyophylli, grains of paradise, all peppers, cubebæ, ana ℥ ii.; of the sap of chelidonia, tapsus, balm, ana lib. ss.; of bean ashes ℥ v. Let all these be mixed and digested together in a pelican for twelve days; then separate, and use for above process.—*Description of the Oil of the Sun.* After the sap has been separated from the gold in the way previously stated, let it boil for fifteen days in the digestive compound which follows. Let it afterwards be separated by the bath, when a thick oil will remain at the bottom, and this is unalloyed gold, which use as above.—*Formula of the Quintessence.* Take as much as you please of gold which has been repurged by royal cement or antimony. Remove its metallic quality, or malleation, by means of the water of salt. Wash away the residue with sweet water. Extract its tincture with spirit of wine. Lastly, elevate the spirit from it

While we thus speak of the quintessence, the difference of one from another should be learnt, and then also what it is in itself. And although we have already sufficiently explained it, nevertheless, practice calls us another way by which also the condition and nature of the quintessence can be found out. For although these do not appear in the form of the quintessence, nor are they produced in the same way, nor do they consist of one element only as the quintessence should, still, none the less, we should judge of the quintessence of these things, which is of more importance than that they should be called a quintessence. It should rather be spoken of as a certain secret and mystery, concerning which more should be written than we have written concerning the quintessence. But since we have made that clear in the books of the Paramirum, we pass it by in this place.* The number of the arcana and of the arts of the mysteries is infinite and inscrutable, and many methods in them are met with worthy the attention of the clearest human intellects. Among these arcana, nevertheless, we here put forward four. Of these arcana the first is the mercury of life, the second is the primal matter, the third is the Philosophers' Stone, and the fourth the tincture. But although these arcana are rather angelical than human to speak of, nevertheless, we shall not shrink from them, but rather we will endeavour to trace out the ways of Nature, and we will arrange that everything which proceeds from Nature

---

and the quintessence will remain at the bottom.—*Construction of the Water of Salt.* Take by itself the purest and whitest pounded salt, which is produced by nature without decoction, boiling, or any of those processes by which salt is usually made. Liquefy it several times, pound it very fine, mix it with the juice of raphanum roots, dissolve and distil it, and when a reddish green appears again distil it five times, combining it in equal weights. Dissolve laminated Sol in this liquor till it becomes powder. Let this powder be washed in most limpid water, and distil a sufficient number of times till the salt shall depart from it, which will take place soon, as it does not penetrate the interior nature. When the corrosive has been removed the gold will be found by itself.—*Extraction of the Spirit of Wine.* Take one measure of the best natural wine, red in preference to white, place in a capacious circulatory vessel for its better rotation, seal it, plunge it into the sea-bath, and let it boil for forty days. Afterwards pour it into a cucurbite, distil by the cold way till the spirit shall have gone and all signs of it shall cease. So finish. (The process varies from that of the *Chirurgia Magna.*) After the sign has been given cease. *What follows is the water of life, not spirit; either is efficacious.* Then the sign is double, one of the spirit and one of the water of life. Pour out this spirit of wine to the dregs, and in such a way that it floats over the surface to the height of six broad fingers. Close all openings completely with glass. Digest for thirty days, during which time the tincture enters the spirit. At the bottom there remains a white powder. Separate by art, and let the powder melt. Hence there is produced an aqueous metal or metallic water. Let the spirit evaporate, as we are taught by alchemy, and a sap resembling liquor remains at the bottom. Graduate this four times in a retort adapted to the quantity of the matter. This is performed by elevation, for it renders such like substances subtle, though it permits not the same to be developed beyond the fifth essence.—*De Contracturis,* Tract II., c. 2. Among other uses of potable gold, it is a tonic for the heart which is so efficacious that it is affirmed by Paracelsus to prevent all injury befalling that organ. Of like virtue is the liquor of gold and the substance of pearls reduced into the form of an oil and balsam. After these are enumerated the essence from the crocus of Mars and corals. The description of the potable gold in this connection is given as follows: Let the gold be calcined into yellowness by the royal cement of *Hell and Malch.* Then let it be separated from its impurity, and afterwards mixed with circulated water. Digest for twenty-four hours in a moderate fire, when the oil will flow forth and will float upon the surface of the water. Collect and drink it mixed with water of life. Proceed in the same way with pearls, only adding calx of chelidonia, and confine by means of distilled vinegar until they pass off into liquor. Proceed also in like manner with corals, dissolving them in vinum ardens mixed with *Hell.* Remove the vinum ardens from the putrefaction and you will have the liquor of coral. There are also other essences, as of crocus, chelidonia, mace, cesium, balm, etc., which are suitable for other complaints and affections of the heart, and some of which will fill the old, infirm, melancholy, and depressed with the greatest joy.—*De Viribus Membrorum,* Lib. II., c. 2.

* The reference here made is too obscure for verification. In the large literature of the Paramirum and its connected treatises, there are many observations upon arcana and mysteries, though few passages deal actually with the quintessence. Whatsoever is of importance has been embodied in this translation as notes to the text. That section of the present volume which is devoted to Alchemy as the fourth column of medicine may also be consulted in this connection, as it is a treatise derived from the Paramirum.

shall be capable of being naturally understood. Concerning the Mercury of Life, therefore, we profess that it is not a quintessence, but an arcanum, because there exist in it may virtues and forces which preserve, restore, and regenerate, as we write in our book on the Arcana.

In like manner, also, the Primal Matter, not only in living things, but also in dead bodies, operates more in the same manner than is thought to be naturally possible. So also the Philosophers' Stone acts, which, tinging the body, frees it from all diseases, so that even metals are purged from their impurities. In like manner the Tincture acts, which, as though it should change the Moon into the Sun, so also changes disease into health. The same things equally the other magisteries and elixirs do, and the aurum potabile, all of which are treated of severally in their own respective books.

Concerning the Extraction of the Quintessence from Metals

We will, then, briefly go through the extraction of the quintessences from metals. For, in our times, many persons have made numerous experiments with these, and copious results followed which obliged them to enter on other different ways. With regard to metals, then, it must be understood that they are divided into two parts, namely, into their quintessence and into their body. Both are liquid and potable, and do not mix, but the impure body ejects the quintessence to its surface, like the cream from the milk. In this way two fatnesses or viscous liquids are formed out of the metals, and these liquids have to be separated. The fatness of the body is always white in all metals; but their quintessence is coloured, as we have before explained concerning the seven metals. All, moreover, have the same process, which is as follows :—

Let the metal be dissolved in water, and afterwards this solution distilled by a bath and drawn off. Let it be putrefied until it is reduced to an oil. Let this oil be distilled from small phials or cucurbites by means of an alembic, and one part of the metal will remain at the bottom. Let this be reduced to an oil as before, and be distilled, until all the metal shall have ascended. Afterwards let it be again putrefied for a month, and at length again distilled with a slow fire. Then the vapours will at first ascend and afterwards fall into the receiver. These vapours remove; and there will ascend two obscure colours, one white, but the other according to the nature and condition of the metal. When they have ascended altogether they become separated in the receiver, so that the quintessence remains at the bottom and the white colour of the body floats at the top. Separate these two by means of a tritorium, and in another phial receive the quintessence, into which pour purified ardent wine and let it remain until the wine is completely acidulated. Afterwards let it be strained or separated from the quintessence, and let more fresh wine be poured on. Do this until you no longer perceive any sharpness. At last, pour on doubly distilled water, so that it may be washed and brought to its proper sweetness, and so keep it. In this way the quintessence of metals is prepared. But if you reduce the white portion, you will have therefrom a malleable, white, and metallic body of which it cannot be known under what species it is embraced. There are many other ways found out for extracting the quintessence. About these we keep silence, for the reason that they are not considered by us to be true extractions of quintessences, but only transmutations in which no extraction is produced or comes to be used.

Concerning the Extraction of the Quintessence from Marchasites

In marchasites also are found various methods for the extraction of their quintessence, and yet we scarcely judge these to be true quintessences. And although they are of greater virtue than their quintessence, as we teach of Arcana, Magisteries, and Elixirs, nevertheless, this our mode and manner of extracting the quintessence from all the metallic marchasites is like the true extractions of the metals. But the following is the reason why we before said that the quintessence is the supreme virtue of things, and now, on the contrary, do here say of Arcana that they are greater than the quintessences themselves : because all the Arcana contain in themselves the quintessences and, moreover, are reduced to such subtlety and acuity that they hence receive a far greater virtue than the quintessence does. The same is laid down, too, concerning their appropriate and specific quality. But the process for extracting the quintessence from marchasites is as follows :—

Take a pound of a marchasite very finely ground, of the eating water two pounds, mix them together in a pelican, let them remain in process of digestion for two or three months, and be reduced to a liquid. Distil this entirely with the fire and it will pass over into an oil which you will putrefy in the dungheap for a month. Afterwards distil it like metals, and in the same way two colours will ascend therefrom, one white, and the other the true colour of the quintessence. Leave the white, unless it be from bismuth or from a white marchasite, in which case you shall know one from the other by the density. Take the lower one, and reduce it to its sweetness as was said above concerning the metals. In this way you have extracted the quintessence from marchasites without any corruption of their powers or virtues.

Concerning the Extraction of the Quintessence from Salts

The method of extracting the quintessence from salts is brought about in a somewhat peculiar way, so that their force may not be diminished, and is as follows :—

Take salts, which you will calcine perfectly ; and if they be volatile you will burn them. Afterwards let them be resolved into tenuity and distilled into water. Put this water in putrefaction for a month, and distil it by means of a bath, when the sweet water will ascend, which throw away. That which has not ascended put again in digestion for another month, and distil as before, repeating the process until no sweetness is any longer perceptible. By this means you now have the quintessence of the salt at the bottom, scarcely two ounces in weight out of a pound of the calcined or burnt salt. Of this salt, thus extracted, though it shall be only common salt, an ounce and a half seasons food more than a pound and a half of the other. For only its quintessence is present, and the body is taken away from it by means of the liquid solution.

In this way is separated the quintessence of all salts; but it is extracted in another way from alum and from vitriol, as follows:—

These will not allow of their being calcined into flux like salts. Therefore, after their calcination it is necessary to burn them, and resolve them according to the usual method. After they have been resolved, pour on them again the waters which have proceeded from them, and go on according to the method prescribed in the case of salts. For much of the essence ascends with the moisture, which again subsides at the bottom in process of composition and putrefaction, and so they meet together in one.

Concerning the Extraction of the Quintessence from Stones, from Gems, and from Pearls

The method of extracting the quintessence out of stones, gems, and pearls, with all of which the process is the same, is of all others the most excellent, and in its operation is very subtle and ingenious. A very small quantity of this quintessence is to be obtained from gems, and the more subtle and pure the gem is, the more minute is the essence. It is scarcely worth the trouble to extract the essence from dense, great, and cheap stones, since little virtue exists in them, whence it happens that very little comes forth from them. The process is of the following kind:—

Take gems, margarites, or pearls, pound them into somewhat large fragments, not into powder, put them into a glass, and pour on them so much radicated vinegar as will exceed the breadth of four or five fingers. Let them be digested for an entire month in a dung-heap, and when this is over the whole substance will appear as a liquid. This you will lighten with other radicated vinegar, and by shaking mix them together. The vinegar then acquires the colour of the stone. Pour the coloured liquid into another glass, again pour on vinegar as before, until the whole has no longer any colour. In that colour is contained the quintessence; the residuum is the body. Then take the colours, suffer them to be cooked to dryness, and afterwards wash often with distilled water until all becomes sweet as above. At length let this dust be dissolved on marble. In this manner you will have the quintessence of the gems and pearls. It must be remarked, however, in the colour of pearls, that they themselves are resolved into the colour of thick milk, and their body is sandy and viscous. In like manner is it to be laid down about the crystal. Its quintessence comes to the top, a certain viscous body remaining, by which the success of this kind of extractions may be known.

Concerning the Extraction of the Quintessence from Burning Things

Those things are called burning which are used neither for food nor for drink, and which of their own nature burn and keep alive the fire in the bodily substance. The method of extracting the quintessence from these is as follows :—Take such a body, very finely pounded, place it in a glazed pot, until it be full, seal it with the seal of wisdom so that it shall not breathe forth, and burn it in a circulatory fire for twenty-four hours, so that it shall remain at an even temperature, while the pot glows like the coals. Then take it out of the fire, let it putrefy in dung for four weeks, and afterwards distil it. Whatever ascends, let this be placed in a *venter equinus*, in order to distil all the moisture from it, and again set it to putrefy until no moisture any longer issues from it ; and then at length the quintessence of the body which thou hadst taken remains at the bottom. In this manner the quintessence is extracted from all things which contain oil in them, or resin, or pitch, or anything of that kind, as out of turpentine, fir, juniper, cypress, and the rest ; and in like manner out of all seeds, fruits, and similar things.

But it must be remarked that many more methods of extracting the quintessence from these are handed down elsewhere, ways and modes by which it comes forth quite odoriferous, subtle, and clear ; but these methods are not extractions of the quintessence; they are rather certain magisteries of these things, by which certain portions of the quintessence mount up at the same time in the process of mixing. They are not, however, perfect quintessences. For the essence of woods is a certain fatness or resin and a thick substance, whence it is not extracted in the form of magisteries. And the cause is this. The quintessence of the turpentine tree heals wounds, but when, in the above-mentioned way, it has been extracted from other magisteries, it does not heal wounds, because it has not in itself the fundamental power of the quintessence. The magisteries, indeed, are separated from the quintessences on this principle, that they only concern the complexions and the four elements, which, however, is not the nature of quintessences. Moreover, these magisteries receive it spiritually and not materially in its proper essence, as is clear from the chapters.

Concerning the Extraction of the Quintessence out of Growing Things

Those things are called growing in this place which fall and grow again, as herbs, leaves, and the like. For extracting the quintessences of these, several methods also have been found out by the addition of other things. But they should be extracted without the admixture of anything, so that they may retain their taste, colour, and odour ; and that these properties may be increased in them, not diminished. Thus, if the quintessence be extracted from musk, amber, and civet, their body afterwards stinks, so that they are no longer of any account in taste, in odour, or in nature. So of other things in this class it must be understood, so far as relates to the extraction of their quintessences. But we do not discuss musk, amber, and civet in this chapter, because we write specially of them elsewhere, and at present we are only treating of growing things, as the lily, spike, leaves ; for the extraction of the quintessence from which the following is the process.

Take growing things, bruised as completely as possible. Put them in some fitting vessel, and set this in a *venter equinus* for four weeks. Afterwards distil them by means of a bath; again let them be placed in horse-dung for eight days, and once more distilled by the bath of Mary.

Thus the quintessence will ascend by the alembic, but the body will remain at the bottom. If any of the quintessence shall have remained at the bottom, putrefy it still further, and proceed as before. Then take at length this distilled water, add it again to this growing thing, and so, by means of a pelican, let them be digested together six days. Thereupon the colour will be dense. Abstract this by means of the *balneum maris*, when the body will disappear, the quintessence remaining at the bottom. Separate this by a retortive process, that is, by pressure, from the dregs, and let this quintessence digest for four days. In this way you will have it perfect in odour, in juiciness, in taste, and in virtue, as well as consisting of a thick substance.

Concerning the Extraction of the Quintessence from Spices

We will now teach the method of extracting the quintessence from spices, as musk, civet, camphor, and the like. First the quintessence ought to be reduced to another form, and at length to be separated therefrom. In that same process of separation the quintessence is found, as follows:—

Take oil of almonds, with which let an aromatic body be mixed, and let them be digested together in a glass vessel in the sun, for the proper time, until they are reduced to a paste. Afterwards let them be pressed out from their dregs. In this way the body is separated from the quintessence, which is thus mixed with oil, from which it is separated in the following way:—

Take rectified ardent wine, into which let the aforesaid oil be poured, and let them be left in process of digestion for six days. Afterwards let them be distilled by ashes. The ardent wine will ascend, and the quintessence with it. The oil will remain at the bottom without any of the quintessence remaining. Afterwards let this wine be distilled by the *balneum maris*, and the quintessence will remain at the bottom in the form of an oil distinct from all similar ones.

Concerning the Extraction of the Quintessence from Eatables and Drinkables

The quintessence of food ought to be none otherwise than in a form similar to that wherewith we should be fed, namely, flesh. And although, as we said before, no quintessence can be extracted from flesh, nevertheless we are easily able to extract from it, so far as it is its own, that which is equivalent to the quintessence, as follows:

Take some eatable thing, cut up, put it into a vessel or jug, carefully luted, and suffer it to boil for three days. Then strain off what is in the pot, and distil it by means of a bath. Thus there will ascend first of all a kind of wateriness, and when this has entirely passed over, the quintessence will be left at the bottom. This is the chief nutriment beyond all others which we could put down or describe. And in respect of its nutritive power it is equal to a quintessence. The quintessence may also be extracted out of drinks in various ways; but this is the true process which we consider chiefly useful and convenient for the work in hand:

Take anything drinkable, enclose it in a pelican, just as it is, with its whole substance. Let it digest for a month in horse-dung. It will be still better to let it stay a year or more, and you will find in the pelican a certain something digested. Separate this by means of the bath, afterwards by ashes, and lastly by fire. In this way you will have three quintessences, which, in like manner, are in all drinks, for many reasons, which we enumerate in our treatise on Generations.

These three quintessences having been extracted, and each put into its proper vessel, the two latter should be further digested and then placed in the *balneum Mariæ*, when there will ascend more of the previous quintessence. Keep on doing this until no more of the former ascends, and in this way they are separated as completely as possible.

What we have so briefly taught about the quintessence of all things, and the short method of its extraction, ought not to rouse wonder in any at the rapid course of our hand and pen. For they are all well and completely handed down, and not so succinctly is the quintessence written of by us but that the work and labour necessary for it are clearly demonstrated. What need is there of much writing, which shall only nauseate ourselves and our readers, if we do not take into consideration that exercise and experience teach everything?

But how wonderful are its virtues and powers if it be extracted in the aforesaid manner we have even now partly taught, and shall make more clear in the last books that which belongs to this part of the Archidoxies; and thus we shall have fully described the quintessence of all things. And although many before us have, in various documents, written great things about quintessences, still we do not think of their writings as a quintessence, the cause whereof we have already sufficiently adduced.

We have even learnt from them that verdigris was accounted the quintessence of Venus, when it is not so, but the crocus of Venus is the quintessence of Venus, which is thus to be understood. The flower of brass is a transmutation with the substance, at once dense and subtle, and extracted from every complexion of Venus. Wherefore it can be no quintessence, but the crocus of Venus, as we have taught, is the true quintessence: for it is a potable thing divided from the body without corrosion or admixture, very subtle, even more so than one cares to write in this place, for the sake of avoiding prolixity. So also, the crocus of Mars and its rust have hitherto been considered a quintessence, but it is not so. But the crocus of Mars is the oil of Mars. Concerning these things we set down more about transmutations in our philosophy.

A quintessence, therefore, is thus to be understood, namely, that it is nothing else than a certain separation of virtues from the body, wherein exists the whole virtue and essence of medicine. But what are the flower of brass, and the flower of Mars, and many similar matters, is handed down in the treatise on Magisteries.

None otherwise must it be judged of vegetables and herbs and such things than concerning the metals. And although we have put forth lofty and numerous virtues of the quintessence, nevertheless, only the smallest part of their forces and qualities has been told. But we have principally made it clear how these things are to be understood ; yet least of all have we been able to tell what and how great are their powers and virtues. From this may be hinted how great is the power which we have in our hands, only let us know how to use it well. Hence, also, is made clear the cause why man was created, and all things on earth were made subject to him ; and also why it is that nothing good or ill happens without a cause, which we set forth more clearly in our book on the Nature of Things.* For that foundation brings with it a faith fixed on the Creator, and a hope of His love towards us, as of an excellent father for his children. So, then, we must not snatch at any shadowy and vain faith ; but it is right that we regard only God and Nature, and the Art of Nature. Wherefore, with good reason, we call only on Him in this life and for evermore, and believe that only which we see to be so, receiving or approving nothing that does not agree with Nature, or which is beyond Nature.

THE END OF THE FOURTH BOOK OF THE ARCHIDOXIES ON THE QUINTESSENCE

---

* In the first book of the treatise entitled *Philosophia Sagax*, man is considered as the quintessence of the macrocosm. This point is frequently a subject of consideration in the transcendental physics of Paracelsus, and some reference has been made to it already in the annotations of the first volume. Man is a certain extract of the whole machine of the world derived no otherwise than the physician extracts the strength and essence from a herb, the result of which process is called a quintessence. That which the despoiled body is to the extracted virtue, so is the world to man. By so much as the body is weaker the quintessence is more efficacious ; the more is extracted the less remains in the original matter. In the case of the macrocosm, however, there has not been so exhaustive an extract as is performed by medicine. That only has been taken which was needed for man, and what remains is sufficient for his nourishment. Now, concerning Adam it must be known that he was made man in the image of God ; his wife, Eve, was made, and not born, out of Adam. It was not the will of God to make a double extract from the world, but one only, which is the quintessence of the Microcosm. He extracted, therefore, the man, not the woman. . . . Man is generated from putrefaction. Putrefaction takes place through the operation of the macrocosm, through the elements, and through the stars, in the father and mother, as by instruments which Nature has bestowed. Not that the exterior world works here, but the microcosm through the quintessence. . . . From the father and mother no intellect, or sapience, is born, but only out of the firmament, by the operation of the quintessence and the microcosmic virtue. . . . The quintessence which is made in man is retained, being ordained for seed, whence children are born. That sperm which is the quintessence retains the nature, essence, and property of the mass and clay of the earth.

THE FIFTH BOOK OF THE ARCHIDOXIES

From the Theophrastia of Paracelsus the Great

Concerning Arcana

HAVING treated of quintessences, we will now turn our attention in due course to write about Arcana, since we understand more about them than about the forces of quintessences; wherein experience teaches us that there is the greatest difference on account of the very powerful operations whereby are demonstrated to us by most evident signs what things are better, more useful, or inferior in their powers. In this way we are able to avail ourselves of one or the other, according to their usefulness in medicine. Indeed, the ancients often thought that the arcana were quintessences, since they saw that they were much more subtle than dense substances, and knew that they operated in a wonderful manner through their subtle nature. But this error of theirs arose, not from reason, but rather had its origin in a lack of practice; since there was among them no knowledge as to the determinate difference of the high degrees, but they esteemed every higher and highest degree as quintessences. This difference, however, ought to be known and defined, not by practice alone, but also, and rather, by the operations of medicine.

Before, then, we treat of arcana, we must see and know why they are so called, and what an arcanum is, since it has so excellent a name, and well deserves to have it, too. That is called an arcanum, then, which is incorporeal, immortal, of perpetual life, intelligible above all Nature and of knowledge more than human.* Compared, indeed, with our corporeal bodies, arcana are to be considered incorporeal and of an essence far more excellent than ours, the difference being as great as between black and white. They have the power of transmuting, altering, and restoring us, as the arcana of God, according to their own induction. And although there is not eternity in our arcana, or that harmony which is celestial, nevertheless, compared with us, they ought to be judged celestial, since they preserve our bodies more than can be done or found out by Nature, and operate upon them in a wonderful way by their virtues. None otherwise, therefore, should these

---

* The arcanum is, as it were, a potent heaven of medicine set within the hand and the will of the physician.—*De Aridura*.

natural arcana be compared to our bodies, so far as medicine is concerned,* than as the secrets of God are. Nor shall we fear to write that these arcana are higher and greater than ourselves, and that they have the greatest power of inaugurating life in us, as witness those four which we shall set down. Neither shall we take any heed of the empty stories told by those slavish dwarf-divines, since we consider that they have no more understanding than a blind man has sight. One arcanum, then, is of a single essence; another is the arcanum of Nature herself: for the arcanum is the whole virtue of a thing, excelling a thousandfold the thing itself. We are able, therefore, fearlessly to assert that the arcanum of a man is every gift and virtue of his which he retains to eternity, as we teach in another book of these Archidoxies. An arcanum, then, comes to be understood in two ways: one is perpetual, the other is quasi-perpetual. This quasi-perpetual arcanum we judge to be like that which is perpetual according to the esteem and predestination of it. Four arcana only have been known to us from our boyish years; with these we will complete this book, and work out for ourselves a sufficiently praiseworthy memorial, so that we may never forget them, praying the Supreme God that of His mercy He will allow our human flesh to attain to many years, so that we may lay up a long and gentle repose for our age, to hope faithfully in Him, and in no way to doubt that He, since He deigned to take our human nature, will allow us to enjoy it, so that we may in no way be disappointed of our hope, as we confidently expect will be the case.

Relying thus upon this hope, we will begin by making clear what is the difference between these four arcana, as to the labour, the art, and lastly, the virtues. For this purpose, a final and conclusive knowledge of the virtues of each is required. Generically, first of all, they keep the body in health, ward off diseases from it and drive them away; they enliven the depressed spirits, freeing them from all sorrow; they protect from all sickness, and happily conduct the body even to its predestined death, which has no end save by a lessening of consumption, as we lay it down in our treatise on Life and Death.

And although we have already made clear their virtues in a general way, and their nature, nevertheless they differ very much in particulars one from the other, so that no one of them operates altogether in the same way as another, or fulfils its virtues, but they are different both in manner and form, each with its own proper and peculiar ways. So, then, the *Prima Materia* is the first Arcanum; the second is the Philosophers' Stone, the third is the *Mercurius Vitæ*, and the fourth is the Tincture, for even thus we will set down the

---

* Now, the difference between arcana and medicines is this, that arcana operate in their own nature, or essence, but medicine in contrary elements. Yet arcana do not prefer themselves to medicine. Medicines are those things wherein it is understood that cold is to be removed by heat and superfluity by purgation. Thus, there are reckoned substances of the arcana which by their natures are directed against the property of their enemy, even as one pugilist is opposed to another. Accordingly, the conflict of Nature is such that craft is circumvented by craft, and all things that we possess naturally in the earth, the same Nature also requires to be preserved in medicine. This is, therefore, the part of the physician, namely, to act not otherwise than as if two enemies opposed one another, who were equally cold or equally glowing with heat, and are both armed with similar weapons. Since the victory lies between these, so must you also understand concerning man, that there are two combatants, soliciting natural aid from one mother, namely, the virtue alone. The arcana also operate with like virtue.—*Paragrinum*, Tract II., *De Astronomia*.

practice of them in order, after we have explained their modes of operation as follows :—

In the beginning, it must be remarked, concerning the Primal Matter, that it puts forth its predestination, to which it is foreordained, entire, and from its first origin to its final end well-defined and exemplified. For as the seed gives of itself the entire herb, with renewal of all its forces and consumption of the old essence, so that the former substance, nature, and essence have no further operation, so do we say of the primal matter, that we are born from one seed like something growing in the field according to its growing nature. According to the aforesaid example, the primal matter introduces new youth into a man, just as a new herb springs forth from a new seed in a new summer and a new year.

The Philosophers' Stone, which is the second arcanum, perfects its operations in another form, namely, this: As the extrinsic fire burning the spotted skin of the salamander, renders it pure as if it were newly born, so, also, the Philosophers' Stone purges the whole body of man, and cleanses it from all its impurities by the introduction of new and more youthful forces which it joins to the nature of man.

The Mercury of Life, which is the third arcanum, gives proofs of its operations like those which the kingfisher displays, which is renewed every year at its annual period and endued with new plumage. Even so this arcanum casts off the nails from the hands and feet of a man, the hair, skin, and everything that belongs to him, makes them grow afresh, and renews the whole body, as was said above concerning the kingfisher.

But the Tincture, which is set down by us as the fourth Arcanum, displays its operations like the Rebis, which makes gold, or the Sun, out of the Moon and other metals. None otherwise does the tincture affect the body of man, and take away from him his corruption and impediments, changing all into the greatest purity, nobility, and permanence.

How, then, can it be that we should withdraw from this noble art of medicine, or from philosophy itself, when we see so clearly their force and power, which alone confirm us, and deservedly so, that we should place the greatest confidence in them? For we have not applied our minds to believe, learn, or imitate things which cannot be proved and attested by most true and certain evidence conjoined with reason. When Christ hung upon the Cross, if the sun and moon had not been affected with a kind of sympathy, so that they were deprived of their light and darkened, and had not the earth been shaken with a fearful tremor, and had not other signs been manifested at His nativity, no one would now believe in Him. They naturally teach us to see and know this, that Jesus Christ is God, and took human nature upon Himself. We may say the same of the Arcana, that they cause and even compel us to believe in them, so that we should not withdraw from them even up to the time of death, but rather strictly and constantly, among many hindrances, every day go on to give thanks to God. So neither the eclipse nor the moon will detach anything from us. We will therefore put forth the practice and elaboration of these four Arcana, by which we are able to drive away the accidents and corruptions of our youth, and to rejoice in them as our eternal Arcanum rejoices in its eternal life.

Concerning the Arcanum of the Primal Matter [*]

Since we have sufficiently pointed out concerning the Primal Matter, whence it proceeds and what it is, we must understand that it is based not only on men, but on all bodily creatures, that is, on everything that is born from any seed. Whence it may be inferred that if it has its operation in any created body and perfects it, as we have before declared, it is able also to preserve trees from corruption, herbs from being dried up, and also metals from rust, concerning which the same thing must be understood of men and of brute beasts. So, then, a tree which is now almost consumed with age, and daily more and more verging on its own corruption, not from defect of root or of nutriment, but of its own proper virtue, can be renewed by its own primal matter just as we said about the skin of the salamander, and so may arrive at another age according to its predestination, nay, even at a third, a fourth, or more. On this principle virtues are to contribute to it, namely, in order that its corruption and destruction may be now and again renewed in a long process of time. No less is this to be understood of herbs, which last only for a single year, because their predestination is no longer. For they, even when they begin to be dried up, are renewed by their primal matter, so that they remain green and fresh for another annual period, or for a third, a fourth, or more. The same thing understand of brute beasts, as, for instance, old sheep and other animals. They can be renewed for a fresh period of life, having received their virtues, such as milk and wool, like young sheep. Equally, too, can man be led on from one age to another, as we have said before. From this it ought be known what the primal matter is according to its essence. In created things, such as have no sensation, it is their seed; but in created beings endowed with sensation, it is their sperm. For it must be known that the primal matter is not to be taken from the thing out of which this created body is produced, but out of the produced and generated material. For the primal matter has in it such virtues that it will not allow the body which is born of it to go into consumption, but abundantly affords whatever is necessary for the supply of every requirement. Indeed, death only arises from the destruction or infection of the living spirit. Now, that spirit grows out

---

[*] According to one of his treatises on turpentine Paracelsus held that the ultimate matter is contained in that seed wherein God has digested each thing, but those more especially which are subject to natural growth, created by Himself. The primal matter is gross seed. It is the ultimate and not the primal matter, which, on the authority of this treatise, is useful to man. According to the tenth chapter of the *Labyrinthus Medicorum*, every growing thing whatsoever is in its first matter without form, or unformed. For example, the fir, the beech, the oak, are all in the beginning seed, wherein there is nothing which ought to be, that is, nothing which we should expect to find, having regard to what they become subsequently. But if such a seed be put into the ground, it is needful that it should first putrefy. Otherwise nothing is produced from it. When it putrefies, it dissolves altogether, and from being a seed it becomes nothing, but, at the same time, the putrefaction is the first matter from which the tree develops.

of the sperm, or out of the seed, and is altogether a spermatic substance; therefore it can be helped by its like. For wherever the like is given as a help, there is introduced a new period of life, for many causes which we do not detail in this place, but make clear in our " Philosophy."* Furthermore, although we did not propose to write in this book anything about the nourishment and renovation of trees, since we undertook to treat only of the medicine of the human body, still, let those facts about the trees and other transmutations of this kind be set down, in order that, parabolically, and by those examples, we may make our meaning easier to be understood. The quintessence of the seed of the nettle (otherwise the lavender) if it be poured on to any root of its own herb, so that this herb may receive its tincture and be affected by it, it remains another year as in the former year, not putrefying until that second year shall have been completed.

In like manner, if the quintessence of the seed of quinces be poured on the root of a quince tree, the tree remains green to the end of another year, and also produces flowers and fruits. In the same way, the quintessence of cherries causes trees to put forth their fruits twice in a year, as if in two summers; one is the middle summer of the seasonable cherry trees, and the other summer is made like the former.

Not only, therefore, is it fit that we should speak of the quintessence of the sperm, but also concerning the arcanum of the sperm, from which proceed far more wonderful things than we have already pointed out.

So then, let us first make clear the process of this practice. In the first place, it is alike among men and animals. Secondly, it is made out of primal matter only in the following way:

Take primal matter, let it digest in a flaccum in a resolutive digestion for a month. To this let there be joined the addition of a monarchy of equal weight. Let them be suffered to digest again from one to two months. Then distil this matter [? by the cloth] and what ascends will be the arcanum of the primal matter, concerning which we are here writing. And let no one wonder at the brevity of this method or process, for complexity is apt to involve much error.

Concerning the Arcanum of the Philosophers' Stone

I am neither the author nor the executor of that Philosophers' Stone, which is differently described by others; still less am I a searcher into it, so that I should speak of it by hearsay, or from having read about it. Therefore, since I have no certainty thereof, I will leave that process and pursue my own, as being that which has been found out by me through use and practical experiment. And I call it the Philosophers' Stone, because it affects the

---

* Paracelsus has bequeathed to his followers a Philosophy of the Four Elements, a Philosophy addressed to the Athenians, an Occult Philosophy, Five Philosophical Tracts, and a vast system under the title of Philosophia Sagax. The connection of the living spirit with the sperm is discussed very largely in most or all of these, as, indeed, elsewhere in his writings. Under these circumstances there is scarcely any need, as, in fact, there would be great difficulty to distinguish the special section to which reference is here made.

bodies of men just as their's does, that is, just as they write of their own. Mine, however, is not prepared according to their process; for that is not what we mean in this place, nor do we even understand it. We do not set down in this our practical treatise the process of the operation, since we have before mentioned it in the beginning of this book when we were writing of its force and its effects, which it has by means of separation.

Concerning the entrance of this penetration, you shall also further note, by which entrance it penetrates the body and all that therein is. For by that penetration it restores and renews it, not that it removes the body altogether, and introduces a new body in its place, or that, like the primal matter, it infuses its spermatic arcanum thereinto, but that it so purges the old whole body as the skin of the salamander is purged, without any injury or defect, and the old skin none the less remains in its essence and form. In like manner, this Philosophers' Stone purifies the heart and all the principal members, as well as the intestines, the marrow, and whatever else is contained in the body. It does not allow any disease to germinate in the body; but the gout, the dropsy, the jaundice, the colic, fly from it, and it expels all the illnesses which proceed from the four humours; at the same time,

purges bodies and renders them just as though they were newly born. It banishes everything that has a tendency to destroy nature, none otherwise than as fire does with worms. Even so, all weaknesses fly before this renovation.

This Philosophers' Stone has forces of this kind, whereby it expels so many and such wonderful diseases, not by its complexion, or its specific form, or its property, or by any accidental quality, but by the powers of a subtle practice, wherewith it is endued by the preparations, the reverberations, the sublimations, the digestions, the distillations, and afterwards by various reductions and resolutions, all which operations of this kind bring the stone to such subtlety and such a point of power as is wonderful. Not that it had those powers originally, but that they are subsequently assigned to it. Something like this is to be understood in the case of honey, which, by its elevations becomes far sharper than any aquafortis or any corrosive, and more penetrating than any sublimate. Such a property of sharpness it has not by nature, but this proceeds solely from the elevation, which changes all the honey into a corrosive. In these effectual Arcana, too, it must be considered that those who use them, as well as the children sprung from them, live afterwards endued with such health that no sickness or ailment, or anything like a flaw, afterwards happens to their bodies, but they are adorned altogether with such a subtle and pure complexion of Nature, that it is impossible a more noble state of the complexion can be induced. For that most choice and excellent medicine effectually renews and purifies, and introduces an incorruptible life, which cannot be contaminated by any kind of life. It suffers nothing to become enfeebled, but secures that men shall live in the highest nobility of Nature, while it advances their offspring to the tenth generation.

This Philosophers' Stone not only transmutes one weight, but this transmutes another, and this again another, and so on, in so far that these mutations might be extended almost endlessly, just as one light kindles a second, and this second a third. So it should be understood of the Philosophers' Stone in relation to health, as out of a good tree good seed and good shoots are born, out of which again good trees are produced. The power and potency of the Philosophic Stone is exalted to so wonderful an extent that it is impossible to trace how it can be naturally brought about; and unless most evident signs lay open to our eyes, it would be incredible that men could perfect and accomplish such wonderful things; since the virtue of that operation passes from generation to generation without any break. On the other hand, by the mercy of God, it exists in one body, and at length, according to their deserts, it is either denied to others or conceded as a special act of grace.

Now let us set down the process of our Philosophic Stone in the following manner.

In the name of the Lord, take Mercury, otherwise the element of mercury, and separate the pure from the impure. Afterwards let it be reverberated even to whiteness, and then sublimate this by sal ammoniac until it is resolved. Let it be calcined and again dissolved, and digested in a pelican for a month. Then let it be coagulated into a body. This body no longer burns nor is consumed in any way, but remains in the same state. Those bodies which it penetrates are permanent in the cineritia, and cannot be reduced to nothing or altered; but the stone takes away every superfluous quality from sensible and insensible things, as we have before related. And although we have set down a very short way, nevertheless it requires a prolix labour, difficult in its adjuncts, and requires an operator who is affected by no weariness, but is in the highest degree active and expert.

Concerning the Mercurius Vitæ

Next in order we wish to write concerning the *Mercurius Vitæ*, the virtue of which far excels the virtues of the two preceding Arcana; for that virtue consists, not in the art or in the operation, but in the *Mercurius Vitæ* itself, like which we have never known any simple anywhere existent; forasmuch as that nature and property is innate therein, not from the virtues of the quintessence or of the elements, but from the specific quality of its predestination. Neither has it only the virtues of transmuting persons and other essential things, but also of renewing every growing thing, and their likes, out of the old quality into a new, after the following manner. The *Mercurius Vitæ* reduces Mars into its primal matter, and again transmutes it into its perfect matter, so that out of it iron is again made. It also renovates gold in the same way, because it reduces it into proper mercury and tincture, and again digests it into gold, so that it becomes a metal as before.

Nor, in truth, does it operate in metals only, but similarly in other bodies, as in herbs. When their roots are suffused with it they will bring forth a second crop of flowers and fruits. If the first seed has fallen off, and they are at that time suffused with the same, they will produce second flowers and fruits, irrespectively of the season.

And none otherwise must it be understood of animals than of men and other things. When this Mercury is applied it renovates all their old and consumed members, and restores the defective and lost powers into a youthful body or abode, so that in old women the menses and the blood flow naturally as in young ones. It also brings back aged women to the same perfection of nature as the younger ones.

Concerning the arcanum of life it is further to be remarked that its forces exist so potentially in its specific form that it separates the old from the new, or age from youth; that it augments the latter, and so renovates the period of life. Whence it may be inferred that youth and its powers do not fail on account of old age, but that these exist equally in the old as in the young. The corruption, however, which grows up with youth is so strengthened that it takes away the powers, whence old age is recognised.

As soon, therefore, as this corruption is separated from youth, that youth again manifests itself without let or hindrance. Now this must be understood thus: When any body or corpse putrefies, the quintessence thereof is not putrefied, but remains fresh and unconsumed, and is separated from the corpse into air, or sometimes it is scattered into the earth or into the water, according to the place whither it goes. For there can no destruction of the quintessence occur, a fact which must be clearly noted and regarded with admiration, as we teach in our treatise on Corruption and Generation.* So, also, a rose putrefied in dung retains its quintessence in itself, even whilst in the dung. Though everything becomes fœtid and putrid, still, in the separation of the pure from the impure, the quintessence lives without spot or blemish, though the bodies are noisome corpses. Thus, therefore, we say of the *Mercurius Vitæ* that it separates corruption, even as in wood it separates that which is decayed. So powerful, also, is it in man, that, after the corruption shall have been separated from him, the quintessence is again excited, and lives as in youth. And this is to be understood thus: Not that the *Mercurius Vitæ* stirs up a new essence, as some persons may malignantly interpret our opinion and experience, but that the essence and the youthful spirit whence proceed the forces of youth remain unconsumed, none the less, though, being oppressed, they are beheld as dead. The *Mercurius Vitæ* removes the impurity, whence it happens that the aged life recovers most effectually its powers as they were before. As we said above when speaking of the kingfisher, it is renewed after death for this reason, that the quintessence does not

---

* The reference does not correspond to the treatise on the Nature of Things, and there is no other extant under this exact title.

withdraw from its abode. But if that dwelling-place be dissolved by putrefaction, then the quintessence is received into that upon which it lies. Whence it happens that there are often found wonderful conditions of Nature in growing things, not naturally existing in them, but by some accident of this kind, as we set it down in our book on Generations. In this way, then, the matter is to be understood. In dung there meet and are accumulated many corruptions, as from herbs, roots, fruits, waters, and other like things. Whence it happens that fields are rendered fat and fertile, not on account of the corruption, but on account of the quintessence existing in the dung, which, betaking itself to the roots, exhibits power in the growing things though the body, that is, the dung vanishes and is reduced into nothingness, its substance being consumed. Wherefore human dung has great virtue in it, because it contains in itself noble essences, as of the food and the drink, concerning which wonderful things might be written. For the body receives from it nothing save nutriment, but not the essence, as we write in our treatise on Nutriments.*

We will come, then, more strictly to the practical view of *Mercurius Vitæ*, which, as we said before, perfects its operations in a wonderful manner, as, for instance, by shedding the nails of the hands and the feet, and plucking out grey hairs by the roots, it strengthens youth, so that corruption cannot demonstrate old age by those signs unless a second old age be again attained.

Now, approaching the practice, let us treat with alchemists in a few words; for there is no need to write much or to preach with prolixity to them. Pretenders and fools we will altogether exclude. Let this, then, be the method of practice:

Take essentialised mercury; separate it from all superfluity, that is, the pure from the impure. Then sublimate it with antimony, so that both may ascend and become one. Then let them be resolved upon marble and coagulated four times. This being done, you will have the *Mercurius Vitæ*, about which we have before spoken, and with which, as with an arcanum, we will console our own old age.

Concerning the Arcanum of the Tincture

None otherwise is the arcanum of the tincture to be understood, namely, that it takes away all the inconvenience from old age, all diseases, and whatever corrupts the health or has an influence contrary to it. This arcanum is a certain tincture of such properties and conditions that it operates and induces health, not in the same way as the preceding three arcana, but according to its own name. The tincture *tinges* the good and the evil, the dense and the subtle. None, otherwise, does this perfect its operations on the body so as to transmute corrupt and ill-disposed complexions into sound ones, just like that

---

* This is another treatise which apparently is wanting.

tincture which makes Luna out of Mercury; it does not separate from it what is evil, but it tinges both the good and the evil, so that in the end they both turn out to be excellent. So, too, this tincture tinges the dropsical and jaundiced body and makes it sound, not because the dropsy is taken away, the origin driven out or separated from the good, but it is transmuted into good, as it should be, and is settled in its highest and best grade, in like manner as the corrupted dung may, by the subtle corruption of art, be converted into an elixir, which drives away all corruption; that corruption not being separated, but the entire substance being transmuted into another quality and nature.

The same is to be thought concerning this tincture, that it tinges the body apart from all separation of the evil from the good, or expulsion of man's original essence, but by his renovation.

Yet it should be known that the tinged body no longer lives in its old form, but, like a metal, is transmuted into another, as into copper or some other. Saturn has not in itself its old quality, but the quality of the tincture itself. None otherwise must it be thought of tinged bodies which have received the tinging of the tincture, that they exist no more in the former life from which they were transmuted by the tincture, but far nobler, better, and more healthful is the condition of the body and the form than its native origin was; and it is like natural gold made out of iron by the tincture, as we have also written concerning Transmutations. If, therefore, this tincture is a transmuter of bodies into better ones, as in the case of metals which so few know or have tried, there will be as many and various corporal as there are metallic tinctures. And as of these one is always better than another, so with the corporal tinctures. It must be considered that some tinge and are tinctures naturally, as the crocus, flower, sulphur; some are made by art, as a stone, realgar, and others. It is highly necessary that these things should be understood, because the beginning and the entrance to these tinctures which they exhibit is not unimportant. Moreover, it is to be remarked that these tinctures ought to be made for the seven principal members, and that to every one of them its own property ought to be assigned and given, as to the heart those that serve the heart, to the brain those which belong to it; and such tinctures should be prepared from metals, herbs, and like things which are proper. So it will happen that by them the whole body will be tinged. Nor will it suffice that it be tinged by one tincture only, but as one tincture tinges only one metal, so must it be judged of these. The practical preparation of the Tincture should be as follows:—

Take the essence of the members, from which you will separate the elements. Afterwards put the fire of them in digestion, and leave it so long until there remains nothing more at the bottom and nothing of the matter appears substantially. Then take the matter, and the glass luted in this way with the lute of Hermes; put it in a moist, cold place, where again it will be resolved into visible matter. That visible matter is the one concerning which we have written. With these few words we conclude: for if we were to write more concerning it, that would be a handle for the derision of the stoics.* From this we would fain be exempt, and speak only to alchemists.

---

* In another place Paracelsus affirms that arcana are not old things but new, not ancient but recent productions. The ancient productions are substances and forms as they exist in the world. And as the form of these things is no good to us, but must be resolved and renovated in order that it may be of use, so there is needed in addition the removal of all properties, whether of heat or cold. That is, unless the solatrum lose its cold it does not become a medicine. Similarly, unless anacardi lose their heat they will in no wise prove a remedy. In short, unless all the old natures pass away and are removed, and are brought over into a new birth, they will never be made into medicines. This removal is the beginning of the separation of the bad from the good. Thus, therefore, the latest medicine, that is, born recently, is left, without any complexion and the like, a pure and absolute arcanum.—*Paramirum*, Lib. II., *De Origine Morborum ex Tribus Primis Substantiis*, c. 2.

THE END OF THE FIFTH BOOK OF THE ARCHIDOXIES FROM THE
THEOPHRASTIA, CONCERNING ARCANA

THE SIXTH BOOK OF THE ARCHIDOXIES

FROM THE THEOPHRASTIA OF PARACELSUS THE GREAT

CONCERNING MAGISTERIES

THE preceding books on most excellent medicines having been finished, we propose to add this one treating of the Magisteries: and first to declare what a magistery is. It is, then, that which, apart from separation or any preparation of the elements, can be extracted out of things; and yet by the addition of something the powers aud virtues of those things are attracted to that material and kept there.

Those virtues do not in any way proceed from Nature, so far as their operation is concerned, nor from a specific virtue, but from a mixture whereby virtues of the same kind are attracted. If vinegar be poured into wine it renders the whole vinegar. Now, this is a magistery. When wine is poured upon honey the wine is so transmuted; so there is no magistery in that case. Those things, therefore, must be considered which relate to a magistery, even as wines do to vinegar. For such as are perfect, or, as should be the case, ill adapted for this purpose, cannot produce a magistery. So, then, the natures of things come to be considered. In the same way, the difference of the extractions of magisteries must be noted, as, for instance, out of metals, marchasites, stones, herbs, and the like matters, by those things which are not metallic, but which become like metals, just as wine is made like vinegar in its powers, its virtues, and its taste. The cause that wine does not appear different from vinegar is, that a nature like vinegar exists in it, whence it happens that the appearance of their natures is the same. If, in like manner, the nature of metals be but pure, it equally appears in their magisteries, but yet it is not of the same property.

Moreover, mention must be made of additions. Those things which are taken for this purpose, though they are not of one and the same complexion, power, or action, nevertheless agree in preparation, since that which results from power of this kind is appropriated and not complexioned. By means of these substances the metals afford their magisteries, which, indeed, are no less to be accounted of than quintessences on account of their virtues. Gold in its magistery lays down all its quality and complexion in one essence. It must not for that reason be supposed that because the body is of no account the rest will therefore be infected with that failing; for in this case its leprosy is in no way injurious, but the thing is entirely good. Sugar, again, is wholly sweet whilst as yet it remains in the body and is not separated, and it can be so prepared, while retaining its sweetness, that it shall turn out by far sweeter and more efficacious than it was before. But the quintessence when extracted is not sweeter than when in or with its own body. Wherefore, this body by no means brings loss to it. But still its virtue separated from the body is less than when prepared in or with it. The magistery amends it more than Nature can be supposed to do.

The same must be understood of stones, which enter into the number of magisteries, and also of their bodies, namely, that what is assumed for their use is not deficient in its virtue, but is a sufficiently powerful magistery. This you may understand thus, as when sulphur is lighted and burnt up; that which burns is the very smallest part; and so likewise stones; for example, the crystal, when it is reduced to a magistery, brings all stones to their first matter, and grinds them in a wonderful manner, in the same way as its essence does, for this reason, that the nature is tinged by the quintessence, and may be held for a quintessence, as in the case of the vinegar and the wine, of which the one makes the other like itself without any defect. In like manner, it is not only in stones that tincture of this kind is made, but also in metals, as the quintessence of gold tinges all its body into a pure quintessence, which light we think great, and even too great, for it is the light of all the secrets in our Archidoxies: wherefore we kindle coals with a light mind, so that we may investigate the final end of these secrets of Nature. We derive our instruction from examples proved experimentally by ourselves, as in magisteries, and chiefly in that of gold, which, containing the body and the quintessence, is drawn therefrom just as the quintessence is. Wherefore our magisteries are known to be dowered with special virtues, and we write them down to our own praise even until death.

We speak also in like manner concerning the magisteries of herbs, which indeed are so efficacious that half an ounce of them operates more than a hundred ounces of their body, because scarcely the hundredth part is quintessence. So, then, the quantity thereof being so very small, a greater mass of it has to be used and administered, which is not required in the case of magisteries; for in these the whole quantity of the herbs is reduced into a magistery, which is not then, on account of its artificial character, to be judged inferior to the true extracted quintessence itself. One part of this being exhibited is of more avail than a hundred parts of a similar body, for this reason, because the magisteries are prepared and rendered acute to the highest degree, and are brought to a quality equal to a quintessence, in which magisteries all the virtues and powers of the body are present, and from these its own helping power arises to it. For in them the penetrability and the power of the whole body exist, from the mixture that is made with it. For the body receives none of those things with any desire except such as are spiritual to it. Whence it happens that it attracts that magistery and mingles itself therewith none otherwise than as gold attracts to itself the *Mercurius Vitæ*, and is mingled therewith. Now, iron does not do this, because they are not, in their composition, so much in agreement with each other. So, then, both the body of it and this magistery are amalgamated, and become one; and hereof many examples may be found which need not be brought forward in this place.

Some of the marchasites, too, perform the office of a medicine in a like way, although with a difference: that is to say, they leave their body, and the best part of them, as the juice, is extracted, and none the less exists as a magistery, though the body be separated from it. This, however, must be understood, that it is not the body of the marchasite, but rather of the earth or the mineral in which the marchasite exists. Its virtue, therefore, is not efficacious of itself, that the earth or the mineral should be separated from it, but it abides therein as in a marchasite, which, indeed, it is thought to be. In order, therefore, to make it clear concerning a marchasite, what it is, whence produced, and with what virtues endowed, we wish to relate its practice in the following treatises, and also to describe the art of its preparation. The process, however, cannot be comprised in one general explanation, but the matter must be treated of particularly, concerning each special case, as of metals by themselves, marchasites, too, in particular, and also stones and herbs.

There is, likewise, reckoned a special magistery of the blood, which is taught in a peculiar form and manner. In it is considered what virtues and forces of man exist, and what its nature contains in itself, in what there is any defect produced, and so on; but still, without diminution of the natural creation, but that it may be considered as a complete and perfect work in all its parts, as a bird with all its feathers.

Concerning the Extraction of the Magistery from Metals

First of all, we will set down the magistery from metals, and make that clear which openly shews itself to be of wonderful virtues, which are to be known according to the tenour of its essence. The process of them must be carried on without any corrosives and all things which are complexioned contrary to the metals. For, from the conjunction or mixture of contrary things, the essences are corrupted, so that, on account of this error, they put forth no virtues, since the one contrary or the other predominates and prevails.

Seeing, therefore, that great account is to be taken of agreement, only the temperate will agree therewith. Consequently you must know what is a temperate thing. A temperate body is one of a certain complexion; this receives another into itself, and is incorporated with that which shall be joined to it, so that it no longer displays its old and special complexion, but only the virtues of that which has been added to it. So, forsooth, *vinum ardens* has in itself a full and perfect complexion; but yet, whatever is put therein, it, so to say, complexionates more abundantly, fulfilling its operation thus, according to the power of that which was put into it.

Since, then, it attracts the virtues of other things and subdues its own, it is on this account said to be temperate, and is deservedly so called. Here some differences call for notice; we understand the elements only, so that we are able to say of oil that it draws to itself foreign natures and hides its own element. From anything like these a magistery ought to be made, so that the virtues of the metals may pass into the same temperate thing, may be cleansed and purified therewith, and may be distilled to their own end. A magistery like this, after it has been purified, shall be called potable, because it can be taken in drink: whence it comes that the magistery of gold is called *aurum potabile*, that of the moon *argentum potabile*, so also iron, lead, quicksilver, etc., will be capable of being made potable, and so called and described according to their complexions, in relation to which processes they have far greater operation than it becomes us to describe. After the following manner, with one temperate thing, and by one process, as also by one practical method, all the magisteries of the seven metals can be made. The practical method is as follows:—

Take circulatum thoroughly purified, even to its highest essence, in which place very thin metallic plates or filings of any metal you please, perfectly and subtly wrought and purified. When these two have been put together in a sufficient weight, let them be circulated for four weeks, and by means of this temperate medium, the leaves will be reduced into an oil, and into the form of a fattiness swimming at the top, coloured according to the condition of the particular metal which you shall separate by means of a silver attractorium from the circulatum. This same is the potable gold or silver. The like is extracted with the other metals, and may be taken as drink, or with food, without any harm.

The Extraction of the Magistery out of Pearls, Corals, and Gems

The Magistery from precious stones is to be understood in the same way as from metals, according to the virtues which each stone has in harmony with its nature. It must be remarked, however, that for stones there is no need of a temperate medium, or of any addition whatsoever, because the solution of them is not the same as the resolution of metals, but the magistery of them is extracted in another way; in the practice of which three processes are understood, one for gems, another for marchasites, and a third for corals, according to which stony growing things may be brought each into its own magistery. In the process of stones, the colours need not be considered, nor their brightness be taken into account, since all their magisteries have a white colour. So also pearls, with the exception of corals, which retain their colours in a remarkable way above others in the magistery. In them, therefore, the colour is chiefly to be noted; for the magistery of them, together with the body and the element and the whole essence, is extracted by means of additions, without any corruption. None the less, however, they may again be restored to their perfection. Wherefore, in respect of their generation and nature, they cannot be compared to stones, and yet they have a stony condition. This, too, may be done: the whole colour may be extracted out of the body of corals into another medium, and afterwards from their body a form may be made as if from clay. After this formation the colour can be again infused, so that it becomes a coral as before. This cannot be done with pearls or gems, which are unable to return to their perfection by a similar process, but they remain in their magistery, their essence not having been corrupted. We have even seen this penetrate glass and instruments, and form them according to its nature. The magistery of the magnet has drawn things towards itself, in the form and after the manner of the material magnet, and afterwards it has fixed itself in the glass and tinged that, so that this also has attracted needles and straws. None otherwise must it be understood of the rest. Therefore they must be kept in gold only. More of these things have occurred to us than is credible, whereof we put together this record, so that from this inducement other persons also may investigate the arts and the magisteries. Since, therefore, they exhibit a demonstration apart from other extractions, we will now teach the practice of them in the following manner. First we will speak of gems.

Take gems, having first bruised them and calcined them according to the reverberatory condition, together with common salt-nitre, of an equal weight, that is to say, one pound. Let these be burnt together into the form of lime, and afterwards washed in *vinum ardens* until no surplus matter can be found. After the washing, calcine this substance again, and cause the whole of it together to pass through the *vinum ardens*. Let this evaporate by boiling; and thus you will have an alkali, which dissolve in water and keep. There is no reason why you should fear to make use of this administration, since, however sharp and calcined it may be, with all its sharpness it only affects that which resists it; and it is so subtle, that one single drop alone tinges the whole body so as to produce a remarkable virtue.

Pearls, too, are to be reduced to a water in the following way, namely: Take corrected vinegar, in which place bruised pearls, and let them digest therein for a month; so that they will be reduced to a water. Then distil it, and separate the *acetum* from it by a bath, having done which, you shall find the pearls in the bottom resolved into water. This is the magistery of pearls or unions. And although the process of this practice is a short and easy way, nevertheless, give credence to one who has tried it; for the operation of these same things is indeed wonderful. This action of their virtues, however, is not produced by art, but is placed in the nature of them, which is hidden in a dense substance, on account of which it cannot operate, just as is the case with a dead body; but by making a resolution its body will be vitalised.* Enough of this. Corals, however, must be ground and calcined from the commencement with salt nitre; then, afterwards, they must be prepared as gems are, and also resolved. So you have the magistery of corals, the virtues whereof I very much and specially wonder at, which God has bestowed on this growing thing, which also do operate powerfully and wonderfully, even as they grow.

Concerning the Extraction of the Magistery out of Marchasites

Concerning the magistery of marchasites thus much must be known; that they are only minerals. So, then, the mineral is not brought to its magistery, but only the true marchasite, as is also ascertained with regard to metals, since these do not pass into their magistery unless they shall have been first separated from their mineral. And although marchasites cannot well be separated from these, still this may none the less be done in the magisteries. There are, indeed, different kinds of marchasites, such as the golden, the silver, the golden talc, the white, the purple, the tin or bismuth, antimony, granite, and others of a like nature, for all of which there is only one appropriate extraction. In like manner, the virtues and powers of them manifest themselves in medicine according to the conditions of metallic operations. And although they do not exist as metals, none the less they have the properties of them. Wherefore, we set down only a few details about them, because we treat them more at length in our book on Extrinsicals.† There is a difference to be noted, according as they come together and agree one with another, as gold and the marchasite, antimony and lead, which, indeed, in their fabrication and constellation are compared one with the other, but are none the less separated in their virtue. For in some marchasites there are more virtues than in the assimilated metals. This we see to be the case with lead and antimony, of which the one cures morphia, alopecia, and the like, and all scabs and scars, the leonine, the elephantine, the Tyrian, etc., which the magistery of lead or Saturn does not affect. So, then, the properties of this kind are to be noted, which are sometimes latent not only in great things but also, and that more abundantly, in smaller ones. Let us, therefore, go

---

* Among other virtues which Paracelsus attributes to the pearl is the increase of milk in women when the supply is deficient. Let the pearls be resolved into a liquor, otherwise an elixir, but in the body a ferment. Whatsoever other things are required for this ferment, these the body supplies. The quantity administered must be regulated by the experience of the physician.—*De Aridura*. Another method for the extraction of a medicament from pearls is as follows: - The matter in the form of a potable sap is extracted from pearls, and this is so powerful that it can scarcely find its equal among sperms. *Formula for compounding the Sap of Pearls.*—Take one pound of purest distilled vinegar, half a pound of circulated water of life, and four ounces of cleansed pearls. Reduce to a fine powder, prepare, digest with others for a month in a circulatorium. The matter of the pearls will then have sunk to the bottom like a thick liquor. Then separate the waters by effusion from the sap. Of this sap administer doses of six grains. There are many other processes for the extraction of this kind of sap, but the above is the most useful of all.—*De Contractiuris*, c. 3.

† The book concerning Extrinsicals, and that also concerning the Generation of Wines, mentioned a little further on, are unknown. A similar remark will apply to the treatise on the Properties of Things, which is cited in the seventh book. There is probably little doubt that many of these apparently missing works are merely the mis-stated titles of others which have come down to us. For example, the treatise on Physics which is quoted in Book VII. may be extant under a different name, or again, the reference may be to a projected treatise which was never accomplished.

on to investigate this reason, why antimony has more virtues than its metal. Its body is not fixed, and not sufficiently digested into its own perfection, as is the case with Saturn, whence it assumes a volatile property. But the very material out of which it is produced is privative, and of a cleansing character, from its own natural property, as we set down in our book concerning Generations. Hence it happens that it purifies gold and silver more than does the fire or any other element. Thence, too, it proceeds that it cleanses and purges the body, even as gold and silver are freed thereby from all their impurities. The magistery of antimony drives out leprosy in a more than credible manner. And so, too, must it be understood of the others.

Let us now approach the practice by which we teach the preparations of the magisteries from all these substances, in the following manner :

Take a marchasite, ground very small, and add to it so much dissolving water as will cover the breadth of six fingers. Let it be dissolved and subsequently putrefied for a month. Afterwards let it be distilled and separated, as we teach concerning the metals, which having done, you will have the magistery of whatever marchasite you selected.

Concerning the Extraction of the Magistery from Fatty Substances

None otherwise must it be thought concerning the extraction of the magistery from fatty substances, as the fat of amber, of resins, of oils, and of other things, even as they appear in like material substances. Concerning these there are three methods of extracting their magisteries. One special method is that of amber, another of resins, and the other of fatty substances, such as oils, tallows, butter, and the like. For amber does not in any way admit of the same process as fatty substances, since its virtues would be lost. Resinous substances, again, do not allow the same practice as amber does, for it would be destructive to them. We will therefore teach the preparations of those magisteries in three ways, since such excellent virtues appear in fatty substances, and these in many forms. In cases where essences produce no effect these fatty substances render aid; for they have that property on account of their specific and appropriate virtue, which is not found to be so perfect in other substances, neither, indeed, is it so. None otherwise than as the difference is between corals and gems when compared must it be understood of these things, since the practice is as follows; and first concerning amber:

Take of amber well ground as much as seems good, and of circulatum a sufficient quantity. Digest them in a flat vessel in ashes for six days. Afterwards let the circulatum be distilled off from thence, and again poured on. Let this be repeated until an oil is found at the bottom; which oil is the magistery of amber. This has revealed to us its wonderful virtues. May it continue so to do!

Resinous bodies are reduced to their magistery by the following method:

Take of turpentine, gum, or resin, as much as you will. Place this in a luted glass vessel and let it be digested by itself for a month in a warm digestion. Afterwards, having mixed it with dissolving water, let it be boiled in *vinum ardens* for half an hour only. Then distil it in a blind alembic and let all stand for one day. Thus you will find certain oils distinct from one another, which separate. Each one of these is a magistery in its own nature.

But the magisteries of oils are made without the addition of dissolving waters; and these in like manner have virtues corresponding with the virtues of their matter.

The Extraction of the Magistery from Growing Things

If we speak of growing things, we understand as growing things those which are green and afterwards wither, and again become green in their season, even as they were before. The magisteries of these are made and extracted from them in different ways, as, for instance, in one way from trees, in another from herbs; the difference whereof consists in this, that the former is wood, the latter putrefiable matter. Therefore the leaves and flowers must be prepared as herbs are. Accordingly, we will set down these Magisteries separately. The preparation of the magistery of woods is as follows:—

Take wood, cut sufficiently small, which put into a glazed pot able to stand the fire, and closely shut up. Let it be burnt in a coal fire for four hours. Then take it out and putrefy it in glass for a month. Afterwards distil it in ashes even to its last spirits, and when these are perceived, presently cease, so that the magistery may not acquire any evil odour from the fire. In this way you will have the magistery of the wood which you undertook to prepare.

So, too, may seeds, roots, barks, and the like, which contain an oil in their material substance, be extracted. There lies hid in these extractions a greater art than is spoken of or understood, though the process is here shewn in its entirety.

Herbs, however, and other things of that kind, must be mixed with *vinum ardens*, and putrefied with it for a month. Then they must be distilled by the *balneum maris*, and that which is distilled must be again poured on. This must be repeated until the whole quantity of *vinum ardens* shall be four times less than the juices of the herbs. Distil this in a pelican with new additions for a month, and then separate it. When you have done this, you will have the magistery of that matter or that herb which you selected.

We now wish to make clear the magistery of wine, which, indeed, appears to be endowed with countless virtues, since it receives a nature of the same kind from many virtues lying hid in the earth, as we set down in our treatise on The Generation of Wine. Now, it should be known that the magisteries of wine are produced in two ways, one of which we pass over in silence, because the practice thereof is common, such as we use in many extractions of wines. Some practical methods for this magistery of wine are here set down. Certain persons seek to extract it while the wine is still new and boiling, during the time of purifying. Some bury it, and leave it so a 100 days (or years). Most people separate it without fire. However, this may be done, I do not wish to write of it here. I will only hand down that method which I have proved experimentally. First of all, it should be known that wine is a very subtle spirit, though small in quantity, and contained in much phlegm. And although this be a quintessence, nevertheless, a magistery can be made from it, but by using a superior practice and process.

It should be known, too, that there are more and greater virtues in that wine which has not yet deposited its tartar, since there are sometimes more virtues in the tartar than in the wine itself. The age of wine, too, is more commendable than its youth or newness, for its spirit is more digested by the lapse of a longer than of a shorter time. Moreover, it must be borne in mind that the wine for this purpose has to be buried in the cold earth, and the vessel containing it must be closely shut below and above without any vent. In this way it is preserved for many hundreds of years without tartar. We are unwilling to speak here of any prolonged time, which would be tedious; but, nevertheless, let this be committed to memory. So, too, that is not a magistery of wine which is extracted out of must, or new wine, but a magistery of must. Nor is it a sign of art to distil it with its own fæces, or its own phlegm, as with *vinum ardens*, because in this way the spirits of the wine, which are in its essence, are destroyed. This should on no account be done. So, the oftener the best vinegar is distilled, the farther does it withdraw from the spirit of wine. Wherefore, the utmost care must be used so as to see that the quintessence shall not be in any way corrupted in the magisteries, but that it shall rather be increased and fortified in its virtues. Again, when it is separated without fire, it cannot in any way be a magistery, since the substantiality is lacking to it. Also, it should be known that the spirit of the wine must be kept along with its substance and not with its phlegm. For in wine two substances are found: the one is vinous, in which is the spirit of the wine, and from which it cannot be separated; the other is phlegmatic, which is mixed with the dregs, and with the sweet water. These should be separated from the true substance, as a metal is from its mineral, or from the earth.

It must be known concerning wine that the dregs and the phlegm are, as it were, the mineral, and that the substance of the wine is the body in which the essence is preserved, even as the essence of gold is latent in gold. According to which we put the practice on record, that so we may not forget it, as follows:—

Take very old wine, the best you can get as to colour and taste, and of the same as much as you please. Pour this into a glass vessel, so that a third part thereof may be full. Close it hermetically, and keep it in horse dung for four months at a continuous heat, which heat do not allow to slacken.

Having done this, then, in the winter season, when the frost and cold are excessive, let it be exposed to them for a month, that it may be frozen. In this way, the cold thrusts the spirit of the wine, together with its substance, into the centre of the wine and separates it from the phlegm. Throw away that which is frozen, but that which is not frozen you must consider to be the spirit with the substance. Having placed this in a pelican with a digestion of sand, not too hot, let it remain there for some time. Afterwards take out the magistery of the wine, concerning which we have spoken. As to those additional processes which are in existence, and are put in practice, we will speak clearly when we shall treat of elixirs. Coming to an end here, we will ignore some of those other processes concerning wine which have little attraction for us.

CONCERNING THE EXTRACTION OF THE MAGISTERY OUT OF BLOOD

We wish also to unfold our opinion concerning the blood, in which there are very many and wonderful virtues, exceeding belief, but still sufficiently evident, chiefly on this account, that the blood exists out of the best root and most potent fountain of the heart, as we make clear in the treatise on The Composition of Man.* In this exists, and can exist, no defect, for it has its conditions according to the nature of the heart, and is a costly treasure of the whole nature, with all that is therein.

Here someone may urge that the blood, when it has flowed out of the veins, will be soon deprived, by the nature of the case, of those virtues which renew and sustain that blood. But this is not so; for it can be preserved in the essence, as we shall point out below. Let us, therefore, consider the small number of men who live with healthy body and blood. Wherefore, care must be taken that men of this kind be brought to a renewed quality and essence by arcana and the quintessence, as we have before mentioned, so that the blood may flow from them incorrupt and healthy. Nor do we speak of man's blood only, but also of the blood of the sperm, which we describe in our secrets, in which blood there is no disease or alteration, but the most wonderful blood of the human sperm, which we intend to take in this place, and this for many causes not here set forth. We speak also of the blood of bread, which is to be taken for the same use and in the same way. For there are in it such virtues as we are scarcely able to scrutinise thoroughly, nor do we undertake the task of doing so to the full. In the same way it may be understood of other nutriments and comestibles, in all of which blood is present, although we do not see it in them any more than in bread; yet, nevertheless, by putrefaction, as in the stomach and the liver, it becomes blood. So also every condiment which is taken therewith is changed after the same fashion as in the body. We are unwilling to write of this blood more largely or at

---

* The *Liber de Generatione Hominis*, to which allusion is apparently made, exists only in the form of a fragment, and is concerned chiefly with the seminal philosophy of Paracelsus.

greater length, especially for the reason that we cannot make the subject agreeable to any but ourselves. We determine, therefore, to take rest in sleep, and then, waking from our pleasant slumber, we will go on to speak further about this blood. Let each thing prevail so far as it can by its own virtues, and according to what it has in it: for out of a good thing much good proceeds, and this always stands forth for our consideration. Nor shall we speak only of the blood of edible things, but also of potable things, which exhibit simply blood to our body. There can also be extracted from blood quintessences as well as arcana; but of these former here we make no mention, having set before us to treat only of magisteries and to comprise each in this one:

Take blood, shut it up in a pelican, and suffer it to rise up so long in a *venter equinus* until the third part of the pelican shall be filled. For all blood in its rectification is dilated according to the quantity, and not according to the weight. When this time is fulfilled you must rectify it by a bath. In this way the phlegms recede, and the magistery remains at the bottom. Having shut this up in a retort, and hermetically sealed it, distil it nine times, as we have taught in the book, "Concerning Preparations." In this way you will arrive at the magistery of blood.*

---

* The treatise Concerning Preparations makes no reference whatever to the extraction of a magistery from blood.

THE END OF THE SIXTH BOOK OF THE ARCHIDOXIES FROM THE
THEOPHRASTIA CONCERNING MAGISTERIES

THE SEVENTH BOOK OF THE ARCHIDOXIES

From the Theophrastia of Paracelsus the Great

Concerning Specifics

IT remains, now, that we speak also concerning specifics, in which reside many marvellous and great virtues, which do not derive their origin from Nature, so far as they are warm or cold; but besides these qualities they have one nature and essence, as we have mentioned in several places. Such a specific takes its nature from externals, as when a piece of wood is thrown into the fire and is burnt; that burning is not part of its peculiar nature, but the being wood is. Specifics, therefore, are produced from a conjunction, as when mastick and colophonia are blended, an attractive substance is generated, which is neither of these; or, when terebinth is coagulated, a stone is produced therefrom which attracts iron to itself like the magnet. There are kindred substances which have such virtues, but from their compositions and from without. Ellebore, too, is compounded from the liquid of stone and earth. From the composition of these two a specific proceeds. So the oil of cherries and acetum after their digestion produce a laxative, though neither of them in its own nature has a laxative property. Wherefore, such specifics are produced out of their own nature by composition of elements and of the proper nature, just as tincture or colour, which is not derived from cold and heat, but from composition, as galls with vitriol produce ink, though neither the one nor the other is black by nature. So, also, sal ammoniac and urine produce a black colour, when both of them are white. It is after the same manner with specifics. They have their origin from externals; but some things which assume such virtues from without may exist in any herb, not in any one kind only, which must be understood in the following way. Wherever the magnet has grown, there a certain attractive power exists, just as colocynth is purgative and the poppy is anodyne. This arises from the composition which exists in them. Hence it is that every magnet is attractive and all colocynth is purgative, But such is not the case with specifics from without, of which the condition is as follows: if one flint has the virtues of a magnet, and another like it has not such virtues, this latter would be an external specific.

But it seldom happens that a peculiar condition is found in one herb and not likewise in another which is similar to it. Then, again, though many similar specifics arise from influences, nevertheless, we will not here discuss at length whether they arise from the same influences or not. We prefer to reserve this discussion for its proper place in our Physics, and to pass it over here. Moreover, many specifics are found—the odoriferous ones, to wit— which derive their origin from composition and digestion, as water of vitriol with sal ammoniac possesses the odour of musk, though neither of the constituents has this by itself. There are many such, which become odoriferous when they were not so before, and acquire a notable fragrance, like a rose or a lily, though they had no odour before, yet by labour, by digestion, and separation, this fragrance is eventually aroused. In like manner, cow-dung is a fœtid excrement, but if it be elevated, it acquires the odour of ambergris, while the residuum which is left at the bottom smells worse than human fæces.

There are some diaphoretic specifics, producing sweat, which acquire a similar virtue by composition, as a burning coal placed on fat earth emits a vapour. So ginger placed on a body burns, and is extinguished like lime when water is poured on it.

This heat accrues to the ginger from the sharpness or asperity which it contains within it, and with a warm element it is coagulated like a stone, which, when placed in the fire, is brought to the same degree of heat. Indeed, every diaphoretic is a calx of the liquid of the earth, as we state in our treatise on Generations. In like manner, purgatives themselves come from composition, as in the case of rhubarb, which is also the calx of a liquid, with a certain difference intervening : for as a Tartarus draught it is resolved into water, and has with it some liquid in itself when it is put in a damp place ; so also rhubarb and other purgatives have a manifold origin, as being a calx of the earth. Some remove the cholera, as rhubarb which is like calcined tartar. Some take away phlegm, as Turbith— so written because the word was wanting in the autograph. With these realgar is dissolved, but nothing else. Others cure melancholy, like senna, which is to be understood after the manner of nitre, which resolves stones as nothing else does. Some act on the blood, as manna, which, like arsenic, resolves sublimates. Thus we must judge concerning the difference of these substances, seeing they are divided one from the other as we have mentioned. Some which are strengthening are derived from composition, as sperm lacking strength, from which, nevertheless, a glandule is produced according to its predestination. So then the strengthening specific is a certain predestinated body by the predestination of its composition. But the carline, which is not produced in this way, attracts to itself the virtues of other herbs, and takes away their strength, which it then possesses solely, just as the sun attracts to itself the moisture from wood, as we declare more at length in the treatise on Generations.

In like manner, also, some mundificatives acquire that virtue by composition, as when the calx of the earth is transmuted into another form by liquid, as resin, honey, gums, pitch, etc. Similar alterations exist in the flowers of Venus which at first are a purgative, as should be the case with a calx. Afterwards they are reduced by fire into a styptic, so that they lose their purgative properties and have a mundificative effect. It is the same with corrosives which are salts, and sometimes calcined in earth, sometimes substantiated (if one may so say) into one matter, as we set it down in the book on the Different Species of Salts. Many and various are the properties of this kind in things, which we handle in our treatise on The Properties of Things, and about which we have said sufficient here. Also why some are hard and others, on the contrary, soft, this we are unwilling to disclose here. Our Archidoxies do not treat of the whole principle, but only the special subject of compounding specifics and bringing them to the highest grade of Nature. With these we will now deal in due order, putting forward our own experience of them, and leaving behind those crooked haranguers who talk about God and understand nothing but hypocritical rites and similar fables, who also are the enemies of such as practise these arts and arcana. They are plunged in worldly glory, romancing and cavilling through their capacity for much speaking. At it they go with both shoulders (as folks say); they flatter and pride themselves upon being wise, when they are only stupid and fatuous, and for filthy lucre's sake they befool their fellow men. Now, then, let us go on to treat concerning specifics according to our usual custom, and leave those University doctors, who only read and think—be their success what it may—to gnash their teeth against us if they will.

Concerning the Odoriferous Specific

Now, then, let us speak of odoriferous specifics, in what way and in what form they are produced. And, first, as concerns their powers. An odoriferous specific is a matter which takes away diseases from patients, just as civet destroys ordure, with its odour. This specific mingles itself with the foul smell of the ordure; and the smell of the filth can no longer hurt or remain. The stench is tinged with fragrance, so that the good odour is as strong as the bad one was before. Since there is nothing which can take away the good odour from civet or musk, it is transmuted, as we prove in many passages. Hence it happens that occasionally some of the excrement is mingled with the musk, because this penetrates more readily than any lily with all its operations. It is well known that more of bad odour than of good is met with; for as tyrus is taken in theriaca in order to penetrate all the limbs at once with another influence, so, in our opinion, does the odoriferous specific act.

By means of the odoriferous specifics diseases are cured in persons who cannot take medicines, as in cases of apoplexy and of epilepsy. Many odours exist which relieve the epileptic; and many, too, which aid the apoplectic. They may not cure them, but they pave the way for a cure. A force of this kind, brought to bear on the body, immediately stirs up the blood, and by forcing this to the heart, revives it to an indescribable degree. We will therefore prescribe an odoriferous specific which shall serve as a foundation for compounding such specifics against all diseases. The process is as follows:—

Take of white lilies, anthos, basilicon (carbon), cardamum, and roses, one handful respectively, with two handfuls of spike. Pound into a coarse paste. Add two quarts of the juice of oranges. Let all these be digested in a pelican for the space of one month. Afterwards separate them from their dregs with the hand, or, what is still better, with a press. Put this again in a pelican, and add one ounce each of mace, caryophylli, and cinnamon; half an ounce of ambergris, two ounces of musk, and one ounce of civet. Having ground these very minutely, put them in the aforesaid pelican, and digest them in dung with the other ingredients. Then add half an ounce of gum arabic in solution, also one ounce of similarly dissolved tragacanth. Place these in a closed glass vessel and indurate them with the clear part from white of eggs. You will presently see the mixture assume the form of glass. Break the vessel, take out the stone, and you will have the odoriferous specific, concerning which enough has now been written. It would, however, be well to add to it aurum potabile.

Concerning the Anodyne Specific

Many causes combine to induce us to write about this anodyne specific. There are some diseases in which all arcana fail us, with the exception of this anodyne specific, which works wonders. Nor need we be surprised at this, when we see water extinguish fire. Just in the same way the anodyne specific extinguishes diseases. There are many reasons why it does this, but we pass them over in silence.

That which rests or sleeps does not, in the course of nature, cause any discomfort. A paroxysm, if it sleeps, is not felt; if it does not sleep, its operation is accomplished. We console ourselves with the reflection that many anxious cares and cases of melancholia are removed by simply sleeping.

And note this: Sleep does not necessarily apply to the sufferer only, but may be predicated of the disease itself. We, therefore, compound a specific which operates on the disease alone, not on the entire patient, as is the case in those applied to fevers. What we put forward would be deadly to the whole man, but is salutary in its application to the disease. We attack the disease itself, and so elaborate our remedy that the ailment shall have, or can have, positively no effect on the body. The following is the formula: Take of Thebaic opium, one ounce; of orange and lemon juice, six ounces; of cinnamon and caryophylli, each half an ounce. Pound all these ingredients carefully together, mix them well, and place them in a glass vessel with its blind covering. Let them be digested in the sun or in dung for a month, and then afterwards pressed out and placed again in the vessel with the following: Half a scruple of musk, four scruples of ambergris, half an ounce of crocus, and one and a half scruple each of the juice of corals and of the magistery of pearls. Mix these, and, after digesting all for a month, add a scruple and a half of the quintessence of gold. When this is mixed with the rest it will be an anodyne specific, capable of removing any diseases, internal or external, so that no member of the body shall be further affected.

Concerning the Diaphoretic Specific

Let us now speak of the diaphoretic specific, whereby every disease is cured which can or should, from its nature and properties, be treated by sudorifics. Such a disease is removed barely better than by any medicament. By means of the diaphoretic, a cold disease grows warm, and is removed by that heat. It has often happened to us that a cassatum of twenty years' standing has been cured by the diaphoretic; and other diseases which they call intercutaneous, or which have their seat in the marrow. On these the quintessence has no effect, and still less the strengthening arcanum; principally for this reason, that there is not sufficient strength in the heart to drive away the cassatum unless this be done by the diaphoretic alone. As the sun warms a frozen stone, and liquefies the hardest ice, just in the same way the diaphoretic exerts its powers on a disease which can be overcome by no other power, however excellent and good. Now, although the flammula be a very warm herb, still its heat cannot be compared in the faintest degree with that of the sun; and so these same warm diaphoretics differ one from the other. Wherefore we put forward at this point a specific which, as it were, summarises the diaphoretic properties. Take one pound of ginger, half an ounce each of long pepper and of black pepper, three drams of cardamum, and one ounce of grains of paradise. Let these be ground to powder, and placed in a glass vessel with an ounce and a half of best powdered camphor and two ounces of dissolving water. Let this mixture remain in a sealed glass vessel in sand so long until its digestion is completed. Then separate therefrom the dissolving water, let it putrefy for one month, and circulate it for one week. Afterwards press and keep it. This is the best and most powerful diaphoretic, acting more vehemently than could be believed both in cassatum and in other diseases. Enough on this point.

Concerning the Purgative Specific

We must also describe the purgative specific. And although we shall have to consider its complexions and the like, still we base our remarks on a more solid foundation, and accept only those specifics which exhibit not one disease or another, but all diseases. Hence it is gathered that whatever in the case of cholera is superfluous, and of no moment, is removed by this specific. Similarly also in cases where the symptoms show the phlegm or the blood at fault, or in melancholia, as also in abscesses and other affections which cannot be cured by mere complexions, nor be resolved by purgatives, as is the case with many diseases.

Our endeavour is solely to remove the peccant matter, whether it be corrupt or not, whether in the form of an abscess, or in complexion, or arising from any admixture. Paying no heed to the prolix and useless discourses of the University doctors, we give our thought simply to sanitation, and, with this end in view, we shall build up our medicaments. Of such a nature, for example, is tartar, which, by its nature and properties, takes away all putrefactions and is not inclined to cholera, to melancholy, to phlegma, nor to the blood, but expels all that is useless from the body, or that can do harm to it. In like manner, vitriol purges every complication from which diseases arise. Coloquinth does not drive out all cholera or purge what arises therefrom; nor does rhubarb effect this. Neither does turbith arrest or evacuate all that proceeds from phlegma. Neither does lapis lazuli remove the impediments produced by melancholia, or manna drive out all bad blood; but these two, besides other of our purgative specifics which we do not mention here, accomplish this, principally in all those putrefactions and superfluous fæces of the body, whencesoever they arise, just as water washes linen rags, and soap clears them from all defilement and impediments of any kind. So also do these act singly on single diseases. Let us then lay down a specific drug which, to our thinking, acts thus in the way just described. Take the magistery of tartar and the magistery of vitriol, mix them together, and afterwards add equal parts of the quintessence of crocus. Put them in a pelican, let them be digested for a month in sand, and let the drug of which I have spoken be carefully preserved. Concerning the other things necessary for this, greater intelligence exists among the erudite than among the ignorant.

And not only human beings may in this way be purged of their diseases and superfluities, but also trees and herbs. For in growing things, just as in persons, infirmities prevail, and their remedies, too, have grown, as we remark elsewhere. The magistery of vitriol heals the defects of anthos, when it cannot vegetate perfectly, and causes it to grow excellently, as we state in our work on Plants.* With these few words, therefore, we now conclude this subject.

Concerning the Attractive Specific

Before we begin to speak of the attractive specific, it should be known that this attractive specific draws to itself everything that is superfluous in the body, and leads out whatever evil may adhere to it. This frequently happens; and the result is proved to demonstration.

Some attractive specifics are so adapted to the flesh that they attract to themselves a hundred pounds of flesh, just as a magnet attracts iron. It has happened in our time that an attractive of this kind has drawn a man's lungs into his mouth and so suffocated him. It has also occurred that, in another case, the pupil of the eye was drawn from its position right down to the nose and could never be removed thence. Attractives are found to act not only on

---

* The *Herbarius Theophrasti* contains no reference to the subject, and selections have already been made from that chapter in *De Naturalibus Rebus* which is devoted to vitriol.

iron, but also on wood, herbs, flesh, and water. We have seen a plaster which attracted so much water that a vessel could be filled from it, and the water ran down from that plaster just as from the roof of a house. In the same way, lead, tin, copper, silver, and gold can be drawn by the composition of attractives. By these same attractives, too, the branches may be torn from trees, a cow lifted into the air, and many other effects produced, which we have detailed in our secret writings as a Thesaurus; so that we may admire and venerate that man alone who has brought it about by his art that so many discoveries should be made; demonstrating as we do these incredible operations which so far excel Nature as it is constituted in itself.

We will, then, lay down certain attractive virtues for the body, whereby what is evil and corrupt may be extracted from it and separated from what is good, so that the attractive may be applied on an emunctory to the particular spot where any defect has shewn itself, as on an ulcer which may be taken on an emunctory, or if a glandular swelling has arisen this must first be opened like an emunctory. We know from experience that by means of an attractive of this kind the pestilence has been extracted in a way that it would not become us to describe here. When this medicament has been used, no sick person has ever perished, however severe the disease with which he was affected. The following is the recipe for the attractive specific: Take of the quintessence of all gums of every kind one quart, a pint of the magistery of the magnet, one pound of the element of *carabis igneus*, of the element of *mastix igneus* and of myrrh, each three pints, and ten ounces of the element of scammony. Of these make an ointment with wax, gum, tragacanth, and turpentine, using it as directed above.

Concerning the Styptic Specific

We now have to speak of the styptic specific, the virtues whereof are very great, more so even than the others. When people see with their own eyes so many wonderful works of Nature, they urge us, in their delight, not to desist from them, but to tax our memory so as to impart what is offered by these arts of ours. And if, perchance, something was discovered by the old physicians and philosophers (which we deem by no means certain), that does not disturb us in any way, since all those things they wrote about were mere blind gropings. We delight in that great Nature which presents itself to our hands, and we rightly pass by the works of these ancients whom we esteem to have been blind, as we mention in many places. Is it not subject for wonder that styptics are so strong and possess a power in their quintessences, arcana, and mysteries, whereby two fragments of iron are so fastened together that afterwards they can only be separated by fire? What is still more remarkable is, that one piece of copper can be so joined to another by similar styptics that they can no more be disjoined either by fire or by water. By a like attractive styptic a heap of stones can be conglutinated into a mass like a rock. In like manner, sand and lime, by means of such styptics, are welded together in perpetual compaction harder than marble. So far we speak of hard material. Let us go on to other kinds. By the same method we have seen leaves joined together so that they were taken for one concrete natural growth; for instance, the leaf of the lily with the leaf of the rose in one combination. Blacksmiths, acting on my advice, have made their weldings as compact as though they had been originally one solid mass.

We have seen, too, by means of these styptic specifics, the lips of a person so firmly drawn together on a single washing, that afterwards they could only be separated forcibly by instruments, the blood flowing freely in the process. Some persons, too, for a joke, closed up another's anus with these styptics so firmly, that when it was necessary to go to stool it had to be opened with a gimlet! So, too, in the case of wounds and in rupture of the bladder, we have seen the compaction made so effectually that no opening occurred in our time or during the life of the patients. Whatever limb this styptic material touched, it so contracted the orifice or the bare flesh that it could only be made smooth with a file. No water softens the force of such styptics, which is really much greater than we have described. So far, however, as relates to medicine, we will detail the styptic specific as follows:—Take quintessence of bolus, that is to say, of iron, and the quintessence of carabis, otherwise called cathebes, one pound of each. Let these be digested in ashes for a month, and afterwards put in a pound and a half of dried tartar. Keep this body until required. This and similar styptics are inscrutable in their own bodies, but when employed on separate substances, they attract in an incredible manner, through the nature and condition of their great dryness. So, then, these styptic specifics may be called styptic beyond all others.

Concerning the Corrosive Specifics

We would now describe in addition the corrosive specific, wherein marvellous powers are naturally present. It has this wonderful property, as compared with the old corrosives, that it completely annihilates metals, so that no body is any longer found in them, just as wood disappears in the fire. It is true that the consumption of metals is brought about by aqua fortis, but still some part of their bulk remains undiminished and unchanged in its essential nature, so that it can be once more brought back by means of fire to its original body and matter. But this is not the case after consumption brought about by the corrosive specific, and for this reason: because no matter is any longer found in this case which can by any means be brought back to a metallic nature, any more than ashes can be again turned to wood. It should be known, too, that this corrosive acts upon flesh in a way to which nothing else is comparable. Its velocity in one moment penetrates the hand like an awl.

Now, we recount this in the interest of medicine, with the following end in view. In the body a good deal of putrid superfluous flesh grows up, in the way of ulcers, like fistulas, scrofulous cancers, etc., all of which can be cured by corrosives of this kind. In this specific a styptic virtue exists of very great power, by which it acquires special curative faculties. It might really be called rather a fire than a medicament, seeing that it consumes iron chains and bars in a way beyond belief or power of description. We give the recipe for it as briefly as possible: Take one pound of aqua fortis rectified from its *caput mortuum*, half a quart of sublimated mercury, two ounces of sal ammoniac. Mix all these and let them be consumed; then add an equal weight of mercurial water. No adamant can resist this corrosive. Although the same is understood concerning the quintessence and the arcanum, namely, that the skin may be taken away by either mode of cautery, and new skin superinduced in its place, as in leprosy, morphea, serpigo, lentigo, pannus, etc., against all these this corrosive specific comes into use; but we omit it on account of its violent action. We choose in preference this mixture, with which the skin should be washed, when it falls off and is laid bare. Afterwards it can be consolidated in due course. The mixture should be as follows:—Take one pound of the juice of the flammula, four and a half of cantharides, and ℨ ij. of the Ignis Gehennæ aforesaid. Mix and use as above.

Concerning the Specific of the Matrix

One would speak of the specific of the matrix for several reasons, specially on account of the affections to which it is liable. But in this place we hesitate to treat of those substances which warm or refrigerate it, since these results can be brought about by magisteries and arcana. Here, however, we set down two specifics, one for suffocation of the matrix, the other for promoting or restraining the menses. Suffocation can only be remedied by the specific, though not elementated or prepared for this object, but in its common form or essence, adapted to the said purpose even as the skin of the fig is. As soon as ever its fume enters by the vulva the disease is expelled. And this is chiefly remarkable on account of its being such a cheap simple; and though it be prepared, the essence of its fume, wherein lies its sole virtue, is not lost. For promoting the menstrual discharge the method is the spleen of a castrated ox reduced to a magistery or quintessence. This is an admirable provocative not only in young, but in old, women.

As to restriction, the best method is to use the quintessence of coral, the oil of iron, or *ferrum potabile*, which restricts more than anything else.* There is no need to describe the properties at greater length, since this would be too prolix for our Archidoxies. It should be remarked that, under the above-mentioned compositions, that is to say, the incarnatives, the conglutinates, and the specifics, are comprised laxatives, mundificatives, and the like, under purgations themselves; and so with others, as the deoppilatives under

---

* By this oil of iron every genus of humid, fluid, and flaccid ulcers, those also which have swollen and ruddy lips, are completely cured, and the older and more deeply rooted the ulcer is, so is the remedy more ready and easy.—*De Tumoribus, etc., Morbi Gallici*, Lib. X.

purgatives and attractives. Here we would conclude our book on Specifics, as an aid to memory, lest we forget them. In like manner the confortatives are also recounted in single chapters.

The End of the Seventh Book of the Archidoxies from the Theophrastia of Paracelsus the Great concerning Specifics

THE EIGHTH BOOK OF THE ARCHIDOXIES

From the Theophrastia of Paracelsus the Great

Concerning Elixirs

AFTER many of the most secret mysteries of Nature, we desire to treat compendiously of Elixirs; and that not in vain, since we see that there is latent in them the chief conservation, which compels us to bend our mind thereto without any rest.* For every elixir is an inward preservative in its essence of that body which shall have taken it, even as the extrinsical balsam is an external preservative of all bodies from putrefaction and corruption, a fact which is evident enough in balsam, that is to say, it preserves bodies so that they abide many hundreds and thousands of years without corruption or change.†

When, therefore, we see a gift like this in balsam, which preserves dead bodies and keeps them in incorruptibility, it is to be understood equally that by this same gift, or mystery, the sound and living body can be far better, more usefully, and more conveniently preserved. But we have not this naturally that these mysteries of Nature and those constituted above Nature, by which we may preserve the body inwardly and outwardly from all contrariety, should be known to us; but among them many things meet us which are most occult.

Concerning elixirs, then, it should be known that these have not their operations from Nature, nor from their complexion, but that they are mysteries rather than specifics, leading us to the very highest admiration of the Creator by many demonstrations. Yet they are implanted in Nature herself, so that they are in her, as is seen in the case of balsam. If, therefore, it be possible to preserve dead bodies, still more so living bodies themselves can be preserved. And there is no reason why we should heed the words or arguments of our adversaries, but rather it is well that we should disclose our own, by which means we desire to guide towards the true foundation of the internal balsam, not heeding their triflings and their useless words who talk of a limit

---

* The *Scholia in Libros Paragraphorum* define the elixir as a ferment, a quintessence, the pure substance separated from the impure.

† Of balsam in general, Paracelsus says that it is a temperate thing, neither sweet nor bitter, nor pontic; it is a liquor of salt and a salt of liquor. Therefore it most powerfully preserves from putrefaction. - *De Tartaro*, Tract. IV., c. 2.

of death and of its predestination, and close it in with determinate points. For God our Father gives us life, and along with it medicaments by which we are able to defend and sustain it. If, therefore, the limit of death were laid down at a precise point, it would of necessity follow that this other theory should be false, which is not the case; but as long as we have power and knowledge, we possess the capacity of sustaining our life. For Adam attained to such an advanced period of life not from the nature or condition of his own properties, but simply from this reason, that he was so learned and wise a physician, who knew all things in Nature herself with which he sustained himself during so long a period. And there were many others, in like manner, who used similar remedies.

There were in the days of Adam many who died without reaching his age, and some did not even attain our limit, as we are now constituted since the Deluge; and these died just as we die, because they were unskilled in those arts which Adam and others understood; and hence it happened that they were deprived of life before their due time, nor did their foods or drinks help them.

Since, therefore, we are enabled from such examples as these to discover naturally that a protracted life proceeds from Nature, we desire to investigate what it is in which Nature and the gift of God consist. Some things preserve a dead body from putrefaction for one year only, as the *oleum laterinum*. Others preserve it for ten years, as the corrected oil of the philosophers. Some preserve it for twenty, as the water of honey. Some for fifty, as the distilled preservative. But some preserve it for ever without end, as balsam. Some preserve it only for eight days, as salt; others for a night, as distilled water. Some preserve it longer, as *vinum ardens*. There are also some others which preserve the body from corruption in a new and strong essence of Nature when a man is confirmed by them, as aloes, citrine, and myrrh. Some preserve bodies from corruption only by their great tincture, which is so powerful that it admits no evil, nor suffers it to grow or to enter secretly, as gold, the sapphire, pearls, arcana, magisteries, and the like, as has been before written about these.

We purpose, therefore, to describe a preservative against all corruptions of the living and the dead body. But it is considered that a preservative of the living body must be taken at the mouth, so that there may be no member of the body which does not receive that same preservation, or which may not be informed by attracting to itself the benefit thereof. Moreover, it must be remarked that the spirits of the excrements existing in the bowels are so strong that they fight against a preservative, for this reason, that nothing which is putrefied can be embalmed or preserved, for it has not in itself any essence as recently dead flesh which is embalmed has. But this agrees with a preservative none otherwise than as worms do with the best herbs, and as a putrefied substance does with one that is incorruptible; since whatever is putrefied cannot be further corrupted or be changed, since there is no movement in it. On the other hand, a preservative cannot in any way be putrefied, for it is like gold, which never rusts. They are mutually separated one from the other, so that each fulfils its own proper function. This, therefore, is said, because, in lapse of time, the excrements are able to overcome the preservatives; but this cannot occur in dead bodies, because they are disembowelled, or, if not yet disembowelled, they are coagulated by death, just as the blood withdrawing from the veins coagulates.

We call this preservative an elixir, as if it were yeast, with which bread is fermented and digested by the body. Its virtue is to preserve the body in that state wherein it finds it, and in that same vigour and essence. Since this is the nature of preservatives, namely, that they defend from corruption, not in any way by purifying, but simply by preserving. The fact that they also take away diseases is due to the subtlety which they possess. So, then, they do not only preserve, but they also conserve. They have a double labour and duty, that is to say, to prevent diseases and to keep the essence itself in its proper condition.

Nor do they do this in human bodies only, but in all bodies possessed of sensation. Thus, also, dead wood can be preserved from corruption, just as any body that is treated with balsam. Herbs, too, can be preserved in their essence none otherwise than a living body can: since those conservations which apply to herbs keep them in the same essence as that wherein they find them, so that they flourish and remain as fresh as in the fields and gardens, or elsewhere, to the fifth or sixth day. If they find them with flowers, they preserve these; if with fruits, these also.

Nor wonder at this, when it is possible for dead wood to live again, and for iron to be so fixed that it shall never rust again, and likewise for sulphur to be made incombustible, all which things are beyond the understanding of a simple man. The cause of all these things we set down more broadly and fundamentally in the treatise on Conservations. Nor must they be considered impossible, since many other things which are deemed impossible can be most easily accomplished. We wish, then, to speak of the conservation of balsam by the distinctions of ages in the following manner.

On Preservation and Conservation by Elixirs

We purpose to write of the first elixir, which conserves the body in that essence where it finds it, and does not suffer it to putrefy or grow weak, but conserves it in the spirit of life, so that no accident can happen to it. It also brings it to a third period of life, or to more. As to its use, the operation on dead bodies differs from that on the living, since the dead must lie in the balsam nights and days, but the living and healthy bodies neither can nor need do this.

Wherefore, this elixir must be understood to be of use only for life, and for the heart and those parts especially in which the life chiefly flourishes, that is, the spirit of life dispersed throughout the entire body. And it preserves the spirit of life in that virtue wherein a dead body or corpse is kept from putrefaction; because, if a wound or an ulcer may be externally preserved from putrefaction and from every mishap, so, also the inner body is arranged so as to be defended from all adversity. Wherefore, we arrange the elixir which is directed to the spirit of life none otherwise than as yeast operates in paste; and with reference to the body, as when a tree is tinged in its root, so that the same colour may never withdraw from it. In this way the whole body is preserved, when the tincture is more or less scattered amongst all the members and penetrates them none otherwise than as the whole metal is tinged into gold, and becomes gold, or is preserved from rust. So, also, is it in the conserved body; there is no member comprised in it which is not full of the elixir.

After that, being dispersed throughout the whole body, it shall have acquired its virtue, and already is exercising its operation by itself, no corruption can happen to it by contact, because the life of every member is full of the elixir as the body is tinged with the balsam. But it should be understood that it is not necessary the whole body should be affected with the balsam by means of the elixir which has been taken; for where the spirit of life has only been surrounded therewith at its root, this suffices for the conservation of the body.

Now we must come to the practical method; and first of all treat concerning the elixir, which by the conditions of the powers of balsam preserves the whole body from decay. Then concerning that which preserves the body by the potential power of salt. Thirdly, concerning the elixir of sweetness, which supports the body in its conservation. Fourthly, we will treat of the elixir which enters the human body by the powers of a quintessence. Fifthly, another shall be appended which is truly noble by the force of its great subtlety: for it resists all the enemies of Nature, by which resistance it never suffers the body to fall into disease. In place of a conclusion, we will add that elixir which, by the forces of its own proper nature, is endowed with similar conservations.

Concerning the First Elixir, that is, of Balsam

Take of the true and very best balsam, well known to us, one pound. Let this be put into a glass covered with a blind alembic, and together with it pour in two ounces of the quintessence of gold and one ounce and a half of the essence of the greater circulatum. Let all these be digested together with a slow fire, so that the vapours may ascend night and day. Afterwards let the fire be increased so that some drops may adhere, and may fall down, drop by drop, for two months. At length let them remain in horse dung for four months, so that they may have their digestion without intermission. When this has been done the elixir is finished. It must be understood that this balsam or elixir becomes a fermentation which is developed and mingled with the root of life, and has the power of ruling the life in a good essence, so that no nature can resist it. None otherwise than as arsenic overcomes Nature for evil does this elixir, on the other hand, overcome it for good, defending the body. The dead body is kept safe by this odour so that it cannot putrefy when it is put in the tomb, and is covered up so that it may not evaporate. Much more do its own virtues remain in a living body, which we hope to have sufficiently explained in this place.

Concerning the Elixir of Salt, by the Force of Which the Body is Conserved

There is no less power and virtue in salt than in balsam, whereof we have spoken; for this reason, that flesh is preserved from putrefaction by salt for days, years, and long periods; and that in different ways, and one way more than another. On the same basis it would be possible to conserve and to preserve the body; not because we mean to use salt in the precise way as with dead flesh, but from it should be compounded the elixir of salt, which penetrates materially the spirit of life, so that it lives by the salt as salted flesh does. For this elixir is so subtle that it can be brought to bear on the spirit of life. Thus these two meet closely in one conjunction, so that the one is tempered by the other to perfection, just in the same way as salt perfects certain food in point of taste, without which it could scarcely in any other way be brought to perfection in respect of unity. And it must be remarked that the elixir of salt is a fermentation in which exists a certain tincture whereby the whole body is penetrated. It is also an inconsumable thing, which is not consumed in the body by digestion along with natural things, but is fixed as glass in the fire, which does not at all perish in the process. This fixed elixir so fixes the body that it becomes permanent in life, none otherwise than as when a metal is fixed, in which case no damp afterwards, no corrosive, or anything of the kind, can injure it, or produce rust in it. So, then, it may be inferred from hence that the elixir is a fixed body, like gold, into which nothing impure can penetrate so as to hurt it. The practical preparation of this elixir of salt we will set down as follows:—

Take salt, prepared in the best possible way, very white and pure. Let this be placed in a pelican with so much dissolving water as may exceed six times its weight. Let them be digested together in horse dung for a month. Afterwards let the dissolving water be separated by distillation, and again poured on and separated as before. Let this be repeated until the salt is converted into an oil, to which let there be added an eighth part of the quintessence of gold. Let them be digested together in a pelican and in horse dung for four months, and afterwards circulated for a month. Let there be added another part of circulated wine; and let them so remain in ascension for a month longer. When that time has elapsed you will have the elixir of salt, concerning which we have made for ourselves a memorial according to rule for the relief of our ancient days.

Concerning the Third Elixir, namely, of Sweetness

We are certainly assured that bodies may be preserved from corruption by sweetnesses; but by what forces this is done we set down in the books of the Generations of Honey,* Sugar, Manna,† Thronus, and the like; and we are unwilling to repeat it in this place on account of the writings of the ancients. We are able to transmute sweets into an elixir, the preparation of which rather conserves the living body in its conserved essence than a languishing body. For it is the property of all specific sweetnesses, that they are neither corrupted nor do they allow this to be corrupted, unless by contrary things they become liable to corruption, as out of honey and bread worms and ants are produced, as also out of sugar and thickened milk. Out of manna and water is produced a corruption like dung. Many more similar compositions may be made, by which the sweets pass into corruption. That this may be obviated, the following is our intention and experience, namely, that in this composition such a thing should be taken as will not prevent the sweetness from remaining in its proper essence, and such as may be without the corruption of other things. In this way, it has the virtues of a balsam for preserving the dead fleshly bodies of corpses or other things. A like sweetness is the balsam of the earth, and some others of the dew, because it derives its origin from them. Now, therefore, we will set down the elixir of thronus, since no sweetness can be compared therewith, and it contains more mysteries than could be believed, as we state in our treatise on Generations. By the preparation of this may be gathered the methods of preparing other sweetnesses. Let that of thronus be as follows:

Take as much thronus as you will, which place in a pelican, and set in the sun for digestion during two months, or, better still, for the whole summer. Afterwards let there be added a fourth part of the quintessence of gold, and so let them be circulated together for two months. Keep this. Although this method is short, nevertheless the elixir made by it is wonderful in the case of very aged persons.

---

* In the description of the essence and property of honey, it must first of all be understood that the prime matter of honey is the sweetness of the earth which resides in naturally growing things, and is extracted out of the property of the same by magnets. Hence you ought to gather that in every prime matter all that concerns growing things is collected, just as when several colours are combined one only appears though all are present. Similarly, also, the seed is the wood, the leaves, and the branches, not actually and in the present, but for the future, if it be brought forth and grow. So, also, in the prime matter of things there is a similar composite from which all growing things attract what they require. . . . Now ye must know that honey in its first matter is a resin of the earth, but resin is a gum not of all flowers and growing things, but of some only. Of some there is a resin produced, of others a sulphur and bitterness, and of yet others something else. That which grows in flowers and locusts is conceived with the form and appearance of honey. It does not yet dwell in its ultimate matter, being perfected by the Sun and the Moon. These two planets, however, can bring nothing to its final perfection without all the assistance of a celestial operation, which in the case of plants, etc., is the summer star. . . . Honey, therefore, is a terrene spirit at first, but if there be applied the influence of summer, it generates a corporal spirit, that is, the spirit which in the sun and moon was terrene becomes a corporal spirit. Bees can remove the same and take it away to their hives. This is the prime materialised matter, for honey and wax are one; if they be separated like chalybs and iron, it is then called a separated corporal from a materialised matter. For as the alchemists, in a circulatory or pelican, circulate the spirit of wine, so, also, the summer star in natural growing things circulates liquid. So honey emerges from the earth into the materialised matter, which is the subject of the bees, and this materialised matter requires further perfecting. . . . There is a threefold kind of honey in all growing things—the lowest, middle, and topmost. It is the last which the bees seek, and they find it in the flowers, where it is purest, the gross being always relegated by Nature to the lowest place.—*De Melle.*

† Manna is the chiefest and most excellent nutriment and the marrow of locusts. It is the highest natural preparation of the star. The food of those bees that are fruitful in honey and wax is this manna and tereniabin.—*Ibid.*

Concerning the Fourth Elixir, which is that of Quintessences

Similarly quintessences can be reduced to an elixir, which, like balsam, conserves living as well as dead bodies. In this place we make very slender mention thereof, because it has been previously demonstrated in the process of quintessences. So, then, advancing a little farther, we will set down such things as we are mindful of as making for preservation and conservation. Afterwards out of the elixirs of those processes we will teach the composition of one elixir equally profitable to the body as the three preceding. It must be understood that this elixir of quintessences has in it a secret virtue which daily tends to restoration, and endeavours to renovate and restore the whole body. So, then, it produces something more than a mere conservation; for it also renovates, not, however, so perfectly as we have described above concerning quintessences and arcana, but with inferior force, because the conservation and the restoration of these cannot co-exist; still, by this method renovation is disposed towards conservation, in the following way:—

Take the quintessence of chelidony and of balm, each two ounces; of quintessence of gold and quintessence of mercury, each half an ounce; of the quintessence of saffron and of all the mirobolanes, each one ounce. Let all be mixed together and remain in digestion of the sun, enclosed in a blind alembic, for two months. Afterwards add one ounce and a half each of the quintessence of wine and the magistery of the same, and let them be again digested as above for a month. Then keep it as a treasure, not only for preservation, but also for restoration.

The Fifth Elixir is called that of Subtlety

Now, we have thought that we ought to set down something concerning the elixir of purity or of subtlety, which conserves by the force of its great purity, as is the case with the corrected oil of the philosophers. This suffers nothing which has been anointed with it to putrefy. The same effect is produced by the corrected laterine oil, and many others, whereof the property, however, is not to preserve from putrefaction, but they acquire this, and take it as their property, from the preparation and the labour bestowed, as distilled or corrected wine does not allow putrefaction any more than digested wine does, and this, moreover, is not changed by the fire.

The water of honey by its preparation resists putrefaction, so far as concerns sensible bodies; though the crude substance thereof does not produce this effect, but is itself liable to putrefaction. Wherefore we set down an elixir of subtlety, since, just as mercury itself, which is volatile, is fixed and becomes permanent through its own water, so the human body also is fixed into consistency and permanence. Now, although this may be done by many other methods than that which we here describe, nevertheless, we mention only those which are experimentally known to ourselves. Not that on this account we wish to detract in any way from the others; only we say that all these things have not yet come to our knowledge and experience. The process of this elixir is as follows :—

Take olive oil, honey, and *vinum ardens*, one pound of each. Distil them all together after the manner of the alchemists, and do this thrice. Afterwards separate all the phlegm from the oils, which are distinguished by their many colours. Put all these oils into a pelican and add to them a third part of the quintessence of balm and of chelidony. Digest for a month. Afterwards keep it for use. No sensible or insensible body can resist it, on account of many causes and properties which we are unwilling to set down in this place.

The Sixth Elixir, which is that of Propriety

Equally from natural objects a perfect elixir can be extracted, as out of myrrh, saffron, and alöepatic. As to what forces it proceeds from, that we set down in our treatise on the Generations thereof. Here we only put forth the process, leaving the origin, which we often treat of elsewhere.

Take of myrrh, of alöepaticus, and of saffron, each a quarter of a pound. Put these all together into a pelican, set them in sand, and let them ascend for a month. Then separate the oil from the dregs by means of an alembic without burning. This oil suffer to digest for a month, together with circulatum of equal weight. Afterwards preserve it. In this elixir are all the virtues of the natural balsam, and, moreover, such a conservative virtue for old persons, more than it seems right to assign to it. For not only one period of life seems to proceed from it, but four, seven, or ten. It is scarcely possible to express its force and natural powers, but, so far as our judgment goes, it has been sufficiently elucidated, nor do we think it requires fuller explanation.

The End of the Eighth Book of the Archidoxies from the Theophrastia of Theophrastus Paracelsus concerning Elixirs

THE NINTH BOOK OF THE ARCHIDOXIES

From the Theophrastia of Paracelsus the Great

IN the preceeding books we have treated of internal diseases. Now we have to write of those which arise without, and to set down remedies for the same. Although we insert nothing in these books as to the origin of diseases, internal or external, we will lay down the origins of the medicaments against them, and afterwards the composition of similar remedies for external ailments. Some refer only to wounds, and by these remedies a wound can be healed in twenty-four hours. This is to be understood in the following manner:—A wound that has been inflicted requires nothing else save that it be again connected or conjoined, just as two planks are connected with glue. You should on no account allow wounds to lie open, but should endeavour to refill them with flesh. This is a matter for the rustic rather than for the physician. Consider that when the lips of the wound are joined, as the planks by glue, they are more than half healed. What has to be done by some kind of medicament is to bring together each side as well and as closely as possible. Thence it follows that when the lips touch, and the compression of the medicament aids Nature, the cure is complete. So that no wound, in which no fracture of a limb is involved, can be so bad as not to be cured in twenty-four hours. Bones cannot be connected in the same way as flesh can; so in this place we do not speak of them. You can understand the matter by an example. When some limb is altogether cut through, before the veins are dead, and while they are still warm and fresh, let them, as soon as possible, be moistened with the medicament, the wound joined together and its two sides connected exactly like two sticks fastened together with glue. Thus they are healed and grow together. This the medicament effects, which Nature takes care of by its power of resiccation, and heals through that power whereof we have spoken above. But it should be understood that the medicament for wounds should not be incarnative, or mundificative, or attractive, because such medicaments draw out the putrid fluxes and cause much matter to be formed. Moreover, the opening or cavity of the wound has to be filled with flesh; and this is done very slowly and, in consequence, with a good deal of peril, and without a magistery. This is also the case with old-standing ulcers, which in course of time have become loaded with discharges, whence it happens that they can only be cured with many accidents and much difficulty, indeed, sometimes can never be cured at all. Wherefore, there is urgent need of some medicament for them, such as we have mentioned, which also with a certain force compresses in like manner the skin and the cavities.

In like manner, it must be remembered in the cure of ulcers that the formation of flesh is necessary, and this cannot be brought about by mere compression. As we said in the case of wounds, so in that of ulcers, fistulas, and the like, all these ailments have to be cured by the force of medicaments. We lay down, therefore, two fundamental rules for this flesh-forming—one being incarnative and the other exsiccative.

And now to speak of other malformations of the skin, such as cicatrices, morphea, serpigines, pannus, spots, leprosy, etc., including all diseases arising from the skin. For these we prescribe the following method of treatment. First we ordain that all the skin shall be stripped off, just as the flesh is stripped from a calf. Afterwards a new skin must be induced by the proper medicament. It will be inferred that the skin must be removed by some medicament, and a new, pure, immaculate skin generated by some other in colour, like that which follows, so that not much of the flesh and the humour may be attracted thither; and thus, as we have said, any spots are removed. The origin of such removal we have not from the beginning mentioned here, because it has been treated of elsewhere, and it does neither good nor harm to our intention and to our present teaching. There are other diseases also, such as cancers, buboes, and the like, which require their special medicament to remove their origin, and to purge away all their impurities. This is best effected by the attractive specific. Then there is need of consolidation, as we show in our treatise on fistulas and the like.

Ruptures of the bones and the like can only be consolidated with an attractive styptic; and this we need not discuss afresh, since we have spoken of it elsewhere. In like manner, many superfluous growths are found, as strumæ, glandular swellings, etc., which ought first to be emptied and afterwards cured.

We will, therefore, divide surgery into three parts or methods of cure: one referring to wounds, a second to ulcers, and a third to spots. Cancer we shall cure only with an attractive specific, and afterwards treat with those medicaments which we shall describe below.

A Remedy for Wounds

If it be necessary to have such a medicament as by its special nature shall connect the lips of wounds, as glue joins two boards, this must be accomplished by its very great dryness and its styptic qualities, which act on the flesh only in the following manner.

Take Samech which has been well burnt and calcined to whiteness. To this add a smaller quantity circulated. Afterwards distil, in order that a very dry *caput mortuum* may remain at the bottom, and that the whole glass may glow. Then pour in fresh matter again, as before, and do this until the circulated substance comes out quite sweet, as it is in itself. Then allow it to be resolved by itself.

That which is so resolved becomes a remedy for a wound; in fact, it might be called "A Balsam for a Wound," because in our common German speech balsam is the same as Baldtzusammen, that is, *mox conjunctum* (soon joined), and the term is not derived from the Latin idiom. We are unwilling to speak in detail as to the virtues of this medicament, but generally, we assert, it has such an effect on all wounds that with one single washing we have cured many hundreds of these in a manner which is not credible, judging by natural methods.

A Remedy for Ulcer

Ulcers, it should be understood, must, in like manner, be compressed by a medicament with the addition of a generative virtue. The writings of the ancients are not worthy of our imitation; they are malicious and wicked. What we have to consider may be expressed thus : " Compel them to come in ;" and we do it in the following way :

Take of the aforesaid balsam for wounds, and of balsam similarly made from rust—of Samech, for instance—one pound each. Mix these together, and add a pound and a half of oil of iron. When all these ingredients are mixed together, let them be placed on the ulcers, which must be washed daily, as shall seem expedient. Let a consolidative plaster be applied such as we prescribe for ulcers. Follow up your ligatures thus to the end, until a cure is effected. It must be noticed that the members have to be compressed with these ligatures, as we have pointed out at sufficient length elsewhere. Let this suffice on the subject of ulcers.

A Remedy against Spots

We have sufficiently explained the removal of the skin by a corrosive specific, and together therewith the cautery—how it is to be produced and adapted for use. After the skin has been removed, and with it the spot, the cure is as follows :—

Take the above-mentioned balsam for ulcers, and add to this washed turpentine, oil of worms, and oil of eggs, equal parts of each. With this wash all the flesh when it has been stripped of its skin. After this treatment nothing more is required. The property of such a medicament is that it induces a new colour along with the new skin, and a natural hardness, so that it can be no longer defaced by the previous spots.

Although it is true that such spots can be removed by many waters, as for instance, the water of bean flowers, of sigillum Mariæ and the like, as well as by human dung, still these do not come within our scope, since their purpose is not uniformly compassed, and the spots are much more effectually removed by the method we have pointed out.

Nor let any be surprised that we set down so few and such concise remedies for the whole range of surgery, and do not follow those surgical methods which the ancients have described and the moderns have adopted, like them, for their own uses. For by that method of medicine, so long as we followed it, we could never find or perceive anything well founded and certain. But we have used our own remedies according to our experience, and in this way we have found out the best medicaments in the whole of surgery, and have comprised them here in three paragraphs only.

And although, no doubt, other diseases can be found besides those which we have mentioned here, such as bullæ, alopecia, etc., they are, nevertheless, comprised under spots and scars, and must be cured according to the treatment prescribed for them, for many causes which are not here brought forward, but left to our own experience; we had not forgotten them. We have had hundreds and thousands of these wounds passing through our hands, and when we saw them cured so quickly and so effectually by these remedies, why should we imitate the tedious and empty processes of the ancients, forgetting those nearest to ourselves? Why, then, we use mundificatives, lotions, sutures, ligatures, corrosives, and the like—which are all effectual against wounds, and destroy them most thoroughly—we have given the cause of all this at sufficient length in our book on wounds. Why do we use different plasters, ointments, unguents, and the like, even in the cure of ulcers? Why ligations, anointings, and the like? Well, to go through all this would be tedious since it only leads us to enter on a prolix, intricate, and foolish course, where we seek mere accidents without finding anything. It is a mere superstition to pin one's faith to antiquity. For in surgery to debate as to the nature of fistula, cancer, or ulcer, or the like, and then to assign to each its special remedy is mere vain talk and waste of writing, which will not repay the outlay in paper; when all can be thoroughly cured and removed by one single remedy—as, for example, external leprosy, alopecia, serpigo, spots, and the like, pustules, the itch, and scars, which can all be thoroughly removed by one medicament and one method of practice, as can also artetic wounds by spears, weapons and bullets. With these few words we would close our treatise on surgery and bring it to an end.

THE END OF THE NINTH BOOK OF THE ARCHIDOXIES FROM THE THEOPHRASTIA CONCERNING EXTERNAL WOUNDS AND THEIR REMEDIES

THE KEY OF THEOPHRASTUS PARACELSUS BOMBAST VON HOHENHEIM

OR

THE TENTH BOOK OF THE ARCHIDOXIES*

From a German manuscript codex of great antiquity

WE had decided to write our Archidoxies, as also other books concerning Medicine, with especial clearness and lucidity, inasmuch as all the highest medical arcana cannot be prepared without true chemical encheiries (undertakings), nor yet can they be speedily exalted in grade, and it is notorious that almost the whole world, through its devotion to riches and earthly wealth, zealously pursues tinctures only and transmutations of metals in order to amass the greatest possible amount of gold and silver, to obtain which they stand in the greatest need of chemical preparations, which also they would like to find in concise form and easily in our Archidoxies. Notwithstanding, it was in consideration of the very great evil which might thence arise, and at the same time to oppose their malice, that we have concealed our doctrine, according to ancient philosophic method and cabalistic practice. I shew this, my doctrine, clearly to the upright and the perfect, yet leave it none the less dark to contemptuous and impious men.

Our writing according to the method of cabalistic philosophy is not due only to the lachrymistæ who gape after gold, but also to the majority of the

---

* The following editorial preface introduces the tenth book of the Archidoxies in the Geneva folio :—Behold, gentle reader, the tenth book of the Archidoxies, which has been long demanded by the wishes of all lovers of chemistry, but till now has not been included among the works of Paracelsus, because, being in the possession of few, it has been held among secret things, as a most precious treasure replete with great arcana. Once, indeed, it was printed in the vernacular tongue of the author at Mentz, but the envy of certain persons, who gazed with a fierce eye at the revelation of the mysteries therein revealed, suppressed the majority of the copies, and compelled very many persons, studious in chemistry and in the writings of Theophrastus, to make transcripts for themselves. We, in order to consult the public good, have thought proper to add to the nine books previously edited, this tenth book, rendered from German into Latin, by one who is skilful in both languages, so that those who are studious in this science may be saved from transcribing, and the number of the Archidoxies may be complete. The author has denied in several places of this work that he would write the said book, lest he cast pearls before swine, and lest these arcana should come to the knowledge of the unworthy and the impious, but he changed his mind at the persuasion and prayers of his friends, who impartially weighed the disadvantages of giving their mysteries to the public with the advantages which could be derived to the human race by their communication. The latter seemed to preponderate by far ; consequently, they at last obtained this concession, that he would entrust the work to certain of his familar acquaintances. It appears from the preface, which he prefixed, with how great an oath he bound each and every person who should obtain an exemplar that he should guard it with the greatest secrecy as a treasure of Nature and Art, and should have nothing to do with avaricious chemists, or with the ambitious and envious followers of Galen. Nor did the author forbid the publication without a reason ; he rightly rejoiced in the title of a key, whereby the doors of the preceding books are opened, and the bars removed, so that

followers of Galen and of Avicenna. The latter would very gladly avail themselves of our medicaments and arcana to repel chronic diseases, which otherwise, using the method of Galen, would be incurable—they would be glad, I say, to use them if they were able to find a short, sure, and easy method of preparing and administering the same, without renouncing the error of the heathen and false Christians, and providing also they could ascribe the honour, fame, and riches which they thence would obtain to the writings of Galen only and to themselves, out of envy ignoring my name and glorious achievements, but claiming the same for their own writings, keeping secret the fact that the whole art proceeds from me. For they themselves, being old doctors, do not wish to appear as if, at their advanced age, they were reduced to be disciples of an unpolished Swiss and younger apostle, and to make public profession of this circumstance, seeing that all detest him for subverting their principles. On account of this, their laziness, ambition, envy, and hatred, as also ingratitude, I have taught and philosophised in my Archidoxies, and in my other books, in the aforesaid manner, as has pleased me, which I will justify before God and my conscience at the last day, in order that those who desire to arrive at the fundamental principle of my Archidoxies, may publicly call themselves Theophrastics, acknowledge me as their monarch, follow me in their labours, frequent my school, and, *vice versa*, discard their old fathers. But although they may chance to secretly obtain some process from a miserable and simple rustic, or from elsewhere, they will not, however, understand the arcana of administering my medicaments, and will thence derive more shame than honour. Wherefore, although it has been shewn to them by Anicula, that the young of swallows, their cranium, and glue of the oak, are a sure remedy for the falling sickness, which is the case, yet even by this you will not effect a cure. Whence is this, and what is the cause? It is for this reason, namely, that you do not understand the mode of administering them and the great Ilech, nor will you learn from Galen unless you shall have frequented my school and learned philosophy according to Christ and not after Mammon.

---

an entrance for all is left open to the sacred and more secret penetralia of the divine chemical art. It is, however, by no means to be thought that anyone who has only slightly occupied himself with the school of Vulcan, who has seen nothing of importance, save obscurely, in this most noble art, or has tasted of its fruits for the first time, should fancy that the things herein contained are for his use, or should deceive himself with a false opinion, persuading himself with excessive credulity that they are intended for him. Let him not apply himself to putting in practice the prescribed formulas of the recipes, unless he wishes to be wise too late, like the Phrygians. No one, indeed, should suppose that food is here set before him which is made all ready for his palate; he who would obtain the kernel must first crack the nut, and then at length he will taste the sweetest fruits of chemistry. Not a raw recruit but a veteran soldier, not one slightly tinged but he who is completely permeated with chemical undertakings, not soiled by scholastic dust but by the smoke of coals and cinders, not accustomed to arguments which savour more of vain subtlety than of real usefulness, but acquainted with true and sound philosophy—those, in a word, who are most skilled in chemical practice--let such gird themselves to the execution of these matters. But let the profane crowd retire as far as possible. If any one doubt that this is the genuine production of the author, the peculiar style and mode of writing affected by Theophrastus should remove his difficulty, while the promises made in the preceding books and here fulfilled should place the matter beyond question. Nor let anyone be surprised that the author declared he would keep the tenth book locked up in his heart, or, to use his own expression, he would retain it in his occiput, whereas the prayers of his friends persuaded him to change his mind. Do you, therefore, gentle reader, being assured by this lawful witness of Theophrastus, enjoy the labour and work of so great a man, and accept with grateful mind our zeal in acquiring and publishing the treatises of our great doctor.—F. B.

Since, therefore, the glue of the oak does not fulfil your expectations, you imagine that it is too weak by itself, wherefore you correct it with other herbs, and make a great composition of sixty or more ingredients, which you digest and purge of dross. You do not, however, expel the disease even by this means, since you do not understand either simples or compounds, nor yet the method of administering them.

Had such persons, indeed, accepted my doctrine in a grateful manner, and had they cast out of doors the red bonnet, or fool's cap, received from Galen, and at the same time submitted themselves to my discipline, I would have put on them a better cap, that of Fortunatus himself, wherein is concealed more art than in all other writings. So, in the presence of none would they need to doff it, but, just as Fortunatus cured the king's daughter, might they cure chronic diseases.

But they are unworthy of better things, and are to be blamed for their mischief, since they are completely ignorant of the great secrets or the mysteries of the sanctuary of Nature, and as much concerning the celestial treasure, which, in these last days of grace, has been freely revealed to me from on high, which, indeed, make a true Adam and paradoxic physician, according to the days of Enoch, in the intellects of a new generation. But these ignorant persons boastfully refuse it. Wherefore, I pity them not, but leave them in their own ignorance.

There is no doubt, that in that very great multitude of men, mentioned in the fourth book of Esdras, the Lord God will reserve for Himself a small number of certain elect persons, who will desire faithfully to pursue my Theophrastic doctrine, to love the truth, and help their neighbours in their destitution and diseases, out of a true and unfeigned Christian love, not for the sake of filthy lucre or ambition, but for pure love of God; and also that out of Nature's light the marvels of God may manifestly appear. At the same time, all are not born under such a constellation as to be able to perceive the sense of our books, however diligently they study, without divine aid. It is on account, therefore, of their sincere intention and love, and that they may understand our excellent writings and arcana of medicines, and may arrive at a blessed end; and also, lest the most precious secret of Nature, divinely revealed to me should altogether be buried with me; that we shall write this book, therein shewing in full light the principle of our Archidoxies and universals, and shall teach the preparation of singular arcana, the quintessence, prime entities, and magisteries.

But, lest this clear light itself should reach the ungrateful and unworthy, I exhort all who have a supply of this book, I bind you by the very great sacrament and oath which you have given to God in baptism, that you shall hide all these things secretly and as the noblest treasure of Nature, lest you admit any unworthy person. Do you rather venerate that treasure as a most blessed talent, and help your neighbour in adversity.

May God grant benediction and favour, that whosoever partakes thereof may rightly use it!

THE TENTH BOOK OF THE ARCHIDOXIES

OF THEOPHRASTUS

Comprised in Ten Separate Chapters.

CHAPTER I

CONCERNING THE SEPARATIONS OF THE ELEMENTS

IN all things four elements are commingled one with the other, but in each thing one of these four is perfect and fixed. That is the predestinated element in which dwells the quintessence, the virtue, and the quality; but the other elements are imperfect, and each an element only, in which is no more virtue than in any other single element. These are all, as it were, the abode of the true, fixed, and perfect element; whence, also, they are called qualified things. Now, the fact is that some persons think the body to be a true element and quality, and that it in some way displays the virtue of a true element. This is because the body, like the three imperfect elements, is tinged and qualified, each according to its own nature, by the fixed, perfect, and predestinated element, as by its indwelling inhabitant.

For example, in some bodies the element of water predominates, in others that of fire excels, in some earth, and in others, again, air. If, then, the predestinated element has to be separated, it is necessary that the house be broken up; and this breaking up or dissolution of the house is brought about in divers ways, as is clearly said in my Metamorphosis concerning the death of things. If the house is dissolved by strong waters, by calcinations, and the like, care must be taken that what is dissolved from that which is fixed must be separated by common distillations. For then the body of the quintessence passes over like phlegm, but the fixed element remains at the bottom. But since we are little concerned about the house or the dwelling, it is necessary to find it in a fixed, predestinated element, and thence to extract it after the mode of a quintessence, so that that fixed element may be dissolved by other stronger artifices than calcinations, sublimations, and so on, and the pure separated from the impure. The pure is the quintessence; but the impure is the superfluous tartar, which is mixed up with every generation, concerning which see the book on Tartareous Diseases.

But since my theory is given at length in other books of Archidoxies, of Metamorphosis, and the Generations of the Paramirum, I am on that account unwilling to cause any weariness, but will briefly point out the practical method. Prepare a metal according to the process in the book on the Death of Things; reduce it to a liquid substance according to the method which we have taught in the book on the Separation of the Elements; separate by continual cohobations and putrefactions the three imperfect elements; and then the fixed element, of whatever kind it may be, remains at the bottom, and in this way these four elements are correctly separated.

CHAPTER II

CONCERNING THE QUINTESSENCE

Abstract the volatile portion, which passes over in the separation of the elements, several times from that which is fixed, so that the quintessence, which partly was raised with the phlegm, may be again conjoined. Take the fixed element that remained after the separation of the three imperfect elements, of whatsoever sort it may be, then dissolve it in its proper water, each according to its nature, as we have said in the Archidoxies concerning the Quintessence. Keep it in the highest state of putrefaction, distil it by cohobation, and the rest by descent. Putrefy still a little, distil, and join all. Then distil it in a *Balneum Mariæ*, even to oiliness. Then break it up with the subtle spirit of wine by boiling; the impure will sink to the bottom and the pure will float on the surface. Separate this by means of a tritorium, and, in order that it may at the same time lose the nature of the aqua fortis, pour on a greater quantity of the spirit of wine, which frequently abstract until the quintessence turns out sweet. Lastly, wash it in common cold water. In the same way, it must be understood of marchasites, stones, resins, herbs, flesh, watery and fixed substances, that first of all, according to the teaching of the book on Separations, the three imperfect elements shall be separated; and that afterwards measures shall be taken with the fixed element according to the instruction given in the book on the Quintessence.

By eating or corroding water, understand *acetum* mixed with spirit of wine, and that which being frequently abstracted from the spirit of salt nitre becomes *acetum*. In this the fixed elements of marchasites should be dissolved, purified, and elevated by means of an alembic; then, lastly, corrupted by spirit of wine, so that the impure may sink to the bottom, and separate itself from the pure.

With regard to the essence of gems, by radicated *acetum*, understand that you have a sharp *acetum* corrected with bricks, a sufficient number of times from the tartarised matrix of *acetum*. Dissolve therein the gems, which have been first calcined by sulphur, putrefy, and then separate the pure from the impure by breaking them up with the spirit of wine.

From fruits, herbs, and roots the essence is easily perfected, so that you dissolve the imperfect elements by the highest secret putrefaction of extreme heat. Then putrefy in dung, drive out by descent all that can go out, and from thence abstract the injurious imperfect body of the moisture by distillation in a bath. Thereupon there will remain at the bottom the predestinated element. Separate this from the impure residuum by corruption with its proper spirit, or with spirit of wine. Abstract this, and you will have the pure quintessence.

The extraction of the quintessence from salts (for example, vitriol, common salt, salt nitre, antimony, etc.) is accomplished in this way: cohobate them frequently with their own proper liquid, or with water, putrefy with phlegm, and abstract the body from thence after the manner of phlegm, even to the fixed spirit. This dissolve in water, or its own proper liquid, and separate in heat with spirit of wine, the pure from the impure.

CHAPTER III

Concerning Magisteries

Magisteries are deservedly to be called mysteries, on account of the great tinctures which they display in an appropriate body, such as *acetum* or wine; and as we mention elsewhere, so here also we teach that the one thing to be considered is with reference to the agreements which are adapted to the extraction of magisteries. For if you take distilled *acetum* you must not tinge water but wine into *acetum* if, indeed, the tincture or the *acetum* was made from wine. If you have well and rightly understood the magistery of *acetum*, you will also sufficiently understand the book of Magisteries. In the magistery of *acetum* it is to be understood that from corrupted wine, by a fermentation which is naturally adapted to it—for example, by tartar—you make, first of all, the tincture, that is, the *acetum*. Then with a small quantity of this same *acetum*, you shall thoroughly tinge a large body of wine, previously corrupted and putrefied, in a short time, into the best *acetum*. If, therefore, you purpose to convert metals into a magistery, and altogether to tinge the whole body into an essence, you must take a principal and an open metal to which all the others in Nature are cognate. That you must corrupt in its own matrix which has been placed in water, and is called the Mother of all Metals, purge it from its superfluous elements and reduce it to its primal liquid entity, that is, the sharpest metallic *acetum*. As often as all the metals are digested therein, they are of necessity transmuted by it into *acetum*, that is to say, into a quintessence. But as wine must be already in some way previously corrupted, if, indeed, good *acetum* is to be quickly prepared from it, so, in like manner, metallic bodies, too, must be previously corrupted, or putrefied and mortified, as is said in the Metamorphosis concerning the Death of Things; and then they are truly called potable.

In this manner, also, the magisteries of marchasites are to be prepared, in which almost more virtue is found than in metals, just as the other magisteries are prepared; and by our dissolving water is to be understood the water of salt.

But the magistery from gems is that you first of all calcine them with sulphur for four hours, then reverberate them, and afterwards burn them with nitre. Then boil them with simple water eight hours, filter, coagulate, and extract them with spirit of wine.

The magistery from gums and resins, for example, from turpentine and amber, is made after the following manner: First boil in spirit of wine, then corrupt in fresh spirit of wine mixed with a dissolving water, of salt, for instance, then distil from it.

The magistery of herbs, in like manner, as also of all spices and fruits, is thus accomplished. First of all, let them be fermented like must. Then extract the spirit, as from the dregs of wine. In that spirit digest the putrefied herb, frequently renewing it with fresh herbs until the spirit shall have become quadrupled in quantity. But since frequent mention is made in our Archidoxies of First Entities, and since the chief foundation is hidden in them, we will here briefly add the preparation of our water of circulated salt, which is here required, but was omitted.

The Preparation of Circulated Salt

In our other books we have sufficiently shewn and made clear that the true element is water, or the sea, as if the true mother of all metals, and from its primal essence it received the sperm of the three principles, of which none before me has made any mention, only they built up their principles from sulphur and mercury, neglecting all mention of the third principle, that is to say, of salt which lies in the sea. But, having been taught by experience, I have in my other books touched upon the fact that the first entity or quintessence of the element of water is the centre of metals and minerals; and I have elsewhere added that every fruit must die in that wherein is its life, so that it may afterwards acquire a new and better life, and thus by laying aside the old body be brought back to the first entity. Wherefore we will add here the extraction of the centre of the water, in which metals ought to lose their body.

Take first the true element of water, or, in place of it, some other salt not yet boiled to dryness, or even purified salt of a gem. Pour two parts of water mixed with a little radish juice; putrefy in an accurate digestion, the longer the better. Afterwards let it congeal; putrefy again for a month, and then distil by a retort. Urge the residuum with a strong fire, so that it may melt. Reverberate it in a retort over a continuous fire. Dissolve it on marble. Pour upon it the water that flows from it, and putrefy it again. Distil it once more even to oiliness; mix it with spirit of wine, and the impure will fall to the bottom, which separate, but the pure will be crystallised in the cold. Pour the distilled matter on it again, and cohobate until a fixed oil remains in the bottom, and nothing sweet afterwards passes over. Digest for a month; and then distil until the arcanum of the salt passes over through the alembic. Do not grudge this protracted labour, for this is the third part of all the arcana which are hidden in metals and minerals, and without it nothing fruitful, nothing perfect, can be brought about.

But although there are several ways for extracting the first entity of salt, this is the most useful and expeditious; and after this is that other way which we have mentioned as the elixir of salt, namely, that fresh salt being mixed with dissolving water, which is the distilled spirit of salt, should be putrefied and distilled until the whole substance of the salt shall be dissolved and reduced to a perpetual oiliness, the body being removed thence as phlegm. In this way it is taught that the arcanum or magistery of vitriol and tartar, and of all other salts, is to be prepared.

CHAPTER IV

Concerning First Entities; and primarily concerning the Extraction of the Quintessence or First Entity of Common Mercury

If the common mercury is to be reduced into its first liquid entity, then it must be previously mortified and brought out from its own proper form. That is done by various sublimations with vitriol and common salt, so that at last it becomes like fixed crystal. Then dissolve it in its own matrix, that is to say, in the first entity of salt. Putrefy for a month; corrupt with fresh arcanum of salt, that the impure part of it may be precipitated to the bottom, but the pure turned into crystals. Sublimate the stones in a closed reverberatory; when it is sublimated, always invert it until it grows to redness. Extract this sublimate with spirit of wine rectified to the highest point. Separate the spirit of wine, dissolve what remains upon marble, and digest it for a month. Pour on fresh spirit of wine, digest for a time, and distil it. Then the arcanum of the first entity of mercury will pass over in a liquid substance, which is called by the philosophers a very sharp metallic *acetum*; and in our Archidoxies the Greater Circulatum. And the same is to be understood concerning antimony, gems, and herbs.

CHAPTER V

Concerning Arcana

What we say concerning arcana is to be thus received: that they are nothing else than a graduated quintessence or first entity. And under the first arcanum of the primal matter, we wish to be understood the first material or first essence of the limbus of man. Also we understand the first matter of the mercury of salt, for that is most closely conformed. Wherefore, according to the process of the first entity, you will reduce all to a liquid substance, then join it again with a monarchy, as if with the living unreduced body of that thing, and so promote it for distillation.

What we think concerning the arcanum of the Stone shall be made clear in the succeeding practice. But by the arcanum of the *Mercurius Vitæ* we intend a living fire, so that the mercury of common life shall be essentialised with the quintessence of salt and be vitalised with the first ens of antimony, as if by a celestial life. But the arcanum of the tincture unfolds itself, wherefore we omit it here.

CHAPTER VI

Concerning the Arcanum of the Stone, or of the Heaven of the Metals

What we have in one place and another theoretically advanced concerning the arcanum of the stone we here pass over; and I say that this arcanum must not be sought in rust, which many have wrongly named "flowers," but in the mercury of antimony. And this mercury of antimony, when it is brought to its perfection, is none other than the heaven of the metals, because its virtue is always vital, and nothing else than a perfect, pure quintessence. Therefore, even in the Deluge no virtue or efficacy was taken away from it; for the heaven, as though it were life itself, can be destroyed by no lesser thing. Its preparation I briefly subjoin here:—

Take antimony, purge it from dross and realgar, in an iron vessel, until the coagulated mercury of the antimony appears white and beautiful. And although it is an element of mercury, and has in itself a true, hidden life, yet all these things are potentially but not actively present.

If, however, you wish to bring it down to activity, it is necessary that you excite that life with what is like itself, such as living fire or metallic *acetum*, with which fire many philosophers have proceeded in different ways. Since, however, they agreed fundamentally, they all arrived at the destined end. One, with much toil, extracted a quintessence out of the coagulated mercury, and led down the mercury of the antimony therewith into activity. Others have discerned that a uniform essence exists in different mineral substances, as, for example, in fixed sulphur of vitriol, in magnetic stone, and thence extracted the same quintessence, and with that same have ripened their mercury or heaven, or have brought it into activity. And since they extracted their quintessence from a stony material, on that account they called that magistery a stone; and, indeed, their opinion is right. Nevertheless, that fire, or corporal life, is found much more perfectly and sublimely in common mercury, which is testified plainly by its flowing, namely, that there is hidden in it a consummate fire and a celestial life. Whoever, therefore, desires to bring his metallic heaven to the highest grade, and to lead it to activity, ought first of all to extract out of the corporal life (the common mercury) the first liquid entity (as if celestial fire), the quintessence of the sun and a very sharp metallic *acetum*, by solution with its mother, that is, to mix it with the arcanum of salt, and with the stomach of Anthion, that is, with the spirit of vitriol, and he should dissolve therein the coagulated mercury of antimony, should digest it, and, lastly, reduce it to crystals, that it may be like a yellowish crystal, concerning which we have made mention in our manual.

CHAPTER VII

CONCERNING THE ARCANUM OF THE MERCURIUS VITÆ

Just as out of herbs, as the vine, a temperate essence is extracted, by which from every kind of herb or root their own essence may be drawn, even so that the mercury of wine does not shew its own peculiar nature, but the nature of that with which it is essentialised, in the same way is it with metals and minerals, for a like mercurius or spirit is extracted from the open and middle metal (mercurius), if the essence be extracted out of the perfect metals with that same spirit. Then, afterwards, that essentialised mercury is joined to the celestial balsam of the quintessence in a closed reverberatory, by means whereof it acquires life, and is on that account called the *Mercurius Vitæ*. The virtues whereof seem to us admirable, and must be kept silent and occult by us, lest they should be despised.

CHAPTER VIII

CONCERNING THE GREAT COMPOSITION, BEING THE CHIEF OF OUR SECRETS IN MEDICINE

In our Paramirical writings it is made sufficiently clear, that is to say, so far as it is necessary for a philosopher or a physician, if it be necessary that the whole human body, not only in its corporeal and earthly mass, but in its celestial balsamic part, should be preserved from and healed of all heavenly and earthly diseases—it is made clear, we say, that in the work of healing such a composition should be made as does not consist of a number of ingredients. For example, if any one should think that by pouring together water and wine a real mixture ensues, this is false, because one part can be separated from the other without injury to either, which is not the case in our great composition. For here is made a uniform and concordant mixture, so that two things, distinct in nature and properties, are united, and neither can be separated from the other without injury on account of their remarkable agreement, as occurs also in the male and female semen. If, therefore, such a composition is to be prepared conformable in its condition to man, through the due proportion of heavenly and earthly things, it is fitting to consider the name of the microcosm and that man is a little world. Wherefore if he is to be cured universally of all diseases, that must necessarily be done by his like. Concerning which Hermes Trismegistus said that it was necessary for him who intends to make this composition to create a new world; and as God created the heaven and the earth, so also the physician must form, separate, and prepare a medicinal world. And in order that he might point out to his disciples with sufficient fidelity from what thing or material this composition should be made, and how, also, the concordances of heavenly virtues are discovered by us in the Valley of the Shadows, he wisely and truly adds, a little after, that what is beneath is as that which is above, and that the inferior and the superior stand related to one another as man and wife; and for the better understanding hereof he teaches that the heaven of itself agrees with the element of water, because it had its first spermatic matter in the water, and that the element of earth, coagulated and changed from its spirituality into corporeality and earthiness, is like the planets and the other stars, which also obtained their spermatic matter by their origin in heaven, and thence by separation passed over or were changed from a heavenly pellucid nature into a dense coagulated body.

In the primal creation, things above and things below, the upper and lower heaven, or water, the upper coagulated nature, or stars, and the lower terrestrial nature, were all mixed together and made one thing. But God separated the subtle from the dense, so that out of one water two were produced. The upper water was subtle, and to be considered as of the masculine sex, compared with the lower, denser, feminine water. But as God divided and separated still further the upper water, so that the subtle, airy part should be appointed for the stars, that thus the celestial bodies or stars may stand related to heaven as sons to a father, so, by parity of reasoning, has God appointed, together with this which is above, a separation also in the denser bodies, that is to say, in the female waters in the Valley of Shadows, and divided them into two parts. The seventh, clearer part He called water, and the other six dried parts, or coagulated portion, He called earth, which comprises in itself all the special fruits and planets which had their prime origin in water as their heaven; as metals, minerals, and gems, which, in respect of the water, are reckoned as daughters in respect of a mother. So the superior heaven has a nature and properties like its own in its feminine nature, that is to say, in the inferior heaven or water; and the superior terrestrial bodies, or stars, like the sons of a father—that is, heaven—have a similar agreement with and relation to their sisters, the earthly bodies. And just as by close relationship the higher heavenly bodies or stars are joined to their father, the heaven, so too, by a like and equal relationship, the lower earthly minerals and metals are connected with their heaven, the water, as with a mother. Whence the truth of the sentiment of Hermes is evident, which we commend to our sons of the doctrine in these words: that indeed the whole microcosm, so far as relates to the comprehensible mass, and to the living, moving, corporeal, generating spirit, ought to be collected and composed of these lower elements and dark waters which are their noblest essences. But as to the mental arcana, wherein consists the sound mind in a sound body, these should be attracted from the superior, heavenly waters, and their astral influences spiritually, in a mental manner, through the mind of the image of the Gamahela; or, if these are not pleasing to us, they should be declined, as in our books on A Long Life we point out these things at length and with clearness. Since, both in other places and especially in the Paramira, we have included the theory of this grand composition, we pause at these words and add the practical method, namely, how the inferior world or heaven should be united and conjoined with its earth, or the sun with its heaven. But because we have premised already the preparation of the heaven and have taught it under the arcanum of the stone, we omit it here. And since of itself alone, like the male semen, it can bring no advantage in the body of man, but only restores the celestial parts, that is to say, the radical moisture, or balsam of life, therefore it ought to be joined to its terrestrial corporeal mass, and be brought into concord therewith, that so also the carnal element in man may be refreshed and restored, and not only one member, but the whole body, be restored to soundness. Therefore let a corporeal mass be taken, which in nature is equal to the sun above, and embraces in itself the properties of all the stars, since it is impossible for all the subterranean stars and the coagulated bodies to be included together in the number of the ingredients. This coagulated essence of heaven, that is, the sun, in its essence and temperate element, is so elevated and graduated that it also fixes with itself its own habitation, that is, the superfluous elements, that it cannot be destroyed by any element, and the inhabitant or corporal balsam hidden in it is able to remain eternal. If, therefore, as was before said, the whole microcosm is to be healed, then the corporeal coagulated balsam should be united with the spiritual celestial balsam and the discord between the elements of the sun should be reconciled, so that the superfluous elements may be separated from the fixed predestinated element and altogether die out and leave the fixed element, as their inhabitant, alone. If this dead body of the sun be afterwards cleansed from superfluities and brought into a volatile spiritual nature, then is perfected the true, sublimated, and resolved mercury of the sun, not that horizontal which many try to prepare with common mercury and sal ammoniac.

CHAPTER IX

Concerning the Corporal Balsam or Mercury of the Sun

In order that you may excite discord among the elements of the sun, or the habitation of gold, you must draw out Sol in a strong solution, by means of the phlegmatic fire, or quintessence of tartar, into its proper heat. By this method the element of air in Sol is very greatly increased, and by the air approaching the fixed element of the sun, as being its proper fire, it is so graduated that it can conquer and destroy the habitation of the other three. Putrefy this destruction with the quintessence of tartar and with struthio. Convert it by a proper sublimation into the matter of mercury; and then the fixed mercurial element of Sol will remain alone without any habitation. But since this is still mixed with its superfluous tartar, therefore this must be removed from it. Dissolve it, then, in the circulated water of salt, corrupt it, and the tartar will be precipitated. Sublime the pure in a closed reverbatory of Athanor, dissolve it upon marble, and putrefy it. Thus is the mercury sublimated, graduated, and dissolved into the first matter of Sol, and is prepared in the highest degree.

CHAPTER X

CONCERNING THE COMPOSITION OF THE SPIRITUAL BALSAM, AND OF THE BALSAM OF THE COAGULATED BODY.

As is remarked in the Manual, this composition is made in the philosophers' egg. And so we put an end to this great work, in the name of God, to His praise and glory.

HERE END THE TEN BOOKS OF THE ARCHIDOXIES

THE MANUAL OR TREATISE
CONCERNING THE MEDICINAL PHILOSOPHIC STONE

PREFACE TO THE READER

READER. God, indeed, permitted the true spirit of Medicine to be brought into operation by Machaon, Podalirius, Hippocrates, and others, so that the true medicine, which shines forth from the clouds (where it cannot be fully and plainly known), should come forth into the light of day, and be manifested to mankind. By that same operation, too, He placed His prohibition on the spirit of darkness, so that it should not altogether overwhelm and extinguish the light of Nature, and that the mighty gifts of God, which lie concealed in Arcana, Quintessences, Magisteries, and Elixirs, should not be altogether unknown. God, therefore, has ordained certain means whereby, moreover, through the ministration of good spirits, research into such arcana and mysteries should be implanted in man, just as certain men have received angelic natures from that heaven which is familiarly acquainted with the angels. Such men have been able afterwards, as being endowed with a perfect intelligence of Nature, to search into Nature and her daily course more profoundly than other people, to compare the pure with the impure, to separate one from the other, and to adapt and modify what is pure in a manner that seems impossible to others. These, as being true and natural physicians, know how to supplement Nature, and by their arts to bring her to perfection. It must be, therefore, that all imperfect and diabolical operations give way before them, as a lie always gives way to what is true and perfect. We must, I assert, speak the plain truth if we would arrive at any happy result. And if it be lawful to grasp the truth by any means, no man ought to be ashamed to seek it in any quarter.

Let none, then, take it in bad part of me, that I myself have loved this truth and pursued it. I was forced to seek it, for it did not seek me. If a man wants to see a foreign city it is no good for him to stay at home with his head on his pillow. He must not roast pears at the fire, for in that way he will never become a doctor. No one will ever get to be a renowned cosmographer by sitting at table. No chiromancist ever became so in his chamber, or geomancist in his cell. So neither can we arrive at the true medicine save by investigation. God makes the true physician, but not without pains on man's own part. He says: "Thou shalt eat the labour of thy hands, and it shall be well with thee." Since, therefore, seeing goes before truth, and the things which are perceived by sight gladden or terrify the heart, I shall not esteem it toilsome, or deem it beneath my dignity, to travel about and join myself to such men as fools despise, in order that I may discover what lies hid in the limbo of earth, and that I may fulfil the duty of a true physician, by exhibiting Medicine according to the ordinance of God and for my neighbour's good; and that it may not do more harm than good. An easy-going man will not take this trouble. Let him who will, then, sit in his cushioned chair. I like to travel about, to see and examine whatever God and the opportunity allow. For the sake of sincere readers, however, who desire to learn, and love the light of Nature, I have written the present treatise in order that they may know the foundation of my true medicine, pass by the absurdities of pseudo-physicians, and be able, in some degree, at least, take my part against them. For those distinguished fellows, of course, know all these things beforehand, and the asinine doctor has them all in his wallet, only he is never able to get at them. For a man must be a good alchemist who wants to understand this treatise, one who is not afraid of the coals, and whom daily smoke does not disgust.* Let those who will take pleasure in these matters, I force myself upon nobody. This alone, I say, the thing will not prove infructuous, however much my enemies, the sham doctors, blame and accuse me.

---

* I praise alchemy, which compounds secret medicines, whereby all hopeless maladies are cured. They who are ignorant of this deserve neither to be called chemists nor physicians. For these remedies lie either in the power of the alchemists or in that of the physicians. If they reside with the latter, the former are ignorant of them. If with the former, the latter have not learnt them. How, therefore, shall those men deserve any praise? I, for my part, have rather judged that such a man shall be highly extolled who is able to bring Nature to such a point that she will lend help, that is, who shall know how after the extraction of the health-giving parts what is useless is to be rejected; who is also acquainted with the efficacy, for he must see that it is impossible that the preparation and the science—in other words, the chemia and the medicine—can be separated from one another, because should anyone attempt to separate them he will introduce more obscurities into medicine, and the result will be absolute folly. By this distinction all the fundamental principles of medicine will be overthrown. I do not think I need labour very hard in order that you may recognise the certainty of my reasons. I give you this one piece of advice: Have regard to the effect of quack remedies. They first destroy wounds which are already aggravated by a succession of processes, miserably torture the patients, and having after all accomplished no good, but removed all chance of recovery, they do the unfortunate to death. . . . I who am an iatro-chemist, that is, one who knows both chemistry and medicine, am in 'virtue hereof in a position to point out errors and to profitably reject all pestiferous remedies, relegating them to their own place. My ardent desires and ready will to be of use prompt me to this. – *Chirurgia Magna*, Pars. I., Tract I., c. 13.

THE MANUAL
CONCERNING THE PHILOSOPHERS' STONE

IN order that the Philosophers' Stone, which, for sufficient reasons, we call a perpetual or perfect balsam, may be made by means of Vulcan, it must first of all be known and considered in what way that Stone may be placed materially before our eyes, and become visible and cognisable by the other senses; and, in like manner, how its fire may be made to come forth and to be recognised. In order, then, that this may be the more clearly set before us, we will take the illustration of common fire, that is to say, we will inquire in what manner its force shews itself and becomes visible; and this is as follows:—First of all, by means of Vulcan, the fire is smitten out of the flint. Now this fire can effect nothing unless it meets with some substance that is congenial to it, and on which it is capable of acting, such as wood, resins, oil, or some other like substance, which, by its nature, readily burns. When, therefore, the fire meets with some such object it goes on forthwith to operate, unless it be extinguished or hindered by something of a contrary nature to itself, or unless the material wherein it should multiply itself be deficient. For if wood or some similar substance be applied, its violence becomes stronger, and operates in the same way until no more fuel is applied. Now, then, as the fire shews its effects in the wood, so is the same thing produced with the Philosophers' Stone, or the Perpetual Balsam acting on the human body. If that Stone be made out of proper material and on a philosophical principle by a careful physician, and due consideration be given to all the surroundings of the man when it is exhibited to him, then it renovates and restores the vital organs just as though logs were put on a fire, which revive the almost extinguished heat and are the cause of a brilliant and clear flame.

Hence it is clear that much depends on the material of this balsam, since it ought to have a special adaptation to the body of the man, and should so exercise its virtue that the human body should be safe from all the accidents which might occur to it from such matter.

Wherefore, not only much depends on the preparation of the Stone or Balsam, but it is of much greater importance, that before all things, the true matter adapted for it should be known, and then that it should be properly prepared, and above all that it should be soberly and prudently used; so that such a medicine should have power to purge away all impurities of the blood and induce soundness in place of disease.

On that account, it becomes the true and honest physician to have good knowledge, and not to regard ambition and pomp, nor to order dubious or contrary things, nor to trust too much to the apothecary, but to make himself well acquainted with the disease and with the sufferer; otherwise you will be constantly treated wrongly, and the only result will be that the sick man is deceived and defrauded, solely by the pride and ignorance of the unfit and unqualified practitioner. For what else is it but a deliberate fraud, when a man asks money and fees for that about which he knows nothing, and tries to lord it, to his own infamy; since many men think nothing of money if they can only get good advice. If this is not given, and they lose both their bodily health and their money, still it is considered quite laudable to demand a fee. Let him who will trust them. I, for my part, would provide marks for such a doctor in quite a different way. It is quite evident that of such doctors, who in their own conceit are most highly learned, there is not a tenth part who possess an adequate knowledge of simples; much less of what they order to be done, or how the medicine is compounded by the apothecary. Hence it often happens that such a doctor orders some kind of simple to be taken, which he himself does not know, and the apothecary knows still less, and has not got in his store. Yet this medicine is called perfect, and is swallowed as such by the sick man, often at a high price. The result the sick man soon feels. Although it furthers his recovery in no way, still it is profitable to the doctor and the apothecary for filling their purses. If the doctor or the apothecary suffered from the same disease, they would not take the same remedy. Hence it can be well understood how nefarious their conduct is, and how highly necessary is it that they should approach the subject in a different way, should amend their errors, and follow a better course of practice. Still I fear that old dogs are very difficult to tame.

But to return to my subject, from which zeal for the suffering and solitary patients has led me astray. In order to do it justice I would say that I do not purpose to romance or to boast about this Philosophers' Stone, but the nature of the case requires that it should be made of the proper material, and that it should be prepared and used with due caution. You must know that many of the ancients in their parabolic writings have sufficiently indicated this material, and described the operation in figurative words, but did not altogether disclose it, so that unqualified persons should abuse it; and yet they took care that it should not be concealed from their own disciples.* When, however, few

---

* The congeries of chemical philosophy, which has been the subject of frequent reference and citation in the first volume, contains excerpts from the *Manual Concerning the Medicinal Stone of the Philosophers*, which offer in some cases considerable variations from the above text. The most important passage is as follows:—Very many of the ancients have with sufficient clearness revealed this matter and its preparation for the ingenious, but still in parabolic and enigmatical words and figures, so that they might drive away the unworthy from so great a mystery at once of Nature and of art. A very few, nevertheless, even of those who are fit for this art, have sought the perpetual balsam of Nature and the perfect Stone, on account of the vast labour and intricate difficulty which meet the investigator at every step. Hence it is that sluggish and slothful dispositions stand aloof from this work. The avaricious, whom the greed

persons followed their meaning, or properly undertook the operation, in course of time their instructions were forgotten, and in their place the nostrums of Galen were introduced. As was the foundation of these nostrums, such was the superstructure, and matters daily became worse. This you see, for instance, in their herbaries, how they fuss over these, and how the Germans mix up Italy in the matter, when it is quite certain that Germany does not lack those imported herbs, but possesses in itself a full supply of perfect medicine. In order, then, that truth may not give place to a lie, and that the obscurities of Galen, with his accomplices, may not quench and suppress the light of Nature in medicine, it is necessary that I, Theophrastus, in this book, should speak, not as a quack, but as a scientist who is not ashamed of his achievements in medicine, who also, by the grace of God assisting him, has had proof of this matter in many cases which you, O Galenist, would not have dared to visit! Tell me, Galenic doctor, whence comes your qualification? Are you not putting the the bridle on the tail of the horse? Have you ever cured the gout? Have you dared to attack leprosy? Have you cured dropsy? I believe you will wisely hold your tongue, and acknowledge that Theophrastus is your master. If, however, you want to learn, learn and note what I here write and say, namely, that the human body does not need your clumsy herbal, especially in chronic or long-standing diseases which you in your ignorance call incurable. Your herbs are too weak for these cases, and by their very nature are impotent to get at the centre of the disease.

You will effect nothing by your pills beyond merely purging the excrements. And even here, on account of their unsuitability, you often expel the good with the bad, and this cannot be done without severe damage to the patient. So, then, these pills must be left off. Furthermore, your draughts do nothing beyond causing nausea in those who swallow them by their foul taste, which irritates the sick person, and by and bye they cause gripings and danger

---

for gold and silver—vile things that they are—has inspired with courage, have persevered most diligently of all in the work, so much so that for it they have neglected their life and substance. But because so much worldly happiness was not intended for them by God, they have wasted their oil and their labour. There is in truth need of the keenest zeal and judgment for this matter, that from various comparisons and similitudes the meaning of authors who write about this art may be detected. In order, therefore, that the more intelligent may comprehend up to a certain point, we will bring forward a fitting similitude by which is prefigured the matter of the perpetual balsam, agreeing with a like balsam in the human body for the purpose of restoring and conserving its highest state of health and driving away disease. Take common natural fire for an example. This is invisible to us, wherefore it must be sought in the air where it is latent, and is to be found by the striking together a flint and steel. Now it does not owe its existence to these, but to the air, and is only retained by some dry object such as firewood. For the dryness immediately takes hold of its heat and that which is like itself, and both operate in the same way in the subject until all the humidity is consumed, and there remains only a dry and ashen body subject to death, being deprived of the fire and food of life. In no other way should philosophers investigate matter wherein the food and fire of life are chiefly present. These should be drawn out by preparations so that they may increase the vitality of human life when it has begun to fail. For as the flame and life of the fire revive the wood all but consumed if there be left only a few bits of coal or sparks of fire, by adding appropriate wood and fuel, so in the human body the perpetual balsam applied even to the smallest remaining atom of life, rouses this into a flame and into its pristine vigour. But a question arises as to the matter in which this rousing fire remains latent. It is not right, or even safe, to speak clearly and openly about this to everybody. Is not the grace of God sufficient for the children of light, whereby the faculty is given to them of bringing light out of the darkness of shadows, figures, and enigmas? The sons of darkness are proved in this way, since for them, even from the light itself, nothing can be elicited save the merest shadows. Nature requires a nature like itself, and takes pleasure in it. As iron, too, is attracted by the magnet, so darkness begets darkness, and light brings forth light. This Stone is hard—who will be able to extract the kernel from it? What is harder than the stone and steel, except it be the diamond? Yet this is worn away. But by what artifice the Philosophers' Stone is to be disclosed, so

through their unnatural action. But I quit the subject of your absurd and useless medicaments, since they are in direct opposition to Nature, and never ought to be used. Since, then, those of which I speak are the true medicines (and no such true medicine can be found in Galen, in Rhasis, or in Mesne, which attacks and purges those diseases I named from the very roots, as the fire purifies the spotted skin of the salamander), it necessarily follows that the curative process of Theophrastus differs altogether from the fancies of Galen, since it emanates from the fountain of Nature. Were it not so, Theophrastus would stand disgraced like the others.

If, then, we are willing to follow Nature, and to use natural medicine, let us see what substances amongst those used in medical art are most adapted to the human body for keeping it in soundness up to the limit of predestined death by means of their virtue and efficacy. If thought be given to the subject, I doubt not but all must confess that metallic substances have the chief adaptation to the human body, and that the perfect metals, in proportion to their degree of perfection, and especially the radical humour of those metals, can produce the greatest effects on the human body. For man partakes of that salt, sulphur, and mercury which, though hidden, enter for the most part into the composition of metals and metallic substances. Thus, like is applied to like, and this process is most serviceable to Nature if only it be dexterously applied. This is the great secret in medicine, worthy to be called its very arcanum. What marvel is it, then, if great, unheard of, and unhoped-for cures follow, such as the ignorant believed impossible? Not to delay longer, I am constrained here to set down what I determined to write in this treatise. I purpose to treat here more clearly than elsewhere of true medicine. First, however, it had to be pointed out how man derived his origin from sulphur, mercury, and salt, regarded as metals. This I have sufficiently indicated in the Paramirum, and it is not necessary to repeat it here.* I will, therefore,

---

that henceforth we have fire and life enough, we do not see. The eyes of the mind must be opened, and the first consideration must be what medicine is by Nature and art congruous with human life before all others, so that it may be conserved and preserved in health to its predestined end, and also may be safe against all corruptions. Nobody—at least, no true physician—will doubt that the metallic essences, especially those of the perfect bodies, are the most durable and least corruptible of all that Nature produces. If, then, life be the fire and heat of the natural form united to the humidity of its own matter by light, as is clear from Genesis, and the light lives more brightly nowhere than in bodies least liable to corruption, what will prevent the heat of the fire and the radical humour in the metals, each being incorrupt, from rousing in the organs joined to human life this vitality that is well-nigh dormant? For these are sleeping in metallic bodies alone, and in a state of repose, as a man overcome with sleep lies as if dead, and is only moved by respiration, but not in his body. As the spirit of metals, if it be liberated from its bodily sleep, will perform movements and actions as if its own in any body that is applied to it, none otherwise must we judge of human bodies. While these are sick the vital spirits in them sleep; they are not able to breathe truly or freely on account of their corrupt domicile. But when the corruptions of darkness are removed from the body, not by an extraneous physician, but by Nature itself, fortified by medical aid, and with an accession of extraneous life, that is to say, of incorruptible metals, the vital spirits in men exercise their movements freely. It is no wonder, then, if miraculous cures are wrought by Spagyric physicians, otherwise impossible by the vulgar medicine of the Greeks, whence it came about that they considered those diseases incurable which they themselves were unable to drive away with their sleepy and, as it were, dead medicaments. Hence the distinction between the Spagyric and the Greek medicine is clearly seen. The latter sleeps with the sleepers; the former, watchful and free from all slumber, rouses the dormant faculties of life. But returning to our investigation of the matter, this cannot be noted better than from the errors of those who inquire into this branch of the subject.

* ON THE ORIGIN OF DISEASES FROM THE THREE PRIMARY SUBSTANCES.—There are three substances which confer its own special body on everything, that is to say, every body consists of three ingredients. The names of these are Sulphur, Mercury, and Salt. When these three are compounded, then they are called a body;

only shew how the Philosophers' Stone may be recognised, and according to what method it is prepared.

Know, then, for a fact, that nothing is so small but that from it anything can be made and can exist without form. For all things are formed, generated, multiplied, and destroyed in their proper agreement, and they shew their origin so that it can be seen what each separate thing was in the beginning, and what it becomes in its ultimate matter, while that which intervenes is a kind of imperfect condition which Nature intermingles in the process of generation. Since, however, these accidents can be separated by the action of Vulcan, so that they shall be rendered inoperative, Nature can, in this instance, be corrected. This is what is done in the Stone. For if you would make it of its proper material, which can be perfectly learnt from the circumstances pointed out, you must remove from it its superfluities, and you must form, multiply, and increase it as a separate thing in its adaptation, which without such adaptation cannot be done. In this instance Nature has left it imperfect, since she has formed, not the Stone, but its materials, which are impeded by accidents, so that it is not able to produce these effects which the Stone, after due prepartion, *is* able to produce. Such material, without preparation, is, so far as regards the Stone, a mere fragmentary and imperfect substance, which has in it no harmony whereby alone it could be called perfect, or serve the human body for healing purposes. You have an illustration of this in the microcosm. See a man who is formed by the Mechanical Power only as a man. He is not an entire and perfect work, since he lacks harmony, but is only fragmentary until the woman is created like him; then the work is entire. Each of these is earth, and the two at last make the entire human being, capable of increasing and growing,

---

and nothing is added to, or coheres with, them save and except the vital principle. Thus, if you take any body in your hands, then you have invisibly three substances under one form. We have now to discuss concerning these three. For these three substances exist under one form, and they give and produce all health. If you hold wood in your hand, then by the testimony of your eyes you have only one body. But it is of no advantage for you to know this. The clowns see and know as much. You should descend and penetrate beneath the surface, when you would learn that you are pressing in your hands Sulphur, Mercury, and Salt. Now if you can detect these three things by looking, touching, and handling them, and perceive them separated each from the other, then at last you have found those eyes with which a physician ought to see. Those eyes ought to see these three constituents as plainly as the clown certainly sees the crude wood. Now this example may make you able to recognise man, too, in these three, no less than wood itself. That is, you have man built up in a similar form. If you see only his bones, you see as the clown sees. But if you have separated his Sulphur, his Mercury, and his Salt, then you see clearly what a bone is, and if that bone be diseased you see where it is faulty, and from what cause, and how it suffers. So, then, the mere looking at externals is a matter for clowns; but the intuition of internals is a secret which belongs to physicians. Now if these visible things must exist, and beyond their mere aspect medicine fails, then their nature must be deduced so that it may lay itself bare and be exhibited. Moreover, see into what ultimate matter these things are resolved, and into how many kinds. You will find these three substances divided from one another into just as many kinds. Now the clown cares nothing about all this; but the physician cares. The mere experimentalist neglects it, but not the physician. The quack thinks nothing of it; but the physician considers a good deal. Before all else these three substances and their properties in the great universe should be understood. Then the investigator will find the same or similar properties in man also: so that he now understands what he has in his hands, and of what he is making himself master.—*Paramirum, Liber I.*

MORBIFIC EFFECTS OF SULPHUR, SALT, AND MERCURY.—With regard to Sulphur, its effect should be thus estimated. It never produces an evil effect by itself, unless it be astral, that is, unless a spark of fire shall have been cast into it. Then is power awakened by that spark. Is not the act of burning a virile one? Without it nothing is produced. So, then, if any disease declare itself from Sulphur, then, first of all, the Sulphur should be properly named. It is essentially a masculine operation. There are many sulphurs, such as resin, gums, botin, axungia, fat, butter, oil, *vinum ardens*, etc. There are some sulphurs of woods, some of animals, some of men, some of metals, as oil of gold, of Luna, of Mars; some of stones, as liquor of marble, of alabaster, etc.; some of seeds, and of all other things, each with their own special names. There is afterwards the fire falling upon each, which alone is their star, and this, too,

and this power is effected by indwelling harmony. So the Philosopher's Stone, which should renovate man no less than metals, if it be freed from its superfluous accidents and established in harmony with itself, performs wonders in all diseases. Unless this be done, all your attempts with it are in vain. But if you wish to establish it in its harmony you must bring it back to its first matter, so that the male may be able to operate on the female, that the outer part may act on the inner, and the inner be turned outwards, and so both seeds, the male and female, may be enclosed in complete concordance; that by the action of Vulcan they may be brought to more than perfection, and be exalted in degree, so that each, as a qualified, tempered, and clarified essence, pours all virtue into the human body as well as into metals. Thus will they render each sound, will drive away defilement by the method of expulsion, and introduce what is good into the human blood by its power of attraction to a due place, so that the microcosm which is situated in the limbus of the earth, and is formed out of the earth, may by this medicine, as by something like itself, be radically, and not in mere imagination, but most surely, led to health and kept therein. This is the Mystery of Nature, and such is the secret which every physician ought to know. And this, too, every one can comprehend who is born of astral medicine. But, that I may more clearly describe the nature and preparation of so excellent a medicine, so that the sons of learning, who love the truth, may be initiated, know ye that Nature has given a certain thing wherein, as in a chest, are enclosed 1, 2, 3, the virtue and power whereof suffice abundantly for preserving the health of the microcosm, so that, after its preparation, it drives away all imperfections, and is the veritable defence against old age. This we name the Balsam.*

---

with its distinguishing name. And this operation is, in one respect, peccant matter. Moreover, with regard to Salt it should be known that it exists of itself as a material humour, and introduces no disease unless it be joined with its star. Its star is resolution, which gives it a masculine power. Salt, no less than the spirit of vitriol, tartar, alum, nitre, etc., exhibits itself tumultuously if it is resolved. Now, whence can such a nature be infused into humours, except by a star? About this physicians have formed a conspiracy of silence. Even if they had been guilty of no other blunder save that in all their causes and cures they had omitted the star, that would have been quite sufficient to prove that they had built their house on a foundation of moss and sand. You should know, also, that salts are manifold. Some are limes, some ashes, some arsenical, some antimonial, some marchasitic, and others of a similar sort. And from all of these are produced and begotten peculiar diseases according to the body of the salt; which diseases thereupon take the name and nature special to each. So, too, understand concerning Mercury. This of itself is not virile unless the star of Sol sublimates it. Otherwise it does not ascend. Its preparations are many, but there is only one body. But the body of this is not like Sulphur or Salt, which have many bodies, from whence come different salts and different sulphurs. This has only one body, but the star changes this in different ways, and into various natures. So, then, from the same source many diseases are produced. Thus its masculine nature comes from the star, and by this nature it has its morbific effect. If, then, all diseases are comprised under these three heads, each with its own name and title, know that you must reduce to Sulphur all that is sulphureous, in the sense that it burns. What is Mercurial should be reduced by sublimation if it be adapted thereto. What is from Salt should be reduced to such salt as is appropriate. These, then, are the three general causes of disease, as we have grouped them together above.—*Ibid.* Some further extensions of the philosophy of Paracelsus concerning the prime principles in their relation to disease will be found in an appendix to this volume.

* We must know before all things that a balsam resides in our body by the virtue of which the body is preserved from putrefaction. That singular balsam is present in all members of the body, for there is one in the blood, another in the marrow, a third in the arteries, bones, etc. While this balsam remains entire and uncorrupted it is impossible for any opening of the skin to take place. But when by means of the preparation of salts it happens to be corrupted, then the principles of corrosion begin to act. For cure in this case the whole attention must be devoted to the restoration of that which has been removed from the balsam by corruption. This is effected by the balsam derived from other elementary bodies. This, I say, is found in external elements, by which also other generated things are preserved from putrefaction. Hence it is reasonably called the mumia of the external body of the elements, that is to say, of their fruits. Such balsam is found congenital in the radical matrices of all growing things. Every body pro-

In what substance Nature has placed such a number you must know beforehand; since I am unable, for many reasons, to describe it more clearly. How it is prepared, Galen, Rhasis, and Mesne were ignorant, and it will not be reached by their followers. For this medicine requires such preparation as mere pill-sellers do not compass, and understand less about it than a Swiss cow. Moreover, it involves, as it were, celestial and special modes of operation. For it purifies and renovates, so to say, by regenerating, as you can read more at length in my Archidoxies, and at the same time study the origin and essence of metals and of metallic substances, together with their powers. He, therefore, who has ears to hear, let him hear, and see whether Theophrastus writes about lies or about the truth, and whether he is uttering mere inanities from the devil, as do you, O Sophist, utter your trifles—you who are from the devil and are surrounded by lies and obscurities, and call nothing good unless it be comprehensible by your fool's head, and that makes for your broth without any previous labour. For you, with your one eye, wander about at random, and cannot get straight to your kitchen window. By all means, twist your intricate thread, and try to find the centre of the labyrinth, by means of an obscure star; this will in no wise offend me. But if you would only use your foresight sometimes, and see whereupon the art of Theophrastus is founded, and how feeble your patchworks are, Theophrastus would not be so opposed to you. But what things I now write briefly, and shall write, so that my astral disciples may be able to perceive them, and enjoy them, and boast of them, these, through the diligence of one who is not ashamed to learn, can be easily understood; since there is nothing so difficult but that it may be understood and learned by labour and study. The practical method of this work, then, is as follows:—

The Preparation of the Matter of the Stone

Take some mineral electrum in filings. Put it in its own sperm (others read, Take the immature mineral electrum, place it in its own sphere), so that all impurities and superfluities may be washed away from it, and purge it as

---

vided with life is vital by reason thereof. The first thing to be considered in its extraction is the quantity present in the given body, for our intention regards not the body, nor the form, but solely the inherent balsam by which it lives. The extraction of these balsams is almost accomplished by separation, that is to say, while the arcanum is being removed from the body which was sustained by the balsam. You must know that elementary balsam is nothing else than that which we are accustomed to call mercurial liquor in the three principles. Whence it follows that all curative efficacy resides in Mercury. Mercury of this kind usually appears most manifestly in thereniabin and nostoch, as also in minerals of water, the fruits of the earth, and in the stars. Hence it is clear why antimony is so efficacious in the cure of ulcers, namely, because it contains mercurial liquor more abundantly than the other species of marcasite. In like manner the curative virtue of gold must be understood, for no body produced from the element of water contains a more copious or more subtle mercurial liquor. Thus also, among things which grow out of the earth, there is not a more arcane remedy than the mercurial liquor of cheirine and realgar. The same is to be understood of chaos and the firmament according to the method of their operations. But for the understanding of every mercurial liquor it is necessary to know that metals can be transmuted into a certain matter of this kind—as iron into crocus of Mars, Venus into flower of copper, tin into spirit of Jupiter. Yet these are not true liquors, but need a more exact separation according to their canons. There are extant various forms and descriptions of balsams according to the categories of the four mineral elements in the writings of the ancients. Some are wont to describe them according to the form of a plaster, others of a powder, others of the oil of water, etc. But I think that these descriptions ought to be completely rejected as purposeless, devoid of skill, plainly repugnant to Nature, and dependent on a dietary regimen which is the foe of Nature. The true and only balsam perfectly purges out and removes that which is separated, restoring the body to its original condition by the accession of fresh balsam.—*Chirurgia Magna*, Pars. III., Lib. V.

completely as possible by means of stibium, after the manner of the alchemists, so that you may suffer no harm from its impurity. Afterwards, resolve it in the stomach of the ostrich, which is born in the earth, and is strengthened in its virtue by the sharpness of the eagle. When the electrum is consumed, and, after its solution has acquired the colour of the calendula, do not forget to reduce it into a spiritual pellucid essence, which, indeed, is like amber. Then add half as much of the spread eagle as the corporal electrum weighed before its preparation. Frequently extract from it the stomach of the ostrich, by which means your electrum becomes continually more and more spiritual. But when the stomach of the ostrich becomes weary with labour, it is necessary to refresh it and always to abstract it. Lastly, when it again loses its sharpness, add tartarised quintessence, but so that at the height of four fingers it may be deprived of its redness, and may pass over together with it. Do this for so long a time and so often until it grows white of itself. When, now, it is enough (for you will see with your eyes how it will gradually fit itself for sublimation), and you have this sign, sublimate. Thus the electrum is turned into the whiteness of the exalted eagle, and with very little labour it is transferred and transmuted. Now, this is what we seek in order to use it in our medical practice. With this you will be able to succeed in many diseases which refuse to yield to vulgar medical treatment. You will also be able to convert this into a water, into an oil, and into a red powder, and to use it for all purposes for which you require it in medicine.

But I say to you in truth that there is no better foundation for all medicine than lies hid in electrum. Although I do not deny, nay, I write it in my other books, that great secrets lie concealed in other mineral bodies as well; yet they require greater and longer labour, and the right use of them is not easy, especially to the unskilled; so that if any one attempts them he causes more harm than good. For this reason it is not advisable that every alchemist should seek to practise the art of medicine when he has no acquaintance with it. Some prohibitory method should be devised which might debar such imaginary physicians. I, indeed, will not take their blame, or recognise them as disciples, since they do not pursue the truth, but I hold them as known deceivers, and lazy vagabonds, who take the bread out of the mouth of true disciples, and do harm to such deliberately, and think nothing of conscience or of art. In the prepared electrum we have described there lies hid so much power of healing men that no surer or more excellent medicine can be found in the whole world. The Galenists, indeed, those drug-selling doctors, call it poison and oppose it, not from any knowledge of it, but from mere pride and folly. I grant, indeed, that there is a poison in its preparation, as great as, or greater, than that of the Tyrian serpent which forms an ingredient in theriaca. But that after its preparation the poison remains has not been proved. For Nature, though by some blockheads this is not understood, always inclines to her own perfection, and much more, therefore, may be brought to perfection by proper methods. What is more, I grant that after its preparation it is a greater and more potent poison than before; but it is such a poison as seeks after its like, to find out fixed and other incurable diseases, and to expel them. It does not suffer the disease to speed its course and do injury, but as if it were an enemy to the disease it attracts the kindred matter to itself, consumes it from the very roots, and washes it as soap washes the spots from a foul rag, along with which spots the soap retires also, leaving the rag pure, uninjured, clean, and fair to look upon. So, then, this poison, as you say, has a far different and better effect than your axungia which you employ to cure the French disease, anointing more frequently than the currier does a hide. This arcanum, which lies hid in this medicament, has within it a well-proportioned, well-prepared, and excellent essence, which can be compared to no poison, unless one understands it, as I said before; and it is as different from your quicksilver, which you use as an ointment, and from your precipitate, in virtue and efficacy, as heaven from earth. It is, therefore, called, and really is, a medicine blessed by God, which has not been revealed to all. It is better compounded than that dirty medicine which the slowly walking doctor has in his gown, or has filtered through his fillet or hood. Moreover, this blessed medicine has thrice greater force and virtue for operating in all diseases, however they may be called, than all the drug shops you ever saw. But I did not find this in idleness, by sitting still, and by sloth, nor did I find it in an urinal, but by travelling, or, as you say, by wandering, and by much diligence and labour it was necessary for me to learn it, so that I might know and not merely think. But you suck your medicine out of the old cushion and bolster on which some old witch has sat who has covered your celestial head with a blue hat for medicine and breathed on you. So, then, I shall never regret my travels, and I shall remain your master and follow in the steps of Machaon, which, indeed, spring from the light of Nature as a flower from the warmth of the sun. But in order that my proposed task may not be delayed and remain imperfect, observe further how the thing must be done, and what power and property of medicine Nature has given to the Philosophers' Stone, and how it leads on to its end.

Here follows the remainder of the Preparation

Having destroyed your electrum, as aforesaid, if you wish to proceed urther, in order that you may come to the desired end, take the electrum which has been destroyed and rendered evanescent, as much of it as you wish to bring to perfection, place it in the Philosophic Egg, and seal it closely so that nothing may evaporate. Stand it in Athanor until, without any addition, it begins of itself to be resolved from above, so that it looks like an island in the midst of that sea, gradually decreasing every day, and at last being changed into the resemblance of blacking. This black substance is the bird which flies by night without wings, which the first dew from heaven, with its constant influence, its ascent and descent, has changed into the blackness of a crow's head. Then it assumes the tail of a peacock, and subsequently acquires the wings of a swan. Lastly, it takes the highest red colour in the whole world, which is the sign of its fiery nature, whereby it drives out all the accidents of the body, and warms again the cold and dead limbs.

Such a preparation, according to the opinion of all philosophers, is made in a single vessel, in a single furnace, in a single fire, the vaporous fire never being allowed to cease.

So this medicine is as if heavenly and perfect, or at least can become more perfect than the moon, by its own flesh and blood, and by the interior fire being turned outwards and prolonged, as was just now said, whereby all the impurities of metals are washed away, and the occult properties of the same are made manifest. For this more than perfect medicine is all-powerful, penetrates all things, and infuses health at the same time as it expels all disease and evil. In this respect no medicine on earth is like it. In this, therefore, exercise yourself and be strenuous, for this will bring you praise and glory, and you will not be a mere quack, but a real scientist, and, moreover, you will be compelled to love your neighbour. For such a divine secret no one can perceive and understand without divine aid, since its virtue is unspeakable and infinite, and by it Almighty God is known.

Know, too, that no solution will take place in your electrum unless it thrice runs perfectly through the sphere of the seven planets. This number is necessary for it, and this it must fulfil. Attend, therefore, to the preparation, which is the cause of the solution, and for your glorified, destroyed, and spiritualised electrum use the tartarised arcanum for washing off superfluities which accrue in the course of preparation, so that your labour may not be in vain. Of the arcanum of the tartar nothing will remain, only you must proceed circularly with it according to the number mentioned. Thus, easily of itself is produced in the philosophic egg and the vapour of fire the philosophic water, which the philosophers call *aqua viscosa;* it will also coagulate itself, and represent all colours, so that at last it is adorned with the deepest red. I am forbidden to write more to you on this mystery, such is the command of the Divine Power. Assuredly is this art the gift of God. On this account it is not all who understand it. God gives it to whom He will, and suffers no one to extort it from Him by violence. He alone will have honour in this work; Whose name be for ever blessed. Amen.

Here follows the Use of the Stone

And now it is necessary that I should write concerning the use of this medicine, and concerning its weight. You must know that the dose of this medicine is so small and so light as is scarcely credible. It should only be taken in wine, or something of that kind, and always in the smallest quantity on account of its celestial power, virtue, and efficacy. For it is only revealed to man for this reason, that nothing imperfect should remain in Nature, and it has been so provided and predestined by God that its virtue and arcanum should be produced by art, in order that to man, made in the image of God, all created things should be made to do service, and that before all else God's almighty power should be known. Whosoever, therefore, has intellect from God, to him this medicine shall be given. That ignorant Galenist, Beanus, will not be able to comprehend it, nay, he will turn from it in disgust. For all his works are works of darkness, while this work takes effect and acts in the light of Nature. And so, now, in brief but true words, you have the root and origin of all true medicine, which no one shall take from me, even though Rhasis with all his base progeny should go mad, and Galen be bitter as gall. Let Avicenna gnash his teeth, and Mesne in a word measure the length and breadth, it will be too high for all of them, and Theophrastus will stand by the truth. On the contrary, the lame works of the ointment-people, and the unctions of the physicians and doctors, with all their pomp and authority, shall go to utter destruction.

There is only one thing more to say, since to many this description of mine will appear obscure. You will say, O my Theophrastus, you speak too briefly and abstrusely to me. I know your discourses, how correctly you impart your subjects and your secrets. Wherefore this description will be of no avail for me. To this I answer that pearls are not for swine, nor a long tail for a goat. Nature has not seen fit to bestow them. Wherefore I say that to whomsoever it has been given by God, he will find sufficiently and above measure, more than he wished for. I write this by way of initiation. Follow with foresight, and avoid not study and labour, nor let the parade of fools lead you astray, nor the diligence which is necessary turn you away. By constant meditation many things are found out, and this is not without its reward. So use to a good purpose what I here give you; let it be to you a fountain, and then you will have no need to drink from the troughs of the pill-sellers, nor will your lot be with the body-snatchers, but you will be able to serve your neighbour well, and to give praise and honour to God.

These things I determined to set down briefly, in my book on the Philosophers' Stone, lest men might think that Theophrastus cured many diseases by diabolical methods of treatment. If you follow me rightly, you will do the same, and your medicine will be like the air which pervades and penetrates all that lies open to it, and is in all things, drives away all fixed diseases, and mingles itself radically with them, so that health takes the place of disease and follows it. From this fountain springs forth the TRUE AURUM POTABILE,* and nothing better can anywhere be found. Take this as faithful advice, and do not annihilate Theophrastus before you know who he is. I am unwilling to set down more in this book, even though it might be necessary to say something, and to philosophise somewhat concerning the *aurum potabile* and the

---

* The lungs, the liver, the spleen, and the reins may all be sustained and nourished by potable gold, which preparation all physicians ought to have by them, because no physician is of any consideration who is without it. I am acquainted with its preparation and am possessed thereof. It would not yet be advantageous to publish it, but time will perhaps reveal it.—*Chirurgia Magna*, Part I., Tract I., c. 17.

liquor solis.* I only wished to note these things here, and if you prepare them rightly they are not to be despised in their powers. But since other of my books treat of these secrets and make them sufficiently clear, namely, what a true physician ought to know, I will let the matter rest, hoping that this book may not be altogether without fruit, nay, that it may be of use to the SONS OF THE DOCTRINE. May God grant His grace, to His own honour and glory! Amen.

---

* According to the treatise entitled *De Male Curatis Aegris Restituendis*, Lib. III., c. 30, Paracelsus in some cases made use of the *Liquor Solis* for the cure of leprosy.

HERE ENDS THE MANUAL CONCERNING THE MEDICINAL STONE OF THE PHILOSOPHERS

A BOOK CONCERNING LONG LIFE*

SEEING that certain medicines are discovered which preserve the human body for a second and subsequent periods of life, which also protect it altogether from diseases, corruptions, superfluities, and other diminutions of its powers; nay, even when these infirmities and corruptions have broken in upon it, take them away, every physician must carefully study these medicines, and learn them from the very foundations. Indeed, numberless wearisome diseases and accidents of all kinds are taken away and extirpated from the very roots by this conservation of life.

So, then, when we propose to write concerning the preparation for a long life it must be understood that we do this in two ways. The one is theoretical, the other practical; for the subject of long life can be understood in both of these ways. And no physician ought to wonder that life can be prolonged—this also for two reasons. One is, that no fixed limit has been laid down for us, so that on some pre-determined day we must die; nor is this left in our own power. The other reason is, that we have medicine from Him who has made us, by means of which we can keep the body in that state of soundness wherein it was created, and expel from it all diseases whatsoever.

Now from this it may be gathered that death brings with it no disease, nor is it the cause of any. On the other hand, no disease causes death. And although the two coexist together, they are still no more to be compared one with the other than fire and water. They are no more akin one to another, nor do they agree better together. Natural sickness abhors death, and every member of the body avoids it. Death, then, is something distinct from disease. So far as relates to our present purpose what we wish is to speak to our own disciples from experience. We would speak to men who know those properties of things which are discovered by the highest art and by daily practice. To the mere pretenders and presumptuous men of medicine all these matters are perfectly occult and unknown. So it is that we write only for our own disciples, and not for those others. It is more

---

* The Geneva folio contains a considerable literature on the subject of long life, and this, as it will be readily understood, is full of repetitions and unimportant variations. At the same time, there is much in it which is only partially represented by the treatise translated in the text, so much, indeed, that it exceeds the reasonable limits of foot-notes, and it has, therefore, been thought better to print the *Book Concerning Long Life*, which pretends to be complete in itself, without annotations of any kind, and to reserve the additional materials to be dealt with in a separate appendix.

certain than certainty itself that the restoration and renovation of the body do take place, and that by these processes the whole frame can be transmuted to something better. In like manner, we see with our eyes that all metals are in their bodies purged so that they are protected from rust, and that even wood and the dead bodies of corpses are embalmed so that they suffer no further decay.

If, therefore, such things as these are possible to Nature, why should any one shrink from our writings, supported, as they are, by examples, although put in a brief space, or stand aloof from them because we draw a comparison between the bodies of metals and those of men? We do not suppose them to be one and the same. We know that they are altogether different; but in both we see the same method of conservation, and this is just what would have been inferred from experience, for if a dead body, by means of embalming, can be conserved, by how much more can a living one be kept from decay?

First, then, it should be known that this conservation, so far as it refers to the body of man, is to be considered in three parts: firstly, with regard to early youth; secondly, with regard to middle age; and thirdly, with regard to old age. Hence it can be understood why we begin severally from youth, from middle age, or from old age, since the end cannot be expected at any determinate period of life. The young life is sometimes destroyed in the mother's womb, sometimes in the cradle, sometimes during the period of growth, and sometimes, again, by the inordinate use of food or drink, whereby the nature being depraved and the strength diminished, it fails to reach the true limit of human existence. In a case of this kind the physician cannot but compare this period of life to old age, because it is equally defective and stunted in its nature. When, therefore, as often happens, children are born diminutive and sickly from their mother's womb, directly they come to the light they ought to be imbued with these conserving substances, by smearing the mother's nipples therewith, as is more fully explained in our practical method. By this method life and strength can be promoted, just as they can in the case of old men who were not so treated in their infancy. But in the matter of bodily power there is considerable resemblance between the old man's last span of life and that early period in which children are sometimes so broken down that they cannot attain to old age unless that first period of existence be fortified for them.

That is considered middle age when the body has ceased to grow and gray hairs begin to appear. At that time should anything occur to weaken or break up this period of existence, such as excessive work or debauchery, a similar method of treatment should be adopted before old age supervenes. If there be too long delay old age may never be reached at all, because that aid came too late.

The final period of life may be said to begin when the hair grows gray, and it lasts up to death. If this stage be complicated with weakness and decay, again the same method must be adopted in good season. But where the powers are sound and strong the method of conservation need only be used in proportion to the requirements. This, then, is the three-fold division of life to which we alluded ; and in whatever period the regimen of conservation shall have begun, in each alike existence starts afresh from youth, and goes through the different stages according to its limits.

Now, although we are supposed to be talking with our own disciples, yet this argument, amongst others, might be brought against us by empirics. Since we have such a means of attaining long life and driving away disease and death, how comes it that so many princes, emperors, kings, and other great persons die premature deaths ? Why do they suffer from infirmities when they could purchase immunity for money, and even ward off impending death ? We answer those who make such enquiries thus : We have never read or heard of a prince or king who used these remedies, with the single exception of Hermes Trismegistus. Many others have existed, as we point out in our book on Restorations, but they are unknown to the dwellers upon earth. And, though we do not consider them as a solution of the difficulty, yet still it must be remembered that the physicians of these emperors and princes understand less about medicine than the clowns who spend their lives in the fields. Hence it happens that they more frequently promote and conduct their princes and heroes to death than to life. Acting upon the advice of their medical men, it would be quite impossible for them to reach their proper period of life, and this on account of the ignorance of the faculty, who are physicians only in name. The argument, we repeat, is not thus answered ; but we might also put forward the irregular life of princes, who often shorten their own existence, and this more as a punishment than through devotion. Then, too, there are many princes to whom these and the like remedies are altogether unknown.

By these three cases we consider that we have answered the argument advanced against us.

In what has been just said, we desire to point out not that this three-fold division of the periods of life is alone to be considered, but also something else which is by far more important, and this in two ways: one, namely, where bodies lay themselves open to diseases by irregular life, whence proceed dropsy, jaundice, gout, the falling sickness, pleurisy, and other like diseases, chronic or acute. Not only the division of the periods of life, but each disease in particular has to be taken into account by itself.

In the other case diseases occur in conjunction with irregular life through the seasons or by means of accidents, as in the instance of pestilence, mania, and other visitations which are destructive of life. These two methods involve the process of long life. In this way we shall have to understand what diseases arise naturally and what from external causes connected with Nature; then those which derive their origin from something beyond Nature, as from incantations and superstitious practices, about which last there will be much more to say. For where such a disease occurs it must often be considered as a punishment, and on this account is not curable. We admonish our disciples to abstain from such practices. But so far as the visitation proceeds from Nature, this we teach how to cure and guard against by natural means.

Although we know and do not forget the effect of rings, images, and the like, which guard life against death, still we set down nothing on that subject here, but pass it by, because it has to do with astronomy. We speak, however, of these matters elsewhere, and proceed by another method, which is well worthy of all attention.

We first set aside diseases which do not arise from the bodily constitution, or from debility, but from some other source, before we speak of means of conservation. So, too, secret diseases will have to be examined. If they were diseases like gout in the feet or the hands, falling sickness, etc., they would not have to be set aside on these grounds, but would be cured by methods of conservation. So, then, there are three different classes of disease: one of a long-standing character, such as fevers, hyposarcæ, jaundices, and the like; another of diseases proceeding from within, as pestilences, pleurisy, abscesses, and so on; and a third of chronic diseases, as gout, falling sickness, and others of like nature. The first we must pass over, for the second prescribe remedies, and in the third case suggest methods of conservation, so that they may be guarded against.

Then, again, we must not omit those diseases which have a mental origin, and arise from the sufferer's own imagination, or from that of another person, as in the case of incantations, superstitions, and the like.

With such a division as this, it must be understood, by way of conservation, that those diseases which have a mental origin are in like manner removed by mental treatment: those which proceed from the patient's own ideas, by some objective mode of treatment; those which arise from imagination, by imaginations; and those which arise from incantations, by counter-incantations; while those which come from superstitions have to be relieved and cured by counter-superstitions. Then, after ailments of this kind have been removed from the body, there follow at once seven means of preservation, one directed to the natural strength or weakness; another against casual or accidental diseases, and the like, with the view of guarding against them for the future; a third against mental attacks; a fourth against incantations; a fifth against imaginations; a sixth against the patient's own notions; and the seventh against superstitions.

Marvel not that no mention is here made of temperaments, because the preservation of the whole body does not depend upon these, or upon their proportions, but rests on the virtue of Nature alone, from which all other excellencies arise. For the sole virtue is that which resuscitates and re-kindles the humours, which are four in number, but of which no account need be taken in medicine. The physician who bases his treatment on the natural temperaments may be fitly compared to a person who extinguishes a fire and leaves coals still burning. It is better worth considering how to preserve the root than the branches of a tree, because from the root the strength of the tree issues forth. If by chance anything happens to the branches—as may occur to the root of life from the temperament and the humours—this is merely an accident which in no way brings health, but rather takes it away. That we said was to be guarded against, when we were speaking of the three classes of disease, with the idea of resuscitating and re-kindling the life of the root, and, together with the branches, deriving its own nutriment thence, according as the complex qualities suggest their own remedies. When we seek to compare a short life with a long one it is necessary to know what life is, where it is, and how it may be diminished or augmented, as also whether it is something analogous to sight, touch, taste and smell. Judging by Nature, it would be impossible for us to know what it is which produces sight, and how it supplies it, as we have set down in our treatise on the Bodily Faculties, or which of those powers possessed by the body can be investigated by natural means. It is not so with life, because life does not derive its origin from natural seed, but has a spiritual source, though its origin is still natural. It is just as when the flint is struck by the steel, the fire leaps forth. There is no fire in either, and yet fire is produced from them. They are not fiery by nature, neither in their elements nor in their combinations are they combustible bodies; they rather resist fire more than other stones and metals do, as is seen in their transmutations, wherein no fire is elicited or can be obtained. None otherwise life takes its origin from something wherein is no life, as from the seed, the root, the sperm, and the like, just as the spark comes from the flint by the force of the *ens*, not of the *esse*. So, also, the life of man proceeds from the *ens*.

And then one has to consider whether this life can be produced, improved, or strengthened, seeing it is an incorporeal, volatile something, like a fire to which the more wood you supply the more fiercely it burns. In like manner, life, the more of the humour of life it has, the more the spirit of life abounds in that life. *How* the fire bursts forth from some substance in which is no fire, it is impossible for us to know. Some think that it is due merely to hardness; but how this can be the cause of a substance supplying that which is not contained in it, is not a subject for discussion in this place, where we treat solely concerning life.

Since, then, we see that life is, as it were, a burning and living fire, in this way it is illustrated to us in a material fashion and in a sense which comes under our ocular observation, that life belongs to the same category as fire, and we know that fire lives in wood, as also in resins and oily substances. None otherwise life according to its goodness or evil is strong or weak. Whatever of good or evil fire there is in it, as in the wood, we are able naturally to amend, or to add that in which it delights, or by which it will be rekindled; we can regulate it at our pleasure, can supply something stronger and stronger when another thing fails, substituting something else in its place exactly as the fire is renovated with fresh wood. Though it had been well nigh extinguished down to its last spark, it is again kindled, and burns more brightly than before. It is increased by those replenishments, or by any other mode of regulation we see fit to adopt.

Nor let us think that we must die on some precise day, sooner or later, or that it is derogatory that a Christian should believe it possible to prolong life by medicaments created by God for that purpose. Nay, it is altogether a mark of idolatry or of idiotcy not to believe that this is put in our power, just as we said above of fire. Our only defect is that we do not know the special kinds of wood by which we can kindle our life. It is not against Nature that we should live until the renovation of the world: it only passes our comprehension. For the most part we lack wisdom, so that we are unable to judge what there is in existence which is useful for us, especially since we altogether ignore our powers in this respect. Adam, whom we think the wisest of mortal men, had perfect knowledge of these matters. Although he was deprived of the tree of life, this is not a matter of theology, but of medicine. There *is* a natural tree of life —the tree of the soul.

Since, then, we are writing and teaching about long life, we should, in due course, at this point consider that kingdoms, districts, states, and valleys more or less contribute to the prolongation of existence. Some are more useful for the purpose than others. They afford more joy, more fresh air, and humours which are healthful to life. Among these exist causes which we ought to understand, connected with the land, the elements, the winds, the stars, and so on, which can make life longer than the common span of human existence. From the earth is produced everything that nourishes and supports our life; and also that which destroys it and makes it perish. Hence we can gather that evil arises to our bodies not only from such things as poison and opiates, but from those things by which life is chiefly supported, as from gold and balm. Nor should it be thought that medicaments were created only for use against disease, as, for example, the Tyrian remedy against poison, nor should it be inferred that these have no operation in healthy bodies. For we should know that to keep the body in health just as many things have grown up as for taking away health. We can, by our daily food and drink, at one time injure our bodies, at another benefit them and keep them in health, according to our use or abuse of these things. Whatever we use of these elementary things becomes useful or hurtful to us according to our reception of them. For whatever the fire in us consumes as it burns, that the fire restores to us when it is extinguished. Whatever loss is brought about by the one or the other the air itself compensates. It never refuses life or deserts us; it is we who do this for ourselves when we withdraw from the air. In this way, whatever one element takes away from us, some other element supplies in the way of conservation; and by an alternation of this kind those elements are regulated, so that they may do us as little harm and as much good as possible.

And although the air may sometimes destroy us by the infection with which it is occasionally contaminated, still we are able to discover this, and even in another direction turn it to our advantage. The air is both particular and universal, good in one place, bad in another, just as one land is better than another. In like manner, it is compared with water, which, like the air, is sometimes to be avoided by us, and the same is the case with fire. For from one kind of wood a better and more healthful flame proceeds than from another. So, also, influences and the stars above ought to be understood, since they have the power of breaking up our health in different ways, and of destroying our health. In like manner, it must be remarked, that among these some have the power of conferring health and of prolonging life. For there is no section, species, or kind so bad but that some good also exists therein, though we are not able to twist and turn this, or to regulate it for our advantage, as, for instance, that we should be able to appropriate to ourselves Jupiter and to reject Mars, or to select for ourselves other useful and convenient stars, as we do with herbs, some of which we choose and others we reject. We have not the same kind of power over superior objects of rejecting the one and selecting another; though even here we have some power when we bring it about through some means that they exert their influence on the inclinations, as when we use the rings or images of the planets, and other methods which we will explain more fully and clearly as we write. Furthermore, too, it must be considered that sometimes our life depends upon the aforesaid causes, and similar ones, either to be diminished or prolonged. We will say at greater length what good or evil can result to us from these, so that we may be able to understand the powers of God and His creation; not that we would presume to scrutinise these from their foundation, or fully describe them from the beginning, but only give some fragmentary thoughts concerning them.

More at length we have to consider in what essential way such things as diet, medicines, places of habitation, and the higher bodies exert their influence over us, so that we may be able to avail ourselves of the virtues contained in them, and to extend our life, and understand what their influence is, as also how they are able to do us good and help us. Their virtue far exceeds ours, because it is less subject to any sensible virtue, and it may help us just as a log is not able to burn unless some oiliness survives in it. The virtues of the things we have spoken about by far surpass ours in their consummate strength, and this in the following way: Their essence and property, their nature and quality, are incorruptible and permanent, since they do not die like mere sensible bodies, or as man dies, who is deprived of life and stripped of all his powers when his body becomes a corpse. Yet herbs and like objects do not perish essentially. For the substance of these things, even when the material body putrefies, remains, an incorrupted essence indeed, in the earth, passing on in turn to its own like until the remaining earth is consumed. They are taken up for its use, and their essence passes through their decaying bodies into the earth. But in the case of man, his virtue, essence, and properties do not remain with his corpse, but withdraw themselves from decay.

Hence it is gathered that we are not able to receive into our bodies the essence of herbs and the like, so that those bodies may be increased and nourished by the virtues of those herbs, simply because we receive the properties of the herbs, so that if one takes a pound of the quintessence of hellebore it relaxes the body as that herb does, or if a man eats gold his body shall become golden. By no means. Here is the difference: the one is appropriated, and the other remains in a material form within the body with all its virtues abiding in itself. This is an essence, not a property, but a certain *esse* from which is produced an essence endowed with all its properties and appropriate natural gifts. Such an essence, when it has entered the body, mingles with the humours for sustaining the spiritual life. This essence is a humour, and constitutes the life of the thing, and for this reason the two humours combine in one, and agree in their intermixture so that the inner receives the *esse* of the outer. The one is united inseparably with the other, just as one wine is mixed with another. Moreover, it must be remarked that consumption goes forth from that body because the humours of man's body are always mortal, and liable to decay; whence it occurs that essences are deteriorated and weakened for the future. There are, then, in this one two contrary things, whereof one is corruptible as the humour of life, but the other, which is fixed, consists of the essences of things, their humours, and liquids. That which is fixed gives assistance until it is consumed, just as talc does, which fire has no power to injure, but, nevertheless, in process of time it is consumed in another way. With regard to long life, we complain bitterly, because we are deprived of the power to understand what we should take when weakness seizes on the fixed substance, and we become weak, like a lion worn out with a long fight, and no more abounding in strength, or when we have no wood to add to a fire that is going out.

We know that there are some medicines and some regions in which there is no death, in others it is slow, and life is very long because the harmonies which prolong life are in these places most abundant. Although there are in these places some who are mortal, still they live a very long time, concerning whom I do not purpose at present to write more fully. We should know also that some persons are immortal, so that we ought to learn what is mortal and what is not, as also what the grades of the universe are, what vast natural powers exist therein, and how supremely man may be beautified by a long life. This can be understood in two ways. That is mortal which cannot remain until the destruction of the world. That is immortal which will await such destruction, though these very things are themselves mortal. This lower world is not going to remain for ever in its substance. What is born therein is divided into mortal and immortal, so far as natural things are concerned only in common speech. For the world is not one and undivided, but partitioned off into regions, namely, Paradise and the outer world. In the latter we live with such hope as we can. But in Paradise, which is the other world, there is no death. This is no miracle, but purely natural, because the nature of that world allows it. Just as our gold preserves one from leprosy, so does Paradise preserve us from death, not by a miracle in either case. Nor must it be understood as though our dead body was to rise, which would be a miracle, as we have said in our book on Heaven and Earth. It is most necessary we should know that in Paradise is born a being immortal in its substance. Moreover, in the essence of the outer world we have nothing but a fixed essence and a corruptible body, which are both preserved in Paradise. But to write much about this belongs not to our experience, beyond what the earthy essence teaches, which affords a centre. Nor ought we to discuss at length about these matters, since they far exceed our imagination, and every faculty which seeks to learn whilst on earth the order of Paradise. We speak of these things after a spiritual manner, rather in a dream than waking, and only for this reason to shew that the life there is enduring, up to the consummation, or perhaps beyond it, but this is to us occult.

Whatever, then, is from Paradise is able to render our life immortal, and not the Nile itself could wash away that attribute or despoil us of that virtue. We are by our materiality, by our place of abode, and by our pursuits deprived of a certain amount of power, and so we will not write further about these matters. Still, we will endeavour to teach in the sequel about a long life, so far as it shall be in our power to look into the matter, leaving those things which we are not able to grasp, and rather considering the predestination of each in particular.

In order to write more fundamentally concerning long life, weighing everything that conserves and defends it, it should be known that mind and being are the two methods by which soundness is either taken away or secured. In these two the health of our whole body, and everything that we do, is established. Out of the mind those matters regulate us which are therefrom produced, which also specially belong to it, as incantation, imagination, thoughts, and influences, all of which operate in the mental sphere. From the being those things govern us which belong to that department, such as the complexions, the qualities, the limbs, and the bones. Furthermore, this must be known and understood, that incantation, estimations, imagination, and influences are submerged in our mind when our thoughts are so plentiful and so powerful as to master our reason and our mind. So, too, from the most powerful forces of imagination, estimation, and incantation, the reason is submerged like the fragrance from the rose, whence it happens that it introduces syncope and madness amongst its influences, and so rules in the mind that it masters it and dominates it, not because it is produced in the mind, as the three preceding things are, but it has such a power of inclination, that it can turn it to good or evil, just as the sun shining through some glass tints according to its own nature whatever is contained in it. No otherwise does our mind exert a transmuting power according to its natural life, like ice melting in the sun. Those things which proceed from the being bring to us bodily infirmities from our complexions and qualities, which complexions originally arise from the being, and this is conserved by the humours, according to the bodily organisation.

It is in no way necessary to consider diseases of this kind, or the origins, beginnings, and essences of the mind and the being, nor again to inquire what the cholera is, or the phlegm, or the blood, or melancholy, but only to proceed generically, as will be shewn below. First, then, we will approach our practical method, because those other matters have no relevance to mental affairs, or mental things to them. So we will assign to bodily things each its own substantial medium, and, while remaining silent about mental things, we will point out the effect which these others have in the curative process. Understand, then, that the bodily substance must be conserved in the humour alone, as being that wherein is the life, and from which the complexions, and so on, are regulated. It is not the complexions or the qualities that have to be conserved or purified, nor need the losses of the liver or the spleen to be taken into consideration. Conservation itself removes all these things, and blots out all defects of a similar kind, together with all that is superfluous in the body, and thus reduces it to a state of equilibrium. Conservation of this kind does not operate thus by means of some grand specific of its own, but by means of its altogether incorruptible essence.

We determine, therefore, to forget for the present higher matters, and to point out the practical method in its due order. Here the regimen must be first of all preserved, afterwards the disposition of the body must be considered, and thirdly, medicine. On these three the conservation is to be established. Regimen, from which we start, as regards regions, scarcely comes into the question amongst other matters, as it needs the least precautions, and only the convenient order and disposition of medicine. We put, therefore, our chief trust in that medicine which not only conserves man in youth, but also affects brutes. The chief essence of the same, which is present in natural objects, kindles the body with so excellent a virtue that no strength or virility can be wanting to it. At the same time no disease can affect it, since, through the conservation of youth, no decay is suffered to take place, no putrefaction can occur, no disease arises. It does not even admit the excrescence of superfluities, whence it follows that no disease proceeds from superabundance, whilst, at the same time, it prevents the body from being corrupted, and so no evil can arise from corruption. Finally, it so protects the body that none of those diseases can enter which flow out from imagination or impression. In like manner, those diseases which proceed from tartarus are unable to adhere to it, and if any has fastened upon it beforehand it is removed. When you have once received what we lay down concerning a long life, there is nothing left which you can wish to understand from common sense, since we disclose our meaning only to those who have a good and extensive knowledge.

Now, then, we will explain our method and practice as briefly as possible, though they will not be intelligible by common persons. We write only for those whose intelligence is above the average.

Its practical use is that one dram of this conservation should be taken once every month in the best wine, if the nature, up to that time, has not become debilitated. If decline has commenced, a dram should be taken every week. In the same way, if ninety or a hundred years have been reached, the same quantity mentioned above should be taken every three days. It should be carefully noted that if any one has been deprived of virile powers, the weight should be increased as necessary. For the natural power of man is virile in its own degree, and more potent for suffering in one than in another. All this must be taken into account in cases of conservation for the reason we have mentioned in the words preceding. Sometimes the constitution of a man is so weak that he cannot be conserved, as in the case of those who from birth have no good foundation or root. As a sponge cannot admit fire in the same way as wood does, so these persons do not admit of conservation in the same way as others do. So, then, though the medicine be perfect, still, on account of defect in the subject, in the way of constitution and perfection as required by Nature, the operation is weakened.

Concerning the female sex, too, it should be understood that, so long as this method of imagination is applied, the menses will not fail for another period of life, nor yet fecundity for conceiving and bearing children ; provided only that the women, by their natural constitution, are qualified for this. For the spirit of life which is in them expels every contrary influence and conforms everything as it ought.

So, too, of the yet unborn fœtus. If at birth it be thus conserved it acquires a thoroughly sound constitution, proof against all diseases and preserved for a long life.

A long life depends on three things : the vital spirit which conserves life in us all : after this arise the different temperaments and qualities, like the trunk and branches from a single root. We place the reception of conservation herein, and endeavour to defend the rational spirit, by our conservation of the nature, from all those accidents which could be occasioned by temperament, such as excessive sorrow or joy, which arise from the humours. These, again, are fourfold, varying with the temperament and the qualities, and so on. For these no special medicine need be used. If the vital spirit only be conserved, these vital humours are conserved also, and if these exist without any defect, no perverted elements, such as temperaments, can arise. So, by this method, the body, and all that is therein, is rendered sound and healthy. Hence, too, it happens that no evil can occur to the rational spirit through the aggravation of the body, but it is conserved in a good essence and nature, just as it can easily be understood that it will be conserved with all its faculties, such as sight, hearing, taste, etc., in due proportion. The process of conservation, then, which we have already mentioned, we describe in two ways, one by means of simples, the other by arcana. Yet, we would not wish to separate altogether simples from arcana, but, on the contrary, join them together, since simples of themselves have such marvellous virtues and powers that some of them conserve for forty years, others for a hundred, and this, so far as conservation is concerned, we place on a par with the powers of the arcana. The essence of such simples which secure so long a life we consider in no way inferior to an arcanum. Take, for instance, the leaves of the daura, which prevent those who use it from dying for a hundred and twenty years. Its virtues we describe in the treatise on the Nature of Things, and here, for the most part, we preserve silence about them on the account of the incredulity of men. In the same way the flower of the secta croa brings a hundred years to those who use it, whether they be of lesser or longer age. There are many more of these no less worthy of regard which we do not set down here. The arcana, it is true, have greater powers of conservation than the simples we have named, because they fortify and nourish the radical humours more than the simples do, just as flesh and herbs nourish men, but in unequal degrees, since one has more power than the other, but Nature derives satisfaction from this rather than from that, and partakes of it with greater advantage. We give the recipe, then, for conservation during two or three periods of existence, for men or for women, as follows :— Take of cut flowers ʒj. ; of leaves of the daura ʒv. ; of essence of gold and of pearls each ʒss. ; of the quintessence of saffron, chelidony, and balm, each ʒv. Mix them all together according to the artistic method, and keep as a compound for use, in glass vessels as above. This medicine is also sufficient to cure and to drive out any accidental diseases which may subsequently occur in the subject.

Now we have to speak of mental affections and assign to each its special remedy. Mental diseases, either those which proceed from the mind or come into the mind from without, may occur in different ways, as we have before said. And though we cannot ascertain exactly the origin and nature of these latter, since they are invisible to our eyes in their seats, and consequently in their powers, still we speak only of their effects which we can see, and of the operations which they shew us. On these we base our practical treatment, saying that their effects are produced in the spirit of the mind (as we have elsewhere remarked), as our mind is merged in them.

We neither wish nor are able to hinder the course of the heavenly bodies, yet we have the power of resisting them, just as a strong wall can be assailed with bombardment and with engines. The sun impresses its influence on a stone. If that stone be thrown into the water, the sun can no longer bring its powers to bear on it, and thus the stone is preserved. Similarly, influence can enter a man in two ways : in one, when it is impressed upon him by ideas. This cannot be hindered, as the grass in the meadows cannot vegetate and grow without the sun. The cause of this we set down in another place. In like manner, there is one kind of influence which conserves and sustains us as nutriment does the body. Such is the influence of the constellations. The other is accidental, and this does us injury by hindering the former so that it cannot exercise its operations upon us. We know this of planets and the like.

We adopt measures of preservation against them so that their effects may have no power upon us. Such methods of conservation against the planets consist of rings and images specially constructed against their influences. What we have said on the subject of imaginations should be carefully noted. The rings of the planets have power of defending us from accidents, and so affect long life. But they are not satisfactory, for several reasons. And here is something to be known, namely, that the mental influence should be directed to some other object. For example, if Mars should be disposed to destroy me, and there be a mental inclination from him in my mind, which might induce mental disease, I construct my homunculus, that the operation of Mars may be directed to this image, and I may get off safely. It is easier to affect the homunculus, and so the planet is able to work its will more gently and without resistance. It takes the easier course, and leaves the more difficult one. The material is the same by means of the opening that has been made. To this material Mars was inclined with the view of injuring my nature; but he begins to operate on that which is easier, is satisfied therewith, and remains therein. Other reasons may be assigned for this, but we do not put them forward, since they have no bearing on the conservation of long life.

In this way there is produced a conservation of the mind, that is to say, the heavenly bodies are distorted towards something less resistent, so that the mind may be freed from the heavy yoke of those heavenly bodies by means of which death is often inflicted.

Against incantations, too, we would prescribe something, so that a long life may not be taken away by means of these. Remedies of this kind we have mentioned in several places, not as conservations, but simply for a cure of incantations. But the same is to be understood concerning them as in the case of the malefic stars. Incantations, that is, are to be guarded against in the same way as mental influences transferred to a homunculus. A similar operation holds good in its reversal to that which binds our minds and mental organs and our beings, the seat whereof is in the mind, as we remark in our treatise on Incantations. It must be directed to some other subject, and not to that which we have from the stars, but to its own incantation in the following manner: I construct a homunculus of wax to serve my purpose, and this I put in its place. Then, whatever attempt is made against me by way of incantation will be fulfilled on this image. For that proceeds from my mind, and the incantation from his mind, so that the minds meet, and on neither side is any harm done, or any effect produced. Under this form we have made clear the mode of resisting and preventing incantations by means of images.

So, too, imagination destroys an imagination directed against us. It may happen that I am being killed by the imagination of another. Such an imagination, which should be fulfilled in me, should be diverted elsewhere, lest health and long life be taken from me thereby. This I illustrate by the following example. If anyone feels great enmity against me, so as continually to wish for my death, and is not an incantator, but still a most persistent enemy, and if I do not know this, but his evil disposition is hidden from me, I, on the other hand, settle my mind in the greatest possible repose, by this method of protection being hostile to none, hurting none. By means of this piety, such great envy which is directed against me cannot be carried out; for piety is the principal and most consummate means of preservation against bad imaginations which can possibly be devised.

Estimation, in like manner, which is in ourselves, may operate in the same way. When I think more of myself than my reason warrants, such estimation is a submerging of my reason, so that I may lose my reason by the inordinate mixture of self-esteem, as we have said elsewhere on the subject of estimation. Against such estimations as these the best method of conservation is not to impose more on the reason than it can bear, which is the way in which such estimations are fulfilled. Here is another method of conservation. For instance, where we make false estimates of God, a subject which surpasses our reason, we betake ourselves to a homunculus directed towards the stars, which homunculus is of the same nature as ourselves, and so our speculation, which might corrupt us, passes to the homunculus. For there is no resistance to such speculations being fulfilled without loss, by a subdivision of the senses existing in me and in the homunculus, with a difference of perfection and imperfection.

By means of estimations and imaginations many superstitions are fulfilled, which are not impressions, or incantations, or estimations, but simply superstitions capable of being understood by a similar example. I suppose myself imbued with the superstition that when I hear crows chatter on my roof I esteem it the sign of some one's death (there are many similar examples which need not be quoted). This superstition may make me ill myself, or kill my sick friend, the cause whereof I set down in my treatise on Superstitions. My method of conservation is this: that I regard such chattering as natural, and that it does not operate according to the foolish idea of men and of myself. By this means my superstition is destroyed, for it is rendered doubtful when I attribute it to my folly. Nothing destroys superstitions more thoroughly than considering them to be follies. Thus all harm is taken out of them. It is the consensus that leads to action, and this consensus is at once removed if I think of my own simplicity and the folly of such empty credulity. And this does not happen only in the case of crows, but with many other things, which it is not necessary, nor would be useful, to rehearse in a treatise on long life. Enough has been said on the subject of conservations.

As to the fact of our being hard to understand, that happens only to those who understand neither us nor Nature herself. Yet is what we say of some moment. For those who do not understand us on the subject of long life, we decline to teach them more plainly. For those who have any depth, we are conscious that we have written of our processes at sufficient length, and have disclosed them with ample clearness.

Concerning the regimen of food and drink, we will only set down so much as bears on long life, reserving the rest to be treated in another place and by another method. First, it should be known that foods ought to be prepared according to a process, just as medicine, so that they may expel all the superfluities of humours which lie hid and grow in the body. These humours, by means of conservation through food and drink, are completely removed. Next, it is necessary to know that the cures of all internal diseases are accelerated and promoted by suitable medicines, and also by Quintessences. As to the origin of diseases and medicines, we say nothing about these matters here, because they are extant in many parts of our Physics. This only we say, that for a long life the following is the best regimen: moderate diet. We do not give this in detail, because it is well known to every physician. One thing we will add, that food ought to be separated from all its impurity, as we teach in our treatise on the separation of the elements. The use of such separated food with moderate diet exhibits a wonderful sustentation and conservation of life; it restores the flesh and blood, so that no disease can do harm to the body through external foods, which is not the case with other foods and drinks. The use of such separation of foods for the foundation of a long life we set down for those who wish to live long; and we do this for many reasons, not necessary to specify here. This we do affirm here, that this is a regimen by which the complexions of the blood and the flesh are nourished. Nevertheless, conservation belongs to the vital spirit and its humour. Thus we believe we have said enough on the regimen necessary for a long life.

We describe the separations of foods and drinks in the treatise in our Archidoxies on the Separations of the Elements, the pure from the impure. In like manner, in the book on Regimen, in the same place, we set down what may be best and most conveniently adopted for a long life. We do not repeat here what we have written at sufficient length elsewhere.

What air, earth, and elements are best, we will indicate in the same manner. In the element of fire no watery disease can arise, and *vice versa*. So, in like manner, we must understand of the earth and the air. Whatever lives in the fire is free from other infirmities except those of fire.

What has to be considered, then, and assumed necessarily for practice, is a certain desired equilibrium of elements, wherein long life consists. This is the work of wisdom alone, in which work are comprised the operations of the elements with the perfection of their forces, so that the fire shut up in man shall avail to take away all diseases, drive them out, and keep them at the greatest possible distance. We look upon this work of wisdom as another earthly Paradise, in which no disease could germinate or remain, no venomous animal could dwell or enter, and health could never be destroyed. This we could wish conceded to us by the Lord God, that we might write freely, without the contempt of idiots, what experience has taught us about this work of wisdom. But, on account of those idiots, one has patiently to hold one's tongue with regard to the miracles and marvels of that work of wisdom, wherein is reserved the earth of the wise. Since, then, I must be silent about this, I determine to describe it only among my secrets, that it may remain buried within me, though without any end of life.

Thus far have I written on the subject of Long Life for our own and other disciples who are endowed with a happy and subtle intelligence.

HERE ENDS THE BOOK ON LONG LIFE

THE BOOK CONCERNING RENOVATION AND RESTORATION

By Theophrastus, Philosopher and Physician, of Germany, called Paracelsus the Great

FIRST of all we must understand what Restoration and Renovation are; what those things are which restore and renovate, and also what that is which can be renewed and restored in the creation of things. All minerals, indeed, are thus brought back to youth, are renovated and restored, so that rusted iron can be again brought back to new iron, and the rust or flower of copper into its copper. So, likewise, minium can be brought into lead, and Saturn into Mercury (or, as others read, the calx of Jupiter into tin). In this place, therefore, renovation and restoration signify that process which brings back a destroyed, or rusted, or consumed substance to its youth and its perfect essence. But still this renovation which we have here introduced cannot be in any way compared to that restoration or renovation which we purpose to expound. For though rust and verdigris are not a metal, still none the less in their essence the essence of the metal has not yet perished or been consumed. So, then, that renovation cannot, in this place, be taken for an explanation of restoration and renovation, because in the human race such rustings and ablutions do not occur; and hence it is that men do not require a renewal of this kind.

In this manner, if a decrease be understood to happen to an old or decrepit man, like a kind of rust in his substance, then, in like manner, his body will be capable of being brought back from its state of decrepitude to youth, and there is a restoration from any kind of disease to health, but concerning this we have no desire to write at present. This, too, may be accounted a kind of restoration when out of salt, sulphur, and mercury a metal is made naturally. When this perfection is completed and brought to the actual metal, that metal may again return to its three primal elements, so that the salt, sulphur, and mercury again appear as they were at its first generation, insomuch that the metallic element altogether passes away, and there is no longer any metal present. So also it may happen that the matter of the three primal elements may return again to a metal as before, namely, if from the three primal elements of copper, copper be again made, and so on. This, too, in the case

The Book concerning Renovation and Restoration 125

of metals is a restoration or renovation, when a kind of regeneration takes place from a metal previously complete into a metal again, perfect and complete. But that is not to be thought of as renovation or restoration if it be referred to man, because we cannot be brought back to our three primal elements, or reduced to our sperm, from which, again, we might be again renovated and restored, as in the case of metals quoted above. For then we should have the power of improving ourselves by a second generation better than the first was; even as iron, which is reduced into its three primal elements, and afterwards to silver or gold, and becomes incorruptible by this very process; or as Saturn, which is again reduced to its mercury, is at length changed to an incorruptible metal. None otherwise should we, too, be able to produce or create from ourselves an incorruptible creature, whereas we have no power to do this. For we lack that primal matter,* and are unable to go back, so as to be constituted an irreducible mass, but we must progress as we have begun, and in no way are we able to recover or to possess that out of which we proceeded.

Restoration, then, or renovation, is twofold. One, as applied to metals, we have introduced and made clear. Another, when an old picture is renewed with fresh colours, so that it appears recent and new as it had been before. But we must not, therefore, understand in this place that a new matter is made out of an old one, but that the old picture is so draped that it may appear new. Wherefore, again, this kind of restoration cannot be cited with reference to the restoration and renovation of man. But restoration and renovation must be understood in this way: that man's radical moisture, acting upon and energizing the spirit of life, shall not be diminished or driven back, but rather shall be increased in its powers and pushed forward, as a tree to which aid is given for the production of its flowers and fruits, so that when these drop off and are done with others are again procreated as before. But, although the example here quoted does not in every detail illustrate our theory, nevertheless, it affords the means for understanding how to promote the radical moisture of life just as we shewed in the case of the tree. We intend renovation and restoration to be understood thus: that they are not produced in the radical moisture,† but in that which is generated from the aforesaid moisture and derives its origin materially and corporeally therefrom. For as a bell made by fusion does not receive its sound from the tone, but from the body, so restoration or renovation does not receive its operation in the spirit of life, but in that which makes this same spirit; that is to say, the one is material, the other substantial. But when all this substance in which the radical moisture is

---

* It should be noted that while all things are constituted in the three prime principles, they cannot be separated without the destruction of the simple matter itself; for in separation the virtue of Mercury, of Sulphur, of Salt, vanishes and goes back into the first matter, as may be seen outside the Microcosm, Mercury being transmuted into soot, Sulphur into oiliness, Salt into alkali, whence it is manifest that the first matter cannot pass into the ultimate matter in the absence of a medium.—*Chirurgia Magna*, Tract III., Lib. III.

† For the conservation of the radical moisture in its proper quality a medicine is required which is also a material humidity, and while this is administered, no disease can be contracted. *De Morbis Metallicis*, Lib. II., Tract IV., c. 5. In the same treatise alum is said to contain an elementary humidity against the fire of the Microcosm.

present shall have been purified, its tone will be also improved, and the better the tone is the better the body will be. And when we say that the radical moisture proceeds from the body and the members, we understand it in this way, that the radical moisture itself, and that which proceeds from it, are just like the root and the tree, of which the one cannot live or subsist without the other.

Equally must it be understood here that these two things are so intimately united and conjoined as to be incapable of being separated. The radical moisture and the spirit of life,* together with the moisture of life, are in bodies and in limbs, just as in metals is the tone, which is not seen, only heard. For the spirit of life and the radical moisture are truly in bodies. It would be idle, therefore, if we endeavoured to purify it or to renovate the body by its means; but it is just that the body and the matter which are born and have their origin therewith should be restored and renewed. Hence, it may be gathered that restoration and renovation are transmutations of members existing superfluously in the body; so that all which proceeds from the body, and not from the radical moisture, falls away, and something new is born in its place, just as we perceive in a tree, from which all the leaves, the flowers, the fruits, and the fungi fall away, and are born again, yet the wood itself is in no respect changed, so as to fall away and be re-born, but it remains. So, too, does the radical moisture remain. This is the life in the body; and when bodies reject from themselves hair, nails, teeth, and such things, these are presently re-born. This is restoration and renovation, whereby that very thing which should be restored and renovated is restored and renovated. For every restoration and renovation occurs in superfluities, and in those things which have their origin and are born from the substance. By what method the body is able to be restored and renewed may be sufficiently understood from the demonstrations which we have made, and from the superfluities which do not form excrescences on material effects, as the hair, the teeth, the skin, and the nails; but are in the body as something in excess, which are not absent from matter or corporeal substances, but remain in their essence as four complexions†

---

* The spirit of life is a spirit situated in all the members of the body, however they may be denominated individually. In all and each of these the said one spirit abides, and it is the sole virtue indifferently of them all. It is that highest and most noble grain whereby all the members have their life. But being extended and diffused it manifests in various ways, according to the variety of its seats. . . . Nevertheless, its potences are one. The virtues which sustain the bones are in no way feebler than those which nourish and fortify the heart, nor do they abound more in the brain than in the marrow, although the opposite may seem to be true. There is the same necessity of the marrow as of the brain, and the virtues of both are the same. A like law prevails through all the members. Some of them may appear to be of greater importance but, nevertheless, one spirit of life is the moderator, virtue, power, and operation of them all. The spirit of life originates from outside causes or generations, not from those which are natural according to the flesh. While the generation of other things is twofold, that of the spirit is simple. – *De Viribus Membrorum*, Lib. I., c. 1.

† Concerning the four complexions—cholera, blood, melancholy, and phlegm, we would by no means be identified with that opinion which asserts that they are or do derive from the stars or the elements. We do not regard this as true even in the least degree. The principle or beginning of cholera is from bitterness; melancholy is acidity; phlegm rises from sweetness for every sweet thing is cold and moist. Blood is from salt; whatsoever is saline is sanguineous, that is, warm and moist. The four complexions, therefore, are acidity, sweetness, bitterness, and saltness. If salt in any man predominate from the ens of the complexion, then is he sanguine; if bitterness, then he is choleric; if acidity, he is melancholic; if sweetness, he is phlegmatic. Thus, therefore, the four complexions exist in the body as in a certain garden, wherein flourish amarissa, polypodium, vitriol, and salt nitre. And all these may coexist in the body, but so, nevertheless, that one alone will prevail. – *Paramirum*, Tract III., c. 10.

The Book concerning Renovation and Restoration

(otherwise humours), whereof one proceeds from coldness and humidity, which is retained in the whole body, and is born, not having any special place, nor any beginning or initial point from whence it proceeds, as is proved concerning the four complexions. Another springs from the exact contraries of the former, namely, from heat and dryness, which, too, are similarly in the body and have no special abiding place or origin, and also produce liquid. The third is cold and dry, deriving its birth in the same way. The fourth is warm and moist, itself also proceeding as the others did.

And here it must be noted as happening that these four humours* do not all exist in all bodies at all times, but sometimes one only, sometimes two, sometimes three, at other times four. Of them it must be remarked, too, that, in the process of renovation and restoration, they are consumed and expelled, for this reason, that Nature and the life of man are able to exist without them, and stand in no sort of need of them, since they exist only as superfluities, like the dregs in wine, or the froth flowing from it at the time of vintage.

Concerning the four complexions thus displaying themselves in man, this, too, is to be noted: that these are not renovated or restored, because they spring from no one of the members, either greater or lesser. Neither are they in the blood, nor in the flesh, nor in anything like these. Nor, again, is it true that the sanguine complexion proceeds from a liver abounding in blood, or the melancholy from the spleen, the choleric from the gall, and the phlegmatic from the brain, and the like; since the aforesaid members do not supply their complexion to man, but those complexions come at birth itself and last right up to death. These points we do not undertake to discuss in this place, because they are too remote from our text concerning renovation and restoration.

Since, then, no one of the four complexions has its place or origin in the bodies we have spoken of, but exists in the spirit of life and in the radical moisture, complexions cannot be renewed or restored. But when the body shall have been clarified, their nature, too, shall be made clear.

In like manner do we point out as foreign to our text the division of complexions according to age, region, and regimen, because no complexions are impressed on a body by these three. It may, indeed, happen that old age brings sadness to bodies, but this is not a complexion. In like manner, the dwelling place may induce phlegm, but the complexion is not on that account phlegmatic. Bile may make one acquire a yellow colour, which need not be discussed here since it is made clear when we treat of the Construction of the Body.† For a division of this kind a special phase of intelligence is required,

---

* There are four humours contained in man—blood in the veins, moisture in the flesh, viscosity in the nerves, grease in the fat. These four have each their natural purpose. – *De l'este, cum additionibus*, Lib. II., Tract III. At the same time, the doctrine of the four humours as commonly expounded at his time was rejected by Paracelsus, because it was a thing hard to believe, founded upon faith only, whereas medicine is established, not in faith but in sight, and nothing in such a matter should be accepted upon faith, except the diseases of the soul and eternal salvation. — *Paramirum*, Lib. I., c. 1.

† Paracelsus has a treatise on jaundice, which will be found in the first volme of the Geneva folio. As in so many other cases, there is no work which precisely corresponds by its title to that named in the text.

since it must be remarked that they are not only humours, but sometimes minerals too, and sometimes corruptions, which all exist as superfluities contrary to Nature and virtue. In like manner it must be said concerning the principal members which resist renovation and restoration, that is to say, in this way, that they do not perceive it, for they do not receive them into themselves, but they take up everything that passes through them, and is prepared with them, just as they take up food, not a medicament. But wherever by any chance humours or superfluities are produced in them, they would be expelled. So, also, of the other members, too, it must be equally understood, namely, of the bones, the marrows, the brain, the heart, the liver, the lungs, the kidneys, the spleen, the stomach, the intestines, the cartilages, the muscles. And of the blood, too, it should be known that corruption or superfluity exists in it, though it be only an accident. And so equally of the flesh. This accident is, as it were, purged away in the process of renovation and restoration. Not, indeed, that another blood is produced, but that what is depraved is removed from it, and the good is preserved and predominates. The same judgment, too, is passed concerning the flesh.

To explain briefly what are those things that can be restored and renovated: leprosy, falling sickness, mania, pustules, gout in the foot, or in the hand, or in the joints, and other like ailments are removed in renovation and restoration, unless, indeed, it be some disease taking its origin from birth. This will not be removed.

But concerning leprosy, or any more severe disease which may exist, it is well to know that it undergoes transmutation in the body, not, indeed, that there is a separation of the pure from the impure, but that the leprosy is converted into health, as copper or iron are transmuted into gold. And no one ought to be staggered by this conversion, for renovation and restoration consume, none otherwise than fire in silver or in gold, its falsities and impurities, and leave it pure. In the same way the falling sickness and gout are taken away; for all things which are in the whole body are forthwith renewed, the blood and the flesh, with the other things which are embraced in it. For, as alkali purifies mercury into the very best silver, so also renovation and restoration transmute the body into a good essence, as has been said above.

Renovation and restoration, then, expel whatever is superfluous and incongruous with Nature in the body, and change all this which Nature does not want, or which was of no account, into something good. In this way it re-establishes everything and causes it to grow again, as we have said above, brings back the entire body to youth, and so on; for this cause, that nothing can resist those things which are in the nature itself.

And now we must see by what method the body is restored and renovated, namely, by the kindling of the renovating and restoring medicament, which it has in the spirit of life and in the radical moisture, by which kindling the operations are prescribed like the burning forces of a nettle. For who is so sagacious that he can investigate rightly forces of this kind when they do not appear to us in natural act, but are sensibly apprehended? In this way, also, renovation and restoration are accessions to Nature produced by forces which we cannot express. This, however, is openly known to us, that every visible thing is cleansed and purified with fire. Nature, indeed, demands that this one process shall be accomplished by fire, and that it shall not be possible by any other means. We understand, therefore, a two-fold fire, a material and an essential fire.* The material fire operates by flame; the essential by means of the essence and the virtues, like cantharides, burning the skin and raising it in pimples as a very violent fire does; yet still it is not fire, nor is felt as fire. A flame and a nettle produce the same effect, as we have often said.

It is in like manner certain that renovation and restoration in this way accomplish their operations when they come into the body or are joined in union within it, because a like operation takes place thereupon as is the operation of Saturn or Mars in Mercury, which are cast into the fire with their realgars, and although neither of them is warm or fiery, they are, nevertheless, burnt up like wood, and in the bottom the perfect metal is found which before appeared altogether leprous.

And, again, who can altogether trace or investigate how it happens that when a migdalio has been powerfully liquefied by means of vitriol, it becomes copper, in all respects and throughout its entire substance like copper, though before it had no likeness to copper at all. None otherwise must renovation and restoration be thought of by us than that they accomplish their operations like lime, which is extinguished by water, and purifies itself, so that all its powers and acridities are taken away and removed by its essential fire.

The renovation and restoration of our nature are none otherwise than in the case of the halcyon, which bird, indeed, is renewed in its own proper nature; and many other like animals are found which have the power of doing this, of which we have made mention in different ways in our Archidoxies, and still more in our Secrets,† from which could be quoted more examples were they not too far removed from our text concerning renovation and restoration, where the demonstrations we make come to be understood equally in this place concerning renovation, while we say again and again that we cannot sufficiently know how the fire operates, though we see it consume the wood, because by its excessive heat it overcomes and consumes everything else. But, leaving this, let us betake ourselves in another direction.

Since, then, we have spoken at sufficient length thus far concerning the beginning of renovation and restoration, let us now point out those things which do renovate and restore. We have, indeed, in our Archidoxies taught

---

* Fire in its nature is four-fold, that is, the sun and moon govern one part in water, the second and the third, which are resident in air and earth, are ruled in like manner by the sun and moon, and there is hence conjoined in all creatures that magnetic virtue concerning which there must be nothing more openly spoken, for herein is the knowledge of the labour of Sophia, the mother and fountain of the Magi; I have said.--*De Pestilitate*, Tract II., s. v. *De vi magnetica mumiæ in homine.*

† The reader of Paracelsus may not unreasonably be disposed to think that his secrets are synonymous with his whole philosophy. At any rate, there is no individual treatise under this title.

in writing how to prepare them, and entitled them with their proper names so that they may be known and marked. Now it is our intention to lay down the composition thereof, and first of all the processes. But while we teach concerning simples and arcana, it must be understood that their operations are brought about in different ways. For some are found which violently purge leprosy by means of renovation and restoration, but beyond that touch no other disease; yet, nevertheless, they are perfect in renovation and restoration. Beyond these, in the distinctions of these kinds of diseases, are the Quintessence, the Magistery, and the Element of Antimony, which purifies the body from leprosy, none otherwise than silver and gold liquefied therein, and in these it leaves no trace of impurity.

In like manner the element of the sun and its quintessence, as also the oil of the same, and aurum potabile, take away leprosy, together with all diseases, renovate and restore. So also the quintessences of hellebore, of chelidony, of balm, of valerian, of saffron, of manna, and of betonia renovate the body, with the exception of those diseases which we have mentioned above, for these they in no wise diminish.

The quintessence of pearls, too, or of unions, of smaragdines, sapphires, rubies, granates, jacinths, renovate also the body and bring it to entire perfection. They take away tartareous diseases, as the calculus, gravel, gout in the hands, feet, and joints, together with congelations and coagulations, and similar diseases which arise from tartar. So also the quintessence and magisteries of minerals and of liquids renovate and restore the whole body from any defect, and free it from the falling sickness, syncope, suffocation, and all diseases which happen with a deprivation of the senses, such as mania, vitista, and the like.

The magisteries and the essence of tartar and alkali also renovate the body with perfect restoration, take away all abscesses, and remedy the putrefactions and the grossness of the humours.

In like manner, the essences, extractions, and magisteries of the greater drugs renovate and restore the whole body, take away fevers, both quotidian, quartan, chronic, and ephemeral. Likewise the first elements of the sub-margarites can renew and restore the whole body, and remove all diseases from women, with their accidents, and also render mankind fruitful, both the husband and the wife. The same arcana in like manner take away all long-standing and incurable diseases by renovation and restoration of the body to its highest powers.

None otherwise, too, the quintessence extracted from balsam renovates and restores the body. It takes away pleurisies and pestilences by its admirable operations and power of perfection. Of this class, too, are many more things which are also enumerated elsewhere, of much greater virtue than can be attributed to them.

In these matters, however, it must be understood that the compositions have to be carefully watched. For though there are many of them, still no one suffices generally to cure all diseases by itself, but such diseases are to be expelled by the medicaments of renovation. Finally, therefore, we purpose to demonstrate the manner and the practice of our intention, though we may not set down all the processes, for this may not be necessary. He who understands us will perceive the drift of our writings. He who does not understand us is not capable of being taught by our writings, yet none the less we will set down all the processes in sufficient detail. In truth it would be a heavy task for us to write those things which have been written by many or are already known beforehand. This doctrine cannot be conveyed better than by the first entity, in which there is a singular nature for operating upon the body and transmuting its essence. For that first entity, indeed, is an imperfect compound predestined to a certain end and to corporeal matter. And because it is not perfect it is able to alter everything with which it has been incorporated as Mercury can, which is like a primal imperfect entity, in respect of its own imperfection. Although this be determined and limited, nevertheless it is not changed from imperfection, but still it is limited thereby.

Mercury, also, has the power of renovating the whole body because there is in it a powerfully laxative force, and an alterative as well, which cannot be sufficiently explored. Yet, nevertheless, it is as a whole imperfect and useless in its operation, because, so to say, it is mercury, and its first entity should not be predestined into another body. For such as it is itself, such is its perfection. Nevertheless, we speak of a first entity, which is perfect for renovating and restoring the whole body, as is the first entity of gold, and for this reason, because it embraces altogether the spirit of gold and is most subtle, far more subtle, indeed, than the true body itself, that is, than gold.

Hence, also, the first entity of the sun, or gold, is penetrable, even as mercury in metals, and does not contain within itself the spirit of salt, whereby it may be coagulated. For the spirit of salt coagulating the first entity takes so much power away that the gold becomes not by a hundredth part so powerful in its effects as its first entity is. In the same way, generous wine coagulated by frost never again returns to its pristine power.

Now, in order that we may speak and write perfectly concerning renova- and restoration, it should be known that the first essence, that is, the first composition of gold which exists as a liquid not yet coagulated, renovates and restores whatever it lays hold upon, not only man, but also all cattle, fruits, herbs, and trees. And this must be understood, just as of the universal form of any metal, which is endowed with far greater virtues than its own metal, seeing that there is in the mineral form the spirit of arsenic and salt of sulphur and of mercury, all of which are lost in the purgation of the metal, the said metal remaining in one essence alone.

The very same thing is, in like manner, to be understood concerning the first entity of marchasites, as of antimony, which ought to be known no less than the first entity of gold. In the first entity of antimony, indeed, there exists such a virtue that of itself, by its own special nature, it transmutes everything of which it takes hold, none otherwise than antimony itself does by fire. For the virtue of that same separates from the body everything which is generated by the radical moisture, and altogether renovates the said body from its very foundation; for its first entity becomes so fixed in that predestination that such an essence proceeds out of it, and goes forth from it as heat does from a fire. No otherwise must it be understood concerning the first entity of resins. The first entity of sulphur is an entire transmutation of the body into certain renovations and restorations; and it is so vehement that it tinges all the first entities of metals into its own essence, takes away their operations, reduces them again to their first matter, and afterwards brings them to a perfect and new body. Indeed, the first entity produced from sulphur has such power over the body of man that it renovates all the radical humours in it, in all its places and parts.

In like manner, also, may we speak concerning the first entities of gems, which, indeed, by their primal essence most powerfully reinstate the whole body in its pristine powers, cleanse it from all its impurities, and renovate and restore it none otherwise than the fire changes lead into purest glass. For the first entity of the emerald regenerates and renews itself since it exists from the first as a perfect body. So, too, green marble, which, from its own predestination, has such a nature that it renovates itself from all uncleanness and impurities, and coagulates a second time until it becomes pure. Sometimes it renews itself thus a third and a fourth time, and returns to its youth; and the oftener it regenerates itself in this way, the purer and more enduring does it become. So far, then, as the virtues of first entities are known to me, these certainly far excel all the rest.

So, also, concerning the first entities of salts it should be remarked that, like their spiritual virtues, they are by far greater than in their perfection. Thus, the first entity of vitriol transmutes all white metals into red, and overcomes and masters all the perfections comprehended in them. It renovates and restores all the imperfect bodies of the metals, as tin into its first entity, and again into tin, in which there are more virtues than in the original tin.

In the same manner, it brings back to the radical moisture whatever proceeds from those radical moistures, and renders that same renovation and restoration more perfect, more plentiful, and more full: for nothing else operates so powerfully on the radical moisture.

In no way different are the first entities of herbs and trees, which, before they have received their body, that is, their stalk or trunk, are a thousand times more powerful than when incorporated. Even so, the first entity of balm renovates and restores the body far more powerfully than seems possible to be done in natural things. It should be known, that the halcyon is not thus renovated or restored by its own nature, but because its nature is such that it should be nourished and live on first entities in this way: When it eats the bodies of herbs, or of seeds and the like, its stomach, by digestion, reduces them to their first entity, and afterwards, by means of that first entity, it perfects the operations of its own renovation and restoration. For its digestion was by Nature predestined only to first entities, whence it happens that it first of all transmutes all its food and drink into a first entity. On which account, also, it eats those bodies only which regenerate and restore, with which from the very time when it is hatched it is furnished and nourished by the parent-birds. In the meantime, this is its own nature, that after death it is renewed and restored; on this account, forsooth, because all these first entities cannot have their progress in the bird whilst it lives, since the life of this bird takes away all the power of those entities by turning them into blood and flesh; but when it is dead it flourishes according to the annual seasons; and just as first entities put themselves forth in the earth, thus, in like manner, do they then put themselves forward in the bird itself, and so renovate and restore the dead flesh, which, indeed, in Nature itself is a wonderful argument for its powers and virtues. If these things did not lie open to our eyes, they would be incredible, though they should be written down by many persons. From this cause, also, it happens that halcyons renovate themselves at irregular times, some sooner, some later, as they have earlier or later eaten the first entities; for some are born and come forth sooner or later than others. Also many worms are renewed and restored for this reason, that they are fed and nourished on first entities while they are still imperfect in the earth. There are many marvellous things which are occult, and far more than are known or openly investigated, concerning which one could write more copiously, were they not too far removed from the text of a book on renovation and restoration.

And although, as we have previously written, we cannot very well take first entities, or have them in the same essence; still it is possible for us to do so. For if we know where the mineral of gold lies hid, we shall also find its first entity in the same place if we come before its perfection. For there are certain signs by which it may be known how the form of the metal is placed. So whilst it is in its first entity it makes trees fruitful, and renders fertile its foundation, that is, the earth. It renovates old trees which for twenty years have borne no fruits; for when the first entity of gold has seized upon them, or upon their roots, they begin again to live and flourish as before. But although there are many more particulars concerning the first entity of gold which are worthy of our admiration besides those we describe, these suffice for a demonstration of the first entity of gold, namely, that it is there.

But when flames and sparks are seen, then it must be inferred and noted that the metal is being produced out of its first entity, which metal has betaken itself to the process of coagulation. These are reckoned as signs concerning the origin of minerals which apply to gold, silver, or the other metals; for the signs of the other first entities as to their origin are like those whereof we have spoken.

When a sign of this kind has been seen or found it must be understood that this same first entity is not gathered up in one heap, as it is when it lies in its perfection, but that it is spread over the whole of the land in that district. Wherefore, this land is in the power of the first entities, and out of it these are drawn, as it is with the celandine when it is not yet composited. Its first entity is in the earth, where it has its position. For this reason similar earth should be taken, and from this at length an extract should be made, as we have pointed out concerning the virtues of celandine. It must be noted that between the first entity and the perfection there is this difference, namely, that the first entity has power to renovate for the reasons before mentioned : but when perfected it has only the virtues of the natures, so that it tends in that direction, but not perfectly. Thus it may be gathered that if it be wished to have from these virtues of the same kind as those of the first entities, it is necessary that they should be removed from their coagulation, and should be corrupted, as is pointed out in arcana and quintessences. But that every thing has greater powers in its first entity, let not philosophers wonder, because even out of any earth in which a herb is produced essentially, before it is incorporated, all the virtues of that herb can be extracted, so that the virtues may be preserved and the earth be put back into its own place, so that for the future it shall be mere earth, and have no fruitfulness in it, because its first entity is taken away, which had lain hid in the earth.

In this way it often happens that the power of a first entity of this kind may be enclosed in a glass, and be brought to such condition that the form of that herb grows of itself without any earth, and even when it has quite grown it has no body, but something shaped like a body, the cause of which is that it has no liquid of the earth. Hence it happens that its stem is nothing more than a mere apparition to the sight, because it can be again rubbed down to a juice by the finger, just like smoke, which shews a substantial form but is not perceptible by any sense of touch. In growing things of this kind the quintesssence is entirely incorrupt and in its highest state of perfection, as in the earth.

Wherefore, from the first entity of gold is produced, in this way, the finished gold, which to the touch is like red water, and is stirred up and exalted like gold.

So far concerning these things. Now, let us next in order betake ourselves to the practice of those things which renovate and restore, if they be prepared by the power and rule of art. These things, although briefly written by us, are sufficiently patent, nevertheless, to intelligent men, that is to say, those who have good instruction in medicine and philosophy.

First of all, then, must we know those things which renovate and restore, as we have pointed out, and their first entity must be extracted, and by that the work of renovation and restoration must be done. As a deduction from this argument we set down four mysteries, that is to say, of minerals, gems, herbs, and liquids, as follows :—

The First Entity of Minerals

Take of mineral gold, or of antimony, very minutely ground, one pound, of circulated salt four pounds. Mix them together, and let them digest four months in horse-dung. Thence will be produced a water, whereof let the pure portion be separated from that which is impure. Coagulate this into a stone, which you will calcine with cenifiated wine, separate again, and dissolve upon marble. Let this water putrefy for a month, and thence will be produced a liquid in which are all the signs as in the first entity of gold or of antimony. Wherefore, with good reason, we call this the first entity of these things. In like manner, it will have to be understood concerning mercury and other things.

The First Entity of Gems

Take of emeralds, well ground, ʒj., calcine them in dissolved salt until they be turned to a white colour. Afterwards let them be dissolved, and enclosed in a phial hermetically sealed, and placed over an open fire. Let the matter be suspended on high in a naked glass vessel, so that it shall not touch the bottom, and let this be continued until, from its spiritual nature and condition it falls down to the bottom into a body like the liquor of honey. This displays the virtues of the emerald. Wherefore, it may rightly be called the first entity of the emerald.

The First Entity of Herbs

Take celandine or balm; beat them into a pulse, shut them up in a glass vessel hermetically sealed, and place in horse dung to be digested for a month. Afterwards separate the pure from the impure, pour the pure into a glass vessel with dissolved salt, and let this, when closed, be exposed to the sun for a month. When this period has elapsed, you will find at the bottom a thick liquid and the salt floating on the surface. When this is separated you will have the virtues of the balm or of the celandine, as they are in their first entity; and these are called, and really are, the first entities of the balm or of the celandine.

The First Entity of Liquids

Take the mineral of sulphur and of dissolved salt; let them be completely resolved into water by themselves, which distil four times. First will ascend a certain whiteness which displays all the virtues of the first essence of sulphur. Therefore with good reason we can have it in the place of the first entity of sulphur, and so name it.

Now that the four first entities have been thus described generally, it must be further remarked in what way they are to be utilised, so that their virtues may be perceived. This is the method. Let either of those first entities be put into good wine, in such quantity that it may be tinged therewith. Having done this, it is prepared for this regimen. Some of this wine must be drunk every day about dawn until first of all the nails fall off from the fingers, afterwards from the feet, then the hair and teeth, and, lastly, the skin be dried up and a new skin be produced.

When all this is done that medicament or potion must be discontinued. And again, new nails, hair, and fresh teeth are produced, as well as the new skin, and all diseases of the body and mind pass away, as was declared above. Herewith we would conclude our little book on renovation and restoration.

HERE ENDS THE BOOK CONCERNING RENOVATION AND RESTORATION

A LITTLE BOOK CONCERNING THE QUINTESSENCE*

By the Great Theophrastus Paracelsus, Most Excellent Philosopher and Doctor of both Faculties

MANY have written concerning the quintessence of those things which either lie hid in the bowels of the earth or grow and sprout therefrom, concerning the quintessence, namely, of metals, salts, saps, stones, trees, herbs, roots, quadrupeds, fishes, and other animals, etc. To few, however, has it been vouchsafed to point out the method or process whereby the quintessence can be extracted from the said things, so that a certain hope might, indeed, be left for the human race, not only that a certain quintessence of this kind resided in things, but also that the glorious and honourable body of man might become partaker of the same essential blessing. Wherefore, I am now about to teach in part, avoiding the method of those loquacious triflers, who merely threaten with railing, but do not strike. It has often been stated that an implanted quintessence inheres and dwells in all things; but I shall, in addition, indicate how it is to be extracted, not only from metals, stones, salts, and saps, etc., but also from roots, herbs, animals, and all other things which have anywhere been formed by the Creator.

All essential and created things contain in themselves water, oil, and salt; hence it will be a matter of very great moment to separate the salt and the oil from the water. During the process of distillation the waters first appear, next the oils, lastly the salt. However, frequently the oil, the water, and the salt remain conjoined together. Thus, if juniper oil be extracted from berries in which salt is present, then the oil is separated by reason of its levity from the water, which is somewhat heavier. No one, notwithstanding, has hitherto succeeded in separating from it the salt, which is the most precious of all. Moreover, I am convinced that it is a true quintessence of juniper.

Further, no fresh water proceeds from sulphur, but only oil. Yet, under that oil lie hidden salt and concealed water. Nor can it be properly designated oil unless the salt and the water have been previously removed and separated

---

* This treatise is not included in the Geneva folio, and is derived from the Basle octavo of 1582. As the fourth book of the Archidoxies has already discussed at considerable length the subject of quintessences, and much further information is scattered through previous books, there is no need for annotation at this point.

from it. Moreover, oil and water are present in common salt, but it can not be properly designated salt until the water and the oil have been separated from it.

In plants and things growing in the earth each has its own water, salt, and oil, yet in a distinct manner. Thus, the warmest things have the most salt, the coldest the most copious supply of water. The lukewarm contain equal proportions of oil, salt, and water; each in its own grade and kind. Thus they differ from one another in colour, odour, and savour. Also the subtlest and purest spirits which are in these things are extracted, and justly and rightly deserve to be described as salutary to the life of man.

Growing metals, as are gold, silver, iron, lead, tin, copper; also arsenic, marcasite, lapis lazuli, rust of copper, green of the mountain, calamine, vitriol, are the first and chief. These are followed by gems, as the emerald, carbuncle, amethyst, which is also called pyrops, cyamus, hyacinth, which is also called chrysolite, corals (white and red), also pearls, etc. And, indeed, the last possess the highest virtues of them all, provided that they be duly prepared; although in addition to the above-mentioned stones various stones are found, as tiles, wherein no small efficacy for preserving the health of the human race is discovered—alabaster, bolus armenus, and others. But before the rest we desire here to speak of the more excellent. In the third place, order demands the mention of vegetables growing out of the earth, as trees, shrubs, herbs, roots, and similar vegetables, their marrow, and other things under which both the saps and the liquors which proceed from growing things are comprehended. Thus, wine, various kinds of oil, as of olives, of nuts, of flax, of the berry of the laurel, of the nutmeg; as also gums distilling out of trees, shrubs, and from the stalks of other vegetables, as Chian turpentine, myrrh, mastic; also from the cherry and heliotrope, from the plum tree and the blackthorn, and others innumerable, each of which has been formed by the Creator of things for the profit and use of mankind. In the fourth place follow the animals, in which no small virtue and healthy operation lie hidden in many ways. Firstly, they are in the blood, which is most certainly the soul of every brute; secondly, in the marrow or fat. Also, there is a force lying hidden in the flesh of animals provided with blood; and equally in fishes, which have it not, from all which things the quintessence can be extracted to the unspeakably great advantage of man. Moreover, that I may at once shew by means of what things and methods this is effected, I will take the essence of the blood of a stork as an instance. This is a most excellent remedy against any poison which has been taken, a result which arises solely from sympathy and compassion. Moreover, the blood of man when distilled abounds in these powers, because if its essence be kindled, it burns without ceasing, as long as the man whose blood it was, lives, and is extinguished when he dies. If those great doctors do not hold it as a nigromantic exorcising, the cause will be that they do not understand the natures of these things. But there exists not a more excellent cordial than that which is extracted from pearls or margarites, which possess such marvellous powers that

A Little Book concerning the Quintessence

by its means I could restore dying men who are already in the agony of death, revive them, and enable them to speak again. It has also given me no small aid in raising up and restoring a man both paralysed and convulsed, who (in the judgment of all the bystanders) was solely in God's hands. The man afterwards spoke, wrote, and begat children. Nor do the other gems possess less efficacy, yet each after their kind. Concerning herbs, roots, and the family of the same, there is no doubt that even with the doctors who follow Galen they possess unspeakably great power and efficacy. What shall I say with regard to the excrement of men and of other animals? If so many virtues reside in this castaway and refuse thing, how much greater virtues must dwell in the noblest metals, in gold, silver, antimony, etc. But how great a virtue must reside also in marcasite, since, without the addition of any other medium, it is a most salutary medicine and purge for all fistulas, for cancer, and similar ulcers? I will say nothing here of vitriol, antimony, and other things which abound in virtues in no wise inferior to the former with reference to diseases of this kind, yet each according to its mode and genus. The same also is (perhaps the essence) of gold thus prepared, whereby I have frequently purged and cured an exceedingly foul leprosy and elephantiasis. I pass over the fact that I have cured gout of forty years' standing, contractions and relaxations (of the sinews) and other (ailments) by means of this and other adjuncts. Nor are marrows and fatnesses to be altogether rejected, especially that of a man, the badger, pigs, rams, bulls, stags, apes, and similar animals, living either in water, air, or land, none of which things has grown or has been formed except as a special boon (to mankind), as is proved by the fact that the minutest Spanish flies, and the dung of the fly and the moth, have their respective properties, all of which matters we have thoroughly enquired into, have examined completely, and have ascertained that nothing remains which has not received its powers and its virtues for the use and profit of man. But to approach our theme.

Concerning the Quintessence out of Oil, also of the Salt of the Same

Take of the purest gold, reduced to minutest grains and lunated, or of gold calcined together with plumbago. Add to this 100 parts of the most excellent white wine discoverable, and 10 parts of the white pine. Leave them to be macerated in a glass for 40 days. Pour out the wine. Pour on the same quantity of fresh wine. Similarly macerate. Do this a third time. Pour these three relays of wine into a glass. Seal well, and distil in a strong fire, so that it may come forth strongly. When it has been distilled, place the glass with water (liquor) upon hot ashes, being well sealed up with a blind alembic, the ashes being too hot for the fingers to bear. Let them stand under till nine parts are consumed or dried up, and the tenth part alone remains. Add the same quantity of the albumen of eggs to the water that remains. Shake together. Distil together, at first slowly; when white water comes forth, separate this. Next distil more strongly until the bocia glows, and the matter comes forth in the form of a somewhat attenuated honey, of a strong (offensive) odour. Retain this, for this is the Quintessence of gold, the royal medicine. Place somewhat whitish and thick water in a glass. Cook until it be consumed. You will then find white and excellent salt. This is salt of the sun. However, the half of your gold will have gone, that is to say, the superior, which enters into the Quintessence. It is permissible to call this same essence the oil of the Sun.

Concerning the Oil or Quintessence of Silver and its Salt

Take silver reduced to the thinnest filings. Cut them up into the smallest parts in the form of a denarius. Do the same with particles of cinnabar. Place together in a glass, layer by layer. Let the glass be half filled and perfectly closed up. Then place it in a cupel for 30 days. Keep in a continuous fire. At last let it cool. It will then have extracted the soul or essence, and the silver will have been reduced to the form of a sponge. Purge the cinnabar with lead, and you will find the soul of silver in the residuum, and this is the most excellent silver that can anywhere be found. Reduce the said silver into the minutest grains, and pour on the strongest vinegar of wine. Let them stand in a glass, and the silver will become blue. Take away the blue and reserve it. Pour on other vinegar. Let it become a second time blue; remove this again, repeating the process of adding blue matters till the silver be totally removed. If there were two ounces of silver, take one ounce of camphor. Let it be dissolved in the said blue silver and vinegar. Distil it as follows, at first slowly, then stronger and stronger, until the water commences to be tinged with a swarthy hue. Then remove the water from its receptacle, place and retain in another glass. After pressing the residuum through a bag, distil with a very strong fire from the fæces until the bocia glows, and you will find a quintessence in the glass, like a dark or coarse beer of somewhat caustic quality. This is that oil and quintessence of Luna. Cook the water retained until it be consumed (as in the case of the salt of the sun), and it will become a green salt, which is salt of silver.

Oil of Mars

Pound up crocus of Mars into a most minute and delicate powder. Wash this with fresh water. Pour out the water of the lotion. Let it stand until it sinks. Then separate the water from that which sinks, and let the crocus be dried up. Take any quantity of this, and make a paste with the yolk of eggs. Let it be again thoroughly dried. At last beat it into a powder, and spread on a smooth glass slab. Place in a wine cellar, and it will be dissolved into a clear oil. This is the oil of Mars, suitable for all external ulcers.

Oil of Saturn

Take viii. oz. of spume of silver. Beat and pound extremely small. Place in a jar, which must lean in the fire to the side, as cinders of lead are usually produced. Stir the spume with an iron spoon. When it is sufficiently heated pour it out into an iron frying-pan. Let two measures of sharp and boiling vinegar be injected. The oil will then separate itself from the spume, when let the oil be again poured into the frying-pan. Let it be consumed until scarce one quarter of one measure remains. This oil has a marvellous sweetness. If you mix it with stale urine it will grow white like ceruse, and if it be boiled in an iron frying-pan it will subside and become like silver. Further, if a small quantity be placed in a vessel and it be left until it dries, there is produced from it a tenacious matter like thin (delicate) gluten. Let the matter which produces the oil be distilled through a well-sealed retort; you will then have the pure and clear oil which is called the Quintessence of Saturn.

Oil and Salt of Jupiter

Let Jupiter be calcined in the following fashion: that is to say, let him be placed in a crucible and cooked by means of a secret fire, that is, by descent, for forty days. You will then have the powder which they call tin cinder. Take of this ʒv., and of juice of lemons ʒlxv. Let these subside for the space of twenty days. Then wash away the ash. Afterwards wash the cinders. Pour on it again the said juice of lemons. Do this a third time. Afterwards let these be distilled, and let them be poured again over the cinder. Let them once more subside twenty days as before. Finally let them be distilled through the alembic, until at length pure water comes forth. If you perceive its colour to be red, apply a strong fire. The water which first comes forth is to be thrown away; there will next issue a water mixed with oil, which is to be separated. You must then rectify the oil in the heat of the sun, and this is the oil of Jupiter. The second water is to be diminished by boiling to the tenth part. This you next allow to subside until it commences to have rays of such an extremely beautiful green colour as to almost surpass the colour of the emerald, and this is the salt of Jupiter, which also I judge as the spirit of Jupiter.

Of the Quintessence of Antimony

This is the most excellent and most sweet matter of all things which has ever existed; it also excels all things proceding from Art or Nature. I except potable gold, because it is surpassed by the quintessence of our Stone, which not unfittingly but truly, we call by its true name, the Stone of the Philosophers. Take, therefore, antimony, which breaks into long and sharp grains which glow. Let it be pounded and sifted until it becomes a powder most thin and most subtle. Let this be imbibed in strong and good wine for thirty days. Let the vessel be well closed. But notice that for a pound of antimony there must be taken two measures of wine. Let these be put in the sun; afterwards let them be distilled in a fire, slow at first, then stronger, until at last the water coming forth commences to redden. Then bring out another receptacle. Keep the water, having sealed it up well in the vessel, sufficiently long to permit it to subside. Then, within the space of nine or ten days, something black will appear at the bottom, from which take the water which swims above it, and for every ʒv. let there be taken an ounce and a half of carline thistle cut up small. Let these again be distilled together. The other water, which is also red, needs in distillation a strong fire so that the lid may glow, whereupon you will see upon the water a red oil, which must be separated from it. In this manner is produced juniper oil. Each must be kept in a separate place. Also in this way are extracted those three waters out of antimony, which contain in themselves the Quintessence. The process for extracting it is as follows : While these three species of waters are subsiding during the period of 30 days, you will find something earthy at the bottom of the glass vessel, just as was done previously with the first water. Then let the waters be strained, and the clear separated from the turbid; the former are to be retained, the latter rejected. Afterwards let them be distilled in the first grade of the fire only for 30 days, among cinders, until the matter be coagulated and become a hard stone, like to granite in colour. Let the stone be beaten up and dissolved in distilled vinegar; afterwards let these be distilled through the alembic. The water which comes forth is to be placed in a glass vessel over cinders in the second grade of the fire. You will then find a red stone like to spinetum, and this is the quintessence of antimony. It was never previously known that it was useful to humanity, but it cures all leprosies and removes all fistulas, also the French disease; indeed, all the incurable diseases of that type, as also all pellicles and small specks in the eyes. It is neither bitter nor acrid, but has a sweet savour; its consistency is like oil, its colour that of red wine. It is the special cure of dropsy, for it quickly consumes (removes) the dropsy if the patient take the quantity of one pea of this medicine in the water of violets. It also heals paralysis, apoplexy, and epilepsy, if with three drops of potable gold the quantity of one third part of a scruple be taken, for this stone is dissolved therein. Lest it should perish, it must be kept in dry places, and preserved in seed of millet; for if it be placed in humid situations it corrupts in four months' time.

Concerning Oil and Salt out of Marcasite

Pound as subtly as possible such marcasite as is found on the mountains. Pour over it strong wine; then, in order that it may overflow, let it be stirred with a stick daily, and after three days let the wine be poured out, fresh being added. Let this be done until to 10lb. of marcasite an addition has been made of 20 measures. Let the wine strained or poured out be distilled with marcasite until water issues forth, which water is to be retained. Next close the aperture of the tube in the glass vessel, in an effectual manner, with lute. Apply a somewhat fierce fire, so that the matter in the upper part of the cucurbite, which resembles pure silver, may attach itself to the sides. Keep this and cast away the dross. Afterwards diminish it by boiling, so that a single measure only may remain. Put the white matter, which was in the upper part of the cucurbite, previously well pounded over a stone, into a cellar, and beneath let there be placed a small vessel into which oil flows. This is a most excellent remedy against fistula, cancer, and other diseases of this kind. That is called the quintessence which is extracted from immature metal. However, again distil the water cooked in this manner. Diminish it by boiling till half has departed, that is to say, till half remains. Let this subside for the space of 30 days. Throw into it straws and pieces of wood cut small, upon which there will collect a salt like to crystal, except as regards colour, which is slightly green. And thus you have oil and salt out of marcasite, which two are by no means the smallest aids to the preservation of the life of the human race and its health.

Oil out of Common Salt

The confection of oil out of salt is not a useless and unproductive labour, since that wherewith it is prepared possesses the greatest diversity in nature from it, not so much in point of provocation of thirst, and removal of putrefaction, as of sweetness of savour. For oil painlessly eats into and consumes in a single hour all things whatsoever which are smeared therewith, and are putrid by nature, whether it comes in contact with them in wounds or other injuries. On the other hand, salt nourishes whatever is putrid, and prevents its consumption, and is sharp. Moreover, salt excites thirst, whereas oil repels it, as may testify several subjects of dropsy, who, having taken some, were free from thirst between nine and ten days. Lastly, as regards the savour, the oil is not sharp as the salt, but forcibly reminds one of the sweetness of honey, or of the juice of wild apples.

Oil and Salt of Coral, also of Crystal

Crystals of the first quality are found in the Helvetic Alps. Let these be reduced by pounding to the smallest powder, over which let there be poured the juice of lemons; let them be put into a cucurbite with a narrow neck on hot cinders as deep as the matter which is in the glass. Let them subside for 40 days. Then the crystals will be dissolved, and from them is produced a gross water. Add to these vinegar, equal in quantity to the juice; let them subside for 20 days again. Afterwards let there be added musk, which is good, in order that the matter may further ascend; then let all be distilled in a well-closed glass vessel. All the water which issues forth must be retained, for no oil comes out, and care must be taken lest the fire be too fierce, otherwise the matter would be burnt up. Next cook the water until half be consumed, and distil until it acquires a gold colour. Afterwards pour it out and distil it in a strong fire until there comes forth pure water of a yellow hue. But if it be turbid let it be preserved apart, and that which is coarse (gross) is to be distilled in a glowing cucurbite, wherein, while it cooks, you will find a yellow oil swimming on the surface of the water. This is the quintessence of crystal. Collect all the waters, the white, also the yellow, and that which comes out last, into a glass vessel. Put thereupon small and thin straws of the length of a finger. Seal up the glass vessel well, place in a cellar, and leave it for 40 days. Then upon the said straws there will grow a matter having rays like salt nitre. Dry these, and you will then have salt of crystal of a marvellous nature, virtue, and efficacy.

Oil and Salt of Pearls

Arabian pearls, and those which are fetched from India, are among others the best. They require to be purged with pure and warm water; next to be dissolved in vinegar nine times distilled. You will be able to renew the vinegar every three days; pour out into a glass vessel, well sealed up, and keep therein. And if they be altogether consumed and dissolved, the vinegar must be abstracted by means of distillation: that which remains at the bottom must be again dissolved into vinegar, and again distilled as before. Then a portion of the pearls or margarites will distil. Keep the distilled water until nothing remains of the pearls. Add to one measure of the same water half an ounce of camphor, which is of so great a virtue that it will not suffer any dross, but renders that which is earthy and heavy light and volatile, so that it can ascend. These are to be distilled again, and the vinegar consumed by boiling to the twentieth part. This must be done in an open glass vessel having a large and wide mouth. Let the remainder be distilled through the alembic until pure water flows. When this has commenced to turn yellow put another glass vessel beneath. Subject it to a fire which increases in vehemence, and then there will proceed a gross matter resembling thin honey. This is the quintessence of pearls, a true crown of human life. But salt is confected from the water which precedes the yellow colour. Let that be cooked until it becomes a salt. This salt is beautiful, white, and soft, yielding the sweetness of camphor.

Of the Essence and Salt of Things growing on the Earth

There exists no better method of extracting the essence of herbs and roots than to cut them up as small as possible and boil them in strong wine in a jar, well closed, lest any of the spirits should evaporate. Let the wine be separated by frequent straining. Also let fresh be poured in again until they lose their strength. Then no further process is needed except to collect all the wines and distil them together through the alembic. This can be done as frequently as you desire, but the aperture of the glass vessel must be effectually sealed up. It is then necessary to wait until the virtue which resides in the wine, and circulates throughout the vessel, collects into one place, for the smaller the quantity of the essence of these things, the better and more subtle the quality. But when at length the waters have boiled after the first and second distillation, salt is discovered at the bottom. Note, however, that in order to the extraction of oil from most herbs it is requisite that both herbs and roots be distilled. But in the case of those which are fat, their leaves alone, when exposed to the sun in a glass vessel, yield an oil, whence ultimately a quintessence can be extracted, and in the same manner also a salt. Wherefore it is impossible to assign to each herb its peculiar operation. Nevertheless, we shall distinguish between those which have oil in them spontaneously, and distil it of their own accord, for instance, nutmegs, the rind of a species of quinces, or golden apples, and the like. Eggs, nutmegs, berries of the laurel, and similar fruits must be melted up and roasted a little in a frying-pan over the fire, so that they may acquire a savour of burning. Afterwards press out the oil. This method of extracting is simple child's play, but there exists scope for skill in extracting from that oil something yet more subtle : just as no small art is required to extract oil from cinnamon, ginger, cloves, and the like, in which the method is quite dissimilar to the above. Indeed, these are distilled through the pores like juniper oil. Consequently, the same amount is not extracted from these as from other matters, although they be more excellent, whereas the berries of the laurel, nutmegs (they are also called *myristicæ*), and similar things out of which an oil proceeds without any process, as aforesaid, yield a larger supply than those things which are light and tender, or are even pressed.

The Method of Separating Oil and Salt from Oil of Olives

Oil which proceeds out of olives also contains salt, though not of special quality. It is extracted more abundantly out of old oil than out of fresh. Out of this is produced a red water, which, marvellous to say, is a most speedy remedy for the stone, whereas it is manifest that no oil contains an aqueous humidity, nor is of such a nature as to break the stone. Wherefore let those who exclaim " How can this or that be done ?" speak with more deliberation and modesty. They bring forward the argument that it is contrary to its nature and property, but they totally ignore the fact that cooking involves so great an alteration as to frequently effect what otherwise lies not in the essence or nature of a thing. Similarly, our silver, although it be not gold, is reduced to such an extent by coction as to arrive at the most excellent gold. Thus also iron is converted into genuine and most excellent copper. The matter of which we are here speaking takes place in the same way.

That Red Water out of Oil of Olives is made in a similar Way

Take lb. iiij. of oil. Place in a kettle of copper. Heat it so that it may commence to smoke; let it cool again, and put in a bocia. Close up its neck most effectually with a sponge. Superimpose a helm well stopped up with lute, lest any vapour should escape. Distil the matter in a slow fire; at length the virtue begins to issue forth. The fire is to be maintained at an equal temperature—not at one moment hot, and at another cold. The water will then become red. If it commence to thicken in the upper part of the receiver, it is necessary to cease operation and proceed no further. And this is the water which is extracted out of oil. It cannot long be kept entire by reason of the putrid nature it contains in itself. Notwithstanding, it dissolves in a moment, as it were, the stone in the bladder. For nine days daily let three spoonfuls be administered, one in the morning, one at noon, and one in the evening. After this the patient must fast a whole hour. The savour of this water is nauseous, and frequently affects a man so much as to give him a headache.

Salt is made in the following Fashion

Distil oil out of which the red water has been made; abstract it through the alembic by means of a moderate fire, and if it flow in excess, remove the fire. If it cannot endure great heat you will find turbid matter at the bottom of the bocia. This is to be removed into a small glass vessel. There must be added to it as much pure fountain water as there is matter. The oil will then float on the top, and must be separated from the water, in which there is no fatness. Distil it again and you will have pure water, which must be cooked in a glass vessel until it be consumed. You will then discover at the bottom a somewhat black salt, most suitable for purging; it is also a strong laxative.

Method of Extracting the Oil and Salt from Pepper

Beat the grains to a fine powder. Add thereto the sap of the alder, sufficient to cover the powder. Let it settle, exposed to a very hot sun, as during the dog-days, for nine days. Afterwards press through a bag. When the water has been distilled pour it again upon the fæces, and let it settle for nine days. Next, when the water commences to recover colour, let it be removed. Put out another vessel in the place of the first. Extract the oil with a strong fire. If any water has been mingled with it, it must be separated by means of a glass funnel. The last water must be added to the first, and again distilled, and consumed, till but a quarter remains, by coction. Let this settle for 30 days in a glass vessel well secured over a capella. A salt will then be found at the bottom, possessing the savour of pepper, most excellent as a pickle for food, by reason of the extremely excellent odour it yields. It is wonderfully hot, and adapted for cold limbs and nerves.

Oil out of Gums

Take myrrh, mastic, and gum, or like things, which must be pure and clear. Let them be pounded and sprinkled over fresh eggs, which have been cooked hard, split through the middle, and placed in sand in a pan, which must be put in a cellar until the powders are dissolved into oil. This cures all wounds and alleviates gout. A man's face is rendered fair if it be anointed therewith after a bath. In like manner it preserves the flesh from putrefaction so that the same can never ensue.

Preparation of Colocynth

Take an apple of colocynth and remove all the seeds. Cut them up small. Let marrow be taken to the weight of three coins of Ravenspurg. Let it be placed in a glass vessel. Add thereto of

Cinnamon	
Cloves	
Ginger	The weight of three coins of Ravenspurg.
Nutmeg	
Mastic	

Pound them all together and place in the glass vessel where the colocynth was put. Pour over five tablespoonfuls of good wine. Let it settle for eight hours. If you desire to use it, it must be prepared about twelve o'clock, mid-day, and must be taken at night. It is next necessary to lie down, then purgation will commence at midnight, or at one o'clock. Note, before taking this medicine, that it must be pressed through the bag, and it must be eaten. But be careful not to press it out too violently. Take only that which penetrates by itself, so to speak, spontaneously, about seven or eight drops, lest it become too violent. When, therefore, the purgation has occurred several times, take, about seven o'clock in the morning, some soup wherein peas have been cooked, but which contains neither salt nor butter, only a slight admixture of flesh. Take it as hot as can be borne, for two hours, without anything else being eaten.

HERE ENDS THE BOOK CONCERNING THE QUINTESSENCE

ALCHEMY: THE THIRD COLUMN OF MEDICINE*

THE third fundamental part, or pillar, of true medicine, is Alchemy. Unless the physician be perfectly acquainted with, and experienced in, this art, everything that he devotes to the rest of his art will be vain and useless. Nature is so keen and subtle in her operations that she cannot be dealt with except by a sublime and accurate mode of treatment. She brings nothing to the light that is at once perfect in itself, but leaves it to be perfected by man. This method of perfection is called Alchemy. For the Alchemist is a baker, in that he bakes bread; a wine merchant, seeing that he prepares wine; a weaver, because he produces cloths. So, whatever is poured forth from the bosom of Nature, he who adapts it to that purpose for which it is destined is an Alchemist. Hence you may understand the difference between this art and all others, from the comparison which has been set before you. For, consider, if any one should put on a rough sheep's skin for garment, how rude and coarse this would be compared with the work of the furrier and clothier. Equally rude and coarse would it be if one had anything taken straight from Nature, and did not prepare it. Nay, this would be much more rude than the former. For this is concerned with the body itself, its life and health. Hence it has to be handled and dealt with much more carefully. Now, these universal methods of treatment have rivalled Nature, and have so mastered her properties that they can express the nature itself in everything, and elicit that

---

* This treatise constitutes the third section of the *Liber Paragranum*, wherein are described the four columns upon which Paracelsus built his system of Medicine. These are Philosophy, Astronomy, Alchemy, and Rectitude. The first distinguishes between the false and adulterated philosophy of Aristotle, and that sure and genuine species expounded by Theophrastus himself. The latter alone enlarges that knowledge whereby the physician is instructed in the matter of all diseases. There is no other way by which the truth concerning the body of man and its nature can possibly be estimated. Outside it there is only pure imposture. Disease itself is of Nature, Nature alone understands and knows disease, and Nature also is the sole medicine of disease. The matter out of which man is made testifies to the physicians concerning that which is produced therefrom. Man, as the exemplar and type of all things, contains within himself all knowledge and wisdom which are required in Medicine. There are two species of philosophy—one is of heaven, the other of earth. The philosopher is he who is acquainted with the things of the lower sphere; the astronomer, on the other hand, is he who is familiar with the things of the sphere above. In their medical aspect philosophy is concerned with the earth and water in man, and astronomy with the air and fire which exist in the same subject. For man is heaven, air, water, earth. As there is a Zodiac in heaven, so is there a Zodiac in man; as there is a firmament in heaven, so is there a firmament in the body. Man has his father in heaven and in the air. He is the son, made and born from air and the firmament. Heaven operates in us. This operation can be understood by no one who is ignorant of the essence and nature of heaven. The star is the basis of celestial knowledge; he who is acquainted therewith is already the disciple of Medicine, and is on the way to understand the heaven in man. So far concerning the first two pillars of Medicine. But the fourth is the probity of the physician, which consists in the certitude of his art, and the rectitude, and sincerity of his faith in God, which forbids him to falsify anything, and makes him a fulfiller of the works of God.

which is the chief feature in each. But in Medicine, where it is most specially necessary, this power exists in the smallest degree; so that, in this respect, Medicine is most rude and unpolished. How can man be more rude than when he eats raw flesh, clothes himself in undressed hides, and has his dwelling in the nearest rocks, or is exposed to the rains? In the same way, how can a physician be more dense, or the preparation of medicine than what ointment-sellers use for decocting substances?* Nothing can be more objectionable than this method for pounding, subduing, and mixing medicines, or for polluting them in any other different way. As its own special art dresses the hide, so so does its own special art prepare the medicament. And since, in this place, the true basis of preparing remedies, in which lies the whole essence of medicine, is laid down and established, be well assured of this, that such a foundation must be extracted from the most secret recesses of Nature, and not from the imaginative brain, as a cook dresses a mess of pottage, according to his own judgment. For in this preparation the extreme and ultimate condition of things is posited. Thus, if philosophy and astrology—that is, the nature of diseases and their remedies, with all their combinations and conjunctions—be understood, it follows next in order, and is chiefly necessary, that you should decide how you will use and avail yourself of your knowledge. Nature, indeed, of herself teaches you on these subjects, and you should give her your chief attention in order that you may reduce your medical science down to practice. As summer puts forth the pears and the grapes, so should you do with your medicine. If you act thus you will assuredly compass the results you covet. And if it so be that as summer puts forth its fruits, so you do with your medicine, be certain that summer does this by means of the stars, and in no case without their aid. Now, if the stars accomplish this, take care so to regulate your preparation in this case also that it shall be duly directed by the stars. These it is which complete and perfect the work of the physician. Now, if the stars have this effect, it is right that medicine should be understood, and naturalised in all respects, with reference to them. Let it not be said: This is cold, this warm, this moist, this dry; but rather let it be said: This is Saturn, this Mars, this Venus, this the Pole-star. In such a way the physician proceeds by a straight road, especially if, beyond this, he also knows how to subject, conjoin, and harmonise the astral Mars with that which is produced from it. For here is situated the covering or nucleus which none of the physicians from first to last before me has ever arrived at. This must be understood, that medicine should be prepared with reference to the stars, so that they exert their astral influences. The higher stars weaken and cause death, but they also heal. If any of these effects is to be produced, it cannot be done without the stars. But if this is to be done by the help of the stars, it will be done after such a manner, and in such a way, that the preparation will be reduced to practice, so that medicine will be compounded and prepared by means of heaven, just

---

* The text at this point is unintelligible, as the comparison is not completed. It has been rendered literally in translation.

as prophecies and other acts are settled by heaven. That is, you see, that the stars presage and prognosticate unfavourable weather, that they foretell diseases, and the deaths of kings and princes; the stars, moreover, portend battles and wars, pestilences, and famine. All these things are indicated by heaven. It makes and produces them. What it produces it naturally predicts. All these effects come from this source, and from it, too, proceed all the branches of this science. So, then, if they are produced by heaven, and come from heaven, they will also be governed by heaven, so that all those matters which have been mentioned and pointed out will be produced at its will and pleasure. The occurrences which have been predicted from heaven come to pass at its pleasure; so that heaven produces and regulates them. Moreover, lay this well to heart. If Medicine is from heaven, without any contradiction it will remain subject to heaven, will accommodate itself thereto, and be regulated according to its will and pleasure. Now, if this be true, it is absolutely necessary that the physician should form an opinion concerning degrees and complexions, humours and qualities, and, whether he will or not, he must learn that Medicine is in the stars; that is, he must judge the nature of Medicine according to the stars, so that he shall understand the superior as well as the inferior stars. Since Medicine is worthless save in so far as it is from heaven, it is necessary that it shall be derived from heaven.* And this bringing down from heaven means neither more nor less than the abolition and elimination of every earthly element which exists in it. Heaven does not rule it except these earthly elements be separated from it. If you have effected this separation, then Medicine is in the power of the stars, and is ruled and protected by them. For instance, everything relating to the brain is led down to the brain by means of Luna. What relates to the spleen flows thither by means of Saturn; all that refers to the heart is carried thither by means of Sol. So, too, the kidneys are governed by Venus, the liver by Jupiter, the bile by Mars. And not only with reference to these, but in all other respects, this must be, in an ineffable manner, perceived. For of what use is the medicine which you exhibit for the matrix of a woman unless it be directed by Venus? What remedy would there be for the brain unless Luna gave it its origin? So judge with regard to the rest. Otherwise all remedies would remain in the bowels, and by-and-bye, being ejected through the intestines, would produce no effect whatever. Thence it happens that, if heaven does not aid your efforts, but refuses to direct your medicaments, you will profit very little. There is need of heaven as the regulator. Herein consists true

---

* The stars, therefore, have generated physicians constituted in the light of Nature, so that they might not deviate from investigating by their own skill the various arts. The first source of their discoveries was in the stars and influences, which, turned into alchemy, in no slight degree enriched the medical art. For alchemy is but a medical pyrotechny, whereby marvellous preparations, transmutations, transubstantiations, etc., of things medical are artificially produced. For such is the affinity of the firmament of the constellation with the nature of terrestrial bodies, that he who is informed with celestial doctrine is not debarred from the desire of the knowledge of terrestrial things, and when these are combined there is joined to them an influence from heaven, and so out of those three, thus united, there arises the true physician. A physician generated after this manner will never lack those remedies which suddenly become necessary for a pressing purpose. —*Chirurgia Magna*, Pars II., Tract I., c. 16.

art, that you should not speak after this fashion: "Melissa is a herb that acts on the matrix, marjoram on the head." Thus speak the inexperienced. The matter rests with Venus and Luna. If you wish to attain the ends you anticipate you must have heaven kindly and benignant to you, otherwise no effect will be produced. This is the source of that error which is so abundant in medicine. "Do you at least drink it. If it does you good it does, and there is an end of it." Any clown can practise this art. There is no need of an Avicenna or a Galen for it. You physicians, who have sprung from them, are wont to say that we must have directories for the head, the brain, the liver, etc. And how can you dare to lay down these directories when you understand nothing about heaven? That is the sole director. Moreover, you forget one thing, which convicts you all of folly. You have, indeed, found out what things direct the brain, the matrix, the anus, the head, the bladder, etc.; but of what things rule disease you are utterly ignorant. Now, if you do not know what rules a disease, you are ignorant as to where the disease is situated. You do just the same with the principal parts which you say are affected as the sacrificers do with their gods. They put the whole of them in heaven, although very many of them lie buried in the infernal regions. So, according to you, all diseases arise from the liver or the lungs, though sometimes they affect the rectum.

This is because heaven rules by means of the stars, but not so the physician. So medicine must be reduced to air, that it may readily be ruled by the stars. Can a stone be lifted up by the stars? No, unless it be volatilised. Hence it is that many, by means of Alchemy, hunt after a fifth essence; which means nothing else than that the four bodies shall be separated from the arcana. Then, what remains is an arcanum. This arcanum, moreover, is a chaos, as easy to be deduced from the stars as a feather to be swayed by the wind. Such, then, should be the preparation of medicine, that the four bodies shall be taken from their arcana. To this should be added the knowledge as to what star is in any arcanum. Then it must be known what is the star of this disease, and what is the star in medicine that operates against this disease. Hence, at length, proceeds a direction. If you drink a medicament, then the belly, which is your alchemist, is compelled to prepare this for you. But if the belly can reduce the medicament to such a condition that it is received by the stars, then that medicament is directed. If not, it remains in the belly and passes off with the excrement. Now, what, I ask, is more worthy of a physician than a knowledge of the concordance existing between each star? In this consists a knowledge of all diseases. In this respect Alchemy is an external bowel, which prepares its own sphere for the star. It is not, as some madly assert, that Alchemy makes gold and silver. Its special work is this— To make arcana, and direct these to disease. To this it must come, and here it is symmetrical. For all these things proceed from the guidance of Nature, and with its sanction. So ought Nature and man to be conjoined, brought together, and estimated one by the other. The whole principle of cure and sanitation rests in this. Alchemy perfects all these processes; and without it not one of them can exist.*

Now if all arcana belong to Medicine, and all medicines are arcana, and, moreover, all arcana are volatile, by what right, I ask, can that sausage-stuffer and that sordid concocter of the pharmacopœia give himself out as a dispenser and a veritable concocter? In undisguised truth he is a dispenser and concocter, but of mere trumpery. How great is the folly of those doctors who trick people by means of such clownish concocters with their electuaries, syrups, pills, and ointments, which are based on no foundation, or art, or medicine, or knowledge! Not one of you, if put on your oath, would dare to examine what works conscientiously and truly. The same, too, is the principle of your Uroscopy. From the urine you divine as to the blue sky, and you persist so strongly in your trifling as to confess that there is nothing but divination and conjecture in the whole matter, nor any coincidence except what occasionally happens by chance. In your surgeries you lie so consummately, and with your washings and your decoctions you assume such a magisterial air, that nobody could think otherwise than that the whole kingdom of heaven is affected by you, whereas you conceal nothing save the mere bottomless pit of the infernal regions. If you would put aside these your incapacities, and would examine arcana, what they are, what director they have, and how the stars rule disease and health, then at the same time you would learn that your whole foundation amounted to nothing but phantasy and private opinion. The ultimate and sole proposition is that the principle of medicine consists of these arcana, and that arcana form the basis of a physician. Now, if the sum total of the matter lies in arcana, it follows

---

* Thirdly and lastly, there is alchemy, wherein the physician should eminently excel. For if he does not thence take his preparation, his practice is nothing worth. Herein consists all the art of preparation. It is also the art which teaches how to separate the stars from the bodies, so that those stars obey the stars and firmament in direction, for the direction is not in the bodies, but only in the firmament. Hence it also follows that everything which the brain produces is a sign to the Moon through its course; that which the spleen produces Saturn attracts to himself; that which the heart produces is attracted by the Sun; and in this manner the external firmaments are the directors of the interior. So do they speak wrongly who say that Melissa is good for the matrix and sage for the cerebrum. For unless Venus and Luna direct them thither, they sink into the stomach and pass out through the intestines. Therefore that which does not separate in medicine is not directed by heaven, that is to say, the course of heaven is absent, and so nothing operates. So has every part its director from stars, and they are called stars. But heaven directs nothing except that which is separated from the body—that is, heaven directs only the arcanum, not the body itself; just as it directs reason to man and then reason directs the body - so heaven directs substances, which, if they be in the stomach are cooked therein, and then it directs one thing—that is, the arcanum. The stomach, indeed, is an alchemist, that is, one who fulfils the function of alchemy; but this takes place much more usefully without, before the substance sinks into the stomach, for then its operation will be much more powerful. Unless this be done, it will be like raw flesh, which the stomach digests much more thoroughly if it be assimilated after cooking. But if so much care be required over the preparation of food, how far more is necessary in the case of medicine. Many have despised alchemy as a senseless search after the confection of gold and silver, but it is not our intention to give a more prolix definition here. I have decided only to deal with the preparation - that is to say, how much virtue and efficacy there is in medicine which is devoid of a body. He, therefore, who contemns alchemy, herein despises that which he does not understand. Although I know well enough that not even apothecaries, barbers, and servants about the baths, will cease from their cookings; nevertheless, if you double the faith which you at present have in your medicines, the congeries of your recipes sufficiently proves that you are nothing but fools. It is evident enough whom ye cure, how, where, and when. Since, therefore, I am decided thus, lastly, to teach of alchemy, which is itself the fountain and pillar of medicine, it must be stated that without this art no one can be a physician, for he who lacks it has the same relation to a physician as his own cook has to that of the prince. All Nature therefore recognises alchemy, and desires that it should be understood by the physician, and that the same, being skilled therein, should not ever be cooking soups and colewort. — *Fragmenta Medica.*

that the foundation of all is Alchemy, by which arcana are prepared.* Know, therefore, that it is arcana alone which are strength and virtues. They are, moreover, volatile substances, without bodies; they are a chaos, clear, pellucid, and in the power of a star. If you know the star, and know the disease, then you clearly understand who is your guide, and wherein your power consists. So, then, these arcana prove that there is nothing in your humours, qualities, and complexions, that such terms as melancholy, phlegma, cholera, and the rest, are falsely imported into the question, and that in place of these should be introduced Mars and Saturn, so that you should say, "This is the arcanum of Mars, this of Saturn." In these, true Physic consists. Who of you, my hearers, will venture to reject and turn away from this foundation? Only your teachers do this: and in this respect they are like the old and case-hardened students.

If, then, it be right for the physician to know such things as these, it will also be convenient that he should ascertain the meaning of calcination and sublimation. And he should not only know this as a matter of handicraft, but as one of transmutation, which is far more important. For by these methods, as they are met with in preparation, there are very often produced such maturations as not even Nature herself is able to bring about. Towards this maturation the physician should direct his art. It is the autumn, the summer, and the star of those things which he ought to bring to perfection. Fire is the earth; man is the order; and the thing operated upon is the seed. And, although all these things are simply understood in the world, they are in result various and manifold. So also are they manifold in the locality of the result. And yet by our process all arcana are born and produced in the fire. That fire is their earth; and this earth is also a sun; and so the earth and the firmament, in this second generation, are one and the same. In this the arcana are decocted; in this they are fermented. And as the seed in the earth putrefies before it is reborn, and fructifies, so here also in the fire a dissolution takes place, wherein the arcana are fermented, lose their bodies, and, by means of ascension, go off into their exaltations, the times of which are calcination, sublimation, reverberation, solution, etc.; and, secondly, into reiteration, that is, into transplantation. Now, all these operations take place by means of motion, which is given by the time. For there is one time of the external world, and another of man. But the operation, or force, of the celestial motion is truly marvellous. And, although the artificer may be disposed to rate highly both himself and his work, still here is the sum of the matter, that heaven, in an equally wonderful way, decocts, digests, imbibes, dissolves, and reverberates, while the alchemist does the same. The motion of heaven, too, teaches the motion and regimen of the fire in the Athanor. So also the virtue which is in the sapphire, heaven draws forth and discloses by means of solution,

---

* Alchemy indeed brings forth many excellent and sublime arcana to the light, which have been accidentally discovered rather than sought for. Wherefore let alchemy be great and venerable in the sight of everyone, for many arcana are in tartar, in juniper, in melissa, in tincture, in vitriol, in salt, in alum, in Luna and in Sol.—*De Caducis*, Par. IV.

coagulation, and fixation. Now, if by these three methods the heaven is thus constituted in its operations, whilst it reduces them to this point, it necessarily follows that the solution of the sapphire shall also, in like manner, be made up of these three points. The solution is of this nature, that bodies are thereby excluded and the arcanum remains. For, hitherto, while the sapphire remained entire, there was no arcanum. But afterwards, analogously to the life in man, so this arcanum has been infused by heaven into this matter. Therefore, the body, which impedes the arcanum, has to be removed, For. as nothing is produced or begotten from the seed, unless it be dissolved, which dissolution is nothing but a putrefaction of the body, not of the arcanum itself, so, in this case, is it with the body of the sapphire, save in so far that it has received an arcanum. But now its dissolution is brought about through the same processes which caused its previous coalescence. The seed corn in the field has no little experience of the industry of Nature during its own progress to the corn-ear. For there is an elixir, and a most consummate fermentation, which is retained in Nature beyond all other places. Afterwards follows digestion ; and hence begins increase of the substance itself. Whoever wishes to become such a natural originator must gain his end in this way. Otherwise, he will be a mere cook, or scullion, or dish-washer. For Nature demands that in all respects the same preparation shall take place in man as in herself ; that is, that we shall follow her as our guide and not the follies of our own brain. But you, doctors and ointment-sellers, what do you ferment, or putrefy, or digest, or exalt ? Nothing, save when you make up some medley of sauces which you serve out and shamefully palm off upon people. Who can praise a physician when he has not learnt the method and principle of Nature ? Who will trust him ? The physician should be nothing but the skilled advocate of Nature, who, in the very first place, knows its being, properties, and condition. If he is thus ignorant about the composition of Nature, what, I conjure you, can he know of its dissolution ? Understand that dissolution is a kind of retrogression. Whatever things Nature has gradually formed by composition, those things you ought to be able to dissolve by a reverse process. As long as you or I shall be inexperienced in, and ignorant of, this solution, so long, at all events, we shall act the part of robbers, murderers, rascals, or simple novices.

What, I would ask, can you produce from alum, in which are latent as many arcana for diseases of the body as for wounds ? According to your method of proceeding, who ever, by following the pharmacopœists, applied it to that use in which it is chiefly powerful ? The same may be said of mumia as of alum. Where will you seek it ? Beyond the ocean, among the heathen ? O, you simpletons, who fetch from afar what is before your very houses, and within your city walls ! Because you are ignorant of Alchemy, you are on that account ignorant also of Nature.* Have you persuaded yourselves that,

---

* I also assign the greatest weight to experience, as most helpful in the attainment of a proposed end, especially in alchemy, by which things unheard of, and indeed scarcely credible, are produced, whence also fertile science and

because you disparage Avicenna, and Savonarola, and Valescus, and Vigo, that you are as capable as these men? These are mere trifles. Apart from this arcanum, nobody can inquire into the true composition of anything in Nature. Bring together into one spot all your doctors and writers, and tell me what corals contain, and what they can do. However much you may chatter, and whatever blatant nonsense you may talk about their powers, directly you begin to reduce it to practice, it is proved that you have not one particle of experience or of knowledge about corals. This is the reason why the process of the arcanum has never been handed down in writing. But if the process is accomplished, then its virtue is ready to hand. So great is your simplicity, however, that most of you think everything consists in pounding, and that it suffices if you write: "Let these things be strained and mixed. Make a powder with sugar." What Pliny and Dioscorides wrote about herbs they did not prove by experience, but gathered from the famous authors who knew many such matters, and then they filled many books with their feminine chatter. Dare to make the experiment for yourselves whether what they hand down is true. Will you never be able to reach the goal of experiment and proof? What do Hermes and Archelaus say about vitriol? They mention its vast virtues, indeed; and these are present in it, but you are ignorant what powers are in it, whether the green or the blue. Can you be masters of natural things and not know this? What you know, you have read, indeed, but you profit nothing and do no good by it. What do the alchemists and other philosophers teach about the potencies of mercury? Their teaching is copious, indeed, and full of truth. That you know truly, but how it is to be verified you know not. Cease, then, to shout. In this respect your academies and yourselves are novices and mere tyros. You skim over all these matters in your reading, and you say, "This property is in one thing, that in another; one is black, another is green. God is my witness, I know no more. So I find it written." So, unless it had been thus written, you would have known nothing about it. Do you think that I am wrong in laying and fixing my foundation in the Alchemical Art?* This reveals to me what is true, and that you are unskilled in proving the truth. Is not such art worthy to come into the light? And is not that deservedly termed the foundation of Medicine, which proves, augments, and establishes the knowledge of the physician? But what is to be thought of

---

notable experience are gathered in the light of Nature. I could state on oath that from such experience the greatest and most notable fundamental principle in medicine has arisen. Who without it will ever be a physician, or know and understand anything?—*De Caducis*, Par. III.

* We assume no person will doubt that the chemical art has been devised to supply the deficiencies of Nature; for although Nature supplies very many most excellent remedies, she has, notwithstanding, produced some which are imperfect and crude; for the perfection of these a separation must be effected, by which the pure is set free from the impure, so that it may at last fully manifest its powers. We desire the surgeon to be versed in this art, without which he does not indeed deserve his name. The preparation of medicaments is of great importance, so that they may be brought to their highest grade of action. God does not will that medicines should ready too easily at hand; He has created the remedies, but has ruled that they should be prepared by ourselves. The chemical art must not therefore be repudiated by the surgeon. As long as physicians are content with the preparations of the pharmacists, they will never accomplish anything worthy of praise. Furthermore, the alchemists themselves, despite the excellence of their remedies, will find their operations barren until the arts of medicine and chemistry are completely united. - *Chirurgia Magna*, Tract II., c. 9.

your judgment when you say "Serapion, Mesne, Rhasis, Pliny, Dioscorides, Macer, report about verbena, that it is useful for this or that purpose"? You cannot prove that what you say is true. What sort of a judgment can even you yourselves think this? Be yourselves the umpire: Is he not more powerful who is able to prove that true which is within? And this you cannot do without the aid of Alchemy. Even though you should read and know twice as much as you do, all your knowledge would be of no avail. Let any one read my work, and then how can he have the face to make it a charge against me that I lay these things before you and explain them to you? You do not reduce to action those powers and virtues which you parade and boast that you possess. Answer me: if the magnet fails to attract, what is the cause of the failure? If hellebore fails to make you vomit, why is this so? You know what causes purging and vomiting; but what are the arcana of healing just now spoken of? In this matter you are the very brother of Ignorance. Tell me in whom confidence should be placed as to the operations and powers of natural things? In those who have only written about such things without ever having tried or used them, or in those who have put them to the proof, though they may not have written about them? Is it not a matter of fact that Pliny has proved none of his assertions? Where, then, is the use of his statements? What has he heard from the Alchemists? And if you know nothing of these you can be at best but a travelling quack-doctor.

If, then, it be of such vast importance that Alchemy shall be thoroughly understood in Medicine, the reason of this importance arises from the great latent virtue which resides in natural things, which also can lie open to none, save in so far as they are revealed by Alchemy.* Otherwise, it is just as if one should see a tree in winter and not recognise it, or be ignorant what was in it until summer puts forth, one after another, now branches, now flowers, now fruits, and whatever else appertains to it. So in these matters there is a latent virtue which is occult to men in general. And unless a man learns and makes proof of these things, which can only be done by an Alchemist, just as by the summer, it is not possible that he can investigate the subject in any other way.

Now, seeing that the Alchemist thus brings forth what is latent in Nature, you should know that there is one kind of virtue in the twigs, another in the leaves, another in the unripe fruit, and yet another in the fruit when ripe; and that the difference between these is so palpable that the later fruit of a tree is altogether unlike the earlier, and this not only in form but in virtues. Whence the knowledge should be of such a kind that it shall extend from the first to the last. This is Nature. And since Nature thus manifests herself, so also does the Alchemist when dealing with those matters which Nature defines for

---

* I include Chemia in the circle of medical perfection for many reasons. It supplies true simples, magnalia, arcana, mysteries, virtues, powers, all things which pertain to the science of remedies, much more perfectly than ordinary pharmacy. But you object that alchemy is universally unpopular. I ask for kind words. Other arts also— astronomy and philosophy—are contemned, but are not the more imperfect for this.—*Fragmenta Medica.*

him. For instance, the genestum keeps the process of its own nature in the hand of the Alchemist. So does thyme, with its flower, and the rest. One thing does not contain a single virtue, but several. You see this in flowers. They have not one single colour, and yet they are in one thing, and are themselves one thing; and every colour is severally graduated to perfection. So is it with the different virtues which are latent in these things. Now the alchemy of colours is so to separate Art and Nature, that this separation shall extend not only to the colours, but to the virtues. As often as a transmutation of colours takes place, so often occurs a transmutation of virtues also. In sulphur, there is yellowness, whiteness, redness, darkness, and blackness. In each colour there is a special power and virtue; and other substances which possess these same colours have not the same, but different, virtues lying hid in these colours. And there is a latent knowledge of colours just as there are latent colours, and a latent cognition of virtues, as there are latent virtues. And the manifestation of virtues is the same as in form and colours, where are first the twigs, then the pith, afterwards fronds, flowers, and leaves, then the beginning of fruits, then their mid-period, and, lastly, their full development. If the virtues ripen by a gradual process of this kind, and thus increase, then the indwelling virtues are changed in degrees and in number every day, nay, every minute. For as time and not matter gives its purgative effect to the elder, so that same time confers its powers on other virtues, some in one way, and some in another. As time again assigns styptic powers to the acacias, which do not naturally arise from time, and as is the case with other wild growths, so time also in this case imparts the middle virtues before the final term. For these signs or intervals must be carefully noted in Alchemy, on account of the knowledge as to the true end of operations, and of the autumnal period by which the time of mature or immature virtue is defined; and the same is carefully attended to in Medicine. So, also, these ripenings are divided into buds, fronds, flowers, pith, liquids, leaves, fruits; and in each of these are their own proper beginnings, mid-periods, and ends, divided into three ways or principles, namely, into laxatives, styptics, and arcana. Those things which loosen and constrict are not arcana. And none of these is conducted at once to the final end, but they exist in the primary and middle virtues. How evident is this illustration in the case of vitriol, which is now everywhere very well known to all, and the virtues whereof are especially patent. Now, I propose in this place not to obscure its virtue, but to unfold and manifest it more widely. First of all, then, vitriol puts forth its laxative virtue, being the chief of all laxatives, and possessing the greatest power for the removal of obstructions. There is not in the body any member, external or internal, which is not penetrated and affected by it. This effect arises from the first time. The second time gives it a constrictive power. As powerfully does it now constrict as in the beginning of the first period it loosened. And still its arcanum is not yet at hand, nor have buds, fronds, and flowers burst forth. If it has sped to fronds, what is more effectual in the falling sickness? If it proceeds to flowers, what is more penetrating? It is like an odour which most readily diffuses itself on all sides. If it issues forth in its fruits, what is more excellent for promoting heat? And there are many other qualities in it, reckoned by its appropriate periods. At all events, so much we have endeavoured to explain, how the arcana in any one thing separate into many parts, and each part is borne on to its own special period, and that, moreover, the limit of periods in things is an arcanum.

So in the first transmutation of tartar; what excels this arcanum in cases of itch and scab, or other similarly disagreeable affections? In its second period, what is more effectual for the removal of obstructions—not in the way of a mere laxative? What, at length, is more powerful in the healing of wounds? Now, it is Alchemy which opens and manifests these qualities. Then, why not raise the foundation of Medicine hereupon?* Learn from this, at all events, and dismiss these dirty ointment-vending quacks, who do not know this process, but, together with their teachers, are double-dyed asses, and so mad as to presume to think everything of this kind false and impossible. They are so ignorant and experienced that they have not learnt even the beginning of a decoction, and yet health and safety for the sick are to be sought from such men as these. What else do you find in them but desire for money and thirst for goods? It is all the same to them whether their medicines do good or harm, whether they remove or increase diseases. Is it not right, then, that ignorance of this kind should be publicly revealed? I do not adopt this course from any hope that they will imitate me. They will feel no shame on this account; but rather hatred and envy will so take possession of them that they will persevere in their ignorance. Yet, notwithstanding, whoever wishes to pursue truth alone, will turn aside to my monarchy, and not to any other.

Mark, I beseech you, my readers and hearers, what a wretched and distorted process is adopted in the falling sickness, not so much by the ancients as by those writers who are contemporary with me; and this to such an extent that they are scarcely able to rescue or to cure a single patient. Do I act unjustly when I despise such writers and such teachers, who demand, as a matter of right, that people should use their remedies, though they are not of the smallest power? On the other hand, if any one investigates another method, by which help can be given to patients, they call him a vagabond, a chatterer, and a fool. What is rather true is that their prescriptions, like their diagnoses in the case of falling sickness and other diseases, are mere lies. This is proved by results. The patients themselves bear witness to it, while the nature of things cries out and proves the foundation on which Medicine must be built up. No one disease can they heal by a well considered and consistent system of

---

* If, therefore, it be the part of the physician to cure, and the foundation must be taken from the four things named above, how shall he conclude? By alchemy alone. What is alchemy? That which prepares Medicine, making a pure and precious remedy, exhibiting it perfect and entire, whereby the knowledge of the physician is completed. If such then be Medicine, and the knowledge of the physician must be acquired in this manner, how, I say, dare those square and knotty doctors and masters, without forfeiting their honour, blame me because I do not deal with trifles, but with truth itself, that I may establish science more fundamentally and exactly, for they adhere to their antiquity. — *Fragmenta Medica*.

Medicine, since God does not call and choose such uncertain and erratic men to be physicians, but rather well assured and experienced men. If He supplies an assured and experienced husbandman or quarryman, much more will He give a physician who is certain about his art and confirmed in its practice, since on him rests more responsibility than on all other men. But they call the foundation itself doubtful, and place it in the hand of God. So, then, the hand of God is stretched like a veil over their imposture and ignorance ; and they justify themselves, but accuse God, when they say that their art, indeed, is perfect, but that God interrupts it and stands in its way. What is impiety and imposture, if not this? But see by what unshaken argument I will establish Alchemy as the foundation of Medicine. I base it on this: that the most severe of all diseases, such as apoplexy, paralysis, lethargy, the falling sickness, mania, frenzy, melancholia or gloom, and other similar ones, can be cured by no concoctions of the ointment-sellers. As meat cannot be boiled with snow, so much less can this kind of medicine be brought to any successful issue by the art of the drug compounders. For as the magistery of each several substance is that to which it specially looks, so it should be noted of these diseases that they have particular arcana. Hence, they require special preparations. What I say of these special preparations I would have to be understood in the sense that particular arcana require particular adminstrations, and different administrations in like manner demand different preparations.* Now, in the surgeries there is no other preparation beyond some kind of kneading and decoction, such as one would see in a cookshop. By this kind of cooking the arcana themselves are stifled and result in no energy whatever. Nature must be kept under proper restraint and management. Thus, you see, there is one kind of preparation required for bread, another for meat, and so on. In the same way is it with herbs. By parity of reasoning it should be inferred that Nature never mixes up in disorderly confusion foods, drinks, meats, and breads in one mass, but deals with each separately and by itself. Now, this arises from no trifling causes, though to recount those causes here would be a work of unnecessary labour. Now, if Nature admonishes us hereby that in all things due order is to be kept, we are also in another way warned by the same mistress how to prepare medicines, and how to adapt the several medicines to their special diseases. The thirsty liver demands wine or water. But think how often wine is pre-

---

* If the seed, that is to say, the matter, be present, it requires preparation. But it is prepared by nothing save alchemy. This is not that which teaches cooking and preparation, which Nature has instituted for the benefit of man. Thus, Nature is acquainted with many virtues in S. John's wort, but for every virtue there is another preparation. Nature orders this to alchemy, saying, as it were: Prepare for one disease thus, and for another after another manner. Then arises the physician, and is famed in the medical art, for he knows the foundation, he knows also what cooking or preparation is here needed. But for what purpose do ye scintillate, ye men of Montpelier, of Leipsic, or of Vienna? Ye must turn hither, hither, ye who would know what is philosophy, and what is alchemy, what preparations Nature institutes, and by what methods she instructs her alchemists. Where, then, will ye remain, ye apothecaries and sordid cooks? For it is a shame and disgrace that your whole business is nothing but sheer fancy and wickedness, opposed to the whole art of Medicine. Strange, if all this become publicly known, that is to say, how many tortures are hidden in your golden gallipots and in your solemn concoctions—how great will be the measure of your opprobrium! Yet the matter must be completely brought into the light. But though you multiply scorpions in my food, the venom will only operate in yourselves, and not in me, and will overflow your wily and mendacious designs, and will break your own necks, not mine.—*Fragmenta Medica*.

pared, and, as it were, reborn, before it satisfies the thirst of the liver. In like manner, the bowels require food. Here, too, notice how variously the food is changed and prepared. Believe it to be the same with diseases. Now, if you are going to undertake a pure and artistic method of cure, you will make no difference, but act just as if apoplexy were a thirst for which there was need of a particular medicine and a special preparation thereof. Or suppose that the bowel is falling, and that it requires for its restoration another preparation, and one which acts on the stomach. Imagine, again, that mania is like the spermatic vessels, and demands that its necessities shall be supplied by other methods. So you will come to the same conclusion as to medicines and their preparation in cases of mania. I admonish you, then, with due cause, that if you have chanced to meet with effectual remedies and arcana against diseases, you should not let them be tampered with and wasted by these mere decocters. Are these things not to be brought into the light? Truly, indeed, they are, in order that such errors may be avoided, and patients may advance to those sure arcana which God has designed for their use and requirement. You will gather from hence how necessary it is to act on my prescription rather than on yours. In this respect you have to follow me, not I to follow you. Though you fulminate ever so much against me, nevertheless, my monarchy shall stand, while yours goes to destruction. It is not in vain that I write at such length concerning Alchemy, but I do it with this purpose and for this end, that you may well and surely know what is latent therein, and how it should be understood. Nor should you be offended hereby because you get no gold or silver by it. Rather its result should be that by means thereof arcana shall be unveiled, and the seductions of the ointment-sellers shall be brought to light, since by these the ignorant folk are deceived, while they sell them for a florin what they would not buy back for a penny. So precious, in this sense, are their secrets!

Who will deny that even in the very best things a poison may lie hid? All must acknowledge this. And if this be true, I would now ask you whether it it not right that the poison should be separated from what is good and useful, that the good should be taken and the evil left. Such should certainly be the case. If so, tell me how it is separated in your surgeries. With you all these elements remain mixed. See your own simplicity, then, if you are forced to confess that a poison lies hid, and are asked how it is to be got rid of. Then you bring forward I know not what correctives, which shall drive out and take away the poison. Comfrey, for example, they use to correct scammony, which is then called diagridium. But what kind of a corrective is this? Does not the poison remain afterwards as before? And yet you boast that you have so corrected it that the poison no longer harms. Whither has it gone? It remains in the diagridium. Try it, if you are wise. Exceed the proper dose, and you will soon see where the poison is. You will not be long in finding out. So you correct turbith also, and then call it diaturbith. These are your methods of correction, fit only for clowns, and useful to drench horses with ! Risk an excessive dose; you will see whether you do not feel the poison. To correct is to take that which has been corrected. A criminal who has broken the law is punished; but his correction is not extended farther than the free will of the culprit lasts. Such are your corrections. The power is in them, not in you. In such a case, all the physician thinks of is how to eliminate the poison. This must be done by separation, For example, a serpent is venomous, and yet it is used for food. If you take away the poison you can eat the flesh without injury. And so it is with all other substances; only a similar separation is absolutely necessary. If this is not brought about, you cannot be sure of your work, unless, indeed, it come to pass that Nature supplies your place, or some special interposition of providence favours you. You have no protection in your art. If a sure foundation be necessary for the extraction of the poison, that is afforded by Alchemy. It must be so arranged that if there be Mars in Sol, Mars must be taken away; or Saturn in Venus, the Saturn must be separated from the Venus. As many ascendants and impressions as there are in natural substances, so many bodies are there in them. But when the bodies are contrary, it is absolutely necessary that one of them should be taken away and removed, so that in this way all contrariety should be eliminated, and so the evil which you are searching for should be separated from the good. As gold is useless except it shall first have passed through the fire, so medicine is much less useful unless it, too, shall have passed through the fire. It is necessary that everything which is to benefit man shall have passed by fire to a second birth. Should not this, then, be deemed the right fundamental principle of every physician? A physician should exhibit not poisons, but arcana. But all the preparations of your surgeries, how many soever they be, do not contribute the smallest tittle of learning. They are employed simply in correcting, which is just as if a dog should break wind in a room and you should kill the stink with fumigations and juniper wood; but does not the smell remain in the room as much after as before that process? Although the smell is not perceptible by the olfactory organs, could anybody say that the stink was separated and no longer remained? It is there, though corrected by the fumigation. The stink and the fumigation enter the nose together. Of such a sort also are the corrections of the drug-vendors who disguise alöepaticum with sugar, that it may not offend the palate. The sugar and the honey form a magistery in this case. So, too, they correct theriacum with gentian. Are not all these operations instances of asinine ignorance? And yet these people boast themselves the physicians of princes, and sell their skill for money! Yet who is so dense as not to scent the fact forthwith that all this is worthless? What else can they trumpet forth about their remedies beyond saying "This electuary is sweet, being compounded of spices, sugar, honey, and other condiments, and is held in very great esteem?" And thus you mock your patient with your medicine, though all you can say in its favour is that it is nice. Just think how idle it is to lay your foundation in compounds of this kind, and to entrust everything to fatuous doctors. This method differs as widely as possible from the true basis of Medicine; and is nothing but a worn-out and ridiculous phantasy.

So, then, up to this point, we have sufficiently discussed this Column of Medicine, that is to say, Alchemy, in which consists the fundamental principle of all Medicine. Whoever is not built up on this foundation is washed away by every wave, the wind blows away his work, the new moon breaks through it. Every new moon destroys that building, or the shower softens it and casts it down. In view of a system of Medicine built up on such a foundation, do you, reader, judge whether I am an irregular doctor of Medicine or a heretic, disregarding truth, and with a mad brain. Do I deservedly, or undeservedly, gird at my opponents? What right have they to rise up against me? Let who will care for their cudgel. When once it has grown warm in their hands they will not readily lay it aside. Any fools can do this; but a wise man should not imitate them. A far-sighted man throws away one cudgel and seeks another. What matters to me their persecution? I shall not try to stop them. What I shall do is to shew them up, because they rely entirely on fraud and impostures, and have no foundation save what is elaborated from their own phantasy and their own brain. Whoever is a trustworthy and honest man to his patients, whoever in his practice tries to imitate Nature, will not avoid me or turn away from my teaching. But those who live in this century do not follow Christ; in fact, they despise Him. Why should I expect such a privilege as not to be despised by any? At first, indeed, I ploughed by no means inactively in the same furrow with them. But when I saw clearly that from such art arose only murders, deaths, paralyses, mutilations, and other forms of destruction, I was compelled to retrace my steps and to follow truth by an altogether different road. Then they complained that I neither followed nor understood Avicenna or Galen, or knew their writings. They boasted that they understood all these things. Out of all this boasting it arose that on every side they injured, tortured, and murdered far more people than I, in my ignorance, did. This is really as much as to say that there is one and the same mode of operation for the one who understands and for the one who does not understand, and that neither the one nor the other is worth anything at all. But the more I contemplated the havoc wrought by them and by myself, the more I began to burn with hatred for the system, and I advanced to such a point as to perceive that it was nothing but a patchwork and a hotchpot mixed up with imposture. And I do not wish the matter to be concluded here; but in all my writings I shall make it clear how, and in what way, all these matters are combined with ignorance. Every day I grasp more and more that not in Medicine alone, but also in Philosophy and Astronomy, these people rest on no good and praiseworthy foundation, as I have already said. A vast tumult will be stirred up against me because I reject those who, for so many ages, have alone occupied the throne of glory and magnificence. But I confidently predict that the time will come when they will be cast down from that throne of glory and magnificence. Their force is nothing but phantasy; and I shall not end with the single expression of this sentiment, but shall assiduously bear the same testimony in all my writings. If the academies do not approve of me, what matters it? They will by-and-bye fall to the ground and be humiliated. Meanwhile, I will expose and oppose your errors with so much severity that to the very end of the world my writings shall be truthful and acceptable. Yours, on the other hand, shall meet with this fate, that, full as they are of bile, and venom, and the poison of asps, they shall be cast out by all like toads, trodden under foot, and shunned. I do not attempt to destroy you and level you with the ground in a single year. It is better that at some future time you shall decay and die by your own infamy and ignominy. I shall judge more of you when I am dead than now I am alive. You may demolish my body, but you will only destroy its refuse. Theophrastus will struggle with you even when he has no body!

But those who shall hereafter be physicians, I would admonish that they deal more cautiously with me than with their own teachers, and that they rather weigh our disagreement with due care and judgment than condemn the other and absolve the other without maturely thinking the matter out. Weigh carefully, I beseech you, with yourselves what it is you would aim at, namely, the healing of the sick. If this be your aim and the subject of your argument, tolerate me as your teacher, since my sole object is to lead you towards this healing of the sick. On what basis, and with how much seriousness, I do so has already been said, and shall every day be more copiously set forth. Let not my writings be an offence to you because I stand alone, because I make a new departure, or because I am a German. By these writings and not by any others, the Art of Medicine must be discovered and learnt. Above all, I would enjoin that you carefully read and consider the works which, by the divine favour, I am to finish. I would name particularly one volume on Medical Philosoph in which all the causes of disease will be investigated. Another will be on Astronomy, with a view to sanitation. A third, and last, will be on Alchemy, that is, the method of preparing medicaments. If you read and understand these three books; even you, who before disagreed with me, will become my followers. Nor shall I fix my limit here, but as long as the divine favour illuminates me I shall go on to regard the Monarchy in certain separate treatises published for this special purpose. Indeed, if I had not been oppressed by the unseemly hatred and envy of certain prominent men in Medicine these treatises would have already for the most part seen the light. I can already easily foresee that the astronomers also, like the philosophers, will noisily set themselves against me. It will be that they fail to understand me. They will raise a precocious clamour against me, but at last they will be silent, and betake themselves to their dens. Let not these things affect you, my hearers. Rather do this—read their writings while they follow mine in full cry. Thus you will find what you seek. At all events, I have set myself to write in what position, and on what basis I build up my system of Medicine, so that you may be able to gather what I wish to erect on this foundation. I lay it before you so clearly that you shall not be able hereafter, with any show of justice, to repudiate me at the suggestion of your fathers and teachers and professors. Take care that you be not led away by vulgar physicians, surgeons, or bath-keepers. These like to look great and powerful, and pour out their long words, which have no science in them, but plenty of ambition and boasting. These are like psalm-singers in a choir, who, indeed, chant the psalter, but understand not its meaning. Such are the physicians who constantly chatter and shout. And just as the nun sometimes understands a single word, but then turns ten pages without comprehending them, so the physicians sometimes make a hit, and then go astray again. Think over these matters with yourselves, and be your own witnesses as to what basis most of these people have for their studies. Even in Medicine it is no new thing for these accusations to damage any one. Medicine in their estimation admits any amount of rascaldom, and is directed only towards persecutions and injuries. All these are signs of doubtful and uncertain art. Those who make such professions give themselves up entirely to wallow in envy and hatred, and wherever one man can stand in another's way, he thinks he has reached the highest point in his practice. So the devil governs them. From him they have derived their discipline; this you cannot doubt. This is attested by their constant rendings and tearings of one another. The hand of God is not the cause of such things as these.

THE "LABYRINTHUS MEDICORUM"

Concerning the Book of Alchemy, without which no one can become a Physician *

ANYONE who would become a physician must learn the book of Alchemy thoroughly by heart. Its name, no doubt, will prevent its being acceptable to many; but why should wise people hate without cause that which some others wantonly misuse? Who hates blue because some clumsy painter uses it badly? Who reviles a stone because it has been broken by the quarryman? In like manner, who will hate Alchemy, which is innocent? He deserves hatred who is guilty, who does not take in the Art, or use it properly. Does anybody hate him who has injured none? Who will blame a dog if he bites anybody who seizes him by the tail? Which would Cæsar order to be crucified, the thief or the thing he had stolen? I trow the thief. No science can be deservedly held in contempt by one who knows nothing about it.

Now, in good sooth, this same Science or Art is of great use and necessity. Into it is dove-tailed the Art of Vulcan, and we know how useful a work Vulcan can accomplish. Alchemy is an Art, and Vulcan is the operator therein. Whoever is a Vulcan, he has power in this Art. Whoever is not a Vulcan has no power herein. In order that you may understand this Art more thoroughly it is necessary to repeat, first of all, that God made all things out of nothing. Out of nothing, I repeat, he made something. Now, this something is the seed which gives the result of its own predestination, its own special office. And, although all things are created from nothing for their own end, there is, nevertheless, nothing which is entirely adapted to its end. That is, it is adapted to its end, but not wholly so adapted; and it is Vulcan

---

* The *Labyrinthus Medicorum Errantium* distinguishes eleven books out of which the physician ought to obtain his art and experience. Of these the fifth is alchemy. The others are wisdom, which is knowledge as opposed to surmise and guess work; the firmament, of which book the stars are the alphabet; the elements, which are all essentially present in man; the greater anatomy, by which the physical body of the microcosm is made known; experience, because the whole of medical science is nothing but a great and certain experience, and whatever acts or operates therein is founded exclusively thereon; the entire natural world, for this is the great storehouse of apothecaries and doctors; theoretic medicine, which must be founded in Nature, even as theoretic theology is founded in God; magic, because medicine should not be constituted in speculation but in manifest revelation, because disease and the medicine thereof are alike hidden, and magic is the science which makes manifest that which is concealed; the book of forms, for all medicines have their forms, of which one is visible and the other invisible, one corporal and elementary, the other spiritual and sidereal; finally, the book of the generation of diseases and their Iliastric and Cagastric seeds.

that must complete the adaptation. All things are created with this view, namely, that they should be placed in our hands, but not altogether perfect. Wood grows to its proper end, but not to coal; clay is created, but a vessel is not formed from it. The same reasoning applies to all growing things. Carefully study, therefore, this Vulcan. We will explain the matter by an illustration. God created iron, but not in the form it should afterwards assume; not as a horse-shoe, a sickle, or a sword. These modifications are entrusted to Vulcan, and so this Art is good. Unless it were good Vulcan would not bring about these adaptations. Hence it follows that iron must be first separated from its ore, and then wrought, for this the artificer requires. Now, this is Alchemy. This is the metal-founder, named Vulcan. What the fire operates is Alchemy, whether in the kitchen or in the furnace. He who tempers the fire is a Vulcan, whether he be cook or heat-producer. And the same is the rule of Medicine. It is created, indeed, by God, but not fully prepared for its final end. It is, so to say, hidden in the ore. Now, the work of Vulcan is to separate the ore from the medicine itself. What you saw about iron is also true of Medicine. That which the eyes perceive in a herb is not Medicine, nor what they see in stones and trees. They see only the ore; but inside the ore the medicine is hidden. First of all, then, the ore has to be removed from the medicine. When this is done, the medicine will be ready to hand. This is Alchemy; this, the special office of Vulcan, who superintends the pharmacopœia, and brings about the elaboration of the medicine. And as it often happens that gold and silver are found in a pure state, so medicine also is sometimes found in a state of purity, and its subsequent separation is then the easier, just as the pure gold needs only fulmination and fusion. When all that is necessary has been done, if in this way, by means of Alchemy, the medicine has been prepared and produced, then it is given to the sick as a remedy, or to the sound as food. Take an illustration from bread. The external Art of Alchemy cannot produce the ultimate material in the furnace, but only the intermediate substance. That is to say, Nature produces the first material up to the time of the harvest. Then Alchemy reaps, grinds, bakes, and cooks this up to the very time when it is taken into the mouth. Thus, the first and the intermediate matters are perfected. Then, at length, the Alchemy of the Microcosm begins. This takes up the first matter in the mouth, that is, it masticates it, which is the primary operation. Then it deals with it in the stomach, which is the second matter. It decocts and digests it until, at length, it becomes flesh and blood. This is the ultimate matter, though afterwards another Alchemy may intervene in the shape of weakness, which is a primary matter. To this succeeds decline, a secondary matter; and at last death, the ultimate matter. Moreover, then ensues putrefaction as a first matter. Next to this is decay; at last, dust and earth. Thus Nature deals with us by means of her creatures. And this makes good my position that nothing is created in a state of perfection for its ultimate matter. All things are created for their first matter. Then Vulcan is applied; and, thanks to the alchemical art, reduces this to its ultimate matter. This is seized upon by the Archeus, or inner Vulcan, who, by circulating and preparing, according to the nature and difference of each separate substance, by sublimation, distillation, and reverberation, puts the finishing stroke to the process. All these arts are prefigured and practised within the body of man, no less than without, in Alchemy. It is here that Vulcan and the Archeus differ. This, indeed, is Alchemy, which directs to its final end everything which has attained some intermediate end; by reducing lead ore to lead, and afterwards shaping lead into whatever it is designed to make. Thus there are Alchemists of metals, Alchemists who work with minerals, who reduce antimony to antimony, sulphur to sulphur, vitriol to vitriol, and salt to salt. Know, then, that this only is Alchemy, which, by preparation through fire, separates what is impure, and draws out what is pure. Though all fires do not actually burn, still they are fires and they remain fires. So, also, there are Alchemists of wood, such as carpenters, who prepare timber for building purposes, or statuaries, who take away from the block of wood whatever does not form part of the contemplated statue. So, too, there are Alchemists of Medicine, who take away from medicine what is not medicine.

Hence, then, it is quite clear what sort of an art Alchemy is, such an art, namely, as separates the useless from the useful, and reduces it to its ultimate matter or nature. The reason why I define these things more at length in this book is because most printed books contain no art at all, but are crammed full of elisions and senseless punctuations, so that swine would rather eat dung than taste such a concoction. Since such ill-digested mixtures are of no use or force, God has put in their place Alchemy, the true and sublime Art of Nature herself. That crass and rude preparation of medicines which the drug-vendors of Montpelier produce is not worthy to be called an art, but is mere cramming, and a most abominable concoction. Yet this is how they make up their syrups and laxatives, or compound other like matters. Those printed books of the pseudo-physicians teach the same artifice, or, at all events, they put it forward. Yet not even syrups or laxatives, such as the practitioners of Montpelier prescribe, should be prepared in this way, but as the science of Alchemy teaches Medicine. So has God appointed and arranged. This should suffice for every physician that, since God has created nothing in its state of ultimate finality, but has committed the finishing stroke to the Vulcans, he, too, should fully perfect his medicines, and not weld the ore with the iron into one mass. Take another illustration. Bread is created and given to us by God, but not in that shape which the baker confers upon it. Those three Vulcans, the farmer, the miller, and the baker, produce from that first matter a second, namely, bread. The same should be done with medicaments, and the same mode of reasoning applies to the Vulcan within us. So, then, the physician should not be ashamed of Alchemy; but in all things proceed according to the method which has been pointed out. Unless he does this he will not be a doctor, but just a freshman dubbed doctor—a doctor only to the same extent as that is a man which is seen reflected in a looking-glass.

HERE ENDS THE BOOK OF ALCHEMY FROM THE LABYRINTHUS MEDICORUM

CONCERNING THE ALCHEMICAL DEGREES AND COMPOSITIONS OF RECIPES AND OF NATURAL THINGS *

Theophrastus Bombast, Eremite of Hohenheim, Doctor and Professor of both Faculties, to those desirous of the Medical Art, health in the Lord

SINCE Medicine alone among all branches of learning is necessarily accorded the commendable title of a divine gift by the suffrage of writers both sacred and profane, and yet very few doctors deal with it felicitously at this day, it has seemed expedient to restore it to its former illustrious dignity, and to purge it as much as possible from the dross of the barbarians, and from the most serious errors. We do not concern ourselves with the precepts of the ancients, but with those things which we have discovered, partly by the indications found in the nature of things, and partly by our own skill, which also we have tested by use and experience. For who does not know that very many doctors at this time, to the great peril of their patients, have disgracefully failed, having blindly adhered to the dicta of Hippocrates, Galen, Avicenna, and others, just as though these proceeded like oracles from the tripod of Apollo, and wherefrom they dared not diverge a

---

* The Geneva folio adds the two other dedications which here follow :—Theophrastus, Eremite of Hohenheim, Doctor of both Faculties, Physician in Ordinary at Basle, to his most famous D. Cristophorus Clauserus, the most learned Doctor of the Physicians and Philosophers of Zurich, Greeting. It is the most excellent and the best sign of a true physician to be acquainted with medical truth, and to know whether he possesses the secret or not, exactly as you, O Cristophorus, most eminent of the physicians of Zurich, do nothing in your medical capacity which is contrary to your judgment and your most tender conscience, to which thousands rightly appeal. But understand this authority which I exercise in this our Monarchia. There is inborn in me a medical virtue derived from the soil of my fatherland. For even as Avicenna was the best physician of the Arabs, Galen of the men of Pergamon, and Marsilius of the Italians, so also, most fortunate Germany has chosen me as her indispensable physician. You know well that experience is the very mother of all physicians, yea, also of our whole Monarchia. But since each country is autonomous and a foreigner cannot be properly identified with it, but on the other hand an alien can well be compared with the man who corresponds to him, so this observe, that you may compare me to Hippocrates and Averroes ; you may compare Rhasis with us three together, each according to his country. Thus the Arabs, the Greeks, and Germans stand on the same level, even as a triple horehound, and they equalize the amber of Germany with the Greek amber, with storax, turpentine, balsam, and mumia. Nor are you unaware that each country contains within itself the matrices of its own element, and produces that which is needful for itself. So amber is amber to its own country, and though perhaps there can be no comparison of the Chaldean rose to that of Arpinûm, what has this to do with diseases, since each rose is for its own country? Exactly in the same way every nation brings forth its proper and peculiar physician ; and that from its own Archeus. For every want gives work to an artificer, and the same necessity is the teacher and parent of every physician. Therefore the Italians can dispense with the Greeks and the Germans with both, since each of these have their own need, and its own minister, one for the nature of each nation. There is no call that any one should copy the mind or morals of the Arabs or the Greeks. If there be error at home, there is arrogance abroad. For this takes place at random, as by a dream, and without any reason—and hence a physician must be generated out of these

finger's breadth. From these authorities, when the gods please, there may indeed be begotten persons of prodigious learning, but by no means physicians. It is not a degree, nor eloquence, nor a faculty for languages, nor the reading of many books, although these are no small adornment, that are required in a physician, but the fullest acquaintance with subjects and with mysteries, which one thing easily supplies the place of all the rest. For it is indeed the part of a rhetorician to discourse learnedly, persuade, and bring over the judge to his opinion, but it behoves the physician to know the genera, causes, and symptoms of affections, to apply his remedies unto the same with sagacity and industry, and to use all according to the best of his ability. But to explain the method of teaching in a few words, I must first speak of myself. I, being invited by an ample salary of the rulers of Basle, for two hours in each day, do publicly interpret the books both of practical and theoretical medicine, physics, and surgery, whereof I myself am author, with the greatest diligence, and to the great profit of my hearers. I have not patched up these books, after the fashion of others, from Hippocrates, Galen, or any one else, but by experience, the great teacher, and by labour, have I composed them. Accordingly, if I wish to prove anything, experiment and reason for me take the place of authorities. Wherefore, most excellent readers, if any one is delighted with the mysteries of this Apollonian art, if any one lives and desires it, if any one longs in a brief space of time to acquire this whole branch

---

things. But he who in spite of this randomness and slumber is raised up as a physician by the need of his country, he at length becomes the perfect physician of his nation, and is plainly its true Hippocrates, Avicenna, and even Lully. However, in this place, I cannot praise the men because they were raised up by this necessity, since their own country will not permit that I should pass over their errors in silence. For how, I ask, did Rhasis benefit Vienna? What good did Savonarola do to Friburg or Arnold to the Swiss? What did Gentilius or the commentaries of Jacob de Partibus and Trusanus to the physicians of Meissen? What did Avicenna confer upon them all, since the health of the sick is the one thing to be considered? This, therefore, is the faculty by which I write, which also my fatherland gave me, and this by that necessity whereby I said that I was born. Hence I dedicate the whole of this book to you. But I am persuaded that some ignorant person will at once reply, and I again shall make answer; so is it manifest and clear on both sides that the whole duty of every physician is concerned with the health of the sick. But those whom I love most dearly will perhaps give interpretations of some obscure places herein, though not my oldest friends of all, namely, the foxes. My crowd of physicians is divided into two parts, the false of tongue, and the false both of heart and tongue. Now you understand what I wish. I will shortly send you some prescriptions, together with my improvement of the oil of colcothar. Act as a friend always, and be careful. Farewell.—*Given at Basle, the fourth day of the Ides of November*, 1526. THEOPHRASTUS OF HOHENHEIM, DOCTOR OF BOTH FACULTIES, AND PHYSICIAN AT BASLE, TO THE MOST EMINENT ASSEMBLY OF THE STUDIOUS AT ZURICH, HEALTH. Alas, how wretched is the estate of mortals, because there is scarcely any joy which is not presently followed by sorrow, a most fine company of helpers! Hitherto I have not fully perceived my blindness, for I did not consider in the present that the wise man must most diligently observe not only those things which are at his feet, but those which are behind him, like a two-headed Janus, and those also which are in all directions around him. The reason is that your most delightful assembly, which I lately enjoyed, and do still recollect with gratitude, had so enchanted my heart and eyes that I forgot all about the future. My mind presaged no disaster; I thought the whole matter was well managed and deemed that joy would be obtained and perfected without the company of grief. Now when I see those things which I ought to have foreseen, how, I say, shall I restrain myself from grief and mourning, since the dearest friend I had at Basle, whom I left in health and strength, has been killed by the accident of a sudden fall from an upper storey, where he was accustomed to sleep? He had been freed by me from the heaviest chains into which he had been thrown by the petty doctors of Italy; by me was he restored to health, of which fact Erasmus of Rotterdam is a witness, with all his family, as the epistle written by his own hand sets forth. Now when I was thus feasting with you, and taking life easily, he died whom I had left in good condition; he, I say, whom I loved as my own eyes; being snatched away by the accident I have mentioned, namely, John Frobenius, the parent and tutor of all learned and good men, being himself also wise and good, the most diligent promoter of all kinds of learning. Wherefore also have I need to fear the same suddenness in death which has overtaken him. What shall I say to myself? Death is common to all. Wherefore be warned. Watch, most excellent fellow-learners, and if to any extent we fail in our office, attribute it to that severe grief wherewith I am now tortured, and can find no relief. Farewell, most sweet companions. Love your Theophrastus.—*Basle, from our Library, the third Ide of November*, 1527.

of learning, let him forthwith betake himself unto us at Basle, and he will attain to far other and greater things than I can describe in a few words. But to make it clearer to the studious, we do not, for instance, shrink from submitting that we in no wise imitate the ancients in the method of complexions and humours. The ancients gave wrong names to almost all the diseases; hence no doctors, or at least very few, at the present day, are fortunate enough to know exactly diseases, their causes, and critical days. Let these proofs be sufficient, notwithstanding their obscurity. I do not permit you to rashly judge of them before you have heard Theophrastus. Farewell. Look favourably on this, an attempt at the restoration of Medicine.
—*Basle, the Nones of June, 1527.*

CONCERNING THE ALCHEMICAL DEGREES AND COMPOSITIONS OF RECIPES AND OF NATURAL THINGS*

By Theophrastus, of Hohenheim, Doctor of both Medicines

BOOK THE FIRST

CHAPTER I

BEFORE I begin to treat of Degrees, two complexions of Nature should be noted: one is hot, the other cold. Moreover, each of these has in itself a certain inborn diathesis: for everything which is hot is dry, and that which is cold is moist, nor can heat or cold be alone. So these two natures, heat and dryness, are one thing, and in like manner, cold and humidity. Hence, therefore, degrees are easily determined, how each and every thing exists in its own degree, and what degree each thing respectively occupies. At this point, no doubt, those who are suffering from cataract and have familiarity with works of darkness will cry out that there are four complexions, hot, cold, dry, and moist, from which they gather that cold is present in moisture and in dryness, and in like manner heat is conjoined with both. According to this opinion they have arranged everything, that is as much as to say that the cold may be dry, and heat may be moist, which is a contradiction of terms. If they had approached more nearly and made a more searching investigation into Nature they would have found our arguments, which here follow, to be nearer the truth. They did not sufficiently understand that these four are two only, and so they falsely ascribe to the four elements those which are nothing less than they are elements, as philosophy clearly demonstrates.

CHAPTER II

But in order to more clearly understand what I have said about the two complexions, take the following. Whatever the elements have produced in

---

* While there is some matter in this treatise which is outside the purpose of the present translation, it has been thought well to include it in the section devoted to Hermetic Medicine because it enters at length into a subject, or, more correctly, a class of subjects, to which there is frequent reference in Paracelsus—that, namely, of degrees and complexions. It is not very clear in itself, and it adopts an arbitrary terminology, which will be dealt with in the Vocabulary at the end, but it will help to illustrate the obscurity of previous references, and may perhaps give a little light indirectly.

the nature of things is either cold or hot. If cold, it has in itself a certain innate individual humidity. Where there is humidity there is cold, and so what is hot is dry, for dryness subsists in heat alone. It cannot come about that the cold is dry and the heat is moist. For these are elemental conjunctions which come from Ares, as is clear from the example of a man and a woman. A man has in himself what is warm and dry; a woman that which is cold and moist; but they contribute to the complexions only according to their several degrees. From the very first, therefore, it must be remarked what is moist, what coagulated, and what, lastly, is resolved dryness. For hence arises a common error which is apt to spread even amongst the chief physicians. For example, take a crystal which appears cold, dry, and arid, since it dries and renders arid, but this appearance is delusive. For the most arid force of the crystal is a coagulated moisture, and in its action it masters everything, transmutes and forcibly changes it into a coagulated moisture, which is finally dissolved like ice. Similarly in petroleum, the dryness is not resolved as it appears to be, for the dryness is resolved in the substance of its own body. Wherefore I lay down this definition in place of an epilogue, that degrees must be observed in a twofold manner, the warm and the cold. Moreover, the dry is double and the moist is double, that is to say, dry *per se* and a resolved dryness, moist *per se* and a congealed moisture. The remainder of what is requisite at this point is contained in the Philosophy itself.

CHAPTER III

Although in this place more was to be said on the subject of degrees than I have so far set down, still, since these matters are well established among those who are any way skilled in medicine, I pass them over in silence here, and speak of those subjects which have up to the present been put forth falsely, and with a certain amount of pervading error. This is what should be accepted. In the first place, it is not only necessary to observe the sum of the elementated degrees, because this only serves in the case of elementated ailments; but attention must also be paid to those things which concern mundificatives, incarnatives, laxatives, constrictives, repercussives, diaphoretics, narcotics, cicatrizers, and other things of this kind. For this purpose it is of prime necessity to acquire a full knowledge of diseases, and, moreover, of the special degrees of each kind of disease. In the case of wounds, one has to know the degrees of incarnation; in hyposarcha, the degrees of drying; in gutta, the degrees of strengthening; in epilepsy, the degrees of specification; in cachexy, those degrees which arise from commixture. If you have thoroughly examined all these matters, then at last approach the composition of recipes.

But we should not omit to mention that for the perfect knowledge both of diseases and of degrees there is required not only the medical but the astrological profession; and, moreover, the Spagyric form. All these require perfect, and, moreover, a prolonged experience; since thereby alone, and not by constant reading, or by a judgment, however exact, the scope of this book is made clear.

Lastly, if you miss anything in this place on the subject of degrees, seek it in daily practice, to which I relegate you all, so as to learn the virtue of anthera, more particularly of tereniabin, which is remarkably ennobled; and, lastly, as respects the flower of cheiri.

CHAPTER IV

Before, however, we come to the degrees themselves, we must observe certain rules of the degrees, by which method the degrees are at one time intensified and at another relaxed. In the first place you will observe this method. Whatever proceeds from the elements of the earth, that occupies the first degree. Of this kind are the lettuce, the violet, the anthos, etc. In like manner, whatever is of the air, such as pestilence, pneumonia, fever, is in the second degree. That which is produced from the element of the water holds the third degree, such as lead, sapphire, topaz, etc. But those things which spring from the element of fire, as ice, crystal, snow, claim the fourth degree, either hot or dry. It must be noticed, therefore, that whatever sensitive thing comes from an element is the same as the element, as the frog whose sperm is in the third degree; in like manner, camphor. That which is of the earth, as man, is in the first degree, as Rebis. What comes from flying things is in the second degree, as vua. What comes forth from the fire, like the Salamander, is in the fourth degree. In what respects one excels another will be made clear in the following chapters.

CHAPTER V

Furthermore, in order that the degrees may be more clearly marked in their points, take the present example. As the degrees of herbs have been divided into four, so all of them, how many soever there may be, are referred to the first degree, but still not all on an equality. For one is sometimes more than another as to the beginnings, middles, and end of both; but still so that whatever descends from the element of earth remains in the first degree, and must not be placed outside that same. Among you the nenuphar occupies the fourth degree, and with you Saturn is allotted the third, though in coldness it exceeds nenuphar by almost eight degrees. So, then, they cannot be arranged in the same degrees. So, too, whatever exists in the second degree, there also the first point excels the fourth degree of the element, which is of the earth, and the fourth point is higher by four degrees than the last point of the first degree. In the third degree the same judgment must be formed, and likewise in the fourth. Thence are gathered sixteen points, which mount as if by stairs to the true degrees, even to the six hundred and sixty-third. Deservedly, then, we say that those have been in error who collected camphor, the

Concerning Degrees and Compositions

sperm of frogs, nenuphar, and alums into one degree; since from these a true and certain degree could not in any way be taken in recipes, as will be shewn in the following modes for compositions.

CHAPTER VI

But in order that you may have in particular both the degrees and the points of those things which induce heat, remark: Whatever makes ashes or lime or glass is in the fourth degree of fire, as fire itself, mercurial water, aqua fortis, etc. So whatever produces a biting effect and brings things to an Ischara, so as to cause putrefaction, occupies the third place. Of this kind are colcothar, arsenic, sal ammoniac, borax, pigment of gold, and others of that kind, as alkali. But as to the virtues of these things in which some things excel others, that is a matter of points, not of degrees. Moreover, whatever produces scars or blisters belongs to the second degree, of which kind are rabeboia, cantharides, flammula, melona, and others of that genus; for although flammula be in the first degree, nevertheless, in another way it affects the second, because the spirit of salt in it reduces the flammula so that it is just comprised in the first point of the second degree. Lastly, whatever warms, but does not attain to the signs above mentioned, such as ginger, cardamum, abrotanum, and other things of that kind, exists in the first degree, together with its higher and lower points. But it is to be observed in this rule that the degrees are not arranged according to the nature and proportion of the elements, but are, independently of them, condensed into the present rule, for this reason, because the present rule is taken from the first three principles and serves for them, namely, those which predominate in salt, mercury, and, lastly, in sulphur, wherefore, in this place care must first be taken not to use the present rule in elementated diseases. For they are only, as it were, gathered from these, and serve for diseases which can be healed by the first three principles.

CHAPTER VII

But in order that you may ascertain the degree of cold, apart from that which belongs to the elements, take the following: Whatever congeals humours belongs to the fourth degree, of which kind are those things which are born of the element of fire. But whatever refrigerates (to use a common expression), yet does not injure the vital spirit when administered as a remedy in a proper dose, as narcotics, anodynes, sleeping draughts, the sperm of frogs, hemlock, belongs to the third degree. Whatever extinguishes unnatural heats and allays paroxysms is in the second degree, and, lastly, whatever prevents a disease from breaking out into a paroxysm is of the first degree. This rule does not differ much from that one which applies to heat, for these offer a direct enantiosis to the aforesaid. But whatever degrees they occupy which are of the elements, that same remains, according to what has before been prescribed, together with the present degree; so that now there is produced a double degree of Nature, and it operates exactly according to the proportion and nature of the elements.

CHAPTER VIII

Moreover, the rule concerning colours must be noticed, since these, at the same time, indicate the nature of the things in which they exist. For instance, the centaury, which is red, is therefore of a warm nature; the lily, which is white, is for that reason of a cold nature. But of colours which are external, nothing certain can be defined, except in this way. The rose is red, yet of a cold nature, on account of the anther lying in it which attracts the heat of the rose. Again, wherever there is any yellow in a red flower, there is the heat, but the redness is judged to be of a cold nature, and so must it be concluded with regard to other flowers in like manner. Moreover, there are flowers which, though by nature they appear warm, are nevertheless cold; among these is the minium. Others, again, seem cold when they are warm, as copper is. In order to ascertain these things, observe the following rule: Whatever is green, as soon as it is gathered from that with which it may be mixed, is warm. So, too, is the body under which these colours lie hidden. Silver is, by its nature, cold, and keeps the colour of a cold body, for finally it passes into the colour of lazurium. Mars is naturally of a cold colour, and admits of being transmuted into a warm nature; but, nevertheless, it preserves the force, and so the universal virtue of its own proper nature. Black colours are of no special nature, for they are nothing but sulphur, which is burned, and nothing underlies this, but it belongs to the elements. Whatever is white, livid, black, and hyacinthine, is cold; the other colours are warm. Whatever is variegated belongs to one nature, presumably that of its principal colour. So, also, in green, though cold be present, yet it is comprised under its own head.

CHAPTER IX

Whatever is fat, and, moreover, moist, is cold, even although it exists in something green; for the greenness is changed into a cold nature. Whatever, on the other hand, is dry, assumes a warm nature. Moreover, whatever comes from Sulphur, Mercury, and Salt falls under each nature, the warm and the cold, and that on account of the three principles. Summarily, whatever burns is sulphur, and of a warm nature, unless it exists in warm colours. But whatever goes into sublimation or calcination admits a warm nature. So whatever resolves itself or is brought to an alkali is warm. Moreover, whatever is austere is cold. Sweet and bitter assume a warm nature unless affected by the former rules. Whatever dries the skin is warm; that which constricts it is cold. If you would judge by the odours of these things you can define nothing accurately, except so far as they retain the same nature as the body. Lastly, there are other rules which are to be admitted, so long as they do not oppose those given above.

CHAPTER X

Moreover, it should be noticed that there are certain things wherein, besides those which are natural, degrees are concealed in two ways, and that according to two bodies, as is the case with metals, gems, and stones. According to this view, mercury is chief among the metals, and embraces in itself a certain peculiar nature, warm and cold, nor can this be taken away from it. Now, if from thence be generated a metal, in iron or lead, beyond that nature it acquires another, and so two natures will be in one substance. Wherefore, from henceforth lead will be in place of mercury, if the leaden nature which it has acquired, together with its own, shall be suitable to your affairs. In the same way it must be judged concerning tin, silver, iron, and copper, because they return into their own body. Thus the liquid in gems remains in its own nature, and that a mercurial nature. If, now, it be congealed into a gem, it equally puts on a two-fold nature, because the constituent parts are reduced into their primal liquid. So, too, must it be judged concerning common stones. In certain herbs, too, a similar nature is present. Wherefore, read, and read again, and finally recall for experiment whatever is committed to your memory concerning the nature of things. And so recall it that you may now not merely think, but know exactly each of these things; for in this lies the essence of a true and sure philosopher.

CONCERNING
DEGREES AND COMPOSITIONS IN ALCHEMY

BOOK THE SECOND

CHAPTER I

ALTHOUGH I have written concerning the relollea of Nature, according to its reason and nature, that it belongs both to the cold and the hot, together with its innate essence, still there are other things which the natural Ares has produced which in many respects excel what I have treated of in my former book. And, to begin from this point, if you wish from the beginning to speak exactly concerning accidental complexions, you will find that in this place the former relollea of Nature are little approved, and for this reason there are two natures universally in things which are both together in one substance, although only one of them appears. These are the innate accident and the elemental accident. Moreover, everything in its own nature is warm. The first matter of things is warm *per se*, nor does it change the innate accident, because all the three principles in the complexion remain even to their ultimate matter; that is, in whatever nature they are found before the relollea in that same nature do they still remain until the relolleum departs. In whatever way, therefore, experience compasses the end, in that same way the beginning is manifested in itself. But before we pass on to those simples which are in the degrees we must observe that neither heat nor cold is an innate accident of them, but rather an elemental and external accident.

CHAPTER II

Nature sends forth absolutely nothing from herself, as the man experienced in medicine easily gathers, but she keeps the innate accident so long as the thing or the body, in which that innate accident is, remains. You have an illustration in fire. In this the heat is an innate accident, and the nature of the three principles, which is evidently hot. Moreover, it cannot be otherwise but that the substance passes away together with the heat, if you wish to confer that heat on something else. Although that heat warms, still it is nothing more than a dead heat, nor does it heal disease or confer any other advantage, but is a certain superfluous heat added from without to the body. In this manner every innate accident puts forth and displays its power without any help to a sick person. Whatever, therefore, is adapted for the healing of disease should be prepared in the following manner : In the first place bring your medicine to him who separates the two essences, the one in the substance, the other in the vital spirit ; for wherever we wish to exhibit medicine, there it is necessary for the vital spirit to depart from the substance and to agree with the offending matter of the disease. Then the medicine will appear alone in its own body, and this in proportion to the three principles. The external elementated accidents go to that part where the disease lurks. And so I gather that in the universal nature of things a two-fold accident exists, an innate and an external. The innate confers little benefit on the health, but only the external. In fire there is no external accident, and therefore I assert that it is an imperfect work of nature.

CHAPTER III

In the beginning, when Nature brings forth in its proper element, the Archeus* prepares the same according to the proportion and nature of its peculiar Iliaster, so that the Ares consists altogether of three things, and generates in the same thing the substance of the body. This generation, *per se*, is, for the sake of the body alone, that it may appear the same with the relolleum. But what is this to the sick man? For the fire is equally a relolleum accident, as is also snow. But they do not heal sicknesses, nor have any power in them for doing so, because they are a relolleum *per se*. Moreover, the external elements make up the cherio of Nature, which, also, you must bring to the relolleum, because, although you take this together with the cherio, it is the cherio that heals all sickness. Remark in this place concerning the cherio that the cherio is nothing but the heat or cold of these things which leaves the body and goes away into Nature. You have an illustration of this in camphor. This has its frigidity from the cherio, and so is a

---

* *The Archeus of the Metals.*—Ares contains within itself the first matter of all the metals, but with regard to the manner in which it distributes that matter over the globe, it must be held that it expels all matters not excocted into metals along a trinal line into the Yliadus, and separates them in division. Thus in one place there are branches of copper, in another branches of tin, and so of the other metals. Further, if thus they are brought from Ares along the triple line, out of some of them there is ejected a metal, such as tin, lead, iron, or copper, etc., before any of the marcasite, bismuth, cachimia, or zinc have been previously purged, or collected into fæces, but while they are all present, and according to their smaller or larger proportion an excellent or base metal is generated. For it is endowed with hardness in the triple line, when Archeus has extracted it out of Ares. For then they are found in Yliadus according to various modes and forms. By that preparation of the Archeus various colours are produced, no one colour being repeated, for just as from all fruit trees no apple or pear is exactly like another, so also these are not alike, as they philosophize concerning Thisma. But silver and gold are frequently found solid and pure, for this reason, that the marcasite, bismuth, and other metallic matters, have been properly separated from the metals, and are sent back along the triple line. Accordingly, when the metal has been made pure, gold and silver are produced. The other extraneous metals have already been expelled ; the rest, therefore, are found pure as Archeus has ordained them. Sometimes, also, spumes are found on the surface of the rocks which look like plates of silver, sometimes, again, in meadows like flames of gold. Then also in Yliadus there are many other forms. They are most frequently found in waters, because pure gold of this kind is compelled by the force of the waves, together with grains of sand to assume the shape of a bolt. It is afterwards deposited in grains, as takes place by the Rhine and other rivers. Cataracts of water, if they pass over the triple line where this kind of gold remains, then the water ejects it. The larger quantity is washed out by violent inundations, etc., on to the beach or coast. It also happens that two, three, or more metals are found mixed, as gold and silver are found in copper. The cause is that by the expulsive operation of Archeus in Ares, two or three are sent in company into Yliadus. This occurs chiefly in the case of cognate metals. While, therefore, they are mixed, and, being mixed, are coagulated, they cannot be separated again, but remain joined together.—*De Elemento Aquæ*, Tract III., c. 10.

most opportune remedy in case of inflations; but in the substance of its first elements it remains warm, as sulphur and the spirit of salt, together with the mercurial. Such, also, are gems and herbs. Moreover, whatever Nature produces has its cherio, that is, its external elemental accident. Wherefore, at this point, on the subject of degrees, I assert that there is a greater portion of cherionic heat, or cherionic cold, in one body than in another. Thus has the Archeus disposed all things, and that for the sake of the microcosm.

CHAPTER IV

But in order that our council concerning the compositions of recipes may be more clearly known, it must be noticed that, as I have before mentioned concerning relolleum and cherio, so, in this place, it is necessary that you again understand this with reference to the body, namely, that those sicknesses which are only of a cherionic nature lurk in the body and descend into the body without involving the destruction of the first three principles of the body wherein they exist. For, just as the Iliaster in the four elements, like a mother, produces the relolleum and cherio, so man exists in the four elements and receives, as it were, hereditarily, the sicknesses which forthwith germinate in the body, so that, eventually, they burst forth into external elementated ailments. Wherefore, for arranging cherionic recipes, it is necessary that the external elementated things should leave their own bodies, together with their substances, and should converge on the vital spirit. Thus, the sick person is set free. It should, therefore, be noticed that death is not cherionic, but relollaceous; although it arises hence that in no direction can death occur. For who can separate what is individual from that in which it lies hid? Here, however, we are speaking of cold and heat in cherionic not in relollaceous matters. The remaining desiderata on this topic you can read in the treatise on the Origin of Diseases.

CHAPTER V

As in the former book, I have conveyed a knowledge of the nature of things, with regard to cold and heat, together with many and various rules, so in this place the present rule must be observed with regard to herbs. Most of them are cold and dry, and these put forth a certain obscure greenness. These, though they are pointed out as hot, are in fact cold, as the verbena and the shepherd's purse. Some are reputed to be cold when they are hot, as the bugloss and the anise. The reason is that the coagulated moisture produces by its congelation great aridity, and the resolved dryness does not become dissolved without some little moisture on account of its cherionic nature. For it is certain that in no other way can anything be produced from the element of earth but it must be hot, nor from the element of water without being cold. This is the rule of Nature. But the reason why nothing of this kind takes place is that the external elementated condition corrupts and breaks through the former nature. Wherefore, it must be dealt with according to its cherionic nature, by the guidance of experience. Moreover, since the same nature, whether it be cold or hot, does not form the body under which it lies hid, there is no need that you should labour for the body, but you may spend all your experimentation on the aforesaid three natures, as we have prescribed in the first book.

CHAPTER VI

Lastly, the physician will have to observe the bodies of those things which lack sensation. For all those bodies in which these things lurk are nothing but a liquid under which is hidden that which is cherionic. But the liquid is congealed like its own element, just as the Iliaster produced it. Wherefore, the separations of Nature once again resolve that which Nature has congealed, and in this resolution the two above-mentioned natures are separated. Hence it is clear that the external elementated things of Nature are the relolleum—accident of nature, and exist apart without any virtue. So,—likewise, it is clear that another nature is fully and perfectly present while the innate property and the accidental property remain each in its own separation. Hence it is gathered that nothing which is cold or hot is congenitally so, and more that whatever is inborn can do neither good nor harm to any person. But there is in addition a certain other nature which does induce heat or cold, and by which we judge the heat or the cold, that is to say, by the cherionic indication. When this interposes, all sickness can be healed. For that same coldness or heat, from the moment of its entrance, turns to the ailment—a thing which the innate property never does. All these matters are contained in the book on the Conjunctions of Things in the properties of the two natures, according to the three principles, and that according to the prescription of philosophy. Moreover, in the following chapters you will see the order of the degrees according to the reason and nature of their elements.

CHAPTER VII

Those things which come forth from the earth have a warm nature in the first degree of heat, and among these are the following :—

Dittany	Gentian	Clary
Lion's-foot	Elceampane	Filla
Anthos, or	Cypress	Calamus
Rosemary flowers	Great Sparge	Hirundinaria
Lacca	Gallingall	Peony
Dodder of Thyme	Philipendula	Ginger
Fig	Bloodwort	Flammula
Broom	Laudanum	Herb of Paradise
Costus	Cloves	Lavender
Pennyroyal	Monk's Rhubarb	Mustard

Humulus	Macropiper	Galbanum
Lencopiper	Fennel	Gamandrea
Hartwort	Grains of Paradise	Liquorice
Cretamus	Citonia	Succory
Scammony	Balm	Cubebs
Teazels	Chamæpitheos	Cardamoms
Basil	Bdellium	Marjoram
Horehound	Fumitory	Mother of Thyme
Sagapin	Thistle	Opopanax
Agrimony	Cheiry	Ammoniacum
	Mellilot	

Things which belong to the air are in the second degree of heat. These are :—

Tereniabin	Clouds	Chaos	Heat

Things which proceed from water are in the third degree of heat, as :—

Vitriol	Granate	Realgar
Sulphur	Red Marcasite	Cachimia of Sulphur
Golden Talc	Congealed Salt	Chimolæa Calcis
Copper	Sal Gemmæ	Jacinth
Topaz	Gold	Chrysolith
Carniola	Smaragdine	Ogorum
Red and White Arsenic	Copperas	Alumen Plumosum
Cachimia of Salt	Liquid Salt	Ruby
	Quicksilver	

Things which come forth from the fire are in the fourth degree of heat, and are these, namely :—

Hot Lightning	Hot Hail	All Ætnean Fires

CHAPTER VIII

The things which are here enumerated are of a cold nature.

Among these those which are produced out of the earth are cold in the first degree.

Dodder	Chestnuts	Pisa
Strawberries	Water Lily	The four greater cold seeds
Comfrey	Lentils	
Branca ursina	Eyebright	Flowers of Mulberry
Mandrake	Bitter Vetch	Ribes
Rose	Mallows	Dates
Acetum	Herb Mercury	Beans
Ciconidion	Pomegranate	Galls
Gourd	Henbane	Crispula
Sanders of all species	Purslane	Ash
Tragacanth	Citron	Darnel

Concerning Degrees and Compositions

Nightshade	Mirabolanes of all species	Lily of the Valley
High Taper	Ripe Apples	Cucumber
Lettuce	The four lesser cold seeds	Greater Arrow-head
Endive	Melon	Fleawort
Gladwin	Snapdragon	Poppies of all species
Bread Flour or Corn		

Things which are produced from the air are cold in the second degree, as Nebulgea.

Things which are produced from the water are cold in the third degree. They include :—

Lead	Antimony	Silver
Camphor	Hematites	Alumen Entali
White Cachimia	The three kinds of Tin	White Talc
Electrum terræ	Alumen de Glacie	The three kinds of Coral
Thalena alterrea	Silver marcasite	Lotho
Thalena frigida	Iron	Aqua Glariona

Things which proceed from the fire are cold in the fourth degree.

Crystal	Cold Lightning	Citrinula
Arles	Citrinæus	Snow
Beryl	Cold Hail	Ice

CHAPTER IX

It is, therefore, to be observed that the law which rules the procedure of each thing from a particular element, rules also that it should possess the same degree. The development of sensitive things from the elements is shewn in the following table.

Those which proceed from the earth occupy the first degree of heat, as:—

Men	The Lion	Rams
Boys	The Horse	The Wolf
The Goat	Oxen	Cocks
The Leopard	The Bear	Foxes, etc.

Those which inhabit the air belong to the second degree of heat.

The Eagle	The Phœnix	The Sparrow
The Ostrich	The Swallow	The Heron

And generally all winged animals which are not referable to water.

Those which relate to the water occupy the third degree of heat, as the Beaver.

Those which inhabit the fire belong to the fourth degree of heat, as the Salamander.

The following are of a cold nature, and, among these, those which proceed from the earth occupy the first degree of cold :—

Women	Cows	All Species of Sperm
Girls	Menstruum	

Those which belong to the air are in the second degree of cold, as doves storks, etc. Those which are referable to water occupy the third degree of cold, as fishes, worms, tortoises, frogs, etc. To the fourth degree of cold are referred those igneous creatures known as Gnavi, or Gnani, and Zonnetti.

CHAPTER X

Moreover, there are certain other simples which, by the intervention of composition, attain to the second grade. These, although they do not acquire their grade altogether according to the manner and nature of the elements, yet such as are in the first grade acquire the second; those which are in the second attain the third; while those of the third, in like manner, acquire the fourth grade, as shewn in the ensuing table.

SIMPLES

Rose	Nenuphar	Flowers of the Centaury
Violet	Camomile	Flowers of the Bullace
Solatrum	Flowers of Tapsus	
Anthera	Flowers of Hypericon	

ADDITION OF COMPOSITIONS

Oil	Vinum Ardens
Crude Vinegar	And all fatty substances
Distilled Vinegar	

Further, although Nature by herself is not so frigid, yet composition effects such a reduction that, by means of addition, there results a certain grade of cold or heat, as is obvious in the case of the oil of roses, the vinegar of roses, and other matters of this kind. There are others which, properly belonging to the third grade, attain the fourth, as camphorated vinegar, oil of lead, etc. Moreover, there are grades which, by means of separation, ascend from the first into the fourth, as also from the third into the fourth grade, as will be seen in the third section of the Grades of the Spagyrists. Again, there are those things which are not intensified at all, of which kinds are snow and ice, by reason of their relolleous nature. Then there are those things which do not manifest their nature unless prepared, as is the case with the sperm of grass, the crystal, and sulphur. There are also those which are reduced from a hot grade into a cold, as gems, and others from a cold into a hot, as camphor, corals, etc. Lastly, there are those which lose their grade in preparation, as those which are congealed or resolved. Item. There are certain things which do not operate in the substance of their body, as oil of Jupiter, and the like. Experience will point out those matters which are omitted in this place.

CONCERNING
DEGREES AND COMPOSITIONS IN ALCHEMY

BOOK THE THIRD

CHAPTER I

AT the beginning of this third book, you are to observe that, besides those essences which I have already mentioned, there is another essence and nature which is called Quintessence, or, as the philosophers say, the Elemental Accident, or again, as ancient physics term it, the Specific Form. It is called quintessence for this reason, that in the first three essences there are four hidden, which in this place is called the quintessence, and is neither warm nor cold, without any complexion in itself. But to make the matter clearer by an example, the quintessence alone infuses robust health, just as the strength or robust health which is in man, without any complexion, is brought to its end. Thus virtue lurks in Nature, for whatever rejects diseases is nothing else than a certain confortative, even as, relying upon your strength, you repel a foe. It is part of the nature of things that there is nothing in the nature of things which is lacking in virtue, unless it be of a laxative quality. The same is the case with quintessence, because this is without complexion. But although coldness elsewhere relaxes, as also sometimes heat, yet it is beyond Nature, and from the virtue of a relollaceous nature. Whatever operates according to Nature possesses a quintessence, for its virtue is so ordered that it removes superfluities from the body, just as incarnatives for curing ulcers in such a manner promote the growth of new flesh, that by the intervention of their virtue the offensive matter is removed. These three are of a triple essence, but there is one virtue, which is justly termed the quintessence.

CHAPTER II

In order to become acquainted with the grades which exist of the quintessence, and specially of those things which are confortative, there are four points to be observed at the outset: firstly, whatsoever is of the earth holds the first grade of health; secondly, whatsoever is from the air is referable to the second grade; thirdly, whatsoever is of water belongs to the third grade; fourthly, whatsoever is produced out of fire holds the fourth grade. But, further, it is labour in vain to seek a quintessence out of earthly things, equal to that which is extracted out of air. In like manner, that which is from air can never be compared with that which derives from water. Judge as follows of the fourth element. For example: To extract the quintessence of chelidonia is nothing else than toiling after the quintessence of the phœnix by that quintessence. Similarly, by the quintessence of the phœnix, the quintessence of gold; by the quintessence of gold, the quintessence of fire; but although in chelidonia, in melissa, and in valerian, there is more of the arcanum than in the rest, yet the grade excels so that by this superiority that arcanum is by far surpassing. Thus in every grade one thing is higher than another. Wherefore, with regard to earthly things, notice whether chelidonia is superior to melissa, as melissa to valerian. Judge in the same way concerning the three other elements.

CHAPTER III

Whatsoever has been dealt with in the former chapters has been with a view to proceeding subsequently to the following signs of the grades, so as to elucidate after what manner the grades stand in the elements. Platearius, Dioscorides, Serapio, and others, their followers, who have written much, but falsely, of the quintessence, do not signally differ from us herein. Yet do you, whoever you be, seek a knowledge of the quintessence from experience, for thus you will understand the grades in their division. That the manner in which diseases are repelled by the quintessence may become clear, we must first diligently notice the concordance of things in diseases. For some virtues contend only in synochia, others in mania, others in aclitis, and yet others in lethargic complaints, as is the case with concordances. In this place I think it worth while to know that which lies hidden in Nature, as in gelutta (carlinum) and melissa, which renew and remove disease without any virtue of grades, namely, in the restoration and repair of youth. The manner and the efficacy by which these things are done are indicated in the treatise, *De Vita Longa*, as certain peculiar mysteries which exist in the nature of things besides arcana. Wherefore, I think proper to pass them by here, and at length continue what I have begun concerning the four grades of the elements. Hence, although there be many and various virtues which cure maladies, some through their aperient nature, others through their narcotic nature, others again by other means, I leave such matters to those who devote their attention to theorems.

CHAPTER IV

Everything which strengthens is tempered. No hindrance will arise from the substance which, although it be cold or hot, will, however, not incommode the Quintessence in its body (*others read*, in its work). Moreover, every specific is a quintessence, without any corruption of its body. Furthermore, nothing is tempered except the Quintessence; all bodies are elementated in nature and their accident.

GRADE OF HEALTH

Those things which proceed out of the earth hold the first grade of health.

- Herbs
- Seeds
- Roots
- Sponges
- Animals
- Flowers
- Barks
- Fruits

of all kinds.

Things which proceed from the air hold the second grade.
These are all kinds of winged creatures.
Things which proceed out of the water hold the third grade, as:—

- Metals
- Marchasites
- Cachimiæ
- Salts
- Minerals
- Resins of Sulphur
- Fishes
- Gems
- Stones

of all kinds.

Things which proceed out of fire hold the fourth grade, as the Tincture and the Philosophical Stone.

However, there are certain other virtues to be noted which are concealed in herbs, but not in winged things, nor in metals, as ursina and white thistle indicate, which, beyond their grade, admit foreign virtues. Among these there is also the emerald, which admits a foreign efficacy into itself, yet such in no wise conduce to health, for they are only external virtues which have no internal effect whatever.

CHAPTER V

Enough having been said of confortatives, we will now turn to laxatives and their grades. Accordingly, first observe that we shall not here make use of that classification whereby the laxatives are divided into four natures. They are described in this fashion according to ancient custom. The coloquintida and scammony purge cholera; turbith and hellebore purge phlegm; manna and capillus veneris purge the blood; while lapis lazuli and black hellebore purge melancholy.

Moreover, there are others also which ward off *cholera vitellina*, others which ward off the rust-coloured and yet others the citrine-coloured water of dropsical subjects, with things of like kind, as elsewhere has been sufficiently described.

CHAPTER VI

As in the former chapter I made mention of the grades of laxatixes, so in this place, to impress it more deeply on the mind, I repeat the same—namely, that laxatives in no wise follow the four grades of the elements, but they have their grades mixed without any respect to the same. Wherefore, more diligent attention must be paid to the nature of the disease, lest you should carelessly misuse the confortatives designed for its cure. These should rather be accommodated so that they may agree with the nature of the disease. The grade and the disease should also be invariably compared. But lest with unwashed hands, as the saying is, we should rush in upon this question of purgations, it is needful to proceed as follows, namely, observing that the functions are sometimes unequal in the same operation in the fourth grade, as hellebore sometimes removes that which sea-lettuce cannot, and in like manner cataputia where both the above would fail. At one time precipitate, at another esula, and at yet another cassia, will prevail in the removal of fistula. Moreover, if we speak of fevers, such a laxative as centaury will occasionally purge febrile humours, and hellebore, another laxative, will be of use in an epileptic complaint. So, also, agaric, and things of this kind, will prevail in the case of worms. The reason is to be sought in Nature, not in the humours, for Nature has been equipped to remove whatever is melancholic, choleric, or phlegmatic, or, indeed, anything which could be mentioned here.

CHAPTER VII

Note the following things for very vehement and very gentle purging.

I	II	
Polypodium	Mountain Osier	Lazuli
Locusta Botim	Cyclamen	Scammony
Hairs of Venus	Turbith	Centaury
Turpentine	Azarabachara	IV
Locusta Sambuci	Hermodactylus	Either Hellebore
Senna	III	Coloquintida
Gamandrea	Rhubarb	Sea-lettuce
Stomachiolum	Esula	Serapinum
Locusta Ebuli	Vitriol	Cataputia
Succory	Diagridium	Præcipitate
Serum of Milk	Agaric	

CHAPTER VIII

Observe the following things concerning incarnatives and consolidatives. They contain in themselves the four grades, while the consolidatives, in the same manner as the laxatives, exclude the elements. In the first place, therefore, we must observe the manner wherein the ailments which we desire to heal stand in their grades. For out of these proceeds the grade of natural things. Now, some heal fractures of bones, others cure wounds, others ordinary ulcers, others eating ulcers (others fleshy). Hence arise four grades in the following fashion.

I

Broken bones are cured by Alchimilla, Periwinkle, Perfoliata, Diapensia, Aristolochia rotunda, Consolida, Serpentina.

II

Wounds are healed by Natural Balsam, Artificial Balsam; the oils of Hypericon, Bullace, Turpentine, Laterinum, Centaury, Anise, Benedictus; apostolic plasters and unguents; apostolic powders; potion for wounds.

III

Imposthumes and common ulcers are cured by Emplastra Gummata, Emplastra Mumiata, Emplastra Apostolica, and Unguenta Apostolica.

IV

Cancrous and eating ulcers are healed by composition of Mercury, of Brassatella, and of Realgar.

CHAPTER IX

There are, moreover, others besides the above which equally possess their own grades, of which kind are poisons, wherein the grades should indeed be specially observed. First, by reason of their elementated nature, they should be admitted into the composition of recipes. At the outset, therefore, have regard to the quantity of the poison; the weight must then be prepared, and that in the following way.

POISONS IN THEIR GRADES

I

Simples by themselves: Colcothar and Alum.

II

Reverberated: Spirit of Jupiter and Spirit of Saturn.

III

Calcinated: Tartar and Scissum.

IV

Sublimated: Arsenic and Mercury.

The other species of poison, as those of the spider, toad, scorpion, lizard, and serpent, as also the small dragon, among many, I pass over because they are not ingredients, except the Tyrian poison, which might be named. There are, moreover, those which provoke the courses in women, which also, being specially adapted for this purpose, may be placed among other recipes, according to the manner of their grades. There are others which suppress tumours, some which provoke the flow of urine; all these and their like are to be sought from experience and concordance. Now for the Grades of the Spagyrists.

CHAPTER X

Out of the spagyric industry four grades precede in the same manner with the four elements, and so surpass the other grades in their quantity. Further, wherever the last grade comes to an end, there the first point in the spagyric grades begins, and after this manner.

I

Oil derived by distillation from all herbs, roots, seeds, resins, gums, fruits, fungi, and tree mosses.

II

Oil of the vulture, the dove, the heron, the crow, and the magpie.

III

Water of vitriol, liquor, mercurial water, oil of quicksilver, viridity of salt, aluminous waters, calcined oils, oils of metals, liquors of gems, potable gold, essence of antimony.

IV

Oil of crystal, oil of beryl, tincture, stone of the philosophers.

All these are hot, for the grades remove that which is elementated, and over that which is element they advance their own grades. Therefore, to become acquainted with those grades there is needed full and perfect experience, so that you may see the preparation of these things which proceed out of the elementated, where they surpass the elementated.

Things which proceed out of the earth occupy the first grade of the Spagyrists, as, for example, out of the seeds of Anise, Juniper, Cardamum, Clove Tree; out of the roots of Jusquiam, Repontic, Angelica, Masterwort; out of the woods of Ebony, Juniper, Sandal.

Things which proceed out of the air occupy the second grade, as, for example, out of the fruits of Nuba, Ilech, Tereniabin; out of winged creatures, as the Phœnix, the Eagle, and the Dove.

Things which proceed out of the water occupy the third grade, as, for example, out of metals, as Gold, Mercury, Silver, Copper, Lothon, Iron, Lead, Tin, Electrum, Sapphire, Smaragdum, Granate; out of gems, as the Topaz, the Ruby, the Hyacinth, the Amethyst, and Corals; out of minerals, as Marcasite, Cachimia, Talk, Realgar, and Vitriol; out of salts and alums.

Things which have their origin from fire hold the fourth grade, as out of the Beryl, the Crystal, and Arles.

And the things which descend from the above-mentioned four elementated substances, as out of the earth, Water of Life, Distilled Balsam, Circulated Waters, Distilled Liquors; out of the air, Distilled Birds, also Tereniabin, Cloud, Ilech, distilled by retort; out of the water, Potable Gold, Sublimates, Resolutes, Liquor of Silver, Calcinates, Congelates, Resolution of Mercury, Reverberates; out of the fire, Liquor of Crystal, Liquor of Beryl, Liquor of Arles.

CONCERNING
DEGREES AND COMPOSITIONS IN ALCHEMY

BOOK THE FOURTH

CHAPTER I

THOSE herbs which are of a frigid nature and from the earth, are not altogether adapted to all diseases which are of a warm nature, nor, again, are those herbs which are of a warm quality adapted in all cases to diseases of a frigid nature. Hence, seven genera of diseases and seven genera of heats and colds are distinguished, and among them those of the heart as well as other members. This difference, therefore, should be very carefully observed, that those things which are wanting to the liver, whether it be warm or cold, may be sought from the same herbs. So, also, those things wherein the cerebrum is deficient require their special herbs. However, although herbs in general are either cold or hot, yet those which are for the spleen are in no wise appropriate for diseases of the reins. Wherefore, after an aquaintance with the grades there is required that of the difference between herbs in the manner following.

CHAPTER II

In the first place, herbs are divided into seven species, together with the rest of the elements, and this on account of the nature of the star, which, equally with these, is divided into seven species. Further, in the same way as they admit of a sevenfold division, the body also is subject to the same classification, and they correspond one to another, so that those things which are under the sun are appropriate to the heart, and these are twofold; while those under the moon are, in like manner, appropriate to the brain, and that in either grade. Those things which are under Venus are heating to the reins; those things which are under Saturn strengthen the spleen; those which are under Mercury defend the liver; those under Jupiter have regard to the lungs; finally, those subject to Mars are considered wholly adapted to the gall. Moreover, though herbs are not regulated together with simples of the planets, nor the planets regulated by them, there does certainly exist a singular supremacy in every element without mixture of another.

CHAPTER III

In order to become acquainted with those elements which pertain to the heart, we must, in the first place, observe that whatever regenerates is akin to the heart, as gold, balm, nuba, etc. Moreover, whatsoever removes phlegm, of which kind are the rose, camphor, musk, amber, etc., are brought to the brain through the medium of their native fragrance. In like manner, whatsoever heats or cools the blood is serviceable to the liver. What provokes the flow of urine or increases the semen benefits the reins. That which preserves long life benefits the spleen; that which purges, the stomach. Experience gives acquaintance with these things, but it is rather the experience which is derived from philosophy than from medicine, from regeneration rather than from disease, even that which is produced from transmutation. For when both medical and philosophical experiments concur there is derived a genuine diathesis of everything.

CHAPTER IV

When, therefore, you have become acquainted with the transmutation which indicates seven species, both of the hot and cold, you must observe that whatever regenerates or expels an evil growth, and purifies or restores the matter to wholeness, and thus to an incorrupt state, comes to be included under the same species, whether it arises from the heat or cold of the elements. Moreover, everything in transmutation consumes superfluous humours, as salt removes leprosy of the moon, whence it is a most speedy remedy for the cerebrum. In this place you may observe that the herbs are not to be administered in this fashion, because they are Lunar, but because they reduce and compel Lunar things into their power. By silver or the Moon the brain is in no wise healed, but by those things which are opposed to these. Moreover, whatsoever prevents rubefaction and putrefaction, and conserves into an essence, as fixed things which obtain in the transmutation of metals, in the same manner preserves the spleen incorrupt. Similarly whatsoever resolves a substance and a body into liquor strengthens the liver and expels that which is repugnant thereto. But whatever dissolves to such an extent that the contraries are separated from one another, is beneficial to the stomach. Of this kind are the alkalies in tin. Finally, whatever prepares things and renders them suitable for the augments of transmutations—of which kind are the conjunctions of arcana—is to be used above all. Seek an experience of these things from the transmutation of Nature, but waste not your whole life in those miserable grades, nor in the profitless catalogues of herbs, which are found in senseless codices. For these are inimical rather than beneficial to the stomach.

CHAPTER V

The following table will indicate the manner in which the seven aforesaid species are distinguished in the four elements, namely, which have their

Concerning Degrees and Compositions

origin from earth, which from air, fire, and water. Hence you may judge concerning the method of composing recipes, as follows :—

Those which come from the earth, and are of a hot nature :—

THE CEREBRUM, *Viriditas Salis*, Liquor of Vitriol, Liquor of Lunaria. THE HEART, Essence of Melissa, Quintessence of Gold. THE REINS, Correction of Sibeta, Essence of Satyrion. THE LIVER, Liquor of Brassatella, Liquor of Manna, Xylo aloes. THE SPLEEN, Mystery of Black Hellebore, Mystery of Valerian, Mystery of Verbena. THE CHEST, *Extractio de Pulmone*, Extract of Tree Moss. THE GALL, Quintessence of Chelidony.

Those which come from the earth, and are of a cold nature :—

THE CEREBRUM, Essence of Geloen, Essence of Anther. THE HEART, Matter of Laudanum, Matter of Pearls, Matter of Sapphires. THE REINS, *Materia Sintocorum*, Matter of Lettuce Seed. THE LIVER, Liquor of Senna, Quintessence of Blood, Quintessence of Gamandrea and Cichorea. THE SPLEEN, *Compositio Candi*, Confection of Dubel Coleph, or Dubelteleph. THE CHEST, Matter of Dew, Matter of Sulphur, Matter of Ologan. THE GALL, Composition of Agresta (verjuice), Composition of Pomegranate Flowers.

CHAPTER VI

Those which come from the air, and are of a hot nature :—

BRAIN AND HEART, Nuba, Symona. REINS AND LIVER, Ilech, Hallereon. SPLEEN, CHEST, AND GALL, Tereniabin.

Those which come from the air, and are of a cold nature :—

BRAIN AND HEART, Halcyon. REINS AND LIVER, Crude Ilech. SPLEEN, CHEST, AND GALL, Crude Arles.

CHAPTER VII

Those which come from the water, and are of a hot nature :—

BRAIN, Oil of Mercury of the Moon, Essence of Silver, Essence of the Sixth, *i.e.*, of Venus. THE HEART, Potable Gold, Liquor of the Sun, Oil of the Seventh, *i.e.*, of Saturn. THE REINS, Essence of Vitriol, Quintessence of Sulphur, Flower of Venus. THE LIVER, Mystery of Mercury, Mystery of Antimony. THE SPLEEN, Mystery of Asphalt, *Rubedo de Nigro*. THE CHEST, Flower of Jupiter, Extract of Tin, Resolved Talc. THE GALL, Crocus of Mars, Topaz from Iron.

Those which come from the water, and are of a cold nature :—

THE BRAIN, Juice of Amethyst, Liquor of Granates, Composition of Gems. THE HEART, Composition of both Marcasites, Composition of White Talc. THE REINS, The Tincture, The Physical Stone. THE LIVER, Spirit of Saturn, Essence of Lead. THE SPLEEN, Mystery of Coagulated Mercury. THE CHEST, Flower of Crude Jupiter. THE GALL, Dust of the fifth metal.

CHAPTER VIII

Things which come from fire, and are of a warm nature :—
 Warm Nostoch.
Things from the same element, but of a cold nature :—
 Arcana of Crystal, Mastery of the Beryl, Citron Liquors.

CHAPTER IX

When you have become acquainted with the grades and their species, you must then advance to the composition of recipes, according to the direction of the rule following. For as there are four elements, so four recipes are to be prepared. There are some kinds of diseases which require earthy remedies, more require atmospheric, others aqueous, and yet others igneous. In the first place, therefore, you must notice the diseases in the seven members, among which the elements are distributed. Hence simples are to be extracted of which you may prepare a composition according to the nature both of the grades and the species thereof. Accordingly, with elementated diseases, as, for example, terrene, a composite is not to be prepared higher than its own grade, but is to be left in that same grade. Similarly, with regard to the atmospheric, nothing out of a foreign element is to be introduced. Judge in this same manner concerning the other elements. This, then, is the crucial point, to accommodate each disease to its proper element, for hence arises that common error which is continually recrudescent in the case of gout, paralytic diseases, and others of this kind, by reason of the preposterous healing method which is adopted by unskilled men. Take epilepsy as an instance ; a species hereof is subject to the element of water, wherefore it is to be healed by means of minerals.

CHAPTER X

Take general rules for the composition of recipes as follows. All those which are prepared for elemental diseases consist of six things—two of which are from the planets, two from the elements, and two from narcotics. For although they can be composed of three things, one out of each being taken, yet, these are too weak for healing purposes. Now, there are two which derive from the planets, because they conciliate and correct medicine ; two derive from the elements, in order that the grade of the disease may be overcome. Lastly, two are from the narcotics, because the four parts already mentioned are too weak of themselves to expel a disease before the crisis. Observe, then, concerning composition, to forestall the critical day. Recipes prepared in this manner are very helpful for diseases in all degrees of acuteness.

CHAPTER XI

Lastly, concerning weights, observe the following rule. Observe the grade, whether it be surpassed by the medicine, or whether the medicine agree with the grade. But in order that the three species may not corrupt one another, dispose the weights as follows: The proportions from the planets should be as four, from the elements as three, and from the narcotics as one.

CONCERNING GRADES AND COMPOSITIONS*

BOOK THE FIFTH

CHAPTER I

IN the prescription of recipes divide the disease into four species, and distribute these among the elements, taking the grade which occurs, and proceeding in the following manner: If the disease passes from one grade into another, take the same grade, for thus are healed diseases of the first grade, which are of the earth; of the second grade, which are of the air; of the third grade, namely, of the water; while that which attains to the fourth grade must be cured by the Tincture alone, for otherwise there is nothing which which can be accommodated to this case. Moreover, although in the fourth book I have prescribed that the recipe should be constituted of six parts, it could, notwithstanding, be confected of three, or the six might be doubled. Again, they might be distributed as follows: Of those which are of the planets take four drachms; of those which are of the elementated take three drachms; of those which are of the narcotics take one drachm. Thus the matter consists in the weight, not the number of the simples.

Again, the strength, and what is more, the effect of those things which are admitted into the description of this recipe, are referred neither to the weight nor to the recipe, but to the dose, so that in those things which are of the planets you obtain greater efficacy than with the elementated things. But this is of the dose, not the weight or the recipe. Wherefore the above method is to be observed. Finally, signal skill in medical matters is herein required, to avoid premature application of the healing process; you must rather proceed so as to purge where there is need of purgation, heal where healing is required, and consolidate where consolidation is necessary.

CHAPTER II

You must know that everything which comes to be tested by Nature pertains to that subject which the physician makes his province. That only which is of ocular demonstration is to be considered in doses. Every dose,

---

* The treatise *Concerning Grades and Compositions in Alchemy* consists of seven books, but of these the last three are so much outside the purpose of this translation that they have been compressed into a very small space. All the information which the Hermetic student is likely to require on the subject of complexions and qualities is embodied in the first four books.

according to its proper arete, is either hot or cold. With regard to the composition of recipes, neither the humid nor the dry is to be considered in doses. As there are only two complexions, so there are only two doses. Whoever is acquainted with the grade of hot or cold, will know that there is joined thereto not only dry and humid, but also a dry resolute and a humid coagulate. No arcanum or aniadus resides in the warm or dry, inasmuch as no disease occurs which only seeks one of these. The chief point is in the hot or cold, for that diathesis dominates either in the hot or the cold. The sole inclination of every disease is that the physician should simply observe whether it be hot or cold. Every grade is the dose of its disease, and from every grade should the dose be taken, as may be understood by the comparison of fire, which has only one grade, and yet it is abundantly sufficient to consume its contrary, which is, indeed, according to heat. So every disease has its own grade, neither more nor less. The dose has a like relation to each and every disease. But natural things are not graded equally as to the disease in the matter of the dose; each has a grade equal to its disease, and that is the grade of the dose.

CHAPTER III

Since there is only one grade, and nothing is graded higher in warm or in cold, equality is the chief help to ascertain the dose. There is one grade in disease and in things of Nature. No disease becomes worse because the grade of its medicament is higher. It becomes worse only according to the capacity of its nature. The extent of the disease regulates the amount of the dose. Wherefore the physician must know with what weight the disease is loaded, for the dose demands for it the same weight of medicine. The weight, therefore, and not the grade, is to be administered. Herein consists the chief principle of administering any dose: this ought to proceed from the number, not the body of those things. The end is that the ares of the microcosmus, and not the medicament, should effect the cure. As soon as the disease has been brought to an equality, Nature herself cures that which is contrary to her. The quantity of the dose must therefore not exceed the number which is taken from the disease. There are twenty-four numbers in Nature, and within this number the medicine should also be confined. In the anatomy of Nature there are twenty-four *minuta* of diseases. Medicine, therefore, has twenty-four lotones. The physician should administer his medicines with reference to these two series, so as to produce the same number of each in the microcosmus. When this is done Nature will cure the sick. The absence of such an equilibrium will sometimes cause death, when the disease itself has run its course.

CHAPTER IV

While any disease in itself is one, it has, as we have indicated, twenty-four numbers, degrees, or minims, and the lotones of the medicine must correspond. The proportion of the dose to the disease in any particular case cannot be learnt from theory, but is gathered from experience only. The anatomy of the dose must correspond to the anatomy of the disease. It must not exceed the number twenty-four. Indeed, the object is to restore health, both to the nature of the Microcosm and to that of the external elements, when these agree in the body. The conjunction is the same as zinobrium, which is graded by minium. In that elementated exaltation, they afford their own exaltations to the virtues of the Microcosm, and so the grain passes into the scruple, the drachm, and loton, some, indeed, into the pound, some also into the kist, and others into the talent.

HERE ENDS THE BOOK CONCERNING DEGREES AND COMPOSITIONS

CONCERNING PREPARATIONS IN ALCHEMICAL MEDICINE*

TREATISE I

CONCERNING ANTIMONY AND MARCASITE OF SILVER.
CONCERNING WHITE AND RED CACHIMIA.
CONCERNING TALC, FLUIDIC AND SOLID.
CONCERNING THUTIA, CALAMINE, AND LITHARGE.

CONCERNING ANTIMONY

The virtues of Antimony obtain in Morphew, Leprosy, Elephantis, Wounds, and Ulcers.

HERE FOLLOWS THE PREPARATION OF ANTIMONY FOR THE SEVERAL SPECIES OF LEPROSY

℞   Of the best pounded Antimony, lb.j.
      Of highly distilled (*sic*), lb.iiij.
      Of crude white Tartar, lb.ss.

Reduce to a fine powder in a phial, distil by retort, and a red oil will result.

The preparations of Antimony vary with the diseases for which it is administered. That which is used for wounds differs from that which is applied in the case of leprosy. And so of the rest. To take the same preparation of Antimony both in wounds and in leprosy would be a serious error.

ELEPHANTIS

The preparation of Antimony for Elephantis is, however, the same as for leprosy.

THE PREPARATION OF ANTIMONY FOR MORPHEW.

℞   Of the best pounded Antimony, lb.ss.
      Of calcined Tartar ⎫
      Of Alum   ⎬ an equal quantity:

Arrange in alternate layers; reduce in a fire of reverberation of the fourth grade; then distil, and a thick red oil will come over. By alternate layers

---

* It is in every way highly desirable that this important collection of Hermetic prescriptions should find place among the Hermetic Medicine of Paracelsus. It is of very considerable value as evidence of the extent to which mineral, and especially metallic, substances were applied by Theophrastus in all varieties of disease. The author's intention seems to have been the compilation of a whole alchemical pharmacy, and there is a small fragment extant of a second book, under the title *De Nascentibus ex Terra*.

understand one layer of the Tartar and Alum, afterwards a layer of the Antimony, and so forward.

The Preparation of Antimony for Wounds

℞ Of Antimony }
Of calcined Tartar } an. lb.ss.
Of Alcool of Wine, 1 kist.

Mix; distil by the alembic till the matter is resolved.

℞ Of the substance dissolved as above, ʒj.
Of Alcool of Wine, ʒiiij.

Dry by coagulation, and reduce into oil on a marble slab.

There is no greater cure for wounds than that which is obtained from Antimony, except in wounds of the head. The Antimony should be distilled upwards till what is below becomes aqueous.

The Preparation of Antimony for Ulcers

℞ Of Antimony }
Of Colcothar } any equal proportion.
Of Flos Aeris }

Reduce S.S.S. to the grade of reverberation. Make an extract with red wine, and reduce into alkali.

The said alkali made into an unguent with olive oil and laid over ulcers is of great healing virtue.

Additions of Antimony for Leprosy

℞ Of the said Antimony, ʒj.
Of Oil of Wine Fæces, ʒj.
Of Oil of Bitter Almonds, to the weight of both the above.

Mix. If there be no hoarseness of the voice, anoint once or twice every seven days. If the voice be hoarse, it is useless.

Addition in Morphew

℞ Of the said Antimony, ʒj., with Kist, *i.e.*, Alcool of Wine.
Of Tragagantum, ʒij.
Of consolidated Royal Mucilage }
Of Psyllus Seed } each ʒij.
Of Gum Arabic }

Make an unguent. The process is the same for Morphew and Alopecia. The unguent is to be applied warm, once or twice in every seven days. By this means the scab will rise up. If it peels off, the ulcer may be healed with the following unguent.

℞ Of Spermiola }
Of Camphor } each ʒj.ss.
Of Oil of Ceruse to the weight of both.

Make an unguent. After the scab has come off the place must be anointed with this unguent every eight days.

Addition for Wounds

℞ Of the said prepared Antimony, ʒvij.
Of the juice of White Tartar ⎫
Of Oil of Myrtles ⎭ each ʒv.

Mix. Apply once every other day, and no mischance need be feared. Note.— In summer, add camphor at pleasure.

Addition for Ulcers

℞ Of the said prepared Antimony, ʒiij.
Of Oil of Colcothar, ʒss.
Of Oil of Mastic (*Oleum Lentiscinum*) to the weight of both.

Make an unguent. Anoint around the ulcer. This does not cure cancer, elephantis, or esthiomena. *Oleum Lentiscinum* is oil from the bark of the mountain osier.

CONCERNING LITHARGE

The virtues of Litharge obtain in Cancer and Fistulas, in Tentigo Prava, Esthiomensis, Red Jaundice, Persian Fire, and Wounds.

Here follows the Preparation of Litharge for Cancer

℞ Of pounded Litharge, lb.ss.
Of Salt Water ⎫
Of Alum ⎭ each lb.j.
Of White Vinegar, lb.iiij.

Reduce over hot coals till their moisture is consumed. Take of the said Litharge with an equal quantity of spring water: reduce *ad colores* for a night, and dry.

The same preparation of Litharge prevails in fistulas.

Preparation for Esthiomensis

℞ Of Litharge, lb.j.
Of Calcined Tartar, lb.ss.
Of Spring Water, or *Aqua Fuliginis*, a sufficient quantity.
Of common melted Salt ⎫
Of Rock Alum ⎭ each ʒvj.

Reduce according to the fourth grade of reverberation with the aforesaid water into an alkali.

Aqua Fuliginis is water from sooty roofs obtained during rain.

Preparation for Red Jaundice

℞ Of Myrrh ⎫
Of Frankincense ⎭ each ʒj.
Of Litharge, ʒiiij.
Of very strong Vinegar, lb.ss.

Reduce into a decoction.

PREPARATION FOR WOUNDS

℞ Of Litharge, with four times Whitened Vinegar, lb.ss.
Of the juice of the herb Pellitory ⎫
Of the lesser Comfrey ⎬ an equal quantity.
Of the round Aristolochy (Birthwort) ⎭

Make a compound with *mucilago lumbricata*.

PREPARATION OF LITHARGE IN TENTIGO PRAVA

℞ Of washed Litharge, 1 lb.
Of Rock Alum, lb.jss.

Mix. Pound well, place in the fourth grade of reverberation for four hours, then extract the alkali with spring water, together with the remaining litharge and rock alum in equal quantity. Pound as above till all the litharge has been used.

The process should be as follows: When the litharge has been placed for four hours in rock alum, take of this distilled alkali, of fountain water, and of soot, each half a pound, and mix together.

PREPARATION OF LITHARGE FOR PERSIAN FIRE

℞ Of Litharge, lb.j.
Of Red Realgar, ℥ij.
Of Sal Ammoniac, ℥ss.

Mix, and place in a sublimatory. This must be done ten or twelve times. Then pour on warm water, and let the litharge be separated.

ADDITION IN TENTIGO PRAVA

℞ Of the said Litharge, ℥j.
Of Common Realgar, ʒj.
Of the juice or water of Chelidony, a sufficient quantity.

Make into an unguent, anoint very thinly, and apply four or five times. The skin turns red, and the rank smell goes away. Then use the following recipe:—

℞ Of the said Litharge, ℥jss.
Of mucilage of Fœnugrek ⎫
Of Lumbrici Nitri ⎬ an equal sufficient quantity.

Make into an unguent. Lumbrici Nitri are worms found in dung.

ADDITION IN WOUNDS

℞ Of the said Litharge, ℥iiij.
Of Oil of Camphor, ℈j.
Of Crocus of Mars, ℈iiij.

Make into an unguent.
Apply to the wound once or twice daily, and rub in well.

ADDITION IN ESTHIOMENSIS

℞ Of the said Litharge, ℥iiij.
Of Powder of Chelidony ⎫
Of Oak Apples ⎬ each ℥ij.

Reduce to a powder.

℞ Of the said Litharge, ʒiij.
Of Mucilage of Consolida ⎫
Of Lumbrici Nitri   ⎬ each a sufficient quantity.
Of Oil of Myrrh   ⎭

Make into an unguent. The disease is cured thereby.

Addition in Cancer

℞ Of the juice of Marrubius ⎫
Of Persicary   ⎬ each ʒj.
Of prepared Litharge, ʒij
Of Oil of the Yolk of Eggs, a sufficient quantity.

Compose an unguent.

Addition in Red Jaundice

℞ Of prepared Litharge, ʒss.
Of Rock Alum, ii.oz.
Of Salt Water, ʒjss.

Addition in Persian Fire

℞ Of Elect Vitriol, ʒiiij.
Of Oak Apples, ʒss.
Of Frankincense, ʒj.
Of prepared Litharge, to the weight of all.
Of Wine and Vinegar, as may be wanted.

When it boils (? ferments) then it is to be used, and the more it boils the better it is.

CONCERNING MARCASITE

Gold or silver Marcasite has angles like tiles. The virtues of Marcasite are in Restriction of the Blood, the Menstrua, and Hemorrhoids.

Preparation of Marcasite

℞ Of Marcasite, ʒiiij.
Of Pitch   ⎫
Of Colophony ⎬ each ʒvi.
Of Resin of the Fir, to the weight of all.

Reduce to calx.

To reduce into calx is to place in a brick kiln and burn till the resin flows out twice or thrice until it glows.

Preparation in Restriction

℞ Of best pounded Marcasite, ʒij.
Of Oil of Flax, ʒvij.

When these two are conjoined and set on fire, the true matter will remain.

Preparation for Hemorrhoids

℞ Of Marcasite, ʒjss.
Of best dried Alcool of Wine, lb.j.

Mix.

Addition in Restriction of the Blood

℞   Of the said Marcasite, ℥j.
      Of Corals, ℥ss.
      Of Plantain Seed, ℥ss.

Reduce to a fine powder.

The powder is to be sprinkled upon the wounds, or mixed with vinegar, and be bound up below the wound; thus it will hold it together. Those who, by reason of an accident, bring up blood, should drink it.

Addition for Menstrua

℞ ℥ss. of this Marcasite, and as much as suffices of Oil of Sandarach. Make into an ointment.

If the menstruum flow to excess, let the umbilicus be anointed twice or thrice.

Addition for Piles

℞   Of the said Marcasite, ℥iij.
      Of Sal Gemmæ ⎫
      Of Mumia   ⎭ ℥j. each.

Make into a powder.

The swollen piles must be cut and then anointed.

CONCERNING CACHIMIA

The virtues of Cachimia obtain in Dysentery, Diarrhœa, and Lienteria.

Preparation for Dysentery

℞   Of Cachimia well ground, ℥vj.
      Of Rust of Iron, ℥ss.

Reduce by the second grade of fire for six or seven hours. Afterwards take out and reduce into alkali.

Preparation for Diarrhœa

℞ Of prepared Cachimia as above, and of Nutmeg what is sufficient for incorporation. Reduce to the second grade in the form of a bolus.

Preparation for Lienteria

℞ Of Cachimia prepared as above, and of Gum Arabic dissolved in plantain water. Make a bolus. Reduce to second grade.

Addition for Dysentery

℞   Of the said Cachimia, ℥ss.
      Of Boiled Dove, a sufficient quantity.

Addition for Diarrhœa

℞   Of the said Cachimia, ℥j.
      Of Theriaca, ℥iij.
      Of Sealed Earth, ℥ss.

Make a bolus, the dose containing from ℥j. even to ℥ij.ss. Administer morning, noon, and evening. Abstain for three days. Afterwards repeat it. Do this thrice.

Addition for Lienteria

℞ Of prepared Cachimia, ʒj.
Of Crocus of Mars, ʒij.
Of Red Corals, ʒss.
Of Theriac, as required.

Make a bolus; the dose is from ʒij. even to iij. or iiij. morning and evening. Let a portion of this be administered daily.

CONCERNING THUTIA

The virtues of Thutia are for spots in the eyes.

Preparation for Spots in the Eyes

℞ Of Thutia, ʒj.
Of White Vitriol } equal quantities.
Of Juice of Eyebright

Make into a bolus with Gum Arabic, and bring to the second grade of fire. It becomes an unguent beneficial to the eyes.

Preparation for White Spots in the Eyes

The Thutia must be extinguished in milk and placed for the night in rose water. This water removes the white speck when applied to it.

For Wens

℞ Of Thutia, ʒiiij.
Of Fused Salt } each ʒvj.
Of Live Calx
S.S.S. ʒvj.

Arrange S.S.S. Apply fourth grade of fire; reduce into alcali.

Addition for the Spots of the Eyes

℞ Of the said Vitriol, ʒss.
Of Spawn of Frogs, ʒij.
Of Laterine Oil, Ɔss.

Make an eye salve. If yellow spots appear in the eye and sparkle, they return.

Addition for Wens

℞ Of the said Thutia, ʒj.
Of White Vitriol, ʒvij.
Of pounded Camphor, Ɔjss.

Make a mixture with water of roses or fennel. This disease attacks all pedal animals. The wen assails the eyes of goats, other animals, and men. In the case of human beings camphor must be administered with it, lest inflammation supervene.

Addition for Wens

℞ Of the said Thutia, ʒj.
Of Sal Anatron, or Gall of Glass } ʒj. each.
Of Fused Salt
Of distilled urine, lb.ss. and ʒiiij.

Mix. The process consists in administering this potion to such as are afflicted with wens, in the morning and in the evening, for three or four weeks. This medicine removes all wens, except *grisonum*.

CONCERNING TALC

The virtues of Talc obtain in ulcers and humid wounds.

Preparation for Wounds.

℞ lb.j. of Talc and an equal weight each of Cinder of Beans and of Oats. Reduce at the fourth fire for a day and night; cleanse and dry. Talc dries the bottom or base of the wound, so that it does not change into a fistula. It also powerfully desiccates ulcers. It must not be used beyond the third day.

Addition for Wounds and Ulcers

℞ Of the said Talc, ℨj.
Of Liquor of Mumia
Of Washed Turpentine } each q.s. for an unguent.

Cures eating, cancerous, and other suppurating ulcers.

CONCERNING CALAMINE

The virtues of Calamine are suitable for plasters, eye salves, and the Persian fire. Add Calamine both for ulcers and for wounds, for plasters where a growth of fresh flesh is required, also for eye salves where neither albugo or spots of the eyes are present, as in the case of red eyes, where the greatest experience is requisite.

Preparation of Calamine for Plasters

℞ Of Washed (that is pure) Calamine, ℨj.
Of Colcothar
Of Live Sulphur } each ℨj.ss.

Arrange in layers and apply the fourth fire for a day and night. Reduce by the second ablution.

Preparation for Unguent

℞ Of the said prepared Calamine, ℨiij.
Of the Oil of the Yolk of Eggs, ℨj.

Make a bolus with gum arabic; reduce at a fire of the second grade for four hours, then reduce by ablution.

For the Eyes

℞ Of the said prepared Calamine, ℨj.
Of Distilled Vinegar, ℨvj.

Make an extraction; then dry it.

Preparation for Persian Fire

℞ Of Crude Calamine, lb.ss.
Of Water of Water Lily, ℨvj.
Of Alumen Plumosum, ℨss.

Reduce these by digestion in a glass for the space of a week, and distil.

It is a recipe of Geber. Petrus de Archilata errs in the process: for the medicine is to be used for pandaricia, but not for combustions.

℞ Of Apostolic Plaster, ʒvj.
Of the said Calamine, ʒss.
Of Camphor, ʒj.

Make a plaster.

For Unguents

℞ Of Agrippine Unguent, ʒiiij.
Of Unguent of the Flower of Copper, ʒss.
Of the said Calamine, ʒx.

Make a mixture. Most excellent for ulcers, itch, and scab.

Addition for Eye Salve

℞ Of the said Calamine, ʒj.
Of Water of Eyebright ⎫
Of Water of Fennel   ⎬ ʒiij.
Of Water of Roses   ⎭

To be applied at night.

Addition for Persian Fire

℞ Of the said Calamine, ʒvj
Of Waters of Vitriol and Oak Apples

Some use cobblers' atrament for Persian fire and red jaundice.

TREATISE II

CONCERNING BLOODSTONE, ARSENIC, SULPHUR, SAXIFRAGE, ORPIMENT

CONCERNING BLOODSTONE

The virtues or chief arcana of Bloodstone are for bloody ulcers, resolved menstrua, premature profluvia of the matrix, lax dysentery, diarrhœa.

PREPARATION FOR BLOODY ULCERS

℞ ʒiij. of Bloodstone, and ʒiij. each of *lutum Lephanteum* (that is, clay from which small cucurbits are made), and of Bolus Armenus. Make a bolus with traganth dissolved in vinegar: Reduce by the fourth grade of reverberation; then extract the alkali.

In the case of wounds, of lupus, and of herpata, bloodstone proves extremely beneficial. It binds the veins so that the flow of blood ceases. Let it be sprinkled upon the parts.

PREPARATION OF BLOODSTONE FOR MENSTRUA

℞   Of Bloodstone, ʒiiij.
     Of Mastic dissolved, ʒx.
     Of Carabe, ʒjss.

Make a mixture with a decoction of water of alum; reduce by ablution.

The flow of menstrua should be checked when it causes pallor in the face. The use of this preparation is safe, and effects a complete cure. After decoction for seven hours lute is produced from bloodstone, out of which trochisks are made for menstrua.

THE PREPARATION FOR IMMATURE MENSTRUA IS AS FOLLOWS

℞   Of Bloodstone, ʒj.
     Of Oil of Nutmeg   ⎫
     Of Oil of Grains of Actis ⎬ ʒiij.
     Of Petroleum   ⎭

Reduce into a composition. The dose is ∋j. It ought to be administered with water of roses, decocted with roots of plantain, or with water of plantain.

Bloodstone stops profluvium *sine torsura*. But if there are gripings, it is the generation of the stone.

PREPARATION FOR LOOSE DYSENTERY

℞   Of Bloodstone ⎫
     Of Red Corals ⎬ v.ss. each.
     Of Spodium   ⎭
     Of Tanacetum, to the weight of all.

Make trochisks with mucilage of the glue of *botin*. The dose is ʒss.

Preparation for Diarrhœa

℞ Of Ice Alum
Of Bloodstone
Of Crocus of Mars } equal quantities.

Make trochisks of gum arabic dissolved in plantain water. The dose is from ʒj. to ʒj.ss.

Water of plantain is to be extracted from the roots and herbs. The cornelian, if carried in the hand, stops blood, but not so bloodstone.

Addition for Bloody Ulcers

℞ Of Prepared Bloodstone, ʒss. (al. ʒj.)
Of Oak Apples, Ɔss.
Of Seraphinus, Ɖj.ss.
Of Oil of Kerua (Keyri) from Violets, sufficient for incorporation.

Let an unguent be made. In cases of acute ulcers, add in place of Oil of Keyri a proportion of the Liquor of Mumia, as in the case of herpeta, and in eating and cancerous ulcers.

Addition for Menstrua

℞ Of Bloodstone, ʒj.
Of Long Pepper
Of Nutmeg } each ℥ss.
Of Cinder of Frogs' Follicles, Ɖiiij.

Make trochisks with Mint Water. The dose if from Ɔss. even to Ɖj.ss.

Addition for Dysentery.

Let the prepared Bloodstone be given in red wine. Therein let iron be extinguished or let it be given with Tyriaca.

Addition for Diarrhœa

℞ Of the said Bloodstone, ʒiij.
Of Pearls, Ɔss.
Of Liquefied Mumia, to the weight of all.

Mix. The dose is from ʒiij. to iij. or iiij.

CONCERNING SAXIFRAGE

By Saxifrage understand any stone which removes growths like tartars, mosses, sand, frost, and hail.

Saxifrage is properly a pale crystal, called also Citrine Stone or Citrinole Stone. Citrine Stone stands between crystal and yellow beryl. Let the liquor be produced after the mode of an alkali. The dose is Ɔss. in good wine.

First Preparation for Sand, Moss, Frost, Hail

℞ Of Saxifrage, ʒj.
Of Borax, ʒij.
Of Salgemmæ, ʒvj.
Of Fused Salt, ʒj.

Reduce S.S.S. at a fire of reverberation through the fourth grade, from the setting of the sun till morning. Reduce into alcali. The dose is ʒss. in white wine.

Second Preparation

Take of the said Saxifrage ʒj. Reduce it by itself to the fourth grade of reverberation. Also take :—

℞ Of Reverberated Saxifrage, ʒj.
Of Cinder of the root of larger Radish, ʒj.
Of Alkali of the roots of Petroselinon, Əj.

Make a mixture by itself. The dose is from Əj. to Əiij. or iiij.

First Addition

℞ Of the said Saxifrage, ʒj.
Of Millet of the Sun, ʒij.
Of White Wine, ʒx.

The dose is from ʒiiij. to vj.

Second Addition

℞ Of the Saxifrage, ʒj.ss.
Of Seed of Parsley } each ʒj.
Of Rocket
Of Clarified *aqua mulsa*, ʒx.

The dose is from ʒiiij. to vj. or vij.

It is necessary to continue this prescription as long as the tartarised urine issues.

CONCERNING ARSENIC

The virtues of Arsenic are for ulcers, wounds, and other openings.

Arsenic is a soot out of metals, and especially from lead, and it is realgar or fulgurr (or soot) out of metals.

The First Preparation is the Reduction of Arsenic into Mumia.

During the preparation the venom must be removed. Nothing cures ulcers and wounds more perfectly than prepared arsenic. It also cures *syrones* and all ulcers, gangrenes, and fistulas. The arsenic from lead is the best. Next, that which exudes from iron and resembles copper.

The Second Preparation of Arsenic is the Reduction of Arsenic into Balm

The Third is the Reduction into Liquor

Arsenic has three preparations, into Mumia, Balsam, and Liquor.

℞ Of White Arsenic, ʒvj.
Of Fused Salt, } ʒj.ss.
Of Colcothar,

Mix and reduce to the second grade of reverberation for three or four hours. Take out.

It must be removed from the top five or six times, pounded, and again prepared as above. This must be repeated five or six times.

PREPARATION OF BALSAM

℞   Of White Arsenic, ʒx.
Of Talc, ʒiiij.
Of Live Calx, ʒxv.

Make a subtle mixture. Reduce to the fourth grade of reverberation for 24 hours. It is like glass and the venom sticks at the bottom of the calx. That which is removed from the top is to be pounded and placed in a glass vessel. Let it be set in a cellar, whereupon an oil or balsam will come forth.

PREPARATION OF LIQUOR

℞   Of Crude or White Arsenic, lb.ss.
Of Saltnitre, lb.j.
Of Salgemmæ, ʒss.

Pound. Reduce in an open reverberatory for twenty-four hours.

If these being united are placed over a fire of reverberation, the Arsenic glows for three hours; afterwards it liquefies. In this condition it is poured into water, and coagulated after the manner of an alkali.

ADDITION FOR BALSAM

℞   Of the said Balsam, ʒiij.
Of Oil of Yolk of Eggs, ʒx.
Of Distilled Turpentine, ʒj.

Mix. Like Mumia, the balsam is to be applied for the space of twelve hours.

ADDITION FOR LIQUOR

℞   Of the said Liquor, ʒxv.
Of Skins of Pomegranate, ʒvj.
Of the Bark of the Frankincense Tree, ʒij.
Of Mucilage of Botin, to the weight of all. Mix.

CONCERNING ORPIMENT

Orpiment is a Yellow Minera like gold.
The virtues of Orpiment obtain in fistulas, cancers, and eating ulcers.

PREPARATION FOR FISTULA

℞   Of Orpiment, ʒj.
Of Calcined Tartar, ʒiij.

Arrange in alternate layers.

Reduce by fourth grade of reverberation for a day and night, that is, for twenty-four hours.

It melts when thus decocted. Let it be removed again, pounded, and poured into water, whereupon a white powder will settle at the bottom, which is the prepared Orpiment.

If it be put into a glass, an oil results, which must be injected into the fistula, or it may be applied by means of a rag. But let ulcers be sprinkled with the powder.

Preparation for Cancer

℞ Of Orpiment, ʒv.
Of Fuligo, ʒss.
Of Sal Ammoniac, ʒiij.

Reduce by the fourth grade of reverberation a day and a night. Reduce into alcali. Alcali is the chief arcanum for cancer.

Preparation for Esthiomenis

℞ Of Orpiment, ʒiij.
Of Calcined Alum, ʒvj.

Reduce by the fire as above with extraction of alcali.

Addition for Fistulas

℞ Of the said prepared Orpiment, ʒss.
Of Resin of Pine, ʒj.
Of Wax, to the weight of all.

Make into a cerotum. It is to be applied for fistulas.

Addition for Cancer

℞ Of prepared Orpiment, ʒv.
Of Cinders of Pigeons' Dung ⎱ q.s. for unguent.
Of Oil of Yolk of Eggs ⎰

This is used for cancer.

Addition for Eating Ulcers

℞ Of Orpiment, ʒv.
Of Liquor of Mumia, ʒij.
Of Oil of Roses, ʒj.ss.
Of Mucilage of the Seed of Fleawort, to the weight of all.

Make an unguent or cataplasm. Should the sick person complain of heat, the ulcer must be anointed with oil of camphor previous to application of the remedy.

CONCERNING SULPHUR

The virtues of Sulphur apply in cases of very acute imposthumes and asthma; they serve to maintain health. Imposthumes include pleurisy, pest, and the like.

The Following is the Preparation for Extremely Acute Ulcers

℞ Of Live Sulphur, lb.j.
Of Colcothar ⎱ each lb.ss.
Of Fused Salt ⎰

Make a fine powder; reduce by sublimation. As soon as it has been sublimated, make an addition again, and sublimate thrice as above. Live Sulphur coheres in fragments, and is not yet dissolved.

Preparation for Asthma

℞ Of Fused Sulphur, lb.j.
Of Shavings of Red Sandal ⎫
Of Cypress Shavings ⎬ each to the weight of the Sulphur.
Of Pine Shavings ⎭

Arrange in layers. Reduce by fire of reverberation, finally into alcali.

℞ Of this Alkali, ʒx.
Of Myrrh, ʒv.

Sublimate

Preparation to Conserve Health

℞ Of Sulphur, ʒiiij.
Of Oriental Crocus ⎫
Of *Myrobalani* ⎪
Of *Chebuli* ⎬ each ʒj.
Of *Bellirici* ⎭
Of Oil of Juniper Seeds, sufficient for incorporation.

Sublime by a very slow fire.

Addition for Very Acute Imposthumes

℞ Of this prepared Sulphur, ʒss.
Of Oil of Nutmeg, ʒj.
Of prepared Aqua Veronica, to the weight of all.

Make a potion.

Addition for Asthma

℞ Of the said Sulphur, ʒss.
Of corrected Thebanus, ℈iij.
Of Tyriaca, q.s.

Make a bolus. The dose is from ℈j. to two or three.

Addition for Conservation of Health

℞ Of the said Sulphur, ʒss.
Of Red Myrrh ⎫
Of Oriental Crocus ⎬ ʒss. each and ℈j.
Of Aloepaticus, to the weight of all.

TREATISE III

CONCERNING GEMS, TRANSPARENT AND OTHERWISE; OF CORALS, THE MAGNET, THE CRYSTAL, RUBIES, GARNETS, SAPPHIRES, EMERALDS, HYACINTHS, &C. EVERY STONE POSSESSED OF MEDICAL VIRTUES IS A GEM

CONCERNING CORALS

The virtues of Coral are for menstruum and profluvium; poison taken internally; thunderings or rumbling of the stomach; charms, if any be enchanted; obsession, if any be mad; nervousness, if any be timorous; melancholy, for those who appear wise in their own eyes but are foolish. The virtue and the substance is one and the same. Virtue is a thing by itself. Coral simples restrict urine and evacuation, and after a long time the menstrua. When prepared their operation is sudden and safe.

PREPARATION OF CORALS TO RESTRAIN MENSTRUUM AND PROFLUVIUM

℞ Of Corals, ℥ss.
Of Oil of Myrtles, ℥j. (*al.* ℥ss.)
Of Olibanum, ℥j ss.
Of Fused Salt, ℥ij.

Make a mixture; reduce by calcination through the fourth grade of reverberation for twelve hours or more; afterwards reduce by ablution with water of plantain. Corals restrict urine but not menstruum.

PREPARATION OF CORALS AGAINST INTERNAL POISONING

℞ Of Corals, thoroughly pounded, ℥ij.
Of Waters of Ligusticum, lb.ss.
Of Sal Gemmæ ⎫
Of White Vitriol ⎭ each ℥ij.

Reduce by digestion in the second grade of fire for a month; take out the red and coagulate. For poisons the medicine must be without a body; for poison is also without a body. The red which settles at the bottom is a good remedy against poison.

PREPARATION FOR RUMBLINGS IN THE STOMACH

℞ Of Corals, ℨvi.
Of Cinders ⎫
Of Roman Cummin ⎬ each ℨiii.
Of Beans ⎭
Of prepared Alum to the weight of each.

Mix these. Reduce by digestion with lb.ss. of desiccated alcool; desiccate. It becomes yellow all over. This preparation of corals ought to be used for *diacymimum*. The colour becomes red when the corals have been prepared. Separate and desiccate.

Preparation of Corals against Charms, Obsession, Nervousness, and Melancholy

℞   Of Corals, ℥iij.
      Of Glue of Oak, ℥iiij.
      Of St. John's Wort, ℥iiij.
      Of *Storax Calamita* ⎫
      Of Laudanum   ⎬ each ʒj.ss.
      Of Gum   ⎭

Of distilled wine add lb.ij. Reduce by decoction in a closed alembic for a day and a night. Distil and pour over again as above.

Corals, if prepared in this manner, become red and exceedingly hard. Consequently, they must first be pounded.

Addition for Menstruum

℞   Of prepared Corals, ʒj.ss.
      Of Tanacetum ⎫ each ℈j.ss.
      Of Plantain ⎭
      Of Long Pepper ⎫ each ℈ss.
      Of Nutmeg   ⎭

Make a powder. The dose is ℈j. in a tempered egg.

When salt is injected into the egg, it must also be eaten, otherwise it does no good.

Addition for Corals against Poison

℞   Of prepared Corals, ℈v.
      Of Theriac, ℥ss.
      Of Root of Larger Lapathius, to the weight of all.
      Of Alcool of desiccated Wine, lb.ss.

Reduce by digestion for a week. The dose is from gr.xv. to ℈ij.

One who has drunk poison should have administered to him ℥ij. of water of bullace or of roses. Let it be repeated several times, so as to produce perspiration, until the evil be felt no longer.

Addition for Rumblings

℞   Of the said Corals, ℥ij.
      Of Species of Diacymini, ℥iiij.
      Of prepared Blood of the Goat, to the weight of all.

Make a physical powder with Saccharum; the dose acts as a sedative.

The Goat of the Spagyrists is the castrated young of a coney. It must be fed with aperient herbs, then it is good.

Addition for the Other Three Species

℞   Of the said prepared Coral, ʒviiij.
      Of Masterwort, ʒj.
      Of Angelica, ʒv.
      Of Glue of Oak, ʒj.ss.

Mix with water of St. John's Wort: the dose is from ℥ss. to ʒvj.

This is the best medicament at the commencement of *tympanis*.

CONCERNING THE MAGNET

It has a virtue for wounds and ulcers *cum flaxis et ramentis*.

PREPARATION FOR THE ABOVE

℞ Of the Magnet, ℥j.
 Of Calx of Eggs, ℥ij.

Make S.S.S. in the fourth grade of the fire of reverberation for a day and a night. Remove the calx of eggs.

ANOTHER PREPARATION

℞ Of Magnet, ℥j.
 Of Calx of Eggs, ℥vj.

Set in layers in a crucible. Place in a fire of reverberation a day and a night. Extract, and it will be prepared.

Otherwise, if not prepared by pounding, it misses its true extractive efficacy. But if previously prepared, then pounded and mixed, the oppodeltoch has an excellent effect.

ADDITION FOR WOUNDS CUM FLAXIS ET RAMENTIS

℞ Of the Magnet, ℥ss.
 Of Carabe, ℥ij.

Make a subtle powder with lb.ss. of the plaster oppodeltoch or plaster of apostolico. Reduce to a wine by shaking.

This plaster when applied extracts splinters of bone, and bullets received from guns, out of ulcers and other wounds. If the magnet be pounded in unprepared condition, it loses its efficacy, but if you pound and mingle with *apostolico*, it extracts by itself. Unprepared it effects nothing.

CONCERNING GEMS

The crystal has the property of producing an abundance of milk if administered internally to women.

PREPARATION OF GEMS BY DIAPHANITAS

The preparation of gems is fourfold. First, by reverberation. Secondly, by calcination. Thirdly, by elevation, and in the fourth place by means of distillation.

A woman requires over lb.j. of crystal before she experiences an increase of milk. Accordingly prepared crystal is necessary.

REVERBERATION OF CRYSTAL

℞ Of Crystal, lb.j.
 Of Water of Entali, lb ij.

Make a mixture by imbibition. Reduce by reverberation for twenty-four hours.

Thus there is left from lb.j. a *verto* (that is, a kind of weight). Dose ℥ij.

Preparations in Alchemical Medicine

CALCINATION OF CRYSTAL

℞ Of Crystal, ʒiiij.
Of Mastic ⎫
Of Colophony ⎬ ʒij. each.
Of Sulphur ⎭

Reduce in an athanor. The dose is ʒj.

An athanor is a furnace in which things are burnt.

ELEVATION OF CRYSTAL

℞ Of Crystal, ʒj.
Of Sal Ammoniac, ʒiij.

Reduce in a sublimatory to powder. The sublimation is to be performed five or six times and the crystal is always to be removed.

℞ Of the said elevated Crystal, ʒss.
Of common distilled Water, ʒiiij.

Reduce to an alkali. Dose ʒss.

DISTILLATION OF CRYSTAL

℞ Of elevated Crystal, ʒj.
Of Water of Nitre, and ⎫
Of Alumen without distillation ⎬ ʒij. each.

Reduce into digestion for three or four days; then distil; coagulate that which is distilled, and dissolve the coagulate.

Coagulation must take place over a slow and small fire. This coagulate, if placed in a cellar, passes into water, which is the last preparation of the crystal. The dose is ℈j. With all other valuable gems the preparation is as with the crystal.

The chief virtue of rubies is for dysentery. The dose is ʒj. if crude, but if reverberated then the dose is ʒij. After calcination the dose is ʒj. After elevation, ʒss. After distillation, ℈j.

Also garnets, thus distilled, constitute a most powerful salve for spots of the eyes.

Emeralds, if prepared by means of distillation, are beneficial to those with bloodshot eyes.

Sapphires, being prepared to the third or fourth preparation, remove trembling of the heart if prepared into distillation. The dose is gr.v. Sapphires dispel *synthena* and palpitation.

The case is the same with the other gems. Bartholemew, the Englishman, has written voluminously concerning gems and precious stones.

ADDITIONS FOR THE GENERATION OF THE MILK OF CRYSTALS

℞ Of the said Crystal prepared, ʒij.
Of Spermaceti ⎫
Of Seed of Lettuce ⎬ ʒiiij. and ss.

Make a powder with administration of water of almonds.

Addition of Garnets for Trembling of the Heart and Bloodshot Eyes

℞ Of Garnets, ʒss.
Of Aloë Epaticus, ʒiii.
Of Prepared Sulphur, ʒj.ss.

Mix with clarified Zuccarum. The dose is from ʒij. to ʒij. This medicine must be diligently administered and continued for five days, although the trembling of the heart may disappear previously.

Addition of Sapphire

℞ Of Sapphire, ʒiij.
Of Dissoved Amber, ʒj.ss.
Of Storax Calamita, Ðj.

Make a mixture. The dose is from Ðj. even to Ðj.ss.

The Emerald strengthens women in labour, and is the sovereign arcanum for their ailments if prepared by distillation, as crystal.

℞ Of the said Emerald prepared, Ðj.
Of Liquor of Melissa, ʒj.
Of Southernwood, ʒij.

Mix. The dose is from three to six drops.

Addition for Prepared Jacinth

℞ Of Prepared Jacinth, Ðj.ss.
Of Laudanum, that is, gum, Ðj.ss.

Mix. This is a chief arcanum for fevers arising from putrefaction of air or of water. Should fevers of this kind be usual with any persons, let them drink, every new moon, four or five drops; thus they will be safe from being attacked a second time, and will be absolutely secure during the new moon.

TREATISE IV

CONCERNING SALTS

Sal Gemmæ, Sal Entali, Sal Peregrinorum, Aluminous Salt, Sal Alcali, Sal Nitri, Sal Anatron, Sal Terræ, Salt from Vitriol

All Salts are from the element of water, as also are all alums.

CONCERNING VITRIOL

The virtues of Vitriol obtain in falling sickness, suffocations of the Matrix, Siphita Stricta, or Noctambulones, Gutta, and Obesity.

The varieties of falling sickness are Analentia, Catalentia, Epilentia, etc. The administration of Vitriol is the same in all.

℞ Of cuprine Vitriol, 1lb. Reduce by separation from the phlegmatic part. Reduce the said phlegmatic part over its colcothar. Distil. Reiterate in the fourth grade of fire. The dose is from ℈ss. to ℈j. before and after the paroxysm.

If the disease arises from the element of Vitriol, it is cured by Vitriol. The disease of falling sickness is occasioned by salt of Vitriol. The medicine is to be administered during the paroxysm and on the day when it is expected. Epilepsy is a mineral disease; its cure is also mineral, that is, by the salts of Vitriol, and by the spirits before and after the paroxysm. Before the paroxysm the body is in great agitation; after the paroxysm the patients sleep. The medicine should be given after the sleep, while the body is still under the excitement thereof. When the body is healthy it should not be administered.

Preparation of Vitriol for Suffocation of the Matrix

℞ Of Vitriol, purged from Phlegma and Colcothar, ʒij.
Of Pennywort, ʒiij.
Of Alcool of Wine, ʒss.

Reduce by distillation. The dose is from ℈ss. to ℈j. This is the most efficacious medicine for the complaint in question.

Preparation for Gutta and Siphita Stricta

In Gutta—
℞ Of the said prepared Vitriol, ʒij.
Of Alcool of Wine, ʒij.
Of Jamen Alum, ʒss.

Reduce into liquids by the fourth grade of fire; applied externally, the dose is ʒss. Applied internally, the dose is from six to nine grains.

Jamen alum is white, like that of Crete, and sweetish.

The outside application is on the place of the Syntheoma, and this is where the disease begins, that is, *in pulsu*, which is the Syntheoma thereof. But if the patient still walks, the medicament is then to be bound about the pulse of the wrist and the neck. For siphita parva castigation is an effectual medicine, but it is of no use for siphita stricta. In gutta the medicament should be placed on the tip of the tongue. The paralysis being arrested, apply to mouth and tongue. It is the best medicament for the complaint.

Addition in Epilentia

℞ Of the said Vitriol, ʒj.
   Of the Viscous Liquor of the Oak with Orisons, each ℈ss. and gran. iij.
                    Mix.

The Syntheoma of Caducus is in the nape of the neck. In the case of young persons to anoint the nape of the neck with castor oil after the paroxysm is an excellent method.

Addition in Siphita Stricta

℞   Of the said prepared Vitriol, ʒj.
      Of Seed of St. John's Wort, ʒss.
      Of Amber, gr.vj.
Mix. Seed of St. John's Wort takes away Siphita Stricta.

Addition in Suffocation of the Matrix

℞   Of the said Liquor of Vitriol, gr.vij.
      Of Grains of Actis, ʒj.
      Of Alcool of Wine, to the weight of both.
Make into a composition.

Unless the spot is on the umbilicus let it be applied thereto. Should there be suffocation attended with vomiting, the medicaments previously mentioned are to be taken internally. The most important preparation of vitriol is to separate it from colcothar. Then add an equal quantity of alcool of wine. This done, place burnt bread (namely, bread made *ex furfure scalino*, which is dried so that it can be pounded in a mortar) in liquor of vitriol. Next set it for a month in horse-dung. Then prepare alcool, by means of a bath of the first grade, from vitriol. If the vitriol has lost its acetivity, its strength is gone.

CONCERNING WHITE VITRIOL

There are external species, as scotomia and spots of the eyes.

The virtues of White Vitriol obtain in affections of the external parts of the eyes, and in Neutha. Sometimes a cuticle covers the eyes or the ears of children at their birth. Neutha are pellicles growing anywhere from time of birth, as on the face, on the mouth, the eyes, ears, etc. White vitriol is a great cure in such cases, and also for exterior complaints of the eyes.

Preparations in Alchemical Medicine

PREPARATION FOR EXTERNAL DISEASES OF THE EYES

℞ Of White Vitriol, ʒv.
Of Oil of Siligo, ʒss.
Of Oil of Camphor, ʒij.

Putrefy for a month by means of horse-dung, and distil by descension.

Oil is produced out of Siligo. The Siligo is placed on a red hot iron plate, when it becomes encircled by a kind of grease, which is the oil in question. If, after birth, a pellicle covers the neighbourhood of the eyes, it must be most carefully treated with water of eyebright, of roses, or of fennel.

PREPARATION OF VITRIOL IN NEUTHA

℞ Of White Vitriol, ʒj.
Of Oil of Tartar, ʒvi.
Of Laterine Oil, ʒv.

Distil together.

Neutha (Teutha) should never be cauterized.

ADDITION IN EXTERNAL COMPLAINTS OF THE EYES

℞ Of prepared Vitriol, ∋j.
Of Liquor of Eyebright, ∋ij.
Of Red Poppy, ʒj.

Make an eye salve.

ROCK ALUM

The virtues of Rock Alum obtain in open ulcers, scab, itch, eating ulcers, putrid, lascivious, and humid ulcers.

FIRST PREPARATION

℞ Of Rock Alum, lb.ij.
Of White Vinegar, lb.ss.
Of Burnt Salt, one verto.

Mix till it passes from ebullition to coagulation. Then distil.

If open (cavernous) ulcers are washed with this water, the result is wonderful. If they are not thus cured nothing else will avail.

ANOTHER PREPARATION

℞ White (? Alum), lb.x.
Of the Juice of Chelidony ⎫
Of the Juice of Plantain ⎬ each lb.j.
Of Pellitory, lb.ss.

Distil. Take of this water lb.j., and of common water lb.x Make a lixivium. Foment warm in the case of alopecia, tinea, and ulcers.

ALUMEN PLUMOSUM

The virtues of Alumen Plumosum obtain in paralysis, lethargy, and benumbed limbs.

Preparation for Paralysis

℞ Of Alumen Plumosum, ʒvi.
   Of resolved Colcothar, ʒiiij.ss.
   Of Sal Ammoniac, ʒiiij.

Resolve. Alumen Plumosum confers strength imperceptibly. Hence it is the best medicament for paralysis.

Addition for Lethargy and Benumbed Limbs

℞ Of the said prepared Alum, ʒj.
   Of Dragon's Blood, ʒiij.
   Of Liquor of Mummy, ʒvij.

Make an unguent. The seat of the disease is in the occiput and in the nape of the neck.

CONCERNING ENTALI

The virtues of Entali are in profluvium and hæmorrhoids.

Preparation

℞ Of common Tartar, } each ʒij.
   Of Entali,
   Of Karabe, Ɔj.
   Of Mastic, ʒij.

Reduce by reverberation to the second grade, and afterwards into alkali.

In profluvium the seat is in the umbilicus; in hæmorrhoids it is in the spine.

Addition in Profluvium

℞ Of the said prepared Entali, ʒj.
   Of burnt Bolus, ʒiij.
   Of corrected Hematite, ʒj.ss.
          Mix.

Addition in Hæmorrhoids

℞ Of the said prepared Entali, ʒiij.
   Of prepared Corals, Ɔiiij.
   Of Oil of Nutmeg, as required.

Make an unguent.

CONCERNING ANACHTHRON

Anachthron is a salt growing in rocks, and is like a moss in appearance. When the said moss is decocted a salt results, that is, glass gall. Its virtues obtain in fistulas, cincilla or cintilla (diarrhœa), and scropulas. Cintilla is from the diaphragm, and it is cured by pure anatron.

℞ Of Anatron, ʒvj. (al. ʒj.)
   Of Bean Ashes, ʒij.

Reduce by the fourth grade of reverberation for twelve hours; extract the alkali.

Anathron with deer grease is good for cincilla. Anathron possesses a volatile Mercury, which must be corrected. It is then an addition in fistulas, cincillas, and scrofulas.

℞ Of the said Anatron, ʒij.
Of crude Butter, ʒiiij.
Of the Fat of Marmots, ʒiij.

Make an unguent. The said unguent is the best for fistulas, cintillas, and scrofulas.

CONCERNING SAL GEMMÆ

Sal Gemmæ is called Sal Granatum by the Spagyrists and Sal Lucidum.

It is a laxative of intense salt, *i.e.*, of cholera, and is like colocinthis. It cures jaundice, yellow dropsy, and sufferings arising from corrupt blood.

PREPARATION IN DROPSY AND JAUNDICE

℞ Sal Gemmæ } each ʒj.
Of Tithymal, *i.e*, Esula Major (al. Minor) }
Of Gum of Cherries, to the weight of both.

Make a bolus. Reduce by the third grade of reverberation for two hours; extract the alkali. The dose is from eight to twelve grains.

You may use it in place of diagridium, and add trochisks of alhandal.

ADDITION

℞ Of Sal Gemmæ, ℈ss.
Of Rebotium (*i.e.* true Mumia), } each ℈iiij.
Of Liquor of Centaury, }

Make a compost. The dose is from four or five to ten or twelve grains in an egg.

PREPARATION IN OTHER DISEASES

℞ Of the said Sal Gemmæ, ʒj.
Of the Juice of Cataputia, ʒij.
Of Ground Flour, to the weight of all.

Make a roasted loaf. The dose is from ʒj to ʒij.

CONCERNING SAL PEREGRINORUM

The virtues of Sal Peregrinorum obtain in fortifying the digestion, also against infection of the air, and against future imposthumes.

PRESCRIPTION OF HERMES

℞ Of Burnt Salt Nitre, } each ʒj.
Of Sal Gemmæ, }
Of Galanga, }
Of Mace, } each ℈j.
Of Cubebæ, }

Make a powder. The dose is three grains in the morning. It prevails against seasickness, and confers long life on old persons.

PREPARATION OF SAL PEREGRINORUM

℞ Of the said Salt, ʒiij.
Of dried Alcool of Wine, lb.ss.

Extract the alkali.

℞ Of the said Alkali, ʒj.
Of the Liquor of Juniper Seeds, one kist.

Make a compost. The dose is one grain.

CONCERNING SAL NITRI

Sal Nitri obtains in pleurisy and open ulcers.

PREPARATION IN PLEURISY

℞ Of Sal Nitri, lb.ss.
Of Crude Tartar, lb.j.

Distil in *sextum alembicum*. The dose is from ℈j. to ℈j.ss. in spring water or good wine, in the morning, at evening, and at midnight. Administer often. It purges through the urine.

PREPARATION IN OPEN ULCERS

℞ Of Alumen,
Of Nitre, } each lb.ss.
Of Spring Water, lb.ij.

Distil into water.

ADDITION AGAINST PLEURISY

℞ Of the said Nitre, ℈ij.
Of Aqua Regis, ℈ss.
Of dried Alcool of Wine, ʒv.

Mix. The dose is ʒss. or ʒi.ss.

ADDITION FOR OPEN WOUNDS

℞ Of Plantain Water,
Of Chelidony, } each lb.j.ss.
Of Oak Leaves,

Use for ulcers of the legs.

TREATISE V

CONCERNING METALS

Concerning Gold, Silver, Tin, Copper, Iron, Lead, Mercury

CONCERNING GOLD

The virtues of Gold obtain in Paralysis, Synthena, Fevers, Palpitation of the Heart, complaints of the Matrix, Ethica, Peri-pneumonia, and in acute diseases generally.

Preparation for Paralysis, Palpitation, and Synthena

℞   Of pure Gold, purged from its alloys, ʒij.

       Of the water of Sal Gemmæ, ʒvj.

Reduce into one by separation with alcool of wine. Then—

℞   Of the Crocus, ʒij.

       Of corrected Alcool, ʒvi.

Mix. The dose is from three or four to six grains.

Preparation in Fevers and Acute Maladies

℞   Of melted Leaves of Gold from the Water of Honey, ʒj.

       Of Alcool of Wine, ʒij.

Reduce by separation from the honey. The dose is from ℈ss. to ℈j.

Preparation for Complaints of the Matrix, Ethica, and Peripneumonia.

℞   Of Gold extinguished in Chelidony Water, ʒxiij.

       Of Indian Myrobalani, ⎫
       Of Chebuli,   ⎬ each ℈j.

Reduce to digestion for a week by separating the superfluous aquosity. The dose is from ℈j. to ʒj.

Addition for Paralysis, Palpitation, and Synthena

℞ Of the said prepared Gold, ℈j.

    Of Lavender Water, with corrected Alcool of Wine, and Spicula, each ʒi.

Dose, ℈j.

Addition in Fevers and Acute Stages

℞   Of the said prepared Gold, ℈iiij.

       Of the juice of Centaury, ⎫
       Of the juice of Sage,   ⎬ ʒij.

Dose, from ℈ss. to ℈i.ss.

Addition in Complaints of the Matrix, Ethica, and Peripneumonia

℞   Of the Oil of Nutmeg, ʒss.

       Of the Oil of Cloves, ʒj.

       Of the said prepared Gold, ℈j.

Dose, from ℈ss. to ℈j.

Process for Water of Sal Gemmæ

℞ Of Sal Gemmæ, lb.ss.
Of Rain Water, lb.j.

Distil by retort till the whole substance of the salt is perfected.

Purgation of Gold

℞ Of Gold, ℨss.
Of Antimony, ℨij. or ℨiij.

Melt into a regulus. By this means the antimony takes up the impure part, and the gold remains at the bottom.

CONCERNING SILVER

The virtues of Silver obtain in complaints of the cerebrum, the spleen, the liver, and in the retention of the profluvium.

Preparation in Complaints of the Cerebrum, Spleen, and Liver

℞ Of Laminated Silver, ℨiij.
Of Sal Gemmæ, ℨvi.

Arrange in layers. Reduce to the fourth grade of reverberation for twenty-four hours, and extract the alkali. The said alkali is placed for three or four days in sublimated wine, when the silver becomes like the wine itself. Let stand. Evaporate The alkali sinks to the bottom, and, being received into a glass, liquefies in a cold place. The dose is from five or six to twelve grains.

Preparation in Profluvium

℞ Of filings of Silver, ℨj.

Reduce into calx by means of Aqua Regis.

℞ Of the said Calx, ℨij.
Of crude Tartar, ℨiiij.

Reduce to the fourth grade of reverberation with extraction of the alkali.

Process for Aqua Regis

℞ Of Alumen, ⎫
Of Vitriol, ⎬ lb.ss.
Of Nitre, ⎭

Distil into sweet water.

Method of the Extraction of Alkali

℞ Of the said Silver, as required.
Of Alcool of Wine, ⎫ each ℨx.
Of Water of Chelidony, ⎭

Reduce as above. The dose is from ℈j. to ℈j.ss. If the red flows forth with the profluvium it is a sign that it is going to be restrained. ℨj. of the said water should then be taken.

Preparation of Silver for all the above Complaints

℞ Of Laminated Silver, ℨj.
Of Purged Sulphur, ℨiiij.
Of Pine Resin, ℨij.

Make a bolus, set alight, and reduce to preparation with spring water. The dose is from ℈j. to ℈j.ss. When thus prepared it is good in all the cases, but the first is more efficacious.

CONCERNING TIN

The virtues of Tin obtain in jaundice, asclitis, and worms.

Preparation for Jaundice

℞ Of Calcined Tin, lb.j.
 Of Salt, ℥v.
 Of Bean Ashes, lb.ss.

Reduce into Litharge by a fire of reverberation.

℞ Of the said Litharge, ℥x.
 Of Alcool of Wine, lb.ss.

After resolution reduce into Alkali. The dose is from six to ten or twelve grains.

Preparation for Asclitis

℞ Of purged Tin, ℥j.
 Of Antimony, ℥ij.
 Of Filings of Cinetus, to the weight of all.

Reduce into calx by reverberation for twenty-four hours. Then

℞ Of the said Calcined Matter, lb.j.
 Of Alcool of Wine, lb.j.ss.

Reduce into alkali. The dose is from ʒj to ʒj.ss.

Preparation for Worms

℞ Of Tin, ℥iij.
 Of Common Salt, ℥iiij.
 Of Asphalt, ℥j.

Make into a powder by burning. The dose is from ʒss. to ʒiij.

Addition in Jaundice

℞ Of the said prepared Tin, ℈iiij.
 Of Alipta Muscata, ℈j.
 Of Bdellium, ℈ij.

The dose is from ℈j. to ℈ij.ss.

Addition in Asclitis

℞ Of the said prepared Tin, ʒss.
 Of Dragon's Blood, ʒij.
 Of Liquor of Tapsus, ℥j.

Mix. The dose is ʒss.

Addition for Worms

℞ Of the said prepared Tin, ℥j.
 Of Colocinth seed, }
 Of Plantain seed, } each ʒvi.

Make into a powder. The dose is from ʒj. to ʒj.ss.

CONCERNING COPPER

The virtues of Copper obtain in ulcers, wounds, worms, and ulcers of the mouth.

Preparation for Ulcers

℞ Of Copper, lb.j.
Of Immature Botrum, lb.v.
Of Vinegar, lb.j.
Of Sal Ammoniac, ʒss.

Digest for a month in a closed vessel, afterwards reduce by ablution, and convert into a salt of alkali, *i.e.*, verdigris in ulcers (*sic*).

Preparation for Wounds

℞ Of Copper, lb.ss.
Of Distilled Turpentine, lb.j.
Of Common Salt, ʒj.
Of Vitriol, ʒij.

Mix in a closed glass vessel for three months. If plates of Copper are taken and thus prepared, the best Balsam results. Afterwards take ʒj. of Flos Æris, and ʒj. of common oil.

Preparation for Worms

℞ Of Calcined Venus, ʒj.
Of Water of St. John's Wort, and of Centaury, each ʒvj.
Of Plantain Water, ⎱ each ʒiiij.
Of Sour Wine, ⎰

Digest for seven or eight days. Reduce into Alkali. The dose is from ℈j. to ℈iiij. or v.

Preparations for Ulcers of the Mouth

℞ Of Venus, laminated or cemented, ʒij.
Of Burnt White and Rock Alum, each ʒvj.
Of Distilled Vinegar, lb.j.

Extract the Alkali for a day and night.

Addition for Ulcers

℞ Of the said Flos Æris, ʒj.
Of Aggripine Ointment, ʒj.
Of Earth Worms, ʒiij.

Make an unguent.

Another Addition in Common Ulcers

℞ Of the said Flos Æris, ʒv.
Of Alum Water, ʒxv.

Make a mixture after the manner of a lotion.

Addition for Wounds

℞ Of the said prepared Flos Æris, ʒj.
Of Oil of Anise, ʒiij.
Of Oil of the Yolks of Eggs, ʒ (imperfect quantity).

Compose an oil.

Another

℞ Of the said prepared Flos Æris, ʒss.
Of the Hepatic Aloe, ʒj.
Of Liquor Consolida, ʒiiij.

Make a Gum.

Addition for Worms

℞ Of the said Flos Æris, ʒj.
Of Zuccarum Taberzet, ⎫
Of Liquorice Juice,   ⎬ each ʒij.ss.

Reduce to a powder. The dose is from ℈ss. to ℈j.

Addition for Ulcers of the Mouth

℞ Of the said Prepared Flos, ʒj.
Of Chelidony Water, ʒiij.
Of Alum Water, ʒj.

Make a gargle or mouth wash.

CONCERNING IRON

Iron has styptic, constrictive, and drying qualities.

Preparation for Styptic Quality

℞ Of Iron Filings, lb.j.
Of Common Salt, lb.v.
Of Spring Water, sufficient for incorporation.

Reduce for the space of a month, and afterwards reverberate into a powder. Incorporation is treatment with water till a pulp is formed.

Constrictive Preparation

℞ Of Iron Filings, lb.ss.
Of Alum Water, lb.j.ss.
Of Distilled Vinegar, lb.ss.

Digest for a month. Reduce by ablution and afterwards by reverberation to a crocus.

Drying Preparation

℞ Of Iron Filings, lb.ij.
Of Water of Vitriol, lb.ss.

Digest for a month and reverberate into a powder.

Styptic virtue is closing and drying to fistulas and cancers. Constrictive obtains in lienteria, dysentery, and diarrhœa. Exsiccative is for phlegmatic complaints.

Addition for a Styptic

℞ Of the said Crocus, ʒj.
Of Burnt Bolus, ʒiij.
Of Sealed Earth, ʒv.

Reduce to a powder. It is an incarnative for ulcers and wounds. If taken internally the dose is ʒj.

Addition for a Constrictive

℞ Of the said Crocus of Mars, ʒj.
Of Myrrh, ʒss.
Of Oriental Crocus, ℈j.

Reduce to powder. The dose is from ℈ij. or iij. to iiij.

Addition for an Exsiccative

℞ Of the said Crocus, ʒiij.
Of Pomegranates, ʒj.
Of the Sap of Acacias, to the weight of all.

Make an electuary.

CONCERNING SATURN

Saturn has an incarnative virtue.

Preparation

℞ Of Lead Ashes.

Decoct with vinegar for three or four hours. This is the first preparation; it cures wounds, and grows solid flesh. Ceruse is also made from lead if washed with water in the sun. Minium is decocted from cerussa in a kettle. All medicaments for wounds and ulcers should be prepared from metals.

CONCERNING MERCURY

Mercury has an incarnative and laxative virtue.

Preparation as an Incarnative

℞ Of Prepared Mercury, powdered, ʒij.
Of Aqua Regalis, ʒx.

Reduce by distillation in a bath several times a day, and convert into an oil. It is a most speedy incarnative for wounds and ulcers. It has two objections: it salivates and produces cerussa. Hence it is generally rejected and disliked. Otherwise it consolidates well and quickly.

Preparation as a Laxative

℞ Of Mercury coagulated by the Albumen of Eggs, ʒj
Of Alum Water, ʒvi.

Distil through ashes and make into a powder. The dose is from three to four or five grains. It is a potent purgative for diseases which originate from leprous humidity, such as paralysis, pustules, the varieties of gutta, and humid dropsy.

Here ends the Book concerning Preparations in Alchemical Medicine

THE ALCHEMICAL PROCESS AND PREPARATION OF THE SPIRIT OF VITRIOL

By which the Four Diseases are cured, namely, Epilepsy, Dropsy, Small Pox,* and Gout

To abolish those errors which are usually committed by Philosophers, Artists, and Physicians

THE spirit is extracted out of vitriol by means of colcothar, which is useless and of no efficacy. That which they call a phlegm is the most noble of all spirits, and all virtues should be ascribed to it.

But although the oil of colcothar is indeed of great efficacy in gravel and stone, as also in alopecia, yet it is of no use for the aforesaid diseases, to which, however, it is commonly applied.

Hippocrates, with whom almost all others agree, hands down certain stages and symptoms, which supervening epilepsy and gout must be reckoned as incurable. But since they had no acquaintance whatever with the spirit of vitriol, let their opinions go to the winds.

In the first place, the extraction of the spirit from vitriol must be effected by means of a powerful fire in an upright cucurbit, so that it may be driven into a fresh alembic, and, remaining in the athanor four days and nights, may be most skilfully passed through the reverberatory. And thus have you prepared this spirit of vitriol.

Afterwards colcothar must be distilled through a phial placed in the athanor for the space of three days over a very fierce fire of wood and coal, until there shall appear in the receiver, out of 1lb. of colcothar, ʒvj., which is tinctured with a scarlet colour.

This having been accomplished, the alkali is to be extracted from the *caput mortuum*, and the same having been resolved four or five times, is to be then coagulated. Thus the three things which exist in vitriol shall be extracted and separated.

Process

Proceed as follows in an epileptic disease. After each paroxysm let ƎJ. be given to the sufferer. Let a dose of four grains of the oil of colcothar be

---

* The term *pustula* also signifies St. Anthony's Fire.

administered morning and evening in peony water. This order is to be followed up to the fifteenth paroxysm. If the paroxysms become less frequent, half the dose is to be taken for the next thirty days.

In gout a daily dose must be taken for a period of thirty days. Afterwards the afflicted part must be anointed with the spirit of vitriol till the pain is removed. If it be gout of long standing, add a fourth part of the liquor of mumia* to the said spirit.

In dropsy half a scruple of the spirit of vitriol must be administered in liquor of Serapinus, the dose being repeated three or four times, according to the stage of the disease. In the absence of the liquor of Serapinus, the liquor of crude tar must be substituted.

In the case of small pox observe this order and method. The whole seat of the disease and the part of the skin affected is to be anointed with the spirit of vitriol for nine days. But if the skin be ulcerated, let the oil of colcothar be applied in combination with its alkali, according to chirurgic method. The bandages are to be loosened after the sixth day.

The regimen and diet must be adapted to the condition of the patient, for the whole circle of disease lies in medicine and not in diet. The medicine, therefore, must be administered sedulously. So are the four diseases here dealt with fundamentally and radically cured.

An Addendum on Vitriol

Alchemy has produced many excellent arts for physicians, whereby admirable cures of various diseases are effected. For this reason, therefore, in the commencement of medicine, alchemical doctors did always labour that it might become the mother and parent of many advantages. These two faculties were long cultivated together as companions, until there arose the triflers and sophistic humourists, who mingled poison with medicine, and rendered it meretricious, which medicine will still continue to remain so long as the humourists survive. I preface this that you may pay more diligent attention to the point, by reason of its great medical utility. But this is to be passed over. Wherever unskilful men rush into any art there they corrupt and defile everything, and out of a pearl make a fetid marsh. A like thing happens with regard to vitriol. At first the spirit of vitriol is taken, and is

---

* A Process for Mumia, opposed to the Errors of those who administer it for Poisons. — Many have laboured in experiments, compositions, and recipes, whereby they might draw forth the poison into the universe, yet have they accomplished nothing. For among all things, both in experiments and recipes, it is only Mumia which brings an immediate remedy against all kinds of poisons. The method of dealing with Mumia is as follows. In the first place, cause the Mumia to putrefy in olive oil, and that for four weeks. Then separate in a retort. To each pound which has proceeded from the separation add one drachm of Alexandrine musk, and of Alexandrine theriac six ounces. Lastly, dissolve the mixture in the Bath of Mary for a whole month. You will then have theriac of Mumia. As regards its administration to the sick, give at first one ounce in oil of almonds, by way of a drink. Next see that the poisoned individual takes to a bed wherein he can sweat well, and the medicine may take effect. By this method any animal or mineral poison is expelled. Moreover, such is the virtue of Mumia, that if it have been administered before the reception of the poison, the latter will work no harm. A single dose ($\mathfrak{Z}$) taken in the morning will obviate the possibility of poisoning during the whole day. In cases of poisoned ulcers, plague, carbuncle, anthrax, and pleurisy, give $\mathfrak{Z}$ j., and repeat the dose at the end of six hours, when, if the sufferer has survived to take the second quantity, his recovery is assured. Lastly, there are innumerable other diseases which by this theriac are completely and perfectly cured. — *De Mumia Libellus*.

usually raised to the highest grade. By this exalted they cure epilepsy, whether it be of recent or long standing, in men and women of whatsoever condition. But here the unskilled workmen rushing in, and about to enter on a better way, have attempted to apply the virtues of vitriol to another purpose, and thus departing from the first method and arcanum, they have suffered it to expire, and then have sought oil in colcothar, which can in no wise be usefully done. For whatsoever is to take away epilepsy must have a subtle, sharp, and penetrating spirit.

For therein consists the faculty of pervading the whole body, and passing over nothing. By such prevading or penetration the disease is attacked on its own ground, for it is certain and beyond all doubt that its seat, centre, or sphere cannot be known; hence it is inferred by the physician that there is need of those remedies which penetrate the whole body. And this is the reason why the mercenary humourists cure none, but prostitute all their learning and profession. I therefore freely affirm that in that oil which these workmen have sought there is no penetrating spirit whatever. They supply, so to speak, a mere earthiness which does not penetrate far, but where it falls there it remains. It is therefore to be regretted that, owing to their ignorance, the true process is prejudiced, and a false one is substituted. For I am persuaded that the devil has devised these things in order that health may not be restored to the sick, and that the sect of the humourists may shortly come into still greater power. To return, however, to the beginning, and to the manner in which the spirit of vitriol was invented. They first distilled the humid spirit of vitriol by itself from colcothar, then they intensified its grade by distilling it and circulating it to the highest point, as the process teaches. In this manner, the water comes to be used for various external and internal diseases, as also for the falling sickness. Thus a marvellous cure is obtained. But in the extraction they were much more diligent, for they took the spirit of vitriol, corrected as above, and distilled it from colcothar eight or ten times over a very strong fire. Thus the dry spirits were completely mingled with the humid. They continued their work until the dry spirits departed, by reason of the uninterrupted and vehement extraction. Afterwards they graded each spirit, both the humid and the dry, received in a phial together, to their terminus. They regarded this medicine as of great efficacy against diseases, and were so successful therewith that they completely confounded the humourists. But there is added unto it a certain correction by the artists by means of sublimated wine, and therefore of greater penetrating power, but it has not attained a higher grade.

But I will communicate to you my process, which I recommend to all physicians, especially for epilepsy, which has its cure in vitriol alone. Wherefore charity towards our neighbour demands that we should take greater care in case of this disease. My process is that the spirit of wine be imbibed by vitriol, and afterwards distilled, as I have said, from dry and humid spirits.

This done, I discovered that the following addition was very useful. Let the spirit of corrected tartar be mingled with the third part of vitriol, and let there be added the spirit of the theriacal water of lavender in the proportion of one-fifth in respect of the vitriolated spirits. Then let it be administered to the sick person before the paroxysm, or several times in the day. This medicine possesses a signal efficacy against the said disease, so that it is not lawful to expect a better one from Nature. Accordingly, the first process was invented and retained by the ancients with the said correction. For thus the heart and the whole virtue of Nature is attained.

But I hope that I shall not be reproved by all good persons who think of the terrible nature of this disease, which ought to move the stones to pity, for whether the vehemence and atrocity of that disease be so great or not, it would be permissible for any one to say, Cursed be all the physicians who, passing by the sick, give them no aid, like the priests and Levites in Jericho, who, deserting the wounded man, left him to be treated by the Samaritan! For they were worthy of the fire of Gehenna, from which there is no redemption, and who will admit anything else than that all these physicians, without exception, look at the disease, and yet pass by the sufferer? Who can say anything else but that they will be judged at the last day? For scarcely would they spend one penny to secure a more certain foundation for the cure of this disease. Did they strive to imitate the Samaritan, God would not then judge them, but, in consideration of their faithfulness, would manifest to them all the secrets of Nature, whereby they might assist the sick, and, although the required properties did not exist in Nature, He would create them afresh. Wherefore I testify to you, men both of high and low degree, that all the doctors have shamefully strayed; whatsoever the seducers Galen or Avicenna have concocted, they adhere to it, and weary themselves with lies. To such an extent are they obsessed by the devil that they cannot exercise charity towards their neighbours, and therefore make of themselves children of condemnation. First of all the Kingdom of God is to be sought, yet not with the Levite or the priest, but with the Samaritan.

If we are merciful, and follow the example of the Samaritan, God is with us, and He will immediately confer upon Nature a remedy not hitherto created. While men have levitical or sacerdotal dealings with the sick, God puts off the medicine, and keeps it to Himself. The sick flee to the Kingdom of God, but the physicians to the abyss of hell. The same place is prepared for both doctors and Levites. Therefore, open your eyes, there are two paths —one taken by the Levites, the other, which leads to heaven, along which the Samaritan proceeds.

That vitriolated extraction is not only excellent for falling sickness, but in the same way for cognate diseases, as swoon and trance, also for constipation and internal imposthumes, etc., and for strangulation and precipitation of the matrix. But far other than the aforesaid virtues would be discovered by diligent inquiries. The devil, whom the false physicians serve, however, obsesses them, and he incites them, so that they cannot endure a lover of the truth. So there is an end to the health of the good.

Further, it is to be known that the aforesaid recipes for making a humid spirit of vitriol cannot be more clearly described. An artist is required to understand it, but sordid cooks can by no means grasp a matter of so much moment. It is from artists, therefore, from alchemists, and from experimentalists, that you are to expect sufficient information on all points. Similarly, we shall, by the same, be more fully instructed in the correction of the spirits of the wine. The doctors of the academies are so ignorant that they can scarcely distinguish between agaric and manna. The art and virtue of all vitriol consists in this, that the spirit of the vitriol should be properly extracted, raised to the highest grade, and by addition should be made potent to enter the penetralia where the centre, root, and seed of the disease can be found. For it is impossible to discover these places so exactly as those doctors assume when discoursing of their humours. The fundamental principle has not yet been discovered as to what makes the disease, or where it is situated, or what it is that throws a man into such a severe paroxysm. Therefore the whole operation must be committed to the arcanum alone which Nature has appointed for the disease. That arcanum will find out the disease just as the sun penetrates all the corners of this world. In short, whoever desires to act as a true physician should first of all study to be a Samaritan, not a priest or Levite. If he be a Samaritan all things of which there is need will be given him, nor will anything be concealed from him.

Of the Oil of Red Vitriol

You must know how a most blood-red and vinegar-like oil is prepared from colcothar by distillation in a retort after the alchemistical fashion. This oil the operators have regarded as more efficacious in the aforesaid diseases than the spirit itself, but erroneously. The process of preparation is well known, and need not be here described. The most important part consists in the manual work, in diligent inspection, and, finally, in suitable instruments. Also you should know concerning its virtues, that, in the first place, it is an acid matter, surpassing all acidity, so that there is nothing more acid. Next, it possesses a corrosive nature, whence it follows that it must be used with caution, as also not by itself, but diluted in such a manner as will be in harmony with the nature of the case. By reason of its acidity it is beneficial to a stomach which is free from cholera and ulcers, but not otherwise, for acidity is aggravating to ulcers. If cholera be present, a continual conflict ensues, even as between aquafortis and tartar. It will also conduce to health in the case of all fevers and loss of appetite, the same method being observed, for many virtues are ascribed to this oil, though few are confirmed by experience. Those, indeed, who have boasted of effecting marvellous cures thereby, have been proved by use to have lied disgracefully. It is useful in stone and gravel, though I know no case which has been cured by it.

In all instances it effects something, but it does not get to the root of any disease.

But with regard to surgery, understand concerning this oil that it creates severe pains, but, nevertheless, brings immediate health even in the most desperate diseases. In certain complaints it is better to distil the red oil of vitriol into a spirit, and thus a minimum quantity becomes sufficient to effect a cure. These things have been taught me by experience; the other decoctions of vitriol are of no importance.

The White and Green Oil of Vitriol

Out of raw vitriol there can be distilled, by descent, an oil, sometimes white and sometimes green, according to the conditions of the vitriol. The same calls for special praise. Because it is prepared out of crude vitriol, it therefore also contains the spirit of the same. It is excellent and advisable for internal diseases. The green oil is better if it be circulated and mingled with a commixture drawn from the spirit of vitriol. Let no one who possesses this despair that he has a certain and undoubted remedy for the falling sickness, and all its varieties. It should be brought to the highest grade, that it may be separated from its earthiness and its fæces by the bath of Mary, and afterwards by fire. Thus, in the bath of Mary the phlegm will be removed, the earthiness will be removed by fire, and the spirit of oil must be then collected alone that it may circulate by itself. Afterwards you may take an addition of the spirit of wine, nor is there need that many things be added. It is taken in peony water before the commencement of the paroxysm. When the spirit of the oil has searched out the centre of the disease, the paroxysm abates, creating a certain vertigo at first, but soon after the patient subsides into gentle sleep, and experiences relief. Yet it is necessary all the same to persevere with the administration of the medicine.

THE ALCHEMIST OF NATURE.*

Being the Spagyric Doctrine Concerning the Entity of Poison

CHAPTER I

CONCERNING the nature and the essence of poison by which our bodies are affected, we would thus establish the foundation and the truth.

It is agreed among all parties that our bodies stand in need of conservation, that is, a certain vehicle, by the aid of which they flourish and are nourished. Wheresoever this is wanting, there life itself departs. But this, however, is equally to be borne in mind, namely, that He who built up or created our bodies, the same, in like manner, procreated the foods thereof, and that with the same facility, though not indeed in an equal perfection. I wish the matter to be understood thus: We are endowed with a body which is devoid of poison. But that which we administer by way of nourishment to our body has poison combined therewith. Thus our body is created perfect, but not also the other. Hence, observe that the other animals and fruits are for us designed as food, and so, also, as poison. They are not in themselves either foods or poisons, but, as regards themselves, and inasmuch as they are creatures, they share their perfection equally with us. When they are taken by us as food they are thus poison to us. Thus a thing becomes poison to us which in itself is by no means a poison.

CHAPTER II

To consider the matter further, everything in itself is perfect, is made good in relation to itself, and according to its own law. But if we have regard to its external uses, it has been formed both good and bad. Understand this as follows: The ox which feeds on grass receives both health and poison, for the grass contains in itself both nourishment and medicament. In the grain itself there is no poison. Thus, also, whatever man eats or drinks is at the

---

* Perhaps the chief utility of this treatise will be the illustration which it affords of the extremely wide sense in which the terms Alchemist and Alchemy were applied by Paracelsus. The little work itself is derived from that portion of the *Paramirum* which is entitled *Textus Parenthesis super Entia Quinque*, and, in addition to the entity of poison, is concerned with the astral entity, the entity of seed, the entity of virtue and quality, the natural entity, and the spiritual entity. Finally, there is the entity of God. The whole constitutes a kind of general introduction to the body of exoteric medicine for which we are indebted to Paracelsus. It is to the several treatises dealing with these subjects that reference is intended in the fifth chapter.

same time both venomous and healthy. Take this statement in two ways, one concerning man himself, excluding the nature of animals and plants, the other concerning the assumption. To impress this more plainly on the mind—the one in man is the great nature, the other is the poison inserted into the nature; and, in order that we may conclude this matter, remember that God has formed all things perfect, in so far as regards their utility to themselves, but imperfect to others. Herein rests the foundation of the entity of poison.

God, indeed, has appointed for man or for creatures no Alchemist for His own sake, but He has destined the Alchemist as one to whom we may betake ourselves if any of those things whereof we have need be imperfect. And He has done so for this purpose, that we may eat the poison which we take under the appearance and in place of healthful food, not as poison, but that we may separate and divide it from such healthful food.

What we tell you about this Alchemist do you regard with the most careful attention.

CHAPTER III

When therefore, anything, which is, in other respects, perfect, assumes at one time the form of poison and at another that of healthful sustenance, we say, proceeding with our subject, that God has appointed an Alchemist for him who eats and uses anything which, when taken, tends as much to destruction as to health; and this Alchemist is such an artist that he can only separate these two elements one from the other by banishing the poison to its own place, and by introducing the food into the body. In this way, as we have said above, we would have our fundamental principle understood and accepted by you. Take an illustration of a different kind. Whoever is a lord or prince is, so far as he himself goes, perfect, as a prince should be. But a prince cannot be without servants who shall minister to him in his princely character. Those servants too, are, so far as concerns themselves, perfect, but not as they stand related to the prince. To him they are as a poison, as a loss; they receive pay from him. So understand the natural Alchemist. God has granted that science shall exist bountifully in him, as in a prince. He teaches him how to separate the poison from his ministrants and to accept the good among them. Here you will find the fundamental principle of our present subject, even though the illustration may not recommend itself to you at first. Its teaching is according to the doctrine of the wise man, where the whole is unfolded. The matter stands thus. Man must eat and drink. For the body of man, which is the temporary abode of his life, absolutely requires food and drink. Man, therefore, is compelled to take into him poison, diseases, and death itself, by means of his food and drink. So then, this argument might be used against Him who endowed us with a body and then added food in order to slay us. Learn, however, that the Creator takes nothing away from the creature, but leaves to each his own perfection. And, although one thing is poison to one and another thing to another, the Creator is not to be accused or blamed for this.

CHAPTER IV

But in this way you will track out the Creator. If all things are perfect in themselves, and this from the settled arrangement of the Creator, according to which one serves for the conservation of the other, as when the grass nourishes the cattle and the cattle nourish man; and if thus the perfection of one thing be to another which partakes of it now an evil, now a benefit, and thus imperfect; then we must assume that the Creator, for the sake of enlarging His creation, arranged matters so for this reason: He determined all things should be so created that in whatever is necessary for some other thing such virtue and efficacy should be latent, that by means thereof the poison should be separated from the good for the health of the body and supply of nutriment, and that this mutual arrangement should be preserved. For example: the peacock devours the snake, the lizard, and the newt. These creatures, so far as they themselves are concerned, are perfect and wholesome, but with reference to other animals, all those mentioned are mere poison, except in the case of the peacock. And hear the reason of this difference. Its Alchemist is so subtle that the Alchemist of no other animal can come up to him. He so thoroughly and purely separates the poison from the good that this diet is innoxious for the peacock. So is it true in other respects also that to every animal is assigned that particular food which is adapted for its preservation, and besides this a special Alchemist is given who separates the nutriment. To the ostrich such an Alchemist is given who separates the iron, that is, the dung, from the nutriment, and it is not possible that this should be done for any other creature. To the salamander is given fire for food, or rather a body of fire. For this purpose it has an Alchemist appointed. The swine feeds on dung, although it is poison; and for that reason it is extruded from the human body by the Alchemist of Nature. Nevertheless, it serves as aliment for the swine, since the Alchemist of swine is much more subtle than the Alchemist of man; for the Alchemist of the swine separates that aliment from dung which the Alchemist of man cannot so segregate. On this account, too, the dung of swine is not eaten by any other animal. There is no other Alchemist more subtle, or who can separate aliment more cleverly, than the Alchemist of the swine. And so of others, which we advisedly omit, to make our discourse the shorter, it must be understood in like manner.

CHAPTER V

We have already said something about the Alchemist; and you must believe him to be appointed by God solely for this reason, that he may separate from the good that which differs from it in the body when, by the divine arrangement, it takes something into itself for the support of life. Recur now to the data elsewhere supplied, namely, that there are five things which have power over man, and whereto man is subject. These are the Astral Entity, which we have dealt with, and next the Entity of Poison. Now even though a man may be in no way affected by the stars, still he is not equally safe and secure from the Entity of Poison, but there is reason to fear lest he may go wrong thereby. These we leave as we described them in the prefaces.* But in order that you may more easily embrace all, observe the initial principle, so that you may more clearly understand in what way the poison can or actually does hurt you. Since, then, we have within us an Alchemist placed by God the Creator in our body for this purpose, that he may separate the poison from the good, and that so we may suffer no harm, it is necessary that we should next discuss about this, namely, what is the principle and what the mode according to which all diseases issue from the Entity of Poison as well as from the other sources. In this disquisition we will pass over all that which brings no injury but some advantage to bodies, as we shall in due course make clear.

CHAPTER VI

But first you must know that in this matter astronomers are deceived, for while pointing out the maladies of our body, they make the body fortunate and the body healthy. But if this be not the case, this one cause remains, that the remaining entities, of which there are still four, do injury to the body, and by no means the stars of themselves. Hence we rightly ridicule and explode their writings, wherein they make such lavish promises of health, and do not at the same time consider this, that there exist four other entities, of equal power with the stars. But we ought to have a game with them, for what is the good of a cat without a mouse, and what is the good of a prince without a fool? Clearly, the physiomantist has also concocted a similar history, yet he will never excite our tears. He promises health, nor does he think of the four entities of which he is ignorant. For he augurs from the sole natural entity, and says nothing about all the rest, which tickles us not a little. It is the part of a well-informed man to declare of many things those which depend upon a course, for of motions or courses there are five, of which man is only one. He who omits some of these motions and proceeds with the rest is truly a vain prophet. To divide and to speak according to division, each according as he has learnt, according to his judgment and opinion, this is extremely praiseworthy. Thus, accordingly, the pyromantic Entista delivers his judgment concerning spirits; similarly, the physiognomical Entista prophesies of the nature of man; the theological Entista of the course of God; the astronomical Entista of the stars. Each by himself is a liar, but they are true and just if they unite in one. We tell you this lest you should proceed to prophesy before you have learned the five entities of the entities. Then indeed we shall repress our laughter.

---

* The *Libellus Prologorum* prefixed to the *Textus Paramiri* observes that there are altogether five modes of cure, which is as much as to say that there are five medicines, or five arts, or five faculties, or five physicians. Each of these faculties, taken separately, is sufficient for the cure of all diseases. The professors of the first are termed natural physicians, for this reason, that they effect cures by the administration of contraries. Those of the second are termed specific physicians; they cure all diseases by means of the *ens specificum*. The professors of the third are called magical physicians; they effect the cure of diseases by the use of magical words and characters. The spiritual physicians are those who understand the nature of the spirits of roots and herbs, and have them under their control. The last are the physicians who heal by faith.

CHAPTER VII.

In order that you may have a fundamental knowledge of the Alchemist, know now that God has dispensed to each creature His own substance and all things which are necessary for this, not for His own regulation, but for the use of those who need those things which are conjoined with poison. That creature has within him, in his own body, one who separates the poison from whatever is applied to the body. This is in very truth an Alchemist, because in his mode of action he makes use of chemical art. He separates the evil from the good. He transmutes the good into tincture. He tinges the body for the sake of its life. He arranges and disposes all that is subject to Nature in it, and tinges it so that it turns into blood and flesh. This Alchemist dwells in the bowel, as in his instrument, with which he decocts and where he operates. Understand the matter thus. Whatever flesh man eats has in it both poison and good. In the act of eating, all things are regarded as healthful and good. Under the good, indeed, poison is latent; but under the bad there is nothing good. As soon as ever food, that is to say, flesh, is taken into the stomach, the Alchemist, immediately fastening upon it, effects a separation. Whatever does not tend to the health of the body he puts aside in its special places; whatever he finds good he also sequestrates into its proper abodes. This is the Divine ordinance. In this way the body is preserved so that it shall not be killed by the poison of what it eats. Now, this separation is made by means of the Alchemist without anything being done on the part of man. And thus it is with the virtue and power of the Alchemist in man.

CHAPTER VIII

Understand, moreover, after the following manner, how, in every single thing which man takes for his use there is a poison hidden under what is good. The essence is that which sustains the man. The poison, on the other hand, is that which destroys him and brings diseases upon him. And this is true of every alimentary substance, without exception, in respect of that animal which uses it. Now, physicians, attend to this! If it be thus with the aliment of the body and the body cannot do without it, but is altogether dependent upon it, then the body simply takes the aliment, such as it is, under the twofold species of good and ill, and delegates to the Alchemist the duty of separation. Now, if the Alchemist be weak, so that with all his care he avails not to separate the poison from that part which is evil, then, from the poison and the good there arises a combined putrefaction and, eventually, a kind of digestion, and this it is which inflicts diseases of humanity upon us. For every disease in man begotten of the Entity of Poison emanates from a putrefied digestion, which ought to take place so gently that the Alchemist should perceive in it no measure of excess. But when digestion is interrupted, then the Alchemist cannot remain perfect in his instrument. So, then, corruption necessarily ensues, and this, in its turn, is the mother of all diseases. This is what physicians ought most carefully to watch, and it should not be involved in any of your intricacies. Corruption defiles the body, and it is produced in this way. Water which is clear and limpid can at pleasure be tinged with any colour. The body is like such water: corruption is the colouring matter; and there is no such colour which does not derive its origin from decay. It is at once the signal and proof of poison.

CHAPTER IX

Learn this with the view of more fully following up the subject, that corruption is produced in two ways, locally and emunctorially, according to the following method: If, as we have said, there be an Alchemist present in digestion, and if in the process of separation, he succumbs to the fault of defective digestion, then in place of him there is generated putridity, which is poison. Everything putrid is poisonous for that place where it is detained, and so becomes the mother of certain and deadly poison. For putridity corrupts that which is good, and if the good be stripped of its virtue, then the evil triumphs over the good, and this good no longer appears otherwise than under the false appearance of that good which is really subject to putridity. And so this becomes the source of diseases which, in their turn, are subject to it. But know that which is produced emunctorially occurs through the failure of the expulsive nature, in the following manner: If the Alchemist expels the poison, he does so in every case by the proper emunctories; the white sulphur by the nostrils, arsenic by the ears, dung through the anus, and so other poisons according as each has its own special emunctory. Now, if one of these poisons he hindered by the weakness of Nature, or by itself and in other ways, it then becomes the mother of the diseases which are subject to it. So, universally in all diseases two sources are patent. We will not say more on this subject.

CHAPTER X

Moreover, as has been said above on the subject of natural Alchemy, that it is situated in every animal, on account of that separation which must take place in the bowel, listen to the following doctrine as to how, in the aforesaid manner, all other diseases also can be investigated and searched out: If the man be well and strong in respect of all entities—if, for example, he has a suitable Alchemist who separates well with appropriate instruments, reservoirs, and emunctories, then it is necessary, in addition to the instruments, to regard many other matters, especially to see that the stars are favourable, and that all the other entities are well disposed. And yet these general entities do not greatly affect us; for assuming that they are all good and effectual, still many accidents happen to the body which either break or spoil, befoul or impede, the reservoirs or emunctories. Fire is contrary to the Nature and to the body. For this in its quality, nature, ardour, dryness, and other force, can so corrupt us that by its presence the instruments of the Alchemist are violated, and so he appears to be of feeble powers afterwards. So, too, water itself is adverse in its nature to our body and its reservoirs, so much so that the instruments are either stopped up or perverted by it, or in some other respect altered. The same is the case with the air and with other necessary things, and also with external accidents, which are of universal power, so that they break asunder, change, and render useless the instruments and emunctories. Then, too, the Alchemist, being weak and dead, proves unequal to the accomplishment of his work.

CHAPTER XI

But it should not escape your notice that the reservoirs, instruments, and emunctories can be corrupted through the mouth by the air, food, drink, and other things of this kind, in the following way : The air which we breathe is not without its poison, and to this we are specially subject. Concerning the quantity of food and drink, and its bad quality, which disagrees with the organism of the body, the truth is that by these means the organs are thrown out of order to an excessive degree, and that, in this way, the Alchemist is clearly disturbed in his operations. Hence ensue digestion, putrefaction, and corruption. And whatever be the properties of the poison which man takes, such a nature the bowel assumes, and, together with the bowel, all the rest of the body. Thereupon this becomes the mother of diseases in that self-same body. Hence, you physicians ought to understand that one poison and not more produces the mother of diseases. Thus, if you eat flesh, herbs, pulse, spices, and from the consumption of these poison is generated in the belly, then it is not all these foods which are the cause, but only one of them, as, for instance, the poison of the herb, or of the pulse, or of the spices. This you should consider a great secret. For if you know this thoroughly, what poison is the mother of diseases, then we will allow you with justice to be called physicians. In this way you will have discovered what remedy you ought to use ; otherwise, you will attempt this in vain. This becomes for you the foundation of the mother of all diseases, whereof indefinite numbers are reckoned up.

CHAPTER XII

We will now communicate to you some brief information on the subject of poisons that you may understand to what and to what kind of poisons we allude. We have pointed out to you that poison exists in all foods. From food, therefore, is educed a certain entity which has power over our bodies. We afterwards explained the Alchemist who is in our bodies, who for the well-being of the body separates the poison from the food by means of his instruments and reservoirs. When this has been done, the essence passes off into a tincture of the body. The poison goes from the body through the emunctories. Whilst each operation proceeds in this way man is sound and healthy by means of the entity itself. At the same time, also, we mentioned the hostile accidents which might impinge upon the entity itself so as to destroy it. In this way it becomes the mother of diseases. Having repeated so much, now let us speak about the different kinds of poisons. I think you now understand what the emunctories are, and how many of them there are. Reasoning from these you can get at a knowledge of poisons. Whatever exudes substantially through the pores of the skin is resolved Mercury. What is excreted through the nostrils is white sulphur. Arsenic is ejected through the ears ; sulphur through the eyes. Through the bladder there is a resolution of salt ; through the anus putrefied sulphur. Possibly your reason may seek to know under what form and appearance each of these can be recognised ; but that is a matter which our present parenthetical treatise does not include. You will gather from our book on the Human Construction of Philosophy the fundamental principles which it is necessary for a physician to know. There, too, are given at some length the appropriate remedies in many cases, such, for example, as in putrefactions, and it will be well for you to read these. In the same treatise, too, you will learn in what way poison is latent in food.*

CHAPTER XIII

We will give you an illustration from which you may briefly learn how poison lurks under aliment ; and how it is that the condition of a substance which is perfect in itself becomes vicious and poisonous in respect of men and animals who use it. The ox is created with all appendages sufficient for its own use, its skin adapted to the accidents of its flesh, its emunctories ready to use for the Alchemist. But this illustration does not seem altogether to square with our purpose. We will give another. The ox is created, with a view to its own requirements, in that form wherein we find it ; not for man by way of nutriment in food. Mark, then ; half of that ox is a poison to man. If the ox had been created for man's sake alone, it would not have needed horns, bones, or hoofs. There is no nutriment in these ; and what is made from them does not come under the category of necessities. You see, therefore, that in respect of itself, the ox is created altogether good, nor is there in itself anything which it could do without or which it has in excess. But now, if that ox is taken for human food then man eats at the same time that which is hurtful to himself, which is, in fact, poison, but which to that ox had not been poison. This, then, must be separated from the nature of man, and the work of separation is undertaken by the Alchemist, in cases where several poisons are generated without any provision for carrying them off. By the operation of the Alchemist each poison is draughted off to its emunctories, and these are filled with such poisons. But every Alchemist among men can perform the same office as the Alchemist in the body ; no art is lacking to him. This should be an example to every one of them that as the Alchemist of Nature works so he himself should endeavour to

---

* This treatise cannot be identified from its title among the extant writings of Paracelsus. The matters referred to are the subject of frequent instruction throughout the medical books,

work. And if poisons are so separated that poison no longer appears, think how even from a caul is produced the most beautiful golden oil, which, nevertheless, is of all oils the most detestable. The mucus of the nostrils, too, is not reckoned a poison, but it is, nevertheless, a most accursed poison, from which arise all the diseases of distillations, and it can be easily recognised from these diseases.

PARTICLE I

We seem to have sufficiently explained the entity of poison, namely, that it proceeds solely from that which we use as food and drink. Hence, further note that digestion is the same as corruption, if it is corrupted. Next, further, know, how every poison is generated in its place; and how in the process of time either diseases or deaths are occasioned from that poison.

PARTICLE II

Notwithstanding, in dealing with the entity in question, we shall not explain to you the manner in which every disease originates from the above-mentioned poisons of food, which poisons are removed to their emunctories. However, to avoid mistakes, pass it over in this parenthesis, and seek it in the Books of the Origin of Diseases. We shall there clearly explain it to you according to this fundamental principle. So ye shall at once understand what are the diseases of arsenic, salt, sulphur, and mercury, according to the distribution of every form and species, just as it is fitted to itself and to the generation of diseases. Thus we desire to conclude these matters with this entity. We wish them to be understood as an introduction to all our other books.

HERE ENDS THE ALCHEMIST OF NATURE

PART III.

HERMETIC PHILOSOPHY.

THE PHILOSOPHY ADDRESSED TO THE ATHENIANS*

BOOK THE FIRST

TEXT I

OF all created things the condition whereof is transitory and frail, there is only one single principle. Included herein were latent all created things which the æther embraces in its scope. This is as much as to say that all created things proceeded from one matter, not each one separately from its own peculiar matter. This common matter of all things is the Great Mystery. Its comprehension could not be prefigured or shaped by any certain essence or idea, neither could it incline to any properties, seeing that it was free at once from colour and from elementary nature. Wherever the æther is diffused, there also the orb of the Great Mystery lies extended. This Great Mystery is the mother of all the elements, and at the same time the spleen of all the stars, trees, and carnal creatures. As children come forth from the mother, so from the Great Mystery are generated all created things, both those endowed with sense and those which are destitute thereof, all things uniformly. So, then, the Great Mystery is the only mother of all ephemeral things, from which these are born and derived, not in order of succession or continuation, but they came forth at one and the same time, in one creation, matter, form, essence, nature, and inclination.

---

* The whole literature of alchemy, so far, at least, as regards the Western world, appeals to the cosmological philosophy contained in the writings attributed to Hermes Trismegistus as to its source and fountain-head, and a right understanding of its mysteries is regarded by its authors as impossible in the absence of a right understanding of that philosophy. At the same time, the cosmology of the Hermetic books has at first sight nothing to do with alchemy. In the same way, the cosmological philosophy of Paracelsus does not at first sight seem to have any distinct bearing upon the alchemy of Paracelsus, and yet it is a complement of his alchemy, and is indispensable to students thereof. The editors of the Geneva folio regarded it in this light, and collected his chemical and philosophical works into a distinct volume. It is therefore advisable that readers of the present translation should have an opportunity of judging after what manner the physician of Hohenheim was accustomed to philosophise hermetically. The two most compendious treatises are here given. The Philosophy to the Athenians, that is, to the followers of Aristotle, deals cosmologically with that separation of the elements which is the subject of such frequent reference in other writings of Paracelsus. The *Interpretatio Alia Astronomiæ* is concerned with man and the sciences in relation to the greater world. It illustrates many matters which have been the subject of reference in the sections devoted to Hermetic Alchemy and Hermetic Medicine. Paracelsus is never characterised by extreme lucidity, and if his alchemy, his medicine, and his philosophy are not illustrated by each other, seeing that he intimately connected them together, their individual difficulties are likely to be increased in proportion. At the same time, it is requisite, for many reasons, that this section should be kept within somewhat narrow limits.

TEXT II

This mystery was such that none other like it ever appeared to any creature; this was the first matter from which all transitory things sprang; and it cannot be better understood than by considering the urine of man. This is produced from water, air, earth, and fire. Of these, no one is like another, and yet all the elements proceed from thence to another generation, and so on to a third generation. And yet, as the urine is only a creature, there may be some difference between this and that. The Great Mystery is uncreated, and was prepared by the Great Artificer Himself. No other will ever be produced like it; neither does it return or is it brought back to itself. For as the cheese does not again become milk, so neither does the generation hereof return to its own primal matter. For though all things may be reduced to their pristine nature and condition, still they do not go back to the Mystery. That which is consumed cannot be brought back again. But it can return to what it was before the Mystery.

TEXT III

Moreover, although the Great Mystery appertains to all created things, sensible and insensible, still, neither growing things, nor animals, nor the like, were created therein; but the truth about it is this, that it left and assigned to all, that is, to men and animals, general mysteries; and to those of each sort it gave the mystery of self-propagation according to their own form, to each its own essence. And so, by a similar example, it conferred on each of the rest alike the special mystery to produce its own shape by itself. From the same origin spring also those mysteries out of which another mystery can be produced, which, also, the primal mystery arranged. For a star (*alias* stercus) is the mystery of scarabæi, flies, and gnats. Milk is the mystery of cheese, butter, and other substances which belong to this class. Cheese is the mystery of worms that grow in it. So, again, in turn, worms are the mystery of their fæces. In this way, therefore, twofold mysteries exist; one the Great Mystery, which is the mystery uncreated. The rest, as if springing out of it, are called special mysteries.

TEXT IV

Since, therefore, it is certain that all perishable things sprang from, and were produced by, the Uncreated Mystery, it should be known that no one thing was created sooner or later than another, or this or that by itself separately, but all were produced at one and the same time, and together. For the supreme arcanum, that is, the goodness of the Creator, created or brought together all things into the uncreated, not, indeed, formally, not essentially, not qualitatively; but each one was latent in the uncreated, as an image or a statue in a block of wood. For as this statue is not seen until the rest of the superfluous wood is cut away, but when this is done, the statue is recognised, so with the Uncreated Mystery; all that is fleshly, whether sensible or insensible, arrived straightway at its form and species by a deliberately planned process of separation. Here no section was lost or passed away, but all found their way to their own form and essence and to their own likes. Nowhere, in any age, could sculptor be found so careful and industrious in the work of separation who by like art could utilise the smallest and most shortlived thing, and shape it into a living being.

TEXT V

In this way it should be understood, not that a house was built out of the Great Mystery, or that all animals were brought together, piled up, and then perfected, and that the same was done with other growing things; not so, but as a physician makes up some compound, of many virtues, though it be but a single matter, in which appear none of those virtues, which lie concealed under that particular species. So must we suppose that creatures of all kinds, which are comprised in the ether, were brought together into the Great Mystery and set in order, not, indeed, perfectly, according to their substance, form, or essence, but according to another subtle standard of perfection, which is hidden from us mortals, and according to which all things are included in one. For all of us are sprung from what is perishable and mortal, and are born just as if by the procreation of Saturn, who in his separation puts forth all sorts of forms and colours, of which not a single one shews visibly in himself. Since, then, the mysteries of Saturn exhibit procreations of this kind, much more surely will the Great Mystery have this miracle in itself; in the separation whereof all foreign and superfluous matter has been cut away; yet nothing has been found so empty and unserviceable as not to produce some growth or useful matter from itself.

TEXT VI

Know, too, that in the cutting or carving of the Great Mystery different fragments fell down, and some went to flesh, of which the species and forms are infinite in number; some to monsters of the sea, also marvellous in their variety; others to herbs; not a few to wood; more still to stones and metals. As to how Almighty God carved these things, there are at least two methods of art in answering such inquiries. The first is that He constantly arranged for life and increase. The second, that it was not a single and everywhere similar matter which fell down. If a statue is carved out of wood, all the chips cut from that block are wooden too. Here, however, this was not the case; but a separate form and motion were given to each.

TEXT VII

In this way, distribution ensued upon the working of the Great Mystery, and the things separated from the superfluous ones shone forth. At the same time, from these same superfluities which were cut off, other and diverse things were produced. For the Great Mystery was not elementary, though the elements themselves were latent therein. Nor was it carnal, though all the races of men were comprehended in it. Neither was it wood or stone; but the matter was such that, whilst it embraced every mortal thing in its undivided essence, it afterwards, in the process of separation, conferred upon each one by itself its own special essence and form. Something of the same kind takes place with regard to food. When a man eats it, flesh is generated therefrom, though the food itself in no wise resembled flesh. If it be allowed to putrefy, then grass grows up from it, though there was nothing in the flesh that was like grass. And this is much more the case with the Great Mystery. For in the Mysteries it is abundantly clear that one has gone off into stones, another into flesh, another into herbs, and so on in different and infinitely varied forms.

TEXT VIII

The separation, then, having been now made, and everything being reduced to its peculiar shape and property, so that each shall subsist by itself, then, at length, the substantial matter can be distinguished. What was fit for compaction has been compacted, the rest (so far as its substance is concerned) remaining empty and thin. For when the compaction first took place, the whole could not be equally compacted, but the greater part remained void. This is clearly shewn in the case of water. If this be coagulated, the mass or quantity of that which is compacted is small. The same takes place in the separation of the elements. The whole compaction took place—stones, metals, wood, flesh, and the like. The rest remained more rare and void, each single thing according to its own nature and the properties of the planets. And so the Great Mystery in its compaction was just like smoke which is diffused far and wide. But it does not contain in itself much substance, save a little soot. The rest of the space which the smoke occupies is pure, clear air, as can be seen in the separation of the smoke from the soot.

TEXT IX

The principle, mother, and begetter of all generation was Separation. It is true that men ought not to philosophise about these things beyond the grasp of human reason; but the following is the method of learning about such things, how they come to pass. If vinegar be mixed with warm milk, there begins a separation of the heterogeneous matters in many ways. The truphat of the minerals brings each metal to its own nature. So was it in the Mystery. Like macerated tincture of silver, so the Great Mystery, by penetrating, reduced every single thing to its own special essence. With wonderful skill it divided and separated everything, so that each substance was assigned to its due form. In truth, that magic which had such an entrance was a special miracle. If it were divinely brought about by Deity, we shall in vain strive to compass it in our philosophy. God has not disclosed Himself to us by means of this. But if that were natural magic, it certainly was very wonderful, marked by intensest penetration, and most rapid separation, the like whereof Nature can never again give or express. For whilst that operation went on, part of the things was cleft asunder to the elements, part went to other invisible things, part went to produce vegetables, and this is deservedly held to be a singular and supreme marvel.

TEXT X

So when the Great Mystery was filled with such essence and deity, with the addition of eternal power, before all creatures were made the work of separation began.* When this had commenced, afterwards every creature emerged and shone forth with its free will; in which state all will afterwards flourish up to the end of all things, that is, until that great harvest in which everything shall be pregnant with its fruits, and those fruits shall be reaped and carried into the barn; for the harvest is the end of its fruit, and signifies nothing else than the corporeal destruction of all things. The number of those fruits is, indeed, almost infinite; but the harvest is one wherein all the fruits of creation shall be cut down and gathered into the barn. No less marvellous will be this harvest, the end of all things, than was stupendous at the beginning that Great Mystery. Although the free will of all things is the cause of their mutual affection and destruction, nothing exists without friendship and enmity; and free will exists and flourishes only in virtues; but it is friendly or adverse in its effects. These belong in no way to separation. This is the great divider, which gives to everything its form and its essence.

TEXT XI

But in the beginning of the Great Mystery of the separation of all things, there went forth first the separation of the elements, so that before all else those elements broke out into action, and each in its own essence. Fire became heaven, and the chest of the firmament. The air was made mere emptiness, where nothing appears or is visible, occupying that place where no substance or corporeal matter was located. This is the chest or cloister of the invisible Fates. The water went off into liquidity, and found a seat for itself around the channels and cavities of the centre, within the other elements and the æther. This is the chest of the nymphs and sea-monsters. The land was coagulated into the earth, which is not sustained by the other elements, but propped up by the columns of Archialtis, which are the mighty marvels of God. The earth is the chest of growing things, which are nourished by the earth. Such separation was the beginning of all creatures and the first distribution both of these and of others.

---

* The Book of the Philosophy of the Celestial Firmament, otherwise the *Philosophia Sagax*, thus explains the *rationale* of that primal creation by which the heaven, the earth, and all creatures were made : The first body ; after the body the Rector or Moderator ; then the sensible body ; after this the King governing it ; lastly, the King ruling men. The first body is that of the upper and lower sphere; the Rector is the living spirit which informs it, and this is the motive power. The Rector of man is the divine reason with which he is endowed,

TEXT XII

The elements having been thus produced according to their essence, and separated from one another, so that each should subsist in its own place, and no one encroach upon another, a second separation ensued upon the first, and this emanated from the elements themselves. Thus, all that was latent in the fire became transformed into the heavens, one part being, as it were, the ark or cloister, while the other developed from it as a flower from its stalk. In this way the stars, the planets, and all that the firmament contains, were produced. But these were begotten from the element, not as a stalk with its flowers grows out of the earth. These, indeed, grow from the earth itself, but the stars are produced from the heavens by separation alone, as the flowers of silver ascend and separate themselves. Thus all the firmaments were separated from the fire. Before the firmament was separated from the fire, this fire had existed as the one universal element. For as a tree in winter exists only as a tree, but when summer comes on, the same tree, if those leaves which have to be separated are removed, still puts forth its flowers and fruits (for this is the time of gathering and of separation), so must be understood as like in all respects that ingathering in the separation of the Great Mystery, which could not any longer restrain itself or be delayed.

TEXT XIII

To the separation of the elements there succeeded another separation from the air, made at the same time with the fire. The whole air was predestinated to all the elements. But it is not in the other elements in the way and manner of mixture. It lays hold of all kinds of things in all the elements, and seizes upon these. Nor does it occupy that which was possessed before. No mixture of the elements remained joined or united, but each separate element withdrew according to its own pleasure, in no way united or conjoined with the rest. Now, after the element had in this way withdrawn from the Great Mystery, there were forthwith distributed from it fates, impressions, incantations, superstitions, evil deeds, dreams, divinations, lots, visions, apparitions, fatacesti, melosiniæ, spirits, diemeæ, durdales, and neufareni. From this separation of the aforesaid, which had now been accomplished, there was assigned to each its own prearranged seat, and its own peculiar essence was predestinated. Thus it befell that things in themselves invisible became to us objects of our perception. No element was created by the supreme arcanum more subtle than the air. The diemeæ dwell in hard rocks. They were so created, together with the air, in vacuity. The durdales withdrew to the trees. A separation of them was made into a substance of this kind. The neufareni dwell in the air of the earth or in the pores of the earth. The melosiniæ took up their abode in human blood. The separation of them was made out of the air into bodies and flesh. The spirits were distributed into the air, which is in chaos. All the rest are, and abide in, special places of the air, each occupying its own definite position, and being separated from the element of air, but still so that it of necessity dwells therein, and cannot change that position.

TEXT XIV

By the separation of the elements the water was segregated to the place predestined for it. In this way, everything that was latent in its elementary virtue and property was more fully segregated by another separation, and the water was divided into many special mysteries, all of which had been moulded from the element of water. A certain part, by means of that separation, was cut off to form fishes of manifold forms and kinds; a certain part went to form fleshy animals; some went to salt, and no small portion to marine plants, as corals, trina, and citrones. Many also went to form marine monsters, contrary to the manner and course of all the elements. Not a few went to nymphs, syrens, dramas, lorinds, and nesder. Some went to rational creatures bearing in their bodies something eternal and propagating their kind; some which finally die out altogether, and some which, in course of time, are at length separated. For the perfect separation of the elements has not yet been fully made. As the great harvest overwhelms us, or draws nigh year by year, new growths may emerge in the element of water. And, indeed, that separation was made at the same moment as the separation of the elements, in one day, and by the motion of sequestration. And, so, by that means everything spending its existence in the water, simultaneously and in a single moment, was created and revealed by the process of separation.

TEXT XV

In like manner, also, when the element of earth had been separated from the rest, a terrestrial separation was made, that is to say, of all things which are, or have been, born from the earth. Four elements, in all respects alike, were latent in the Great Mystery at the time of the first creation. The same, in like manner, were also divided at one and the same time. Moreover, these were also, in like manner, divided from one another in the second separation, which is called elementary. By such elementary separation there were divided out of the element of earth, severally, sensible and insensible things, eternal and non-eternal, each being allotted its own essence and free will. Whatever was of a wooden nature therein was made wood. A second went off into metallic minerals, a third into marcasite, talc, bismuth, granate, metallic cobalt, pyrites, and many other substances; a fourth into gems of manifold forms and kinds, and to stones, sands, and chalk; a fifth into fruits, flowers, herbs, and seeds; a sixth into sensible animals, of whom one part are partakers of eternity, as men, the others cut off from it, as cattle and the rest. Very many species and differences can be enumerated; for in the terrestrial element far more species have been separated than in any other of the rest. For by means of the semen, and of the congress of two, the father and mother, to wit, all things were propagated, and this was not arranged and predestinated after the same fashion in the other elements. Here are the gnomes and the sylvesters and the lemures, of which the one are destined for the mountains, others for the woods, and the rest only for the night. Moreover, the giants were separated to the third generation. Great essences were also distributed, forming, as it were, stupendous miracles among men, cattle, and growing things. This is a matter hard to be understood by any philosophy and so is esteemed to be brought about contrary to the serial order and the customary method of Nature.

TEXT XVI

After that, as has been said, the four elements of things were in the beginning severally separated from one single matter, in which, however, their complexion and essence were not present—those complexions and natures emerged by that process of separation. The warm and the dry withdrew to the heavens and the firmament, each falling according to its own property. The warm and moist withdrew to the air, whereupon the warm and the moist were invisibly separated. The cold and moist was cleft asunder to the sea, and places bordering thereupon. The cold and dry degenerated to the earth and all terrestrial things. But contrarieties were originated from this separation of the elements, which appear in no respect like their elements. Among these is reckoned lime, which in respect of its nature is not fire, but originates in fire. The cause of this is, that in the course of the separation of the element, its dissolution departed too far from the fiery nature. Fire contains in itself both coldness and humidity. In fact, fire is fourfold. So, again, the colours which proceed from fire are not always alike. One fire makes a white and a lazurium colour. A dry fire makes red and green. A humid fire makes ashen and black. A cold fire makes saffron and red. For this reason one procreation is warmer than another, because the one fire was more or less graduated than the other. For fire was not only one simple thing, but some hundred kinds existed of which no one had precisely the same grade as another. There was a procreation of each one, as if it were made a certain pre-ordained mystery according to its definite subject.

TEXT XVII

Moreover, neither did the water itself obtain the complexion of a simple species. An infinite number of waters were latent in that element, yet all were veritable waters. It is not matter of discovery for a philosopher that the element of water is only cold and moist of itself. It is a hundred times colder, and not more moist, and, moreover, this is to be referred not so much to its warmth as to its coldness. The element of water does not live or flourish only in cold and moisture of one degree; nay, indeed, it does not even exist of one degree. Some waters are springs, and these are manifold. Some are seas, which again are very numerous, and very different one from another. Some are streams and rivers, whereof no single one is precisely like any other. Some aqueous elements were appointed to stones, as in the case of the beryl, the crystal, the chalcedony, the amethyst. Some went to plants, as the coral and amber; some to chime, as the liquid of life. Not a few went to the earth, as the liquid of the earth. Such are the elements of water, manifold in species. Those things, for instance, which grow out of the ground from seed scattered there are also referred to the element of water. So, too, those which are fleshy, as the nymphs, are referred also to the element of water. For although, in this instance, the element of water must be understood to be transmuted into another complexion, it never lays aside or oversteps the nature of the element itself, from which it proceeds. Whatever is of water, that again becomes water. So, too, whatever comes from fire again becomes fire, what comes from earth becomes earth, and what from air becomes air.

TEXT XVIII

By parity of reasoning, too, it is plain that all things which are constituted from earth retain the nature of the same. Although mineral liquids are taken to be fire, yet they are not really so. Not even sulphur burns so much as to make it a fiery element. Indeed, the cold as well as the warm burns. That which in burning produces an ash is not the element of fire, but a fire of the earth, and that fire must not be esteemed an element. It is not an element, but only a consumption of the earth, or of its substance. Water itself can be quite as easily made to burn and to cause a conflagration. If this be so, then this is aqueous fire. Besides, the mere fact that the fire of earth burns and flames does not justify our considering it to be igneous, though it be true that it is in some respects like fire. That philosopher is at once simple and sensuous who names an element from what he perceives. The element is something widely different from such a fire as this. Why do we say so? Everything that moistens is not the element of water. The element of earth can be reduced to water, yet it still always remains earth. In the same way, whatever is in earth belongs to the element of earth. For it exists and is known by the properties of that from which it proceeded, or to which it appears like. The hard flint and the chalcedony alike emit fire from themselves. That fire, however, is not an elemental fire, but a strong expression in great hardness.

TEXT XIX

The element of air contains within itself a large number of procreations, which, nevertheless, are all merely air. Every philosopher should know that an element does not procreate anything else save of the same kind as it is *per se*. Like is constantly produced from like. So, therefore, since the air is invisible, it cannot beget any visible thing of itself. In the same way, since it is impalpable, it produces nothing which is palpable. Thus it melosiniates

—if I may so say. If the melosinia be from the air it is air, nothing else. Still, a conjunction takes place with some other element, which, in this case, is the earth. For here a conjunction can be made from the air to form a human being, just as in all evil deeds and incantations it happens by means of spirits. The truth is the same as with regard to nymphs, who, though they are ranked under the element of water and are nothing else, can still freely have commerce with terrestrials and generate with them. A similar compaction is made from the air. It is visible and tangible, not, however, as a procreation of the primal separation, but only a sequel of the same. As the scarabæus is produced from dung, so a monster can assume bodily shape from the aërial element with aërial speech, thoughts, and action through commixture with the terrestrial. But miracles and consequences of this kind return again to air, just as nymphs return to water, and as man by decay is reduced to earth and consumed because he was born from the earth.

TEXT XX

In this way, then, by means of the great separation, procreations are produced, the one following from the other. But from these procreations other generations have emerged, which have their own mystery in these procreations, not, indeed, as a separation in the exact form of those before mentioned, but as an error, or an abortion, or an excrescence. Thunder arises from the procreations of the firmament, and this itself exists from the element of fire. Thunder is, as it were, the harvest of a star, at that precise point of time in which the thunder has grown mature for acting according to its nature. Magical storms arise from the air and end in the air again, not because the element of air produces them, but rather the spirit of the air. Some are conceived corporeally from fire, as gnomes are from the earth. Just in the same way, dung is produced by men and cattle, not from the earth. The lorind emerges from the naturals of the water, and is not from the water itself. From that abundance, or that error, or that harvest, many other things, too, are begotten. By impressions are born deformed men, women, and other generations like these. From fatal storms arise infections of particular districts, pestilences, and dearness of the market (famine, etc.). From dung are born scarabæi, erucæ, and dalni. From the lorind is collected or understood prophecy of that particular region, and this is in a certain way a presage of future things, stupendous, rare, and never heard of before.

TEXT XXI

We have seen a threefold separation into threefold forms already made from the mystery. There remains, in like manner, a fourth and ultimate separation. It will be the last of all, none will come after it. Then all the others will perish, and not even the Mystery will remain. By that separation all things are reduced to their supreme principle, and that only remains which existed before the Great Mystery, and is eternal. That, however, should not be accepted in the sense that I am to be turned into *something*, or that, in the ultimate separation *something* will be produced out of me, except by death. I am reduced to nothing, as if being produced by reason, I came forth from nothing. And when the sum total of things shall return to their principle it is well that we should know how it takes place. When they are turned to nothing, then they exist in their prime. That prime must be sought in the beginning. What that means—the going into nothingness—must be accounted one of the secrets. The soul which dwells in me was made of *something*. Therefore it does not pass into nothingness, because it is built up of this *something*. Of nothing nothing comes—nothing is generated. A figure painted into a picture, when it is there, has been certainly made of *something*. But we are not thus constituted in the æther out of something—like a picture. Why? Why, because we came forth from the Great Mystery, not from anything procreated therefrom. So we return to nothing. If the figure be blotted out with a sponge, it leaves nothing behind it, and the picture returns to its former shape. So, assuredly, all creatures will be reduced to their primeval state, that is, to nothingness. And if we would know *why* all things must be reduced to nothingness, let us learn that it is on account of the eternity that stirs within rational corporeities. Such an ultimate separation is the ultimate matter. There will be centred all the numerous procreations, permixtions, conversions, transmutations, alterations, and the like; to know the sum total whereof surpasses the scope of the human intelligence.

TEXT XXII

Moreover, as it is certain from philosophy that all those things which minister as auxiliaries to what is perishable and mortal are themselves equally mortal with that to which they subserve, and that what is divided cannot be again conjoined—as coagulated milk is not restored to its freshness—so we must philosophise that the Great Mystery does not return to that from which it proceeded. Hence it should be realised that all creatures are a picture of the supreme arcanum, and so nothing more than, as it were, the colouring spread over a wall. So we spend our time beneath the æther. One or another may be destroyed and turned to nothingness. As a picture is liable to destruction or conflagration, so is the Great Mystery, and we with it. All creatures, together with the Great Mystery itself, perish, are wiped out, and reduced like some wood which burns down to a small heap of ashes. But out of those ashes is made a little glass; the glass is made into a small beryl, and the beryl passes away into wind. In the same way shall we be consumed, passing from one thing to another, until nothing more of us any longer survives. Such as was the beginning of creatures, such shall be their end. If from a little seed a cypress tree can be produced, surely it can be brought back again into as small a compass as that first little seed was. The seed and the beryl are alike ; and as the beginning is from the seed, so in the beryl will be the end. The separation having been thus made, and every single thing brought back to its nature or first principle, that is, *to nothing*, then within the æther there will be nothing that is not eternal, but all will be without end. That from which the non-eternal came into existence will flourish far more widely than before the beginning of creatures. It has no frailty in it, no mortality. And as glass cannot be consumed by a creature, so neither can that eternal essence ever be reduced to nothingness.

TEXT XXIII

When, therefore, the last separation shall be the dissolution of all created things, and one after another is consumed and perishes, from that circumstance the time of those things is recognised. For after the generation of created things there is in them no passing away. The seed of the old supplies the place of that which perishes. In this way, something eternal shews itself in mortals without any distinction, by the renewal of other seed. This the philosopher ignores. He neither admits nor takes count of any eternal seed. And yet he admits putrefaction ; in course of which that which is eternal is taken back again into the eternal. In this respect, man alone of all created things comprises in himself something eternal joined to that which is mortal. Since, therefore, on the aforesaid reasoning, the mortal and the eternal are combined, it should be known that the mortal prepares an essence in the stomach, and sustains the merit of the body. And this is so because the eternal part of man lives for ever, but the mortal part feebly dies out. And as with the body, so with that eternal part which proceeds from that body. This is the great marvel of philosophy, that the mortal part dominates and sways with its nod the eternal part itself, and this depends upon the man himself. He, therefore, in this way, is more the partner of his eternal portion than is that one from whom emanated both the mortal and the eternal parts of man. Hence it can be gathered that the mortal parts of all creatures dwell together—that is to say, the rational and the irrational parts—the one subserving the use of the other in succession, and that an eternal element is inserted into a mortal one, both dwelling together simultaneously. Whence philosophy teaches that all those things which dwell together without disagreement, without deceit or fraud, without good or evil, cannot be destroyed and consumed. But this would be the case if one were opposed to the other. In those, however, wherein the eternal does not dwell no indication is found. But those where the eternal is will not lack such indication. Now, if discord ensues in this way, the one eternal is compelled to give account to the other, and to pay what is due for injuries inflicted by the one upon the other. And when compensation regards the eternal it is not undertaken by that which is mortal ; for though bodies conciliate one another, still, if anything remains over and above here, that is eternal. So, then, is that which alone is eternal judged in us. And whilst one demands account of the other, all the mortal parts which bear the eternal within them are compelled to die out, so that the eternal may alone survive without the concurrence of the body. So is the judgment completed. For that alone is eternal, nor is there more of it; in the final passing away all is mortal.* If, then, that which they embraced within themselves of an eternal nature thus perished, nothing remains but that it was *per se* eternal, and that it nourished and increased the mortal portion. That which profits not does not remain in the creature. All the rest is present only for the sake of the eternal. Hence it happens that, together with that which the mortal holds within itself of eternity, all those elements which sustained it perish and die out. So, then, it is clear, that the end of human affairs is in nothingness, whereto they all tend. From their essence they are separated into nothingness—that is, from something to nothing. In man, however, a perfect separation, that is, of the eternal from the mortal, is yet wanting. For here the judgment is clear which, in all things within the æther, denounces frailty. And if no reason could co-exist with frailty, there would be no passing away in creatures; all would be eternal. There is one single cause for all this, namely, that we mortals do not dwell in justice, nor do we give just judgments, nay, we have not received the faculty of judging what is eternal. This power belongs to the eternal part. And in order that we may gain this power, it is necessary that we should all be collected and come together. And so the passing away of all things is desired.

TEXT XXIV

Since it has been said, then, that all things were created from the Great Primal Mystery, and that they pass away in like manner, it follows as an evident consequence that some such Great Mystery must exist. This is nothing more than saying that a house has been built by a word. This is an attribute of the eternal, just as it is possible for man to elicit fire where there is no fire, and from that which is not fire. The flint has no fire, though it emits fire from itself. So in the Great Mystery all Primal Mysteries were existent in a latent form and after a threefold manner, in respect of vegetative, elementary, and sensible things. The vegetative things were many hundreds, nay, many thousands. Every kind in the Great Mystery had its own specialty. With regard to the elements, there were only four. These four only had their principles. But men were innumerable. One kind were the loripedes, another the cyclopes, another the giants, another the mechili. In like manner, there are those who dwell in the earth, in the air, in the water, in the fire. So, also, every kind of growing thing had its own proper mystery in the Great Mystery, and from thence emerged all the multifarious created things. There are as many mysteries as there are trees or men. The eternal alone dominates in

---

* The text at this point is scarcely capable of an intelligible rendering.

man, and in the whole of his mystery, not in one more than in another. In the Great Mystery was no kind which was not infinitely formed and coloured, one differently from another. Now, all these things are to perish. What then happens we forbear to say, but—A new Mysterium Magnum is not possible. That would be a greater miracle than we are able even to speculate about.

THE PHILOSOPHY ADDRESSED TO THE ATHENIANS

BOOK THE SECOND

TEXT I

SINCE, then, there was a something by which, in the course of separation, all created things issued forth, in the very beginning we must hold that there is some difference of gods, and that of the following kind : Since created things are divided into eternal and mortal, the reason of this is that there existed another creator of mysteries who was not the supreme or the most powerful. For that Supreme One should be the judge and the chastiser of all creatures, and should know how much had been conceded to them, seeing that they might do either good or evil, although not produced by himself. Moreover, created things are continually stirred up and instigated to evil rather than good; they are driven by the stars, by fate, and by the infernal power; when, if created things had issued forth from the Supreme himself, it would be impossible that we should be forced to these properties of evil or of goodness, but we should enjoy free will in all respects without any such impulsion; whereas the creature has not wisdom enough to know good or evil, or to distinguish between what is eternal and what is mortal. For many are foolish and destitute of mind, scarce one in a thousand is wise. The greater part are false prophets, teachers of lies, masters of ignorance. They are openly esteemed foremost persons, though they are in no sense such. The cause is obvious. We are creatures who do not receive what is good and perfect from our masters, but we are chiefly built up by the mortal gods, who in the Great Mystery had indeed some power, but, nevertheless, were placed by the Eternal for judgment, both to themselves and to us.

TEXT II

Now if, as we know by separation, the created universe consisted of four elements only, and proceeded therefrom, these four will be literally the matrices of all creatures, and are called the elements. And although every created thing is so far an element, or has a portion of an element, it is not really like an element, but like the spirit of an element. But then, too, nothing can subsist without an element. Moreover, even the elements themselves cannot coexist. Indeed, there is nothing which consists of four, three, or two elements. Each single element exists apart; and every created thing has only one element. It is mere blindness on the part of those who set down humidity for an element of water, or burning for an element of fire. An element is not to be defined according to body, or substance, or quality. What is visible to the eyes is only the subject or receptacle. But an element is spirit, and lives and flourishes in those things as the soul in the body. This primal matter of the elements is invisible and impalpable, but present in all. For the first matter of the elements is nothing else than life, which all created things possess. What is dead subsists no longer in any element but in ultimate matter wherein flourishes neither any taste, nor virtue, nor force.

TEXT III

All created things in the universe, then, were born of four mothers, that is to say, of four elements. These four elements, it should be remarked, exactly sufficed for the creation of all things. Neither more nor fewer were requisite. In mortal things more than four natures cannot subsist. In immortal natures, it is true, temperaments can subsist, not elements. Whatever "elementure" (if I may use that term) exists, is dissoluble; but, on the other hand, temperature does not involve in itself the idea of dissolution. Its condition is such that nothing can be added to it or taken away from it, nothing decays, nothing perishes. This being the mortal condition, therefore, as has been said, it can be understood that all things subsist in four natures, and each nature has the name of its element. The warm is the element of fire. The cold is the element of earth. The moist is the element of water. The dry is the element of air. The next thing to be considered is how each of the aforesaid natures is what it is peculiarly or separately. Fire is only warm, not dry, not moist. Earth is only cold, not dry, not moist. Water is only moist, and not warm, not cold. Air is only dry, and not warm, not cold. Thus it is they are called elements. They are of one single and simple nature, not of a twofold one. This declaration on the subject holds good for all created things, that an element is that which subsists with body and substance, and thus operates. The chief point to be understood on the subject of elements is that they each and all have only a single and simple nature, moist, dry, cold, or hot. This is the condition of spirits. Every spirit is simple, not double, in its nature; and this is the case with the elements.

TEXT IV

Now, if it should be that composites do exist in us mortals, this scarcely coincides with the opinions of the ancients. The colic comes from the element of fire. It is not compounded of heat and dryness, but it is only hot. In like manner with the other complexions. And so, if any disease be detected where heat and dryness are combined, one can only come to the conclusion that the two elements exist, one in the spleen and the other in other members. Two elements cannot be in the same member. It is certain that in every member dwells some one element which is peculiar to it; a subject which we leave for physicians to define. This, however, cannot consistently be said, that two elements exist at the same time in the same place, or that one and the same element is both warm and moist. Such a composite is not forthcoming. The elements know nothing of a compound, for the reason already given. Where heat is there cold cannot be, nor dryness, nor humidity. So, in the same way, where cold is, there none of the others is present. The same is the case with moisture and dryness. Every element is *per se* simple and solitary, not twofold in its composition. Whatever philosophy may concede as to the possibility of conjunction of elements—as of heat with humidity—all this goes for nothing. No element of water bears heat. No heat can subsist in humidity. Every element stands by itself alone. Cold cannot of itself tolerate dryness. That subsists by itself uncontaminated. This may be said and understood of the special essence of the elements. All dryness is a dissolution of cold. Just as humidity and dryness cannot be conjoined, so much the less can coldness and dryness and humidity, or warmth or dryness meet together and co-exist. As heat and cold are contraries, so heat and cold stand in the relation of contrariety to moisture and dryness.

TEXT V

It would be erroneous if we wished to assert that because all things are made up of four elements, they therefore necessarily exist in conjunction. Every conjunction is a composition. Since they are compounded, they cannot be the Great Mystery. Every Mystery is simple and a single element. There is a similar difference between elements and compounds. An element, as likewise a mystery, can generate a divertallum. A compound can generate nothing except what is like itself, as men make men. But a mystery does not produce a mystery like itself. It produces something different as a divertallum. The element of fire is the producer of stars, planets, and the whole firmament, and yet none of these is formed or constituted like fire. The element of water has constituted water, which is altogether contrary to the element of water. The element of water *per se* is not at all moist. It is true the element itself has this kind of moisture, that it softens stones and hard metals. But this remarkable power of softening is taken from it by substantial water, so that its virtue is not perfect. The element of air is so dry that in one moment it dries all waters. But this power is taken away and destroyed by substantial air. The element of earth is so cold that it reduces all creatures to their ultimate matter, as, for example, water to crystal and * to Dualech, animals to marble, and trees to giants. The foundation of the knowledge of the elements is to understand that they are of such remarkable and prompt activity and efficacy that nothing else like them can be found, or even conceived. The things in which they reside are attracted and assumed by them as if it were by a fate which is corporeal, and without these has not the smallest amount of virtue.

TEXT VI

In order more fully to understand what an element is, it should be realised that an element is really neither more nor less than a soul. Not, indeed, that it is of precisely the same essence as a soul, but it corresponds with a certain degree of resemblance. There is a difference between the elemental soul and the eternal soul. The soul of the elements is the life of all created things. Fire which burns is not the element of fire, as we see it; but its soul, invisible to us, is the element and the life of fire. The element of fire can be present in green wood no less than in fire. But the life itself is not equally present as it is in fire. There is again a difference between the soul and the life. Fire, if it lives, burns. But if it be in its soul, that is, in its element, it lacks all power of burning. It does not follow that a cold thing must have proceeded from a cold element. It may often proceed from a warm one. Many cold things issue forth from the element of fire. Whatever grows is of the element of fire, but in another shape. Whatever is fixed is from the element of earth. Whatever nourishes is from the element of air; and whatever consumes is from the element of water. Growth belongs only to the element of fire. Where that element fails there is no increment. Except the element of earth supplied it there would be no end to growth. This fixes it; that is to say, it supplies a terminus for the element of fire. So, also, unless the element of air were to act, no nutrition could be brought about. By the air alone are all things nourished. Again, nothing can be dissolved or consumed unless the element of water be the cause. By it all things are mortified and reduced to nothing.

TEXT VII

But although the elements are thus in other respects altogether invisible and impalpable, and are thus hidden from us, they have, nevertheless, the power of putting forth their own mysteries. Thus, the element of fire gives forth from itself the firmament; not, indeed, by any corporeal method, but on the scale of the elementary essence. The sun has a body different from that which he has received from the element of fire. Nevertheless, in this same he exists essentially with heat. For his heat does not arise from his motion or his rotation; it is from himself. Though the sun stood still and never moved, he would still put forth his heat and brightness. The crystal has given the sun from the element of fire, though it has no other body than that which it has received from the element of the fire.* For thence the elements are (if I may so say) incorporated. The sun and the other stars, in like manner,

---

* This literal rendering seems to be without a reasonable meaning.

derived their origin from the element of fire, but only in a red colour, in which no heat or glow, but only a kind of dead splendour, inheres. And although signs appear in the sky differing in form and species—which we do not enumerate here—still the same form is to be understood as with ourselves on earth. It is not one only but multifarious, whether known or unknown to us. When the mystery of fire was divided off, a certain something was produced, such as we see. So the stars are daughters of the element of fire, and the sky is nothing but chaos, that is, vapour exhaled from the firmament, and so fervid as to be, in this respect, indescribable. That fervour or heat it is which gives coruscations, and colours, and appearances; for in that region is the pure element of fire, as shall be pointed out at greater length in its proper place.

TEXT VIII

In the same way as the fire put forth its different species and essences, so in like manner did the element of air. And yet there is a certain difference in the four elements as to those things which are procreated from them. Each of them produces after his kind. The firmament is like neither of the other three. Fate is from the air, and is not like those other three. Those signed by the earth cannot be in any way compared to the other three. In the same way are marine monsters related to the rest. For each single element has procreated in itself both rational and irrational things. The sky has in its firmaments rational creatures just as much as the element of earth has. In the same way, too, the fate of the air is distinct as to its signature in the matter of rationality and brutality. And the same is true of earth and of water. Who is he, then, who shall assure us as to the truth about these signed elements? Who are they that have the true faith handed down and entrusted to them, and the right way of salvation, or who alone shall possess eternity—matters which we pass by for the present? It cannot be but that human beings dwell in all four of these just as in one of them, that is, the earth. And on the subject of fate, it should be made known that its generation from the element is manifold, and yet without any body or substance, according to the property of the air (which is incorporeal), together with its habitation. For some things have it corporeally, others incorporeally, as the thing is understood.

TEXT IX

It is perfectly well known that from one seed emerges a root, divided into many filaments. Afterwards rises a stem, then branches are suspended there; flower, fruit, and seed come forth. Exactly similar is the method of manifold procreations from the four elements. Thus, all those which are from one element cohere, as the herb is produced from one seed. Yet all these growths are not exactly like their own seed. Those things procreated from the water are partly men, partly animals, and partly what these feed upon. One element has left its signature as both its necessity and its sustentation; then it also gives sign of its course and its advent, which are easily recognised by the stars, not because these rule or influence us, but only because they run concurrently with us, and imitate the inner movement of our body And no less in the element of water all those things are produced which are produced in the element of earth. The lorind is a commotion and change of this element of water. This moves itself; and the lorind is like a comet. A monster in the water should be accepted in just the same way as an error of the firmament. So in the water a special world is to be recognised, together with its mystery, even to the end of the world. There is no beginning in these save that which is in the other elements; nor is there any other end than is found in the other elements. The only difference is of forms, essence, and natures, accruing to them, with their signatures and elements. Thus we must understand the four worlds according to the four elements and their primary habitations. With regard to justice there is only one eternal, which is to be recognised alike in all four.

TEXT X

In the element of earth there is much intelligence for us, because we ourselves came forth thence. Each like thing understands its like. But the understanding of the four elements flows out from philosophy. This, again, is like, flowing out as it does from the same source as that from which philosophy afterwards rises. Nevertheless, as the element of earth has procreated its signature, so have also the other elements procreated theirs. Just as we have stones, so are they not lacking in the other elements, though those stones may not be made according to the same form as ours, but in their own peculiar form and with their own special properties. The other elements have also their minerals no less than we have. The celestial firmament puts forth its floral growths as well as its mineral products, and these we reckon among miracles. Here, however, we make the greatest possible mistake, because we reckon natural processes among prodigies, and give it out like prophets that this or that appearance of the firmament portends something special, when we ought to understand that such things happen only in the ordinary course of events. But if such thing does occasionally happen we ought to believe that such was our course or condition. In the meantime, if anything suffers from the error of the elements, the other things grow uncertain too. All ought to proceed with a perfect and unimpeded motion, and though the other three elements serve us for nutriment, just in the same way do their other three serve the firmament, the air, the water, and those who dwell in them. One derives its nutriment from the other, just like many trees in a garden. And the defects and errors of the firmament can be observed by us no less than the firmament observes our defects. And the same judgment may be passed as to the rest.

TEXT XI

That philosophy, then, is foolish and vain which leads us to assign all happiness and eternity to our element alone, that is, the earth. And that is a fool's maxim which boasts that we are the noblest of creatures. There are many worlds: and we are not the only beings in our own world. And the ignorance becomes more marked when we fail to recognise those human beings who spring from our own element, as the nocturnals, the gnomes, etc. Though they do not live in the open light of day, nor use the brightness of the firmament, but hate what we enjoy and enjoy what we hate, and, moreover, are like us neither in shape, nor in essence, nor in their sustenance; still, that is not a subject for our wonder. They were made up as they are in the Great Mystery. We are not the only beings made; there were many more whom we do not know. We ought to conclude, then, that not one simple single body but many bodies were included in the Great Mystery, though there existed only in general the eternal and the mortal. But in how many forms and species the elements produced all things cannot be fully told. But let all doubt be removed that eternity belongs to all these. In this way, certainly, many things which are now unknown will be investigated and in different ways become cognisable by us, not concerning those existences only which have the eternal element in them, but of those which sustain and nourish that eternity. For eternity may be understood in two ways. One is that of kingdom and domination; the other that of ornamentation and decoration. It is opposed to all true philosophy to say that flowers lack their own eternity. They may perish and die here; but they will re-appear in the restitution of all things. Nothing has been created out of the Great Mystery which will not inhabit a form beyond the æther.

TEXT XII

The universal procreations indicate that all things must have four mothers, neither more nor fewer. Not, however, that this can be understood from the basis of the general demonstration concerning the present Great Mystery as it appears in the beginning according to its properties; but rather the Great Mystery is known and understood by the ultimate mysteries, and by the procreations which have gone forth having their origin in the first. It is not the beginning but the end which makes either a philosopher or a magistrate. The knowledge of a thing according to its perfect nature is only found at the end of its essence. Perhaps there might have been more elements than have been given to us. In the latest knowledge of all things, however, only four are found. And although we can, indeed, suppose that it would have been easy for God, who only created four, to have created more by the same operation; still, when we see all mortal things made up of these four only, we shall not do ill to believe that more than these would not have been well. It is reasonable to think, perhaps, that after the destruction of the four elements spoken of, certain others will come into existence essentially unlike those before mentioned; or, that after the passing away of the present creation a new Mysterium Magnum may supervene, and we have a better and fuller cognition of it than we had of that which shall then have passed away. We base nothing on this idea, but to every one who would know the origin of the universe we say, that he must understand the world to have sprung from the elements; and as there are four elements, so there are four worlds, and in each exists a peculiar race, each with its own necessities.

TEXT XIII

Though all things exist in the four elements whereof we have spoken, we could not, however, insist that there are four elements in all things, or that four elements abide in all things. The reason is this. The world which is separated and procreated from the element of fire needs neither air, nor water, nor earth. By the same argument, the world of air needs neither of the other three. The same is equally true of the earth and the air (*sic* ? water). For the doctrine of the elements does not lay it down that the world must be sustained by the four elements, but rather that everything is conserved by one element, namely, by that from which it is sprung. And although you may not deny that the firmament nourishes the world by means of its elementary forces, which are all igneous, as they descend to the earth, still that nutrition is not a matter of necessity. Neither is the world going to perish by itself, but suffices for its own sustentation, just as also any other world nourishes itself without the help of the earth. For example, the earth confers nothing on the water as respects its own proper essence, nor, on the other hand, the water on the earth. The same is the case with the air. Yet still we do not take the course hereof to be that each world exists solely, or *per se*, in its own element, but rather that the light from heaven is, as it were, something drawn from the four elements, and most noble in its full and perfect properties. But let no one on this account imagine that the sun received his brightness and his motion from the element of fire, since neither the planets did this, but rather from an arcanum. The splendour of the firmament irradiating the world emanated, not from the element of fire, but from the arcanum. The earth of itself gives Tronum, the water Turas, and the air Samies. These proceed, not from the element, but from the arcanum, and are present in the element. In the arcanum the four worlds agree in this way, that they are useful to each other in turn, afford mutual nutriment, and sustain one another; not, however, from the nature of the elements, for they are themselves elements.

TEXT XIV

That man lives, sees, hears, and the rest, is due not to the elements, but rather to the arcana, and especially to the monarchy. Such is also the case with other creatures. The sole entertainment and nutrition is elementary. Understand that whatever is eternal proceeds from the arcanum, and is the arcanum. Dogs die, and their arcanum remains. A man dies. His arcanum remains, and still more, his soul, which causes him to be so many degrees more worthy than the dog. The same is, in like manner, the case with all growing things. Hence has sprung up an error that in the new *Mysterium Magnum* all created things which have ever existed, will ultimately appear, not essentially, now, but in an arcane manner. We do not say that the arcanum is an essence like the immortal, but that it *is* this in its perfection. The element of fire has an arcanum in itself. From this there are added, or there flow out to the other three, light, brightness, influence, increase, but not from the element. These arcana could exist without the element itself, just as the element without the arcanum. But mark, moreover, that the element of air has in itself an arcanum from which nutriment is supplied to all the other three and to itself, not as an element from itself, but as an arcanum by means of the element. The element of earth has in itself the arcanum of permanence and fixation, which imparts to the others the virtues of duration and generation, so that nothing may perish. The element of water has the arcanum of sustentation for all the elements, and conserves whatever is in them, so that it escapes destruction. In this respect, too, there is a difference between an element and an arcanum. The one is mortal or perishable from the elements, the other permanent in the ultimate *Mysterium Magnum*, wherein all things will be renewed, yet no other things will be produced save what have been.

TEXT XV

We come to the conclusion, then, that all the elements are not joined together, but that they are altogether aërial, or igneous, or terrestrial, or aqueous, solely and without admixture. This, also, is settled, that every element nourishes itself, or that does which is in it or its world. For this reason a medicine drawn from the element of water is of no avail for those who are from the element of earth, or any other, but only nymphs, sirens, and the like. So, also, a terrene medicine does not benefit the three other worlds, but only the animals of its own world. The same is to be said of the air; for in the air there are diseases and physicians—learned and unlearned—just as in our own world, each having its own peculiar mode. So with the fire. If it sometimes happens that nymphs copulate with terrestrials, and children are born, it is open to question whether this can happen from the possibility of ravishment. By this power other aërial beings, such as melosinæ, have intercourse with terrestrials. So, too, the trifertes are borne from the fire to the earth. Now, if three of these unfamiliar beings should come from their world to ours, as we have said, they would be looked upon as gods compared with us, on account of the vast distance between us, and the entirely foreign essence which they possess. But if a man can be carried to them, there is a corresponding power of being borne to them (*sic*). So, then, the elements have no need one of the other; only one is the ark or receptacle of another. As water and earth are mutually separated, so, in like manner, the air and the fire each occupies a special position without any contact of the four elements, save that which is like the party-wall of a house.

TEXT XVI

If there is at length to follow a conjunction and a gathering together in which all things shall return to their primal essence, then that will be an arcanum, and, indeed, according to the aspect or appearance of the elements. For there nothing corporeal can appear from generation, but the appearance and present exhibition of all the generations contained will supply that place, and so all those things will be known to everyone which have before been made, or shall afterwards be made, which also will be as well known as if they were seen before the very eyes; yet, still herein is a hidden perception of the ultimate Great Mystery. And that will be known or made certain not by Nature, but from a knowledge as to the causes of the final separation of the elements and of all created things: where each one will give an account of his own death. That is the cause of the perishable, of the living, and of the permanent. For there will be one Judge who has power from the eternal, and from age to age there has been this one Judge. This, also, is the cause or origin of religions, and of the different mystæ who serve the gods, all of which customs are false and erroneous. For no other judge has ever existed except the one God, who has been the Judge from eternity. It is too impious a piece of folly to wish to worship one who is mortal, perishable, and liable to decay, in place of the Author of creation and the Ruler of the eternal. Whatever dies has no power of governing and of ruling. There is only one way and one religion, nor should others be rashly adopted.

TEXT XVII

If, therefore, whatsoever things are created return to that unto which they were predestinated from the beginning, in that place an arcanum will be produced. For predestination is, as it were, the ultimate matter, which will be without element, and without present essence; but that which is temporated and that which is incorrupt will the more certainly follow. For these things are not understood by the spirit, but from Nature, with this evidence that the eternal follows the mortal. For, if an insensible plant perishes, then something eternal succeeds to its place. There is no frail or fading thing in the whole world which does not substitute in its place something which is eternal. Nothing has been created void, nothing that is mortal without a succession of what is eternal. After the end of all created objects these eternal things will meet together, and be collected not only as nutriments, but as a magistery of Nature, both in the perishable and in the eternal. And thus the eternal is a sign of the dissolution of Nature, and not the beginning of created things, and the end in all things which no nature is without. And although, indeed, the fatal beings, such as melosinæ and nymphs, are to have something eternal behind them, yet we will not here treat of their corruption. For a manifold decay is here involved, inasmuch as there are four worlds. For decay is terrestrial, aërial, igneous, and aqueous. Each of these, with those created with it, is turned and led downwards to decay with the eternal which is left. Nevertheless, these four decays will reduce their eternal portion to one similitude, notably and visibly, not with their works, but with their essence. For each eternal thing is a single habitation, prevailing, however, in many differences.

TEXT XVIII

We must not pass by in silence the evestrum, according to its essence; it is either mortal or immortal. The evestrum is like a shadow on a wall. The shadow grows and originates with the body, and remains with it up to its ultimate matter. The evestrum has its origin with the first generation of everything. Everything animate and inanimate, sensible and insensible, has conjoined with itself an evestrum, just as everything casts a shadow. The trarames is understood to be, as it were, a shadow of the invisible essence. It is born with the reason and the imagination of both intelligent animals and of brute beasts. To philosophise about the evestrum and the trarames belongs to the very highest philosophy. The evestrum gives prophecy, and the trarames gives acumen. To prophecy about what shall happen to a man or animal, or even to a piece of wood, belongs to the shadowed evestrum. What will be the method comes from the trarames. So the evestra either have a beginning or have it not. Those which have a beginning carry dissolution with them, together with something eternal surviving. That which has no beginning possesses the power of sharpening the traramium in the intellect. The mortal evestrum knows the eternal. This knowledge is the mother of prophecy. For the foundation of all intellect is from the evestrum as it is extracted or elicited by the light of Nature. Thus the prophet "evestrates," that is, he vaticinates by means of the evestrum. If the spirit prophecies, that is without the light of Nature. So it is for us fallacious, liable to imposture and uncertain, as also certain and true. And in the same way the trarames also, as the shadow of reason, may be divided.

TEXT XIX

Moreover, in the dissolution of all things the evestrum also and the trarames will be dissolved, but not without some relics of the eternal. So, then, the evestrum is exactly as it were the firmament in the four worlds. The firmament is fourfold, according to the four worlds, divided into four perfect essences, each world perfectly regarding its own creation, because it is itself of that nature: one in the earth from the firmament, one in the water, air, and fire respectively. But the firmament which is in the evestrum is dispersed, not stars which are visible, but these are firmaments to the nymphs which are not stars. Neither do they use stars, but they have a peculiar and proper firmament as fates; the igneous have each their private heaven, earth, abode, dwelling, firmament, stars, planets, and other like things, which are not in the least the one like the other. As water and fire, the substantial and the impalpable, the visible and the invisible, stand mutually related to one another, so is it with matters of this kind. In these the evestrum is divided with respect to fatal things, and remains a shadow of itself after its essence from dissolution; and the evestrum after fire adheres to the igneous man, as another to the aqueous, and another to the terrene. This evestrum falsifies and deludes the world, shadowing itself by fraud from one world to another, shewing visions, splendours, signs, forms, and appearances. Hence arise the evestrum of comets, the evestrum of impressions, the evestrum of miracles. These three are the prophetic evestra and the shadowed evestra.

TEXT XX

It is, however, firstly and chiefly, necessary that we should know this prophet evestrum. For turban is the great essence of this kind, presaging all things which are in the four worlds. For whatever is about to happen by way of prodigy, and against Nature, or against common opinion and life, is known by the prophet evestrum, which is taken, and shadows itself forth from the great turban. It is most necessary that the prophet should know the great turban. It is most difficult to be understood, and is united with reason. It is, therefore, possible for a mortal man to know the great turban even to its extreme resolution. By this all the prophets spoke; for in it are all the signs of the world. From it all evestra are born, from it the comets are shadowed, prodigious stars which arise contrary to the usual course of the heavens. From turban all impressions get their beginning, not from the firmament or from the stars. As often as something unheard of and rare is about to happen, outsiders sally forth and messengers in advance, by whom the coming misfortune is announced to the people. These presages are not natural, but they come from the prophetic evestrum. All epidemics, wars, seditions, have their presages, which arise from turban. Whoever has knowledge of the properties of evestra, he is a prophet and a seer of futurity. For the Most High does not parley with men, nor does he send down angels from His throne to announce these things in advance. They are fore-known and understood from the great turban which many pagans and Jews have worshipped as God, being blinded in their senses and intellect as to who the true God was.

TEXT XXI

Since, then, a shadowed evester is born with every created thing, and arises therefrom, it is well to know that by means thereof may be prognosticated the fortune and life of that thing whose evestrum it is. For example, when an infant is born there is born at the same time with it an evestrum constantly expressed in that same child, so that from the cradle to the hour of death it prophecies concerning the child, and points out what is going to happen. If it is to die, death does not arrive until the evestrum has announced it beforehand, either by knocking, shaking, or falling, or by some other warning of this kind. Whence, if the evestrum be understood, it can be presaged that this is a sign of death. Moreover, the evestrum is united with the eternal. After the death of a man his evestrum remains upon earth and gives signs whether the man is in happiness or misery.

Nor should one say, as simple folks do, that this is the spirit or soul of the man, or that the dead himself is walking about. This is the evestrum of the dead man which does not yield its place until the final end, in which all things will meet. This evestrum works signs. The gods, by their evestrum alone, have wrought miracles; and as the sun in its splendour puts forth his heat, and his nature, and his essence, so it is in our case with the presaging and prophetic evestra, in which our trust should be placed. These evestra regulate sleep and sleeplessness, the prefigurations of future events, the nature of things, reason, concupiscence, and thought.

TEXT XXII

In this way when future events are predicted in the elements by that in which evestra dwell, there will be evestra in water, some in mirrors, some in crystals, and some in polished surfaces; some will be understood by the motion of waters, some by songs and animum; for all these things can—so to say—"evestrate." The great and good God has a mysterious evestrum, in which are seen His essence and attributes. By the mysterial evestrum are known all that is good and all that is illuminated. On the other hand, the Damned Spirit himself has an evestrum in the world by which evil is known, together with all which violates and corrupts the law of Nature. But though these two evestrate, still they do not not in any way affect our life. Only by our evestrum shall we know ourselves. In all creatures are evestra, which are in all respects prophets, whether rational or irrational, whether gifted with sense or destitute of it. The evestrum is the spirit which teaches astronomy (astrology). Not that it is known by prognostications and nativities drawn from the stars; but its essence, if I may so say, is in the evestra, and its being is in these, as an image in a mirror, or a shadow in water or on the ground. And, just as growing things increase or diminish, so is it with the stars; not because from their own nature they have such a course, and damp and cold arise from the earth, but because the essence of the earth is such. So it is shadowed forth in heaven and over different parts according to the evestrum, and not as a power.

TEXT XXIII

Evestra of this kind will be corrupted, yet, nevertheless, they will not perish without the eternal. Nor will the evestra themselves be regarded, but they will dwell near those to whom they belong. Whence each will be its own adviser, so that before all else it may exhort and know itself. The nature and number of evestra is infinite. These guide the sleepers, foreshadow good and evil, search into the thoughts, and accomplish the labours or the works without any bodily motion. So, then, the evestrum is a marvellous matter, and the mother of all things in prophets, in astronomers (astrologers), and in physicians. Unless the understanding thereof be sought through the evestrum there is no knowledge of Nature. As theft betokens punishment, as a cloud foretells rain, and lotium bespeaks disease, so the evestrum shews everything without any exception. By it the sibyls and prophets spoke, but, as it were, in a kind of sleep-wakefulness. In this way the evestra are in the four worlds, the one ever communicating to the other presage, image, and miracle; and by their regeneration they will be found much more wonderful. Nor let us here omit to say that the evestrum is a remnant of the eternal, the sustainer of religions, and the mode of operation for celestial things. For happiness alone and blessedness, and the chief good, and the last judgment, move us, and instigate us the more keenly and more profoundly to search and to investigate what difference there is between these two, the true and the false; and this must be weighed and learnt not spiritually, but naturally.

THE PHILOSOPHY ADDRESSED TO THE ATHENIANS

BOOK THE THIRD

TEXT I

EVERYTHING in existence necessarily has a body. The mode and manner may be understood as being like a smoky spirit, which indeed has substance, but is not a body, nor is it tangible. But, though this be the case, still both bodies and substances can be produced from it. This may be understood from fuming arsenic; since after the generation of a body, nothing more is seen of the fume of the spirit, just as if it had been all reduced to a body. But this is not so. Something of a most subtle nature still remains in that place of generation. Thus by the process of separation there are produced something visible and something invisible. By this method and in this mode all things are propagated. Wood has still surviving the spirit from which it has been separated. So have stones, and so all things, without any exception. Their essence still survives just as it was separated from it. Man, in like manner, is nothing else than a relic and a survival from the separated fume. But mark this, that a certain spirit existed, and from that man is made up, and is most subtle in spirit. This spirit is the index of a twofold eternity, one being the caleruthum, and the other the meritorium. The caleruthum is the indication in the first eternity. It seeks or makes for the other, that is, God. That is a natural cause because all things affect or tend towards that from which they proceeded, or those natures which have been in contact with it; for whatever anything when building up used in the process, that the thing when it is built up desires after and pursues. And this should be understood, that a thing which has been built up does, by Nature, or by natural instinct, tend not towards its builder, but towards that from which it has proceeded. So the human body longs for the matter from which it has been separated, and not for God, since it was not taken out of Him. And that matter is the life and the habitation in which the eternal meritorium abides. So everything returns to its essence.

TEXT II

Since, then, everything has an appetency for its source, that is, for the mystery whence it sprang, it must now be further understood that this is eternal life, and that what comes forth from thence is mortal. None the less, however, there remains in the mortal something eternal, that is, the soul, as may be learnt elsewhere. And if a perishable thing is to return to its pristine condition, that can only be done by a conjunction of what is permanent; then at length there is a collocation and a union of things. The form and substance, however, both of transitory and of non-transitory things, proceeds from that spirit of fume, just as hail or lightning emerges from a cloud. These are corporeal, and that matter from which they have proceeded remains invisible. So, then, it may be laid down that all things spring from the invisible, yet, without its suffering loss, for the matter has always the power of regenerating and recuperating that loss. Hence, also, it happens that the whole world will pass away like snow and return to the same essence of the spirit of smoke, and then will come together or coalesce apart from all tangible essence. In this way, it can be again re-born as at first. Hence, also, it is known that no created thing exists which has been born, but only as it has been built up or created. So, the chief good is constituted in the beginning of all things that anything shall thus proceed from the invisible and become corporeal, and then shall afterwards be separated again from its body, and once more become invisible. Then all things are again joined together and united and reduced to their primal matter. And although, indeed, they may be united, yet still they involve some distinction and difference one from the other. One is the abode of the other; that other is the inmate of the abode. For that is the habitation of all things; sensible and insensible alike must all return to that condition and to that place. For, whether rational or irrational, nothing is free from this change, but will return to its habitation, from which it has been separated, and there appear.

TEXT III

So, too, every body, or every tangible substance, is nothing else but coagulated smoke. Hence it may be assumed that such coagulation is manifold. One kind refers to wood, another to stone, a third to metal. But the body itself is none other than smoke, breathing forth from the matter or the matrix in which it is present. What grows from the ground is a smoke brought forth from the liquid of Mercury, which is various, and emits a manifold smoke for herbs, trees, and the like. But that smoke, if it issues forth from its primal, or as soon as it expires from the matrix touches foreign air, is thereupon coagulated. So this smoke constantly and persistently evaporates. As long, therefore, as it is driven and disturbed, so long the thing grows, but when the ebullition ceases the smoking also ceases. This terminates, too, both the coagulation and the growth. Wood is the smoke from Derses. Therein is latent a specific from which wood is produced. Nor is it only produced from this smoke; it may be produced also from other dersic matter. In like manner, leffas is boiling matter, from the smoke of which all herbs are gendered. For the only predestination of herbs is leffa;

there is no other. God is more wonderful in specifics than in all other natures. Stannar is the mother of metals, furnishing the first matter for metals by its fume. Metals, in fact, are nothing but the coagulated fume from stannar. Enur is the fume of stones. In fine, whatever is corporeal is nothing but coagulated smoke, in which there is latent a specific predestination. All things, too, will ultimately be resolved like smoke; for the specific which coagulates has no power save for a definite time. The same may also be said of coagulation. All bodies will at last pass away and vanish in smoke, and will be terminated only in smoke. This is the consumption of everything corporeal, both living and dead.

TEXT IV

Man is coagulated smoke. Only from the boiling vapours and spermatic members of the body is the coagulation of spermatic matter produced. Man, too, will be resolved into a vapour of this kind; so that death may be like birth. Moreover, we see in ourselves nothing else than that man is coagulated smoke formed by human predestination. Whatsoever, too, is taken or given forth is merely the coagulated fume from liquids. And so whatever is injected is consumed by the life on the same principle, so that the coagulation may be again dissolved and liquefied, as ice is liquefied by the sun, that it may afterwards vanish into the air like smoke. Life consumes everything; for it is the spirit of consumption in all corporeities and substances. Here, too, attention must be given to the preparation of the digested mystery; for if everything is due to return to that state from which it originated, and so anything is given forth, then it is consumed together with the life. This, however, happens only in those things which are not transmuted. Transmutation is not driven back or repressed; and some transmutation is produced by means of life. Thus, then, is transmutation altered into the frailty of the body; but, nevertheless, it is again separated from the body. For in its putrefaction, transmutation has no further power, and in putrefaction the digested mystery ensues as a consequence. In the meantime, there are mutually separated all the properties which man had in himself from herbs and other things, each returning to its own essence. Separation is, in fact, like that process by which, if ten or twelve things are mixed they are again dissevered, so that each regains its own special essence. Thus, eating is nothing else than a dissolution of bodies. Hence the materials of bodies are separated in vomitings and dejections from the bowels, which are simply fœtid smoke mixed with good. Nature, indeed, seeks only the subtle, avoiding what is dense. Stones, metals, and earths—in a word, all things—are dissolved by life; nor is there any other dissolution of them by the body than that which is brought about by its life.

TEXT V

Moreover, it is equally necessary to understand the process by which each separate thing regains its own essence. This cannot be more fitly compared to anything than to fire, which is elicited from a hard flint, flaming and burning beyond all natural knowledge. For, as that hidden fire takes its origin and proceeds to work its effects, in the same form and appearance also is the essence led to its nature. And here reflect that in the beginning there existed only one thing, without any inclination or speciality, and from this afterwards all things issued forth. This origin exactly resembles some well-tempered colour, purple for instance, having in itself no inclination to any other colour, but conspicuous in its just temperature Yet, notwithstanding, in that colour all colours are existent. For the other colours cannot be separated from it—red, green, blue, clay colour, white, black. Each of these colours, again, brings forth other blind colours, while yet each one is by itself entirely and properly tinted. And although many and various colours are latent in them, still, nevertheless, they are all hidden under one colour. In the same manner, too, everything had its essence in the Great Mystery, which the Supreme Architect afterwards separated. The crystal emits fire, not from a fiery nature, but on account of its hardness and solidity. It also hides in itself other elements, not essentially, but materially, ardent fire, blowing air, moistening water, and earth which is black and dry. Besides all these things it possesses in the composition of its qualities all colours, but hidden within itself, as the fire lies hid in the steel, betraying its presence neither by burning, nor by shining, nor by casting a colour. In this respect, all colours and all elements are present in everything. But how all things arrive at and penetrate to all other things, if anyone cares to know, let him believe that all these matters are brought about and cared for by Him alone, who is the Maker and Architect of all things.

TEXT VI

Although, as has been said, Nature lies invisibly in bodies and in substances; nevertheless, that invisibility is led to visibility by means of those bodies themselves. According as the essence of each is situated, so is it seen visibly in its virtues and in its colours. Invisible bodies, however, have no other method than this corporeal one. So mark, then, that the invisibles contain within themselves all the elements, and operate in every element. They can send forth from themselves fire and the virtue of its element; and so, too, do they send forth air, as a man sends forth his breath, or water, as a man sends forth urine. They are also of the nature of earth, and sprung from the earth. Know this, too, that the liquid of the earth always boils, and sends forth on high, beyond itself, the subtle spirit which it contains in itself. From this are nourished the invisibles and the firmament itself, and this could not be done without vapour. Incorporeal as well as corporeal things need food and drink. For this reason stones come forth from the earth from a like spirit of their nature. Each one attracts its own to itself. From the same source come spectres, fiery dragons, and the like. If, therefore, invisible as well as visible are each in its own essence, this is due to the nature of the

Great Mystery, as wood acquires ignition from a light or a taper, though this suffers no loss. And though, indeed, it be not corporeal, still it needs something corporeal in order to escape death, which is produced by the wood. In the same way, all invisibles need to be sustained, nourished, and increased by some visible thing. With these, indeed, they will at length perish and come to an end, still, however, having their activity in them without any waste or loss to other things, that is to say, to the corporeal and the visible, although this is brought about by the invisible, and apprehended by the visible.

The rest (for doubtless the author advanced further) has not come down to us.

HERMETIC ASTRONOMY

PREFACE TO THE INTERPRETATION OF THE STARS

Written by the Doctor Theophrastus of Hohenheim

I have accounted it a thing at once convenient and reasonable that, seeing there is an order of the whole of astronomy, myself to explain the method by which it may be taught and known. It is the more needful that I should come forward because all science is completely corrupted and polluted by shameful notions, and imperatively calls for illumination by its true and genuine sense. Yet, having regard to the present times, I do not doubt that my labour will appear absurd and useless to many. I write these things by reason of those very persons—for the detection of their ignorant and random judgment. Such is the blindness of the world that it invariably prefers the name rather than that which the name signifies.

Astronomy contains in itself seven faculties or religions; he who does not know them all is unworthy to be called an astronomer. Let each man remain in that religion or faculty wherein he occupies himself; let the astologer deal with astrology; he who treats of the religion of magic, let him remain exclusively a magus; he who is concerned with divination, let him remain a diviner; he who regards nigromancy, a nigromantic; he who studies signatures, a signator; he who is devoted to the uncertain arts, an incertus; he who investigates matter, a physicist. Let not the astronomer deny that magic is astronomy, nor yet refuse the name to divination, nigromancy, and the rest. All these things are comprehended under astronomy as much as is astrology itself. They are natural and essential sciences of the stars, and he who is acquainted with them all, he is worthy to be called an astronomer. But albeit these sciences are sisters, they have heretofore been ignorant of their relationship, which it is important to recognise, so that one may not be despised by the other.

As, however, astronomy does not lead to the life which is eternal, though it may be called the highest wisdom of mortals in the light of Nature, it is not the highest wisdom of men. Beyond this wisdom there is another given from on high, which transcends the created and surpasses by far all mortal sapience. But, you will reply, the Father made the Light of Nature and also man himself. Blessed is the man who walks in that light for which he was formed, seeing that it derives from the Father! But what follows? The Son gave the Light of Eternal Wisdom to man that he might also walk therein. Can one contaminate the other? Happy is he that walks in the Father, happy is he that walks in the Son! It is right to live in both—in one to that which is mortal, in the other to that which is eternal. For whereas the Father is not angry with the Son, neither the Son angry with the Father, how then can the Light of Nature be separated from that which is eternal? One remains in the other. Nevertheless, these twain are separated by him who hands over to another what does not belong to himself, and speaks from his mouth that which he knows not in his heart. Each one on earth has his special predestinated gift, and it is lawful for him to work therewith. This gift in the light of Nature has regard to the neighbour. But besides this there are the gifts of the Holy Spirit—that is to say, prophetic and apostolic. Those who hunger after the goods of their neighbour possess not the divine gift, and those who speak hypocrisy have not the gifts of the Holy Spirit.

I freely confess that I have seen no prophets or apostles. I have, however, seen their writings, which dictate an eternal wisdom, and for this reason I by no means prefer the light of Nature thereto, but tread it under my feet, for the prophets have prophesied such things as no astronomer could have done. The apostles have healed sufferers whom medicine could never have restored. Therefore the relation of the physician to the apostle is the same as that which obtains between the astronomer and the prophet. What physician can restore the dead to life? Can any astronomer prophesy as David? Medicine is fallible, not so the apostles. But I only teach concerning ourselves as mortals in the light of Nature, with this limitation, that the wisdom of God is before all. The astronomer is acquainted with the figure, form, appearance, and essence of the heaven. The magus operates on the old and new heaven. The diviner speaks from the stars. The nigromancer controls sidereal bodies. The signator is versed in the microcosmic constellation. The adept in uncertain arts rules the imagination. The physicist composes. Now, those who give light on earth as torches in the Light of Nature shall shine, through Christ, as stars for ever. Wherefore let every one so consider these my sayings that he may gain the more from them than is written. The seed is cast into the earth, but it is another who giveth the increase. I, indeed, offer you the seed; He who brings forth seed-time and harvest, may He so conduct you to the end that ye may rejoice in an abundant vine!

THE INTERPRETATION OF THE STARS

IN order that, before advancing to definitions and proofs, I may communicate to you my full scheme, I would have you know somewhat concerning the stars. The whole machinery of the universe is divided into two parts, a visible body and an invisible body. The visible and tangible is the body of the universe, consisting of three primals, Sulphur, Mercury, and Salt. This is the elemental body of the universe, and the elements themselves are that body. The body which is not tangible, but impalpable and invisible, is the sidereal heaven or firmament. The firmament which we see is corporeal, visible, and material. This, however, is not the firmament itself, but its body. The firmament no one has ever seen, but only its body, just as the soul of man is not visible. The whole universe is thus divided into two parts, into body and firmament. Moreover, the firmament consists of two parts. One is in heaven among the stars; the other in the globe of the earth. Hence two essences of the firmament are built up. One is peculiar to the firmament of heaven, and the other peculiar to the element of this globe and sphere. The firmament of the globe or sphere is of such a nature that out of it grows whatever the body of the earth or of the elements gives or appoints. Thus from the ground the firmament of the globe brings forth the fruits, which could not be accomplished without the firmament. And the same is the case with all things that are produced from the ground. The other firmament has its special operation in heaven, that is, it relates solely to man. Now, although both star-systems, the upper and the lower, are linked together, conjoined, united, and run one with the other, still there is this difference, that the upper stars govern the higher senses, and the lower govern growing things; that is, the upper system arranges the animal intellect, and the lower those things which grow, springing forth from the sphere itself.

Beyond what has been already said, I shall enter upon no discussion as to the firmament of the globe, save so far as concerns its fruits and its growths. These are its philosophy. I shall put forward so much, however, with regard to the sense-producing star as will enable us to know that man is divided in himself; namely, into the body of the globe and the body of the senses, that is into a visible, palpable body and a body that is invisible and impalpable; or, in other words, into an elementary body of the three primals, Salt, Sulphur, Mercury, and into a sidereal body. So far as relates to the body of man, he is merely flesh and blood. That which is impalpable in him is called spirit. Thus man is made up of flesh, blood, and spirit. Moreover, the flesh and blood are not the man, but the spirit existing in himself. The spirit of man is wisdom, sense, intellect; and these are the man. The body is mere brute matter. The spirit is subjected to the stars, and the body is subjected to the spirit. So the star governs the man in his spirit, and the spirit governs the body in the flesh and blood. That spirit, however, is mortal, since it is not the soul. The soul is supernatural, and I do not speak of that here, but only of that which, being created in Adam, trenches on Nature, that is to say, flesh, blood, and this spirit. Whoever, therefore, is not reborn dies, and cannot sustain that spirit, but is carried off to death. So, then, there is a certain conjunction of the star and the man, of the elements and the man. It is a single conjunction, and a single alliance, of such a nature that no partition or separation can occur. All which can happen is that the soul departs, and is separated, that it leaves what is produced by the machinery of the world, and takes to itself what is eternal. This I point out in order that the star may be rightly comprehended among the things that are above and in the globe of the world; whilst, at the same time, it may be duly understood how each constituent part has been united in man. In this way the agreement and the operation of the star, one with another, will be understood, and we shall have ascertained what effect the external stars can have on man, and also what those things which are in man do in external things. For it is true that the external stars affect the man, and the internal stars in man affect outward things, in fact and in operation, the one on the other. For what Mars is able to effect in us, that also can the man effect in himself if he restrain himself in his manly operations. Thus are the double stars related one to the other. Man can affect heaven no less than heaven affects man.

And now we have to discuss the medium between the principal stars and the body. There is one star which governs all things; in man the animal intelligence, in brutes sensation, in the elements their operation. The star is the one supreme thing created from destruction or dissolution; and it is that in Olympus which has all these things under itself. Its office is to operate in man, to operate in elements, to operate in animals, to turn and to change their senses and their mind. Now, it is impossible to do this without a medium. This same medium is and must be a star situated in those things where the supreme operates. By this medium is produced an effect on the substance and on the body. Let us illustrate the matter by an example. If Mars is to act on a man, that cannot be done without a medium, which shall serve as the material star. By means of this Mars acts. Thus, if the higher star is to act on a parrot, it is necessary that there should be in the parrot a star as a medium by which the superior star acts. Hence it is clear that there is some star in man, in birds, and in all animals; and whatever these do, they do by the impulse of the higher influence which is received from the constellation, and regulates the unequal concordance.

Moreover, there is a similar star also in the elements, as in the earth, and that an efficacious one. That star receives an impression from the higher star, and then of itself acts on the earth, so that there is drawn forth from the earth whatever exists or lies hid in it. The same is the case with the element of water and the rest. So a person is first of all an astrologer from the higher star, and another from the star of men. There is an astrologer from the star of the elements, and there is an astrologer from the star of animals. In this way there are four astrologers of the elements, two of the stars of men and animals respectively, which make six; and then one of the superior star, which is the seventh. Besides these there remains yet another astrology born of the imagination in man, superior to all the rest, and standing eighth in order.

This, like the others, has been neglected and passed over by astrologers; but whoever would be accounted an astrologer must have a perfect knowledge of all the eight. But, although those who are skilled in particular departments ought not to be despised, yet they cannot act universally. The star is divided into eight parts; one is effective, six are subject to it; the eighth is in itself effective and like the first, nay, in some respects it is superior to, and more excellent than, the first, as will hereafter be more clearly shewn, when we speak of the new heaven and firmament. But it is only right that the celestial astronomer should know also about the rest. Now, these intermediate stars act, follow one another, and agree, so that nothing shall be predicted from the higher star, with the accomplishment of which the lower star interferes and produces something else, be it better or worse.

Hence, it is clear that astronomy was always highly valued by the ancients from the time of the Deluge up to the birth of Christ.

All species of Astronomy, it is well known, were highly cultivated. In the time of Christ, however, this ceased; and it is (? not) matter for regret that under Christ this should have occurred, because the Father had determined by His mighty love that men should confine their thoughts to Him, and that useless things should be omitted; and yet many of these things originated and grew up under Christ and afterwards inundated the whole world. In order to understand this, I would have you know that Christ taught eternal wisdom, and took care for the soul, not without a purpose, but that the image of God might be promoted to the kingdom of its Creator, and so the lower wisdom might be neglected while the higher might be more actively cultivated. Although, therefore, in this book I write these things like a heathen man, I profess myself a Christian. The heathen, however, can rejoice in the Father who is not opposed to the Son; and he is not really a heathen if he walks in the light of Nature. The wisdom of Christ is better than all the wisdom of Nature. I myself avow this, that one prophet in a single hour speaks more certainly and more truly than all the astrologers in many years, and one apostle far excels in truth all the magicians.

What could resist the school of those who spoke with tongues of fire? And yet, though these gifts were possessed, a certain sect rose up in the time of Christ speciously boasting an eternal wisdom which they did not possess, though it grew and spread abroad. This sect cut itself off from the noble science of astronomy, and took its place much as dung might take the place of fruit. Hence it occurred that, in course of time, not only the foundation and light of eternal wisdom, but also the true astronomy itself was obliterated, and the entire light of Nature at length corrupted and obscured. This was a lamentable evil and sin. Woe to those who sin against the Holy Spirit as do these of whom I complain! I confess that it is better to speak from God than from astronomy; it is better to heal from God than by means of herbs; better to preach from God than from false prophets, which is the sin against the Holy Ghost. What comes from God is not halt or maimed, but has, as they say, hands and feet. What comes from Nature is, for the most part, worm-eaten and decayed. All things are not from God; but some are from Nature. All are not from Nature, but some from God. If the magi, the astronomers, the signators, the necromancers, the incerti, the diviners, should give up their science and follow the prophets, the apostles, and especially Christ Himself, who could impute it to them for a fault that they aspired to the greater from the less, from Nature to Christ? Yet we cannot but lament that they did not penetrate to that school which spoke with the tongues of fire, though they had almost lost the light of Nature. Hence it happens that they thoroughly detest both kinds of wisdom.

Concerning the stars, I lay it down that they it is which confer all animal intelligence. As the body is conferred by the globe, in the same way is the intellect conferred by the star. One cannot exist without the other. I am forced to admit that it repented the Father that He had made man, whom the Son regenerates. It is therefore wiser to be in communion with the Son than with the Father, though the light of the Father must not be abandoned. For the Father is not opposed to the Son, nor the Son to the Father. Woe to him who sins against the Holy Spirit! I acknowledge that man is dust; for he was taken from the elements. What are the elements? Nothing. What is man? Nothing. Better is it then to follow, not that which is nothing, but that which is something. But when it comes to recognising the wondrous works of God, it cannot be but that I shall feel a difficulty. For the gifts of God are given to the Prophets, are given to Apostles, and to Saints. But so also those gifts are bestowed on astronomers and on physicians. All are by God and from God. Whatever is pre-destined to Prophets and Apostles will succeed. May that too succeed which is pre-destined to astronomy and to medicine, but all by means of God and of His operations! It is not everything that regards what is eternal or that regards even Nature; everything looks to its own. What I have to say of man, of animal, of elementary body, or of wisdom from the stars, is strictly true; and since man remains as he was formed at the beginning, I describe him as such, making in this place no mention of the new birth. Still, if the old birth and the new could not coexist, I would not describe the one or the other, but would vote all things vain. In the meantime, as to the accusation that I, being a Christian, treat a heathen topic, if the Father and the Son be agreed, and the one exist in the other, I would hope that this fact need cause no strife with any person; and unless, indeed, opposition were raised by that sect which has darkened the light of Christ as well as the light of Nature, and so brought it about that between two stools one comes to the ground, any one would readily undertake to write on these matters. In the meantime, if the renovation of the world takes place, then will be brought to pass the saying that was uttered by the eternal Virgin, "He has filled the hungry with good things; and the rich He has sent empty away."

THE END OF THE BIRTH, AND THE CONSIDERATION OF THE STARS

By Dr. Theophrastus Hohenheim

Concerning the Mass and the Matter out of which Man was Made

IT follows next in order to consider how it comes about that external causes are so powerful in man.

It must be realised, first of all, that God created all things in heaven and on earth—day and night, all elements, and all animals. When all these were created, God then made man. And here, on the subject of creation, two remarks have to be made. First, all things were made of nothing, by a word only, save man alone. God made man out of *something*, that is to say, from a mass, which was a body, a substance—a *something*. What it was—this mass —we will briefly enquire.

God took the body out of which He built up man from those things which He created from nothingness into something. That mass was the extract of all creatures in heaven and earth, just as if one should extract the soul or spirit, and should take that spirit or that body. For example, man consists of flesh and blood, and besides that of a soul, which is the man, much more subtle than the former. In this manner, from all creatures, all elements, all stars in heaven and earth, all properties, essences, and natures, that was extracted which was most subtle and most excellent in all, and this was united into one mass. From this mass man was afterwards made. Hence man is now a microcosm, or a little world, because he is an extract from all the stars and planets of the whole firmament, from the earth and the elements; and so he is their quintessence. The four elements are the universal world, and from these man is constituted. In number, therefore, he is fifth, that is, the fifth or quint-essence, beyond the four elements out of which he has been extracted as a nucleus. But between the macrocosm and the microcosm this difference occurs, that the form, image, species, and substance of man are diverse therefrom. In man the earth is flesh, the water is blood, fire is the heat thereof, and air is the balsam. These properties have not been changed, but only the substance of the body. So man is man, not a world, yet made from the world, made in the likeness, not of the world, but of God. Yet man comprises in himself all the qualities of the world. Whence the Scripture rightly says we are dust and ashes, and into ashes we shall return; that is, although man, indeed, is made in the image of God, and has flesh and blood, and is not like the world, but more than the world, still, nevertheless, he is earth and dust and ashes. And he should lay this well to heart lest from his figure he should suffer himself to be led astray; but he should think what he has been, what he now is, and what hereafter he shall be.

Attend, therefore, to these examples. Since man is nothing else than what he was, and out of which he was made, let him not, even in imagination, be led astray. The knowledge of the fact tends to force upon him the confession that he is nothing but a mass drawn forth from the great universe. This being the case, he must know that he cannot be sustained and nourished therefrom. His body is from the world, and therefore must be fed and nourished by that world from which he has sprung. So it is that his food and his drink and all his aliment grow from the ground. The great universe contributes less to his food and nourishment. If man were not from the great world but from heaven, then he would take celestial bread from heaven along with the angels. He has been taken from the earth and from the elements, and therefore must be nourished by these. Without the great world he could not live, but would be dead, and so he is like the dust and ashes of the great world. It is settled, then, that man is sustained from the four elements, and that he takes from the earth his food, from the water his drink, from the fire his heat, and from the air his breath. But these all make for the sustentation of the body only, of the flesh and the blood.

Now, man is not only flesh and blood, but there is within him the intellect which does not, like the complexion, come from the elements, but from the stars. And the condition of the stars is this, that all the wisdom, intelligence, industry of the animal, and all the arts peculiar to man are contained in them. From the stars man has these same things, and that is called the light of Nature; in fact, it is whatever man has found by the light of Nature. Let us illustrate our position by an example. The body of man takes its food from the earth, to which food it is destined by its conception and natural agreement. This is the reason why one person likes one kind of food, and another likes another, each deriving his pleasure from the earth. Animals do the same, hunting out the food and drink for their bodies which has been implanted in the earth. Now as there is in man a special faculty for sustaining his body, that is, his flesh and blood, so is it with his intellect. He ought equally to sustain that with its own familiar food and drink, though not from the elements, since the senses are not corporal but are of the spirit as the stars are of the spirit. He then attracts by the spirit of his star, in whom that spirit is conceived and born. For the spirit in man is nourished just as much as the body. This special feature was engrafted on man at his creation, that although he shares the divine image, still he is not nourished by divine food, but by elemental. He is divided into two parts; into an elemental body, that is, into flesh and blood, whence that body must be nourished; and into spirit, whence he is compelled to sustain his spirit from the spirit of the star. Man himself is dust and ashes of the earth. Such, then, is the condition of man, that, out of the great universe he needs both elements and stars, seeing that he himself is constituted in that way.

And now we must speak of the conception of man, how he is begotten and made. The first man was made from the mass, extracted from the machinery of the whole universe. Then there was built up from him a woman, who corresponds to him in his likeness to the universe. For the future, there proceeds from the man and the woman the generation of all children, of all men. Moreover, the hand of God made the first man after God's own image in a wonderful manner, but still composed of flesh and blood, that he may be very man. Afterwards the first man and his wife were subjected to Nature, and so far separated from the hand of God that man was no longer built up miraculously by God's hand, but by Nature. The generation of man, therefore, has been entrusted to Nature and conferred on one mass from which he had proceeded. That mass in Nature is called semen. Most certain it is, however, that a man and woman only cannot beget a man, but along with those two, the elements also and the spirit of the stars. These four make up the man. The semen is not in the man, save in so far as it enters into him elementarily. When, in the act of conception, the elements do not operate, no body is begotten. Where the star does not operate, no spirit is produced. Whatever is produced without the elements and the spirit of the stars is a monster, a mola, an abortion contrary to Nature. As God took the mass and infused life into it, so must the composition perpetually proceed from those four and from God, in whose hand all things are placed. The body and the spirit must be there. These two constituents make up the man—the human being, that is, the man with the woman, and the semen, which comes from without, and is, as it were, an aliment, something which the man has not within himself, but attracts from without, just as though it were a potion. Such as the principle of food and drink is, such is also that of the sperm, which the elements from without contribute to the body as a mass. The star, by means of its spirit, confers the senses. The father and mother are the instruments of the externals by which these are perfected. In order to make this intelligible, I will adduce an example: In the earth nothing grows unless the higher stars contribute their powers. What are these powers? They are such that one cannot exist without the other, but of necessity one must act in conjunction with the other. As those without are, such are those within, so far as man is concerned. Hence it is inferred that the first man was miraculously made, and so existed as the work of God. After that, man was subjugated to Nature, so that he should beget children in connection with her. Now, Nature means the external world in the elements and in the stars. Now it is evident from this that those elements have their prescribed course and mode of operation, just as the stars, too, have their daily course. They proceed in their daily agreement, and at particular epochs Nature puts forth new ones. Now, if this form of operation—if the father and mother—with this concordance meet together for the work of conception, then the fœtus is allotted the Nature of those from whom it is born, namely, of the four parents—the father, the mother, the elements, the stars. From the father and mother proceed a like image and essence of flesh and blood. Besides this, from their imagination, which is the human star, there is allotted the intellect, in proportion wherein the concordance and constellation have exhibited themselves. So, too, from the elements there is allotted the complexion and the quality of the nature. So, too, from the external stars their intelligence. As these meet, the influence which is stronger than the others, preponderates in the fœtus, or else there is a mutual commingling of all. Thus man becomes a microcosm. The father and mother are made from the universe, and the universe is constantly contributing to the generation of man. In this way, there is constituted a single body, but a double nature, a single spirit, but a twofold sense. At length the body returns to its primal body, and the senses to the primal sense. They die, pass away, and depart, never to return. The ashes cannot again be made wood, neither can man from that state in which he is ashes be brought back so as to be man again.

Now we have traced the generation of man to this point as a general and universal probation of the whole of astronomy, in order that it might be understood from thence why the astronomer studies and gets to know men by the stars, namely, because man is from the stars. As every son is known by his father, so is it here; and this science is very useful if a man knows who is from heaven, from the elements, from father and mother. The knowledge of the father and mother lies at the root. The knowledge of the elements pertains to medicine. The knowledge of the stars is astrological. There are many reasons why these cognitions are useful and good. Many men are mere brutes, and yet make themselves out angels. Many speak from their mother, calling themselves Samuels or Maccabees. Many in their earthly complexion fast and pray, and call themselves divine. Many handle those things which are not really what they are said to be. Anyone who is an astrologer knows what that spirit is which speaks and is seen. It is matter for regret that many hesitate between the two lights, culling and stealing from each in order to make themselves conspicuous. The spirits are known, indeed, to each, but in a different way, and this should not be so. But though things are thus, man is the work of God, but one only is His very son, that is, Adam. Others are sons of Nature, as Luke in his genealogy recounts of Joseph, that he was the Son of Helus, which Helus was the son of Mathat, which Mathat was the son of Levi, and so on back to Adam; yet there is no mention of the son of God. Thus man is a son in Nature, and does not desert his race, but follows the nature of his parents, the stars. Now, he who knows the father and mother of the stars and of the elements, and also the father and mother of the flesh and blood, he is in a position to discuss concerning that offspring, concerning its nature, essence, properties, in a word, concerning its whole condition. And as a physician compounds all simples into one, preparing a single remedy out of all, which cannot be made up without these numerous ingredients, so God performs His much more notable miracle by concocting man into one compound of all the elements and stars, so that man becomes heaven, firmament, elements, in a word, the nature of the whole universe, shut up and concealed in a slender body. And though God could have made man out of nothing by His one word " Fiat," He was pleased rather to build man up in Nature and to subject him to Nature as its son, but still so that He also subjected Nature to man, though still Nature was man's father. Hence it results that the astronomer knows man's conception by man's parentage. This is the reason why man can be healed by Nature through the agency of a physician, just as a father helps a son who has fallen into a pit. In this way Nature is subjected to man as to its own flesh and blood, its own son, its own fruit produced from itself; in the body of the elements wherein diseases exist; in the body of the spirit, where flourish the intelligence and reason; and the elements, indeed, by means of medicine, but the stars by their own knowledge and wisdom. Now, this wisdom in the sight of God is nothing; but the Divine wisdom is preeminent above all.* So the names of wisdom differ. That wisdom which comes from Nature is called animal, because it is mortal. That which comes from God is named eternal, because it is free from mortality. These two parts, therefore, seemed to me necessary to be treated before I commented upon astronomy itself, so that from these universal proofs the whole foundation might be the more easily gathered.

The following are the numbers, religions, and faculties of the whole of astrology, which are treated naturally and artificially. Neither more nor fewer than these exist essentially and spiritually. Their names and differences are as follows:—i., Astrology; ii., Magic; iii., Divination; iv., Nigromancy; v.,

---

* In an exceedingly abstruse treatise on The Foundation of Wisdom and the Sciences, Paracelsus thus delivers himself upon the subject of sapience and knowledge: Whosoever undertakes a treatise on the foundation of wisdom, the same before all things should admonish or teach the reader concerning the origin of the sciences and of sapience. So also the physician who has decided to write concerning diseases must first explain from what foundation he writes, and the author whom he follows, as also how he teaches, in order that the same may be proved in the case of diseases. For out of these the probity and truth of his doctrine and science can be judged. Similarly, we who are about to treat of the foundation of the sciences and the arts must necessarily teach their origin, whence they have proceeded, and whence they are to be learned. Having done this, we must proceed to explain the matter itself. This book treats not of corporal matters, but of things invisible, that is, of reason itself. Hereof I have been impelled to write, seeing that many persons before me have supposed many kinds of wisdom, whereas, so far as regards man, there is only one wisdom; for how can one carpenter differ from another carpenter when both construct the same house, both use the same axe, and both have one method of building? There is one compass. Concerning this, it must be stated in a treatise on science that two compasses are not to be used, for the compass is one and not two. In the same fashion, the carpenter, the quarryman, and the bricklayer can use no compass different from this. Thus the house of wisdom neither can nor ought to be established except upon one foundation. And as the builder's art is defined by one circle, one number, one line, one square, so also through all methods there is one wisdom; and as the distribution takes place from one circle into a triangle, quadrangle, etc., which, however, are all one circle, the same is to be understood concerning the distribution of sapience. So also, as heaven, earth, air, and water use one and the same line, thus according to one line all wisdom is educed and extended. And as all men and all things are numbered by one number, so there is one number of wisdom itself, nor is any other beyond it to be taken. But inasmuch as a line is drawn by a free hand, yet by no means a correct one, or a circle, also by no means accurate, is described by hand, etc., so also lines, circles, and other figures proceed from sapience, but are by no means correct. For such sapience is by no means a true circle, quadrangle, or line. Hence I have determined to treat shortly of the foundation of science and wisdom, whence they proceed, what they are, and who bestows or imparts them. Artificers are greatly in need of the knowledge

Signature; vi., Uncertain Arts; vii., Manual Arts. What will be handled in each religion, and what the religion itself is, seek in the sequel.

ASTROLOGY.—This science teaches and treats concerning the whole firmament, how it stands with the earth and with man according to the primæval order, and what is the connection between man, the earth, and the stars.

MAGIC.—This science brings down and compels heaven from above to stones, herbs, words, etc. It teaches also the change of one thing into another, as well as the knowledge of the supernatural stars, comets, etc., and what their signification is.

DIVINATION.—This science is from heaven to man without any formal institution, so that he knows how to speak of things future, present, and past, though he has never looked into those things himself, and speaks nothing save what heaven impresses upon him. This science is most of all seen among simple persons.

NIGROMANCY.—This treats of sidereal bodies, which are without actual body, flesh, and blood. This operation stands related to the necromancer as a servant to his master, the latter commanding the former.

SIGNATURE. —This science teaches one to know the stars, what the heaven of each may be, how the heaven has produced man at his conception, and in the same way constellated him.

UNCERTAIN ARTS.—These sciences are without any principles on which they rest, or from which they proceed, and are ruled by the imagination, offering a new spirit and a new firmament by which they work.

MANUAL ART.—This science teaches the preparation of instruments for all astronomy, and with slender material expresses or comprises the form of the stars, and brings heaven and earth into one figure.

For the sake of fuller understanding I will add how many species each religion has, in this way:—

---

of such foundation, for what else is sapience but an art or science which one man has derived from another. He who by counsel flourishes in prudence, what has such a man learned except the art of provident wisdom, which another man does not know? Thus, the artificer has a skill in regulating the fire in which the tailor is completely wanting. This, therefore, is his art. Hence in different persons there are different arts. What, therefore, is science save art? It becomes this art to proceed out of a circle, out of a line, out of a number, for this gives a mode, and thus a mode obtains in arts. From his individual art the turner derives his special mode, and this also is true—the line, the circle, etc., give a mode of wisdom, and the said mode is wisdom itself. Further, sciences are distributed after many methods, nor could they all consist in one. One man knows one thing, and one another. No person can know and accomplish everything. Who is familiar with, who performs all things? And as no one perfects two labours in one labour, but it is necessary that one thing only should be performed at one time, so the case is the same with the arts and sciences. For the sciences are so extensive and so profound that they cannot be contained and held by one brain. One part is given to one, a second to another, etc. For as in one street of a city there are many gold beaters, in another many shoemakers, and in a third many tailors, so there is an analogous distribution of the sciences. Now, all undertakings proceed from one fount ; from one fount flow all works and all sciences. Their ramifications are distributed like fruit on a tree, none of which can so separate itself as to deny that it was produced with the others from the same tree. If a guide and teacher be needed in this kind of wisdom, so that the source of a writer's instruction may be known, it becomes right and just that I should divide this discourse into two parts. The sapience of man is twofold—one relating to the soul, the other to the body. Having made this distinction, it is necessary that we should understand the animal, that is, the corporeal, and also, so to speak, the mental, that is, the eternal. But as we have elsewhere described the origin of the animal man, so here we shall chiefly concern ourselves with the man which is spiritual, and here is explained the invisible source of science and wisdom.—The rest of the work, which contains several treatises, and is at the same time only of a fragmentary nature, attributes, as might be expected, all wisdom to God, for man has nothing of himself. Man, however, is the heir of heaven, and the world exists solely for his benefit. It further divides wisdom into the animal and the angelical, and affirms that men themselves are angels, and are before all heaven and all angels.

Astrology has three species, as referring to man, the inferior bodies, animals.

Magic has six species. It belongs to comets, images, gamahei, characters, spectres, incantations.

Divination has five species, dreams, brutes, the mind, speculation, phantasy.

Nigromancy, of which there are three species, material visions or spectres, astral spirits, and inanimate, phantastical bodies, that is, those assumed by the dead or by lifeless things.

Signature has three species, chiromancy, physiognomy, and proportion.

Uncertain Arts, of which there are four species, geomancy, pyromancy, hydromancy, ventinina.

Manual Art has five species, arithmetic, geometry, cosmography, instrument, sphere (a mathematical instrument).

The Interpretation of the Species according to each religion.

Astrology—The First Religion

This science embraces three species in which it is occupied. It operates against man, against elements, against animals. For since heaven and the lower bodies are mutually connected, the heaven teaches us to know the lower bodies by means of a figure which represents the whole heaven. From this figure is inferred the property of the inferior bodies, and what effect heaven produces in those inferior bodies.

Magic

Of Comets. This species teaches us to recognise all these supernatural signs in the sky, and to understand what they signify. Of this class are comets, halos, and the other figures of the sky. This science is founded on the Apocalypse, on dreams, and on the saying of Christ, "There shall be signs in the sun, moon, and stars." Since all these signs are supernatural, they refer not to astrology, but to magic.

Of Images. This science represents the properties of heaven and impresses them on images, so that an image of great efficacy is compounded, moving itself and significant. Images of this kind cure exceptional diseases, and avert many remarkable accidents, such as wounds caused by cutting or by puncturing. A like virtue is not found in any herbs.

Of Gamahei. These are stones graven according to the face of heaven. Thus prepared they are useful against wounds, poisons, and incantations. They render persons invisible, and display other qualities which, without this science, Nature of herself cannot exhibit.

Of Characters. These species are words which are either spoken or written. They have power against all diseases, which they also avert. They divert misfortune and all accidents, they set free prisoners so that they are loosed from their chains, and produce those effects which Nature itself is not able to bring about, but only magical science can accomplish.

OF SPECTRES. This species exhibits the likenesses of men, so that something appears which is not really present. These visions with their signs are produced by night, not by day, and lack the body, blood, flesh, soul, and spirit of man.

OF INCANTATIONS. This species teaches how to turn men into dogs, cats, etc. It teaches a man how to convert himself into all kinds of appearances and forms. It renders people invisible, changes the minds of men at the will of the artificer, impels, leads, and directs impressions and generations according to his pleasure.

DIVINATION

DREAMS. If anything is presented to a person by means of a dream, be it present, future, or past, be it knowledge, a treasure, or any other secret, it bears reference to this art. It can direct the stars to a dream, so that anything may be thereby revealed.

BRUTES. This species teaches us to distinguish the prophecies which come from animals, so that man may see and understand what the heaven does or is about to do. It operates also in fools, in animals, and in other simple beings.

THE SOUL. This species refers only to the mind of man, so that by chance and not by premeditation it is suggested to the mind of man what he ought to do. This species is of great importance, and should be studied among the very first by man, so that he may know what the mind suggests to him from its true foundation.

SPECULATION. If any one carefully weighs and speculates, and, by means of a strong imagination, finds what he seeks, it ought to be referred to this species. It arises from the stars, which are occupied about man and teach him.

PHANTASY. If any one in mere sport finds out or learns anything, this also is from the star, when it is matured. This often reveals many things such as treasures, mines, and others which are hidden, operating without any previous knowledge or investigation, and benefiting him who does not seek it.

NIGROMANCY

VISIONS. This species sees in crystals, mirrors, polished surfaces, and the like, things that are hidden, secret, present or future, which are present just as though they appeared in bodily presence.

ASTRAL SPIRITS. This species teaches how to deal with sidereal spirits separated from the body, so that they may be compelled to serve men like slaves.

Inanimates are men without a soul produced by the stars, dwelling and conversing with men and doing the same as they.

SIGNATURE

CHIROMANCY, by which the star is exhibited in man with that appearance in which the heaven was at the time of his nativity. It appears in the hands, feet, and other lines and veins of the body, shewing themselves differently in different bodies.

PHYSIOGNOMY. This species teaches how to know a man by his countenance, manners, and gestures. This also has for its cause the hour of birth, which signs a man, and by those signs forms his nature.

PROPORTION. This species judges the properties from the general habit of a man, whether he be lame, too tall, too short, etc.

UNCERTAIN ARTS

GEOMANCY. This science is practised with a free mind without foundation or certain knowledge or signs (tesseræ). It agrees with astrology.

PYROMANCY. This species is fortune telling by fire. By fire is seen what is the motion of the heaven, what its nature and condition. In this the moon is principally consulted.

HYDROMANCY. This species teaches how to see in water certain secret and hidden things, closed and sealed letters, and persons who are travelling in distant countries, whether they are living or dead. This operation proceeds from the constellation of the new firmament, by means of imagination.

VENTININA. This teaches how to determine from the wind what the future state of the heaven will be as regards man, whether good or bad, fruitful or sterile, and other similar things in the future which cannot be determined by Nature.

MANUAL ART

ARITHMETIC. This species teaches how to find the number of heaven and earth in the stars and the like.

GEOMETRY teaches how to measure the height of heaven and earth, and of the things contained in them.

COSMOGRAPHY teaches the situation and distance of all things, the manners and nature of peoples.

INSTRUMENTATION. This species teaches how to make instruments, with which is known how heaven and earth are connected.

SPHERE. This species teaches how to learn by means of an instrument what is the knowledge and correspondence of heaven and earth.

With this brief discourse I have endeavoured to describe the different species of religions that astronomy itself may thus be more rightly understood. All these make up astronomy. But how each one may be proved is afterwards described, with this view, that it may be clear that astronomy is no inconsistent or mendacious science, but that it is based on a solid foundation drawn from the light of Nature itself; which, indeed, is necessary for establishing all truth and knowledge.

PROOF IN ASTROLOGICAL SCIENCE *

Having treated of the generation of man we must now deal with his sustenance, and in this way astrology will be sufficiently proved. There is a certain congenital virtue in man which attracts into man from the external

---

* The proofs in astrological science, magic, and divination are wanting in the treatise, and the deficiency has been supplied from another work, the *Explicatio totius Astronomiæ*, which duplicates the Hermetic Astronomy (*Interpretatio alia Astronomiæ*). It has been thought advisable at this point to compress somewhat the prolixity which further obscures the original.

sphere. Now, from that which is attracted man is sustained, and he is well and ill according to that which he has attracted. The attractive virtue is twofold, one of the elementary body, the other of the sidereal body. The desire of man for sustentation is to be understood as follows: The rays of the external sphere penetrate to us; the internal economy of man accomplishes the rest. Thus the sphere extends its fruits from the radix even to the outward locust. Hence it follows that there is a certain nature, namely, hunger and thirst, which is implanted in us and compels us to eat those fruits. So do the rays of this sphere enter us. Now, even as the food of the physical body comes to us from the elements, so is the sidereal body supplied by the constellation with all science, all arts, all prudence. Man is formed in such a manner that he should derive all his knowledge in the same way as he gathers fruit from a tree. Thus originates music, the metallic art, medicine, agriculture: whatsoever the earthly body requires, that he finds in the wisdom of the stars, and all wisdom, whether good or bad, is derived to him from the stars. Two things only, namely, justice and holy scripture, proceed immediately from the Holy Spirit. In the stars then is the whole light of Nature founded. For as man seeks food from the earth in which he was born, so also does he seek it from the stars in which he is likewise born. Thus the wisdom to which he is born is twofold—one is animal—but of the other Christ said, "For this I was born," as if He had affirmed "I was born in the Eternal Wisdom." The wisdom of earth should be employed only over carnal matters; the other and higher wisdom should be learnt and employed according to the words of Christ.

Now, the sidereal wisdom is foolishness before God, whence comes that saying: The wise man rules the stars, in the sense that eternal wisdom governs the animal. Thus natural wisdom is given to the body and not to the soul. Those things, therefore, which concern the soul are by no means to be polluted by the light of Nature. This must only be used with Nature. By the light of Nature all arts and operations have been invented. In the mansions of the planets there are workmen who have taught all other workmen, and they, indeed, are the best of all, for they have their arts implanted from birth. These, were they men, would everywhere forge iron and handle it as if it were wax. Mortals as yet have not learnt this arcanum, but they would do so did they drink from the true fountain. So also masons dwell in the habitations of the planets, from whom all other masons learn, and if they did this fully all matter would be plastic in their hands. Thus the firmament formed by God is our perfect instructor in all the arts if we refer to their true source. Thus, too, the palmary physician is in the firmament, who is acquainted with all diseases, and even sees those things which are hidden from our eyes. God created him such that he might beget physicians on earth. Now, concerning evil sources, there are unskilled artificers in heaven even as on earth. This ignorance and clumsiness may be discerned even by the animal wisdom, which is given for this end, that the good and not the bad may be chosen. So does the natural light lead up to the higher light. Further, Lucifer in heaven made himself other than he was created, together with his companions, and the same thing can also take place in the stars. Hence contrary conceptions may arise, adverse and perverse arts. We must not, therefore, believe every spirit, since of spirits there are two kinds, even as there are two kinds of angelic intelligences—those who remained as they were and those who fell from their first estate. Astronomy is important, in that it teaches us to discern between these two kinds of spirits. This same science also contains a great arcanum, nor can anything be learnt without it. Wisdom is eternal and natural. The eternal is immutable and constant; the natural, from its mutable conception, generates a false spirit which misinterprets scripture. But if astronomy is acquainted with this, and if, indeed, nothing is so hidden as not to be revealed thereby, who shall not extol it with the highest praises?

It has already been shewn after what manner man was made, how he possesses hunger and thirst, an elementary and sidereal body, to produce an appetite for nourishment, and finally that he tends to that which was implanted in him at conception. Hence it follows that such virtue, nature, property, and condition, and finally all the concordance and constellation, can be described by the astronomer, for in this way various nativities are constituted, and hidden things are prognosticated. In all who live according to Nature nothing is hidden from the astrologer, and thus for the generation of man a figure of heaven is erected, in order to know the properties of the stars, as also the particular mode. Understand, therefore, concerning astrology that it knows the whole nature, wisdom, and science of the stars, according as they perfect their own operation in conception and constitute an animal man. The astrologer can easily describe a man or an animal by reason of such a conjunction and concordance. But if astrology be fundamentally and properly known, and the nativities of infants be erected rightly according to the mode of the influence, may evils will be avoided which would otherwise be occasioned by the unpropitious constellations.

Proof in the Science of Magic

In the first place let us define the nature of Magic. It is that which brings celestial virtue into the medium, and thence is able to perfect its own operation. The medium is the centre. The centre is man. By means of man, therefore, the celestial force can be transmitted into man so that in man may be found such an operation as the constellation itself can produce. Moreover, in magic there is a further operation which it performs itself while exercising its art, that is to say, while the nature itself of the constellation does that which the magus ought to do. If the magus be himself the medium and centre, and, what is more, be capable of performing the operation of the constellation for man by means of man, it is in addition given to this art to produce another medium which is to be understood as a subject, by which subject that operation is just as well performed as by man, who is the true medium. Thus in magical science there exist two operations, one which Nature herself produces, selecting man as the instrument, and as the recipient of her influence, whether bad or good, the other operates by means of arbitrary instruments, such as statues, stones, herbs, words, also comets, similitudes, halos, and any other supernatural generation of the constellation. Thus Nature herself is able to prepare her magical powers and perform her own operations by their means, as, for example, when something extraordinary takes place amidst a rude populace and is referred to miraculous agency, whereas it is only Nature who has worked magically.

Whatsoever Nature is able to accomplish in a foreign body, the same also can man accomplish, if he direct his operation so that conception can be attained, namely, the image, having neither flesh nor blood, and being like to the comet, so that the words and characters possess their own virtues equally with medicaments. It is, in like manner, possible to bring about such a condition in herbs and gamahei, that they become like to the planets and the dwellers therein. Now, it is no matter for astonishment that man accomplishes such things, for if it be true, as the scripture says, that ye are gods, we shall certainly be superior to the stars. If the stars as a fact are found to govern the majority of men, that is because men have abdicated their power as gods; few, indeed, are those who have exercised gifts such as those of the apostles and saints. The difference between the saint and the magus is this, that one operates by means of God and the other by means of Nature. Magic is a sublime science, and by reason of its operations is very hard of attainment. We must have regard to the word of Christ, which passes not away, when He said; "If ye believe, ye shall accomplish more things than these." Now, if we can exceed that which is accomplished by Christ, we can also exceed that which Nature accomplishes, seeing that she was created on our account and is therefore in our power. The wise man rules Nature, not Nature the wise man. For the same reason we can accomplish more than the stars. In us, then, should abound so great a wisdom that we shall thereby control all things, not only firmamental virtues, but also living animals which yet are much stronger than man. The will of man extends over the depth of the sea and the height of the firmament.

Nature herself is a magus. If about to announce anything, she creates for herself messengers, such as comets and other celestial signs. The magus man is comparable to the physician. The physician knows the hidden virtues of herbs, but the magus the hidden potencies of the stars. The physician extracts the virtues of herbs, and produces a remedy which is small in weight but represents the powers contained in a whole field of vegetation. The magus can transfer the powers of a whole celestial field into a small stone, which is called the gamaheus. As the physician infuses herbal virtues into the sick man, and so heals his disease, so the magus infuses into man the heavenly virtues just as he has extracted them. Medicines are renewed yearly, but the stars have their exaltations in place of a summer. The sun is the highest grade of diurnal light, plus the congenital heat which belongs to it. How shall this light and heat be brought downward by means of man into a subject, so that its light will be intolerable to the eyes and sense shall scarcely be able to endure its heat? This takes place in the sphere of the crystal, which then is termed Beryl.

If the Magus can draw down virtues from heaven and infuse them into a subject, why should we be unable to make images conducive to health or disease? If poison, and the rest, can arise from earth, it can issue also from heaven. But why should not similar things take place in the case above, whether the subject be images, herbs, stones, or woods? The birth corresponds to what is sown in the constellation, and it is not man alone who operates such things; Nature also variously exercises herself. But if it be possible to Nature, why not also to man? Let Nature be an example to us. As she works we must follow in imitation. Herein lie hidden medical science, all artifices, all arts, all animal industries. It frequently happens that Nature advances some person beyond the knowledge he can derive from man, who also by skill and industry surpasses all the rest. Such a man is born like the comet, which differs from other stars. Thus it becomes possible that the Magus also, by means of magic science, may produce such an industrious man like a comet. These are the mysteries and the great things of God. The firmament, by means of the magi, exhibits the glory of God. By means of the magi out of Satra and Tharsis, by the ascendant of Christ in Bethlehem, is made manifest whatsoever the firmament and heaven do reveal in the Arcana of God.

Proof in the Science of Divination

Astronomy creates herself, and from herself performs astronomical operations which do not require art and industry. This mostly takes place among those who are of a good and honest disposition, as also temperate. The ancients preserved both their bodies and souls from pollution, so that they might more successfully perform operations of this kind in themselves. This is divination. When men, having no knowledge of astronomy, perform such operations, they are considered miraculous, and the operators are regarded as gods. The operation is revealed by dreams, by the soul, by speculation, and by animals. Divination was of much importance among the ancients. It is a part of astronomy, but it is not a science, for the operation occurs spontaneously. It is often said in common parlance: "My angel told me this." Here the operation is called an angel, as if it took place by God; it is ascribed to the angel, as if to a medium between God and the man. At the same time, the whole operation is merely celestial. Now this is the origin of divination. Man possesses a sidereal body united with an external constellation. These two communicate when the sidereal body is not affected by the elementary. In sleep, when the elementary body is quiescent, the sidereal body performs its functions. Hence arise *insomnia*, according as the constellation operates them, and as the constellations are badly or well disposed, so also are the *insomnia*. When the constellation and the sidereal body are favourably co-ordinated, future things are truly predicted. In this manner, also, many remedies have been discovered which prevail over different diseases, also hidden treasures and other concealed things, so that scarcely anything can be compared to this very great science. The firmament foreknows all future things, nor does anything escape its knowledge, whether of things past or things present. If a sidereal body of this kind be found suitable by the constellations, and if the constellations be prepared, many marvels are manifested, both present and past. In this manner old men and women, unendowed by any knowledge, as it were by their simplicity and fatuity, have often made prophecies which the event marvellously verified.

In the same way, also, many have become learned men, who, having attained a suitable sidereal body, have sedulously exercised themselves in their native influence. Hence it happens that they at last draw down upon themselves the influence of their native constellation, just as rays from the sun. So an admirable science, doctrine, and wisdom are discovered, yet is the whole animal alone, not from on high, but taken from the stars alone. Heaven being thus constituted, and producing for itself a sidereal body, there arise many great minds, many writers, doctors, interpreters of Scriptures, and philosophers, according as each is formed from its constellation. Their writings and doctrines are not to be considered sacred, although they have a certain singular authority, given by the constellation and influence, by the spirits of Nature, not of God. Operations of this kind sometimes proceed from the mind of man in a stupendous manner, when men, changing their heart and soul, would make themselves like to the saints, being made such by a drunken star; whereas wine changes man, so also these are changed. It is, therefore, worth while to understand this sort of astronomy. Intoxicated writers of this kind lead many astray; they are wanderers in the Spirit of God as well as in the Light of Nature, flitting about like dreams. Many things are done by these which yet are of no moment, nor can be understood by others.

The force and efficacy of the constellations impresses itself upon brute animals, for whatsoever lives contains in itself the sidereal spirit, and wherever the operations are, there they are manifested. So the clamour of peacocks presages the death of their owners. For no man dies without the previous indication of portents. When a man is about to die the constellation within him loses its operation, and this loss takes place by means of a sign or a great mutation. So the stars shuddered at the death of Christ. From motions taking place in Nature, the death of every man can be prognosticated. Knockings in houses will sometimes precede the death of some occupant, yet these are not the work of spectres, but are natural operations, which in this manner are accomplished in men by means of the stars. The stars singularly sympathise with man, for man has been so formed by God that the whole firmament is consensitive with man, and out of compassion gives its presages to his grief.

Proofs in Nigromancy

Regard, in due order, nigromancy, so that it may be possible to learn and judge sidereal spirits and those who have no soul. The judgment is directed to that whereof we proceed to speak. The man who buries a treasure in the earth and hides it has all his mind intent upon that treasure. If he dies, his elementated body is buried, his sidereal body withdraws from it, and walks about on the earth up to the time when its decay is complete. This body carries about with it the thoughts and the heart of the dead man. Hence, as may be inferred, it keeps itself in the neighbourhood of that place where the treasure has been buried, about which the heart of the dead man was anxious. Such sidereal spirits are constantly seen at or near the place where such treasure is. The same thing occurs in other matters about which anybody has been anxious with the whole desire of his heart, whether it has been food, or drink, or debauchery, gambling, or hunting. In all these things the spirit acts for the imagination of that heart, and it does the same thing in a shadowy way after death, until the star consumes that spirit also, as the elements have consumed the body. Hence it follows that the necromancers get to know these sidereal spirits and to ascertain for what reason they are walking about in one place or another. In the same way, they explain the nature of lemures, giants, and gnomes. Nigromancy is the philosophy of spiritual sidereal bodies, and of inanimate beings who are, nevertheless, human, as onagri, nymphs, lemures, etc. The man who busies himself about these is a necromancer. The same is the case with exorcists who adjure bodies and inanimate beings of this kind. They differ from necromancers in this respect, that the exorcists are occupied with bodies obsessed by the devil, while the necromancers find their occupation, both naturally and philosophically, with those who are not obsessed. The ignorance of men has confounded exorcism with necromancy, and taken them to be one and the same. However, their distinction has now been settled. Moreover, I have determined to say nothing about exorcists here; it will be better to relegate them to the devil, whose servants they are. But I would wish to commend necromancy to you as a remarkable natural science, which produces some marvellous effects, since by means of sidereal spirits are laid bare the very hearts of men, shewing how they are inclined, what they long for, and what their ambitions are.

It is, moreover, pleasant and delightful to rightly understand nigromancy. The knowledge of nymphs, also the discovery of lemures, gnomes, and giants, is very subtle and ingenious. Indeed, the philosophy of these four inanimate generations is a truly noble one, which many babblers oppose and prefer their own nonsense to it. Since God is wonderful in all His works, it is more than likely that, one of these days, the temerity of these people will be brought out into the light of day, and, in God's own good time, branded openly. Moreover, necromancers use beryls because, in respect of astral spirits, they have some familiarity with magicians in the way of visions, but the magi do not admit these. The causes of this fact will be noticed in these treatises. In the present discourse it has been made sufficiently clear what nigromancy is, and what is its subject-matter.

Proof in the Science of Signature

God has enriched the light of Nature with such ample gifts that even one who is not addicted to the light can know all things that are therein. Is not this a great thing which external signs offer to man's knowledge? And God has arranged it so. Possibly you wonder how this can be done. Let the following example put an end to your wonder. The carpenter is the seed of his house. Whatever he is, such will be his house. It is his imagination which makes the house, and his hand which perfects it. The house is like the imagination. Now, if such be the property of imagination that it makes a house, Nature also will be an imagination making a son, and making him according to its imagination. So the form and the essence are one thing.

Whatever anything is useful for, to that it is assumed and adapted. So if Nature makes a man, it adapts him to its design. And here our foundation is laid. For everything that is duly signed its own place should properly be left; for Nature adapts everything to its duty.

If any lord or prince builds a city he so builds and arranges all the walls, towers, citadels, and the rest, that they shall as closely as possible suit his design. If man does this, how much more shall Nature, which is higher than man? It makes one man lame, because it is going to use such an one for lame purposes. It makes another blind, he being destined for blind purposes. In one word, whatever it requires any one to be, such an one it produces.

This, then, being the custom of Nature, that it produces such a man as it wishes, those vestiges will be clear and plain in the man. By these vestiges is meant whatever Nature is going to use such a man for.

Since Nature, therefore, works thus openly and puts forth its work in public, it is right and convenient that some one should be met who sees what sort of a person Nature has in each case prepared and produced, that is, how it opposes a rascal to an honest man, and sets a man-wolf over against a shepherd.

A signature, then, is that which has to do with the signs to be taken into consideration, whereby one may know another—what there is in him. There is nothing hidden which Nature has not revealed and put plainly forward.

Rightly, therefore, should its proper place be given to signature, because it is a part of astronomy, for this reason, that the star builds the man up at its own pleasure, with the marks belonging to him. What is going to be tinged with black Nature makes black, what blue, it makes blue; that which is going to sting is made a nettle, and what is to purge is made an equisetum, what is to be used for smoothing and polishing is made a smiris. In fine, to everything is assigned its own form, by which it may be known for what purpose that thing is made by Nature.

Whatever is in anything according to its properties, quality, form, appearance, etc., is revealed in herbs, seeds, stones, roots, and the rest. All things are known by their signature. By the signature those who are instructed trace what lies hid in herbs, seeds, stones. But when the signature is obliterated and trifles are substituted for it, then it is all over with everything, even philosophy and medicine being at fault.

The cry goes everywhere that I burn with hatred of learned men, doctors, magistrates, bachelors, senators, consuls, and the like. What is the reason? Nature has signed them too clearly. I can see what they are made of; and I hate every house that lets in the rain.

In like manner, I am accused of disliking physicians and surgeons. Why? Just because they are not signed for their profession, but as rogues and impostors. The same is true of others also. I know plenty of them, if it were only safe to speak out.

How can I favour a man who is branded with so many stigmata and disgraceful marks as the Consul of Astorza, Niger, and of Nuremberg, Muffel? And how many others are there like them? Of course they detest this art, because it too clearly betrays bad men.

Proof in Uncertain Arts

Nature puts forward a way and clear order in which man should consider what belongs to Nature and its properties. Thus astrology teaches us to know the nature of the sun by the accustomed order of the stars. So what the moon is and what her nature, the astrologer learns from her course, which he sees to be regular. The same judgment is to be passed on the other stars. In like manner, philosophy is learnt from that which appears, how Nature stands related to the earth; hence it is ascertained that the method of philosophy ought to be the same. Thus all things have their own proof and comprehension. And so all arts, such as medicine and the rest, are conceived in a natural order. Without this order nothing can be done or brought to a perfect end.

Moreover, the uncertain arts, of which four are shewn in the table, have not this order and process, which can be materially proved and demonstrated; but this differs from the order spoken of. With regard to this it ought to be understood that there are many things which do not indeed square with the same order, but still are not opposed to Nature. They only differ from the order of material nature, as God has settled it. But what there is besides in this order ought to be understood from the Uncertain Arts in the following manner. The firmament and new heaven are constituted by the imagination; and it should be known that this imagination is effective, and produces many things, being marvellous in its operations. It often happens that the imagination of the parents, father and mother, confers on the offspring born in that creation a different heaven, another figure, another ascendant besides that which astrology gives. Thus it often happens that an offspring is begotten contrary to the star, and arranged otherwise than the figure of the heavens dictates. By the force of this imagination many learned men are often born.

Nothing, therefore, ought to be accepted beforehand in the way of proof for these uncertain arts short of the operation which takes place through the imagination by means of a new heaven, new ascendant, and firmament. In proportion as this is good, strong and just in operation, so the judgments fall. Let us take an example. Speculation is the wishing to know this or that thing. This speculation produces imagination; imagination begets operation; and operation leads to judgment and opinion. Now imagination is concerned, not with the flesh and blood, but with the spirit of the star which exists in every man. This spirit knows many things: future, present, and past, all arts and sciences. But flesh and blood are crude and imperfect, so that they cannot of themselves effect what the spirit wishes. But if flesh and blood are subject to the senses, and are purged by them, then the spirit acts thereby, if only the body be consentient. These senses are supreme in the uncertain arts. It is for this reason they are called uncertain arts; for who can know what imagination is in them? What does the spirit which is given to them imagine and effect? Yet, nevertheless, the art itself is certain. But the artist who uses it may be unfit for the creation of new heavens and the generation of a firmament. Because, therefore, there is the element of doubt on these points, credence cannot be given to opinion, but one has to wait for the issues. At last, however, the force and efficacy of these things are discovered. Moreover, it is not to fight against God if the future is explored apart from him whom God has set over the nature of the firmament. Suppose, for example, that someone is going to be stabbed with a dagger. Let this be foretold to him by some other person. Premonitions of this kind have often been found true, though there might have been strong opposition. Now, if this happens by uncertain arts God himself suggests the prophecy, the prediction, and the premonition in a manifold way. So many prophets have predicted such things by dreams. It was by a dream Joseph was admonished about Mary. And since these things did not seem natural to flesh and blood, they were thought nothing of until, the event corresponding, they were believed. Now the uncertain arts, just like dreams and other revelations, are intelligible. God chooses to appear wonderful in His works. For this reason the uncertain arts are by no means to be despised, because they eventually become known by the result. God does not intend that we should always foreknow the future for certain, as can be done by the order of Nature. He wishes us to know, indeed, but sometimes to doubt; that seeing we may not see, as Christ Himself also was known, yet not known by the Jews, seen yet not seen, heard yet not heard.

It has been said above concerning the imagination that it draws the star to itself and rules it, so that from the imagination the operation itself may be found in the star. Just as a man with his imagination cultivates the earth according to his judgment, so by his imagination he builds up a heaven in his star. The imagination of the artist in uncertain arts is the chief art and head of all. But in addition to this, imagination is strengthened and perfected by faith, so that it becomes reality. All doubt destroys the work and renders it imperfect in the spirit of Nature. Faith, therefore, ought to strengthen the imagination. Faith bounds the will.

Now, faith is threefold. There is faith in God. This produces what it believes. By faith mountains are moved, the dead are restored to life, sight is given back to the blind, the lame walk. What marvels faith produces if imagination looks to God with full faith which is unbroken and unmutilated. We find an example of this in the Saints of the Old as well as the New Testament, who, according to their belief, were made to obtain their wish, so that nothing was wanting to them. There is another faith in the Devil and his powers. Whoever has this faith, to him it happens as he believes, if only it be possible for the Devil to fulfil it. Lastly, there is also a faith in Nature, that is, in the light of Nature. He who believes in this obtains from Nature as much as he believes. Now more cannot be obtained from Nature than is given to it and conferred upon it by God. It is, then, imagination by which one thinks in proportion as he fixes his mind on God, or on Nature, or on the Devil. This imagination requires faith. Thus the work is concluded and perfected. That which imagination conceives is brought into operation.

Note an example of this. Medicine uses imagination strongly fixed on the nature of herbs and on healing. Here is need of faith that such imagination may act in the physician. If this is present, imagination conceives and brings forth spirit. The physician is spirit, not body. Hence infer that the same fact holds good in all arts. Moreover, there are physicians without imagination, without faith, who are called phantastics. Phantasy is not imagination, but the frontier of folly. These work for any result, but they do not study in that school where they ought. He who is born in imagination finds out the latent forces of Nature, which the body with its mere phantasy cannot find; for imagination and phantasy differ the one from the other. Imagination exists in the perfect spirit, while phantasy exists in the body without the perfect spirit. He who imagines compels herbs to put forth their hidden nature. So also imagination in the uncertain arts compels the stars to do the pleasure of him who imagines, believes, and operates. But because man does not always imagine or believe perfectly, therefore these arts are called uncertain, though they are certain and can give true results. The other sciences of astronomy hold their own even without faith or imagination, just as a mechanic who, if he follows his own order in working, has no need of imagination or consideration, and yet finishes his work.

But, it should be remarked, that by faith water can be crossed over without drowning or wetting; and a man without faith can do the same thing if he crosses the water by a bridge or in a ship. So also healing the sick is accomplished by means of medicine without faith; but by means of faith it is found out what medicine is. Imagination takes precedence of all. What this discovers and gives, the other, who acts phantastically, uses.

Man is not body, but the heart is man; and the heart is an entire star out of which it is built up. If, therefore, a man is perfect in his heart, nothing in the whole light of Nature is hidden from him. Thus from one point in Geomancy his whole will is accomplished. So, too, in Austrimancy, Pyromancy, and Hydromancy. The newly-born and self-begotten spirit shadows forth its knowledge and intelligence, in a figure and by a figure, as the man imagines, and remains firm therein without any dissolution. It is in this way the spirit of those sciences is begotten which at last operates and perfects that which is sought. The first step, therefore, in these sciences is to beget the spirit from the star by means of imagination, so that it may be present in its perfection. After that perfection is present even in uncertain arts. But where that spirit is not, there neither judgment nor perfect science will be present. Hence wonderful things are now found out in future and occult things, which are laughed at and despised by the inexperienced, who never realise in themselves what is the power of Nature in their spirit, that spirit, I mean, which is born in the manner described, and given and assigned by God for this special purpose.

To believe in the Devil leads to doubtful results, and the thing is mixed up with fraud. The reason for this is to be sought from God, who has determined that all who believe in the Devil should be or become liars like himself. But this faith in God is perfect and free from all defect. It is in Nature such as its power is. So, then, the uncertain arts are sciences, but with this condition added, that a new generation of the prophetic and Sibylline spirit shall take place by which the art and hand may be ruled and guided. Who was the inventor of these uncertain arts, I have not been able to ascertain. I know this, that these arts are very old, were held in great esseem by the ancients, hidden and handed down as special secrets. They spent their time on imagination and faith, by which they tracked out and demonstrated many consummate results. At present, so much imagination and faith do not exist; but most men fix their minds on those things which minister to the pleasures of flesh and blood. These they follow; to these they give their attention. These arts, therefore, even on this account, are uncertain, because man within himself is so doubtful. He who is doubtful can accomplish nothing certain; he who hesitates can bring nothing to perfection; he who pampers the body can attain to nothing solid in the spirit. Everyone should be perfect in that which he undertakes. So the spirit will be entire, and will conquer the body, which is nothing worth. The spirit is fruitful. This a man should have perfect within him, and put aside flesh and blood.

THE END OF THE PROOF IN UNCERTAIN ARTS

Imagination has impression, and impression makes imagination. Therefore from impression descends imagination. Hence, it follows that whatever be the impression, influence, constellation, star—such is the imagination.

Hence, too, it ensues that imagination brings forth a new heaven above impression, and as the imagination, such is the figure of the heaven.

Proof in Manual Mathematical Science

Though everything in the whole of astronomy be seen and discovered, yet there must be respectively numeration, dimension, occasion, and instrument. These are the principles of all sciences, that is to say, they are those things which concur with all sciences.

It is difficult to understand how numeration can be brought to bear in the case of stars, on account of their infinite number. The greatest part of them is never seen, or seen with difficulty, yet all of these must be reckoned in their number. It is, however, impossible for a man who only uses his eyes to count these. He who uses more than his eyes can count them, but not that other one.

The same is the case with geometry, for the measuring of height, depth, breadth, etc., is much too difficult to be undertaken by all. That is not geometry which is handed down among the seven liberal arts. Our geometry is astral, not terrestrial, and is known only to him who makes his measurement magically, not elementarily, but beyond the elements.

In like manner, the work of cosmography is material. The invention of the art itself is material, not elemental, but rather connected with nigromancy and divination. They who practise it examine the state of all things in heaven and earth, in what position they are placed and constituted, and with what conjunction they are connected. These matters are found out with so much subtlety that they will be described by-and-by with reference to the globe, instrument, and sphere.

Now, if a manual mathematician be so skilled a numberer in arithmetic, a measurer in geometry, an explorer in cosmography, an experimenter with instruments, then he may with the utmost propriety give himself out as a mathematician. Of these three departments does mathematical science consist, and these four make up mathematics. In this way the invisible body of astrology, which is known to the wise men, can be deduced.

But there are other mathematics, which only concern the Magi. They are very apt at making magical instruments, such as gamahei, images, characters. For these things, too, are instruments. The art of making them has to be sought in magic. Their preparation is part of mathematics. It is necessary, therefore, that these persons should be certain and well constellated, fit for preparing these things and disposing them in their place. That is, they must be virgins.

So, also, in Nigromancy. It is mathematical so far as making its preparation goes. Divination and signature need no mathematics. In nigromancy, however, it is necessary that an instrument of certitude, as also one regalia, and other defensives be used; for spirits are very prone to obsession. It is, therefore, necessary that all should fortify themselves well against them, since the danger is imminent. But where that kind is (if I may use the expression) obsessible, it is worth while to know.

And so with regard to the mathematics required, as has been said, for the science of astronomy, let this be settled and determined, that herein is need for the most consummate prudence and intelligence. Nothing will be done by the common method. It is requisite that a man should be one who discovers these things in a more sublime way than by the ordinary and earthly light of Nature. There is need, I say, of a higher light, that is, of one that is above the artificial.

In this way, the mathematics in astronomy are proved by means of their own instruments, which agree with the great world. These instruments are so connected and bound up with the elements and the stars that they assume the form of a microcosm, which is itself made from the greater world, but consists of a smaller body, yet one which contains the universal world in itself like a quintessence extracted from it.

Here follow certain fragments aud schedules on the same matter as the preceding.

CONCERNING THE KNOWLEDGE OF STARS

Before all else you should be taught about the stars, what they are; for the astronomer is directed to the stars only and to nothing else. It should be known, however, in this place that elementary bodies are not concerned; also, that flesh and blood effect nothing, but only stars. In order that you may thoroughly understand this, I would have you know that man's senses are apart from his body. Whatever is not corporeal is either star or ether. But of those things which have not body there are many species in man. However this may be, man's sensation is certainly not flesh and blood. The body, therefore, is one thing and the sense another. The body is flesh and blood. The sense is soul. The soul, not the body, is the subject of astronomy. But the body is ruled by the soul. So, then, the body, too, is the subject of astronomy, because the body underlies the soul, is obedient to it, and ruled by it. Moreover, the soul is not something eternal in man; it is not the *summum bonum*, but is something mortal existing in man; it is the man built up in Adam. Since, then, the soul is subject to astronomy, and astronomy acknowledges the star alone as its lord, know that in the star there are many essences, that is, not one star, but many. It is known, also, that one star exists higher than all the rest. This is the Apocalyptic star. The second star is that of the ascendant. The third is that of the elements, and of these there are four; so that six stars are established. Besides these there is still another star, imagination, which begets a new star and a new heaven.

But although, as is now understood, there are seven stars, still the astrologer is not so conditioned as to act the astrologer in these seven. One is an astronomer of supernatural astrology; another over the ascendant;

another over the four elements, and yet another over the imagination. Each discourses of his own astrology, each one is an astrologer, and each sufficient in himself. Now, he who is an astrologer does not rest in one thing, but is conversant with all, if he does not expound his own species with which he is conversant. But that is an intolerable error which, neglecting the different kinds of stars, deals only with the horoscope, the ascendant, and the figure of the heaven. But though the rest of the horoscope should not be understood by the astrologer, this would matter little if he only confessed that these other parts were good and belonged to astrology. For there is a star of the firmament, that is, fire, which has nothing to do with the horoscope. There is a star of the earth, because the earth, no less than the heaven, has its astrologer. So water and ether equally have their own star, and in like manner the air. Let no one think there is only one star. There are more; but beyond all that have been mentioned there is one. Beyond the fifth, again, there is another supernatural one; and beyond this sixth, one which is hidden in man himself, making the seventh.

I speak of the seven kinds of astronomy which make up the entire man, as has been before pointed out. Moreover, these seven kinds are not under seven stars. But I say this, that astrology alone embraces these seven in itself, and hence it is necessary to understand how these seven stars are essentially conditioned. In this way a perfect astrological judgment issues forth, which can be obtained in no other way. For there must be a medium, by which the last operates, and another after the first of the four, add also in the last. I add this with the view of making quite clear what is not sufficiently insisted upon in astrology, that Mars in the sky must be thoroughly understood, which looks there like a live coal. For besides this many another exists, and, moreover, four others in the four elements, and, lastly, one in the imagination. What sort of smith would he be who could forge a horse-shoe but not a nail? What sort of a carpenter who could only cut his wood, and not join it? Science ought to be perfect in all particulars, without exception. What things should be joined, let them be joined.

In this place it should be specially considered that before the Deluge our ancestors, up to the birth of Christ, devoted themselves with constant zeal and unwearied labours to the discovery of wisdom; and now, since the advent of Christ, all this has perished and become extinct, so that it is difficult to find any of it anywhere surviving. I will tell you the cause of this. Christ offered eternal wisdom to the world. When this was offered it was only right and just that the inferior wisdom should be repudiated, and the higher acknowledged. In this respect I confess that I write like a heathen, though I am a Christian. For by right the lower wisdom gives place to the higher. The wisdom of Christ is better than the wisdom of Nature. A prophet or an apostle is better than an astronomer or a physician. A prediction from God is better than one from astronomy. A cure wrought by God is better than one by herbs. Prophets speak infallibly. The sick are healed and the dead raised by apostles; nor is there any deceit about these things.

Although, therefore, astronomy with its light was obliterated by Christ, who will impeach that light? And thus much farther I am commanded to say. The sick have need of a physician, but not all of them need apostles. So predictions need an astronomer, but not all need a prophet. Distribution being made, one part goes to the prophets, another to the astronomers; one part to the apostles, another to the physicians. Each has his own limitations. And so, indeed, astronomy is not taken away from or interdicted to us Christians, but we are commanded to use it in a Christian way. We are created by the Father for the light of Nature, and it is only right that we should know and practise this. We are called by the Son for eternity, whence this, too, should be known. So, therefore, the light is transferred to us from the Father as if by inheritance, and the light from God the Son here in this world to eternity. Neither hinders the other—the Son the Father, or the Father the Son. By this means man is able in both ways to learn, to know, and to work.

Having made this excursus, I end my treatise on astronomy, that you may know what the stars are, and what power the astronomer or astrologer has, in what respect the one differs from the other, and how the stars are situated. I have made mention of seven, not for the moment taking thought of one other star, which is the Signed Star of the Microcosm, so that really they should be reckoned as eight. In the following explanation and proofs all these things will be found connected together so that you will understand them.

I could wish, indeed, that those who put themselves in the place of Christ would shew themselves His real disciples. Then the light of Nature would be more rightly understood, that is, the miracles of God would be more carefully looked into. As it is, mere trifles and deceits are obtruded, in which there is no juice, no marrow, no wisdom. If folly and wickedness like this are allowed to succeed, what success can there be for the noble wisdom of Nature? In this way no consideration is given either to the wisdom of Nature or to eternal wisdom, but both lose esteem together. It is the way of the world to oppose every kind of wisdom. This being so, I thought it best not to refrain from writing, but by all means to go on. For the renovation of the world will be upon us; and then at length will be found that which is now sought after; and it will be so put before us that nothing of it will perish. It is a good thing to keep for our heirs a treasure predestinated to its special purpose. This is a real treasure, which is dug up with that end in view. Let no one think that I mean here to treat of anything save of the stars, and these are sufficiently explained, that being added which so far ends our knowledge. What we have deemed necessary we have linked together. And it should be known that the medium must be rightly nnderstood; for without this nothing is done; this is so.

The higher star governs all lower things. Now, if there is no star in the earth, the higher star will affect nothing. But the star of the earth conceives the power of the higher star, and is capable of containing it, which else would not be the case. It is so also with the water and the rest, as we have said above.

Another Schedule

I. This threefold operation of astrology has one mode in a figure of the heaven. By this it is understood how the heaven stands related to lower things, so that a perfect judgment is able to be made.

II. COMETS. What we understand as such are newly-begotten stars, not produced at the first creation, but freshly exhibited by God. Such were the star of Christ and others like it.

IMAGES are made from terrestrial things endued with celestial powers, by means whereof they heal diseases and turn aside wounds in the case of those whom they mark.

CHARACTERS are words which heal diseases and act like images. They are drawn from the higher stars and are artificially assumed by the lower.

GROWING THINGS OF THE EARTH are like Characters and Images. Sometimes trees are brought to such a state as to put forth flowers. Sometimes these growing things are changed into frogs, serpents, owls, scarabæi, dragons, etc.

SPECTRES are visions which sometimes appear to men. They portend wars and other future evils, like comets. They should be explained magically.

DREAMS occur if the heaven and its sidereal spirit sport and joke with men, concerning the past, present, and future.

BRUTES are used when heaven works by them and foreshadows the future, so that by them we can be informed of some impending evil and misfortune.

III. THE MIND. This is when the mind itself within man expects something, good or evil. The origin of this is from heaven, which thus sways the mind, and impresses on it its good or evil fortune.

SPECULATION is when a man speculates and imagines within himself, and thereby his imagination is united with heaven, and heaven operates so within him that more is discovered than would seem possible by merely human methods.

PHANTASY is when a fool or silly person speculates, and heaven is at the same time in connection, and so operates by him that from the phantasy of a fool heavenly influence is recognised.

IV. VISIONS are apparitions artificially produced in mirrors, crystals, nails, etc.

ASTRAL SPIRITS are those which dwell in man on the earth, separable from man, and serving him, as long as they exist.

INANIMATES are men who are produced without the seed of Adam by the operation of Nature, such as are giants, lemures, nymphs, gnomes, etc.

CHIROMANCY is a science pointing out the stars by the lines in the hands, feet and other parts of the body, as we have said above in this treatise.

SCHEDULE CONCERNING THE PROOF OF MAGIC

From what source magic proceeds, and how it interprets new signs. "There shall be signs."

Besides, how impressions from above impinge upon lower bodies. Moreover, what effect heaven has with its signs, as earth with its medicines. In order that you may understand this source from which magic draws its interpretation, attend. All sciences, all branches of human knowledge, are from God. These sciences either come from the light of Nature, or are learnt by instruction, or are secretly instilled by God.

The first mode is that in which man learns by himself without the instruction of man. For magic is not learnt by its interpretation, unless it be spoken from on high.

The magician is born, as all arts are born, as is the case with those who find out new arts, as letters, or Montanica:—

The Magi have the new spirit, not created by this or that man. That spirit is born by asking, by searching, by knocking, out of the heart and by the spirit.

NOTE.—Whatever God says, He adds an interpreter thereto. Let none, therefore, ask, whence is this? It is from God. He, for example, has said, "There shall be signs in the sun and in the moon." This needs interpretation. It cannot be explained by Nature, because it transcends the limits of Nature. The spirit must concur with what is said, and he who interprets this is a magician. The spirits are in the stars.

HERE ENDS THE TREATISE ON HERMETIC ASTRONOMY

APPENDICES

APPENDIX I

CONCERNING THE THREE PRIME ESSENCES*

CHAPTER I

EVERY thing which is generated and produced of its elements is divided into three, namely, into Salt, Sulphur, Mercury. Out of these a conjunction takes place, which constitues one body and an united essence. This does not concern the body in its outward aspect, but only the internal nature of the body.

Its operation is threefold. One of these is the operation of Salt. This works by purging, cleansing, balsaming, and by other ways, and rules over that which goes off in putrefaction. The second is the operation of Sulphur. Now, sulphur either governs the excess which arises from the two others, or it is dissolved. The third is of Mercury, and it removes that which changes into consumption. Learn the form which is peculiar to these three. One is liquor, and this is the form of mercury; one is oiliness, which is the form of sulphur; one is alcali, and this is from salt. Mercury is without sulphur and salt; sulphur is devoid of salt and mercury; salt is without mercury or sulphur. In this manner each persists in its own potency.

But concerning the operations which are observed to take place in complicated maladies, notice that the separation of things is not perfect, but two are conjoined in one, as in dropsy and other similar complaints. For those are mixed diseases which transcend their sap and tempered moisture. Thus, mercury and sulphur sometimes remove paralysis, because the bodily sulphur unites therewith, or because there is some lesion in the immediate neighbourhood. Observe, consequently, that every disease may exist in a double or triple form. This is the mixture, or complication, of disease. Hence the physician must consider, if he deals with a given simple, what is its grade in liquor, in oil, in salt, and how along with the disease it reaches the borders of the lesion. According to the grade, so must the liquor, salt, and sulphur be extracted and administered, as is required. The following short rule must be observed: Give one medicine to the lesion, another to the disease.

---

* The doctrine of the three prime principles being the foundation of the physics and philosophy of Paracelsus, it is the intention of this brief Appendix to exhibit that doctrine in connection with the origin of diseases.

CHAPTER II

Salts purify, but after various manners, some by secession, and of these there are two kinds—one the salt of the thing, which digests things till they separate—the other the salt of Nature, which expels. Thus, without salt, no excretion can take place. Hence it follows that the salt of the vulgar assists the salts of Nature. Certain salts purge by means of vomiting. Salts of this kind are exceedingly gross, and, if they do not pass off in digestion, will produce strangulation in the stomach. Some salts purge by means of perspiration. Such is that most subtle salt which unites with the blood. Now, salts which produce evacuation and vomiting do not unite with the blood, and, consequently, produce no perspiration. Then it is the salt only which separates. Other salts purge through the urine, and urine itself is nothing but a superfluous salt, even as dung is superfluous sulphur. No liquor superfluously departs from the body, for the same remains within. Such are all the evacuations of the body, moisture expelled by salt through the nostrils, the ears, the eyes, and other ways. This is understood to take place by means of the Archeus from these evacuations. Now, as out of the Archeus a laxative salt comes forth, of which one kind purges the stomach because it proceeds from the stomach of the Archeus, so another purges the spleen because it comes from the spleen of the Archeus; and it is in like manner with the brain, the liver, the lungs, and other members, every member of the Archeus acting upon the corresponding member of the Microcosmus.

The species of salt are various. One is sweet as cassia, and this is a separated salt which is called antimony among minerals. Another is like vinegar, as sal gemmæ; yet another is acid, as ginger. Another is bitter, as in rhubarb or colocynth. So, also, with alkali; there is some that is generated, as harmel; some extracted, as scammony; some coagulated, as absinth. In the same way, certain salts purge by perspiration alone, certain others by consuming alone, and so on. Wherever there is a peculiar savour, there is also a peculiar operation and expulsion. The operation is of two kinds—that which belongs to the thing and the extinct operation.

CHAPTER III

Sulphur operates by drying and consuming that which is superfluous. Whether this proceeds from itself or from others, it must be completely consumed by means of sulphur, if it be not subject to salts. Thus, a medicine for dropsy is made of the salts produced out of the liver of the Archeus to consume the putrefied and corrupt. But to remove the disease itself the strength of sulphur is necessary, to which diseases of this kind are subjected in virtue of their origin. Yet, it is not every kind of sulphur which will effect this purpose, and so it results from the nature of the element that every sickness produced by the nature of the body has its contrary from the nature of the element. This takes place both universally and particularly, and, consequently, from the genera of an element the genera of diseases may be recognised. One is always the sign and proof of the other.

The same sign occurs in the case of mercury; it assumes that which separates from salt and sulphur. Hence are produced diseases of the ligaments, arteries, joints, limbs, and the like. Hence in these diseases we must simply remove the liquor of mercury. But the ailments themselves ought to be removed by those things which are favourable and conducive to them, when proof has been obtained of the speciality of the thing in Nature.

CHAPTER IV

The physician should understand the three genera of all diseases as follows. One genus is of salt, one of sulphur, and one of mercury. Every relaxing disease is generated from salt, as dysentery, diarrhœa, lienteria, etc. Every expulsion is caused by salt, which remains in its place, whether in a healthy or suffering subject. The salt in the one case is, however, that of Nature, while in the other it is corrupted and dissolved. Cure must be accomplished by means of the same salts from which the disease had origin, even as fresh salt will rectify and purify dissolved salt. The sulphureous cure follows as a certain confirmation of the operation in salt.

All diseases of the arteries, ligaments, bones, nerves, etc., arise from mercury. In the rest of the body the substance of corporeal mercury does not dominate. It prevails only in the external members. Sulphur softens and nourishes the internal organs, as the heart, brain, and reins, and their diseases also may be termed sulphureous, for a sulphureous substance is present in them. Let us take colic as an example. Salt is the cause of this, because this predominates in the intestines. In its dissolved state it produces one kind of colic, and when it is excessively hard it produces another kind; for when it passes from its own temperature it becomes excessively humid or excessively dry. In the cure of colic by elemented salts the human salt must be rectified. But if a salt other than from sulphur be applied, you must regard it as a submersion of salt and not a cure of colic. Similarly, in the case of mercurial and sulphureous diseases, each must be administered to its counterpart, not a contrary to a contrary. The cold does not subdue the hot, nor vice versa, in congenital diseases. The cure proceeds from the same source as the disease, and has generated the place thereof.

CHAPTER V

The genera of diseases are also divided into various branches, locusts, and leaves. Yet is there one cure. The mercurial disease is an instance, for mercurial liquor separates into many branches, locusts, and leaves. So all varieties of pustules are subject to Mercury, because the disease is mercurial. But some are subject to common and others to metallic mercury, some to mercury xylohebenus, some to mercury of antimony. It is necessary, therefore, to know that liquor of mercury, which cures that which the salt of mercury dissolves, and it has also an incarnative virtue. For mercury is multiplex. In metals the liquor of mercury is like a metal; in juniper and ebony it is like wood; in marcasite, talc, and cachimia, it is like a mineral; in brassatella, persicaria, and serpentina, it is like grass, and yet there is but one mercury variously manifested. What has been said of pustules must be understood also of ulcers, of which some are cured by the mercury of persicaria, some by the mercury of arsenic, and some by the mercury of xylon guaico. Consequently, the physician should know the tree of diseases and the tree of natural substances, but of these there are indeed many. There is the tree of salt, which is twofold, namely, of rebis and of the element. There is also the tree of sulphur and there is the tree of mercury. Accordingly, the physician must guard against inserting two trees into one cure; he must remember that Mercury must be administered for mercurial, salt for saline, and sulphur for sulphurous diseases. To each malady let the corresponding remedy be applied. So are there only three medicines as there are three forms of diseases.

CHAPTER VI

In fine, the physician should classify diseases under the name of their medicine. It is opposed to the usage of art to say that a complaint is, for example, jaundice; any rustic knows this. Let him say rather: This is the disease of Leseolus. Thus, in one word, you comprehensively express the cure, property, name, quality, disposition, art, and science thereof. For Leseolus cures jaundice and nothing else. I would persuade every one to become accurately acquainted with the trees, for he who knows not their seed is involved in fundamental errors. So also we must say that this or that is a disease of gold, and not that it is leprosy. In like manner let him speak of disease of the tincture, whence it will be evident that the complaint is one which belongs to age, for the tincture regenerates age. So also we shall have a disease of vitriol, and this is epilepsy, which is cured by the oil or spirits of vitriol. I have comprehended these matters under a theory because of the special mode from which it is first deduced and the mysteries of Nature which were hidden by alchemical authors. From these I prove my theory of the elementary in its production and the annual in its generation. Let us instance the operation and virtue of Mercury. There are many of these operations and virtues both in the elementated and the annual, which experience teaches to those who know in what things Mercury and in what things other spirits lie hidden. They also will know how to prepare that Mercury, and how to form one kind into a topaz, another into crocus Sandalius, a certain other into a spirit, and any they choose into the exaltation which best suits it. The power of flesh astringents and flesh formers for wounds proceeds from Mercury alone, in which there is no sulphur and no salt, and the same is extracted and produced into its own pure liquor. But some Mercury is quicker in operation than others, as the Mercury in resin, which is quicker than that in mumia or tartar. The same process must be followed with sulphur and salt; the physician must understand their exaltations if he would cure his patients. I know perfectly well that Porphyrius would marvel were he to hear that Mercury becomes sapphire and a noble jasper because he has not seen it or handled it.

CHAPTER VII

Ginger is diaphoretic by reason of the salt out of the body whereof it is made. But that virtue belongs to the fire, through which generations boil up, as is held in philosophy, and by reason of this boiling up it removes obstacles, and reduces or elevates the humours of sulphur, salt, and Mercury to the second, third, and fourth grade of ebullition. And as it is constituted out of the igneous nature of salt, so it also ascends a grade, by which grade the humidities distil through the pores and guttas. Thus purifiers perform their work by the sole force of salt, as, for example, honey. The balm of salt is situated in honey, which, consequently, does not putrefy. For balm is the most noble salt which Nature has produced.

Attractive force is of sulphurous nature or essence. Mastic is a sulphur thus produced, and so also opopanax, galbanum, and others. Nor must we accept the axiom of physicians, that it is the property of heat to attract. We should rather say that it is the property of sulphurs to attract. Hot things only attract in so far as they burn, but that which burns is sulphur, which is not fixed, and hence evaporates like gums. Laxatives also attract like a magnet from those places where they are not. But the reason why salts attract is that salt is impressed upon sulphur and coagulated by means of the spirit of sulphur. Therefore, it attracts from places more distant than itself. Thus there are aperients of sulphur, whether cold, or green, or purple red, of any fashion whatever. For it is the nature of aperient sulphur to operate and to drive before it every moveable thing which it reaches. Nor is it true, as physicians say, that it is the nature of cold to cause evacuation.

CHAPTER VIII

What we should know about tonics is explained by the Archeus, who is like man, and remains hidden in the four elements—being one Archeus indeed, but divided into four parts. He therefore is the great cosmos and the small man, and one is like to the other. From Archeus proceeds the force of tonics. That from the heart of Archeus acts as a tonic to the heart, as gold, emeralds, corals, and the like. That which proceeds from the liver of Archeus strengthens the liver of the lesser world. Thus, neither Mercury, sulphur, nor salt, bring out this kind of healing virtue. But the heart of the elements sends it forth; from this does that flow. In the elements there is a force and potency which produce the tree from the seed; thence it derives the strength by which it bcomes erect and stands fast. So, also, by an external strength which the eyes see do hay and straw grow up. There is a like strength in animals, by virtue of which they stand and move. Moreover, there is another strength which is not visible to the eye, but is inherent and is the principle of health in the subject wherein it abides. This is the spirit of Nature, which if a thing have not, it perishes. This spirit remains fixed in its own body. The same strengthens man. In this manner does the strength in all the limbs of Archeus flow down into the lower world by means of vegetables.

APPENDIX II

A BOOK CONCERNING LONG LIFE[*]

BOOK THE FIRST

CHAPTER I

SINCE it is becoming to Theophrastus that he should philosophize further concerning long life, it is necessary, in the first place, and worthy to be known, in my judgment, what life is, especially immortal life, which subject the ancients completely passed over, as I believe, either because it was by them unknown, or was not sufficiently understood. Hence it is that so far they have made provision only for the mortal life. Now, I will straightway define what life is. Life, by Hercules, is nothing else than a balsamite mumia, preserving the mortal body from mortal worms and from dead flesh, together with the infused addition of the liquor of salts. Moreover, our life is long, for neither spirits nor the light of Nature affirm that it is short. The life of the ignorant is short, with art it is long. What is shorter than art? What is longer than life, at least among those who are not superstitious? Further, what is more durable, more healthful, and more vital than balsam? What is more transient, more weak, and more mortal than the physical body? Its measure varies between long and short. Why, therefore, is life long and why is it short? But that life which is of the supercelestial-physical is outside our rules. The pomp of our authority extends only over the mortal body, and is regulated, so to speak, by art unto the third terminus, unto the fourth, even unto the fifth. So much for the living body. What, then, about death, and what is death? Certainly nothing else than the dominion of balsam, the destruction of mumia, the ultimate matter of salts. The separation of immortal from mortal things produces a dissolution of the mortal

---

[*] This Appendix may be regarded as serving two purposes. The subject of long life is, of course, a highly important branch of the Hermetic Mystery, and whatever Paracelsus wrote concerning it should be included in a collected edition of his Hermetic writings. But the alternative treatise entitled *De Vita Longa* shews Paracelsus at his darkest, and, it may be added, at his worst. From beginning to end it is not only unintelligible, but almost incapable of translation. It is well that one specimen of his really arcane manner should be given to the reader, so that he may regard more hopefully the difficulties which encompass the comparatively lucid works which have preceded. The present version has been reasonably compressed, but it can only be affirmed that it interprets the original about as accurately as can be expected.

members. But this is accordingly called long life from the beginning. This also is short life, that is to say, it is said of death. Now, death is not life, but art is longer than this death. These are the dissolutions of life, also digested separations of that which is pure, of life long and healthy, both mortal and immortal, which the day of birth has united and conjoined, and that from both bodies. For every conjunction of perishable things of a diverse kind brings about dissolution, and how much more then will the conjunction of things natural with things which are beyond Nature be also followed by dissolution? For the cause of death is an empirical war, scarcely different from a duel taking place between the mortal and immortal. Disease may be compared to the javelin, and the Anthos to the breast-plate. What else is there over and above the struggle which these carry on? Herein is the fountain and origin of the generation of disease which presently death follows. Hence we understand what life is, both the mortal and immortal.

CHAPTER II

In order to the clear understanding of what has been already said, I consider that I should next speak of the physical body. The end of the physical body is the sustentation of all those things of which we have been treating. Herein we should divide our examination after the following manner. In the first place, let its parts be considered, not only the special organs but those which are distributed over the whole body, including a right understanding of the marrows, the conditions, the uses of tendons, the forms of bones and cartilages, the nerves, the properties of flesh, the virtues of the seven chief members, bearing throughout this rule in mind. In the first place, we must thoroughly know the whole *rationale* and nature both of the physical body and the physical life. Now, the body and the life of the physical body are alike mortal. But from that which is mortal nothing can be elicited in the direction of long life, and thus with regard to the arcanum and the elixir, in this our Monarchia, neither the body nor the mortal life ought to be considered. For long life is a thing outside the body, is preserved apart from the body, and the body is inferior thereto. Moreover, when the body intervenes, a dissolution of either life takes place. On this subject the Empiric Muse and the medical sophists, following the method of the Spagyrites, preserved the body as a balsam to avoid occasioning death, whereas the balsam is the mumia of life, not of the body, forgetting, meanwhile, that death was not in life, for the death of life is nothing else than a certain dissolution of the body from the immortal. When this takes place, then the body dies. This was the mistake of Hippocrates throughout all his prescriptions, namely, that he administered to the body instead of to the soul, and that he proposed to preserve the mortal by means of the mortal. The body is a creature, but not so the life, and it is indeed nothing but the daughter of death. Therefore, from Archa descended that which is immortal.

But you will say that the Hippocratic Muse is not altogether to be referred to death. Be it so, but you will find a much easier way to health, since the Magnale has descended from above. For God gave unto Hippocrates only those things which are creatures, and among these even the chief mysteries were not imparted in their fulness. To this body God has added another body which is to be regarded as celestial, that, namely, which exists in the body of life. Hereof I, Theophrastus, affirm that this is the work and this the labour, namely, lest it collapse into the dissolution which is of mortal things and belongs to this body alone. And although dissolution can take place in that perishable body, and from this dissolution may be gathered a loss of the heavenly body, yet it cannot stand in the way of long life, by reason of the restoration which must shortly take place, so that the body may be altogether without any defect, for as fire continues to live so long as wood is present, even is it the same in the case of long life, so long as the body out of Archa is present, because the body as a body is to be preserved by the intervention of a body, which extrinsically grows strong. By means of this it is preserved. For the body is nothing else than the subject wherein the long life of the eternal body flourishes. So much for the physical body.

CHAPTER III

It is needful that we should now state after what manner the matter of the same is to be preserved from all corruption. In the first place, whatsoever the body corrupts in itself the same is to be restored by a foreign body, that is to say, as the Monarchia of the Spagyrists does not admit, by the common nature of balsam which labours to preserve the body. For as it is impossible that wood should not be consumed by fire, so it is impossible for the body not to be at length corrupted by life. Therefore, those who are skilled in essential things, who think they can attain long life as a balsam, are the less to be heeded, since the nature of the balsam is rather to preserve the body from corruption lest there should be a vacuum in the body. For every vacuum is a disease of that place, or a sickness in the body, or, again, it is a certain atrophy of long life. For long life has place in the perfect body; in the imperfect it continually fails until it is dissolved in death. We know that the physical body can be sustained from death, and that by virtue of its innate mumia. These things have reference to a healthy life, not to a long one, which is a terminus for the physical body. But it will be worth while before we explain long life to first exhibit and ensure a healthy life. There are certain things which ward off diseases. It is to be observed, therefore, that corruption is to be removed from the body, and that which blazes forth in long life is to be again returned to the refrigerium. Wherefore, in this place the specifics of Nature which are prepared for this purpose must necessarily defend the body wasted by any disease, which is the duty of the physicians, but in long life nothing of the kind is required. And now concerning short life. The specifics for given diseases have nothing to do with long life; they are used solely to fortify the body. It makes little difference as to long life whether provision is made against fevers, etc. For as long as the spirit of Nature remains it preserves the celestial body, and the long life remains, together with the torture of diseases. Death here is not ready, for as long as the body is committed to the care of the physician there is death, but not the celestial body, for out of the body flows a poison into life which so inflames it that it bursts forth altogether into bad flesh, and seeing that death takes its origin from corrosives, and a certain arsenical realgar is for all, therefore it does not cease from the nature of a poison until it has satisfied its nature and consumed the body, converting it into incinerated eschara. Nor after the end will it cease from its malice therein. Therefore, a double praxis is to be begun—one to preserve life, and the other to repress the body and to alleviate it from day to day by reason of the corruption which takes place daily.

CHAPTER IV

Whereas, by the nature of its creation, the body, and its physical life, passes as one part into the composition of the form, and because the physical body is the half, being the whole with the celestial, the physician ought to give more consideration to the question how the major physical man is to be preserved. For in the major life consists long life, but in the minor is the subject of mortality, and this is implanted according to predestination, both the body and the celestial life in the physical body, which, as an individual companion, accompanies this conjunction. Now, it is to be understood concerning predestination that there are some things which are free therefrom, and are therefore disposed according to the Divine Will, without the violation of any law. Further, when another conjunction of these two forms has been produced, of the natural, that is to say, and of that which is beyond Nature, into the form of Nature, and when it is completely elicited in the matrix that there are two fathers and one son, two mothers and one daughter, and that these four persons generate, this celestial seed produces, together with the mortal, that animal which appears when born, the elemental seed, and also the celestial at the same time. For in this place the corporal seed truly works, which also ought to be preserved in the predestination of the natural channel, for that which is beyond Nature ought to be considered in the first place, in order that by these things even those which are beyond Nature may be preserved. Thus a boy is designated as the heir of two inheritances—of the nature and of the essence which exists from Nature, according to the decree of the creation which comes forth at the same time, and of that also from him who is the parent beyond Nature, which parent rules the body and governs it. Out of the two, that is to say, from these parents, arises a conjunction of matrimony from on high. For Adam obtained nothing from his creation, nor was he made subject to ascending signs or to any other matters.

Nothing, therefore, out of the four stars can participate with man. For the stars and the homuncula are not divided, but man has received long life from that which is beyond Nature. In the first place, the physician is to be admonished in order that he may possess and use the truth, not following everywhere the figments of the unskilled, who have written most frigidly on the matter, that he should pay more attention to the things which are beyond Nature than to those which are according to Nature. Next, that he should be fairly acquainted with predestination. From this, as from a source, proceeds that monarchia which is beyond Nature rather than the specific and qualitative. Hence is spread abroad that error wherewith not a few are imbued, so that they determine to study the body, and attribute many more things thereto than of right belong to it. Wheresoever present life exists, it is not in its fulness, and therefore it exists without force. This, although it be dead, because it does not operate, is yet implanted in the body. Thus, in the case of one who holds a knife to his throat, a blow takes away long life. Whatsoever further life arises in the body is the congenital life of Nature. To this, however, attention need not be paid, but to that only which revivifies the body. Moreover, long life exists as a man with us, even as fire put among wood, whereby a man recuperates himself.

CHAPTER V

We will discourse briefly of all these things, that it may become more clear as to what has been said concerning the parent which is beyond Nature, and is engrafted definitely on the natural, also the cause of the dual life of the natural body and the parent of that which is beyond Nature. From this it is clear that man is born of a double seed. For now from the time of Adam his complexion has changed with the nature of the generations in the flesh, by means of the importunate and unseasonable operations of persons in their contrary nature. It is clear that neither the sanguine, the melancholy, the choleric, or the phlegmatic temperament is born with us. From none of these has a complete temperament ever arisen. The physicians should now, therefore, pay no heed to the four complexions, for they did not exist in Adam, and much less then in his progeny, nor can four such diverse things co-exist. For first, as by the intervention of an unseasonable birth, the respective temperaments are corrupted, and this not without loss of children (for what is the temperament? It is the nature of the parent, and that without hot or cold, black or white); thus, also, in the body which is beyond Nature there exists a certain hereditary seed, and if two human beings of the same temperament unite, yet the supernatural semen under which both wisdom and life are hidden is never truly conjoined. There is therefore a dual marriage—one which human reason counselled, the other which is the conjunction of God. The former is not properly marriage, except as the eyes in the senses of Nature permit. In this, although a man in every way considers how he may excuse himself and his children, and require the Spirit of God to unite them, and profess to be honourable, there is nothing but hypocrisy. The new change of locality is a proof. But the conjunction which is of God is properly marriage, and belongs to long life by reason of divorce which in this place cannot take place, a fact which none can understand without the intervention of children. For this reason many are sanctified in the womb of their mother; these are they whom God has joined, as she who was once wife of Uriah and afterwards of David, although this in all human judgment was diametrically contrary to a just and legitimate marriage. But God effected this union because either had attained a long life beyond that which is of Nature, as by heredity, on account of Solomon, who could not otherwise be born except from Bathsheba, by a meretricious power, with David. Therefore whatsoever is beyond Nature is as a treasure committed to God, a fact well known to those acquainted with the Spagyric art, and marvellously conducive to long life.

CHAPTER VI

The practice, therefore, being divided into two, one for physical and one for long life, the physician will diagnose from the end, so far as the use of either shall come in. But of this life which is beyond Nature, whereof we are at present speaking, it must be ascertained whether it may be possible by any means to attain it in the physical life, since its sphere is beyond the powers which are accorded to Nature, and under it lie hidden the arcana of long life. For in this place the impressions which are beyond Nature are openly produced; they flow together into the supernatural life, even as the firmament passes by influx into the body which is according to Nature, and although supernatural impressions appear, yet the knowledge of them is obscure. Hence it is that they received the name of impressions from some, from others that of incantations, from others that of superstitions, while yet further names were bestowed on them according to the rules of magical art. From these proceeds that which the Greeks term Magiria, treating exclusively of impressions, which they call incantations and superstitions, which also belong to the supernatural body. It is important to treat of the supernatural body in relation to its impressions, because the whole of Magia has been perverted to a foreign use by astronomers; it has been wrongly called superstitious, and a certain medical sorcery. After the same manner they referred necromancy and nigromancy to the same source, so that each might be regarded as an idolatry, which things, unless an influence intervenes, would come at once to silence, for although the manes may answer on every side, nevertheless this does not happen without the influence which is beyond Nature, a thing which is wrongly believed to be an imposture of Satan because it is impossible to man, whereas it can be easily produced, as you see in the case of the exorcisms of fantastic spirits. The whole of cabalistical magic is contained in the separation of the body which is according to Nature from the body which is beyond Nature, and is implanted in us as an image, to be sustained and administered, so that although absent it may establish communication between those who are widely separated, and may manifest unknown thoughts. This at the same time may be very difficult for those who are uninstructed in the cabalistic art, seeing that a great mistake has been made even by its professors, a fact which is indicated by their translations out of the Hebrew and the Canons of the Spagyrists. Hence we conclude concerning long life as follows : that out of the supernatural influence not only incantations but the arts of images and gamahei have proceeded. Philosophasters have referred this influence to the stars of the firmament, and out of the coals of heaven have feigned a Mars and Jupiter to govern that body which is beyond Nature, whereas this does not pertain to them except in regard to mortal things which have nothing to do with long life. Hence the things which are to be used for long life are to be extracted from supernatural and not from natural bodies, for the whole of that supernatural force is magic, and every magus obtains the influence which is beyond Nature, together with the body in which it inheres. The body which man bears about within him is invisible to man, as is the case with generation, etc.

CHAPTER VII

But that you may rightly understand after what manner incantations or manes came to be considered superstitions, and how they since came to be abused, so that they ought neither to be called manes nor superstitions, know that the beginning of these things was from the Protoplast, who united a supercelestial and mortal body in his own long life. Now, every phantasy and imagination is a principle and special thing in supercelestial bodies. As the mortal body preserves itself in its own special substance, so does that supernal body in the imaginative, as phantasy is of that body, and is indeed itself a body. Whosoever would, in any sense, control a supercelestial body of this kind, must be thoroughly acquainted with the method of resisting the imagination, for the more frequently that body has intercourse with mortals, with the more peril do these things accompany the body. The protoplasts clearly overcame this, but their posterity, having no solid and perfect knowledge hereof, deceived themselves like madmen and fools, nor are they unjustly considered such. Moreover, that supercelestial body is in no wise dissimilar to the stars out of a certain fire, out of which invisible things there arises a visible cloud. Such also is the property and nature of supercelestial bodies that out of nothing they clearly constitute a corporeal imagination, so as to be thought a solid body. Of this kind is Ares. Those who ignorantly perverted, and knew nothing of the foundation upon which this art is built, feigned that there were manes, which originally were called Fate, and afterwards superstitions and incantations. Out of these supercelestial bodies both nigromancy and necromancy originated, and so also geomancy, hydromancy, pyromancy, and lastly also the arts of mirrors, the divining rod, divination by key, and innumerable other things which are classed among superstitions.

CHAPTER VIII

But, that the physician may know all things fully, let us remark the examples of the elders who laboured very greatly in the said magic, that they might obtain long life, without any mixture of the Hermetic rejuvenescence, and without the art of Spagyric experience, being of the body alone. We see the age of Adam and Methuselah, with whom the art of magic began. It is vulgarly thought that the Protoplast was predestined to attain the greatest possible age, but that the smallest measure of years is allotted to ordinary men. The latter point is much insisted on in the schools, but is by no means to be approved. The source of Adam's longevity was magic, by means of which influence he always lived. The death of Adam is ever to be deplored by posterity, not so much because of the fall, but of the science which died with him, who alone retained the spirit of the highest life beyond that which was of Nature. Understand the same of Methuselah, who was next to Adam. There have been other men, indeed, not unworthy mention, who surpassed the ordinary length of human life, as Moses, who completed one hundred and twenty years, yet not according to the method of magic, but rather of physical life, to whom was joined so strong a nature that it attained a great age without difficulty. Like instances occur in our own days, and will be found occasionally to the end of the world. Some, again, by the help of magic, have lived to a century and a half, and yet some have attained to a life of several centuries, and that by the adjoined force of Nature, which exists fully in metals and in other things which they call minerals. This force lifts up and preserves the body above its complexion and inborn quality. Of this kind are the Tincture and the Stone of the Philosophers, because they are elicited from antimony, and, similarly, the quintessence. These and other numerous arcana of the Spagyric art are met with, which in all manners restore the body exhausted by age, return it to its former youth, and free it from all sickness, a fact which is well known to all acquainted with this monarchia.

CHAPTER IX

There is also another way of preserving long life, which Mahomet prescribed to his disciple according to magic, and endowed him with many years ; nor did he do this from God, but from the influence which is beyond Nature. Because Mahomet, as a magus, exercised this method for the unskilled population, not for himself, he has won an immortal name. Archeus preserved his life for several years beyond a century, a thing which was laid to his discredit, and was referred to idolatry. He was equally skilled in cabalistic art with those three Sabean magi who came, not by natural magic, but by the force of horses, to the Bethlehemites, and was acquainted not only with that which was of long life, but that which is of the intellect beyond Nature. All these things proceed from supernatural influence, which rules and governs the body. These magi were afterwards followed by those who falsely claimed for themselves this almost divine name, among whom was Hippocrates, who preferred rather than that his daughter should remain in her actual form, to transform her outward natural influence into a body alien from all Nature—an evident proof of the power of incantation. In the same way Serellus attained long life and studied the metamorphosis of Nature. In the conservation of that body which is beyond Nature the most part were equal to Methuselah, but they made great errors in the transformations; their operation passed into a fantastic body, by reason of their ignorance of physical things. There are many, indeed, whose length of life will persist up to the last day. Such metamorphoses, however, take place without long life, as we see in the case of sea-wolves, who, if restored to their pristine form, again become subject to mortality. Judge also in like manner concerning the fantastic body, on the intervention of food or the osculum of man. All these things are subject to the deltic impression, but before they pass into the deltic impression death is not present, except as far as a mixed fantastic body is admitted, which produces a narcotic form, preserving even to the last day. Moreover, many have lived upon the life of another, and that according to the rule of the Deltic Nature, among whom was Styrus, who when struggling for life is said to have attracted to himself the strength and nature of a robust young man, who chanced to stand by, so that he succeeded in transferring to himself his senses, thoughts, and even the mind itself. By this imagination Archasius is said to have attracted to himself the science and prudence of every wise and prudent man. Such is the strength of mind in which that supernatural vigour exists, that it sometimes satisfies a glowing and, what is more, a ravenous concupiscence. Hence arises that contempt of images and gamahei among those who abuse this image even to destruction. Hence are those words, characters, signs, forms, and figures of hands, imprecations and orations, which are the principal cause of incantation, and, what is more, of words which are commonly applied to wounds and other diseases. Finally, whatsoever can change into this form does so by the force of that body which, beyond Nature, is implanted in us. Further, out of those impressions which are beyond Nature arise the stars of the firmament, Venus and Saturn, and other planets, so that that influence which is beyond Nature rules and governs inferior things. Whatsoever, therefore, takes place in gamahea and imaginations, by the accession of planets and signs, all this can be transferred to the superior signs. Wherefore those bodies which are perishable can easily be set free from death by that supernatural force. Moreover, Venus and Saturn, Mars and Mercury, exercising their force in the superior firmament, have endowed the most part of mortals with immortality, and that without any human operation, by the accession of imaginations, of whom not a few exist, visible and invisible, both on earth and in the sea. Some of these have attained this point by means of Deltical impressions, not, however, the nymphs, as is the case with animal generations.

A BOOK CONCERNING LONG LIFE

BOOK THE SECOND

CHAPTER I

HAVING spoken of the several arcana which restore to its pristine health a body affected by diseases, we will begin where we last left off. To finish what we handed down in former books, and to shew how the physical body may be preserved like a balsam, the particular arcana and the matter of this second book are referred to the same body. Although, then, one and the same preparation holds good, still the practice comprised in this elixir differs from that special mode of healing. In this second book the first places are held by Flos Cheiry* and Anthos. In this is comprised the arcanum of elixirs, and that by the force and virtue of the whole quintessence. At the outset, therefore, in order that each may be the more clearly noted, I will, with this view, point out in a few words what the quintessence is. Nature procreates the four elements, from which a certain tempered essence is prepared by the spagyrist, as is expressed by the Flos Cheyri.*

Now, here I think it matters little what the art of Lully teaches on this matter, since he wanders more than sixteen feet from that universal Monarchia which the Archidoxies prescribe. One thing is Extraction, another Confortation, another Melioration, to adopt the terminology of these men, of which Raymond makes mention in that treatise which is entitled "The Art of Lully," and from these he has made a false estimate of the quintessence. Since these are mere trifles rather than truths, we will pass them by in silence. But the Flower of Gold, the Flower of Amethyst, and lastly, whatever is of a transparent nature, pearls, sulphurous bodies, cachymiæ, and whatever belongs to the aluminous zerebothini, including all the genus of other things which the water produces, such as carabæ and corals,—these, I say, are all capable of forming quintessences according to the rate of temperation which is wont to be produced by the spagyrist through the intervention of a corruption of the elements.

CHAPTER II

Moreover, the sum total of the whole matter lies in this (since what is said in the book on the Elixir must each and all be referred to the subject of long

---

* Thus differently spelt in the original.

life), that universal Nature is reduced to the spagyric mixture, or temperation, which is nothing else than the goodness of Nature, in which is nothing that is corruptible, nothing of an adverse character. And yet, by another method and a different one, the same goodness of Nature is found in the tincture, according to the prescription of Nature, which exists in the Philosophers' Stone, in antimony according to the Nature of the crow, in sulphur according to the effect of the Lunary, and in the same way in other cases. Nevertheless, in all these there is one and the same temperation which among metals lurks under Mercury (I mention Mercury, which is in all metals), among gems under the crystal, among stones under the zelotus, among liquids under carabe, among herbs under valerian, among roots under sulphur-wort, among bitters under vitriol, among flints (say rather among marcasites) under antimony. Moreover, as Mercury is in all metals, so is antimony in all flints (or rather marcasites), vitriol in salt, and melissa in herbs. These are names of the tempered elixir. It should be remarked, too, that in elixirs following upon the sulphur of those substances which certain people call minerals, there is a quintessence, the Mercury of the Metals, from which is extracted the nature of the body. Cheyri prevails in Venus, Anthos in Mars; and the force and nature of these are not only that they drive away diseases, but that they preserve that body for a long life which is dependent upon the lower influence. With this view we will further say that the Elixirs of Long Life shall be embraced under many and various names, since the force of them all is one and the same. To us (if, perchance, you wonder at this mode of treatment) it has seemed good in the meantime to play with words.

CHAPTER III

Of all elixirs, the highest and most potent is gold. We will, therefore, treat of this first. If you understand the principle of this, you will understand that of other substances which are separated from their bodies. The rest, which are not separated from the body, will be indicated below when we come to mention wine. Concerning the Elixir of Gold, then, so far as relates to practice, act thus: Resolve gold, together with all the substance of gold, as a corrosive, and continue this until it becomes identical with the corrosive. Nor let the mind revolt from this method of treatment; for the corrosive excels gold, so far as it is gold, and without the corrosive it is dead. The quintessence of gold, therefore, without the corrosive, we assert to be useless. It follows, then, that the resolution must be renewed anew by means of putrefaction, although the corrosive adheres somewhat closely. For if the force of gold is so great that it preserves the body and renders it free from all sickness, nor allows it to be corrupted, how much more itself, and that without any infection? It corrects and purifies everything that is not pure. The corrosive, therefore, in the case of gold, ought not really to be called a corrosive at all. For the force of the arcanum overcomes all poison. All realgar dies in the elixir of gold, and goes off to the tincture which excels in medicine. And thus it is in this way that Potable Gold is produced after putrefaction. The common practice of the Spagyrists prescribes this dose, or rather a certain harmony. Lastly, you will notice about the elixir that wherever an elixir is brought to bear on anything, it so transmutes it that it remains fixed in a form similar to itself.

CHAPTER IV

Concerning Pearls

Now, in order to give greater clearness to what we have said about the quintessence, it should be remarked that nothing is nearer to gold than pearls. You must, therefore, reduce to temperation the four elements which are in pearls, whereby exists a quintessence without any loss of substances. Moreover, if you wish to transmute pearls into a quintessence, according to prescribed rule, act after the method of a quintessence. Do not change anything except the principle, in which it is necessary there should be joined the ultimate matter which exists as first matter in finishing the quintessence of Sol. This is extracted by a prescription of the following kind: First of all reduce to liquid a lemon newly re-elevated, in which pearls have been calcined, dried, and resolved; this serves for a resolution into the element, in which resolution is no complexion whatever. There is herein an universal force like a quintessence. I cannot in this place advise you to admit that method of extraction which Archelaus prescribes, nor any other spagyric separations of that kind. The mode of transmutation given above not only restores to their former power those members which are weak, but also keeps in the same vigour those which are strong and robust. So there is much more in pearls than in other sperms; and among these I consider the most excellent are those which come from the oysters.

In this place, too, the homunculus treated of in the Archidoxies bears no small part. The necromancers call it the Abreo; the philosophers name such creatures naturals, and they are commonly called Mandragoræ. Still, error prevails on this subject through the chaos in which certain persons have involved the true use of the homunculus. Its origin is in the sperm. By means of complete digestion, which takes place in a venter equinus, a homunculus is generated like in all respects, in body, blood, principal and inferior members, to him from whom it issued. We will, however, in this place pass by its virtues, because the subject has not been dealt with, as that of pearls has, by those who are acquainted with this matter.

CHAPTER V

Concerning the Extraction of the Quintessence from Herbs

We have made mention above of that Quintessence which should be produced without any extraction, and it is necessary to regard this subject in connection with our present opinion. The quintessence cannot be got from herbs without extraction, on account of the diversity of those essences which are included under one substance. These must be separated, so that the herb shall remain a herb and the quintessence a quintessence. Although in every herb there are four duplex elements, still the quintessence is not duplicated, but one part only. The other part, which belongs to the substance, we relegate to those arts that are special, and will treat of what belongs to the elixir. This is made quite clear by the example of melissa. Digest melissa for a philosophic month in an athanor; then separate it so that the duplicated elements appear separately, and immediately there will shine forth the quintessence, which is the Elixir of Life. Such is the case, too, with generous wine, and differently in other instances. In nepita it is bitter; in the tare, like clay; in tincium, blackish; in the hop-plant, slender and white; in the Cuscuta, harsh. In other cases it must be judged in like manner according to the prescriptions of experience.

Moreover, when this spirit has been extracted and separated from the other, behold the wine of Health! The philosophers have strenuously tried for ages to attain this; but they have never succeeded. A good part of them, followers of Raymund, have emptied several casks, in order to extract the quintessence of wine, but they arrived at nothing save burnt wine, which they erroneously used for spirit of wine. All that is necessary on this subject will be found elsewhere, in the "Philosophy of Generations." Enough to have warned the Spagyrist under what form the quintessence exists in herbs, and what it is worth while to investigate in them.

CHAPTER VI

Concerning Antimony

As antimony refines gold, so, in the same way, and under the same form, it refines the body. There is in it an essence which allows no impurity to be mixed up with that which is pure. No one, even though he be skilled in the Spagyric Art, can apprehend to the full extent the power and virtue of antimony. In the beginning of things antimony was developed, and was so related to the metals, which were produced by the water, that, when the Deluge was over, its genuine force and virtue remained after such a manner that it directs itself under the form of influence, and has never lost anything of power or virtue. With due cause, therefore, we assign to this alone everything which is attributable to minerals, whereof antimony includes within itself the chief and most potent arcanum. It purifies itself, as well as other things which are impure. Nay, more, if there be nothing wholesome present, it still transforms an impure into a pure body. This has been dealt with in the exposition of leprosy; and spagyric practice makes everything clear and comprehensible. But, not to digress at undue length, let us come at once to the mode of preparing the virtue of antimony (one jot or tittle of which is better than all the texts in your possession). First of all, take care that the antimony be not corrupted, but that the total, whatever it be, remains entire, without any loss of form, for, under this form lurks the arcanum of antimony, which should be impelled through the retort without any *caput mortuum*, and be reduced anew in a third cohobation to the third nature. Then the dose will be four grains of it given in the quintessence of melissa. To this the Archeus of the earth assigns nothing further.

CHAPTER VII

Concerning Sulphur

It is specially difficult, yet worthy of all celebrity, to realise the power and nature of the earth which procreates balsam, the characteristic whereof is that it suffers nothing to putrefy. But think of the resins, whereof the principal ingredient is sulphur, and there is nothing which deserves greater praise. In sulphur there is a balsam which none who study the different arts should fail to remember. In it are the balsamic liquids which do not allow wine or anything dead to putrefy, but do so conserve the body that there can attach to it no evil influence, natural corruption or any impressed on it from without. None need be surprised that so great a power is in resins, or that we speak its praises beyond the balsam which grows on the earth, and but, as it were, illustrates the force and virtue of this balsam. In those which are occult much more is found than in those which are manifest. And so, too, much more is found in sulphur than in the other departments of resins. In the case of sulphur, in order that we may arrive at the method of treating it, proceed thus: Elevate sulphur by colcothar in the spagyric manner. Do this so long as the fire does not get the mastery, as colcothar is wont to do in the case of sulphur. This same fixed spirit is the balsam of the earth, concerning which we write very little in this treatise. Its virtue is made clear by experience; and, though certain gums and resins, and other substances of this class, have the same nature as balsam, still, I think that among these sulphur is the first and the best.

CHAPTER VIII

Concerning Mercury

The Elixir of Mercury, prepared in the same way as that in which it is used for transmuting metals, avails in the very highest degree for driving away disease. Its rust, which the followers of Lully falsely call its flower, is nothing but death. As death consumes and wears away the body, so does rust affect the metal. In whatever way, then, this tincture affects it, the result will be that it ministers to long life, and the more efficaciously and powerfully in proportion as (let the expression be allowed) it reaches the grade of a poison, and the more actively and subtly its preparation has been repeated. Let no one be alarmed by those fables of Rupescissa, who, as his custom is, has written at once rashly and frigidly on this subject, namely, that, in the tincture of the body, you should altogether avoid gold and substances of that kind, which belong to Mercury, and, lastly, whatever is prepared from the spirit of salt or of arsenic. Albertus and Thomas have approached more nearly to the tincture of mercury (the virtue whereof is of subtle sharpness, though it derives its nature from the Archeus), but in their excessive coagulation, and also in the degree of repetition wherein they have overwhelmed the whole affair, they are entirely wrong. In preparing the tincture they verge on the true tincture, as in the following opinion: As metals are transmuted and fully fixed, so also is the body in the following manner: Reduce mercury in elevation until it assumes the form of a fixed crystal; then digest it to the point of resolution and coagulation; join it with gold so that this shall produce its ferment. Then proceed according to the prescript of Hermes, and continue to the completion of the stone. The dose thereof is one grain. Its power and virtue preserve the whole body in its entirety.

CHAPTER IX

Concerning the Spirit of Wine

When I mentioned the essence of herbs above, I pointed out that it is nothing but wine, which I would have you thus understand. The spirit of wine proceeds from its substance. Wine is a subjection of this just as marrubium is of proper and native wine. In order, therefore, to get the spirit of wine as an essence, which is truly an elixir, understand thus: As a pound of persicaria sends forth ℈ij. of wine, so a pound of wine takes not more than one scruple. The rest is the phlegma of wine which has no bearing on the present elixir. Let the preparation of this essence proceed in the following manner: Digest in horse dung wine which has been poured into a pelican. Continue this for a period of two months, and you will see a thin, pure substance, like a sort of fat, which is the spirit of wine, spontaneously evolved on the surface. Whatever is below this is a phlegma possessing none of the nature of wine. The fat, put by itself in a phial, and separately digested, is of the utmost power for long life. And not only does it avail for long life, but this preparation can also be adapted to other purposes by the intervention of cinnamon, xylobalsamum, myrobolani, and other things of this kind, in the following manner: Mix, and by the use of digestion so join these ingredients that with the addition of the above-mentioned elixir and of gold, a medicine shall be prepared which removes all contractions and gives free play to the limbs.

CHAPTER X

The Extraction of Mumia

The extraction of the virtues out of mumia is made magisterially (if I may use that expression) by its mixture with the essence of wine taken from chelidony. Digest it for ten days, and distil for five. Moreover, let it be once more digested afresh until the mumia turns into a liquid. When this takes place above as well as below, these portions being separated from the middle, add the sixteenth part of balsam from woods, and a twelfth in weight of the sealed earth of Pauludadum with the same quantity of liquor Horizontis. Digest this for its month, then shut it up and reverberate it. In this way it ascends to its highest degree. Of all those preparations which are dominated by poisons, this is the most powerful and efficacious.

CHAPTER XI

The Extraction of Satyrion

Whatever has to be extracted from satyrion must be procured by means of separation. In satyrion lurks a Saturnian power which, as it were, secretly steals away and weakens the virtue which satyrion possesses, and so its exaltation reduces it by thirty grains. Hence it not unfrequently happens that when satyrion is used it fails in its effects. It is worth while, therefore, to consider how not its form but only its virtue shall be separated. This must be done in the following way: Let satyrion be digested with panis siliginis in a venter equinus for a month. When this is over, take it away from the bread, and throw away the dregs. Then let the blood of the satyrion be digested thoroughly and allowed to effervesce. When this effervescence has subsided you have obtained a medicine which leaves far behind all others for every purpose which relates to conception.

CHAPTER XII

The Extraction of the First Metal

The most complete and perfect conservation of the body is attained by the First Metal; and this is so efficacious, not by the nature of its own strengthening power, but rather by virtue of the minerals which it contains. For, in order to conserve long life, it is necessary to use the prince of minerals, since minerals make up the physical body. This is the temperament which singly and alone resists corrosives, and Ares operates as much chemically as by means of the Archeus. Moreover, it blends the strongest and the weakest body in one degree. Strength, indeed, is that which exceeds the strength of Ares, and weakness is that which falls below it. That which is taken away from the stronger is conferred upon the weaker, and so each is reduced to a mean. It is done in the following way: Take the liquor of coral, in its most purely transparent form, to which add a fifth part of vitriol, which is from Venus. Let these be digested in a bath of Mars for a month. In this way the wine of the First Metal separates itself to the surface, and the vitriol of Venus lays hold of whatever dregs there may be. Thus the First Metal becomes a clear, transparent, and ruby-red wine, whereof the special virtue and power is that of all the minerals over the whole physical body.

A BOOK CONCERNING LONG LIFE

BOOK THE THIRD

CHAPTER I

LEST anything should be omitted which concerns Long Life, it is proper to observe that within the testa and over and above that quintessence, there is enclosed something out of which a certain conjunction, both of the corporal and of that which is beyond the body, outside of that quintum, produces the body into long life. Concerning this understand that it is absolutely nothing and invisible. But in the body there is something exquisite which not only confers long life upon the microcosmic body, but even preserves Dardo itself whole even to the thirtieth year, and guards the anthos and the great cheyri up to the third age. This microsmic thing sustains both the anthers and the leaves which ought to remain in their own conservation throughout the whole anatomy of the four elements. Wherefore at this point the physician must note that the whole anatomy of the four elements can be contracted into a single anatomy of the microcosm, yet not out of the corporal, but from that rather which preserves the corporal. Indeed, the superquintessence sustains the quintessence itself as well as the other four. If it be proper to give it a just and true nomenclature, I may rightly call it the balsam itself out of which life is preserved, which rightly separates itself from the balsam of the body, and is such a balsam as to surpass Nature herself. This surpassing of Nature is by a corporal operation.

CHAPTER II

But of that balsam whereof we have now spoken, which ought to produce long life, a declaration takes place in two ways—one which is secret and happens by accident, whence it follows that long life is dispensed to the majority, who yet have no idea what it is in itself. But the other mode takes place by arts, that is, by those who are able to obtain that conjunction, nor can it take place without a medium. For herein is situated the point of the matter, because in the Iliaster both long and short life are found. For that which is adjoined to herbs has its terminus; similarly, also, there is a terminus to that which is of the water of minerals; in the same way tereniabin, and so also nostoch. Besides all these things such and so great is the strength and power of the conjunction itself, that everything which is produced out of the four elements is conserved above its first terminus, and that is the terminus of Iliaster, by which, indeed, we wish overcome that subtle man who says that a terminus cannot be crossed over, which, if it does not take place, and passes over, is for this and the other reason. For there are two of them, one of which cannot pass over, because the terminus is placed in the nature of the microcosm; one is in the nature of the elements, the other in that of the quintessence, moreover, also, the other is out of the last Iliaster. For these termini consist in the power of the physician, who in these can change what he wishes according to his will, except only the fixed, where he ought to expect the end together with the mutation of himself.

CHAPTER III

Understand this Iliaster as follows, since here three virtues are found besides the quintessence. For there is the Iliaster of sanctity, the Iliaster of the *Paratetus*, and finally that great Iliaster. Of the first understand that such sanctity imparts long life, according to the industry of him who uses it; the second dispenses it by favour; the third, being bruised, consists without harm in long life. Hence consider the Iliaster comprehended in long life. All three are together subject to the microcosm, so that it may reduce them into one gamonynum; but the other is in no wise controlled, for it is acquired according to favour. With the third the case is exactly as with the Enochdiani and the Heliezati, just as it is clearly the case with Aquaster. In the first place, therefore, it has its origin from the elements, as the testa shuts up, and the superquintessence is attributed to the arcana themselves. The second is ascribed to the Magnalia, the third is out of its own specifics. Hence it follows that the dwellers in the earth, the nymphs, the undines, and the salamanders receive their long life in an alien essence. For there is a death, a time, and a will of that third Iliaster, and he it is who grants to the ear of corn that it should bear more than it would by Nature, as also the fragrance of myrrh, and the strength in Leris. This being so, the physician must consider that a conjunction of this kind takes place in a similar manner as the tree of the sea when once fixed and reduced to bondage thereby, can also become an approved and constant cheyri; so by a similar conjunction in the microcosm the same thing comes to pass.

CHAPTER IV

But concerning that first Iliaster, understand that it exceeds a thousand species, not that one excels another, but rather for this reason, that every microcosmus has its peculiar and, what is more, perfect conjunction and virtue: so great is the virtue and potency of Iliaster that by it a dead body is preserved alive, for this reason, because that first terminus is transmuted.

This conservation of long life transcends our powers, but not those of the higher powers. Its sustentation takes place as follows—that it confers long life, yet without the expulsion of the disease. Life, indeed, it affords, but not good health, yet sometimes it affords both, being long life of the kind which proceeds from that Iliaster. It lasts for years and is extended, as, for example, to the tenth year in the case of one who ought to have died in the fifth—a thing which takes place both by reason of the superior and inferior conjunctions.

CHAPTER V

With regard to the true Iliaster, the fact is that nothing of the kind can be reduced without signification or necessity towards the greater Iliaster, which you are to understand as follows. The greater Iliaster which is to extend long life of this kind can by no means do so where there is no place for long life. Hence observe that such a thing cannot take place without transmutation of the place as well as of the elements. That is to say, as the four mortal elements are in the testa, every moment producing a new generation, they lead forth the same to death. In another direction there is a fixed (generation) in its firmament which remains unshaken, neither causing disease nor death. Such fixed spirits suffer nothing to perish altogether, whose long life is immovable and firm even to their transmutation again into the first. The similitude is, indeed, taken out of the text because according to the fixation, as I have said, of the firmament, long life is more prolix in one confirmamentum than in another, although each arrive at their first terminus only. However, some few inhabitants accompany this kind of transmutation, so it is permissible to call it, into the tenth or even into the twelfth, whose death follows on the destruction of that great firmament, where bodies, both celestial and terrestrial, shall be shaken, yea, the supercelestial also. Notwithstanding, this takes place without any distinction of Nature, for in the first moment when such mutations happen the putrefaction of Nature commences, and that is with a still living body.

CHAPTER VI

Now, concerning Iliaster, it is necessary, in the first place, that the impure animate should be depurated without separation of the elements; this takes place without any corporal and mechanical labour, which disposition arises according as man grades himself in mind that he may be rendered like to the Enochdiani, not that he desires the Enochdianian life, for in his mind he differs diametrically from it. Wherefore it is necessary for the microcosmus in its interior anatomy to reverberate it with a supreme reverberation. Thereby the impure consumes itself, but the fixed which is separated from the impure remains without rust. Nor yet is it a fire wherein Salamandrine essence or Melosinic or Ares could be present, but rather a retorted distillation from the middle of the centre, above all coal fire. This reverberation thus being made, in its last terminus it exhibits the physical fulmen, just as the fulmen of Saturn and of the Sun separate from each other. Accordingly, whatsoever advances by this fulmen of long life pertains to that great Iliaster, and this fulmination and preceding reverberation in no wise remove the weight, but rather the turbulence of the body, and that by the method of diaphanous colours.

CHAPTER VII

Moreover, from that Iliaster of the first power long life of this kind does not result, for it affords an inferior grade. Yet he it is, however, who separates in that place, and exactly as a fixed thing can preserve a thing which is not fixed, defends the microcosmus from death, seeing that its operation is not to separate, but rather by means of those perspicuous arcana it should conquer that which is undigested, lest its perdition should follow. Just so mumia, which, together with the body, proceeds from the birth, being itself good, but the body is bad and putrid. Whatsoever life, therefore, the body lives, mumia lives also from it alone, for it is its property and nature to putrefy and revert to dung, of which it is a member, and this is its continual desire. But not so the celestial mumia, for it breaks the worthless part and guards the same by its own will, lest it should ever effect that which it attempts. Wherefore the following is the tenor of the recipe: that the supercelestial mumia sustains the microcosm more than its own mumia. For as often as there is a mumia there is also another terminus. Yet neither time nor number are found in these termini, for they continue to endure till they can no longer escape a second generation. The physician must be perfectly acquainted with the fact that every first matter expels the last. Hence the generation of worms begins where the ultimate matter of the physical body shews itself. Observe, therefore, this Iliaster, that it not only does not destroy the generation of worms, but when their matter is present it does not even impede their generation—a thing which mumia should prevent.

CHAPTER VIII

The natural mumia should be compounded out of three chief antimonies so that the foreign microcosm should govern the physical body, whether by means of the element of water or by means of its metals, salts, etc., or otherwise by means of the element of earth, as by its herbs and boleti, or in tereniabin or nostoch. For all these are mansions of the supercelestial things. Wherefore let no one be surprised that the great virtues of melissa are described everywhere. Seeing that in this a supercelestial conjunction takes place, who shall deny to it a most excellent virtue? These are the magnalia which the Bamahemi contain, and this is plainly Ilech, who, being composed out of the true Aniadus, can in no wise be removed from that elementated thing—a thing which takes place with exaltations of either world, exactly as exaltations of the nettle burn, and the colour of the flammula radiates. Yet in exaltations of this kind their virtue can be reduced into another. Therefore learn to diagnose their exaltations as follows. They are far more potent than the nettle, and also ye may collect the same in the true May when the exaltations of Aniadus commence. For exaltations of the virtues are not only situated in the matrices, but also in supercelestial things. That were a common Idæus and of no importance who knows how to fabricate a single thing subject only to the vision, nor besides tangible things can create greater things still, but he has constructed another May where supercelestial flowers attain their exaltation, in which Anachmus ought to be extracted and preserved, even as the virtue of gold lurks in laudanum. Such, indeed, are the virtues of Anachmus: then will you truly be able to enjoy long life.

A BOOK CONCERNING LONG LIFE

BOOK THE FOURTH

CHAPTER I

WE will in this place complete what has been said previously on the foundation of life, and on the life which is beyond Nature. In the first place, we exhibit to all Spagyrists the age of Adam and Methuselah, after speaking exactly of that long life which is in the hands of the highest Iliaster, according to the manner of magnalia, where more facts are to be dealt with concerning free will than we can administer out of the elements. To make these things understood more clearly we must revert to the Enochdiani. A comprehension of the nature of their influence will enable us to get at the principle of long life, even without any trouble, as was the case with S. John, whose nature comprehended not merely one age or one century. Lest I should give an opportunity to the libellous who wrest the scriptures, we will define nothing certainly in this chapter concerning the life of the highest Iliaster, whether this be present in corporal elements, or whether it lives in the quintessence where no body occurs, and where not only those live whom we have mentioned, but also those whom we thought buried in sleep. All these things I leave to be considered slowly by sublime spirits, while we have descended to these. If that highest Iliaster be impelled, or at least, if it have need of anything, he will easily attain to whatever is Enochdianic, where all our long life is collocated in its proper places in ether and in the clouds. But once for all Iliaster has satiated himself, so that henceforth he lacks nothing.

CHAPTER II

The end of long life is contained within the limit of six or nine hundred years. Concerning the source of this life which is beyond Nature, understand as follows. There are two forces in the power of man—one natural, the other of the air, wherein is nothing corporal. Having treated sufficiently of the natural, the incorporal force shall close our little book. Miserable in this respect are mortals to whom Nature has denied her first and best treasure (which the monarchy of Nature contains), to wit, the Light of Nature. But herein let us not labour vainly, but in the case of philosophy since it departs and diverges from Nature, we will remember the Aniadus, nor will we make further mention of philosophy. Having, therefore, dismissed natural things, and all which has been treated of concerning things out of the elements, as also those which are latent beneath the chaos, that is, the great Iliaster, we will have recourse to what was mentioned in the first book. In order, therefore, that we may arrive at the year of Aniadin, or even further, the following rules are to be observed. Let not what we are about to say of the nymphs offend any one. Here also shall be indicated the force and Nature of the Guarini, the Saldini, and the Salamandrini, and whatsoever can be known concerning Melosina.

CHAPTER III

But in order to make clear at the same time both the place and the body in these things, which have to be ordained and disposed according to a certain harmony, we must observe the nature of Iliaster. It preserves to a period of three hundred, or even six hundred years. Further, whatsoever out of its own nature admits also the nature of the place is brought to one conclusion, like the former century, but where they unite the nature both of the place and the body at the same time, they arrive there, and without any trouble, to the six hundredth year. Some who have reached that age might be enumerated, did not my pen hasten in another direction. There are, moreover, those who for a long time secretly and furtively are preserved to a long life, an account of whom may be omitted because they have given nothing except to Iliaster. Whatsoever does not pertain to aërial life is passed over in this place. Those, therefore, follow who have lived an aërial life, of whom some have arrived at their six hundredth, thousandth, or eleven hundredth year, a fact which can be easily understood according to the precept of the magnalia. Compare Aniadus, and that by means of the air alone, whose force is so great that the terminus of life has nothing in common with it. Further, if the said air be wanting, that which lies hidden in the capsule bursts forth. If the same shall have been filled by that which recently returns, and then is brought forward into the middle, that is to say, outside that under which it lay hidden, it still is so far hidden that as a tranquil thing it is completely unheard by anything corporal, so that there only resound Aniadus, Adech, and Edochinum. These three, and that which verges into these three, are not four but one. You will attain a very long hidden life. Such is the nature of that Aquaster, which is born beyond Nature. But if it has not been able to attain that which was latent, yet here it occasions that which was extrinsically Iliaster, etc.

CHAPTER IV

The monarchia remain, and to this we are recalled by the great Zenio, for there is a life far different, whereunto we are constrained. When all things have passed away the oppressor and the oppressed remain, a fact not sufficiently understood up to this present by the Aliani. Yet a time comes when all these things which we have investigated together shall pass away, from the first even to the last. As to whether a healthy life can be conjoined with a long life, note that there is a double essence, in one of which health resides, and this essence is fixed; in the other disease is centred, and this is similarly fixed. As to the place and the mutations of these things, let us not change anything. What, however, is the use of vainly lingering among those things which the light of Nature has refused to us? Wherefore he who guides us out of the desire of the mind, does not leave us gaping at what he points out. Let us then pass over what is beyond us, namely, certain creatures of a marvellously long life, and proceed to those which have no death, among whom are the Laureus, Siconius, Hildonius, and many others, whose nativity or natural death no one hitherto has attained or heard of. Add the nature and essence of those things, and how many will you find who have written anything at all about them?

CHAPTER V

I make no account of him who by the arts of Lully vaunts himself as a Necrolicus, and inveighs against what is contained in the four Scaiolæ, announcing me as the highest Scaiolus, in order that I may commence Necroleous arts according to the manner of the cedurini. But I envy even the hydra together with the envious Scaiolæ. What shall I say in this place of those things which the sagacious muse embraces in her canons together with the matrix of the four Scaiolæ, which sleep in you, and render your temples anodynic? I occasion so great an astonishment in you that you shall come even to take heed of a poppy. But I confine myself to the cosmographic life, where both the place and the body of Jesihach appear. Further, the things I prescribe I do prescribe beyond the forces of the body and the place. Whosoever understands these things the same has a lawful claim upon the title of a spagyrist. There is no mortality in the Scaiolæ. He who lives according to their manner, he is immortal; this I prove by means of the Enochdiani and their followers. Aquaster will not invade this place. But if I be inserted among the Scaiolæ according to the manner of the Necrolii, there will be something that I might take out and lead, a thing which the Great Adech antiverts, and leads out our proposition but not the mode, a thing I leave to theoretical discussions. And in this manner Melusine departs from the nymphadidic nature, by the intervention of the Scaiolæ, to remain in another transmutation, if that reluctant Adech permit, who is both the death and the life of the Scaiolæ. Moreover, he permits the first times, but at the end changes himself, from which I gather that supermonic figments in Cyphanta open the window. But the doings of Melusine prevent these being fixed, which, being of this kind, we dismiss. But as for the nymphadidic nature, in order that it may be conceived in ourselves, and that we may thus arrive immortal at the year of Aniadus, we take the characters of Venus. If ye recognise these things, nevertheless ye have put them to little use. But we have completed it, so that we may securely attain this life in which Aniadus dominates and reigns, and remains with that at which we ever do assist. These and other arcana are absolutely in need of nothing. In this fashion we leave and conclude long life.

HERE ENDS THE TREATISE ON LONG LIFE

APPENDIX III

A SHORT LEXICON OF ALCHEMY

Explaining the Chief Terms used by Paracelsus and other Hermetic Philosophers

THE materials for the following vocabulary are derived partly from Paracelsus himself, that is, from other writings not included in the present translation; in part from such alchemical authors as Arnoldus de Villa Nova, Eugenius Philalethes, Ferrarius, Raymund Lully, and Cornelius Agrippa, who, however, was more magician than alchemist; and, finally, from the following sources:—

A. T. Pernety. Dictionnaire Mytho-Hermétique; Paris, 1781, 8vo.

William Salmon, M.D. Dictionnaire Hermétique; London, 1695, 12mo.

William Johnson. Lexicon Chymicum (*editio ultima*). Two parts. Frankfort, 1678, 8vo.

Dictionarium Theophrasti Paracelsi; Frankfort, 1583, 8vo.

Rochus le Baillif. Dictionariolum, a brief supplement to the Geneva folio.

Martinus Rulandus. Lexicon Alchemiæ; Frankfort, 1612, 4to.

Michael Toxites. Onomasticon sive Dictionarium Philosophicum, Medicum, et Synonymum; Argentorati, 1574, 8vo.

Gaston le Doux. Dictionnaire Hermétique; published under the pseudonym of a lover of Hermetic Truth.

The vocabulary has not only been compiled, but has, for the most part, been literally translated from these authorities, and where it speaks positively upon the mysteries of Hermetic science it must not be understood that the editor himself is speaking. While the information it contains may perhaps claim to be regarded as reasonably full, the reader must not expect to find a satisfactory explanation of all strange, *bizarre*, and unaccountable terms which are to be met with in the text. Some are peculiar to Paracelsus, and outside the sage of Hohenheim himself it would be useless to look for information. Such words as Deneas, Magdalion, and Censeturis belong to this category.

Others, less apparently recondite, are notwithstanding wholly unknown to the editor, though he may claim a wide acquaintance with Hermetic literature and its curious recipes. The Stomach of Anthion and the Aqua Caudi Magnæ Mirandæ are of this class. Some expressions and some names of substances, such as the Quintessence of Gold and of the Sun, though common to all alchemy, are not capable of more lucid interpretation by recourse to other writers, and the Index which follows this Lexicon, by collecting the references in the text, will supply all the knowledge that is likely to be gleaned concerning them. In such cases the terms have been excluded from the Lexicon, as it is obviously useless to re-dress the materials supplied by the translated treatises as lights for a vocabulary. It is useless, for example, to say that the Balm of Sulphur is the Radical Moisture of metals, unless, indeed, it were possible to explain more fully than the alchemists have elected to do what they mean by their radical moisture, especially in the metallic kingdom. It is useless to look for any real explanation of such terms; they are part and parcel of the philosophical mystery, and unfortunately their number is somewhat formidable. The First Entity of Antimony, the First Entity of Mercury, the First Entity of Salt, the special alchemical significance attaching to *Liquor Solis*, Macerated Tincture of Silver, Mercury of Life, Oil of the Sun, Philosophic Water, Precipitated Gold, Sphere of Saturn, are all of this kind, and all these have been omitted. But wherever the peculiar use which Paracelsus makes of a term is modified or illustrated in the writings of other adepts, there the term has been included; and wherever any commentator has volunteered any explanation of a word which is peculiar to Paracelsus, then that explanation has been given. A careful reader will, however, find that Paracelsus is in most cases his own best interpreter, and much of his coined or mysterious phraseology is explained by the passages which contain it. For example, there is only one reference to *Argentum Potabile*, but the term is accompanied by an explanation concerning it, which makes its presence in a vocabulary unnecessary. A similar remark will apply to Astral Gold, mentioned in the *Catechism of Alchemy*; to Abrissach, which occurs in the *Philosophy Concerning the Generation of the Elements*. In a few cases, mostly regarding unknown or recondite substances, such as *Emplastra Apostolica*, *Mucilago Lumbricata*, *Compositio Caudi*, etc., the interest attaching to the information is not strong enough to warrant the research which would have been necessary to provide it. In some instances the herbs mentioned by Paracelsus have been briefly referred to, with a view to save reference unnecessarily to other sources of information. But these are generally so accessible, and the catalogue in a Herbary of Theophrastus would be so large, that those who wish to become acquainted with the virtues anciently attributed to Gladwin, Gentian, Ginger, Gallingall, Fumitory, Fennel, Dittany, Dodder, Cummin, Comfrey, and a hundred others, must have recourse to Gerard's *Herbal*, or some similar storehouse. That Clary is a plant of the sage genus requires no more erudition to announce than can be derived from a popular dictionary of the English language. In like manner it is a matter of common knowledge that the term Crocus signifies any metal calcined to a deep yellow or red colour, and it would be childish to include information of this kind in a word-book of alchemical technology. In like manner, it may be taken for granted that the readers of Paracelsus are acquainted with the nature of ordinary chemical vessels, such as the Alembic, the Retort, Cupel, etc.

ABROTANUM, *i.e.*, Artemisia Abrotanum, the herb Southernwood, an aromatic plant.

ACETUM. Other meanings are attached to this term in alchemy in addition to the ordinary significance of vinegar. It is the mercurial water of the Sages, or their universal dissolvent, their virgin's milk, their pontic water; it is the vinegar of Nature, and although its elements are various they all come from one root.

ACTIS, possibly Actæa, the wall-wort, or shrubby elder of Pliny.

ADAM. The formation of Adam by God out of the earth, as described in Genesis, is counted, by the alchemists, among the great mysteries. The material was no common potters' clay, but another, and one of a far higher nature. He who knows this knows also the subject of the philosophical medicine, and, by consequence, what destroys or preserves the temperament of man. It contains principles which are homogeneous with man's life, are potent to restore his decaying virtues, and can reduce his disorders to harmony. Arias Montanus calls this matter "the unique particle of the multiplex earth."

ADAMANT. The gem Adamas takes origin from the element of water. As the stone called Lasurius is a transplanted silver, or a transplanted extraction of silver, as the topaz is an extraction from the minera of Mars, and is a transplanted iron, as the sapphire is a quintransplantation of Lasurius, so the Adamas is a second transplantation of Saturn. And they are all extractions out of the fruits of the element of water. The term also was used by the Greeks to signify the hardest metal known to them, probably steel, and also a compound of gold and steel.

ADAMITÆ. These are white stones of an exceedingly hard quality. There is also Adamitus, which properly is the stone in the bladder. It is sometimes written Adamitum.

ADECH. Hermetic Philosophers apply this name to that portion of the human body which is commonly termed the groin; sometimes also it signifies the mind creating conceptions of things with a view to their manual imitation. The invisible and interior man.

ÆTNEAN. Some alchemists apply this name to their fire because it is concentrated and natural, acts perpetually, and is not always manifest.

A Short Lexicon of Alchemy

ÆTNÆI. These igneous spirits appear to be identical with the Salamanders which inhabit the fiery region of Nature. Paracelsus considers that the names which have been given to the Elementaries, and by which they are most commonly known, are not their true designations, which, under the circumstances, is exceedingly probable.

AGRESTA. A beverage made from apples or barberries. Paracelsus forbids it to be used by persons who are afflicted with diseases of a tartareous nature, for every acid is a resolved tartar produced out of a cold coagulate, or, otherwise, from the salts of a vitriolated minera combined with alums. Agresta is verjuice.

AGARIC is described as a medicine made from flies; it destroys worms in the body, and acts as a tonic to the system in certain forms of plague. Agaricon is also a mushroom growing upon high trees; it is of a white colour, and is good for purging phlegm. It was used anciently for tinder.

AGRIMONY, called also liver-wort, once used in the preparation of a medicine which was held to be a valuable tonic.

AGRIPPINE UNGUENT, probably a prepararation of the Agrippum, a wild olive.

AIR. The philosophy of Paracelsus on this subject may be compared with that of other alchemists. For example, Eugenius Philalethes says that air is not an element, but a certain miraculous hermaphrodite, the cement of two worlds, and a medley of extremes. It is the sea of things invisible, and retains the species of all things whatsoever. It is also the envelope of the life of our sensitive spirit. The First Matter of the philosophers is compared to air because of its restlessness.

ALCHEMY. The following remarkable passage occurs in the *Anima Magica Abscondita* of Thomas Vaughan. It has often been made use of as evidence that the adepts had a higher object than the transmutation of ordinary metals :—Question not those impostors who tell you of a *sulphur tingens*, and I know not what fables; who pin also that narrow name of *Chemia* on a science both ancient and infinite. It is the Light only that can be truly multiplied, for this ascends to, and descends from, the first fountain of multiplication and generation. If to animals, it exalts animals; if to vegetables, vegetables; if to minerals, it refines minerals, and translates them from the worst to the best condition.

ALCHEMY. The monk Ferarius defines it to be the science of the four elements, which are to be found in all created substances, but are not of the vulgar kind. The whole practice of the art is simply the conversion of these elements into one another.

ALCHIMILLA, a herb, otherwise called Lion's-foot.

ALCOHOPH is possibly *Sal Alacoph, i.e.*, sal ammoniac.

ALCOL. Some chemists have given this name to vinegar.

ALEMBIC. The name of this alchemical vessel has been sometimes applied to Mercury, because by its means the philosophers perform their distillations, sublimations, etc.

ALKALI. In addition to its chemical meaning, this term signifies the Vessel of the Philosophers.

ALÖEPATICUM, a medicine used in complaints of the liver.

ALOPECIA is a species of scab or mange arising in those portions of the body which are referable to Jupiter, *i.e.*, the cranium. Paracelsus ascribes the disease to the presence of the spirit of Jupiter. When that spirit is separated from its natural humours and passes into its own minera, the result in the metallic kingdom is cachimia; in trees, fungus; and in the human body alopecia. The fish known as sea-fox was called Alopecias by Pliny.

ALUM. Common alum is distinct from the alum of the adepts; the latter is their salt, which is a basis of alum, all salts, and all minerals and metals.

ALUMEN ENTALI is identical with Alumen de Pluma or Alumen Scariola. It is said to be Gypsum and Asbestos.

ALUMEN PLUMOSUM, *i.e.*, Alumen de Pluma.

ALUMEN SACCHARINUM is Zaccharine Alum.

AMALGAM. The Amalgam of the philosophers is properly the union of philosophic Mercury with the sulphur or gold of the Sages. This does not take place after the fashion of ordinary chemistry, by pounding in a mortar, or otherwise, a solid and a liquid substance. It is the conduct of the fire of the philosophers according to the proper regimen; that is to say, it is the perfecting of the work by continuous coction or digestion at an equal fire—sulphureous, covered in, and non-combustive.

AMETHYST. In addition to the precious stone which bears this name, there was a grape so called by the Greeks of which the juice was said to be non-intoxicant. There is also a herb called amethyst by Pliny, having leaves of a red-wine colour.

ANACHMUS, unknown.

ANGELICA. Paracelsus considers that Spica and Angelica are not in their origin the result of a natural generation, but of that which he terms transplantation. In other words, they are hybrids, after the manner of the mule. The Angelica Ursina, Cardopatia, or Carlina, has in its roots the peculiar quality of depriving persons of their virile strength.

ANIADUS. A term of spagyric philosophy which signifies the powers and virtues of the stars, from which we receive celestial influences by the medium of fantasy and imagination.

ANTHERA. A medical extraction made from hyacinths. Paracelsus affirms that the special astrological influences which reside in the sign of Scorpio are resisted by anthera.

ANTHOS. In old botany this term signifies the flower of rosemary; in alchemy it is the quintessence of the philosophers and their aurific elixir.

ANTIMONY. The antimony of the vulgar is to be distinguished from the antimony of the wise. They have applied this name to the sulphureous mercurial matter which forms part of the philosophical composition. By means of their antimonial vinegar an incombustible quicksilver is extracted from the body of Magnesia. This philosophical antimony is identical with the permanent water and the celestial water, in a word, with philosophical Mercury. It cleanses, purifies, and washes philosophical gold after the same manner that common antimony purifies common gold.

AQUA METALLORUM. Trevisan explains that this is Mercury, and Braccesco states that it is a single substance.

AQUA PERMANENS. According to one interpretation, this is the Catholic Magnesia, or sperm of the world. It is also the Mercury of the Philosophers, and the water of the Sun and Moon.

AQUARIUS is Salt Nitre. It is also the alchemical symbol of dissolution and disintegration.

AQUA VISCOSA. The generation of a metallic sperm is the chief object of those who wish to perform transmutation in the metallic kingdom, and this is done by first of all converting the whole body into a thick or viscous water. Indeed, Rupescissa declares that the matter of the stone itself is a viscous water which is to be found everywhere, but if the stone itself should be openly named the whole world would be revolutionised.

ARCANUM. This term is understood by Paracelsus to signify an incorporeal, immortal substance which in its nature is far above the understanding and experience of man. Its incorporeal quality is, however, only relative and by comparison with our own bodies. From the medicinal standpoint its excellence far exceeds that of any element which enters into our own constitution. The term is applied also to every species of tincture, whether metallic, vegetable, or animal. In general Hermetic science it signifies viscous mercurial matter, or Mercury animated by reunion with philosophic sulphur.

ARCANUM OF HUMAN BLOOD. Many alchemists, both before and after Paracelsus, experimented with the blood of animals, and this not so much with a view to medicine as with the hope of discovering therein that matter of which the philosophers form their magistery. This magistery, indeed, passes sometimes under the name of Human Blood. According to Philalethes the reference is to the matter at the black stage. The name is really applied to philosophical Mercury. Even as the blood of animals nourishes their whole body, and is the principle of their physical constitution, so is this Mercury the base and principle of metals. Thus the blood of the little children who were slaughtered by Herod is pictured in the hieroglyph of Abraham the Jew and is a type of the radical humidity of metals extracted from the minera of the philosophers, which is symbolised by the children, this matter being still crude and left by Nature only on the way to perfection. The Sun and Moon come to bathe in this blood, because it is the fountain of the philosophers in which their king and queen lave themselves. Flamel, foreseeing that his allegory might receive a literal interpretation from some, warns his readers not to mistake actual human blood for the material of the stone, as it would be a foolish and abominable thing.

ARCHÆUS. According to one interpretation, this name is applied by the Spagyrists to the Universal Agent specialised in each individual; it is that which sets all Nature in motion, and disposes the germs and seeds of all sublunar beings to produce and multiply their species.

ARCHIALTIS, or Archaltes, Archates, Archallem, the secret power of God by which the earth is held up in its place.

ARES. The occult dispenser of Nature in the three prime principles. It is that which gives form and difference to species. The word is derived from the Greek, ‘αρης, Mars.

ARGENTUM VIVUM. The transmutation of metals, says the *Clavicula* of Raymond Lully, depends upon their previous reduction into volatile sophic argent vive. This is drier, hotter, and more digested than common Mercury.

ARLES CRUDUM. Certain little drops which fall in the month of June. Called also Hydatis.

ARISTOLOCHY. This herb, which is corruptly called Birthwort, was made into a decoction with wine and applied by Paracelsus as a healing plaster for fractures. The variety which he calls Aristolochia Acuta was used as an ingredient in the composition of an unguent for corroding ulcers. Aristolochia is supposed to promote child-birth.

ARSANECH is sublimed arsenic.

ARSENIC. The arcane sense of this term refers it to the Mercury of the philosophers, and at times to the matter of the philosophers when in the stage of putrefaction. It is stated, or supposed to be stated, in one of the Sibylline verses, that the name of the matter whence philosophical Mercury is extracted consists of nine letters. Of these four are vowels and the rest consonants. One of the syllables is composed of three letters, the rest are of two. Hence it was concluded that *Arsenicum* was the name in question, more especially as the philosophers affirm that their matter is a deadly poison. However, the matter of the stone, according to other authorities, is not arsenic, though it is the matter of which arsenic and all mixed bodies are formed. Nor can the Mercury of the Sages be extracted from arsenic, for arsenic is sold by apothecaries and the minera of Mercury is found everywhere. The name has been given by some other writers to the matter in putrefaction, because it is then a most subtle and violent poison. Sometimes it refers to the volatile principle of the sages, which performs the office of female. It is their Mercury, their Moon, their Venus, their vegetable Saturn, their green Lion, etc. The arsenic of the philosophers whitens gold, even as the common arsenic whitens copper.

ARTETIC, Arthetica, a disease which contracts the nerves, tendons, ligaments, etc., and is very enervating and prostrating.

ASCLITIS is dropsy of the stomach.

ASPHALT is, according to Paracelsus, an extraction of black succinum. It is bitumen, also a kind of petroleum, or rock oil.

ASTRUM. This term in Alchemy signifies the fixed and igneous substance, the principle of multiplication, extension, and all generation. It tends of itself to generation invariably; but it acts only in so far as it is excited by the celestial heat which is diffused everywhere. The term also represents the highest virtue, power, and property acquired by the preparation of a given substance. The Astrum of Sulphur is sulphur reduced to an oil which far exceeds the virtues of natural sulphur. The Astrum of the Sun is the salt of the Sun reduced to an oil or water. The Astrum of Mercury is sublimated Mercury. The name is also given to the alcools or quintessences of things.

ATHANOR. In exoteric chemistry this is a square or oblong furnace communicating on one side with a tower. The tower is filled with coals; when these are lighted heat is communicated to the furnace by a funnel. The same name is applied analogically to the secret furnace of the philosophers, wherein a fire is always maintained at the same grade. It is unlike the vulgar athanor; it is actually the matter itself animated by a sophic fire which is innate therein, and is developed by art.

ATRAMENT. At the period of Paracelsus this term seems to have included all varieties of vitriol, chalcanthus, flower of copper, chalcitis, misy, sory, melanteria, etc.

AURATA. The fish called gilt-head by Quintillian.

AURICHALCUM. Brass, copper, ore, etc., *æs montanum*, extracted from cuprine stone. More correctly, Orichalcum.

AURIPIGMENTUM. The body of orpiment is composed of Sulphur, its coagulation is from Salt, its brilliance from Mercury. It is of a petrine and metallic nature, yet it is neither a metal nor a stone. Paracelsus, *De Elemento Aquæ*.

AURUM POTABILE is either the Oil of Gold, or gold reduced to a liquor without a corrosive. The Golden Calf, which was ground to powder, sprinkled upon the waters, and given to the children of Israel to drink, has been regarded as an allegory of the philosophers' potable gold.

AUSTROMANCY is a method of divination by the winds. It is, apparently, a branch of the science of Aeromancy, which, says Agrippa, divines by aërial impressions, by the blowing of the winds, by rainbows, by circles about the moon and stars, by mists and clouds, and by imaginations in clouds and visions in the air.

AXUNGIA is the fat of animals which was made into an oil and used as an unguent for wounds. The axungia of human beings is said to have been most efficacious, and next thereto that of the cock and capon. The only useful axungia from fishes was that obtained from the Thymallus.

AZOC. Mercury of the Philosopers, not vulgar and crude quicksilver, simply extracted from the mine, but a Mercury extracted from bodies by means of argent vive. It is an exceedingly ripe Mercury. It is with this substance that the philosophers wash their Laton; it is this which purifies impure bodies with the help of fire. By means of this azoc there is perfected that medicine which cures all diseases in the three kingdoms of Nature. It is made of the Elixir.

AZOTH also signifies Mercury. When the philosophers say that fire and Azoth suffice for the Great Work, they mean that Mercury prepared and well purified, or philosophical Mercury, are enough for the beginning and the completion of the whole labour, but the Mercury should be extracted from its minera by an ingenious artifice. Bernard Trevisan says that everyone beholds how this minera is changed into a white and dry matter, having the appearance of a stone, from which philosophical argent vive and sulphur are extracted by a strong ignition. Azoth has many names:—Astral Quintessence, Flying Slave, Animated Spirit, Ethelia, Auraric, etc. It will be seen that Azoth and Azoch are terms used interchangeably. They also stand for the universal medicine, which is arcane to the world, the sole true remedy, the physical stone, and also, according to some writers, they signify the Mercury of any metallic body.

BALM. The quintessence of Mercury is called the external balm of the elements.

BALSAM. Paracelsus affirms that the composition of balsams was first discovered by the alchemists, and he refers the name itself to the artifice which is required in its preparation. Every balsam, but more especially for wounds, should be of a sweet nature, not corrosive, not attractive, but possessing a consolidative quality.

BALNEUM MARIÆ. The furnace of the Sages, the secret furnace, to be distinguished from that of ordinary chemistry. Sometimes the name is applied to philosophical Mercury. The term Bath is also given to a matter which is reduced into the form of a liquor. For example, when it is desired to make projection upon a metal, it is said that it must be in the bath, that is, in a state of fusion.

BALNEUM MARIS In alchemy this term seems to apply both to the vessel which holds the sea-water and to the dissolution that takes place in the vessel. The *Balneum Mariæ* is a bath of warm water, and the name is still applied to a large cooking pan. The *Balneum Roris* seems an interchangeable term for the sea-bath. The term bath is also applied to the matter itself when it is in a liquid state. Circulation in the philosophical egg is also called the Bath of the Philosophers.

BALNEUM NATURÆ. Eugenius Philalethes says that this is really the philosophical fire.

BALNEUM RORIS, see Balneum Maris.

BAMAHEMI, unknown.

BASILISK. No one, according to Paracelsus, has any conception of the form or appearance of this animal, for the simple reason that no one can look at it without dying. According to one of his explanations it is a calf generated without a cow, that is, born of the male animal.

BDELLIUM. A plant, and also the fragrant gum which exudes from it.

BEANI. This word, which is rendered *novices* on p. 154 of Vol. II., is said to have originated in the following acrostic : *Beanus est animal nesciens vitam studiosorum.*

BENEDICTA CARYOPHYLLATA, that is, Caryophyllum, ther herb Benet or Blessed Avens. Paracelsus made use of it in tatareous complaints.

BERILLISTIC ART, *i.e.*, divination in the beryl. Paracelsus regards the the stone itself as a formation out of ice, by means of the glacial stars, which have a very great power of congelation.

BITUMEN, *i.e.*, Asphalt.

BISMUTH. As zinc, according to Paracelsus, is for the most part a spurious offspring of copper, so is Bismuth of tin. It is partly fluidic and partly ductile.

BLOOD. Some alchemists have pretended, as previously stated, that human blood is the true subject and matter of the philosophers, but others say that this is not to be literally understood, and that those who experiment with this substance in athanors will have reason to deplore their error. Blood and Human Blood are, however, by no means uncommon designations for the arcane substance of the Magistery. Philalethes says that it is applied to the matter. It is an analogical term, the allusion being to the fact that the blood in animals carries the nourishment to every part of the body, and is the principle of their physical constitution. Mercury fulfils the same function in the mineral kingdom, for it is the foundation and principle of metals.

BOLETI. Boletus is a mushroom. Bolitus is the same as Bolbiton, *i.e.*, the excrement of oxen. These explanations will, perhaps, not throw much light on the use of the term by Paracelsus.

BRASSATELLA, *i.e.*, ophioglossum.

BOCIA. Bocium, according to Paracelsus, originates out of menstrua and hæmorrhoids, in the same place in which both of these fluxes join, producing a third which is peculiar to itself.

BOLUS ARMENUS on account of its great dryness was used by Paracelsus to heal wounds.

BORAX. This is said to have been melted with natural or artificial chrysocolla, and was then used as a dissolvent and purifier of metals, especially gold and silver. The best borax was supposed to come from Alexandria.

BOTIN is Turpentine.

BUFONARIA, probably Buphonon, a herb mentioned by Pliny, that is, the toadstool.

BUGLOSSUM, *i.e.*, Bugloss, or Borage.

BULLÆ. } This causes wind, says Paracelsus, and its presence often
BULLA. } indicates the beginning of colic, while in women it occasions fall of the womb. It is otherwise called Bleb, a little vesicle, or blister. Bulla is also a genus of mollusca.

CABALA. The alchemists recognise a twofold Cabala. There is that which ends always in the letter where it begins, and this is the alphabetical system, the name and not the thing; the shadow, not the substance; the mere type of the inner Cabala. Within this there is the true, ancient, physical tradition, and this also has its complement in a metaphysical part.

The greatest mystery of the Cabala, according to one account by an alchemist, is Jacob's ladder. It is affirmed further that Jacob was not asleep during his vision, except in a mystical sense, for he had passed through death. It was, however, the cabalistical *Mors osculi*, or death of the kiss, of which those who know it must not speak one syllable. The false, grammatical Cabala consists only of alphabetical rotations, and a metathesis of letters in the text, by which Scripture is wrested. The book of Abraham the Jew, which was discovered by Flamel the alchemist, is supposed to prove that the true Cabala was chemical.

CACHEXIA, a disease which, according to Paracelsus, usually supervenes upon other complaints, and in which the nutriment turns to evil humours.

CACHIMIÆ. This term is loosely applied to a variety of substances. It seems to have most generally signified the dross of metals, or an undigested metallic matter.

CAGASTRIC. There are two seeds of disease, the Iliastric and the Cagastric. The first is in the substance from the beginning, the second is generated out of putrefaction. Dropsy and gout are Iliastric; plagues, fevers, pleurisy, etc., are of cagastric origin.

CALAMUS, a sweet cane, growing in Arabia, India, and Syria. The root of the sweet flag. The resin called Dragon's Blood is obtained from a palm of the genus Calamus.

CALAMINE, a stone used in the composition of brass. Also an ore of Zinc, Cadmia.

CATAPUTIA, a plant.

CALCANTHUS and Chalcanthus, see Atrament. It is copperas, vitriol, shoemaker's black, the water of copper.

CALCATRIPPA, or Consolida Regalis, a herb used by Paracelsus for the cure of ulcers.

CALCINATION is a pulverisation and purification of bodies by means of an exterior fire, either for effecting a disunion between their component parts, or for evaporating the humidity which combines them into a solid body. Calcination, corruption, and putrefaction are sometimes used interchangeably by the Spagyric philosophers, but calcination is most commonly that process which follows the rubefaction of the Stone. *Philosophic Calcination* is performed by the moist fire, or pontic water, of the Sages, which reduces bodies to their first principles without destroying their seminal and germinative virtues. These, on the other hand, perish under the calcination which is performed by a vulgar fire, and this is philosophically termed the Tyrant of Nature. There are two kinds of *vulgar calcination*; the one is accomplished by an open fire, as over ashes, the other in a sealed vessel. In the first the sulphureous volatile parts evaporate, and the salts are deprived of a power and virtue which are conserved in the second. All salts extracted from such ashes crystallize.

CALENDULA, the herb marigold.

CALERUTHUM. The reversion of any substance towards its first matter.

CALX. In the language of the adepts this name is applied to all kinds of bodies when reduced to an impalpable powder, whether by the action of fire or of corrosive waters. Some say that it should only be applied to the ashes of metallic or mineral bodies, and that others should be called cinders.

CALX LUNÆ, *i.e.*, Calcined Silver, or Blue Flower of Silver.

CALX OF LEAD, *i.e.*, Red Lead.

CANTHARIDES, a herb praised by Paracelsus because it draws out the humours from ulcers. The term stands for Spanish fly, a beetle which infests corn, and a kind of fish.

CAPELLA, a young goat or kid.

CAPILLUS VENERIS, literally, the hair of Venus, a herb mentioned by Pliny.

CAPUT MORTUUM. The philosophers describe the so-called element of earth as an impure, sulphureous subsidence, or *caput mortuum* of the creation.

CARABA, see Carabe.

CARABE is Succinum, and is said by Paracelsus to be an extract of the resin of its element. In another place he observes that it is a resolved petroleum, for wherever the latter is found, there also is Carabe. This and other stones void excrement like animals. If it be cast into water which is isolated from the air, its evacuations will coagulate the water, and worms will be produced.

CARABIS IGNEUS, Cathabis, or Cathebis, the stag beetle, a prickly kind of crab.

CARBUNCLE, is called also jaspis by Paracelsus, and he describes it as a golden stone, that is, of an aureate nature.

CARDAMOMUM, an Indian spice, of an aromatic, pungent, and medicinal quality.

CARDUUS ANGELICUS, a thistle, which, according to the signatory art, has the magical marks indicating that it is a cure for pleurisy. Paracelsus says that it is to be reduced into ashes and then made into a lute.

CARDAMUM, garden cress, that is, nasturtium, especially its seed, which was bruised and eaten by the ancients, and above all, the Persians, after the manner of our mustard. *See* Cardamomum.

CARNIOLA. The virtues of this plant are regarded by Paracelsus as of a celestial rather than an earthly origin.

CASSATUM, an unhealthy or dead blood in the veins.

CEDUSINI. One of the arcane names of air is *Cedue*, and the cedusini are probably a class of aerial intelligences, or sylphs.

CELLA, the inner chamber of a bath.

CENIFICATED WINE, *i.e.*, Calcined Wine.

CENTAURY. The virtues of this plant prevail in complaints of the liver; it strengthens and cleanses the separative force of this organ. Its liquor destroys serpents.

CERATION. This name is given to that stage of the philosophical process during which the matter passes from the black to the grey, and, ultimately, to the white colour. This is performed solely by digestion and coction without any addition whatsoever.

CEREVISSIA, that is, ale or beer.

CERUSSA, *i.e.*, White Lead.

CHALCEDONY, a stone which is said by Paracelsus to colour the water in which it is steeped. It is classed as an extract from salt.

CHAMEPITIS, the herb ground-pine, or St. John's Wort.

CHAOMANCY is substantially identical with aëromancy. It is a revelation of the stars of the air, and prognostication by means of the air.

CHAOS. Many other definitions of the Chaos are given by Paracelsus besides those which find a place among the treatises translated in these volumes. The term is also applied to the air which contains the cause of corruption. There is also a chaos in the human body, which is the motive force of all its interior operations. The universal world is in man, and that not analogically, but actually. It is also the chaos which conserves the body. Empty, immeasurable space. The rude, unformed mass out of which the world was created. The atmosphere. Finally, the term chaos is applied analogically to the matter of the work in putrefaction, because at that period the elements or principles of the stone are in such a state of confusion that they cannot be distinguished from one another.

CHEIRI is a term of various meanings. Sometimes it stands for Mercury. Flos Cheiri is the Elixir at the white, though also the essence of gold. Flos Anthos is the Red Elixir of Gold. Generally it is the flower of any vegetable or plant. Paracelsus says of it:—There is no more powerful medicine for the liver. Its blackness is sublimed away and a white substance is left, which must be drunk mixed with wine of life. It removes all hepatic corruptions. Iliastric mysteries are contained therein. Here the reference is uncertain; it may be to the narcissus, the violet, or the yellow gilliflower.

CHELIDONY. Paracelsus calls chelidonia a constellated remedy, and names it as a powerful specific for certain ulcers which he also calls constellated. There was, moreover, a philosophic salt of chelidonia, which was medically applied by the alchemists. The same herb was regarded as a preventive of plague, and it was used as a remedy for jaundice. The plant is better known as swallowort or celandine. Chelidonia is also said to be a secret name of gold.

CHERIO. This is described as an accident of the external elements, whether cold or hot, through which all diseases are healed. Cherionium is an unalterable nature, such as indurated crystal, which it is not possible to dissolve.

CHIMOLÆA CALCIS. This should probably read Chimoleæ Calx. Cymolea or Chymolea is a Hermetic name for sedge or reeds. Cymolia is white waste ore, white silver litharge, marl, fuller's earth.

CHIROMANCY is the beginning of Magic; it is to that science what the alphabet is to writing. It is acquired easily, but at the same time is a most useful and illustrious art. It is the star of natural things. For example, there is a chiromancy of the lily, and in the lily a star abides which corresponds to the nature of that flower.

CHRYSOCOLLA, gold solder, borax. Sometimes the term is used in an arcane sense and interchangeably with Argot, Rebis, etc.

CICHOREA, *i.e.*, Succory. To this plant Paracelsus ascribes a certain peculiar congenital influence which it derives from the Sun, towards which its blossoms turn invariably, and when that luminary is absent it has very little virtue. At the proper moment the congenital influence can be extracted from the flowers. After seven years its root becomes transformed into a bird.

CINERITIUM. This term has two meanings. In the first place it is an ash-pan often mentioned by Paracelsus; secondly, it is an equivalent of Regale, an amalgam of gold and silver.

CINETUS, the thickness of clouds.

CINNABAR. Further information concerning this substance is scattered through the chirurgical and medical works of Paracelsus, in one of which he seems to regard it as a form of Mercury.

CIRCULATED WINE, *i.e.*, the extracted spirit of wine.

CITRINUS is a stone which occupies a middle position between the crystal and the beryl; it is of yellow colour.

CITRINÆUS and CITRONES. See Citrinus. It is a pellucid variety of quartz.

CITRINULA, see Flammula.

CLAVELLATED. The *herba clavellata* is the herb-trinity, or heart's ease.

CLISSUS, the entire essence of a substance amalgamated into one composition. Clissus is also an occult force going and returning from one place to another, as the virtue of a root which first passes to the stem and then returns to the root. (Rochus le Baillif, in his Spagyric Dictionary.) Petrus Poterius, in the Spagyric Pharmacopœia, says, " Clissus is a certain union of all virtues in any plant, which virtues consist of the three primary substances, sulphur, salt, and mercury, such substances being severally educed from the single parts of plants."

COAGULATION. The bond of union in composites, and the common attraction between their parts. It is the rudiment of fixation. There are two kinds, even as there are two solutions. One is performed by cold, the other by heat, and each of these again is duplex, the one permanent, the other transitory. The first is fixation, the other coagulation simply: metals are an example of the first, and salts of the second.

COLCOTHAR is defined by Paracelsus as a salt. He affirms that it is a signal specific for obstruction of the fluxes.

COMPLEXIONS. The ancient opinion on this subject was most explicitly rejected by Paracelsus. In man he admits only one complexion, which has a dual mode, namely, as hot or cold. He affirms that the theory which was current in his own day deserves no consideration in diseases. At the same time in his *Explanatio Totius Astronomiæ* he admits four complexions, not, however, in the body, but in the essence.

COLOCYNTH. Whosoever swallows two grains of the essence of colocynth shall be free from every impression of the moon and from all noxious properties present in the air. It was also recommended by Paracelsus as a cure for worms.

COLOQUINTH, *idem*.

COMPOSITION. It is held by some of the adepts that there is no perfect specifical nature which is simple and void of composition save God alone. Thus the soul of man itself was compounded of an excessively tenuous fire and of the most uncompounded form of light.

COPPER. In alchemy an alternative name with Laton, signifying the matter at the black.

COPPER GREEN, *i.e.*, Verdigris.

COPULATION. In alchemical terminology this is the union of the philosophical male and female, the fixed and the volatile.

CORAL. This was used by Paracelsus as a remedy for the plague, the falling sickness, and against poison.

CORPORAL MERCURY. The innumerable recipes for corporal Mercury found in the *Manual* of Paracelsus may be illustrated by a classification which is given in his chirurgical works. He observes that there are three bodies in Mercury—that out of which it is generated, before it has become perfect; that by virtue of which it is that which it is; and that which it becomes when it is prepared by art. According to Arnoldus de Villa Nova corporal Mercury is another name for the Mercury of the Philosophers, which see.

COSTUS. The plant usually known as *herba Maria*, that is, Zedoary, passes under this name.

CRISPULA, *i.e.*, cristula, crispa, the herb called cock's-spur.

CROCUS. Hermetic chemists have sometimes given the name of Crocus or Saffron to their fixed matter when it has attained the colour of red orange.

CRUCIBLE. The alchemical crucible is described as a clay melting vessel, capable of withstanding a severe degree of heat. It had a narrow base and widened out into a round and triangular body. The cupel was a species of crucible.

CUBEBÆ, a drug so-called. A small, spicy berry, something like pepper.

CUBEBS, see Cubebæ.

CYANUM and Cyanus. According to some this is a kind of blue jasper; others say that it is a turquoise or lazule. It is also the flower commonly called bluebottle. A rock bird. A blue dye or lacquer.

CUPELLA, see Crucible.

CYCLAMEN, Cyclaminus, and Cyclaminum, the herb sow-bread; a tuberous rooted plant used for garlands.

CYROGLOSSUM, hound's-tongue, a plant. The liquor of Cyroglossum is an arcanum for falling of the womb.

CYPHANTIC, unexplained.

DARDO, a term of unknown meaning which occurs only in the treatise *Concerning Long Life*.

DAURA is the same as hellebore and winter aconite. Sometimes also it is supposed to signify foliated gold.

DEALBATION, the washing of the Laton, that is, the coction of the matter until all its blackness departs and the substance remains white. It is the removal of all impurities.

DELTIC. The term Deltic impression, made use of by Paracelsus in his work *Concerning Long Life*, is another mystery in the terminology of that strange treatise. The conjecture may be hazarded that it is derived from the Greek word *Deltos*, a writing tablet, when the Deltic impression would be analogical to that made upon a wax tablet by a stylus.

DENARIUS, properly the Roman denier, seven of which were at one time equal to a Troy ounce. Afterwards there were eight to the ounce. In the Lower Empire a silver denier scarcely weighed half so much.

DENARY, see Ternarius. Agrippa calls the number ten a manifold religion and power applied to the purging of souls. It possesses a divine quality; there is no real number beyond it; just as it flows back into unity so everything returns to its proper source, even the spirit unto God who gave it.

DENTARIA, toothwort.

DERSES, a certain arcane smoke or terrene vapour, which is the principle of vegetable birth and growth.

DIAGRIDIUM, a preparation of scammony and quince-juice.

DIAPENSIA, the plant alchimilla, which see.

DIATHESIS, an innate art or nature. A physical predisposition towards a given disease.

DIEMEÆ or Dienez, said to be spiritual essences which inhabit large stones.

DISSOLUTION, according to Trevisan, is the whole mystery of the art, and it is to be accomplished not, as some have thought, by means of fire, but in a wholly abstruse manner, by the help of Mercury.

DISTILLATION. Later writers agree with Paracelsus as to the variety of complicated processes in the operation of the great work. Those who would be accounted wise must labour to find out the Mercury, so that they may reduce things to their mean spermatic chaos, and may avoid broiling destruction. Futile distillation, in particular, will be dispensed with by one who remembers that sperms are not made by separation, but by composition of elements; to bring a body into a sperm is not to distil it, but to reduce the whole into one thick water, keeping all the parts thereof in their first natural union.

DIVERTALLUM, a generation of the elements, otherwise a production from metals.

DRACHUM, *i.e.*, a drachm.

DRACANCULUS, dragon's wort, or wild adder's tongue. There is a shell-fish also so called by Pliny.

DRAGON'S BLOOD. Alchemically speaking, this is the Tincture of Antimony.

DUBELCOLEPH, or Dubelteleph, a composition of white coral and amber.

DUELECH, a dangerous and painful tartareous and porous stone which forms in the human body, especially in the bladder.

DURDALES, wood nymphs, spirits of the trees, etc.

EAGLE. This name has been applied by the philosophers to their Mercury after sublimation, firstly, on account of its volatility, and, secondly, because even as the eagle devours other birds, so does the Mercury of the Sages destroy, consume, and reduce even gold itself to its first matter. The term is also applied to sal ammoniac and to sublimated Mercury, because of their facility in subliming. Yet the reference is not to the vulgar substances, but to those of the philosophers. The *Eagle which devours the Lion* signifies the volatilization of the fixed by the volatile, or of the sulphur by the Mercury of the Sages.

EDOCHINUM, unknown.

ELECTRUM is gold, according to one interpretation, but it was not used in this sense by Paracelsus. Sometimes it is a conjunction of seven metals into one composition according to the conjunction of the planets. It is also the middle substance between ore and metal, neither wholly perfect nor altogether imperfect. It is, indeed, on the way to perfection, but Nature, having encountered hindrances, has left it. Hence the philosophers say that we must begin where she leaves off. It is called, says one account, Electrum because it is composed of two substances, and immature because it must be perfected by the operations of the artist. Properly it is the Moon of the Philosophers, sometimes called Water, sometimes Plant, Tree, Dragon, Green Lion, Shadow of the Sun, etc. Electrum is also one of the names which have been given to the Magistery at the white.

ELECTUARIES, medicinal confections.

ELEMENTS. Some of the adepts enumerate only two elements, earth, the residence of the matrix, and water, which is the mother of all things visible. Every element is, however, threefold, this triplicity being the express image of their Author, and the seal He has set upon His creatures. There is no created thing too simple, vile, or abject in the sight of man, but that it bears witness of God in respect of that abstruse mystery, His unity and trinity. Every compound whatsoever is three in one and one in three.

ELIXIR. Avicenna speaks of a duplex elixir, and the first matter from which it is produced is also duplex. The Elixir is nothing else, according to Trevisan, than the reduction of the body into mercurial water, from which water the Elixir is extracted, that is, an animated spirit. The Elixir is the second part or operation in the achievement of the sages, as Rebis is the first and the Tincture the third. There are three species of Elixirs in the

Magistery. The first is that which the ancients called the Elixir of Bodies. It is that which is performed by the first rotation. The second is performed by seven imbibitions, even to the white and the red. The third is the Elixir of Spirits and is attained by fermentation; it is also called the Elixir of Fire, and with this multiplication is accomplished.

EMBRYONATED, *i.e.*, fecundated, implanted with seed. Thus Embryonated Sulphur is a term used by the alchemists to distinguish one of their principles from the ordinary substance of sulphur.

EMUNCTORY, an opening for the escape of corrupt matter. Thus, an ulcer is an emunctorium, and so are the ordinary channels of expurgation.

ENOCHDIANI. It may be conjectured that these beings are of the race of Enoch and Elias, who were wrapt away into another world wherein they are still supposed to retain their mortal bodies. The prophets mentioned are according to mystic speculation by no means the only persons who have thus been "caught up to heaven," *i.e.*, into the unseen, without dying in the flesh.

ENS, ENTIA. The first extract of mineral natures is the *primum ens* of its kingdom. The *primum ens* of the animal world is in the blood or in the ova. It is the first matter, the seat of life and motion.

ENUR, the hidden vapour of water, out of which stones are generated.

EQUISETUM, horse tail.

ERUCÆ, a palmer, or canker-worm, also the herb rocket.

ESCHARA is dead flesh.

ESSENCE OF THE GREATER CIRCULATUM, unknown.

ESTHIOMENSIS is Lupus, St. Anthony's fire.

ESULA, the herb tithymallus, or spurge.

ETHICA, a kind of fever.

EUPHRASIA, the herb eyebright, once regarded as valuable in diseases of the eyes.

EVE. When used in a purely alchemical sense, this name signifies the mastery of the philosophers at the white stage. Adam, similarly, is the Magisterium at the red.

EYEBRIGHT, see Euphrasia.

EVESTRUM, the eternal substance of heaven. Also a prophetic spirit, which interprets the signs of coming events. Finally the double, living phantasm, or sidereal body of man.

FEL VITRI, glass gall, *i.e.*, Sandiver, a whitish salt scum cast up from glass in a state of fusion.

FERMENT. The sophic Mercury which, together with the sophic sulphur, is said by Avicenna to be the original substance from which all metals were created, is also declared to be a ferment for every body with which it is united chemically. It is the universal vivific spirit which penetrates, exalts, and develops everything. It is the grand metallic elixir, and its potencies are educed by the operation of fire. Though found in all minerals it is really a terrestrial matter, which possesses lucidity, fluidity, and a silverine colour.

Another account says that it is the fixed matter which, combined with Mercury, causes it to ferment and communicates to it the nature which it needs.

FILLA, possibly Fella, *i.e.*, a name sometimes applied to sulphur-water.

FIRE. It is said that the common chemist works with common fire, using no medium, and so he generates nothing, not working, as God does, for preservation, but for destruction. Hence he always ends in ashes. The disciple of the philosophers should use it with an intermediate phlegma, so that his materials shall rest in a third element, where the violence of fire cannot reach, but its soul only.

FIRE OF THE PHILOSOPHERS. Some adepts affirm this to be the greatest crux of the art. It is a close, aërial, circular, bright fire, which the philosophers call their sun. It causes a certain vapour to arise in the glass which contains the matter, and digests the latter by a still, piercing, vital heat. It is continuous, producing at length an alteration and corruption of the philosophical chaos. Its proportion and regimen are very scrupulous. To understand the proper degree of this fire, the generation of man or some other animal should be considered.

FIRMAMENT, a name of Lazurium. Also the upper part of the Hermetic vessel.

FIRST MATTER. The first matter, by the universal agreement of all alchemists, was existent before man and before all other creatures; it was, indeed, the mother of them all. The philosoper, in seeking it for the special purpose of his art, must avoid all common salts, stones, minerals, vegetables, and animals. It is totally impossible to reduce any particular to the first matter, or to a sperm, without philosophical Mercury, and being so reduced, it is not universal, but the particular sperm of its own species, and does not work any effects save such as are agreeable to the nature of that species.

FIXATION is a process by which a naturally volatile substance is rendered fixed. The principle of fixation is a fixed salt, and digestion at a suitable fire. Hermetic chemists say that the perfection of fixation can only be obtained by the operation and processes of the Stone of the Philosophers, that their Matter is also susceptible to it, and that the state is attained when it is brought by coction to the ruby-red colour. It is performed by a philosophical fire of the third degree.

FLAMMULA, or Citrinula, is crow's-foot.

FLOS. This term frequently occurs in Paracelsus, and it has many meanings and many variations in alchemy. It is given to the spirits which are enclosed in the philosophical matter. These spirits are so lively that it is always recommended that the fire should be applied gently, as otherwise they may burst the vessels. The name of Flowers is also applied to the different colours which appear in the matter during the process of the work. The *Flower of the Sun* is the citrine redness which precedes the ruby redness. The *Flower of Lily* is the white colour which goes before the citrine. The *Flower of the Salt of the Philosophers* is the perfection of the stone. The *Flower of Gold* is sometimes the Mercury of the philosophers and sometime the citrine colour. The *Flower of Wisdom* is the Elixir perfect at the red. The *Flower of the Air* is dew. The *Flower of Water* is *flos salis*. The *Flower of the Earth* is the dew and the flowers. The *Flower of Heaven* is a species of manna from which an admirable liquor is extracted. Some have erroneously regarded it as the true philosophical matter. The *Flower of the Wall* is saltpetre. Philosophical *Flos æris* is the matter of the work towards the end of putrefaction, and when it begins to grow white. The *Flower of Cheiri* is essence of gold. The *Flower of Sapience* is the Elixir perfect at the red. There are also other explanations of the *Flower of the Sun and of Gold*. It is the sparkling whiteness, more brilliant than that of snow itself, which characterizes the matter at the white. It is the fixed body of the Mastery, which is not to be understood of any flowers or tinctures extracted from common gold, but of the philosophical gold only, and of the fixed portion of the composition of the Mastery by means of which the volatile part is also fixed, according to the regimen of a prudent heat and the governance of a perfect coction.

FLYING EAGLE, the Mercury of the philosophers.

FŒNO GRÆCUM, *i.e.*, Fœnum Græcum, a leguminous plant of the clover family.

FOLIATED EARTH. This is said to be the Mercurial Water in which gold is sown.

FOUR ELEMENTS. The development of things out of the four elements occurs, according to Agrippa, by the way of transmutation. Each element has two specific qualities, one of which is proper to itself, and the other is a mean by which it corresponds to that which comes after it, as, for example, fire is hot and dry, earth dry and cold. He also recognizes that each element is threefold, that so, he says, the number four may make up the number twelve; and by passing the number seven into the number ten, there may be a progress to the supreme unity, upon which all virtue and wonderful operation depend. Of the first order are the pure elements, which are neither compounded nor changed, nor admit of mixture, but are incorruptible. Through these the virtues of all natural things are brought into act. No man can declare their virtues, because they can do all things upon all things. He who is ignorant of them will never bring to pass any wonderful matter. Of the second order are the elements that are compounded, changeable, and impure, yet such as may by art be reduced to their simplicity; their virtue, thus reduced, perfects above all things all occult and common operations of Nature, and these are the foundation of all natural magic. Of the third order are those elements which originally and of themselves are not elements, but are twice compounded, various, and interchangeable. They are the infallible medium, the middle nature, or soul of the middle nature. Very few understand the deep mysteries thereof. By means of certain numbers, degrees, and orders, they contain the perfection of every effect in what thing soever, natural, celestial, or supercelestial; they are full of wonders and mysteries, and are operative both in natural and divine magic. From and through these proceed the bindings, loosings, and transmutations of all things, the knowing and foretelling of things to come, the expulsion of evil and the attracting of good spirits. Let no man, therefore, without these three sorts of elements, and the knowledge thereof, be confident that he is able to work anything in the occult sciences of Magic and Nature. But whosoever shall know how to reduce those of one order into those of another, impure into pure, compounded into simple, and shall understand distinctly their nature, power, and virtue in number, degrees, and order, without dividing the substance, shall easily attain to the knowledge and perfect operation of all natural things and celestial secrets.

FULIGO MERCURII. The *fuligo Metallorum* is properly arsenic in alchemical symbolism, but it often stands for Mercury.

FULMINATION is the graduated depuration of metals. It is so called because the metals become brilliant and diffuse radiance from time to time during the process. A red pellicle forms above, and when it disappears little sparkles are manifested at intervals.

FUMUS is Fimus, *i.e.*, dung.

GALANGA, *i.e.*, Galingale or Galangal, an Asiatic plant; the roots have a hot and spice-like flavour, accompanied by an aromatic smell.

GALBANUM, a gum or liquor having a very strong smell. Derived from an umbelliferous plant.

GAMALEI, or Gemetrei, Gamathei, etc., certain natural stones which, owing to some powerful astrological influence, receive extraordinary magical impressions. There are also artificial Gamathei engraved with magical figures, and used for talismans.

GAMONYMUM, unknown.

GARYOPHYLLON, *i.e.*, Caryophyllum, the clove gilli-flower.

GEOMANCY. According to Cornelius Agrippa, who preceded Paracelsus, and was, like Paracelsus himself, a disciple of Trithemius, geomancy is an art of divination whereby a judgment may be given by lot, or destiny, to every question whatsoever. It consists in the use of certain points, arranged in figures which are in harmony with celestial figures. It can, however, shew forth no truth unless it be founded in some divine virtue. It is supposed that the hand of the operator is directed by the spirits of the earth, so that incantations and other magical rites are resorted to in connection with geomancy. The soul itself of the operator also enters actively into the process.

GLUTEN, in addition to its ordinary significance, means ox-gall. It is also the sinonium of Paracelsus which resembles the white of egg.

GLUTEN OF SULPHUR. In the three prime principles Paracelsus appears to have recognised a certain superincession, and as gluten was alchemically sometimes a name of salt, the reference may be to the innate salt which was supposed to exist in sulphur. Gluten is also Blood, Lime, etc.

GOLD, the most pure and perfect of all metals, has been called by the adepts the Sun, Apollo, Phœbus, and other names, especially when it has been considered philosophically. Gold, as applied to the ordinary purposes of society, they term Dead Gold. *Etherised Gold* is philosophic gold. *Base Gold* is sometimes a deceptive description of the living gold of the sages. *White Gold* is the Magistery of the Philosophers attained to the white grade from which subsequently develops their orange gold and perfect redness, which alone is their true gold, their ferment, and their red smoke. *Gold in Spirit* is the gold of the sages reduced into its first matter, which they also term gold reincruded and volatilised by Mercury. When the sages tell the student to take gold, the reference is not to the vulgar metal but to the fixed matter of the Great Work, wherein their living gold is concealed as in a prison. Their 24-carat gold is their gold pure and unmixed with any foreign elements. *Volatile Gold* is the fulminating gold of Crollius. *Gold of Coral* is the matter fixed at the red. *Gold of Gum* is the fixed matter of the philosophers. *Exalted, Multiplied, or Sublimated Gold* is the powder of projection. *Vivified Gold* is gold reincruded and volatilised. *Gold of Alchemy* is the sulphur of the philosophers. *Foliated Gold* is the sulphur of the philosophers in dissolution.

GRANATE, *i.e.*, Granatum, a pomegranate.

GRAND MAGISTERIUM. The operation of the *Magnum Opus*, the separation of the pure from the impure, the volatilisation of the fixed and the fixation of the volatile one by the other, because no artist will succeed by operating separately on either. The philosophers say that the principle of their magistery is one, four, three, two, and one. The first unit is the first matter whence all has been made; the number four represents the four elements which are formed of this matter; the number three represents Sulphur, Salt, and Mercury; the number two is Rebis, the volatile and the fixed; the final unit is the stone, or that which is the result of the process and the fruit of all the Hermetic labours. In exoteric chemistry there are three kinds of magisteries, of which one has reference to the quality of the composites, the second to their substance, the third to their colour, odour, etc.

GREAT ARCANUM. Roger Bacon, or, more correctly, a treatise attributed to him, affirms that the great and supreme arcanum of Hermetic philosophy is hidden in the four elements. Espagnet says that its production requires a perfect knowledge of all Nature and art concerning the realm of metals, which is obtained by analysis of metallic principles.

GREEN LION. Philosophical chemists frequently make use of this term to signify one of the matters which enter into the composition of the Magistery. Usually it is their male or their Sun, both before and after the confection of their animated Mercury. Before confection it is the fixed part, or matter capable of resisting the action of fire. After confection it is still the fixed matter, but more perfect than before. In the first instance it is the Lion simply; it becomes the Red Lion by preparation. The Mercury is made with the first, and the stone, or elixir, with the second. The Old Lion is the fixed part of the stone, so called because it is the principle of the whole. The

Green Lion is the matter made use of for the Magistery. It is certainly mineral and derived from the mineral kingdom. It is the base of all the menstrua of which the philosophers have spoken. Of this they have composed their mineral dissolvent.

GRILLUS, or Grilla, a name of Venus, or Copper. It is also a mild species of chalcanthus used as a purgative.

GUARINI. Men who derive their life from the influence of heaven.

GUTTA ROSACEA is a red breaking out in the face.

HALCYON, the kingfisher. The fabulous bird of that name, and also the foam of the sea.

HALLEREON, the Eagle of the Philosophers.

HARMEL, the seed of the herb Rue, but it is impossible to affirm the special significance which Paracelsus may have attached to the term in his supplementary treatise *Concerning Long Life*.

HELAZATI, unknown.

HELIOTROPE, the Melissa of Theophrastus. Heliotrope, a plant which follows the sun with its leaves and flowers, *herba solstitialis*. A gem used as a lens to look at the sun.

HELLEBORE. A plant used by the ancients as a specific for many illnesses, especially for madness.

HERMES. Albertus Magnus affirms that this mysterious adept was the first who discovered the grand magisterium.

HERMODACTYLUS, the wild saffron, or, according to some, dog's-bane.

HILDONIUS, an unknown hierarchy of elementary spirits.

HIPPURIS, the herb horse-tail, or shave-grass. The hipparus is a kind of lobster.

HIPPUS, or Hippeus, is a kind of crab-fish, the sea horseman.

HIRUNDINARIA, or Hirundinina, swallow-wort.

HONEY. A name given to the philosophical dissolvent.

HOP-PLANT, that is, Lupulus.

HUMIDITY OF THE PHILOSOPHERS. Arnoldus de Villa Nova says: I tell thee further that we could not possibly find, neither could the philosophers before us, anything that would persist in the fire, save the unctuous humidity. A watery humidity easily vapours away, its earth remains behind, and its parts are separated because their composition is not natural. Unctuous and viscous humidities are hardly separable from those parts which are natural to them. A treatise attributed to Albertus Magnus assumes that all metals are composed of an unctuous and subtle humidity, intimately incorporated with a subtle and perfect matter.

HUMULUS, a plant of the hop genus. See Hop-plant.

HYDROMANCY, according to Cornelius Agrippa, performs its presages by the impressions of water, its ebbing and flowing, increase, depressions, tempests, colours, etc. To this must be added the visions that are seen in the water.

HYDROMEL. A kind of meal. Honey diluted with water.

HYPOSARCHA, *i.e.*, Hypersarcosis, fungous or proud flesh.

ILECH. This terms seems identical with Ileias and Ileadus. The *primum Ilech* is the beginning. The supernatural Ilech is the supercelestial conjunction and union of the stars of the firmament with the stars of things below. The great Ilech is the star of medicine. When we take medicine we assimilate this star. Crude Ilech is a composition of the three principles, Salt, Sulphur, and Mercury. It seems also to be elemental air.

ILIASTER, Illeias, Eliaster, Iliadum, etc., the first chaos of the universal matter, constituted of Sulphur, Salt, and Mercury. The prime principle. Called also Ilion.

ISOPIC ART, see Ysopus, etc.

JACINTH; from this stone a medicine called Antera was extracted.

JACULATION, a darting or casting.

JESICHACH, or Jesahach, is supernatural.

KING. A name used in two different senses by the philosophers. Most commonly it is the sulphur of the sages, or philosophical gold, being an allusion to vulgar gold, which is called the king of metals. Sometimes it is a name of the matter which enters at the beginning into the composition of Mercury, and is the first fire thereof, that fixed grain which has to overcome the mercurial cold and volatility. It seems to be used in both these senses by Basil Valentine in his Twelve Keys. Subsequently he applies the name of King to perfect sulphur and even to the powder of projection. The King is also identical with the Lion. When referring to the powder of projection it is said that the King so loves his brethren that he gives them his own flesh to eat, and thus makes them all kings like himself, that is to say, gold.

KIST, an uncertain weight. It may be of fifteen grains, or four pounds. It also signifies two measures of wine, or half a gallon.

KYBRICK, or Kibric, a name given to the stone. It signifies also the father and first matter of Mercury and all fluids.

LAC VIRGINIS. A name given to the Mercury of the philosophers. It is also mercurial water, and the vinegar of the philosophers. The Mercury of the philosophers, under the form of a milky water in the humid way. It has further been applied sometimes in the dry way when the matter has been brought to the white.

LAPIS LAZULI, *i.e.*, lazulus, the azure or lazule stone.

LATERINE OIL, an oil obtained from bricks.

LATON, copper, or brass, but also a certain state of the philosophical matter when the red colour has appeared, but is not yet permanently acquired. It is, further, a name of electrum. The blood of the Laton is that dry water which is extracted from the virgin earth of the sages.

LAUDANUM. This name is said to be used by Paracelsus in another than the ordinary sense, and to signify a specific for fevers which was a compost of gold, corals, pearls, etc. Another writer says that it is a medicine beyond all praise, made out of two substances, than which nothing more excellent can be found in all the world, and whereby almost every disease is cured.

LAUREUS. This, apparently, is a race of wood spirits. Compare *Laurus*, a laurel.

LEFFAS, the occult vapour of the earth, the principle of vegetable life and growth, that is, the predestination of plants, the sap of plants. See Derses.

LENTIGO, a freckle, pimple, or spot on the face.

LENTOR, a clammy or gluish humour.

LIGUSTICUM, lovage of Lombardy.

LILI. One author says that this is in general any matter from which a good tincture can be made, as, for example, antimony. The *lilium* of Paracelsus is the extraction of a tincture from metals.

LIMBUS. The spagyric philosophers for the most part identify the Limbus with ancient chaos. It has been described as a huddle of matter wherein all things were strangely contained. Some regard it as uncreated, but if otherwise, then it was an effect of the Divine Imagination, acting beyond itself in contemplation of that which was to come, and producing its passive darkness for a subject on which to operate. It also stands for the universal world, with the four elements thereof.

LIMBUS OF MAN. At death the earthly parts of man return to the earth, but the celestial portion departs to a superior, heavenly Limbus, while the spirit goes back to God who gave it.

LIME, see Calx.

LION'S-FOOT, see Alchimilla.

LIQUOR HORIZONTIS, probably liquor of Mercury.

LITHARGE, *i.e.*, Litharge of Silver, is the matter of the work when it has arrived at the white colour under the process of the Sages. Litharge of Gold is the Stone at the red, otherwise, the Sulphur of the Philosophers.

LIXIVIUM, *i.e.*, lye.

LOCCA, or Lorcha, the essential sweetness of the locusts of trees.

LOCUSTS. This name is applied to several different plants and trees. Paracelsus seems to have used it to signify the young shoots of any vegetation.

LORINDS, water spirits.

LORIPIDES, *i.e.*, bow-legged.

LUNARIA, moon-wort, an ingredient for love-potions.

MACHAON, possibly Macha, a flying worm.

MAGISTERY. The strength of the perfect magisterium, according to Avicenna, is one part upon a thousand, that is to say, it will transmute a thousand times its own quantity of base metal into gold. This tinging power, however, varies, and even larger quantities are mentioned by other adepts. See Grand Magisterium.

MAGNALIA, great and divine works.

MAGNESIA. This term, which is occasionally used by Paracelsus in its alchemical, as distinct from its chemical sense, has received many explanations from the adepts. It is the matter of the stone, which the philosophers sometimes call their red, and sometimes their white magnesia. In the first preparation the chaos is blood-red, because the central sulphur is stirred up and discovered by the philosophical fire. In the second it is exceedingly white and transparent like the heavens. It is something like common quicksilver, but of such a celestial and transcendent brightness, that nothing on earth can be compared to it. It is the child of the elements, a pure virgin, from whom nothing has been generated as yet. When she breeds, it is by the fire of Nature, which is her husband. She is neither animal, vegetable, nor mineral, nor is she an extraction from these; she is pre-existent to them all, and is their mother. She is a pure simple substance, yielding to nothing but love, because generation is her aim, and that is never accomplished by violence. She produces from her heart a thick, heavy, snow-white water, which is the *Lac Virginis*, and afterwards blood from her heart. Lastly she presents a secret crystal. She is one and three, but at the same time she is four and five. She is the Catholic Magnesia, the Sperm of the World, out of which all natural things are generated. Her body is in a sense incorruptible; the common elements will not destroy it, neither does she mix with them essentially. Outwardly she resembles a stone, and yet she is no stone. The philosophers call her their white gum, water of their sea, water of life, most pure and blessed water; she is a thick, permanent, saltish water, which does not wet the hand, a dry water, viscous, slimy, and generated from the saline fatness of the earth. Fire cannot destroy her, for she is herself fire, having within her a portion of the universal fire of Nature, and a secret, celestial spirit, animated and quickened by God. She is a middle nature between thick and thin, not altogether earthly, not wholly igneous, but a mean aërial substance, to be found everywhere and at all seasons.

MAGNET. Alchemically speaking, this is the dew of the philosophers.

MAN AND WIFE. Most philosophers have compared the confection of the Magistery to the generation of humanity. They have, therefore, personified the two parts or ingredients of the work, namely, the fixed and the volatile, as the male and female, man and wife, etc.

MANNA is, in some instances, the Mercury of the philosophers, called also divine manna, because they affirm that the secret of its extraction is a gift of God, as also the knowledge of that Mercury which is the minera whence it is derived.

MARCASITE. Many species were known to the old chemists, for all stones which contained any proportion of metal were so called, and even sulphureous stones, vitriolic stones, etc., were included under the same term.

MARRUBIUM, *i.e.*, the herb horehound.

MARTAGON is Silphium, a kind of lily.

MATTER OF THE PHILOSOPHERS. This has been compared to the sperm of an animal, for it is a most delicate substance, almost a living thing; indeed, it possesses some portion of life, and Nature produces certain animals out of it. For this reason the least violence destroys it, preventing all generation. If it be overheated for even a few minutes, the white and red sulphurs will never essentially unite and coagulate. On the other hand, however well the work has been begun, should it grow cold for so short a space as half an hour, it will come to no good end afterwards.

MELIGIA, probably Melisea, *i.e.*, motherwort, a kind of manna or balsam, obtained by a magisterial process upon vegetables. Meliagris is frittany.

MELLILOT, the *dulcis et mellea lotus* of Ovid, called *Sertula campana* by Pliny. The herb mellilot.

MELISSA, see Meligia.

MELONA, ? Melina, a drink made of honey.

MENSTRUUM. This term is used in a very arcane manner by some alchemists, who speak of the menstruum or matrix of the world, wherein all things are framed and preserved. It is a certain oleaginous and ethereal water.

MERCURIAL LIQUOR, a balsam of things in which all health resides. It is most potent in tereniabin and nostoch.

MERCURY. According to Arnoldus de Villa Nova, the Mercury of the philosophers is an aqueous, cold, and moist element. This is their permanent water, the spirit of the body, unctuous vapour, blessed water, virtuous water, water of the wise, philosophers' vinegar, mineral water, dew of heavenly grace, virgin's milk, corporal mercury, etc. All these names signify one only thing; out of this all the virtue of the art is extracted, and, according to its nature, the tincture, both the red and the white. Mercury, by another account, is composed of a metallic and fluidic earth. There are as many Mercuries as metals, which can all combine with the fluidic earth, and this in such a manner that separation is impossible. The whole secret of Hermetic philosophy and the Great Work consists in the wonderful sympathy between those Mercuries and this earth. *Dissolving Mercury*, used by spagyric philosophers for the reduction of metals, minerals, vegetables, and all bodies to their first matter, is of three kinds: Simple dissolving Mercury; composite dissolving Mercury, which is properly their true Mercury, and common Mercury, or that which is extracted from metals. Simple Mercury is a water extracted according to the principles of their art from a matter, the true name of which they have carefully concealed under an infinity of false descriptions. *White Mercury of the Sages* is the stone at the white. *Red Mercury* is the magistery perfect at the red. *Universal Mercury* is the animating spirit diffused throughout the universe. *Crude Mercury* is the dissolvent of the sages, not that vulgar quicksilver which is called crude mercury by chemists. *Preparing Mercury* is this same dissolvent, which prepares soluble bodies for their development into the perfection of the magistery.

MERCURY OF LUNA. Flamel directs that the Mercury of the Moon should be taken with that of the Sun, and cherished over the fire in an alembic. It must not be a fire of coals or wood, but a bright, shining fire, like the sun itself; its heat must never be excessive, but always of one and the same degree. Thus is performed the plantation of one into the other.

MERCURY OF SOL. The unctuous humidity of the philosophers is called water of silver and water of the moon, but it is really, say others, the Mercury of the Sun, and partly that of Saturn. Indeed, it is an extract of the three metals, and without these it can never be made.

MERITORIUM, a tavern.

MICROCOSM. Sir George Ripley, a celebrated English alchemist, describes the stone as a triune microcosm. Proportion must be studied in its composition. The matter itself is found everywhere. It flies with fowls in the air, swims with fishes in the sea, it is discerned by the reason of angels, and it governs man and woman.

MINERA. The alchemists mention a minera of man as well as of metallic things. It is also the First Matter of the Philosophers' Stone. It is water and not water, earth and not earth, the Damascene earth, the subject of art, and that also of which God made use in Nature.

MITHRIDATIC, pertaining to mithridate, an old antidote against poison.

MOREE, Moro, Mores, a kind of abscess.

MORPHEA, see Morphew.

MORPHEW, or Morphea, a species of leprosy.

MOTHER OF METALS, *argentum vive*, *i.e.*, live silver.

MUMIA. Whatsoever when killed has the power of healing diseases. Hence Mumia of the elements is the balsam of the external elements. *Mumia transmarina* is manna. *Mumia versa*, according to some opinions, is liquor of Mumia.

MURIA, brine or salt water. Also stinking menstruum.

MYROBOLANUM, the Egyptian bean, a fruit used by apothecaries in the manufacture of precious ointments; an oil is obtained from the kernel.

MYSTERIUM MAGNUM. This is used by the alchemical philosophers in another sense than that of Paracelsus. The operation or confection of the great work is called a mystery of the philosophers, because they discover it only to their most intimate friends Some also refer under this name to the first matter of the work, because that is the most concealed portion of all their writings.

NECROLICUS. As Necrolia and Necrolica are medicaments which prevent death and preserve life, so it is reasonable to suppose that a Necrolicus is the administer of their medicines. But he is also a person who has written learnedly on any subject.

NENUPHAR, a generic name of the elementary spirits of the air. It also signifies a water lily.

NEPITA, the herb net, or catmint.

NEUFARINIS, a generic name of elementary spirits.

NOSTOCH, the efflux of a certain star, according to Paracelsus. It is deposited on the earth in summer, appearing like a yellow fungus, but in consistency it is like a jelly. Some say it is wax.

OGERTUM or Ogertinum, *i.e.*, Orpiment.

OIL OF THE PHILOSOPHERS. Though oil simply so called has nothing to do with the Great Work, and must not be used in confection, the name has been applied to the matter during that period when it has assumed an oleaginous colour and viscosity. This is when it is undergoing putrefaction in the philosophical egg. Oil of the philosophers also signifies the Secret Fire of the Sages. What is termed by some alchemists *Blessed Oil* is incombustible, philosophical oil, the sophic sulphur, also the stone perfect at the white, or red, because it runs and melts before the fire like butter. *Oil of Nature* is the prime salt which is the basis of all salts. It is called oil because it is unctuous, melting, and penetrating, and oil of Nature because it is the base of all the individual substances in the three kingdoms, and is their material conserver and restorer. It is the best, the most noble, the most fixed, and at the same time the most volatile before its preparation. When it is sought to be employed by art, it must be brought from the fixed to the volatile, and from the volatile to the fixed: to solve and to coagulate is the whole work. *Essential Oil* is the volatile sulphur of the philosophic metals; it is their soul, the male, the Sun, the gold of the sages. *Oil of Saturn* is the matter of the philosophers at the black stage, because their matter in putrefaction is called lead. *Oil of Sulphur* is also sometimes the matter at the black stage. *Incombustible Oil* is the mastery at the red, so called because of its fixation. *Live Oil* is the mastery at the white. *Vegetable Oil* is oil of philosophical, not vulgar, tartar.

OIL OF IRON. The Sulphur of the Philosophers at the red.

OLIBANUM, an inspissated sap, or gum-resin.

ONAGIR, *i.e.*, Onager, the wild ass.

OPOPONAX, the juice of the panax, or herb all-heal.

PANIS SILIGINIS, or Panis Siligineus, white bread, or fine manchet.

PANNUS, a cloth or bag.

PARATELUS, of unknown signification, but the text of the treatise *Concerning Long Life* makes it barely possible that it is a corruption of Paraclete.

PART WITH PART, a composition of equal parts of gold and silver.

PEACOCK'S TAIL. The matter of the work at that moment when the colours in the tail of the peacock manifest on the surface.

PELICAN. An alchemical vessel, so called after the bird of that name. The body tapers towards the neck, which is bent round, and the tube returns into the body.

PENTAPHYLLUM, the herb cinquefoil.

PERI-PNEUMONIA, inflammation of the lungs.

PERSIAN FIRE, a scorching and spreading ulcer.

PER SCOBAM, possibly per scobem, from *scobs*, any powder that comes of sawing, filing, etc.

PERSICARIA, the herb culerage or peachwort.

PETROSELINAR, a kind of parsley growing among rocks.

PHILOSOPHERS' STONE. This is one, says Arnold de Villa Nova, and he adds that it can be extracted from all bodies, including common quicksilver. The first physical work is the dissolution of the stone in its own Mercury, so as to reduce it to its first matter. Jacob Böhme describes the philosophers' stone as dark, disesteemed, and grey in colour.

PHILOSOPHICAL EGG. While the majority of chemists have supposed that the sages applied the name of Philosophical Egg to the vessel in which they enclose their matter for coction, and have consequently shaped it like an egg, this does not represent the idea or the sense of the sages, although at the same time such a shape is the most convenient for circulation. The egg of the philosophers is not that which contains; it is that which is contained; that is, the true vessel of Nature. Therein the philosophic chicken is concealed, which the internal fire of the egg, excited by the warmth of the hen, vivifies by degrees, and gives life to that matter of which it is the root, whence there is at length born the philosophical infant which perfects and enriches his brethren. Again: the egg signifies most commonly the actual matter of the mastery, which contains the Mercury, Sulphur, and Salt, even as the ordinary egg is composed of the white, the yolk, and the pellicle.

PHILOSOPHICAL MONTH, *i.e.*, the period required for chemical digestion.

PHLEGM. When this term is made use of in connection with the operations of alchemy, it signifies a certain aquosity or vapour given off from the matter of the work, becoming white in the process of distilling. It is for this reason that the same name is applied to the stone and to philosophical Mercury in the grade of the white.

PHŒNIX, the fire in the quintessence, the physical stone.

PISA, a mortar.

POCALE, a measure of wine.

PRUNELLA, *i.e.*, Prunellus, the bullace-tree.

PUTREFACTION. The corruption of the moist substance of bodies by a defect of heat, or, otherwise by the action of a foreign fire on the matter. It is in this sense that the spagyric philosophers say that the matter of their stone is in putrefaction when the heat of the extrinsic fire setting into action the internal fire of the said matter, the two act in concert thereupon, separating the humidity which binds its parts, and after several circulations in the hermetically sealed aludel, reducing the matter into dust.

PYROMANCY is divination by the impressions of fire, by comets, by fiery colours, by visions and imaginations in the fire. Capnomancy, or divination by smoke, is an art mentioned by Agrippa as allied to Pyromancy.

QUARTA, probably a measure containing five ounces of wine and four and a half ounces of oil.

QUARTATION, an old method of testing gold. In melting, nine parts of silver were added to one of gold, and both were resolved by aquafortis, which held the silver in solution, and the gold settled at the bottom.

QUINTESSENCE. Paracelsus, though possibly the first, was by no means the last of the alchemists to describe the term quintessence as a misnomer. Eugenius Philalethes affirms that there is no quintessence, no fifth principle, except Almighty God. There is a quartessence which is a moist and silent fire, which passes through all things in the world and is Nature's chariot, the mask and screen of the Almighty. Wheresoever God is there this train of fire attends Him. It was this fire which was manifested to Moses on Mount Sinai. According to another authority, the terms quintessence, specific, magnetism, bond, seed of the pure elements, etc., are all synonyms of one substance, a subject wherein the form abides. It is a material essence, which encloses an operative and celestial spirit. Others say that it is the fifth principle of composites, comprising the finest portion of the four elements. The *Quintessence of the Elements* is the Mercury of the Philosophers.

RABELOIA, the roots of the larger ranunculus.

RAINWATER. This is the common soft water, mentioned sometimes by Paracelsus.

REALGAR is red orpiment.

REBIS. Among other meanings attached to this curious term, it is said to be the last matter of things, and this may be understood either literally, as excrement, or philosophically, as opposed to the first matter. Alchemically, it is the fixed and the volatile.

RECTA CROA, *i.e.*, flos sectæ croæ, flower of the crocus, an extract of chelidony leaves. Also flower of the nutmeg.

RED LION. The spagyric philosophers have given this name to the terrestrial and mineral matter which remains at the bottom of the vessel after sublimation of the spirits, called eagles. This Red Lion is also called Laton.

REDUCTION. This is the retrogradation of a substance which has reached a certain degree of perfection to a degree of a lower order. The reduction of metals into their first matter is their philosophic, not the vulgar, retrogradation into their proper seed, that is to say, into a Hermetic Mercury. It is also called reincrudation, and is performed by the dissolution of the fixed by its proper volatile, from which it has been made.

REGULUS, literally, the little king.

RELOLLEA, arcane virtues, called also relolleum.

RUBEDO DE NIGRO, a red substance, extracted by art from black lime.

RUBIFICATION. The process under which the philosophical matter passes from the white to the red stage.

RUBY. With the philosophers this sometimes means the magistery at the red. So also their precious ruby is the powder of projection.

SAGÆ. The saga is properly a witch, sorceress, or wise woman. The sagani are the elementary spirits.

SAGAPIN, a Persian gum-resin.

SALAMANDER. In alchemical terminology this name is sometimes applied to the philosophical matter, and then the term Blood of the Salamander signifies the red state. More usually it is the redness which appears in the recipient in the distillation of nitre and vitriol.

SALT. This substance was supposed by the alchemists to consist of a small quantity of sulphureous earth and a large proportion of mercurial water. It was the substantial matter of bodies, of which the form was Sulphur. Three chief species were distinguished, the nitrous, the marine, and the vitriolated, to which some add the tartareous. The marine was regarded as the most important. From this salt volatilised comes Nitre, from Nitre comes Tartar, and from Tartar, when cooked and digested, comes Vitriol. There was also another classification as volatile, mean, and fixed Salt. *Fusible Salt* is the Matter of the Sages cooked and perfect at the white; it was so called because it is actually a salt which melts like wax when placed upon a plate of red hot metal. *Salt of Metals.* Some investigators, taking this expression literally, have imagined that the Matter of the Philosophers was a metallic substance reduced into a salt or a vitriol, but the reference is to the Magistery at the white, because even as salt is the principle of the vulgar metals, so that of the Sages is the root of the first matter of philosophical metals. *Red Salt* is the red Sulphur of the Sages. *Salt Alacoph* is Sal Ammoniac. *Bitter Salt* is Alkali. *Salt of Greece* is Alum. *Indian Salt* is the Mercury of the Sages. *Salt of Bread* is marine or common salt. *Salt of the Sages* is Sal Ammoniac, the natural kind, or philosophic Nitre, but the true Salt of the Sages is their Matter at the white. *Acid Salt* is philosophical Mercury. *Salt Adram* is Sal Gemmæ, also called *Salt of Cappadocia*. *Burnt Salt* is the Matter at the black stage. *Salt of the Earth* is the Mercury of the Sages. *Salt in Flower* is the Mercury of the Sages, or the Dry Water of the Sages. *Salt of Glass* is the Mercury of the Sages. *Fixed Salt* is the Sulphur of the Sages. *Honoured Salt* is the Matter of which the Hermetic Mercury is made. *Salt of Saturn* is lead reduced to a salt. *Sea Salt* is the Mercury of the Sages. *Spiritualised Salt* is the philosophers' spirit of salt. It is their Mercury prepared by Hermetic sublimation. *Universal Salt* is the Mercury of the Sages.

SAL ANATRON, salt nitre, red salt of the Indies. It is also called *Sal Andaron*.

SALDINI, also called Rolamandri, are igneous men, otherwise essences of the race of the Salamander.

SAL GEMMÆ is Hungarian salt. It is also called *Sal Nominis* and Salt of Hungary.

SALT NITRE, see Salt.

SAL PEREGRINUM is very probably another name for the alchemical Salt of Tartar.

SALT OF TARTAR, white calcined tartar.

SAMECH, tartar, or salt of tartar; also the healing power of all wounds.

SANDARAC, a bright red colour used by painters; it is found in mines of gold and silver; some call it red arsenic. There is another species made of burnt ceruse; also red lead. Another name is cerinth.

SAPPHIRE. The philosophers have given this name to their mercurial water.

SATURN, *i.e.*, Lead. This metal is said by Isaac the Hollander to be the first matter of the philosophers. It is also a name for the philosophical matter in putrefaction. It is further the Adrop of the Sages or the Azotified Vitriol of Raymond Lully. Finally, it is a name sometimes confusedly applied to ordinary copper.

SATYRION, or Orchis, the herb ragwort, or priest-pintle.

SCARIOLÆ. The fourfold spiritual powers of the mind, corresponding to the four elements and the four cabalistical wheels of fire of that chariot which took up the prophet Elias. They are soul emanations or faculties, fancy, imagination, speculation, etc. These under religious sanctification become instruments for the attainment, not merely of long, but of eternal life.

SCORIÆ. Impurities separated from minerals and metals during fusion.

SEALED EARTH, a red earth.

SEED. The seed, or first matter of the Stone, says the Persian alchemist, Rachaidibi, is outwardly cold and moist, but inwardly hot and dry. Rhodion, the instructor of Calid, says that it is white and liquid, but afterwards it becomes red. It is the flying stone, aërial and volatile. It has one virtue inwardly and another outwardly. See *Sperm*, First Matter, etc.

SEPARATION. This process, so often referred to by Paracelsus, is also mentioned by pseudo-Hermes, who has these significant words: Know, ye that are children of the wise, the separation of the ancient philosophers was performed upon water, which separation divides the water into other four substances.

SEPULCHRE, the glass vessel which contains the matter of the philosophical work; the dissolvent of the sages; the black stage of the matter.

SERPIGO, a tetter.

SERUM, whey, buttermilk.

SICONIUS, unknown, a race of elementaries.

SILIGO, a species of corn.

SMAGMA, called also both Smegma and Smigma, is a detersive substance.

SMIRIS, Smyris, a stone with which glaziers cut glass, and with which lapidaries polish gems.

SPAGYRIC SCIENCE is that which teaches the division and resolution of bodies, with the separation of their principles, either by natural or violent means. Its object is the alteration, purification, and perfection of bodies, that is to say, their generation and their medicine. It is attained by solution; success is impossible if their construction and principles are ignored, because these serve for dissolution. The heterogeneous and accidental parts are separated with a view to the intimate reunion of the homogeneous portions.

Spagyric Philosophy, properly so-called, is the same as Hermetic Philosophy.

SPATHUS, or Spatha, a pounding instrument.

SPATULA, or Spathula, a broad and flat instrument, like a slice. It is used in certain furnaces for regulating the heat. A damper.

SPELT, *i.e.*, German grain, a kind of wheat.

SPERM OF THE WORLD, that is, catholic magnesia. It is said to be very salt, extremely soft, somewhat thin and fluid, having no relation to ordinary salts, a sperm that Nature herself draws out of the elements without the help of art. Man may find it where Nature leaves it, it is not his office to make the sperm, nor to extract it; it is already made, and wants nothing but a matrix, with heat convenient for generation.

SPHERE OF THE SUN, the dual matter of the stone, the heaven, *i.e.*, the quintessence.

SPODIUM, ash of gold, but sometimes applied to pompholix and to a kind of tutty.

SPREAD EAGLE. In common chemistry this represented sublimed sal ammoniac, and, in the Hermetic sense, it was the volatilisation of the matter.

STEEL. The steel of the philosophers is a subject of frequent reference, and it has caused many persons to seek the philosophic stone in this substance, but in vain. The steel of the sages is the wine of their philosophic gold, a spirit pure above all, an infernal and secret fire, most volatile in its nature, and the receptacle of superior and inferior virtues.

STELLIO, an eft or newt.

STIBIUM. In the first place, black lead and its derivatives; secondly, a name of antimony; thirdly, a certain stone found in silver mines.

STOMACH OF THE OSTRICH. The chemical philosophers give this name to their dissolvent or philosophical Mercury.

STONE OF THE PHILOSOPHERS. The Rosicrucian philosophers say that in the impregnable fortress of truth is contained the true and undoubted Philosophers' Stone, that treasure which, uneaten by moths and unstolen by thieves, remaineth to eternity, though all things else dissolve, set up for the ruin of many and the salvation of some. To the crowd this matter is vile, exceedingly contemptible and odious, but to the philosophers it is more precious than gems or gold. It loves all, yet it is well-nigh an enemy to all; it is to be found everywhere, yet scarcely anyone has discovered it. It is the one thing proclaimed by veritable philosophers, which overcomes all, is itself overcome by nothing, searches heart and body, penetrates everything stony and solid, strengthens all things delicate, and establishes its own power on the opposition of that which is most hard. It is the way of truth, and there is no other path to life. It is the true medicine, rectifying and transmuting that which is no more into that which it was before corruption, even into something better, and that which is not into that which it ought to be. The gold of the philosophers with which the wise are enriched is not that gold which is coined.

STORAX CALAMITA. Storax is a sweet incense or gum.

STRUTHIO, an ostrich. Soap-wort, used for cleaning wool, a chaplet of this flower.

SUBLIMATION is the purification of the Matter by means of dissolution and reduction of the same into its constituents. It is not the forcing of the Matter to the top of the vessel, and then maintaining it separated from its *caput mortuum*, but its subtilization and purification from all earthly and heterogeneous parts, imparting to it a degree of perfection not previously possessed, or, more correctly, its deliverance from the bonds which bind it, and hinder its operation.

SULPHUR VIVE. This substance, which is one of the profound mysteries of alchemy, is identical with the red sulphur of the philosophers. Some persons, led astray by Hermetic symbolism, have gone to work on native sulphur, assuming it to be the true matter of the art, but though this name is applied to it, the reference is really to that substance when it has attained the perfection of the red or white. It is then the true philosophical sulphur, and, according to Raymond Lully, is not to be sensibly distinguished from the true Mercury, which has no connection whatsoever with common sulphur. Mention is also made of a white sulphur, which some say is Mercury educed from potentiality into activity by the operations of the Magistery and following the principles of the Medicine of the First Order. It may be noted in this connection that Sulphur of Vitriol is the Soul of Vitriol, and so also with Mars, Sol, etc. Black Sulphur is Antimony; unctuous Sulphur is another name for the Sulphur of the philosophers. Narcotic Sulphur of Vitriol is an extract of Vitriol which is described by Beguin; Paracelsus himself regarded Vitriol as the best of all anodynes. Ambrosial Sulphur is a natural red Sulphur, very transparent and met with in large lumps. Green Sulphur is the Sulphur of the Sages. What the alchemists call True Sulphur is the fixed grain of their matter, the true internal agent which acts upon, digests, and cocts its own proper mercurial matter wherein it is enclosed. Sulphur Zarnel is another name for philosophical Sulphur. The Sulphur of Nature is yet another. Some, however, apply it to the matter at the white. One author says that it is the essential menstruum made from Mercury and spirit of wine seven times rectified, which dissolves the calx of the Sun and Moon, or at least, extracts their tincture, with which gold is nourished by a simple yet secret series of operations. But there is some doubt on this point. Universal Sulphur is affirmed to be the Light from which all particular sulphurs proceed.

SULPHUR-WORT, *i.e.*, Peucedanus.

SUN AND MOON. The influence of Paracelsus on the later alchemists is shewn very clearly in their speculations concerning Luna and Sol; as for example: The Sun and Moon are two magical principles, the one active and masculine, the other passive and feminine. As they move so move the wheels of corruption and generation. They naturally dissolve and compound, but properly the Moon is the instrument of the transmutation of the inferior nature. There is no compound in Nature which has not a little sun and a little moon, a Son of the heavenly Sun, and a Daughter of the celestial Moon. What is performed by the great luminaries for the conservation of the great world, is analogically accomplished for the microcosm by the lesser luminaries. The little Moon is our incombustible, eternal oil, the receptacle of the little Sun, which is a fire. These are the Sol and Luna of the philosophers, not gold and silver.

SUPERMONIC, *i.e.*, enigmatical.

SYLVESTERS, the fauns of classical mythology.

SYPHITA PRAVA, St. Vitus's Dance.

SYPHITA STRICTA, the fantastic spirit of sleep-walkers.

SYRONES, pimples or boils on the hands.

TALC. The older alchemists have often made reference to what they term an Oil of Talc, to which they have attributed so many virtues that subsequently chemists have exerted all their power to compose it. They have calcined, purified, and sublimed the matter in question, but have met with no success. The reason is that the term was used allegorically, and that the reference was to the Oil of the Philosophers, the elixir at the white.

TANACETUM, the herb tansey.

TAPSUS, *i.e.*, Tapsos, a kind of herb.

TARTAR, a name applied to the mastery at the white stage.

TAURUS ♉. This zodiacal sign represents asphalt or bitumen in alchemical symbolism.

TENTIGO PRAVA, ? a severe stiffness, or contraction.

TERENIABIN. A variety of manna.

TERNARY. The author of *Anima Magica Abscondita* observes that some philosophers who, by the special mercy of God, attained to the Ternarius, could never, notwithstanding, obtain the perfect medicine. Elsewhere in the same treatise he says that this Ternarius, being reduced by the Quaternary, ascends to the magical decad, which is the exceeding single monad, in which state it can perform whatsoever it pleases, for it is united thus, face to face, with the first eternal, spiritual Unity.

TESTA, potsherd, tile, brick; also a metallurgical term, signifying bloom. But it seems to have had another meaning in the treatise *Concerning Long Life*. It is said to be the skin of man's body.

TESTÆ OVORUM, egg shells.

THERIAC. Any antidote to poison.

THRONUS, *i.e.*, Thronum, a flower or herb used as drugs or charms.

THUCIA, tutty.

TIBIA, the shank or shinbone.

TIGILLUM, a funnel, a crucible.

TITHYMAL, the sea lettuce, wolf's milk, or milk thistle. Also spurge, euphorbia; many kinds were known to the ancients. Physicians used the juice or berries as a purgative.

TRAGACANTH, probably Trajanthes, a species of Artemisia. Also gum dragant, a low shrub. The astragalus, whence the gum tragacanth.

TRANSMUTATION. The soul of man is affirmed by the alchemists to possess an absolute power in miraculous, that is, in more than natural transmutation.

TRARAMES. Spirits who are not seen but heard only, as rapping or throwing spirits.

TRIFERTES, the same as salamanders.

TRIPLICITY. Besides the triplicity, which exists in the elements and in all created subjects, there is another more obscure and mystical triplicity, which is recognised by adepts. Without this the former cannot be attained. These three principles are the key of all Nature. The first is one in one and one from one; it is a pure white virgin, and next to that which is most pure and simple. It is the first created unity, the bride of God and of the stars, through which as a medium all things were made, and are still made, both in things natural and in things of art. The second principle does not differ from the first in substance or dignity, but only in complexion and order. The third principle is not, properly speaking, a true principle, but rather a product of art; it is a various nature, consisting of superior and inferior powers. It is the magician's fire, the Mercury of the philosophers, the microcosmus and the Adam.

TRINES, possibly Triones, the constellation called Charles's Wain.

TRIPOLIS. Tripolium is the herb called turbit or blue daisy.

TRITORIUM, a vessel for grinding, mortar, or pestle.

TRONOSIA, that is, Tronossa, honey-dew.

TRONUS, same as Tronosia.

TRUPHIT, the principle of development in every metal, a secret virtue of minerals.

TURBITH, a root much used in medicine to purge off phlegm. Mineral Turbith is sweet, non-corrosive, precipitated Mercury.

TUTIA, see Thutia.

UNIONS. The term *unio* or union was applied to the pearl, because though many are found in one shell, not one is exactly like another.

UNIVERSAL MEDICINE. Most recipes of the adepts given for the preparation of the absolute elixir, are more obscure even than Paracelsus, as, for an example: Take ten parts of celestial slime; separate the male from the female, and each afterwards from its earth, but physically, mark you, and with no violence. Conjoin after separation in due, harmonic, vital proportion. Straightway the soul, descending from the pyroplastic sphere, shall restore, by a vivific embrace, its dead and deserted body. The conjoined substances shall be warmed by a natural fire in a perfect marriage of spirit and body. Proceed according to the Vulcanico-magical theory till they are exalted into the fifth metaphysical rota. This is that world-renowned medicine of which so many have scribbled, and yet so few have known.

VALERIAN, the great set wall.

VENTER EQUINUS. According to an explanation of Paracelsus the Venter Equinus is the digestive power.

VESSEL OF THE PHILOSOPHERS. Geber describes this instrument as a round glass vessel with a flat round bottom. This and all other simple explanations concerning it are supposed to be wholly deceptive, and it is regarded as one of the most profound mysteries of all alchemical art. A treatise attributed to S. Thomas Aquinas says that there is but one vase, one substance, one way, and one only operation.

VERBENA, the herb Vervain.

VERTO, a weight equivalent to the fourth part of a pound.

VESSEL OF HERMES, see Vessel of the Philosophers.

VICTORIALIS is probably Victoriola, the Alexandrian laurel, or tongue-laurel. Paracelsus refers to its magical powers, and says that the lorinds can be made to manifest by means of it.

VINUM ARDENS. According to Paracelsus the correction of Vinum Ardens is by distillation as long as it will ascend, for the removal of the aqueous part.

VINUM ESSATUM, wine impregnated with the virtues of herbs or other substances.

VIRGIN MERCURY and Virgin Sulphur. According to one explanation these are the heaven and earth of Moses.

VITISTA, *i.e.*, St. Vitus's dance. Paracelsus terms it a disease of the imagination.

VITRIOL. The philosophers apply this name to their green or crude matter. Their White Vitriol is the magistery at the white. Their Red Vitriol is their sulphur perfect at the red. Metallic Vitriol is the salts of metals.

VUA, that is, Uva. This is firstly the grape, and stands, by an extension of meaning, for the vine itself. Analogically it is any bunch or cluster. There is also a swelling of the uvula which passes under this name. Finally, there is *uva quercina*, a concretion occasionally found at the roots of oak-trees, possessing medicinal properties, especially in dysentery.

VULCAN is said by Paracelsus to be the master of the alchemists and spagyrists.

WATER. So early as the days of the Greek author Theophrastus, the origin of metals is said to have been referred to water. The same writer mentions that Callias, an Athenian, endeavouring to make gold, brought his materials into cinnabar.

XYLOBALSAMUM, the wood of the balsam-tree.

YLE, wood; timber; the matter or stuff of which a thing is made; the raw unwrought material, whether wood, stone, or metal. In chemical signification, a simple substance, a base matter; as a principle of being, first found in Aristotle, and frequently later in philosophical writers,—usually as opposed to the intelligent principle.

YSOPUS.
YSOPAIC ART. } The art of smelting and fusing.

YLIADUM, Yliadus, Yleidus, etc. The interior spirit which informs the members of every body. Outwardly it generates health, but inwardly disease in humanity. It also leads on to the crisis in diseases. Disease is the resolution of the Yliadus. The reason of this seems to be that the interior spirit contains many species of salts. The resolution of arsenic in the body causes plague; the resolution of ogertinum, or orpiment, causes pleurisy; the resolution of vitriolated salt causes another disorder which Paracelsus describes as a disease characterized by a paroxysm, by swellings of the throat, and ulceration of the tongue. Elsewhere Paracelsus says that Yleidus is elemental air, and that its obstruction in any part of the body occasions disease. There is an Yliadus of the elements and one also of man.

YRCUS. The male coney, or rabbit. The blood of this animal was supposed to soften glass and flints when pounded and made into a paste therewith. It had also a similar effect upon other crystalline substances. It is written Hircus by the ignorant. It should be noted in this connection that *cuniculus*, that is, the word which usually signifies coney, also stands for the long pipe attached to a still or a furnace.

ZELOTUM is Petrine Mercury.

ZINCTUM, or Zinetum. Paracelsus describes this as a little known metal, of a peculiar nature and seed. Many metals are adulterated therein. It is of itself fluxile, for it is generated from the flux of the three prime principles. It is fusible but not malleable. He states that he is unacquainted with its ultimate matter. It is akin to quicksilver, and it does not admit of combination with other metals.

ZINIAT. Ferment.

ZONNETI GNOMI. These are fantastic bodies.

ZWITTER, a species of Marcasite; also roasted ore.

NOTE.—*Some explanations contained in the foregoing Vocabulary concerning known substances are not in correspondence with modern knowledge, and it should be understood that they represent chemical science during the period of Paracelsus and of Alchemy.*

INDEX

Abraham, a Vulcanic philosopher, i. 48.
Abreo, ii. 334.
Abrissach, i. 231.
Acetum, i. 197, 199; ii. 52, 85, 89.
Adam, the inventor of Arts, i. 48; Hermaphroditic Adam, 66; a name of the Stone, *ib.*, 161 *n.*; the Protoplast, 188; the first Signator, *ib.*, 201 *n.*; ii. 70. *See also* ii. 113, 285, 327, 329.
Adamant, a black crystal, i. 17; how dissolved, *ib.*; its generation, 17 *n.*; white adamant, *ib.*
Adamic Matter, i. 66.
Adamic Flesh, i. 94 *n.*
Adech, i. 219, 223; ii. 345, 346.
Aeromancy, ii. 296.
Aes Ustum, i. 141, 142.
Aetnean Fire, i. 159.
Agaric, ii. 188.
Agrippa, i. 150.
Air, element of, i. 206, 208 *n.*, 210, 211; separation of elements from, ii. 20; Air of the Philosophers, i. 298; Air the cloister of the Invisible Fates, ii. 253.
Alabaster, i. 158.
Albertus Magnus, i. 57, 59, 67, 95, 243.
Alchemical Phœnix, i. 40.
Alchemist, one who brings forth that which is latent in Nature, ii. 156.
Alchemy, cause of its difficulty, i. 4; futility of many recipes and processes, 1, 3; actual nature of the art, *ib.*; what alchemy is, 16; what it teaches, 15; materials and instruments required in it, *ib.*; not affected by the aspects and courses of the planets, 16; alchemy as an inferior heaven, 16 *n.*; alchemy from God, 73; comprised in facts, not words, 89; praise of the art, ii. 95 *n.*; Tincture of Alchemy, i. 251; medical pyrotechny, ii. 150; an external bowel, 151; experience in alchemy, 154 *n.*; alchemy of colours, 157; external alchemy, 166.
Alcohol, i. 156.
Aliani, ii. 346.

Alkali, magistery and essence of, ii. 130.
Almadir, i. 66, 69, 70.
Alopecia, ii. 53.
Aloth, Sol converted into, i. 321.
Alum, i. 153, 195, 197, 224, 237, 266; medical virtues of rock alum, ii. 221; medical virtues of alumen plumosum, 221, 222; medical virtues of alumen entali, 222.
Amarissima, i. 230.
Amber, i. 158; ii. 54.
Ambergris, i. 137; ii. 60.
Amethyst, its colour, i. 18; whence extracted, 18 *n.*
Anacardus, ii. 23, 47.
Anachmus, ii 343.
Anachthron, medical uses of, ii. 222, 223.
Anaxagoras, i. 68.
Aniadus, i. 227, 228, 230; ii. 197, 342, 345, 346.
Animal Stone, i. 56.
Animation, i. 64.
Anismut, i. 145.
Anthion, ii. 89.
Anthos, i. 227; quintessence of, 11, 26; ii. 61, 324, 331, 333.
Anthrax, i. 269.
Antimoniac, i. 110; antimoniac minerals, 311.
Antimony, importance of, i. 31; the true bath of gold, 62; the white lead of the wise, *ib.*; supreme arcanum of, 63; oil of, *ib.*; purple liquor of, 63 *n.*; fixation of, 313; antimony in Saturn, 8; *see also* 266, 275; quintessence of, ii. 26; element of, 130; medical uses of, 199, 200, 201.
Apollo, i. 29.
Aquafortis, i. 136, 278.
Aqua Regis, i. 37, 38.
Aquaster, ii. 340, 345, 346.
Aquinas, S. Thomas, i. 67, 69, 117.
Arcanum, what it is, i. 21 *n.*, ii. 37; arcana derived from the firmament, i. 74; oil of arcanum, 333; arcana and quintessences, ii. 37; arcanum of the Stone, i. 53, 71, ii. 89; *see also* ii. 87, 119, 151.

Archa, ii. 324, 325.
Archelaus, i. 19; ii. 155.
Archeus, i. 92, 97; identical with Nature, 97 n., 125, 127 n., 129, 160 n.; Archeus and his signatures, 171; see also 184, 190, 233 n., 240, 247, 291, 292, 293; ii. 167, 179, 180, 318, 321, 322.
Archialtis, ii. 253.
Ares, i. 101 n., 127 n., 233 n.; ii. 179, 329, 341.
Argentum Vivum, i. 148.
Aristotle, i. 20, 26, 73, 95, 99, 244; fatuity of his philosophy, 28; confuted by sulphur, 264, 265; Aristotle the alchemist, i. 68.
Arithmetic, ii. 297.
Arnold de Villa Nova, i. 52, 69, 71, 299.
Arsenic, fixation of, i. 58 n.; false process for, i. 59; three spirits of, ib.; varieties of, 59 n.; alchemical virtues in, 107 n.; see also 288; medical virtues of, ii. 210, 211.
Arsenicals, i. 110.
Art, how to learn it, i. 72; arcanum of, 74 n.; to whom made known, 88; 254, 294.
Artemisia, i. 230.
Art of Vulcan, ii. 165; special office of, 166.
Ascension, i. 153.
Astral Gold, i. 301.
Astral Spirits, ii. 296, 313.
Astral Science, i. 171.
Astrology and Alchemy, i. 16; science of, ii. 294, 295, 297-299, 313.
Astronomic Magic, i. 51.
Astronomy, agreement with Alchemy, i. 23; seven faculties of, ii. 282, 293.
Astrum, i. 37.
Athanor, i. 284; ii. 153.
Atrament, i. 166.
Augmentation of Sol, i. 142.
Aurelius Augurellus, i. 87.
Aurichalcum, i. 164.
Auripigment, i. 110, 184, 268.
Aurum Musicum, i. 149.
Aurum Potabile, i. 38; how gold is made potable, 38 n.; 100; ii. 130.
Auster, i. 218.
Avicenna, i. 20, 36, 39, 73, 95, 243; ii. 82, 155, 162, 169.
Azoc, i. 69.

Bacon, Roger, i. 50.
Balm, ii. 130.
Balneum Mariæ, i. 168; ii. 35, 85.
Balneum Maris, i. 58, 75; ii. 55.
Balneum Roris, i. 157.
Balm of Sulphur, i. 290.

Balsam, i. 227, 245, 257, 258, 259, 261; ii. 69, 71; Balsam of the Stars, i. 38; congenital Balsam, 38 n.; confection of, ib.; Balsam of Mars, i. 136; Balsam the first elixir, ii. 72; corporal Balsam, 92; spiritual, 93; Balsam of the coagulated body, ib.; perpetual Balsam, 96; 101.
Balsamites, ii. 25.
Bamahemi, ii. 342.
Basilicon, ii. 61.
Basilisk, generation of, i. 55, 123.
Bath of the Philosophers, i. 298.
Bdellium, ii. 182.
Bean flowers, ii. 79.
Berberis, i. 229.
Berillistic Art, i. 171.
Beryl, i. 215.
Betonia, ii. 130.
Bishop of Strasburg, recipe of, i. 322.
Bismuth, i. 167, 266; bitumen and antimony, i. 8.
Bitumen, i. 243.
Blessed Blood of Rosy Colour, i. 55.
Blood, arcanum of, i. 131.
Bloodstone, medical uses of, ii. 208, 209.
Botin, i. 227.
Buglossum, i. 189.
Burning Star, i. 288, 301.

Cabalists and the vessel of alchemy, i. 68; cabalistic art, i. 161; cabalistic cosmography, 161 n.
Cachimiæ, i. 223, 237, 255; medical uses of, ii. 204.
Calamine, medical uses of, ii. 206, 207.
Calcatrippa, i. 189.
Calcination, i. 139, 151.
Calcined Blood, i. 59 n.
Calculus, ii. 130.
Caleruthum, ii. 277.
Calid, i. 56.
Cancer, i. 269.
Caput Mortuum, i. 23, 32, 33, 35, 104, 141, 154, 197, 314.
Caraba, i. 137.
Carbuncle, a solar stone, i. 17; of what formed, 17 n.; 276; how made, 318, 319.
Cardamum, ii. 61.
Carraways, i. 229.
Catholicon of the Philosophers, i. 29.
Celandine, a coagulation of Mercury, i. 54; quintessence of, ii. 27.
Celestial powers, the elements of the quintessence, i. 5.
Cement in Sol, i. 312.
Cementation, i. 151.

Cenuphar, i. 230.
Cerussa, i. 143.
Chalybs, i. 199.
Chaomancy, i. 190, 193.
Chalcanthum, in Venus, i. 38.
Chalcedony, the lowest of precious stones, i. 18; extracted from salt through the amethyst, 18 *n.*; 224.
Characters, ii. 295, 313.
Chaos, i. 207, 231, 295, 297, 301, 304.
Chelidony, ii. 130.
Chemical pathway of Paracelsus, the, i. 298.
Cherio, ii. 179.
Cheyri, the Great, ii. 339, 340.
Chiromancy, i. 179-181, etc.; ii. 313.
Chiseta, i. 109.
Christ the first spagyrist, i. 28.
Chrysocolla, i. 162, 183, 184.
Cinnabar, i. 159, 184.
Cinnamon, i. 229.
Circulated Salt, ii. 87.
Circulation, i. 153.
Circulatum, ii. 51, 54.
Citrines, i. 225.
Civet, i. 137; ii. 61.
Coagulation, i. 154; coagulation and resolution, i. 53 *n.*
Cobalt, i. 254.
Cobleta, i. 109.
Cœlum Philosophorum, ii. 22.
Cœruleum, i. 162.
Cohobation, i. 153.
Colcothar of Vitriol as a coagulate of Mercury, i. 54.
Colic, i. 264; ii. 264, 319.
Colocinth, i. 229; ii. 59, 146.
Colophonia, ii. 59.
Comets, ii. 295, 313.
Combustion of Minerals, i. 311.
Complexions, ii. 117, 150, 172.
Conservation, ii. 118.
Copper, i. 224, 278, 336; medical virtues of, ii. 228, 229.
Coral, nature of, i. 17; species of, 17 *n.*; quintessence of, ii. 24; magistery of, 53; medical uses of, ii. 214, 215.
Corporal Resin, ii. 17.
Cosmography, ii. 297.
Crocus of Copper, i. 142.
Crocus of Iron, i. 140, 200 *n.*, 274; ii. 35.
Crocus of Mercury, i. 200 *n.*
Crocus of Sol, i. 144.
Crow's Head, i. 68.
Crystal, whence extracted, i. 18; to what referable. 18 *n.*; coagulation of, *ib.*; theory of conjuration in, 14, 225; medical uses of, ii. 216, 217.

Crystalline, i. 268.
Cyamus, ii. 138.
Cyanean Stone, i. 158.
Cyclopes, ii. 261.
Cyphanta, ii. 346.
Cyroglossum, i. 189.

Dalni, ii. 258.
Danewort, i. 230.
Dardo, ii. 339.
Daura, ii. 119.
Death, the Mother of Tinctures, i. 138; death twofold, 146; death not caused by disease, ii. 108.
Degrees, two only, ii. 172; elementated degrees, *ib.*; rules of, 174; degrees of herbs, *ib.*; of heat and cold, 175; concealed degrees, 176; first degree of heat, 181, 183; second degree of heat, 182; first degree of cold, 182; second degree of cold, 183, 184; third and fourth degrees of heat, *ib.*; grade of health, 187; grades of spagyric industry, 190; degrees in relation to disease, 192, 193, 194; grades of disease, 197.
Deneas, i. 218.
Deltic Nature, ii. 331.
Dentaria, i. 189.
Derses, ii. 278.
Destruction a perfecting agent, i. 4.
Diana of the Wise, i. 298.
Diemeæ, ii. 254.
Digestion, i. 153.
Diocletian, destroyer of Spagyric books, i. 19.
Dioscorides, ii. 155, 156.
Distillation, i. 151.
Divertallum, ii. 265.
Divination, i. 185; stars of, 191; science of, ii. 294, 295, 296, 301, 302.
Divining Rod, i. 185.
Domor, i. 202, 203.
Doses, ii. 196, 197, 198.
Drachum, i. 233.
Dramas, ii. 255.
Dreams, ii. 296, 313.
Dropsy, i. 197 *n.*
Dualech, ii. 265.
Durdales, ii. 254.
Dust of Metals, i. 308.

Earth, element of, i. 221; philosophy of, 228; separation of, ii. 21.
Edochinum, ii. 345.
Electrum, i. 107, 108, 114; ii. 103, etc.
Elementary Gold, i. 301.
Element, one only, ii. 11: long life consists in equilibrium of elements, ii. 122.

Element of the Sun, ii. 130.
Elias the Artist, i. 27.
Eliphas Levi, i. 288.
Elixir, notable, 335, 336; elixir at the white, *ib.*; elixir from Luna, *ib.*; what elixir is, ii. 99; a ferment, 69 *n.*; conservation by, 71; quintessence the fourth elixir, 75; subtlety the fifth elixir, *ib.*; propriety the sixth elixir, 76; elixir of Nature, 154; elixir of gold, ii. 333.
Ellebore, ii. 59.
Emerald, a defence of chastity, i. 17; whence derived, 17 *n.*, 224, 243; quintessence of, ii. 24, 130.
Emerald Table of Hermes, i. 19.
Enigmas of alchemy, ii. 97 *n.*
Enochdiani, ii. 340, 341, 344.
Ens, ii. 112.
Enur, ii. 279.
Erucæ, ii. 258.
Estimation. i. 173; ii. 120.
Eternal Mystery, ii. 4.
Euphrasia, i. 189.
Eurus, i. 218.
Evestra, i. 193; ii. 273, 274, 275.
Exaltation, i. 152 *n.*

Fæces, i. 297.
Faith and Divination, i. 185 *n.*
Falling Sickness, ii. 111, 128, 158.
False Processes for the Stone, i. 55.
Fatacesti, ii. 254.
Fel Vitri, i. 163.
Ferments, secret of, i. 70.
Fire, purges imperfections, i. 4; especially in the prime substances, 4 *n.*; generation of metals by fire, 5 *n.*; virtues of, 42 *n.*, i. 211; the subject of art, 74; not an element, 74 *n.*; manifold nature of, 75; ultimate and primal matter of everything, 90; separation of elements from, ii. 20; its fourfold nature, 129; a twofold fire, 129.
First Entities, ii. 84, 131, 132, 133, 134, 135, 136.
First Principle and ultimate matter, i. 91.
Fishes, separation of elements from, ii. 18.
Fistula Cassiæ, i. 137.
Fixation, i. 153.
Fixation of spirits, i. 309, 310.
Fixed Augment, i. 326.
Fixed Oil, i. 313.
Fixed Substances, separation of elements from, ii. 20.
Flaccum, ii. 41.
Fleshly Substances, separation of elements from, ii. 18.

Flos Æris, i. 199.
Flos Cheiri, ii. 332, 333.
Flying Eagle, i. 40, 286.
Fortunatus, ii. 83.
Foundation of Philosophers, i. 321.
Four Complexions, ii. 126, 127.
Four Elements, ii. 84.
Four Humours, ii. 127 *n.*
François de Nation, i. 299.
Freising, i. 322.
Fuligo Mercurii, i. 139, 149.
Fulmination, i. 274.

Galen, i. 20, 36, 73; ii. 81 *n.*, 82, 98, 99, 162, 169.
Gall, ii. 23.
Gamahea, i. 51; varieties of, 51 *n.*; ii. 295.
Gamandria, i. 230.
Gamonynum, ii. 340.
Garyophyllon, i. 134.
Geber, i. 57, 59, 67, 298.
Gems, manner of their origin, i. 16 *n.*; generation of, 126; transmutation of, 158; 233, 243, 255; sulphur of, 269; medical uses of, ii. 216, 217.
Generation, two-fold, i. 120; ii. 6 *n.*; how originated, i. 296.
Genestum, ii. 157.
Gentian, i. 229.
Geomancy, i. 171, 189; stars of, 190; ii. 297.
Geometry, ii. 297.
Giants, i. 125, 162; ii. 261.
Glass, i. 162.
Gnomes, ii. 258.
Gold, how generated, i. 111; growth of, 129; sulphur of, 266: calx of, 276; Mercury of, 278, 292; astral, elementary, and vulgar, 301; solution of, 313; separation from the cup, 316; quintessence of, ii. 8, 26, 28; difference between quintessence and potable gold, ii. 28 *n.*; magistery of, 48; fifth essence of, 74; first entity of, 131; medical uses of, 225, 226.
Golden Herb, i. 129.
Gout, ii. 111, 128, 130.
Gradation, meaning of the term, i. 31 *n.*; number of gradations, 31.
Granate, i. 255; ii. 130.
Grasses, separation of elements from, ii. 20.
Gravel, ii. 130.
Great Architect, i. 295.
Great Artificer, ii. 250.
Great Composition, ii. 90.
Great Mystery, ii. 249, 250, 251, 252, 253, 254, 255, 258, 259, 261, 262, 265, 269, 271, 272, 280.
Green Lion, i. 38.

Gyphus, i. 162.

Halcyon, ii. 129, 132.
Hamuel, son of, i. 65.
Hare's-foot, i. 143.
Heaven of the Metals, ii. 89.
Heliazati, ii. 340.
Heliotrope, i. 189, 249.
Hellebore, ii. 130.
Hepatica, i. 189.
Herbs, separation of elements from, ii. 18; seven species of, 191.
Hermes, i. 19, 21, 65, 70, 85, 125, 148, 284, 298; ii. 110, 155; Apocalypse of, i. 23.
Hildonius, ii. 346.
Hippocrates, i. 29, 73; ii. 169, 324.
Hippuris, i. 189.
Homunculus, i. 124; ii. 120, 334.
Honey, i. 134; ii. 74.
Hyacinth, ii 138.
Hydromancy, i. 171, 190, 193; ii. 297.
Hydromel, i. 134.
Hyle, i. 297.
Hypericon, i. 134.
Hyposarchæ, ii. 111.

Iliaste, see Iliaster.
Iliaster, i. 201, 203, 204, 205, 206, 210, 216, 226, 297; ii. 179, 180, 339, 340, 341, 342, 344, 345.
Images, ii. 295.
Imagination, i. 122, 123, 173; ii. 7, 117, 120.
Imbibition, i. 153.
Impressions, i. 174; ii. 117.
Inanimates, ii. 313.
Incantation, ii. 119, 296.
Instrumentation, ii. 297.
Iron transmuted into copper, i. 28; to cut iron with iron, 309; to harden iron, 316; oil of iron, ii. 67; medical virtues of, 229, 230. *See also* i. 188.

Jacinth, i. 17; akin to the carbuncle, 17 *n.*; ii. 130.
Jaspis, sulphur of, i. 266.
Jaundice, ii. 111, 127.
Jean d'Espagnet, i. 299.
Jeshihach, ii. 346.
John Fabricius, i. 299.
John de Rupescissa, i. 57, 67, 117, 195.
Juniper, essence of, ii. 26.
Jupiter, not a quintessence, i. 6; elements of, 6 *n.*; Jupiter and Cinnabar produce Sol, 15; Mercury of, 280, 281.

Lac Virginis, i. 54, 55.

Laton, i. 164.
Laureus, ii. 346.
Lavation, i. 153.
Lavender, quintessence of, ii. 26.
Lazuleum, i. 158.
Lazurium, i. 143, 184; sapphires of, 224.
Lead, properties when molten, i. 77 *n.*; test of, 313. *See* Saturn.
Lemures, i. 193.
Leprosy, ii. 116, 128, 130.
Leris, ii. 340.
Leseolus, ii. 320.
Life, a spiritual essence, i. 134; a veil of the three principles, 134 *n.*; twofold in man, *ib.*; definition of, ii. 323.
Light without fire, i. 316.
Lili of Alchemy, i. 22; putrefaction of, 23; the subject of the Tincture, 24; desiccation of, 27; phenomena of, *ib.*
Limaria, i. 249.
Limbus of the Microcosm, i. 66.
Litharge, i. 162; litharge of gold, 184; medical virtues of, ii. 201, 202, 203.
Lixivium, i. 308, 310.
Lizards, a process for Sol, i. 309.
Long life, what it depends on, ii. 118.
Lorinds, ii. 255, 258, 268.
Loripides, ii. 261.
Lotones, ii. 197.
Luna and her properties, i. 8; constituents of, 9 *n.*; Luna out of Venus, 27; Mercury of, 278; projection of, 313; digestion of, 316; gradation of, 319; Luna developed into Sol, *ib.*; augmentation of, *ib.*; Luna and Venus, 335.
Lunaria, i. 53.
Lute of Hermes, ii. 46.
Lydian Stone, i. 277.

Mace, 1. 229.
Machaon, i. 29.
Macrocosm, quintessence of, ii. 36; 289.
Magdalion, i. 215.
Magi, who they were, i. 49; classes of, 50 *n.*; 51, 116.
Magic, celestial and malefic, i. 50 *n.*; its origin, 52; instruction of, ii. 294, 295, 296, 299-301, 313, 314.
Magiria, ii. 328.
Magistery, what it is, ii. 48; magistery and quintessence, 48, 49; Magistery of the Blood, 50; extraction of from metals, *ib.*; from pearls, corals, and gems, 51; from Marchasites, 53; from fatty substances, 54; from growing things, 55; from blood, 57; 86.
Magnalia of Nature, i. 29; ii. 340.

Magnesia of the Philosophers, i. 117; Magnesium and Saturn produce Luna, i. 15; Magnesia and Antimony, i. 8.

Magnet, i. 17; its power over diseases, 17 *n.*; phenomena of, 132; life of, 136; mortification of, 145; attractive virtues, ii. 59; medical uses of, 216.

Magnum Opus, i. 201 *n.*

Man the true quintessence, i. 52 *n.*; created from the Limbus, 66 *n.*; exemplar and type of all things, ii. 148; conception of, 291.

Mandragora, ii. 334.

Mania, ii. 128.

Manna, i. 137, 193; ii. 74, 130.

Manual Art, ii. 294, 295, 297, 309, 310.

Marble, Sulphur of, i. 266.

Marcasites, lixivium of, transmuted into Venus, i. 28; their generation, 109; 223, 256, 266; separation of elements from, ii. 16, 50; magistery of, 53; medical uses of, 203, 204.

Mars, his properties, i. 7; component elements of, 7 *n.*; sulphur of, 266; Mercury of, 280.

Martagon, i. 53.

Mastic, ii. 59.

Matrix, i. 233.

Matter of the Philosophers compared to a golden tree, i. 54.

Matter of the Stone, its colours, i. 54; 64; rule of its discovery, 65; Sol and Luna the Matter of the Stone, *ib.*; ii. 112.

Matter, ultimate, i. 237, 240.

Mechili, ii. 261.

Medicine of Metals, i. 28.

Meligia, i. 227.

Melosiniæ, ii. 254, 258, 271, 273, 346.

Menstruum, i. 301.

Mercurialis, i. 249.

Mercurius the wisest of the Philosophers, i. 66.

Mercury, two genera, i. 5 *n.*; congelation of, 12; warm nature of, 14; extraction of heat of, *ib.*; coldness of, *ib.*; Mercury the body of every precious stone, 17 *n.*; coagulation of, 53, 54; conversion into Luna, *ib.*; Mercury in metallic counterfeits, 57; fixed oil from, *ib.*; tincture of a supreme secret, 81 *n.*; mortification of, 142; precipitate of, *ib.*; calcination of, 143; dual Mercury, 297; fire of, 301; Mercury of the Philosophers, 119, 250, 298; sulphur of, 107; Mercury of gold, 196; the sperm of Sol and Luna, 65; body, spirit, and soul of, 66; coagulation by lead, 45 *n.*; Mercurius vivus, 12, 117, 118, 119, 126, 149, 164, 165, 308; corporal Mercury, 187, 320, 321, 323, 326; Mercury transmuted into Luna, 307; into Sol, 307, 308; into a stone, 308; coagulation of, 311; calcination of, 314; preparation with marcasites, *ib.*; Mercury from Jupiter and Saturn, 320; Mercury of Saturn, 320, 324, 328; Mercury of all metals, 320; Mercury from all bodies, 320, 328; Mercury of the body, 323, 324, 325, 326, 329, 330, 331, 332; Mercury of Jupiter, 326, 333; Mercury of Luna, 328, 329, 333; metals connected with Mercury, 318; Mercury of Sol or Luna, 330; extraction of Mercury of the Moon, 332; water of Mercury, 334; malleable Mercury, 336; fixation of, *ib.*; Mercury of life, ii. 29, 30, 50, 90; the third arcanum, 38, 43; medical uses of, 230. *See also* i. 195, i. 36, i. 288, i. 293; ii. 125 *n.*, 336; Mercurial water the first beginning of metals, i. 90.

Meritorium, ii. 277.

Mesne, ii. 99, 156.

Metals, not devoid of life, i. 117 *n.*; generation of, 125, 248; regeneration of, 126; radical moisture of, 290; transmuted into stones, i. 16.

Microcosmus, consists of four elements,, ii. 3 *n.*; contains all minerals, *ib. See also* ii. 289, 318.

Migdalio, ii. 129.

Mineral Menstruum, i. 299.

Minerals, method of seeking, i. 15; ultimate matter of, 90; minerals out of arsenic, 315; augmentation of, *ib.*; antimoniacal minerals, *ib.*; reduction of the minera, *ib.*

Minium, i. 143; ii. 124.

Mixture, spiritual mixture and communion of the metals, i. 9.

Monstrous Births, i. 121, 122.

Monstrous growths, i. 173, etc.

Moon, digestion of, i. 315; quintessence of, ii. 26.

Morphea, ii. 53.

Morsus Diaboli, i. 189.

Mucilage, i. 243.

Multiplication, i. 317.

Mumia, i. 131; mumia and man, 131 *n.*; mumia and balsam, *ib.*; mumia and sweet Mercury, 169; extraction of mumia, ii. 337.

Muria, i. 160.

Must, magistery of, ii. 56.

Myrrh, i. 272.

Mysterium Magnum, i. 201 *n.*

Narcissus, i. 130.

Natural Magic, ii. 252.
Necrolicus, ii. 346.
Necromancers, i. 185.
Necromancy, i. 171; necromancy and nigromancy, 171 *n.*, 190, 193.
Nedeon, i. 233.
Nenuphar, i. 230; ii. 174.
Nesdar, ii. 255.
Nesva, i. 20.
Neufarini, ii. 254.
New Birth, necessity of, i. 20.
Nigromancy, ii. 294, 295, 296, 303.
Nitre, i. 100, 197.
Nymphs, ii. 253, 255, 258, 271, 273, 274, 340.

Oil of Coral, ii. 143.
Oil of Jupiter, ii. 141.
Oil of Marchasite, ii. 142.
Oil of Mars, ii. 35, 140.
Oil and Salt of Pearls, ii. 144.
Oil of the Philosophers, i. 314.
Oil of Salt, ii. 143.
Oil of Saturn, ii. 140.
Oil of Silver, ii. 140.
Oil of the Sun, i. 38; ii. 28.
Oleaginous substances, separation of elements from, ii. 17.
Oleum Laterinum, ii. 70.
Olympian Spirit, i. 116.
Ophioglossum, i. 189.
Orchis, *see* Satyrion.
Orpiment, medical uses of, ii. 211, 212.

Paracelsus, sprung from Helvetia, i. 19; a prince of philosophy and medicine, *ib.*; chosen by God, 20; monarch of arcana, 21; his experiments in Istria, 29; born of Nature, 39; his academia, 96; *see also* i. 285; ii. 163, 170.
Paradise, fruits of, i. 228; no death in Paradise, ii. 115, 116.
Paratetus, ii. 340.
Parsnip, i. 229.
Passage of the Red Sea, i. 298.
Part with Part, i. 55, 59, 164, 166, 312.
Pearl, not a stone, i. 17; produces milk in women, 17 *n.*, 156; special virtues of, ii. 53; magistery of, 52.
Pelican, ii. 74.
Penates, i. 93.
Pentaphyllon, i. 189.
Perforata, i. 189.
Perpetual Water, i. 319.
Persica, i. 230.
Persian Fire, i. 259, 269.
Petfoliata, ii. 189.
Petroleum, ii. 173.

Phantasy, i. 173; ii. 296, 313.
Philosophers' Egg, ii. 93.
Philosophers' Stone, ii. 29; the second arcanum, ii. 40, 100.
Philosophers' Tree, i. 129.
Philosophic Man, i. 66.
Philosophical Stone, i. 294.
Philosophic Sun, i. 71.
Phœnix and Iliaster, i. 40 *n.*, 286.
Physical Stone, i. 301, 303.
Physiognomy, ii. 297.
Pigmies, i. 125.
Plato, i. 56; ii. 155.
Pleurisy, ii. 130.
Poison, concealed in all things, ii. 160; grades of poison, ii. 189; nature and essence of, 237; alchemy of poison, 238; entity of poison, 239; the natural alchemist as a separator of poison, 241.
Polydorus, i. 67.
Ponderations, magical art of, i. 71.
Porphyrius, ii. 321.
Potable Gold, ii. 28, 51, 106, 130.
Potable Silver, ii. 51.
Potherbs, i. 229.
Precious Water, i. 319.
Primal Matter, ii. 29; the first arcanum, 39, 40, 125; ii. 125.
Promised Land, entrance into, i. 298.
Proportion, ii. 297.
Protoplast, *see* Adam.
Prunella, i. 189.
Purging, simples for, ii. 188.
Pustules, ii. 128.
Putrefaction, the beginning of generation, i. 120; many kinds of, 121, 153; putrefaction and transmutation, ii. 279.
Pyrites, i. 167.
Pyromancy, i. 171, 190, 193; ii. 297.
Pyxis, i. 336.

Quicksilver, generation of, i. 44 *n.*; how rendered ductile, *ib.*
Quintessence, i. 5; its principle, 5 *n.*; the divine quintessence, 52; correction of, ii. 22; formula of, 28; quintessence from metals, 30; from marcasites, 31; from salts, *ib.*; from stones, gems, and pearls, 32; from growing things, 33, 334; from spices, 34; from eatables and drinkables, 35, 85; quintessence of oil, 139; of silver, 140; of antimony, 141.
Quince, quintessence of, ii. 41.
Quintessence of the Sun, i. 199; ii. 12.
Quintessences, ii. 120.

Radical Moisture, i. 290; ii. 125, 126, 128, 132.

Raymond Lully, i. 53, 67, 299; ii. 335.
Realgar, i. 136.
Recipes, composition of, ii. 194; prescription of, 196.
Rector of Man, ii. 253 *n.*
Red Lion, i. 22; short way to, 23; rose-coloured blood of, 25; to be sought in the East, 27; process on, 37.
Red Oil for fixing Luna and Sol, i. 318.
Reduction, i. 167, 326.
Relollea, ii. 176, 179, 181.
Resin of the Earth, i. 265.
Resolution, i. 167.
Rhabarbarus, i. 229.
Rhasis, i. 95; ii. 156.
Richard of England, i. 67.
Roger Bacon, i. 299.
Royal Cement, i. 42; ii. 29 *n.*
Ruby, i. 17, 224; ii. 130.
Rude Stone, i. 297.

Saffron, quintessence of, ii. 127 *n.*, 130.
Salmanazar, i. 70.
Sal Ammoniac, i. 195, 199; ii. 59.
Sal Fluxum, i. 163.
Sal Gemmæ, i. 261; ii. 223.
Sal Nitri, ii. 224.
Sal Palla, i. 215.
Sal Peregrinorum, ii. 223.
Sal Petræ, i. 261.
Sal Stiriatus, i. 261.
Sal Terræ, i. 261.
Salt, corrects Luna, i. 43; extracted out of urine, 55, 261; duplex, 260; elixir of, ii. 73; fixation of salts, i. 310; *see also* i. 197, 224; ii. 132.
Salt, Sulphur, and Mercury, found in all metals, i. 90; many varieties of, 96; source of all metals, 125; as spirit, soul, and body, 125. *See also* i. 201 *n.*, 204, 206, 208, 209, 216, 217, 221, 223, 224, 225, 226, 227 *n.*, 238, 239, 245, 247, 248, 251, 256, 273, 305; ii. 124, 125.
Salt Nitre, i. 195, 278.
Saltpetre, water of, i. 140.
Samies, ii. 270.
Sand, i. 162.
Sandarach, i. 183, 184.
Sapphire, heavenly nature of, i. 17; whence generated, 17 *n.*; varieties of, *ib.*, 224, 245; ii. 130.
Saturn, unspiritual nature of, i. 7; water of Saturn, 8; nature and elements of, 8 *n.*; spirit of, 79; blackest and densest body of all metals, 110; spirit of, 203 *n.*; mercury of, 281, 292; magistery of, ii. 53; medical uses of, 230; procreation of, 251.

Satyrion, i. 189; ii. 338.
Savonarola, ii. 155.
Saxifrage, medical uses of, ii. 209, 210.
Scarabæi, ii. 250, 258.
Scariolus, ii. 346.
Sciomancy, i. 155.
Scariæ, i. 162.
Scrofulary, i. 189.
Separation, ground of, i. 160 *n.*; grades of, 163; separation of Luna from Venus, 311; philosophy of, ii. 10; separation of elements from metals, 15; from marcasites, 16; from stones, 17; from oleaginous substances, *ib.*; in corporeal resins, *ib.*; from herbs, 18; from flesh, *ib.*; from fishes, *ib.*; from watery substances, 19; from glass, 20; from fixed substances, *ib.*; from the four elements, 20, 21; separation the begetter of generation, 252; separation of the elements, 254. *See also* ii. 279.
Serapio, i. 26; ii. 156.
Shadows, art of, i. 185.
Siconius, ii. 346.
Sigillum Bariæ, ii. 79.
Sign, what it is, i. 37.
Signator, i. 190.
Signature, ii. 268, 294, 295, 296, 304.
Silver, sulphur of, i. 266; medical virtues of, ii. 226, 227.
Sleep, i. 186 *n.*
Sol, a royal metal, i. 9; pure fire, 10; liquefaction of, *ib.*; purity of fire of, 11; incorruptible in fire, *ib.*; threefold essence of, *ib.*; gold and the Sun, *ib.*; Sol out of Luna, 27; earth of red Sol, 39; sulphur of Sol, *ib.*; spirit or tincture of, 75; Sol the body of Mercury, 80; fixes Mercury, *ib.*; resin of, 144; calx of, 147; quintessence of, ii. 130; oil of, *ib.*
Sol and Luna, the matter of the Stone, i. 65.
Solatrum, ii. 47.
Solution, i. 51.
Somnambulism, i. 186 *n.*
Sons of spagyric art, i. 20; sons of the Doctrine, ii. 107.
Sophia and Philosophy, i. 95.
Southern Star, i. 36.
Sovereign Artist, i. 295.
Spagyria, i. 17 *n.*
Spagyric Art, i. 157.
Spagyric Books, i. 19.
Spagyric Magistery, i. 28; spagyric generation, i. 284.
Spathus, i. 162.
Spectres, ii. 296, 313.
Speculation, ii. 296, 313.

Specifics, how produced, ii. 59; odoriferous, 60; anodyne, 62; diaphoretic, 63; attractive, 64; styptic, 65; corrosive, 66; specific of the matrix, 67.
Sperm of the Three Principles, ii. 87.
Sphere of Saturn, i. 298.
Sphere of the Sun, i. 36.
Spirit of God, i. 295
Spirit of Life, ii. 126, 127 *n.*
Spirits, dissolution of, i. 310; tincture of, *ib.*; spirit of wine, ii. 29 *n.*
Spirit Voice, i. 186 *n.*
Spots, remedy against, ii. 79.
Stannar, ii. 279.
Stars and their influence, i. 174, 175; ii. 149, 150, 285, 286, 287.
Stars of metals, i. 186, 187.
Star of the Microcosm, i. 116.
Stelliones, i. 308.
Stone of the Philosophers, i. 301; separation of the Stone, ii. 104; use of, 105; a perfect balsam, 96.
Stones, separation of elements from, ii. 17.
Sublimation, i. 151, 197.
Sulphur, its varieties, i. 5 *n.*; fire of, 27; not the beginning of metals, 90; a resin of the earth, 105; animated sulphur, *ib.*; embryonated, *ib.*; quintessence of, 126; reverberated, *ib.*; sulphur in wood, 270; crude sulphur, 272; arsenic in sulphur, 273; sulphur of the philosophers, 249, 291; sulphur of Nature, 297; sulphur vive, i. 33; sulphur of Sol, i. 39; sulphur of cinnabar, 40; red sulphur of the philosophers, 51; sulphur transmuted into Sol, 308; how reddened, 311; note concerning sulphur, 236; sulphureous minerals, 311; first entity of, ii. 132; medical virtues of, 212, 213. *See also* ii. 137, 336.
Supercelestial Marriage of the Soul, i. 61.
Superstitions, ii. 121.
Sweetness, elixir of, ii. 73, 74.
Sydenia, i. 189.
Sylphs, i. 125.
Sylvans, i. 193.
Syrens, ii. 255, 271.

Talc, i. 136, 246, 275; medical uses of, ii. 206.
Tartar, oil of, 310, 330; dissolution of, 310; magistery and essence of, ii. 130; transmutation of, 158.
Temperaments, ii. 111.
Tereniabin, i. 193.
Theriaca, ii. 103.
Thronus, ii. 74.

Time of the Work, i. 40.
Tin, medical virtues of, ii. 227.
Tincture, matter of, i. 22; early process of, 24; magistery of, 25; method of attaining, 26; a universal medicine, 29; as a medicine, 41; tincture of gold, 76 *n.*; how tincture is defined, 155; two tinctures in man, 155 *n.*; how tinctures operate, 156; tincture of the philosophers, 186; tinctures the fourth arcanum, ii. 39, 45. *See also* i. 276.
Tithymal, a coagulation of Mercury, i. 54.
Topaz, i. 17.
Transmutation, i. 28; ii. 279.
Transmutation of metals, time of, i. 118.
Traramium, ii. 273.
Trevisan, Bernard, i. 299.
Trifertes, ii. 271.
Trines, ii. 255.
Tripolis, i. 162.
Trithemius, i. 50.
Tronosia, i. 193.
Tronum, ii. 270
Truras, ii. 270.
Turba Philosophorum, i. 279.
Turban, ii. 274.
Turbith, i. 229; ii. 60.
Turpentine, i. 137.
Tuthia, preparation of, i. 309; medical uses of, ii. 205.
Tyrian remedy, ii. 113.

Ulcer, remedy for, ii. 79.
Uncertain Arts, ii. 294, 295, 297, 305-308.
Undines, i. 193.
Unions, *see* Pearl.
Urine, a resolved salt, i. 100, 189, 199, 260, 262, 327, 328; ii. 59.
Uroscopy, ii. 152.
Uterus, Spagyric, i. 284.

Valerian, ii. 130.
Valescus, ii. 155.
Vase of the Philosophers, i. 298.
Venitian Art, i. 55.
Venter Equinus, i. 75, 121, 124, 148, 150, 157; ii. 58, 334.
Ventinina, ii. 297.
Venus, properties of, i. 7; the first metal generated by Archeus, 7 *n.*; Venus out of Saturn, 27; experiment of Paracelsus, 29; counterfeits of, 58; many tinctures from, 78 *n.*; mercury of, 279; quintessence of, ii. 26, 35; crocus of, 35; flowers of, 60; *see also* i. 292.
Verdigris from copper, i. 141.
Vessel of Philosophers, i. 68, 69.

Victorialis, i. 189.
Vigo, ii. 155.
Vinum Aceti, i. 142.
Vinum Ardens, i. 136, 156; ii. 50, 52, 55, 70.
Virgil, i. 148; Virgil's Bell, 116.
Visions, ii. 296, 313.
Vitriol, varieties of, i. 38 *n.*; nobility of, 60; arcanum of, 60 *n.*; imperfection of, 61; viridity of, 61 *n.*; diagnosis of, *ib.*; species of, 102; tests of, 103; vitriol grillus, 104; vitriol from Venus, *ib.*; spagyric vitriol in verdigris, *ib.*; green vitriol, 278; vitriol of Nature, 300. *See also* i. 195, 199, 224, 266. Medical uses of, ii. 219, 220, 221, 231, 232; oil of red vitriol, 235; white and green oil of vitriol, 236.
Vitriolated Oil in alchemy, i. 101.
Vulcan, art of, i. 22; office of, 22 *n.*, 263, 266, 277; ii. 96.
Vulgar Gold, i. 301.

Water, an element, i. 91; mother, seed, and root of all metals, 92; the first matter of minerals, 95; contains only Salt, Sulphur, and Mercury, 231, 243; water giving weight to Sol and Luna, 313; water fixing Mercury, 316; separation of elements from water, ii. 19, 21; water of life, 28; water of salt, 29 *n.*
Weight, in projection, i. 63; weight of the ferment, 70; artificial weight, 71.
White Eagle, i. 22; gluten of, 25; to be sought in the south, 27.
White and Red, process for, i. 316, 317.
Whitening, mode of, i. 313.
Winckelstein, John, of Friburg, i. 194.
Wine, Magistery of, ii. 55, 56.
Wounds, remedy for, ii. 78.

Xylobalsamum, ii. 337.
Xylohebenus, ii. 320.

Yle, i. 234.
Yliadus, } i. 161 *n.*, 232, 234.
Yliadum, }
Ysopaic Art, i. 266.

Zachary, i. 299.
Zedoch, i. 217, 218, 219. 220.
Zenio, ii. 345.
Zinc, i. 136, 254, 314.
Zodiac, celestial and human, ii. 148 *n.*
Zoroaster, i. 49.
Zwitter, i. 136.